HOLT SCIENCE & TECHNOLOGY

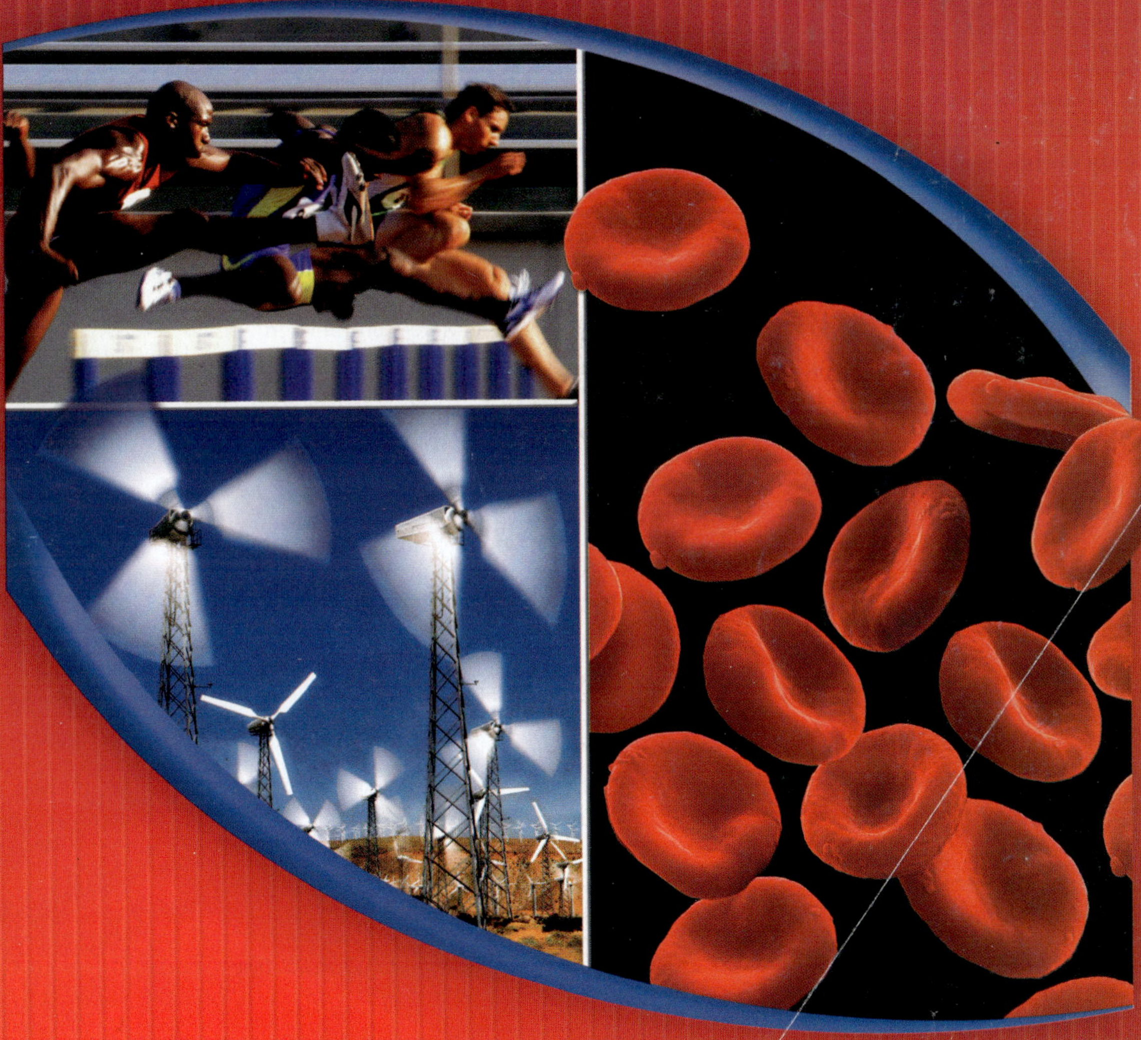

HOLT, RINEHART AND WINSTON
A Harcourt Education Company
Orlando • **Austin** • New York • San Diego • Toronto • London

Acknowledgments

Contributing Authors

Katy Z. Allen
Science Writer
Wayland, Massachusetts

Kathleen Meehan Berry
Science Chairman
Canon-McMillan School District
Canonsburg, Pennsylvania

Linda Ruth Berg, Ph.D.
Adjunct Professor
Natural Sciences
St. Petersburg College
St. Petersburg, Florida

Christie Borgford, Ph.D.
Assistant Professor of Chemistry
Department of Chemistry
The University of Alabama
Birmingham, Alabama

Barbara Christopher
Science Writer and Editor
Austin, Texas

Leila Dumas
Former Physics Teacher
Austin, Texas

Jennie Dusheck
Science Writer
Santa Cruz, California

Robert H. Fronk, Ph.D.
Professor
Science and Mathematics Education Department
Florida Institute of Technology
Melbourne, Florida

Mary Kay Hemenway, Ph.D.
Research Associate and Senior Lecturer
Department of Astronomy
The University of Texas at Austin
Austin, Texas

Kathleen Kaska
Former Life and Earth Science Teacher and Science Department Chair

William G. Lamb, Ph.D.
Winningstad Chair in the Physical Sciences
Oregon Episcopal School
Portland, Oregon

Peter E. Malin, Ph.D.
Professor of Geology
Division of Earth and Ocean Sciences
Duke University
Durham, North Carolina

Karen J. Meech, Ph.D.
Astronomer
Institute for Astronomy
University of Hawaii
Honolulu, Hawaii

Robert J. Sager, M.S., J.D., L.G.
Coordinator and Professor of Earth Science
Pierce College
Lakewood, Washington

Mark F. Taylor, Ph.D.
Associate Professor of Biology
Biology Department
Baylor University
Waco, Texas

Indiana Teacher Consultants

Amy Bethke
Science Teacher
Baker Middle School
Michigan City, Indiana

Kathy Bartley
High School Math and Science Teacher
Highland High School
Highland, Indiana

Richard Bishop
Science Teacher
Clifford Pierce Middle School
Merrillville, Indiana

Jayme Herbert
Seventh and Eighth Grade Teacher
South Dearborn Middle School
Aurora, Indiana

Christina L. Hilton
Science Teacher
Science Department Curriculum Coordinator
Greenfield-Central High School
Greenfield, Indiana

Bruce A. Starek
Department Chairperson, Science Teacher
Baker Middle School
Michigan City, Indiana

Inclusion Specialists

Karen Clay
Inclusion Specialist Consultant
Boston, Massachusetts

Ellen McPeek Glisan
Special Needs Consultant
San Antonio, Texas

Safety Reviewer

Jack Gerlovich, Ph.D.
Associate Professor
School of Education
Drake University
Des Moines, Iowa

Academic Reviewers

Glenn Adelson
Instructor
Biology Undergraduate Program
Harvard University
Cambridge, Massachusetts

David M. Armstrong, Ph.D.
Professor
Ecology and Evolutionary Biology
University of Colorado
Boulder, Colorado

Acknowledgments
continued on page 845

Printed in the United States of America

ISBN 0-03-038143-6

2 3 4 5 6 7 048 08 07 06 05

Contents in Brief

Contents

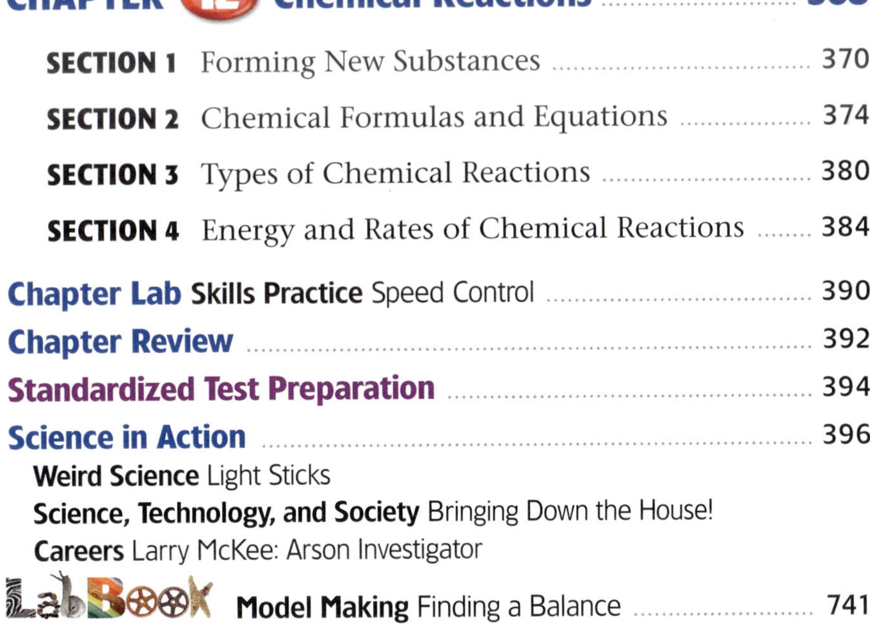

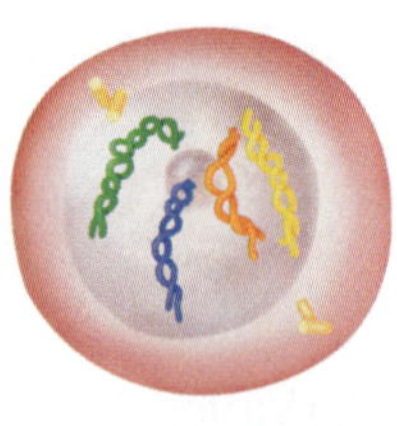
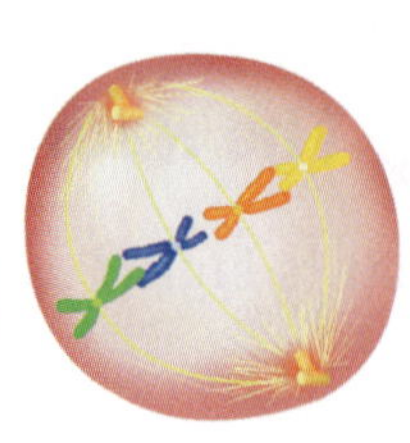
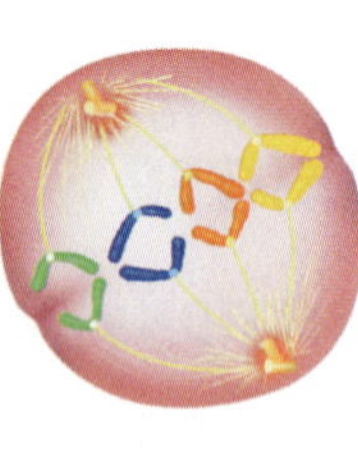
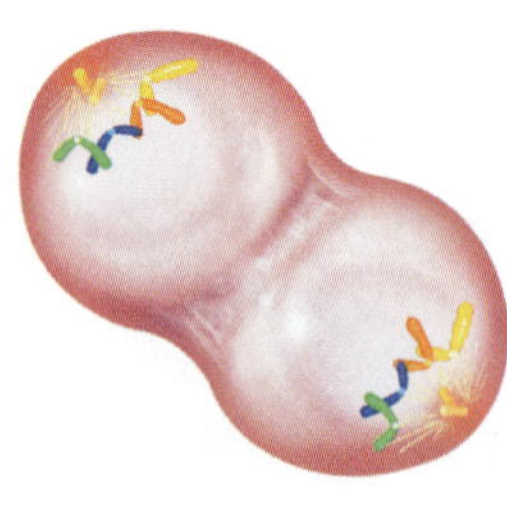

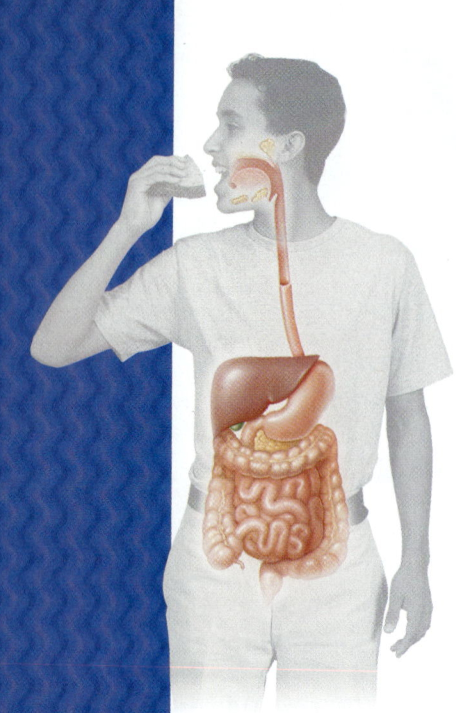

Chapter Labs and LabBook

The more labs, the better!

Take a minute to browse the variety of exciting **labs** in this textbook. Labs appear within the chapters and in a special LabBook in the back of the textbook. All labs are designed to help you experience science firsthand. But please don't forget to be safe. Read the Safety First! section before starting any of the labs.

Start your engines with an activity!

Get motivated to learn by doing the two activities at the beginning of each chapter. The **Pre-Reading Activity** helps you organize information as you read the chapter. The **Start-up Activity** helps you gain scientific understanding of the topic through hands-on experience.

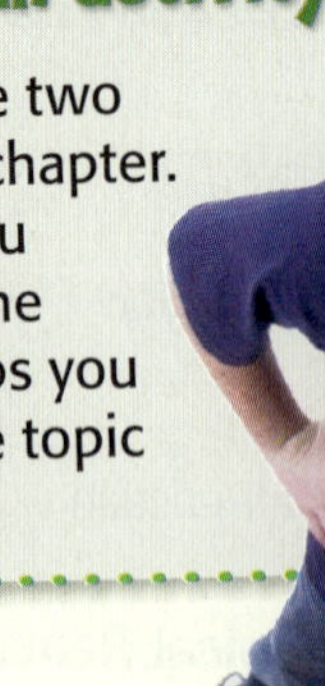

PRE-READING ACTIVITY

FOLDNOTES

Graphic Organizer

START-UP ACTIVITY

READING STRATEGY

Remembering what you read doesn't have to be hard!

A **Reading Strategy** at the beginning of every section provides tips to help you remember and/or organize the information covered in the section.

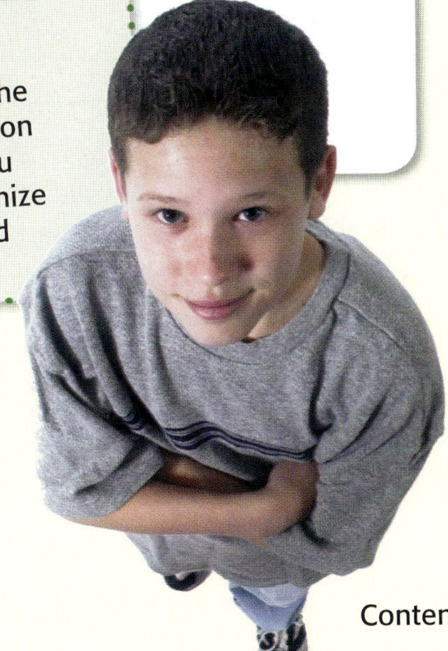

Quick Lab

School to Home

Activity

Science brings you closer together!

Bring science into your home by doing **School-to-Home Activities** with a parent or another adult in your household.

INTERNET ACTIVITY

Get caught in the Web!

Go to **go.hrw.com** for **Internet Activities** related to each chapter. To find the Internet Activity for a particular chapter, just type in the keyword listed below.

MATH FOCUS

Science and math go hand in hand.

The **Math Focus** and **Math Practice** items show you many ways that math applies directly to science and vice versa.

Connection to...

One subject leads to another.

You may not realize it at first, but different subjects are related to each other in many ways. Each **Connection** explores a topic from the viewpoint of another discipline. In this way, all of the subjects you learn about in school merge to improve your understanding of the world around you.

Science in Action

Science moves beyond the classroom!

Read **Science in Action** articles to learn more about science in the real world. These articles will give you an idea of how interesting, strange, helpful, and action-packed science is. At the end of each chapter, you will find three short articles. And if your thirst is still not quenched, go to **go.hrw.com** for in-depth coverage.

Careers

People in Science

Science, Technology, and Society

Scientific Debate

Scientific Discoveries

Weird Science

Science Fiction

How to Use Your Textbook

Your Roadmap for Success with Holt Science and Technology

Reading Warm-Up

A Reading Warm-Up at the beginning of every section provides you with the section's objectives and key terms. The objectives tell you what you'll need to know after you finish reading the section.

Key terms are listed for each section. Learn the definitions of these terms because you will most likely be tested on them. Each key term is highlighted in the text and is defined at point of use and in the margin. You can also use the glossary to locate definitions quickly.

STUDY TIP Reread the objectives and the definitions to the key terms when studying for a test to be sure you know the material.

SECTION 3

READING WARM-UP

Objectives

- Explain how viruses are similar to and different from living things.
- List the four major virus shapes.
- Describe the two kinds of viral reproduction.

Terms to Learn

virus
host

READING STRATEGY

Discussion Read this section silently. Write down questions that you have about this section. Discuss your questions in a small group.

virus a microscopic particle that gets inside a cell and often destroys the cell

host an organism from which a parasite takes food or shelter

Viruses

One day, you discover red spots on your skin. More and more spots appear, and they begin turning into itchy blisters. What do you have?

The spots could be chickenpox. Chickenpox is a disease caused by a virus. A **virus** is a microscopic particle that gets inside a cell and often destroys the cell. Many viruses cause diseases, such as the common cold, flu, and acquired immune deficiency syndrome (AIDS).

It's a Small World

Viruses are tiny. They are smaller than the smallest bacteria. About 5 billion virus particles could fit in a single drop of blood. Viruses can change rapidly. So, a virus's effect on living things can also change. Because viruses are so small and change so often, scientists don't know exactly how many types exist. These properties also make them difficult to fight.

Are Viruses Living?

Like living things, viruses contain protein and genetic material. But viruses, such as the ones shown in **Figure 1**, don't act like living things. They can't eat, grow, break down food, or use oxygen. In fact, a virus cannot function on its own. A virus can reproduce only inside a living cell that serves as a host. A **host** is a living thing that a virus or parasite lives on or in. Using a host's cell as a tiny factory, the virus forces the host to make viruses rather than healthy new cells.

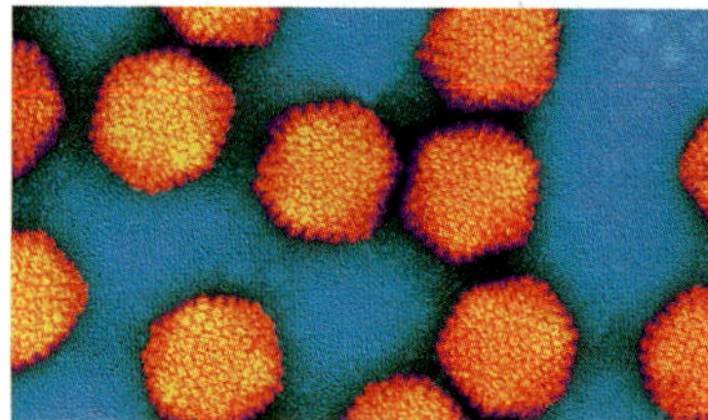

Figure 1 *Viruses are not cells. They do not have cytoplasm or organelles.*

572 Chapter 18 Bacteria and Viruses

Get Organized

A Reading Strategy at the beginning of every section provides tips to help you organize and remember the information covered in the section. Keep a science notebook so that you are ready to take notes when your teacher reviews the material in class. Keep your assignments in this notebook so that you can review them when studying for the chapter test.

Be Resourceful—Use the Web

SCILINKS®

Internet Connect boxes in your textbook take you to resources that you can use for science projects, reports, and research papers. Go to scilinks.org, and type in the SciLinks code to get information on a topic.

Visit go.hrw.com

Find worksheets, **Current Science®** magazine articles online, and other materials that go with your textbook at **go.hrw.com.** Click on the textbook icon and the table of contents to see all of the resources for each chapter.

Figure 2 The Basic Shapes of Viruses

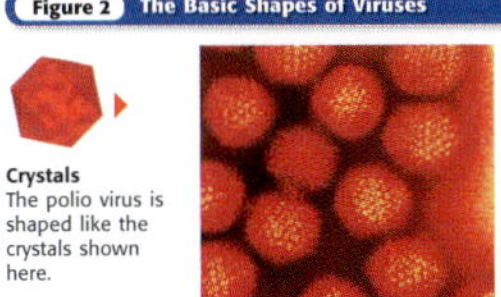

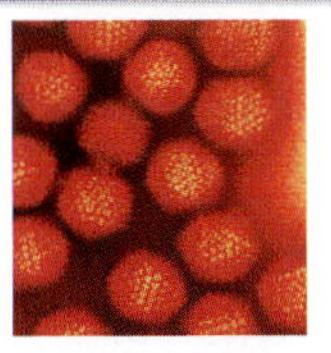

Crystals The polio virus is shaped like the crystals shown here.

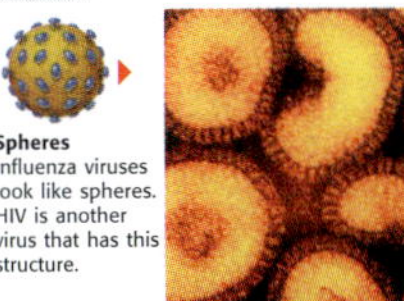

Spheres Influenza viruses look like spheres. HIV is another virus that has this structure.

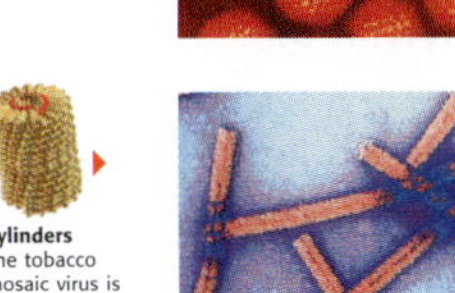

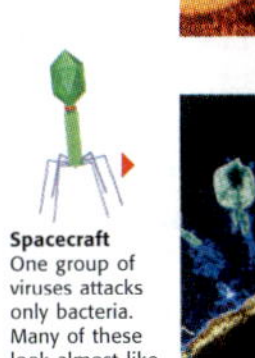

Cylinders The tobacco mosaic virus is shaped like a cylinder and attacks tobacco plants.

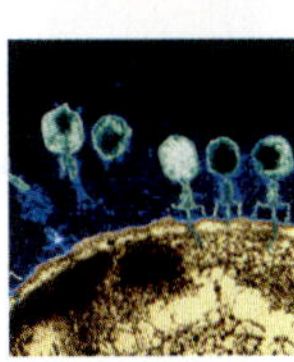

Spacecraft One group of viruses attacks only bacteria. Many of these look almost like spacecraft.

Classifying Viruses

Viruses can be grouped by their shape, the type of disease they cause, their life cycle, or the kind of genetic material they contain. The four main shapes of viruses are shown in **Figure 2.** Every virus is made up of genetic material inside a protein coat. The protein coat protects the genetic material and helps a virus enter a host cell. Many viruses have a protein coat that ma

The genet RNA is made made up of contain info cause warts cause colds a AIDS, which also contains

Reading Ch classified? (See

MATH PRACTICE

Sizing Up a Virus

Treating a Virus

Antibiotics do not kill viruses. But scientists have recently developed antiviral (AN tie VIE ruhl) medications. Many of these medicines stop viruses from reproducing. Because many viral diseases do not have cures, it is best to prevent a viral infection from happening in the first place. Childhood vaccinations give your immune system a head start in fighting off viruses. Having current vaccinations can prevent you from getting a viral infection. It is also a good practice to wash your hands often and never to touch wild animals. If you do get sick from a virus, like the boy in **Figure 4,** it is often best to rest and drink extra fluids. As with any sickness, you should tell your parents or a doctor.

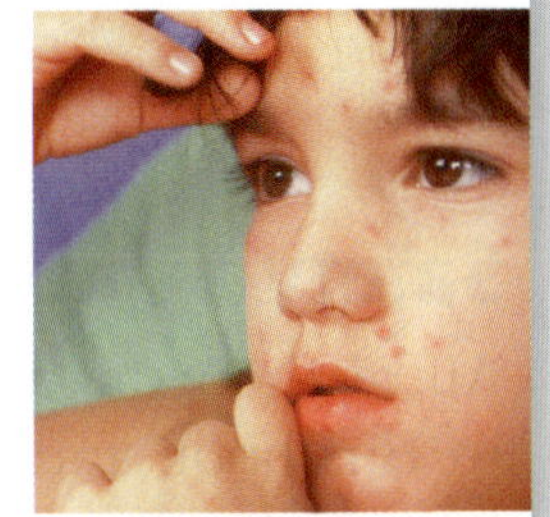

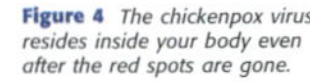

Figure 4 *The chickenpox virus resides inside your body even after the red spots are gone.*

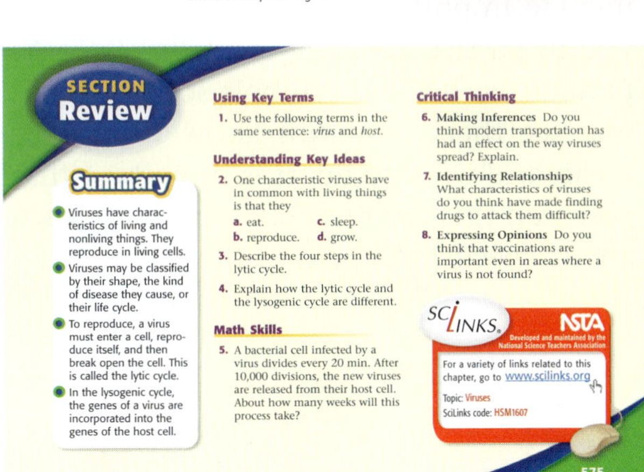

SECTION Review

Summary

- Viruses have characteristics of living and nonliving things. They reproduce in living cells.
- Viruses may be classified by their shape, the kind of disease they cause, or their life cycle.
- To reproduce, a virus must enter a cell, reproduce itself, and then break open the cell. This is called the lytic cycle.
- In the lysogenic cycle, the genes of a virus are incorporated into the genes of the host cell.

Using Key Terms

1. Use the following terms in the same sentence: *virus* and *host.*

Understanding Key Ideas

2. One characteristic viruses have in common with living things is that they
   a. eat.
   b. reproduce.
   c. sleep.
   d. grow.
3. Describe the four steps in the lytic cycle.
4. Explain how the lytic cycle and the lysogenic cycle are different.

Math Skills

5. A bacterial cell infected by a virus divides every 20 min. After 10,000 divisions, the new viruses are released from their host cell. About how many weeks will this process take?

Critical Thinking

6. **Making Inferences** Do you think modern transportation has had an effect on the way viruses spread? Explain.
7. **Identifying Relationships** What characteristics of viruses do you think have made finding drugs to attack them difficult?
8. **Expressing Opinions** Do you think that vaccinations are important even in areas where a virus is not found?

SCILINKS. NSTA Developed and maintained by the National Science Teachers Association

For a variety of links related to this chapter, go to www.scilinks.org

Topic: Viruses
SciLinks code: HSM1607

575

Use the Illustrations and Photos

Art shows complex ideas and processes. Learn to analyze the art so that you better understand the material you read in the text.

Tables and graphs display important information in an organized way to help you see relationships.

A picture is worth a thousand words. Look at the photographs to see relevant examples of science concepts that you are reading about.

Answer the Section Reviews

Section Reviews test your knowledge of the main points of the section. Critical Thinking items challenge you to think about the material in greater depth and to find connections that you infer from the text.

STUDY TIP When you can't answer a question, reread the section. The answer is usually there.

Do Your Homework

Your teacher may assign worksheets to help you understand and remember the material in the chapter.

STUDY TIP Don't try to answer the questions without reading the text and reviewing your class notes. A little preparation up front will make your homework assignments a lot easier. Answering the items in the Chapter Review will help prepare you for the chapter test.

Visit Holt Online Learning

If your teacher gives you a special password to log onto the Holt Online Learning site, you'll find your complete textbook on the Web. In addition, you'll find some great learning tools and practice quizzes. You'll be able to see how well you know the material from your textbook.

Visit CNN Student News

You'll find up-to-date events in science at **cnnstudentnews.com.**

Exploring, inventing, and investigating are essential to the study of science. However, these activities can also be dangerous. To make sure that your experiments and explorations are safe, you must be aware of a variety of safety guidelines. You have probably heard of the saying, "It is better to be safe than sorry." This is particularly true in a science classroom where experiments and explorations are being performed. Being uninformed and careless can result in serious injuries. Don't take chances with your own safety or with anyone else's.

The following pages describe important guidelines for staying safe in the science classroom. Your teacher may also have safety guidelines and tips that are specific to your classroom and laboratory. Take the time to be safe.

Safety Rules!

Start Out Right

Always get your teacher's permission before attempting any laboratory exploration. Read the procedures carefully, and pay particular attention to safety information and caution statements. If you are unsure about what a safety symbol means, look it up or ask your teacher. You cannot be too careful when it comes to safety. If an accident does occur, inform your teacher immediately regardless of how minor you think the accident is.

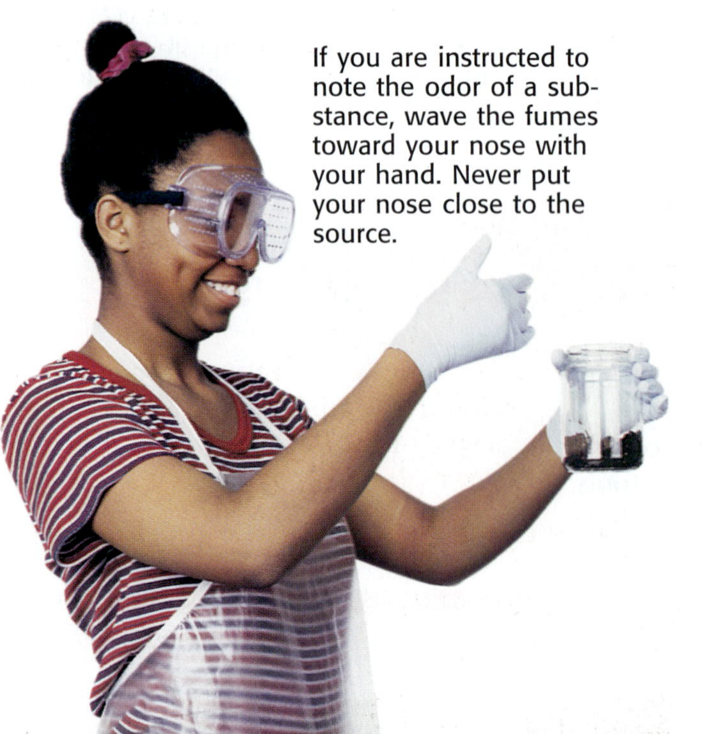

If you are instructed to note the odor of a substance, wave the fumes toward your nose with your hand. Never put your nose close to the source.

Safety Symbols

All of the experiments and investigations in this book and their related worksheets include important safety symbols to alert you to particular safety concerns. Become familiar with these symbols so that when you see them, you will know what they mean and what to do. It is important that you read this entire safety section to learn about specific dangers in the laboratory.

Eye protection

Clothing protection

Hand safety

Heating safety

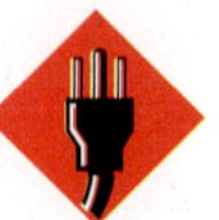

Electric safety

Chemical safety

Animal safety

Sharp object

Plant safety

Eye Safety 

Wear safety goggles when working around chemicals, acids, bases, or any type of flame or heating device. Wear safety goggles any time there is even the slightest chance that harm could come to your eyes. If any substance gets into your eyes, notify your teacher immediately and flush your eyes with running water for at least 15 minutes. Treat any unknown chemical as if it were a dangerous chemical. Never look directly into the sun. Doing so could cause permanent blindness.

Avoid wearing contact lenses in a laboratory situation. Even if you are wearing safety goggles, chemicals can get between the contact lenses and your eyes. If your doctor requires that you wear contact lenses instead of glasses, wear eye-cup safety goggles in the lab.

Safety Equipment

Know the locations of the nearest fire alarms and any other safety equipment, such as fire blankets and eyewash fountains, as identified by your teacher, and know the procedures for using the equipment.

Neatness

Keep your work area free of all unnecessary books and papers. Tie back long hair, and secure loose sleeves or other loose articles of clothing, such as ties and bows. Remove dangling jewelry. Don't wear open-toed shoes or sandals in the laboratory. Never eat, drink, or apply cosmetics in a laboratory setting. Food, drink, and cosmetics can easily become contaminated with dangerous materials.

Certain hair products (such as aerosol hair spray) are flammable and should not be worn while working near an open flame. Avoid wearing hair spray or hair gel on lab days.

Sharp/Pointed Objects 

Use knives and other sharp instruments with extreme care. Never cut objects while holding them in your hands. Place objects on a suitable work surface for cutting.

Be extra careful when using any glassware. When adding a heavy object to a graduated cylinder, tilt the cylinder so that the object slides slowly to the bottom.

Chemicals    

Wear safety goggles when handling any potentially dangerous chemicals, acids, or bases. If a chemical is unknown, handle it as you would a dangerous chemical. Wear an apron and protective gloves when you work with acids or bases or whenever you are told to do so. If a spill gets on your skin or clothing, rinse it off immediately with water for at least 5 minutes while calling to your teacher.

Never mix chemicals unless your teacher tells you to do so. Never taste, touch, or smell chemicals unless you are specifically directed to do so. Before working with a flammable liquid or gas, check for the presence of any source of flame, spark, or heat.

Heat  

Wear safety goggles when using a heating device or a flame. Whenever possible, use an electric hot plate as a heat source instead of using an open flame. When heating materials in a test tube, always angle the test tube away from yourself and others. To avoid burns, wear heat-resistant gloves whenever instructed to do so.

Electricity 

Be careful with electrical cords. When using a microscope with a lamp, do not place the cord where it could trip someone. Do not let cords hang over a table edge in a way that could cause equipment to fall if the cord is accidentally pulled. Do not use equipment with damaged cords. Be sure that your hands are dry and that the electrical equipment is in the "off" position before plugging it in. Turn off and unplug electrical equipment when you are finished.

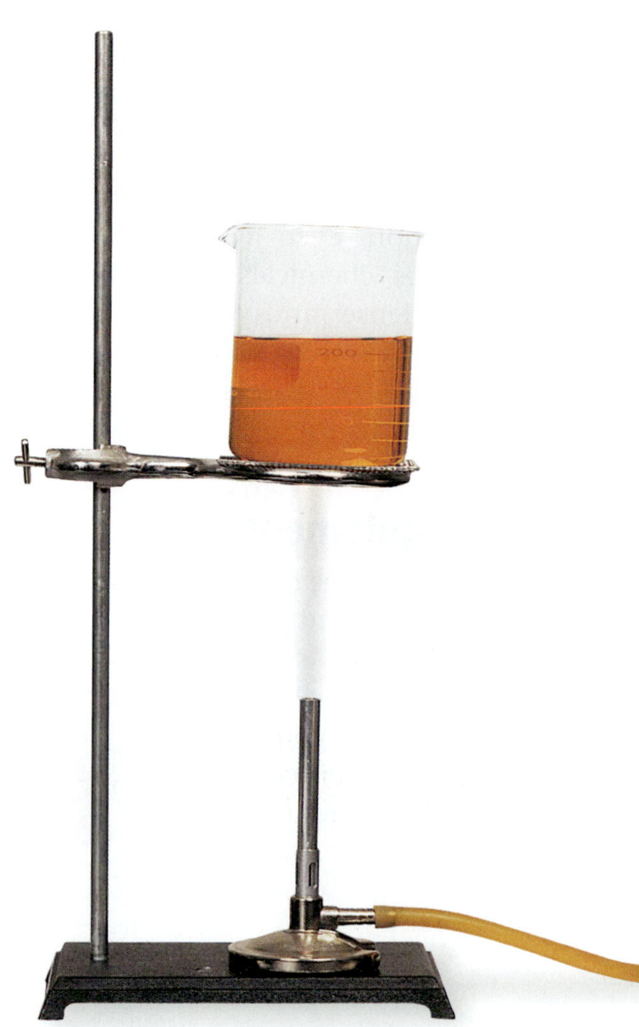

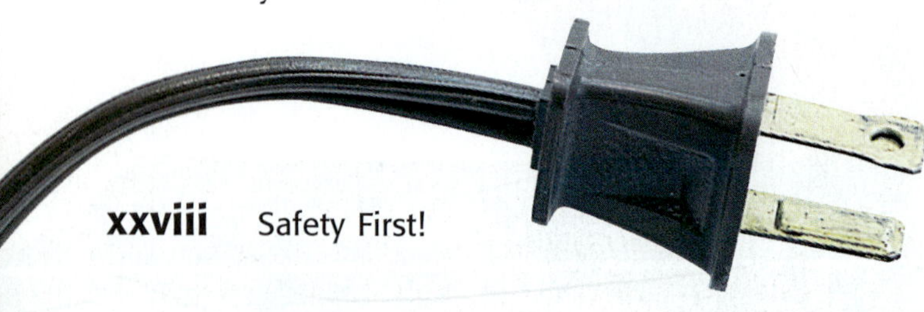

Animal Safety

Always obtain your teacher's permission before bringing any animal into the school building. Handle animals only as your teacher directs. Always treat animals carefully and respectfully. Wash your hands thoroughly after handling any animal.

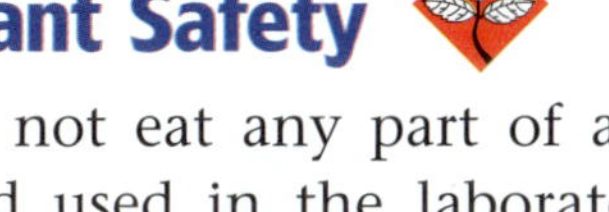

Plant Safety

Do not eat any part of a plant or plant seed used in the laboratory. Wash your hands thoroughly after handling any part of a plant. When in nature, do not pick any wild plants unless your teacher instructs you to do so.

Glassware

Examine all glassware before use. Be sure that glassware is clean and free of chips and cracks. Report damaged glassware to your teacher. Glass containers used for heating should be made of heat-resistant glass.

1

Science in Our World

About the PHOTO

What happened to the legs of these frogs? Science can help answer this question. Deformed frogs, such as the ones in this photo, have been found in the northern United States and southern Canada. Scientists and students like you have been using science to find out how frogs may develop deformities.

PRE-READING ACTIVITY

FOLDNOTES **Layered Book** Before you read the chapter, create the FoldNote entitled "Layered Book" described in the **Study Skills** section of the Appendix. Label the tabs of the layered book with "Examples of life scientists," "Scientific methods," "Scientific models," and "Tools, measurement, and safety." As you read the chapter, write information you learn about each category under the appropriate tab.

START-UP ACTIVITY

A Little Bit of Science

In this activity, you'll find out that you can learn about the unknown without having to see it.

Procedure

1. Your teacher will give you a **coffee can** to which a **sock** has been attached. Do not look into the can.
2. Reach through the opening in the sock. You will feel **several objects** inside the can.
3. Record observations you make about the objects by feeling them, shaking the can, and so on.
4. What do you think is in the can? List your guesses. State some reasons for your guesses.
5. Pour the contents of the can onto your desk. Compare your list with what was in the can.

Analysis

1. Did you guess the contents of the can correctly? What might have caused you to guess wrongly?
2. What observations did you make about each of the objects while they were in the can? Which of your senses did you use?

SECTION 1

Science and Scientists

You are enjoying a picnic on a summer day. Crumbs from your sandwich fall to the ground, and ants carry the crumbs away. You wonder, Why do ants show up at picnics?

Congratulations! You just took one of the first steps of being a scientist. How did you do it? You observed the world around you. Then, you asked a question about your observations. And asking a question is part of what science is all about.

READING WARM-UP

Objectives

- Describe three methods of investigation.
- Identify benefits of science in the world around you.
- Describe five jobs that use science.

Terms to Learn

science

READING STRATEGY

Prediction Guide Before reading this section, write the title of each heading in this section. Next, under each heading, write what you think you will learn.

Start with a Question

The world around you is full of amazing things. Single-celled algae float unseen in ponds. Volcanoes erupt with explosive force. Mars may have had water in the past. And 40-ton whales glide through the oceans. These things or others, such as those shown in **Figure 1,** may lead you to ask a question. A question is the beginning of science. **Science** is the knowledge obtained by observing the natural world in order to discover facts and to formulate laws and principles that can be verified or tested.

science the knowledge obtained by observing natural events and conditions in order to discover facts and formulate laws or principles that can be verified or tested

Reading Check What is science? (*See the Appendix for answers to Reading Checks.*)

In Your Own Neighborhood

Take a look around your home, school, and neighborhood. Often, you take things that you use or see every day for granted. But one day you might look at something in a new way. That's when a question hits you! The student in **Figure 1** didn't have to look very far to realize that he had some questions to ask.

The World and Beyond

You don't have to stop at questions about things in your neighborhood. Ask questions about atoms or galaxies, pandas and bamboo, or earthquakes. A variety of plants and animals live in a variety of places. And each place has a unique combination of rocks, soil, and water.

You can even ask questions about places other than those on Earth. Look outward to the moon, the sun, and the planets in our solar system. Beyond that, you have the rest of the universe! There are enough questions to keep scientists busy for a long time.

Figure 1 *Part of science is asking questions about the world around you.*

Investigation: The Search for Answers

After you ask a question, it's time to look for an answer. But how do you start your investigation? Several methods may be used.

Research

You can find answers to some of your questions by doing research, as **Figure 2** shows. You can ask someone who knows a lot about the subject of your question. You can find information in textbooks, encyclopedias, and magazines. You can search on the Internet. You might read a report of an experiment that someone did. But be sure to think about the sources of your information. Use information only from reliable sources.

Figure 2 *A library is a good place to begin your search for answers.*

Observation

You can also find answers to questions by making careful observations. For example, if you want to know how spiders spin their webs, look for a web. When you find one, return to observe the spider as it spins. But be careful in making observations. Sometimes, what people expect to observe affects what they do observe. For example, most plants need light to grow. Does that mean that all plants need bright sunlight? Do some plants prefer shade? Some people might "observe" that bright light is the only answer. To test an observation, you may have to do an experiment.

Ask a Question

The next time you're outside, look carefully around you. Think of a science-related question that you would like to answer. Write the question in your **science journal.** Discuss with a parent which methods of investigation would be most likely to help you answer your question.

Experimentation

You might answer the question about light and shade by doing a simple experiment, such as the one shown in **Figure 3.** Your research and your observations can help you plan your experiment. What should you do if your experiment needs something that is hard to get? For example, what do you do if you want to know whether a certain plant grows in space? Don't give up! Try to find results from someone else's experiment!

Figure 3 *This student is doing an experiment to find out whether this type of plant grows better in shade or in direct sunlight.*

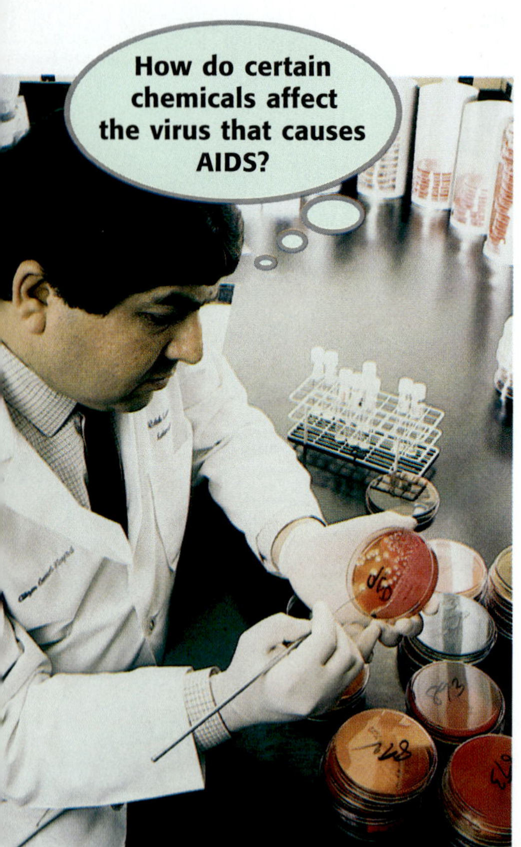

Figure 4 *Abdul Lakhani studies AIDS to find a cure for the disease.*

Why Ask Questions?

Although scientists cannot answer every question immediately, they do find some interesting answers. Do any of the answers really matter? Absolutely! As you study science, you will see how science affects you and society around you.

Fighting Diseases

Polio is a disease that can cause paralysis by affecting the brain and nerves. Do you know anyone who has had polio? You probably don't. But in 1952, polio infected 58,000 Americans. Fortunately, vaccines developed in 1955 and 1956 have eliminated polio in the United States. In fact, the virus that causes polio has been wiped out in most of the world.

Today, scientists are searching for cures for diseases such as mad cow disease, tuberculosis, and acquired immune deficiency syndrome (AIDS). The scientist in **Figure 4** is learning more about AIDS, which kills millions of people every year.

Saving Resources

Science also helps answer the question, How can we make resources last longer? Recycling is one answer. Think about the last time that you recycled an aluminum can. By recycling that can, you saved more than just the aluminum, as **Figure 5** shows. Using science, people have developed more-efficient methods and better equipment for recycling aluminum, paper, steel, glass, and even some plastics. In this way, science helps make resources last longer.

Figure 5 **Resources Saved Through Recycling**

Compared with producing aluminum from its ore, recycling 1 metric ton (1.1 tons) of aluminum:

produces 95% less air pollution

saves 4 metric tons (4.4 tons) of ore

produces 4 metric tons (4.4 tons) fewer chemical products

uses 14,000 kWh less energy

produces 97% less water pollution

Answering Society's Questions

Sometimes, society faces a question that does not seem to have an immediate answer. For example, at one time, the question of how to reduce air pollution did not have any obvious, reasonable answers. The millions of people who depended on their cars could not just stop driving. As the problem of air pollution became more important to people, scientists developed different technologies to address it. For example, one source of air pollution is exhaust from cars. Through science, people have developed cleaner-burning gasoline. People have even developed new ways to clean up exhaust before it leaves the tailpipe of a car!

Reading Check **How can society influence the types of technologies that are developed?**

CONNECTION TO Biology

Technology and Aging People are living longer than ever before. Research some of the health problems that elderly people may face, such as heart or vision problems. Then, research how these health problems are influencing the development of new technologies. Make a poster illustrating one of these problems and one or more technologies being developed to address it.

ACTIVITY

Scientists All Around You

Believe it or not, scientists work in many different places. If you think about it, any person who asks questions and looks for answers could be called a scientist! Keep reading to learn about just a few people who use science in their jobs.

Zoologist

A *zoologist* (zoh AHL uh jist) is a person who studies the lives of animals. Dale Miquelle, shown in **Figure 6,** is part of a team of Russian and American zoologists studying the Siberian tiger. The tigers have almost become extinct after being hunted and losing their homes. By learning about the tigers' living space and food needs, zoologists hope to make a plan that will help the tigers survive better in the wild.

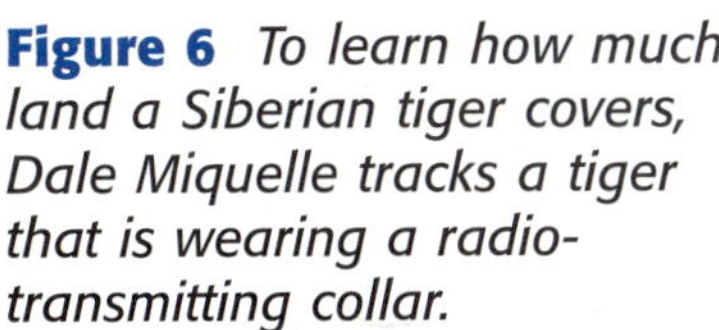

Figure 6 *To learn how much land a Siberian tiger covers, Dale Miquelle tracks a tiger that is wearing a radio-transmitting collar.*

Figure 7 *This geochemist may work outdoors when collecting rock samples from the field. Then, he may work indoors as he analyzes the samples in his laboratory.*

Geochemist

Some scientists work outdoors most of the time. Other scientists spend much of their time in the laboratory. A geochemist (JEE oh KEM ist), such as the one shown in **Figure 7,** may work in both places. A *geochemist* is a person who specializes in the chemistry of rocks, minerals, and soil.

Geochemists determine the economic value of these materials. They also try to find out what the environment was like when these materials formed and what has happened to the materials since they first formed.

Mechanic

Do you have a machine that needs repairs? Call a mechanic, such as Gene Webb in **Figure 8.** Mechanics work on everything from cars to the space shuttle. Mechanics use science to solve problems. Mechanics must find answers to questions about why a machine is not working. Then, they must find a way to make it work. Mechanics also think of ways to improve the machine, to make it work faster or more efficiently.

Figure 8 *A mechanic can help keep a car's engine running smoothly.*

Oceanographer

An *oceanographer* studies the ocean. Some oceanographers study waves and ocean currents. Others study plants and animals that live in the ocean. Still others study the ocean floor, including how it forms.

While studying the ocean floor, oceanographers discovered black smokers. Black smokers are cracks where hot water (around 300°C!) from beneath Earth's surface comes up. These vents in the ocean floor are home to some strange animals, including red-tipped tube worms and blind white crabs.

Volcanologist

If black smokers aren't hot enough for you, perhaps you would like to become a volcanologist (VAHL kuh NAHL uh jist). A *volcanologist* studies one of Earth's most interesting processes—volcanoes. The volcanologist shown in **Figure 9** is photographing lava flowing from Mt. Etna, a volcano in Italy. Mt. Etna's lava may reach temperatures of 1,050°C. By learning more about volcanoes, volcanologists hope to get better at predicting when a volcano will erupt. Being able to predict eruptions would help save lives.

Reading Check What does a volcanologist do?

Figure 9 *Volcanologists gain a better understanding of the inside of the Earth by studying the makeup of lava.*

SECTION Review

Summary

- Science is a process of gathering knowledge about the natural world by making observations and asking questions.
- Science begins by asking a question.
- Even if science cannot answer the question right away, the answers that scientists find may be very important.
- A question may lead to a scientific investigation, including research, observations, and experimentation.
- Science can help save lives, fight diseases, save resources, and protect the environment.
- A variety of people may become scientists for a variety of reasons.

Using Key Terms

1. In your own words, write a definition for the term *science*.

Understanding Key Ideas

2. A zoologist might study any of the following EXCEPT
   a. shellfish living in ponds.
   b. the reason that mole rats live in large groups underground.
   c. environmental threats to sea turtles.
   d. rocks and minerals in the Painted Desert.
3. Describe five careers that use science.
4. How are observation and experimentation different?
5. How may what people expect to observe affect what they do observe? How can people avoid this problem?

Math Skills

6. Students in a science class collected 50 frogs from a pond. They found that 15 of the frogs had serious deformities. What percentage of the frogs had deformities?

Critical Thinking

7. **Making Inferences** An ad for deluxe garbage bags says that the bags are 30% stronger than regular garbage bags. Describe how science can help you find out if this claim is true.
8. **Identifying Relationships** Make a list of three things that you consider to be a problem in society. Give an example of how new technology might solve these problems.
9. **Applying Concepts** Look at **Figure 9.** Write five questions about what you see. Describe how science might help you answer your questions. Share your questions with your classmates.

SECTION 2

Scientific Methods

Imagine that your class is on a field trip to a wildlife refuge. You discover several deformed frogs. You wonder what could be causing the frogs' deformities.

A group of students from Le Sueur, Minnesota, actually made this discovery! By making observations and asking questions about the observations, the students used scientific methods.

READING WARM-UP

Objectives

- Explain why scientists use scientific methods.
- Determine the appropriate design of a controlled experiment.
- Use information in tables and graphs to analyze experimental results.
- Explain how scientific knowledge can change.

Terms to Learn

scientific methods
hypothesis
controlled experiment
variable

READING STRATEGY

Reading Organizer As you read this section, make a flowchart of the possible steps in scientific methods.

What Are Scientific Methods?

When scientists observe the natural world, they often think of a question or problem. But scientists don't just guess at answers. They use scientific methods. **Scientific methods** are the ways in which scientists follow steps to answer questions and solve problems. The steps used for all investigations are the same. But the order in which the steps are followed may vary, as shown in **Figure 1.** Scientists may use all of the steps or just some of the steps during an investigation. They may even repeat some of the steps. The order depends on what works best to answer the question. No matter where scientists work or what questions they try to answer, all scientists have two things in common. They are curious about the natural world, and they use similar methods to investigate it.

Reading Check **What are scientific methods?** (*See the Appendix for answers to Reading Checks.*)

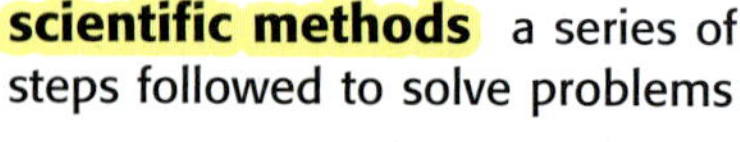

scientific methods a series of steps followed to solve problems

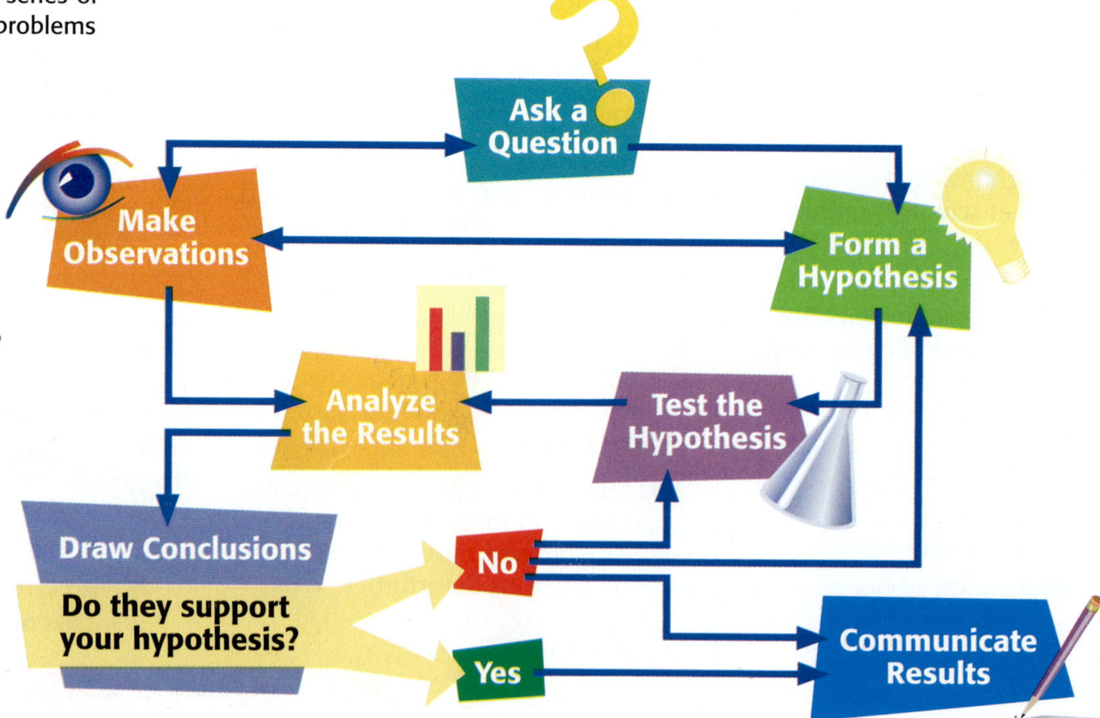

Figure 1 *Scientific methods often include the same steps, but the steps may not be used in the same order every time.*

Ask a Question

Have you ever observed something out of the ordinary or difficult to explain? Such an observation usually raises questions. For example, about the deformed frogs you might ask, "Could something in the water be causing the frog deformities?" Looking for answers may include making more observations.

Figure 2 *Making careful observations is often the first step in an investigation.*

Make Observations

After the students in Minnesota realized something was wrong with the frogs, they decided to make additional, careful observations, as shown in **Figure 2.** They counted the number of deformed frogs and the number of normal frogs they caught. They also photographed the frogs, took measurements, and wrote a thorough description of each frog.

In addition, the students collected data on other organisms living in the pond. They also conducted many tests on the pond water and measured things such as the level of acidity. The students carefully recorded their data and observations.

Accurate Observations

Any information that you gather through your senses is an observation. Observations can take many forms. They may be measurements of length, volume, time, speed, or loudness. They may describe the color or shape of an organism. Or they may record the behavior of organisms in an area. The range of observations that a scientist can make is endless. But no matter what observations reveal, they are useful only if they are accurately made and recorded. Scientists use many standard tools and methods to make and record observations. Examples of these tools are shown in **Figure 3.**

Figure 3 *Microscopes, rulers, and thermometers are some of the many tools scientists use to collect information.*

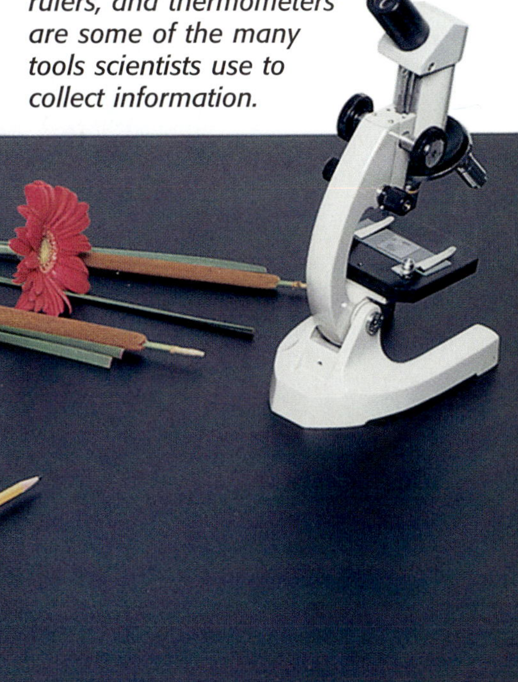

Form a Hypothesis

After asking questions and making observations, scientists may form a hypothesis. A **hypothesis** (hie PAHTH uh sis) is a possible explanation or answer to a question. A good hypothesis is based on observation and can be tested. When scientists form hypotheses, they think logically and creatively and consider what they already know.

To be useful, a hypothesis must be testable. A hypothesis is testable if an experiment can be designed to test the hypothesis. Yet if a hypothesis is not testable, it is not always wrong. An untestable hypothesis is simply one that cannot be supported or disproved. Sometimes, it may be impossible to gather enough observations to test a hypothesis.

Scientists may form different hypotheses for the same problem. In the case of the Minnesota frogs, scientists formed the hypotheses shown in **Figure 4.** Were any of these explanations correct? To find out, scientists had to test each hypothesis.

Reading Check What makes a hypothesis testable?

CONNECTION TO Environmental Science

WRITING SKILL **Minnesota's Deformed Frogs**

Deformed frogs were first noticed in Minnesota in 1995. In 1996, the Minnesota Pollution Control Agency (MPCA) began studying the problem. It funded and coordinated studies searching for the causes of the deformities. Find out what the MPCA is doing about the deformed frogs today, and write a short summary of what the MPCA has discovered.

hypothesis an explanation that is based on prior scientific research or observations and that can be tested

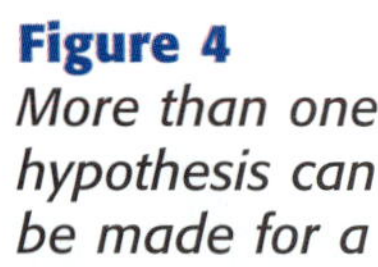

Figure 4 *More than one hypothesis can be made for a single question.*

Predictions

Before scientists can test a hypothesis, they must first make predictions. A prediction is a statement of cause and effect that can be used to set up a test for a hypothesis. Predictions are usually stated in an if-then format, as shown in **Figure 5.**

Figure 5 lists the predictions made for the hypotheses shown in **Figure 4.** More than one prediction may be made for each hypothesis. After predictions are made, scientists can conduct experiments to see which predictions, if any, prove to be true and support the hypotheses.

Figure 5 *More than one prediction may be made for a single hypothesis.*

Hypothesis 1:

Prediction: If a substance in the pond water is causing the deformities, then the water from ponds that have deformed frogs will be different from the water from ponds in which no abnormal frogs have been found.

Prediction: If a substance in the pond water is causing the deformities, then some tadpoles will develop deformities when they are raised in pond water collected from ponds that have deformed frogs.

Hypothesis 2:

Prediction: If a parasite is causing the deformities, then this parasite will be found more often in frogs that have deformities.

Hypothesis 3:

Prediction: If an increase in exposure to ultraviolet light is causing the deformities, then some frog eggs exposed to ultraviolet light in a laboratory will develop into deformed frogs.

CONNECTION TO Language Arts

WRITING SKILL **"Leading doctors say . . ."** Suppose that you and a friend see an ad for a cold remedy on TV. According to the ad, "Leading doctors recommend this product for their patients." Then, a famous actor comes on and says that he or she uses the product, too. Write a dialogue of the debate you might have with your friend about whether these claims are believable.

Test the Hypothesis

After scientists make a prediction, they test the hypothesis. Scientists try to design experiments that will clearly show whether a particular factor caused an observed outcome. In an experiment, a *factor* is anything that can influence the experiment's outcome. Factors can be anything from temperature to the type of organism being studied.

Figure 6 *Many factors affect this tadpole in the wild. These factors include chemicals, light, temperature, and parasites.*

Under Control

Scientists studying the frogs in Minnesota observed many factors that affect the development of frogs in the wild, as shown in **Figure 6.** But it was hard to tell which factor could be causing the deformities. To sort factors out, scientists perform controlled experiments. A **controlled experiment** tests only one factor at a time and consists of a control group and one or more experimental groups. All of the factors for the control group and the experimental groups are the same except for one. The one factor that differs is called the **variable.** Because only the variable differs between the control group and the experimental groups, any differences observed in the outcome of the experiment are probably caused by the variable.

controlled experiment an experiment that tests only one factor at a time by using a comparison of a control group with an experimental group

variable a factor that changes in an experiment in order to test a hypothesis

Reading Check **How many factors should an experiment test?**

Designing an Experiment

Designing a good experiment requires planning. Every factor should be considered. Examine the prediction for Hypothesis 3: *If an increase in exposure to ultraviolet light is causing the deformities, then some frog eggs exposed to ultraviolet light in a laboratory will develop into deformed frogs.* An experiment to test this hypothesis is summarized in **Table 1.** In this case, the variable is the length of time the eggs are exposed to ultraviolet (UV) light. All other factors, such as the temperature of the water, are the same in the control group and in the experimental groups.

Table 1 Experiment to Test Effect of UV Light on Frogs

	Control factors			Variable
Group	Kind of frog	Number of Eggs	Temperature of water	UV light exposure
#1 (control)	leopard frog	100	25°C	0 days
#2 (experimental)	leopard frog	100	25°C	15 days
#3 (experimental)	leopard frog	100	25°C	24 days

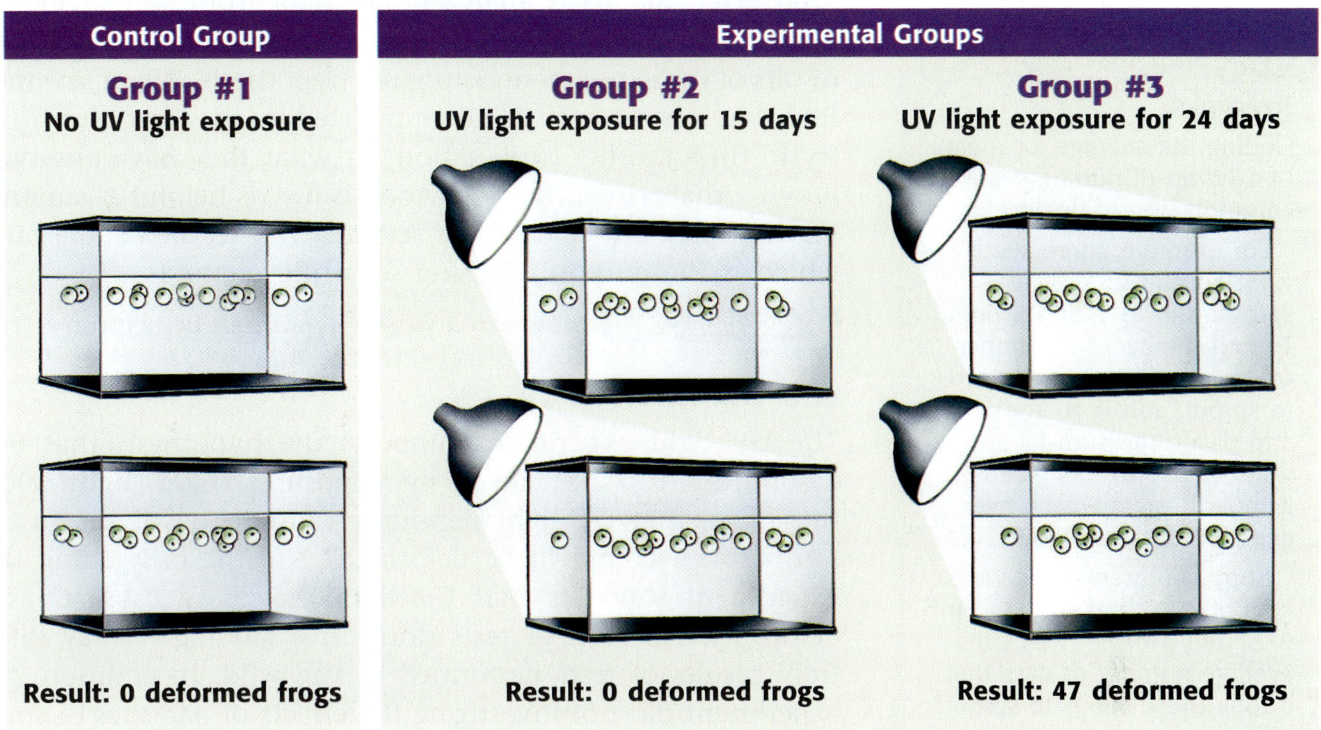

Collecting Data

Figure 7 shows the experimental setup to test Hypothesis 3. As **Table 1** shows, there are 100 eggs in each group. Scientists always try to test many individuals. They want to be sure that differences between control and experimental groups are caused by the variable and not by differences between individuals. The larger the groups are, the smaller the effect of a difference between individual frogs will be. The larger the groups are, the more likely it is that the variable is responsible for any changes and the more accurate the data collected are likely to be.

Scientists test a result by repeating the experiment. If an experiment gives the same results each time, scientists are more certain about the variable's effect on the outcome. Scientists keep clear, accurate, honest records of their data so that other scientists can repeat the experiment and verify the results.

Analyze the Results

After scientists finish their tests, they must organize their data and analyze the results. Scientists may organize data in a table or a graph. The data collected from the UV light experiment are shown in the bar graph in **Figure 8.** Analyzing the results helps scientists explain and focus on the effect of the variable. For example, the graph shows that the length of UV exposure has an effect on the development of frog deformities.

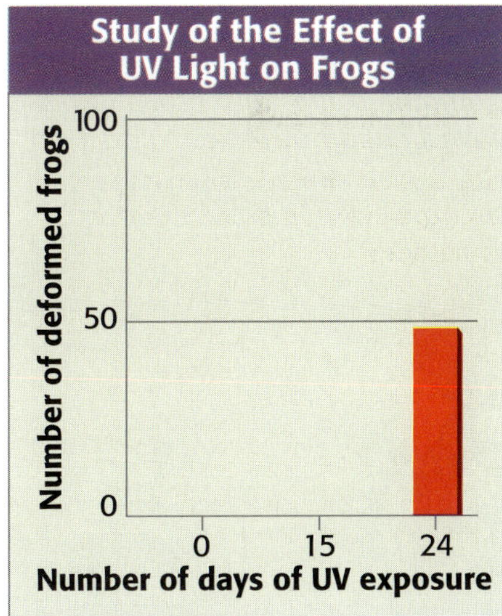

Figure 8 *This graph shows that 24 days of UV exposure had an effect on frog deformities, while less exposure had no effect.*

Draw Conclusions

After scientists have analyzed the data from several experiments, they can draw conclusions. They decide whether the results of the experiments support a hypothesis. When scientists find that a hypothesis is not supported by the tests, they must try to find another explanation for what they have observed. Proving that a hypothesis is wrong is just as helpful as supporting it. Why? Either way, the scientist has learned something, which is the purpose of using scientific methods.

Averages

Finding the average, or mean, of a group of numbers is a common way to analyze data.

For example, three seeds were kept at 25°C and sprouted in 8, 8, and 5 days. To find the average number of days that it took the seeds to sprout, add 8, 8, and 5 and divide the sum by 3, the number of subjects (seeds). It took these seeds an average of 7 days to sprout.

Suppose three seeds were kept at 30°C and sprouted in 6, 5, and 4 days. What's the average number of days that it took these seeds to sprout?

Reading Check How can a wrong hypothesis be helpful?

Is It the Answer?

The UV light experiment supports the hypothesis that frog deformities can be caused by exposure to UV light. Does this mean that UV light definitely caused frogs living in the Minnesota wetland to be deformed? No, the only thing this experiment shows is that UV light may be a cause of frog deformities. Results of tests done in a laboratory may differ from results of tests performed in the wild. In addition, the experiment did not investigate the effects of parasites or some other substance on the frogs. In fact, many scientists now think that more than one factor could be causing the deformities.

Sometimes, similar investigations or experiments give different results. For example, another research team may have had results that did not support the UV light hypothesis. In such a case, scientists must work together to decide if the differences in the results are scientifically significant. Often, making that decision takes more experiments and more evidence.

Figure 9 *This student scientist is communicating the results of his investigation at a science fair.*

Communicate Results

Scientists form a global community. After scientists complete their investigations, they communicate their results to other scientists. The student in **Figure 9** is explaining the results of a science project.

Scientists regularly share their results for several reasons. First, other scientists may then repeat the experiments to see if they get the same results. Second, the information can be considered by other scientists with similar interests. The scientists can then compare hypotheses and form consistent explanations. New data may strengthen existing hypotheses or show that the hypotheses need to be altered. There are many paths from observations and questions to communicating results.

SECTION Review

- Scientific methods are the ways in which scientists follow steps to answer questions and solve problems.
- Any information you gather through your senses is an observation. Observations often lead to the formation of questions and hypotheses.
- A hypothesis is a possible explanation or answer to a question. A well-formed hypothesis is testable by experiment.
- A controlled experiment tests only one factor at a time and consists of a control group and one or more experimental groups.
- After testing a hypothesis, scientists analyze the results and draw conclusions about whether the hypothesis is supported.
- Communicating results allows others to check the results, add to their knowledge, and design new experiments.

Using Key Terms

1. Use the following terms in the same sentence: *hypothesis, controlled experiment,* and *variable.*

Understanding Key Ideas

2. The steps of scientific methods
   a. are exactly the same in every investigation.
   b. must be used in the same order every time.
   c. are not used in the same order every time.
   d. always end with a conclusion.
3. What is the appropriate design of a controlled experiment?
4. What causes scientific knowledge to change?

Math Skills

5. Calculate the average of the following values: 4, 5, 6, 6, and 9.

Critical Thinking

6. **Analyzing Methods** Why was UV light chosen to be the variable in the frog experiment?
7. **Analyzing Processes** Why are there many ways to follow the steps of scientific methods?
8. **Making Inferences** Why might two scientists working on the same problem draw different conclusions?
9. **Making Inferences** Why do scientists use scientific methods?

Interpreting Graphics

10. The table below shows how long it takes for bacteria to double. Plot the information on a graph. Put temperature on the *x*-axis and the time to double on the *y*-axis. Do not graph values for which there is no growth. At what temperature do the bacteria multiply the fastest?

Temperature (°C)	Time to double (min)
10	130
20	60
25	40
30	29
37	17
40	19
45	32
50	no growth

11. What would happen if you changed the scale of the graph by using values of 0 to 300 minutes on the *y*-axis? How might that change affect your interpretation of the data?

SECTION 3

Scientific Models

How can you see the parts of a cell? Unless you had superhuman eyesight, you couldn't see inside most cells without using a microscope.

What would you do if you didn't have a microscope? Looking at a model of a cell would help! A model of a cell can help you understand what the parts of a cell look like.

READING WARM-UP

Objectives

- Give examples of three types of models.
- Identify the benefits and limitations of models.
- Compare the ways that scientists use hypotheses, theories, and laws.

Terms to Learn

model
theory
law

READING STRATEGY

Reading Organizer As you read this section, create an outline of the section. Use the headings from the section in your outline.

Types of Scientific Models

A **model** is a representation of an object or system. Scientific models are used to help explain how something works or to describe the structure of something. A model may be used to predict future events. However, models have limitations. A model is never exactly like the real thing. If it were, it would not be a model. Three major kinds of scientific models are physical, mathematical, and conceptual models.

model a pattern, plan, representation, or description designed to show the structure or workings of an object, system, or concept

Physical Models

A model volcano and a miniature steam engine are examples of physical models. Some physical models, such as a model of a cell, look like the thing that they model. But a limitation of the model of a cell is that the model is not alive and doesn't act exactly like a cell. Other physical models, such as the model of a skyscraper in **Figure 1,** look and act at least somewhat like the thing that they model. Scientists often use the model that is simplest to use but that still serves their purpose.

Figure 1 *The model of the skyscraper doesn't act like the real building in every way, which is both a benefit and a limitation of the model.*

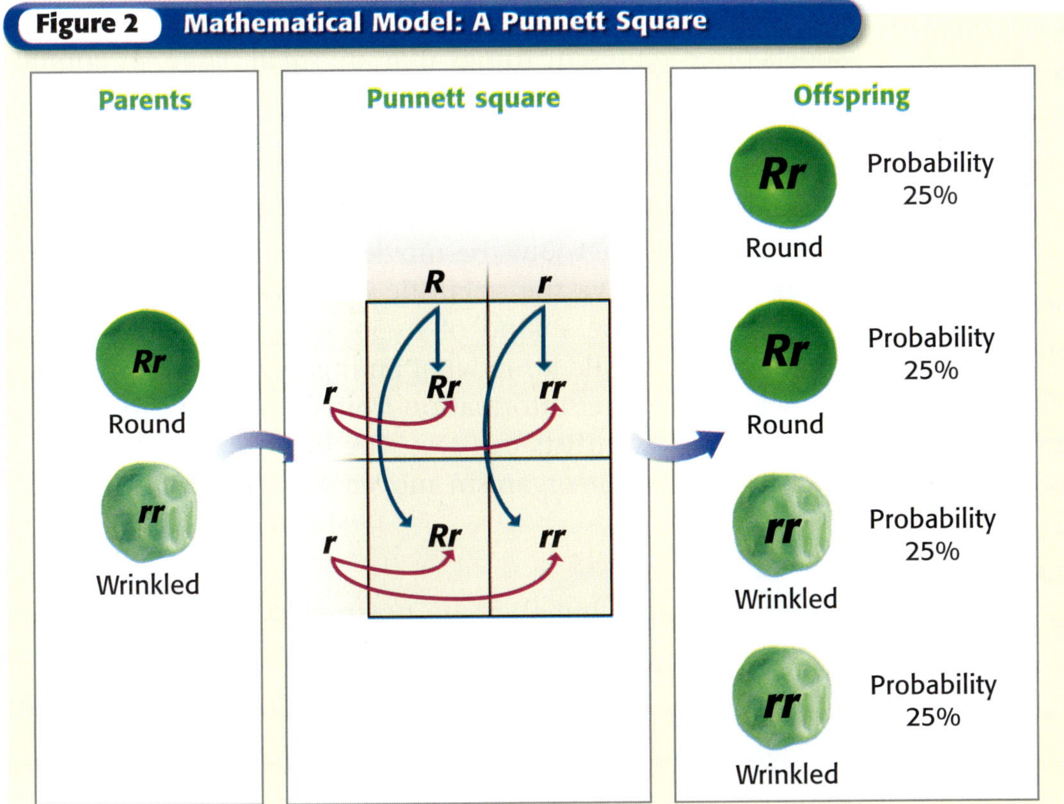

Mathematical Models

A mathematical model may be made up of numbers, equations, and other forms of data. Some mathematical models are simple and can be used easily. The Punnett square shown in **Figure 2** helps scientists study the passing of traits from parents to offspring. Using this model, scientists can predict how often certain traits will appear in the offspring of certain parents.

Computers are useful for creating and manipulating mathematical models. They make fewer mistakes and can keep track of more variables than a person can. But a computer model can also be incorrect in many ways. The more complex a model is, the more carefully scientists must build and test the model.

Reading Check **What type of model is a Punnett square?** (*See the Appendix for answers to Reading Checks.*)

Conceptual Models

The third type of model is the conceptual model. Some conceptual models are systems of ideas. Others compare unfamiliar things with familiar things to help explain unfamiliar ideas. The idea that the solar system formed from a spinning disk of gas is a conceptual model. Scientists also use conceptual models to classify behaviors of animals. Scientists can then predict how an animal might respond to a certain action based on behaviors that have already been observed.

CONNECTION TO Environmental Science

Samples Scientists studying deformed frogs in Minnesota wanted to know at what stage in the frogs' development the deformities happened. So, the scientists collected a large sample of frogs in all stages of development.

The larger a sample is, the more accurately it represents the whole population. If, for example, a sample of frogs is too small, one unusual frog may make the results of the study inaccurate. If the sample has too many old frogs or too many tadpoles, the sample is unrepresentative of the whole population. Give an example of an unrepresentative sample. Make a poster describing how that sample might make the experimental results inaccurate.

ACTIVITY

Figure 3 *This computer-generated model doesn't just look like a dinosaur. It may also open and close its jaws in much the same way that a dinosaur does.*

Benefits of Models

Models often represent things that are small, large, or complicated. Models can also represent things that do not exist. For example, **Figure 3** is a model of one type of dinosaur. Dinosaurs died out millions of years ago. Some popular movies about dinosaurs have used computer models like this one because dinosaurs are extinct. But the movies would not be as realistic if they did not have the scientific models.

A model can be a kind of hypothesis, and scientists can test a model. To build a model of an organism, even an extinct one, scientists gather information from fossils and other observations. Then, scientists can test whether the model fits their ideas about how an organism moved or what it ate.

Limits of Models

Models are useful, but they are not perfect. For example, the model in **Figure 3** gives scientists an idea of how the dinosaur looked. But to find out how strong the dinosaur's jaws were, scientists might build a physical model that has pressure sensors in the jaw. That model would provide data about bite strength. Scientists may use different models to represent the same thing, such as the dinosaur's jaw. But the kind of model and the model's complexity depend on the model's purpose.

Even a model jaw that has pressure sensors is not perfect. Scientists can compare the dinosaur bite with the bite of a crocodile. Next, scientists use their model to conduct tests. Scientists might then estimate how hard the dinosaur could bite. But without a live dinosaur, the result is still a hypothesis.

Building Scientific Knowledge

Sometimes, scientists draw different conclusions from the same data. Other times, new results show that old conclusions are wrong. Scientists are always asking new questions or looking at old questions from a different angle. As scientists find new answers, scientific knowledge continues to grow and change.

Scientific Theories

For every hypothesis, more than one prediction can be made. Each time another prediction is proven true, the hypothesis gains more support. Over time, scientists tie together everything that they have learned. An explanation that ties together many related observations, facts, and tested hypotheses is called a **theory.** Theories are conceptual models that help organize scientific thinking. Theories are used to explain observations and to predict what might happen in the future.

theory an explanation that ties together many hypotheses and observations

law a summary of many experimental results and observations; a law tells how things work

Reading Check How do scientists use theories?

Scientific Laws

What happens when a theory and its models correctly predict the results of many experiments? A scientific law may be formed. In science, a **law** is a summary of many experimental results and observations. A scientific law is a statement of what *will* happen in a specific situation. A law tells you how things work.

Scientific laws are at work around you every day. For example, the law of gravity states that objects will always fall toward the center of Earth. And inside your cells, many laws of chemistry are at work to keep you alive.

Scientific Change

New scientific ideas may take time to be accepted as facts, scientific theories, or scientific laws. Scientists should be open to new ideas but should always test new ideas by using scientific methods. If new evidence challenges an accepted idea, scientists must reexamine the old evidence and reevaluate the old idea. In this way, the process of building scientific knowledge never ends.

CONNECTION TO Chemistry

Model Cocaine in the Brain
Analyze and evaluate information from a scientifically literate viewpoint by reading scientific texts, magazine articles, and newspaper articles about how drugs, such as cocaine, affect brain chemistry. Create a model to show what you have learned. Use your model to describe possible treatments for drug addiction.

ACTIVITY

SECTION Review

Summary

- Models represent objects or systems. Often, they use familiar things to represent unfamiliar things. Three main types of models are physical, mathematical, and conceptual models. Models have limitations but are useful and can be changed based on new evidence.
- Scientific knowledge is built as scientists form and revise scientific hypotheses, models, theories, and laws.

Using Key Terms

In each of the following sentences, replace the incorrect term with the correct term from the word bank.

theory law hypothesis

1. A conclusion is an explanation that matches many hypotheses but may still change.
2. A model tells you exactly what to expect in certain situations.

Understanding Key Ideas

3. A limitation of models is that
   a. they are large enough to see.
   b. they do not act exactly like the things that they model.
   c. they are smaller than the things that they model.
   d. they model unfamiliar things.
4. What type of model would you use to test the hypothesis that global warming is causing polar icecaps to melt? Explain.

Math Skills

5. If Jerry is 2.1 m tall, how tall is a scale model of Jerry that is 10% of his size?

Critical Thinking

6. **Applying Concepts** You want to make a model of an extinct plant. What are two kinds of models that you might use? Describe the advantages and disadvantages of each type of model.

SECTION 4

Tools, Measurement, and Safety

READING WARM-UP

Objectives

- Collect, record, and analyze information by using various tools.
- Explain the importance of the International System of Units.
- Calculate area and density.
- Identify lab safety symbols, and demonstrate safe practices during lab investigations.

Terms to Learn

meter	volume
area	temperature
mass	density

READING STRATEGY

Reading Organizer As you read this section, make a concept map by using the terms above.

Would you use a hammer to tighten a bolt on a bicycle? No, you wouldn't. You need the right tools to fix a bike.

Scientists use a variety of tools in their experiments. A tool is anything that helps you do a task.

Tools for Measuring

You might remember that one way to collect data is to take measurements. To get the best measurements, you need the proper tools. Stopwatches, metersticks, and balances are tools that you can use to make measurements. Thermometers, spring scales, and graduated cylinders are also helpful tools. Some of the uses of these tools are shown in **Figure 1.**

Reading Check Name six tools used for taking measurements. (*See the Appendix for answers to Reading Checks.*)

Tools for Analyzing

After you collect data, you need to analyze them. Perhaps you need to find the average of your data. Calculators are handy tools to help you do calculations quickly. Or you might show your data in a graph or a figure. A computer that has the correct software can help you make neat, colorful figures. Of course, even a pencil and graph paper are tools that you can use to graph your data.

Figure 1 Measurement Tools

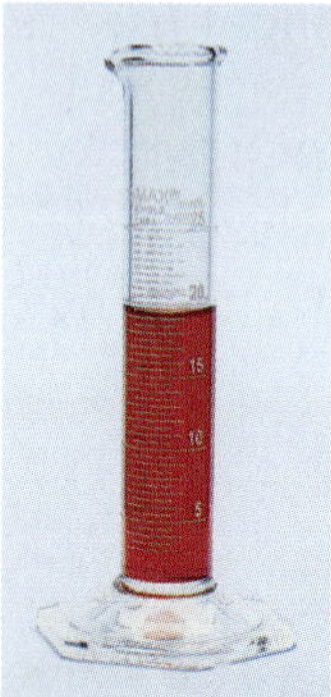

You can use a **graduated cylinder** to measure volume.

You can use a **thermometer** to measure temperature.

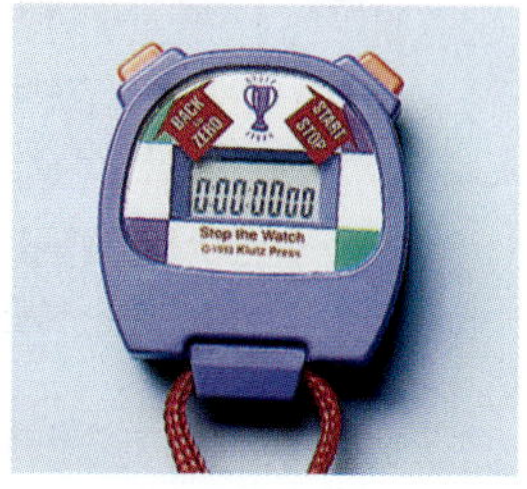

You can use a **stopwatch** to measure time.

You can use a **meterstick** to measure length.

You can use a **balance** to measure mass.

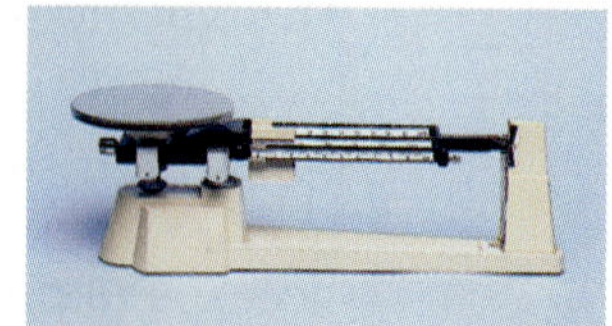

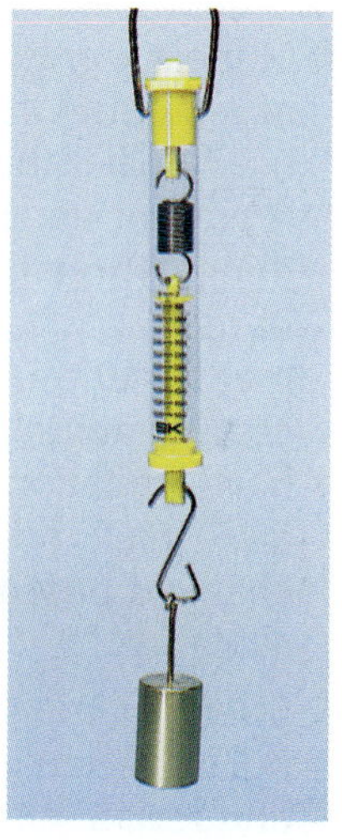

You can use a **spring scale** to measure force.

Units of Measurement

The ability to make accurate and reliable measurements is an important skill in science. Many systems of measurement are used throughout the world. At one time in England, the standard for an inch was three grains of barley placed end to end. Other modern standardized units were originally based on parts of the body, such as the foot. Such systems were not very reliable. Their units were based on objects that had different sizes.

No Rulers Allowed

1. Measure the width of your desk, but don't use a ruler.
2. Select another object to use as your unit of measurement.
3. Compare your measurement with those of your classmates.
4. Explain why it is important to use standard units of measurement.

The International System of Units

In the late 1700s, the French Academy of Sciences began to form a global measurement system now known as the *International System of Units,* or SI. Today, most scientists and almost all countries use this system. One advantage of using SI measurements is that doing so helps scientists share and compare their observations and results.

Another advantage of SI units is that all units are based on the number 10, which makes conversions from one unit to another easy. The table in **Table 1** contains commonly used SI units for length, volume, mass, and temperature.

Table 1 **Common SI Units and Conversions**

Length	**meter (m)** kilometer (km) decimeter (dm) centimeter (cm) millimeter (mm) micrometer (μm) nanometer (nm)	1 km = 1,000 m 1 dm = 0.1 m 1 cm = 0.01 m 1 mm = 0.001 m 1 μm = 0.000001 m 1 nm = 0.000000001 m
Volume	**cubic meter (m^3)** cubic centimeter (cm^3) liter (L) milliliter (mL)	1 cm^3 = 0.000001 m^3 1 L = 1 dm^3 = 0.001 m^3 1 mL = 0.001 L = 1 cm^3
Mass	**kilogram (kg)** gram (g) milligram (mg)	1 g = 0.001 kg 1 mg = 0.000001 kg
Temperature	**Kelvin (K)** **Celsius (°C)**	0°C = 273 K 100°C = 373 K

Figure 2 *This scientist is using a metric ruler to measure a lizard's length. The unit chosen to describe an object, such as this lizard, depends on the size of the object being measured.*

Measurement

Scientists report measured quantities in a way that shows the precision of the measurement. To do so, they use significant figures. *Significant figures* are the digits in a measurement that are known with certainty. The MathFocus below will help you understand significant figures and will teach you how to use the correct number of digits. Now that you have a standardized system of units for measuring things, you can use the system to measure length, area, mass, volume, and temperature.

Length

How long is a lizard? Well, a **meter** (m) is the basic SI unit of length. However, a scientist, such as the one in **Figure 2,** would use centimeters (cm) to describe a small lizard's length. If you divide 1 m into 100 parts, each part equals 1 cm. So, 1 cm is one-hundredth of a meter. Even though 1 cm seems small, some things are even smaller. Scientists describe the length of very small objects in micrometers (μm) or nanometers (nm). To see these small objects, scientists use powerful microscopes.

meter the basic unit of length in the SI (symbol, m)

area a measure of the size of a surface or a region

Area

How much paper would you need to cover the top of your desk? To answer this question, you must find the area of the desk. **Area** is a measure of the size of the surface of an object. To calculate the area of a square or a rectangle, measure the length and width. Then, use the following equation:

$$area = length \times width$$

Units for area are square units, such as square meters (m^2), square centimeters (cm^2), and square kilometers (km^2).

MATH FOCUS

Significant Figures Calculate the area of a carpet that is 3.145 m long (four significant figures) and 5.75 m (three significant figures) wide. (Hint: In multiplication and division problems, the answer cannot have more significant figures than the measurement that has the smallest number of significant figures does.)

Step 1: Write the equation for area.

$$area = length \times width$$

Step 2: Replace *length* and *width* with the measurements given, and solve.

$$area = 3.125 \text{ m} \times 5.75 \text{ m} = 18.08375 \text{ m}^2$$

Step 3: Round the answer to get the correct number of significant figures. Here, the correct number of significant figures is three, because the value with the smallest number of significant figures has three significant figures.

$$area = 18.1 \text{ m}^2$$

Now Its Your Turn

1. Use a calculator to perform the following calculation: 125.5 km × 8.225 km. Write the answer with the correct number of significant figures.

Figure 3 *Adding the rock changes the water level from 70 mL to 80 mL. So, the rock displaces 10 mL of water. Because 1 mL = 1 cm^3, the volume of the rock is 10 cm^3.*

Mass

How large a rock can a rushing stream move? The answer depends on the energy of the stream and the mass of the rock. **Mass** is a measure of the amount of matter in an object. The kilogram (kg) is the basic unit for mass in the SI. Kilograms are used to describe the mass of a large rock. Grams are used to measure the mass of smaller objects. One thousand grams equals 1 kg. For example, a medium-sized apple has a mass of about 100 g. Masses of very large objects are given in metric tons. A metric ton equals 1,000 kg.

Reading Check **What is the basic SI unit for mass?**

Volume

Think about moving some magnets to a laboratory. How many magnets will fit into a box? The answer depends on the volume of the box and the volume of each magnet. **Volume** is a measure of the size of a body in three-dimensional space. In this case, you need the volumes of the box and of the magnets.

The volume of a liquid is often given in liters (L). Liters are based on the meter. A cubic meter (1 m^3) is equal to 1,000 L. So, 1,000 L will fit into a box measuring 1 m on each side. A milliliter (mL) will fit into a box measuring 1 cm on each side. So, 1 mL = 1 cm^3. Graduated cylinders are used to measure the volume of liquids.

The volume of a large, solid object is given in cubic meters (m^3). The volumes of smaller objects can be given in cubic centimeters (cm^3) or cubic millimeters (mm^3). The volume of a box can be calculated by multiplying the object's length, width, and height. The volume of an irregularly shaped object can be found by measuring the volume of liquid that the object displaces. You can see how this works in **Figure 3.**

mass a measure of the amount of matter in an object

volume a measure of the size of a body or region in three-dimensional space

Figure 4 *This thermometer shows the relationship between degrees Fahrenheit and degrees Celsius.*

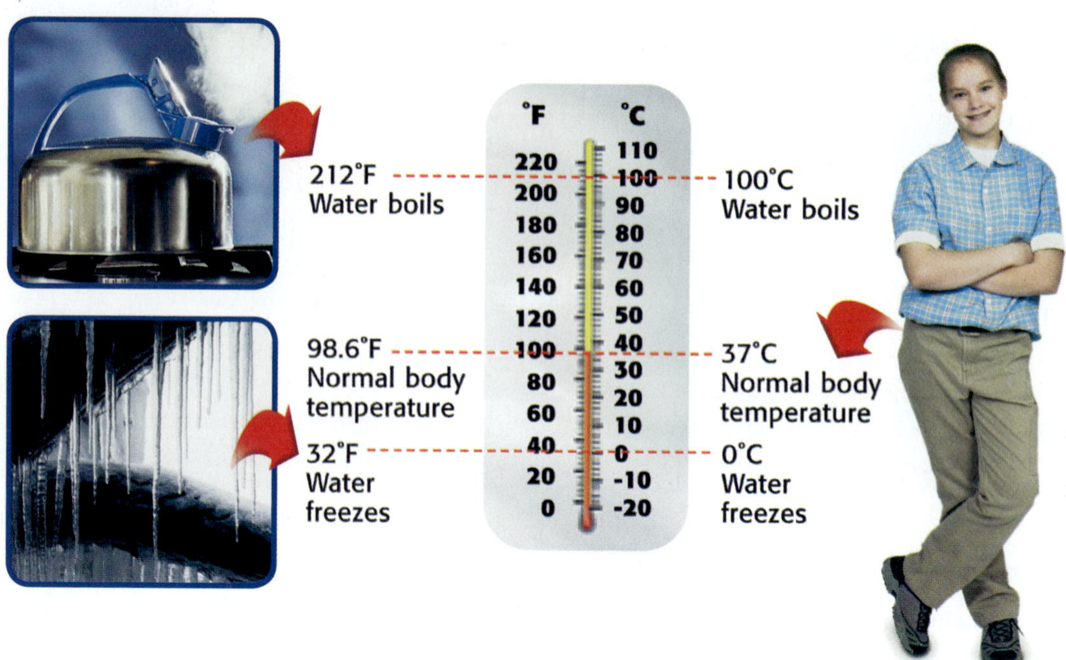

temperature the measure of how hot (or cold) something is

density the ratio of the mass of a substance to the volume of the substance

Temperature

How hot is melted iron? To answer this question, a scientist would measure the temperature of the liquid metal. **Temperature** is a measure of how hot or cold something is. You probably use degrees Fahrenheit (°F) to describe temperature. Scientists commonly use degrees Celsius (°C), although the kelvin (K) is the official SI base unit for temperature. You will use degrees Celsius in this book. The thermometer in **Figure 4** compares the Fahrenheit and Celsius scales.

Density

If you measure the mass and volume of an object, you have the measurements that you need to find the density of the object. **Density** is the amount of matter in a given volume. You cannot measure density directly. But after you have measured the mass and the volume, you can use the following equation to calculate density:

$$density = \frac{mass}{volume}$$

Density is the ratio of mass to volume, so units often used for density are grams per milliliter (g/mL) and grams per cubic centimeter (g/cm^3). Density may be difficult to understand. Think of a table-tennis ball and a golf ball. They have similar volumes. But a golf ball has more mass than a table-tennis ball does. So the golf ball has a greater density.

CONNECTION TO Social Studies

Archimedes (287 BCE–212 BCE) Archimedes was a Greek mathematician. He was probably the greatest mathematician and scientist that classical Greek civilization produced and is considered to be one of the greatest mathematicians of all time. Archimedes was very interested in putting his theoretical discoveries to practical use. Use the library or Internet to research Archimedes. Make a poster that illustrates one of his scientific or mathematical discoveries.

Safety Rules!

Science is exciting and fun, but it can also be dangerous. Don't take any chances! Always follow your teacher's instructions. Don't take shortcuts—even when you think there is no danger. Before starting an experiment, get your teacher's permission. Read the lab procedures carefully. Pay special attention to safety information and caution statements. **Figure 5** shows the safety symbols used in this book. Get to know these symbols and their meanings. Do so by reading the safety information in the front of this book. **This is important!** If you are still unsure about what a safety symbol means, ask your teacher.

Reading Check **Why are safety symbols important?**

Figure 5 **Safety Symbols**

Eye Protection

Clothing Protection

Hand Safety

Heating Safety

Electric Safety

Sharp Object

Chemical Safety

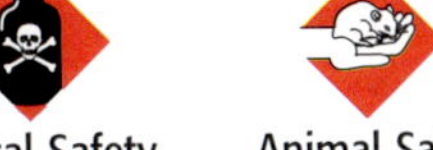
Animal Safety

Plant Safety

SECTION Review

Summary

- Scientists use a variety of tools to measure and analyze the world around them.
- The International System of Units (SI) is a simple, reliable, and uniform system of measurement that is used by most scientists.
- The basic units of measurement in the SI are the meter (for length), the kilogram (for mass), and the Kelvin (for temperature).
- Before starting any science activity or science lab, review the safety symbols and the safety rules for that activity or lab. Don't take chances with your health and safety.

Using Key Terms

Complete each of the following sentences by choosing the correct term from the word bank.

mass	area
volume	temperature

1. A measure of the size of a surface or a region is called ___.

2. Scientists use kilograms when measuring an object's ___.

3. The ___ of a liquid is usually described in liters.

Understanding Key Ideas

4. SI units are
- **a.** based on standardized measurements of body parts.
- **b.** almost always based on the number 10.
- **c.** used to measure only length.
- **d.** used only in France.

5. What is temperature?

6. If you wanted to measure the mass of a fly, which SI unit would be most appropriate?

Math Skills

7. What is the area of a soccer field that is 110 m long and 85 m wide?

8. What is the density of silver if a 6 cm^3 piece of silver has a mass of 63 g?

Critical Thinking

9. Applying Concepts Some people are thinking about sending humans to the moon and then to the planet Mars. Why is it important for scientists around the world to use the International System of Units as they make these plans?

10. Making Inferences Give an example of something that can happen if you do not follow safety rules.

11. Applying Concepts What tool would you use to measure the mass of the air in a basketball?

Using Scientific Methods

Skills Practice Lab

OBJECTIVES

Apply scientific methods to predict, measure, and observe the mixing of two unknown liquids.

MATERIALS

- beakers, 100 mL (2)
- Celsius thermometer
- glass-labeling marker
- graduated cylinders, 50 mL (3)
- liquid A, 75 mL
- liquid B, 75 mL
- protective gloves

SAFETY

Does It All Add Up?

Your math teacher won't tell you this, but did you know that sometimes 2 + 2 does not appear to equal 4?! In this experiment, you will use scientific methods to predict, measure, and observe the mixing of two unknown liquids. You will learn that a scientist does not set out to prove a hypothesis but to test it and that sometimes the results just don't seem to add up!

Make Observations

1. Put on your safety goggles, gloves, and lab apron. Examine the beakers of liquids A and B provided by your teacher. Write down as many observations as you can about each liquid. **Caution:** Do not taste, touch, or smell the liquids.

2. Pour exactly 25 mL of liquid A from the beaker into each of two 50 mL graduated cylinders. Combine these samples in one of the graduated cylinders. Record the final volume. Pour the liquid back into the beaker of liquid A. Rinse the graduated cylinders. Repeat this step for liquid B.

Form a Hypothesis

3. Based on your observations and on prior experience, formulate a testable hypothesis that states what you expect the volume to be when you combine 25 mL of liquid A with 25 mL of liquid B.

4. Make a prediction based on your hypothesis. Use an if-then format. Explain the basis for your prediction.

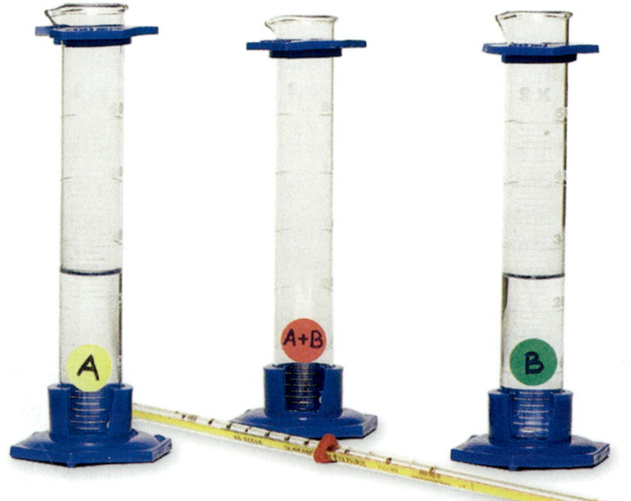

Data Table				
	Contents of cylinder A	Contents of cylinder B	Mixing results: predictions	Mixing results: observations
Volume				
Appearance				
Temperature				

Test the Hypothesis

5. Make a data table like the one above.
6. Mark one graduated cylinder "A." Carefully pour exactly 25 mL of liquid A into this cylinder. In your data table, record its volume, appearance, and temperature.
7. Mark another graduated cylinder "B." Carefully pour exactly 25 mL of liquid B into this cylinder. Record its volume, appearance, and temperature in your data table.
8. Mark the empty third cylinder "A + B."
9. In the "Mixing results: predictions" column in your table, record the prediction you made earlier. Each classmate may have made a different prediction.
10. Carefully pour the contents of both cylinders into the third graduated cylinder.
11. Observe and record the total volume, appearance, and temperature in the "Mixing results: observations" column of your table.

Analyze the Results

1. **Analyzing Data** Discuss your predictions as a class. How many different predictions were there? Which predictions were supported by testing? Did any measurements surprise you?

Draw Conclusions

2. **Drawing Conclusions** Was your hypothesis supported or disproven? Either way, explain your thinking. Describe everything that you think you learned from this experiment.
3. **Analyzing Methods** Explain the value of incorrect predictions.

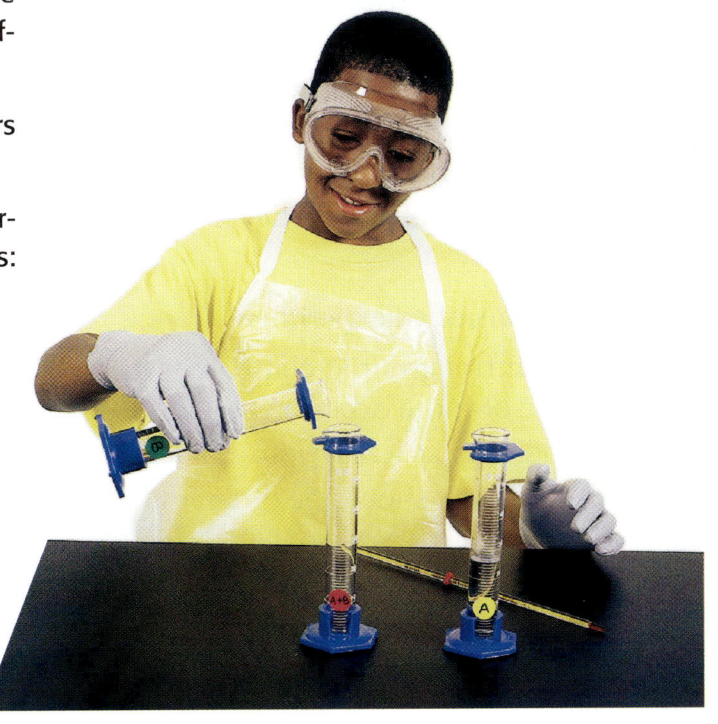

Chapter Review

USING KEY TERMS

1. Use the following terms in the same sentence: *science* and *scientific methods*.

2. Use the following terms in the same sentence: *controlled experiment* and *variable*.

For each pair of terms, explain how the meanings of the terms differ.

3. *theory* and *hypothesis*

4. *controlled experiment* and *variable*

5. *area* and *volume*

6. *physical model* and *conceptual model*

UNDERSTANDING KEY IDEAS

Multiple Choice

7. The steps of scientific methods
   - **a.** must all be used in every scientific investigation.
   - **b.** must always be used in the same order.
   - **c.** often start with a question.
   - **d.** always result in the development of a theory.

8. In a controlled experiment,
   - **a.** a control group is compared with one or more experimental groups.
   - **b.** there are at least two variables.
   - **c.** all factors should be different.
   - **d.** a variable is not needed.

9. Which of the following tools is best for measuring 100 mL of water?
   - **a.** 10 mL graduated cylinder
   - **b.** 150 mL graduated cylinder
   - **c.** 250 mL beaker
   - **d.** 500 mL beaker

10. Which of the following is NOT an SI unit?
    - **a.** meter
    - **b.** foot
    - **c.** liter
    - **d.** kilogram

11. A pencil is 14 cm long. How many millimeters long is it?
    - **a.** 1.4 mm
    - **b.** 140 mm
    - **c.** 1,400 mm
    - **d.** 1,400,000 mm

12. The directions for a lab include the safety icons shown below. These icons mean that

    - **a.** you should be careful.
    - **b.** you are going into the laboratory.
    - **c.** you should wash your hands first.
    - **d.** you should wear safety goggles, a lab apron, and gloves during the lab.

Short Answer

13. List three ways that science is beneficial to living things.

14. Why do hypotheses need to be testable?

15. Give an example of how a scientist might use computers and technology.

16. List three types of models, and give an example of each.

17. What are some advantages and limitations of models?

18. Which SI units can be used to describe the volume of an object? Which SI units can be used to describe the mass of an object?

19. In a controlled experiment, why should there be several individuals in the control group and in each of the experimental groups?

CRITICAL THINKING

20. **Concept Mapping** Use the following terms to create a concept map: *observations, predictions, questions, controlled experiments, variable,* and *hypothesis*.

21. **Making Inferences** Investigations often begin with observation. What limits the observations that scientists can make?

22. **Forming Hypotheses** A scientist who studies mice makes the following observation: on the day the mice are fed vitamins with their meals, they perform better in mazes. What hypothesis would you form to explain this phenomenon? Write a testable prediction based on your hypothesis.

INTERPRETING GRAPHICS

The pictures below show how an egg can be measured by using a beaker and water. Use the pictures below to answer the questions that follow.

Before: 125 mL After: 200 mL

23. What kind of measurement is being taken?
    - **a.** area
    - **b.** length
    - **c.** mass
    - **d.** volume

24. Which of the following is an accurate measurement of the egg in the picture?
    - **a.** 75 cm^3
    - **b.** 125 cm^3
    - **c.** 125 mL
    - **d.** 200 mL

25. Make a double line graph using the data in the table below.

Number of Frogs		
Date	**Normal**	**Deformed**
1995	25	0
1996	21	0
1997	19	1
1998	20	2
1999	17	3
2000	20	5

Standardized Test Preparation

READING

Read each of the passages below. Then, answer the questions that follow the passage.

Passage 1 Zoology is the study of animals. Zoology dates back more than 2,300 years, to ancient Greece. There, the philosopher Aristotle observed and theorized about animal behavior. About 200 years later, Galen, a Greek physician, began dissecting and experimenting with animals. However, there were few advances in zoology until the 1700s and 1800s. During this period, the Swedish naturalist Carolus Linnaeus developed a classification system for plants and animals, and British naturalist Charles Darwin published his theory of evolution by natural selection.

1. According to the passage, when did major advances in Zoology begin?

- **A** About 2,300 years ago
- **B** About 2,100 years ago
- **C** During the 1700s and 1800s
- **D** Only during recent history

2. Which of the following is a possible meaning of the word *naturalist,* as used in the passage?

- **F** a scientist who studies plants and animals
- **G** a scientist who studies animals
- **H** a scientist who studies theory
- **I** a scientist who studies animal behavior

3. Which of the following is the **best** title for this passage?

- **A** Greek Zoology
- **B** Modern Zoology
- **C** The Origins of Zoology
- **D** Zoology in the 1700s and 1800s

Passage 2 When looking for answers to a problem, scientists build on existing knowledge. For example, scientists have wondered if there is some relationship between Earth's core and Earth's magnetic field. To form a hypothesis, scientists started with what they knew: Earth has a dense, solid inner core and a molten outer core. Scientists then created a computer model to simulate how Earth's magnetic field might be generated.

They tried different things with their model until the model produced a magnetic field that matched that of the real Earth. The model predicted that Earth's inner core spins in the same direction as the rest of the Earth, but the inner core spins slightly faster than Earth's surface. If the hypothesis is correct, it might explain how Earth's magnetic field is produced. Although scientists cannot reach the Earth's core to examine it directly, they can test whether other observations match what is predicted by their hypothesis.

1. What does the word *model* refer to in this passage?

- **A** a giant plastic globe
- **B** a representation of the Earth created on a computer
- **C** a computer terminal
- **D** a technology used to drill into the Earth's core

2. Which of the following is the **best** summary of the passage?

- **F** Scientists can use models to help them answer difficult and complex questions.
- **G** Scientists have discovered the source of Earth's magnetic field.
- **H** The spinning of Earth's molten inner core causes Earth's magnetic field.
- **I** Scientists make a model of a problem and then ask questions about the problem.

INTERPRETING GRAPHICS

The table below shows the plans for an experiment in which bees will be observed visiting flowers. Use the table to answer the questions that follow.

Bee Experiment				
Group	Type of bee	Time of day	Type of plant	Flower color
#1	Honey-bee	9:00 A.M.–10:00 A.M.	Portland rose	red
#2	Honey-bee	9:00 A.M.–10:00 A.M.	Portland rose	yellow
#3	Honey-bee	9:00 A.M.–10:00 A.M.	Portland rose	white
#4	Honey-bee	9:00 A.M.–10:00 A.M.	Portland rose	pink

1. Which factor is the variable in this experiment?

A the type of bee
B the time of day
C the type of plant
D the color of the flowers

2. Which of the following hypotheses could be tested by this experiment?

F Honeybees prefer to visit rose plants.
G Honeybees prefer to visit red flowers.
H Honeybees prefer to visit flowers in the morning.
I Honey bees prefer to visit Portland rose flowers between 9 and 10 A.M.

3. Which of the following is the **best** reason why the Portland rose plant is included in all of the groups to be studied?

A The type of plant is a control factor; any type of flowering plant could be used as long as all plants were of the same type.
B The experiment will test whether bees prefer the Portland rose over other flowers.
C An experiment should always have more than one variable.
D The Portland rose is a very common plant.

MATH

Read each question below, and choose the best answer.

1. A survey of students was conducted to find out how many people were in each student's family. The replies from five students were as follows: 3, 3, 4, 4, and 6. What was the average family size?

A 3
B 3.5
C 4
D 5

2. In the survey above, if one more student were surveyed, which reply would make the average lower?

F 3
G 4
H 5
I 6

3. If an object that is 5 µm long were magnified by 1,000, how long would that object then appear?

A 5 µm
B 5 mm
C 1,000 µm
D 5,000 mm

4. How many meters are in 50 km?

F 50 m
G 500 m
H 5,000 m
I 50,000 m

5. What is the area of a square whose sides measure 4 m each?

A 16 m
B 16 m^2
C 32 m
D 32 m^2

Science in Action

Scientific Debate

Should We Stop All Forest Fires?

Since 1972, the policy of the National Park Service has been to manage the national parks as naturally as possible. Because fire is a natural event in forests, this policy includes allowing most fires caused by lightning to burn. The only lightning-caused fires that are put out are those that threaten lives, property, uniquely scenic areas, or endangered species. All human-caused fires are put out. However, this policy has caused some controversy. Some people want this policy followed in all public forests and even grasslands. Others think that all fires should be put out.

Social Studies ACTIVITY

WRITING SKILL Research a location where there is a debate about controlling forest fires. You might look into national forests or parks. Write a newspaper article about the issue. Be sure to present all sides of the debate.

Science Fiction

"The Homesick Chicken" by Edward D. Hoch

Why did the chicken cross the road? You think you know the answer to this old riddle, don't you? But "The Homesick Chicken," by Edward D. Hoch, may surprise you. That old chicken may not be exactly what it seems.

You see, one of the chickens at the high-tech Tangaway Research Farms has escaped. Then, it was found in a vacant lot across the highway from Tangaway, pecking away contentedly. Why did it bother to escape? Barnabus Rex, a specialist in solving scientific riddles, is called in to work on this mystery. As he investigates, he finds clues and forms a hypothesis. Read the story, and see if you can explain the mystery before Mr. Rex does.

Language Arts ACTIVITY

WRITING SKILL Write your own short story about a chicken crossing a road for a mysterious reason. Give clues (evidence) to the reader about the mysterious reason but do not reveal the truth until the end of the story. Be sure the story makes sense scientifically.

People in Science

Matthew Henson

Arctic Explorer Matthew Henson was born in Maryland in 1866. His parents were freeborn sharecroppers. When Henson was a young boy, his parents died. He then went to look for work as a cabin boy on a ship. Several years later, Henson had traveled around the world and had become educated in the areas of geography, history, and mathematics. In 1898, Henson met U.S. Naval Lieutenant Robert E. Peary. Peary was the leader of Arctic expeditions between 1886 and 1909.

Peary asked Henson to accompany him as a navigator on several trips, including trips to Central America and Greenland. One of Peary's passions was to be the first person to reach the North Pole. It was Henson's vast knowledge of mathematics and carpentry that made Peary's trek to the North Pole possible. In 1909, Henson was the first person to reach the North Pole. Part of Henson's job as navigator was to drive ahead of the party and blaze the first trail. As a result, he often arrived ahead of everyone else. On April 6, 1909, Henson reached the approximate North Pole 45 minutes ahead of Peary. Upon his arrival, he exclaimed, "I think I'm the first man to sit on top of the world!"

Math ACTIVITY

On the last leg of their journey, Henson and Peary traveled 664.5 km in 16 days! On average, how far did Henson and Peary travel each day?

To learn more about these Science in Action topics, visit go.hrw.com and type in the keyword HZ5SW7F.

Current Science

Check out Current Science® articles related to this chapter by visiting go.hrw.com. Just type in the keyword HZ5CS01.

TIMELINE

Earth Science

In this unit, you will learn about the basic components of the solid Earth—rocks and the minerals from which they are made. You will also learn about other resources the Earth contains. The ground beneath your feet is a treasure-trove of interesting materials, some of which are very valuable. Secrets of Earth's history are also hidden within the ground's depths. This timeline shows some of the events that have occurred through human history as scientists have come to understand more about our planet.

1543
Nicolaus Copernicus argues that the sun rather than the Earth is the center of the universe.

1860
Fossil remains of *Archaeopteryx*, a species that may link reptiles and birds, are discovered in Germany.

1936
Hoover Dam is completed. This massive hydroelectric dam, standing more than 221 m, required 3.25 million cubic yards of concrete to build.

1975
Tabei Junko of Japan becomes the first woman to successfully climb Mount Everest, 22 years after Edmund Hillary and Tenzing Norgay first conquered the mountain in 1953.

1681

The dodo, a flightless bird, is driven to extinction by the actions of humans.

1739

Georg Brandt identifies a new element and names it cobalt.

1848

James Marshall discovers gold at Sutter's Mill, in California, beginning the California gold rush. Prospectors during the gold rush of the following year are referred to as "forty-niners."

1947

Willard F. Libby develops a method of dating prehistoric objects by using radioactive carbon.

1955

Using 1 million pounds of pressure per square inch and temperatures of more than 1,700°C, General Electric creates the first artificial diamonds from graphite.

1969

Apollo 11 astronauts Neil Armstrong and Edwin "Buzz" Aldrin bring 22 kg of moon rocks and soil back to the Earth.

1989

Russian engineers drill a borehole 12 km into the Earth's crust. The borehole is more than 3 times deeper than the deepest mine shaft.

1997

Sojourner, a roving probe on Mars, investigates a Martian boulder nicknamed Yogi.

1999

A Japanese automaker introduces the first hybrid car into the U.S. market.

Rocks: Mineral Mixtures

About the 

Irish legend claims that the mythical hero Finn MacCool built the Giant's Causeway, shown here. But this rock formation is the result of the cooling of huge amounts of molten rock. As the molten rock cooled, it formed tall pillars separated by cracks called *columnar joints*.

PRE-READING ACTIVITY

Graphic Organizer

Spider Map Before you read the chapter, create the graphic organizer entitled "Spider Map" described in the **Study Skills** section of the Appendix. Label the circle "Rock." Create a leg for each of the sections in this chapter. As you read the chapter, fill in the map with details about the material presented in each section of the chapter.

START-UP ACTIVITY

Classifying Objects

Scientists use the physical and chemical properties of rocks to classify rocks. Classifying objects such as rocks requires looking at many properties. Do this exercise for some classification practice.

Procedure

1. Your teacher will give you a **bag** containing **several objects.** Examine the objects, and note features such as size, color, shape, texture, smell, and any unique properties.
2. Develop three different ways to sort these objects.
3. Create a chart that organizes objects by properties.

Analysis

1. What properties did you use to sort the items?
2. Were there any objects that could fit into more than one group? How did you solve this problem?
3. Which properties might you use to classify rocks? Explain your answer.

SECTION 1

The Rock Cycle

You know that paper, plastic, and aluminum can be recycled. But did you know that the Earth also recycles? And one of the things that Earth recycles is rock.

READING WARM-UP

Objectives

- Describe two ways rocks have been used by humans.
- Describe four processes that shape Earth's features.
- Describe how each type of rock changes into another type as it moves through the rock cycle.
- List two characteristics of rock that are used to help classify it.

Terms to Learn

rock cycle	deposition
rock	composition
erosion	texture

READING STRATEGY

Reading Organizer As you read this section, make a flowchart of the steps of the rock cycle.

Scientists define **rock** as a naturally occurring solid mixture of one or more minerals and organic matter. It may be hard to believe, but rocks are always changing. The continual process by which new rock forms from old rock material is called the **rock cycle.**

The Value of Rock

Rock has been an important natural resource as long as humans have existed. Early humans used rocks as hammers to make other tools. They discovered that they could make arrowheads, spear points, knives, and scrapers by carefully shaping rocks such as chert and obsidian.

Rock has also been used for centuries to make buildings, monuments, and roads. **Figure 1** shows how rock has been used as a construction material by both ancient and modern civilizations. Buildings have been made out of granite, limestone, marble, sandstone, slate, and other rocks. Modern buildings also contain concrete and plaster, in which rock is an important ingredient.

Reading Check **Name some types of rock that have been used to construct buildings.** (*See the Appendix for answers to Reading Checks.*)

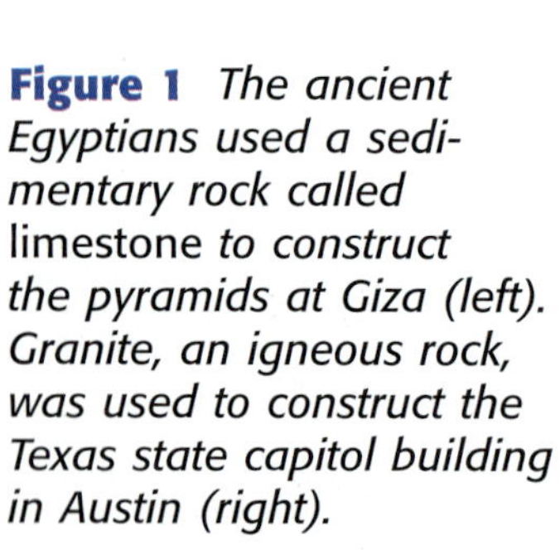

Figure 1 *The ancient Egyptians used a sedimentary rock called* limestone *to construct the pyramids at Giza (left). Granite, an igneous rock, was used to construct the Texas state capitol building in Austin (right).*

Processes That Shape the Earth

Certain geological processes make and destroy rock. These processes shape the features of our planet. These processes also influence the type of rock that is found in a certain area of Earth's surface.

rock a naturally occurring solid mixture of one or more minerals or organic matter

rock cycle the series of processes in which a rock forms, changes from one type to another, is destroyed, and forms again by geological processes

erosion the process by which wind, water, ice, or gravity transports soil and sediment from one location to another

deposition the process in which material is laid down

Weathering, Erosion, and Deposition

The process in which water, wind, ice, and heat break down rock is called *weathering*. Weathering is important because it breaks down rock into fragments. These rock and mineral fragments are the sediment of which much sedimentary rock is made.

The process by which sediment is removed from its source is called **erosion.** Water, wind, ice, and gravity can erode and move sediments and cause them to collect. **Figure 2** shows an example of the way land looks after weathering and erosion.

The process in which sediment moved by erosion is dropped and comes to rest is called **deposition.** Sediment is deposited in bodies of water and other low-lying areas. In those places, sediment may be pressed and cemented together by minerals dissolved in water to form sedimentary rock.

Heat and Pressure

Sedimentary rock made of sediment can also form when buried sediment is squeezed by the weight of overlying layers of sediment. If the temperature and pressure are high enough at the bottom of the sediment, the rock can change into metamorphic rock. In some cases, the rock gets hot enough to melt. This melting creates the magma that eventually cools to form igneous rock.

How the Cycle Continues

Buried rock is exposed at the Earth's surface by a combination of uplift and erosion. *Uplift* is movement within the Earth that causes rocks inside the Earth to be moved to the Earth's surface. When uplifted rock reaches the Earth's surface, weathering, erosion, and deposition begin.

Figure 2 *Bryce Canyon, in Utah, is an excellent example of how the processes of weathering and erosion shape the face of our planet.*

Illustrating the Rock Cycle

You have learned about various geological processes, such as weathering, erosion, heat, and pressure, that create and destroy rock. The diagram on these two pages illustrates one way that sand grains can change as different geological processes act on them. In the following steps, you will see how these processes change the original sand grains into sedimentary rock, metamorphic rock, and igneous rock.

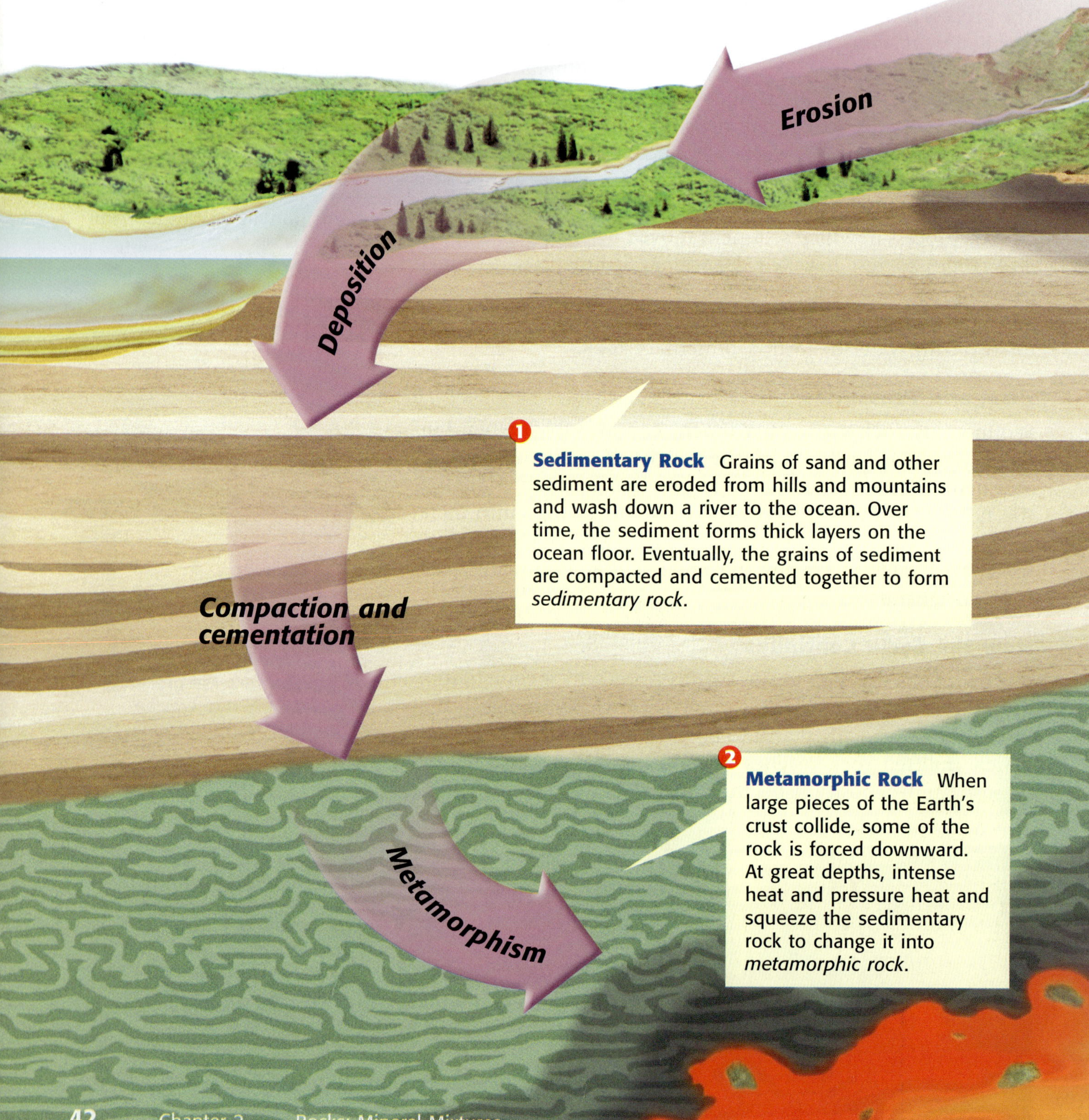

1 Sedimentary Rock Grains of sand and other sediment are eroded from hills and mountains and wash down a river to the ocean. Over time, the sediment forms thick layers on the ocean floor. Eventually, the grains of sediment are compacted and cemented together to form *sedimentary rock*.

2 Metamorphic Rock When large pieces of the Earth's crust collide, some of the rock is forced downward. At great depths, intense heat and pressure heat and squeeze the sedimentary rock to change it into *metamorphic rock*.

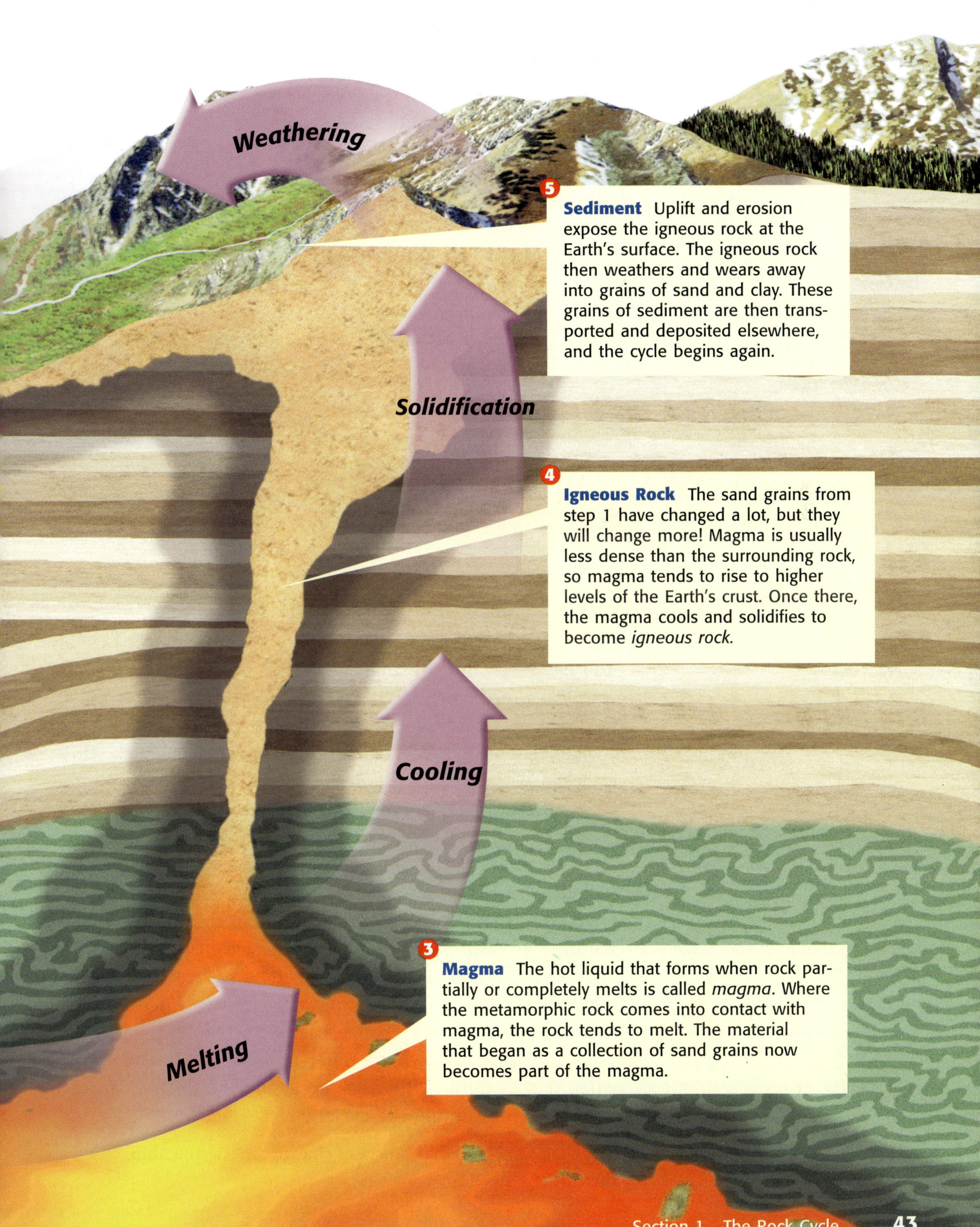
Weathering
Sediment Uplift and erosion expose the igneous rock at the Earth's surface. The igneous rock then weathers and wears away into grains of sand and clay. These grains of sediment are then transported and deposited elsewhere, and the cycle begins again.
Solidification
Igneous Rock The sand grains from step 1 have changed a lot, but they will change more! Magma is usually less dense than the surrounding rock, so magma tends to rise to higher levels of the Earth's crust. Once there, the magma cools and solidifies to become igneous rock.
Cooling
Magma The hot liquid that forms when rock partially or completely melts is called magma. Where the metamorphic rock comes into contact with magma, the rock tends to melt. The material that began as a collection of sand grains now becomes part of the magma.
Melting

Figure 3 The Rock Cycle

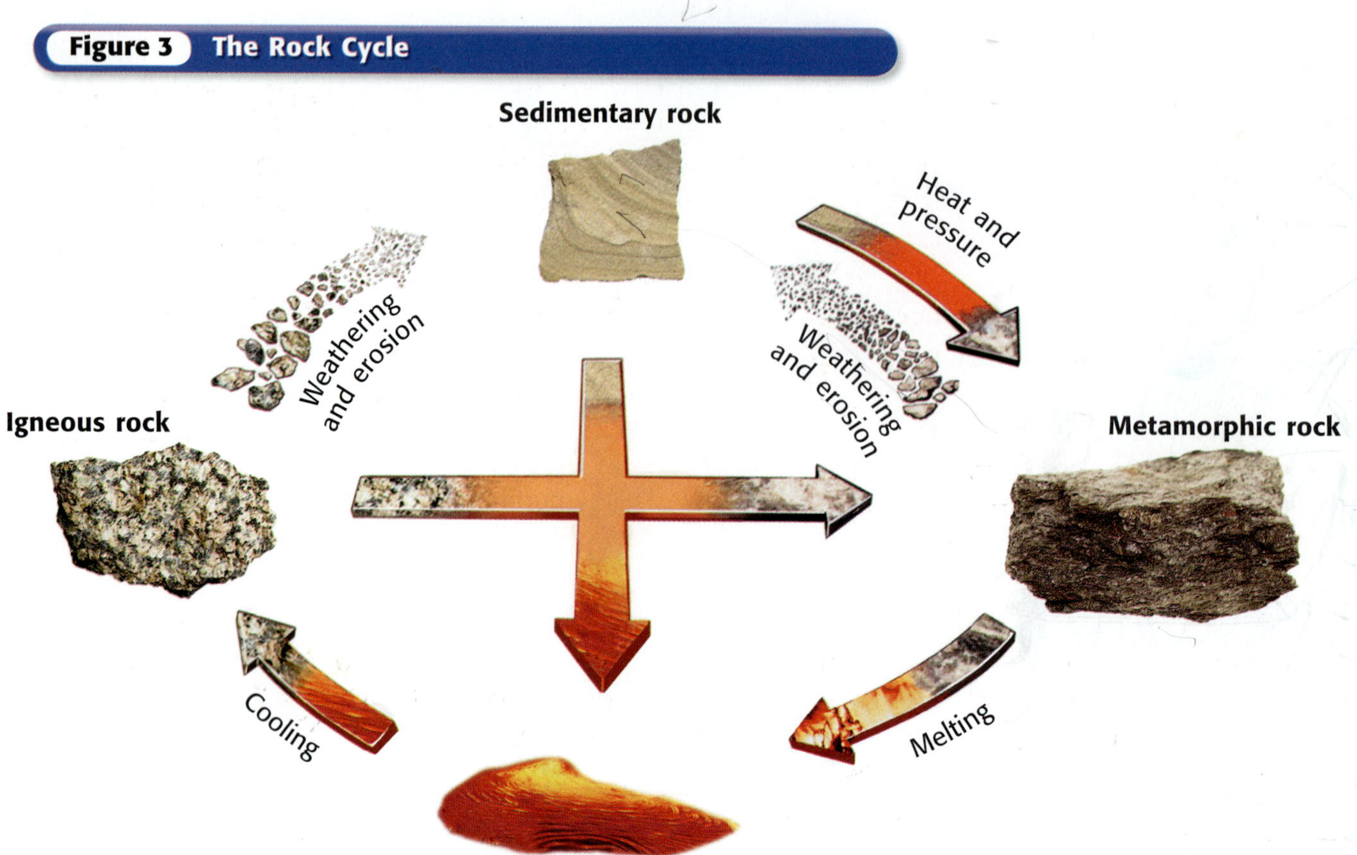

Round and Round It Goes

You have seen how different geological processes can change rock. Each rock type can change into one of the three types of rock. For example, igneous rock can change into sedimentary rock, metamorphic rock, or even back into igneous rock. This cycle, in which rock is changed by geological processes into different types of rock, is known as the rock cycle.

Rocks may follow various pathways in the rock cycle. As one rock type is changed to another type, several variables, including time, heat, pressure, weathering, and erosion may alter a rock's identity. The location of a rock determines which natural forces will have the biggest impact on the process of change. For example, rock at the Earth's surface is primarily affected by forces of weathering and erosion, whereas deep inside the Earth, rocks change because of extreme heat and pressure. **Figure 3** shows the different ways rock may change when it goes through the rock cycle and the different forces that affect rock during the cycle.

Reading Check What processes change rock deep within the Earth?

Rock Classification

You have already learned that scientists divide all rock into three main classes based on how the rock formed: igneous, sedimentary, and metamorphic. But did you know that each class of rock can be divided further? These divisions are also based on differences in the way rocks form. For example, all igneous rock forms when magma cools and solidifies. But some igneous rocks form when magma cools *on* the Earth's surface, and others form when magma cools deep *beneath* the surface. Therefore, igneous rock can be divided again based on how and where it forms. Sedimentary and metamorphic rocks are also divided into groups. How do scientists know how to classify rocks? They study rocks in detail using two important criteria—composition and texture.

What's in It?

Assume that a granite sample you are studying is made of 30% quartz and 55% feldspar by volume. The rest is made of biotite mica. What percentage of the sample is biotite mica?

Composition

The minerals a rock contains determine the **composition** of that rock, as shown in **Figure 4.** For example, a rock made of mostly the mineral quartz will have a composition very similar to that of quartz. But a rock made of 50% quartz and 50% feldspar will have a very different composition than quartz does.

composition the chemical makeup of a rock; describes either the minerals or other materials in the rock

Reading Check What determines a rock's composition?

Figure 4 Two Examples of Rock Composition

The composition of a rock depends on the minerals the rock contains.

Figure 5 Three Examples of Sedimentary Rock Texture

Fine-grained

Siltstone

Medium-grained

Sandstone

Coarse-grained

Conglomerate

Texture

texture the quality of a rock that is based on the sizes, shapes, and positions of the rock's grains

The size, shape, and positions of the grains that make up a rock determine a rock's **texture.** Sedimentary rock can have a fine-grained, medium-grained, or coarse-grained texture, depending on the size of the grains that make up the rock. Three samples of textures are shown in **Figure 5.** The texture of igneous rock can be fine-grained or coarse-grained, depending on how much time magma has to cool. Based on the degree of temperature and pressure a rock is exposed to, metamorphic rock can also have a fine-grained or coarse-grained texture.

The texture of a rock can provide clues as to how and where the rock formed. Look at the rocks shown in **Figure 6.** The rocks look different because they formed in very different ways. The texture of a rock can reveal the process that formed it.

Reading Check **Give three examples of sedimentary rock textures.**

Figure 6 Texture and Rock Formation

Basalt, a fine-grained igneous rock, forms when lava that erupts onto Earth's surface cools rapidly.

Sandstone, a medium-grained sedimentary rock, forms when sand grains deposited in dunes, on beaches, or on the ocean floor are buried and cemented.

SECTION Review

Summary

- Rock has been an important natural resource for as long as humans have existed. Early humans used rock to make tools. Ancient and modern civilizations have used rock as a construction material.
- Weathering, erosion, deposition, and uplift are all processes that shape the surface features of the Earth.
- The rock cycle is the continual process by which new rock forms from old rock material.
- The sequence of events in the rock cycle depends on processes, such as weathering, erosion, deposition, pressure, and heat, that change the rock material.
- Composition and texture are two characteristics that scientists use to classify rocks.
- The composition of a rock is determined by the minerals that make up the rock.
- The texture of a rock is determined by the size, shape, and positions of the grains that make up the rock.

Using Key Terms

Complete each of the following sentences by choosing the correct term from the word bank.

rock	composition
rock cycle	texture

1. The minerals that a rock is made of determine the ___ of that rock.

2. ___ is a naturally occurring, solid mixture of crystals of one or more minerals.

Understanding Key Ideas

3. Sediments are transported or moved from their original source by a process called

- **a.** deposition.
- **b.** erosion.
- **c.** uplift.
- **d.** weathering.

4. Describe two ways that rocks have been used by humans.

5. Name four processes that change rock inside the Earth.

6. Describe four processes that shape Earth's surface.

7. Give an example of how texture can provide clues as to how and where a rock formed.

Critical Thinking

8. Making Comparisons Explain the difference between texture and composition.

9. Analyzing Processes Explain how rock is continually recycled in the rock cycle.

Interpreting Graphics

10. Look at the table below. Sandstone is a type of sedimentary rock. If you had a sample of sandstone that had an average particle size of 2 mm, what texture would your sandstone have?

Classification of Clastic Sedimentary Rocks

Texture	Particle size
coarse grained	> 2 mm
medium grained	0.06 to 2 mm
fine grained	< 0.06 mm

SECTION 2

Igneous Rock

Where do igneous rocks come from? Here's a hint: The word igneous *comes from a Latin word that means "fire."*

READING WARM-UP

Objectives

- Describe three ways that igneous rock forms.
- Explain how the cooling rate of magma affects the texture of igneous rock.
- Distinguish between igneous rock that cools within Earth's crust and igneous rock that cools at Earth's surface.

Terms to Learn

intrusive igneous rock
extrusive igneous rock

READING STRATEGY

Reading Organizer As you read this section, make a table comparing intrusive rock and extrusive rock.

Igneous rock forms when hot, liquid rock, or *magma,* cools and solidifies. The type of igneous rock that forms depends on the composition of the magma and the amount of time it takes the magma to cool.

Origins of Igneous Rock

Igneous rock begins as magma. As shown in **Figure 1,** there are three ways magma can form: when rock is heated, when pressure is released, or when rock changes composition.

When magma cools enough, it solidifies to form igneous rock. Magma solidifies in much the same way that water freezes. But there are also differences between the way magma freezes and the way water freezes. One main difference is that water freezes at 0°C. Magma freezes between 700°C and 1,250°C. Also, liquid magma is a complex mixture containing many melted minerals. Because these minerals have different melting points, some minerals in the magma will freeze or become solid before other minerals do.

Figure 1 The Formation of Magma

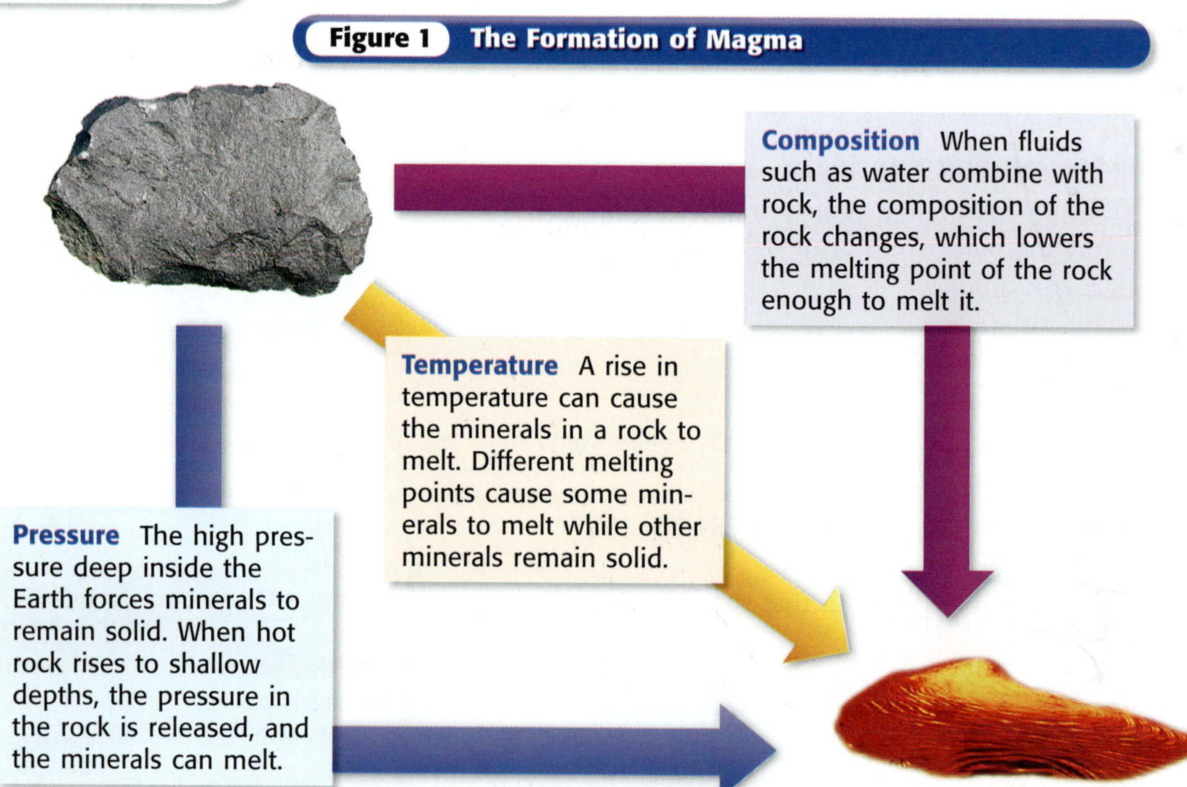

Figure 2 **Igneous Rock Texture**

	Coarse-grained	Fine-grained
Felsic	 Granite	 Rhyolite
Mafic	 Gabbro	 Basalt

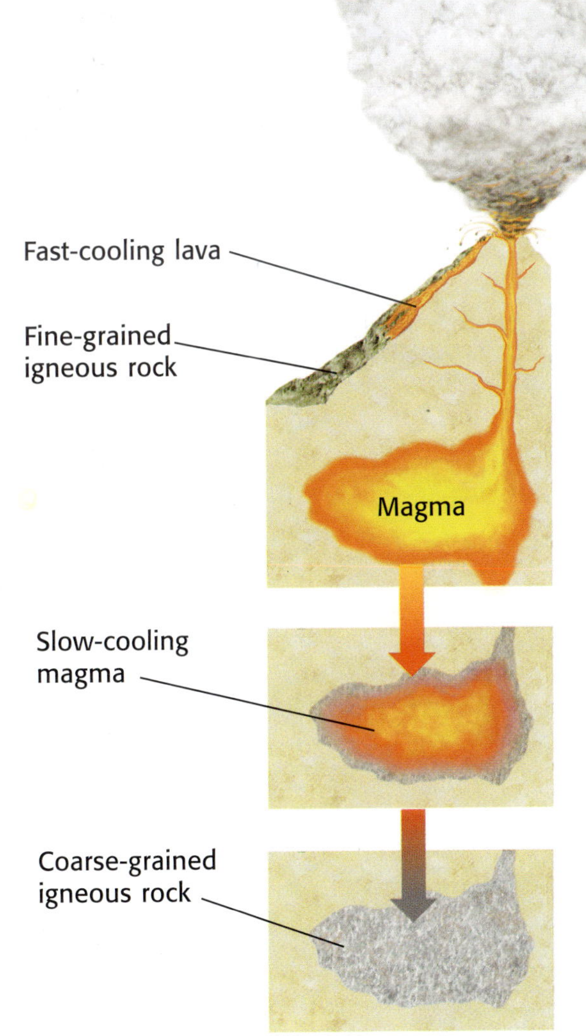

Figure 3 *The amount of time it takes for magma or lava to cool determines the texture of igneous rock.*

Composition and Texture of Igneous Rock

Look at the rocks in **Figure 2.** All of the rocks are igneous rocks even though they look different from one another. These rocks differ from one another in what they are made of and how fast they cooled.

The light-colored rocks are less dense than the dark-colored rocks are. The light-colored rocks are rich in elements such as aluminum, potassium, silicon, and sodium. These rocks are called *felsic rocks*. The dark-colored rocks, called *mafic rocks,* are rich in calcium, iron, and magnesium, and poor in silicon.

Figure 3 shows what happens to magma when it cools at different rates. The longer it takes for the magma or lava to cool, the more time mineral crystals have to grow. The more time the crystals have to grow, the larger the crystals are and the coarser the texture of the resulting igneous rock is.

In contrast, the less time magma takes to cool, the less time crystals have to grow. Therefore, the rock that is formed will be fine grained. Fine-grained igneous rock contains very small crystals, or if the cooling is very rapid, it contains no crystals.

✓ Reading Check **Explain the difference between felsic rock and mafic rock.** (*See the Appendix for answers to Reading Checks.*)

INTERNET ACTIVITY

For another activity related to this chapter, go to **go.hrw.com** and type in the keyword **HZ5RCKW.**

Igneous Rock Formations

Igneous rock formations are located above and below the surface of the Earth. You may be familiar with igneous rock formations that were caused by lava cooling on the Earth's surface, such as volcanoes. But not all magma reaches the surface. Some magma cools and solidifies deep within the Earth's crust.

Intrusive Igneous Rock

intrusive igneous rock rock formed from the cooling and solidification of magma beneath the Earth's surface

When magma *intrudes,* or pushes, into surrounding rock below the Earth's surface and cools, the rock that forms is called **intrusive igneous rock.** Intrusive igneous rock usually has a coarse-grained texture because it is well insulated by surrounding rock and cools very slowly. The minerals that form are large, visible crystals.

Masses of intrusive igneous rock are named for their size and shape. Common intrusive shapes are shown in **Figure 4.** *Plutons* are large, irregular-shaped intrusive bodies. The largest of all igneous intrusions are *batholiths*. *Stocks* are intrusive bodies that are exposed over smaller areas than batholiths. Sheetlike intrusions that cut across previous rock units are called *dikes*, whereas *sills* are sheetlike intrusions that are oriented parallel to previous rock units.

Figure 4 *Igneous intrusive bodies have different shapes and sizes.*

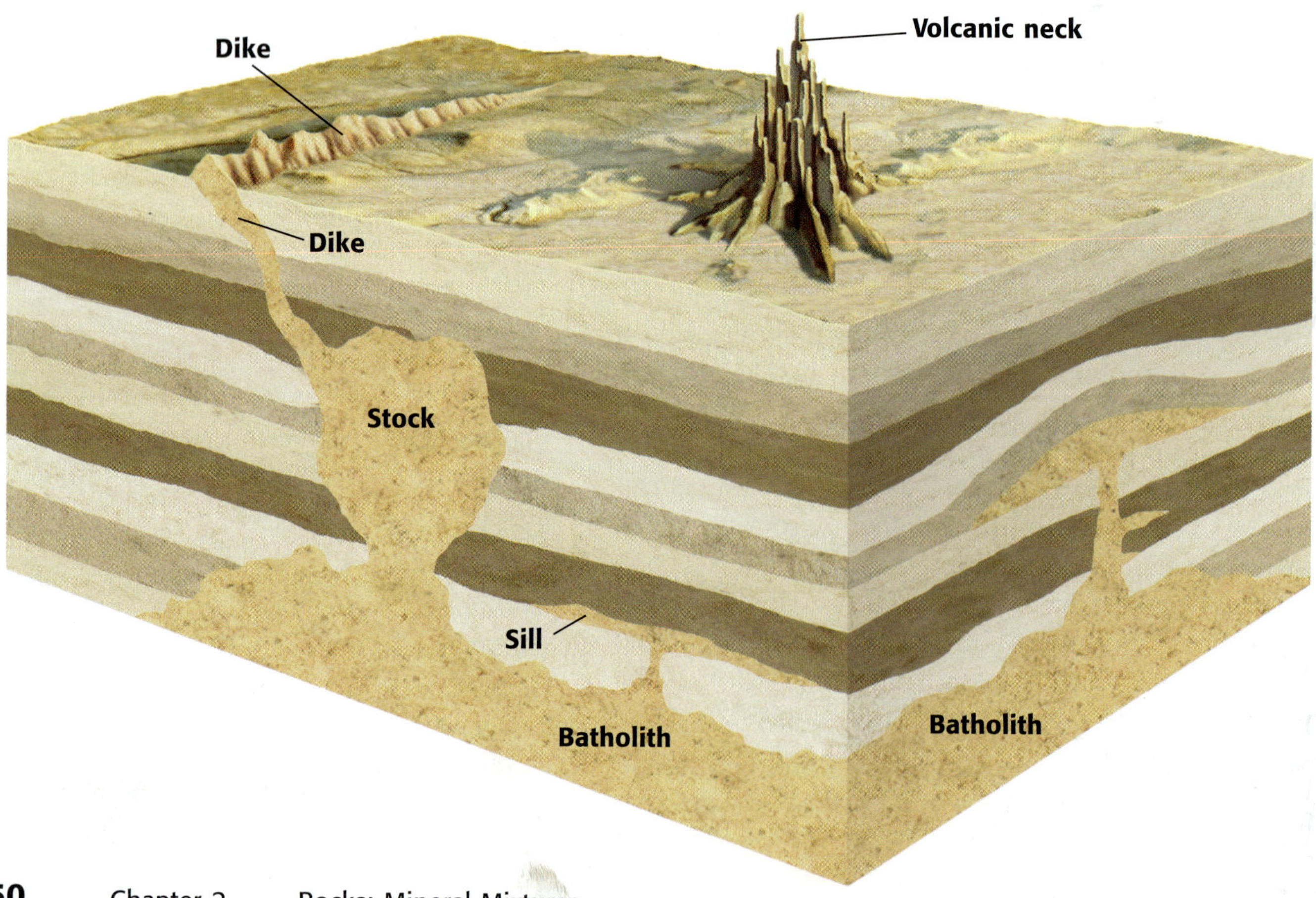

Extrusive Igneous Rock

Igneous rock that forms from magma that erupts, or extrudes, onto the Earth's surface is called **extrusive igneous rock.** Extrusive rock is common around volcanoes. It cools quickly on the surface and contains very small crystals or no crystals.

When lava erupts from a volcano, a *lava flow* forms. **Figure 5** shows an active lava flow. Lava does not always flow from volcanoes. Sometimes lava erupts and flows from long cracks in the Earth's crust called *fissures*. Lava flows from fissures on the ocean floor at places where tension is causing the ocean floor to be pulled apart. This lava cools to form new ocean floor. When a large amount of lava flows out of fissures onto land, the lava can cover a large area and form a plain called a *lava plateau*. Pre-existing landforms are often buried by these lava flows.

Reading Check How does new ocean floor form?

Figure 5 *An active lava flow is shown in this photo. When exposed to Earth's surface conditions, lava quickly cools and solidifies to form a fine-grained igneous rock.*

extrusive igneous rock rock that forms as a result of volcanic activity at or near the Earth's surface

SECTION Review

Summary

- Igneous rock forms when magma cools and hardens.
- The texture of igneous rock is determined by the rate at which the rock cools.
- Igneous rock that solidifies at Earth's surface is extrusive. Igneous rock that solidifies within Earth's surface is intrusive.
- Shapes of common igneous intrusive bodies include batholiths, stocks, sills, and dikes.

Using Key Terms

1. In your own words, write a definition for each of the following terms: *intrusive igneous rock* and *extrusive igneous rock*.

Understanding Key Ideas

2. ___ is an example of a coarse-grained, felsic, igneous rock.
   a. Basalt
   b. Gabbro
   c. Granite
   d. Rhyolite
3. Explain three ways in which magma can form.
4. What determines the texture of igneous rocks?

Math Skills

5. The summit of a granite batholith has an elevation of 1,825 ft. What is the height of the batholith in meters?

Critical Thinking

6. **Making Comparisons** Dikes and sills are both types of igneous intrusive bodies. What is the difference between a dike and a sill?
7. **Predicting Consequences** An igneous rock forms from slow-cooling magma deep beneath the surface of the Earth. What type of texture is this rock most likely to have? Explain.

SECTION 3

Sedimentary Rock

Have you ever tried to build a sand castle at the beach? Did you ever wonder where the sand came from?

Sand is a product of weathering, which breaks rock into pieces. Over time, sand grains may be compacted, or compressed, and then cemented together to form a rock called *sandstone*. Sandstone is just one of many types of sedimentary rock.

READING WARM-UP

Objectives

- Describe the origin of sedimentary rock.
- Describe the three main categories of sedimentary rock.
- Describe three types of sedimentary structures.

Terms to Learn

strata
stratification

READING STRATEGY

Reading Organizer As you read this section, create an outline of this section. Use the headings from the section in your outline.

Origins of Sedimentary Rock

Wind, water, ice, sunlight, and gravity all cause rock to physically weather into fragments. Through the process of erosion, these rock and mineral fragments, called *sediment*, are moved from one place to another. Eventually, the sediment is deposited in layers. As new layers of sediment are deposited, they cover older layers. Older layers become compacted. Dissolved minerals, such as calcite and quartz, separate from water that passes through the sediment to form a natural cement that binds the rock and mineral fragments together into sedimentary rock.

Sedimentary rock forms at or near the Earth's surface. It forms without the heat and pressure that are involved in the formation of igneous and metamorphic rocks.

The most noticeable feature of sedimentary rock is its layers, or **strata.** A single, horizontal layer of rock is sometimes visible for many miles. Road cuts are good places to observe strata. **Figure 1** shows the spectacular views that sedimentary rock formations carved by erosion can provide.

Figure 1 *The red sandstone "monuments" for which Monument Valley in Arizona has been named are the products of millions of years of erosion.*

Composition of Sedimentary Rock

Sedimentary rock is classified by the way it forms. *Clastic sedimentary rock* forms when rock or mineral fragments, called *clasts,* are cemented together. *Chemical sedimentary rock* forms when minerals crystallize out of a solution, such as sea water, to become rock. *Organic sedimentary rock* forms from the remains of once-living plants and animals.

strata layers of rock (singular, *stratum*)

Clastic Sedimentary Rock

Clastic sedimentary rock is made of fragments of rocks cemented together by a mineral such as calcite or quartz. **Figure 2** shows how clastic sedimentary rock is classified according to the size of the fragments from which the rock is made. Clastic sedimentary rocks can have coarse-grained, medium-grained, or fine-grained textures.

Chemical Sedimentary Rock

Chemical sedimentary rock forms from solutions of dissolved minerals and water. As rainwater slowly makes its way to the ocean, it dissolves some of the rock material it passes through. Some of this dissolved material eventually crystallizes and forms the minerals that make up chemical sedimentary rock. Halite, one type of chemical sedimentary rock, is made of sodium chloride, NaCl, or table salt. Halite forms when sodium ions and chlorine ions in shallow bodies of water become so concentrated that halite crystallizes from solution.

Reading Check **How does a chemical sedimentary rock such as halite form?** (*See the Appendix for answers to Reading Checks.*)

WRITING SKILL **Salty Expressions** The word salt is used in many expressions in the English language. Some common examples include "the salt of the earth," "taken with a grain of salt," not worth his salt," "the salt of truth," "rubbing salt into a wound," and "old salt." Use the Internet or another source to research one these expressions. In your research, attempt to find the origin of the expression. Write a short paragraph that summarizes what you found.

Organic Sedimentary Rock

Figure 3 *Ocean animals called* coral *create huge deposits of limestone. As they die, their skeletons collect on the ocean floor.*

Most limestone forms from the remains, or *fossils,* of animals that once lived in the ocean. For example, some limestone is made of the skeletons of tiny organisms called *coral.* Coral are very small, but they live in huge colonies called *reefs,* shown in **Figure 3.** Over time, the skeletons of these sea animals, which are made of calcium carbonate, collect on the ocean floor. These animal remains eventually become cemented together to form *fossiliferous limestone* (FAH suhl IF uhr uhs LIEM STOHN).

Corals are not the only animals whose remains are found in fossiliferous limestone. The shells of mollusks, such as clams and oysters, commonly form fossiliferous limestone. An example of fossiliferous limestone that contains mollusks is shown in **Figure 4.**

Another type of organic sedimentary rock is *coal.* Coal forms underground when partially decomposed plant material is buried beneath sediment and is changed into coal by increasing heat and pressure. This process occurs over millions of years.

Figure 4 The Formation of Organic Sedimentary Rock

Marine organisms, such as brachiopods, get the calcium carbonate for their shells from ocean water. When these organisms die, their shells collect on the ocean floor and eventually form fossiliferous limestone (inset). Over time, huge rock formations that contain the remains of large numbers of organisms, such as brachiopods, form.

Sedimentary Rock Structures

Many features can tell you about the way sedimentary rock formed. The most important feature of sedimentary rock is stratification. **Stratification** is the process in which sedimentary rocks are arranged in layers. Strata differ from one another depending on the kind, size, and color of their sediment.

Sedimentary rocks sometimes record the motion of wind and water waves on lakes, oceans, rivers, and sand dunes in features called *ripple marks*, as shown in **Figure 5.** Structures called *mud cracks* form when fine-grained sediments at the bottom of a shallow body of water are exposed to the air and dry out. Mud cracks indicate the location of an ancient lake, stream, or ocean shoreline. Even raindrop impressions can be preserved in fine-grained sediments, as small pits with raised rims.

Reading Check What are ripple marks?

Figure 5 *These ripple marks were made by flowing water and were preserved when the sediments became sedimentary rock. Ripple marks can also form from the action of wind.*

stratification the process in which sedimentary rocks are arranged in layers

SECTION Review

Summary

- Sedimentary rock forms at or near the Earth's surface.
- Clastic sedimentary rock forms when rock or mineral fragments are cemented together.
- Chemical sedimentary rock forms from solutions of dissolved minerals and water.
- Organic limestone forms from the remains of plants and animals.
- Sedimentary structures include ripple marks, mud cracks, and raindrop impressions.

Using Key Terms

1. In your own words, write a definition for each of the following terms: *strata* and *stratification.*

Understanding Key Ideas

2. Which of the following is an organic sedimentary rock?
   a. chemical limestone
   b. shale
   c. fossiliferous limestone
   d. conglomerate
3. Explain the process by which clastic sedimentary rock forms.
4. Describe the three main categories of sedimentary rock.

Math Skills

5. A layer of a sedimentary rock is 2 m thick. How many years did it take for this layer to form if an average of 4 mm of sediment accumulated per year?

Critical Thinking

6. **Identifying Relationships** Rocks are classified based on texture and composition. Which of these two properties would be more important for classifying clastic sedimentary rock?
7. **Analyzing Processes** Why do you think raindrop impressions are more likely to be preserved in fine-grained sedimentary rock rather than in coarse-grained sedimentary rock?

SECTION 4

Metamorphic Rock

Have you ever watched a caterpillar change into a butterfly? Some caterpillars go through a biological process called metamorphosis in which they completely change their shape.

READING WARM-UP

Objectives

- Describe two ways a rock can undergo metamorphism.
- Explain how the mineral composition of rocks changes as the rocks undergo metamorphism.
- Describe the difference between foliated and nonfoliated metamorphic rock.
- Explain how metamorphic rock structures are related to deformation.

Terms to Learn

foliated
nonfoliated

READING STRATEGY

Discussion Read this section silently. Write down questions that you have about this section. Discuss your questions in a small group.

Rocks can also go through a process called *metamorphism*. The word *metamorphism* comes from the Greek words *meta,* which means "changed," and *morphos,* which means "shape." Metamorphic rocks are rocks in which the structure, texture, or composition of the rock have changed. All three types of rock can be changed by heat, pressure, or a combination of both.

Origins of Metamorphic Rock

The texture or mineral composition of a rock can change when its surroundings change. If the temperature or pressure of the new environment is different from the one in which the rock formed, the rock will undergo metamorphism.

The temperature at which most metamorphism occurs ranges from 50°C to 1,000°C. However, the metamorphism of some rocks takes place at temperatures above 1,000°C. It seems that at these temperatures the rock would melt, but this is not true of metamorphic rock. It is the depth and pressure at which metamorphic rocks form that allows the rock to heat to this temperature and maintain its solid nature. Most metamorphic change takes place at depths greater than 2 km. But at depths greater than 16 km, the pressure can be 4,000 times greater than the pressure of the atmosphere at Earth's surface.

Large movements within the crust of the Earth cause additional pressure to be exerted on a rock during metamorphism. This pressure can cause the mineral grains in rock to align themselves in certain directions. The alignment of mineral grains into parallel bands is shown in the metamorphic rock in **Figure 1.**

Figure 1 *This metamorphic rock is an example of how mineral grains were aligned into distinct bands when the rock underwent metamorphism.*

Figure 2 *Metamorphism occurs over small areas, such as next to bodies of magma, and over large areas, such as mountain ranges.*

Contact metamorphism

Sedimentary rock

Magma

Regional metamorphism

Contact Metamorphism

One way rock can undergo metamorphism is by being heated by nearby magma. When magma moves through the crust, the magma heats the surrounding rock and changes it. Some minerals in the surrounding rock are changed into other minerals by this increase in temperature. The greatest change takes place where magma comes into direct contact with the surrounding rock. The effect of heat on rock gradually decreases as the rock's distance from the magma increases and as temperature decreases. *Contact metamorphism* occurs near igneous intrusions, as shown in **Figure 2.**

Regional Metamorphism

When pressure builds up in rock that is buried deep below other rock formations or when large pieces of the Earth's crust collide with each other, *regional metamorphism* occurs. The increased pressure and temperature causes rock to become deformed and chemically changed. Unlike contact metamorphism, which happens near bodies of magma, regional metamorphism occurs over thousands of cubic kilometers deep within Earth's crust. Rocks that have undergone regional metamorphism are found beneath most continental rock formations.

Reading Check **Explain how and where regional metamorphism takes place.** (*See the Appendix for answers to Reading Checks.*)

Stretching Out

1. Sketch the crystals in granite rock on a **piece of paper** with a **black-ink pen.** Be sure to include the outline of the rock, and fill it in with different crystal shapes.
2. Flatten some **plastic play putty** over your drawing, and slowly peel it off.
3. After making sure that the outline of your granite has been transferred to the putty, squeeze and stretch the putty. What happened to the crystals in the granite? What happened to the granite?

Figure 3 *The minerals calcite, quartz, and hematite combine and recrystallize to form the metamorphic mineral garnet.*

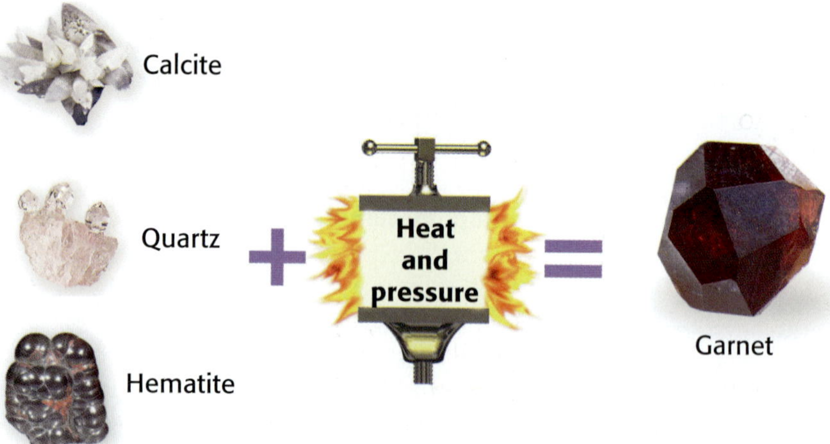

Making a Rock Collection

With a parent, try to collect a sample of each class of rock described in this chapter. You may wish to collect rocks from road cuts or simply collect pebbles from your garden or driveway. Try to collect samples that show the composition and texture of each rock. Classify the rocks in your collection, and bring it to class. With other members of the class, discuss your rock samples and see if they are accurately identified.

Composition of Metamorphic Rock

Metamorphism occurs when temperature and pressure inside the Earth's crust change. Minerals that were present in the rock when it formed may not be stable in the new temperature and pressure conditions. The original minerals change into minerals that are more stable in these new conditions. Look at **Figure 3** to see an example of how this change happens.

Many of these new minerals form only in metamorphic rock. As shown in **Figure 4,** some metamorphic minerals form only at certain temperatures and pressures. These minerals, known as *index minerals,* are used to estimate the temperature, depth, and pressure at which a rock undergoes metamorphism. Index minerals include biotite mica, chlorite, garnet, kyanite, muscovite mica, sillimanite, and staurolite.

Reading Check What is an index mineral?

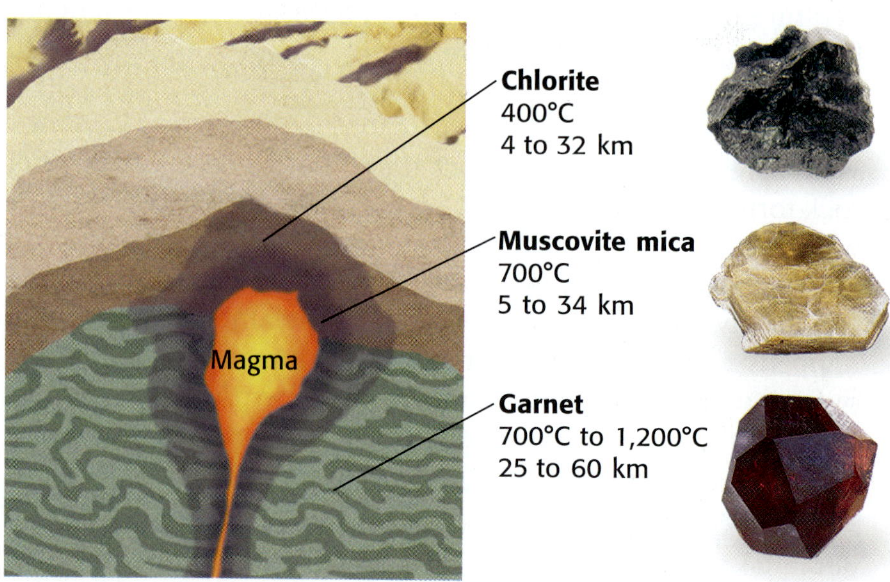

Figure 4 *Scientists can understand a metamorphic rock's history by observing the minerals the rock contains. For example, a metamorphic rock that contains garnet formed at a greater depth and under greater heat and pressure than a rock that contains only chlorite.*

Textures of Metamorphic Rock

You have learned that texture helps scientists classify igneous and sedimentary rock. The same is true of metamorphic rock. All metamorphic rock has one of two textures—foliated or nonfoliated. Take a closer look at each of these types of metamorphic rock to find out how each type forms.

foliated the texture of metamorphic rock in which the mineral grains are arranged in planes or bands

Foliated Metamorphic Rock

The texture of metamorphic rock in which the mineral grains are arranged in planes or bands is called **foliated.** Foliated metamorphic rock usually contains aligned grains of flat minerals, such as biotite mica or chlorite. Look at **Figure 5.** Shale is a sedimentary rock made of layers of clay minerals. When shale is exposed to slight heat and pressure, the clay minerals change into mica minerals. The shale becomes a foliated metamorphic rock called *slate.*

Metamorphic rocks can become other metamorphic rocks if the environment changes again. If slate is exposed to more heat and pressure, the slate can change into rock called *phyllite.* When phyllite is exposed to heat and pressure, it can change into *schist.*

If metamorphism continues, the arrangement of minerals in the rock changes. More heat and pressure cause minerals to separate into distinct bands in a metamorphic rock called *gneiss* (NIES).

Figure 5 *The effects of metamorphism depend on the heat and pressure applied to the rock. Here you can see what happens to shale, a sedimentary rock, when it is exposed to more and more heat and pressure.*

Nonfoliated Metamorphic Rock

The texture of metamorphic rock in which the mineral grains are not arranged in planes or bands is called **nonfoliated.** Notice that the rocks shown in **Figure 6** do not have mineral grains that are aligned. This lack of aligned mineral grains is the reason these rocks are called *nonfoliated rocks*.

Nonfoliated rocks are commonly made of one or only a few minerals. During metamorphism, the crystals of these minerals may change in size or the mineral may change in composition in a process called *recrystallization*. The quartzite and marble shown in **Figure 6** are examples of sedimentary rocks that have recrystallized during metamorphism.

Quartz sandstone is a sedimentary rock made of quartz sand grains that have been cemented together. When quartz sandstone is exposed to the heat and pressure, the spaces between the sand grains disappear as the grains recrystallize to form quartzite. Quartzite has a shiny, glittery appearance. Like quartz sandstone, it is made of quartz. But during recrystallization, the mineral grains have grown larger than the original grains in the sandstone.

When limestone undergoes metamorphism, the same process that happened to the quartz happens to the calcite, and the limestone becomes marble. The calcite crystals in the marble are larger than the calcite grains in the original limestone.

CONNECTION TO Biology

WRITING SKILL **Metamorphosis** The term *metamorphosis* means "change in form." When some animals undergo a dramatic change in the shape of their body, they are said to have undergone a metamorphosis. As part of their natural life cycle, moths and butterflies go through four stages. After they hatch from an egg, they are in the larval stage in the form of a caterpillar. In the next stage, they build a cocoon or become a chrysalis. This stage is called the *pupal stage*. They finally emerge into the adult stage of their life, in which they have wings, antennae, and legs! Research other animals that undergo a metamorphosis, and summarize your findings in a short essay.

nonfoliated the texture of metamorphic rock in which the mineral grains are not arranged in planes or bands

Figure 6 **Two Examples of Nonfoliated Metamorphic Rock**

Marble and quartzite are nonfoliated metamorphic rocks. As you can see in the views through a microscope, the mineral crystals are not well aligned.

Marble

Quartzite

Metamorphic Rock Structures

Like igneous and sedimentary rock, metamorphic rock also has features that tell you about its history. In metamorphic rocks, these features are caused by deformation. *Deformation* is a change in the shape of a rock caused by a force placed on it. These forces may cause a rock to be squeezed or stretched.

Folds, or bends, in metamorphic rock are structures that indicate that a rock has been deformed. Some folds are not visible to the naked eye. But, as shown in **Figure 7,** some folds may be kilometers or even hundreds of kilometers in size.

Reading Check How are metamorphic rock structures related to deformation?

Figure 7 *These large folds occur in metamorphosed sedimentary rock along Saglet Fiord in Labrador, Canada.*

SECTION Review

Summary

- Metamorphic rocks are rocks in which the structure, texture, or composition has changed.
- Two ways rocks can undergo metamorphism are by contact metamorphism and regional metamorphism.
- As rocks undergo metamorphism, the original minerals in a rock change into new minerals that are more stable in new pressure and temperature conditions.
- Foliated metamorphic rock has mineral crystals aligned in planes or bands, whereas nonfoliated rocks have unaligned mineral crystals.
- Metamorphic rock structures are caused by deformation.

Using Key Terms

1. In your own words, define the following terms: *foliated* and *nonfoliated.*

Understanding Key Ideas

2. Which of the following is not a type of foliated metamorphic rock?
   a. gneiss
   b. slate
   c. marble
   d. schist
3. Explain the difference between contact metamorphism and regional metamorphism.
4. Explain how index minerals allow a scientist to understand the history of a metamorphic rock.

Math Skills

5. For every 3.3 km a rock is buried, the pressure placed upon it increases 0.1 gigapascal (100 million pascals). If rock undergoing metamorphosis is buried at 16 km, what is the pressure placed on that rock? (Hint: The pressure at Earth's surface is .101 gigapascal.)

Critical Thinking

6. **Making Inferences** If you had two metamorphic rocks, one that has garnet crystals and the other that has chlorite crystals, which one could have formed at a deeper level in the Earth's crust? Explain your answer.
7. **Applying Concepts** Which do you think would be easier to break, a foliated rock, such as slate, or a nonfoliated rock, such as quartzite? Explain.
8. **Analyzing Processes** A mountain range is located at a boundary where two tectonic plates are colliding. Would most of the metamorphic rock in the mountain range be a product of contact metamorphism or regional metamorphism? Explain.

Skills Practice Lab

Let's Get Sedimental

How do we determine if sedimentary rock layers are undisturbed? The best way to do this is to be sure that fine-grained sediments near the top of a layer lie above coarse-grained sediments near the bottom of the layer. This lab activity will show you how to read rock features that will help you distinguish individual sedimentary rock layers. Then, you can look for the features in real rock layers.

OBJECTIVES

Model the process of sedimentation.

Determine whether sedimentary rock layers are undisturbed.

MATERIALS

- clay
- dropper pipet
- gravel
- magnifying lens
- mixing bowl, 2 qt
- sand
- scissors
- soda bottle with a cap, plastic, 2 L
- soil, clay rich, if available
- water

SAFETY

Procedure

1. In a mixing bowl, thoroughly mix the sand, gravel, and soil. Fill the soda bottle about one-third full of the mixture.
2. Add water to the soda bottle until the bottle is two-thirds full. Twist the cap back onto the bottle, and shake the bottle vigorously until all of the sediment is mixed in the rapidly moving water.
3. Place the bottle on a tabletop. Using the scissors, carefully cut the top off the bottle a few centimeters above the water, as shown. The open bottle will allow water to evaporate.
4. Immediately after you set the bottle on the tabletop, describe what you see from above and through the sides of the bottle.
5. Do not disturb the container. Allow the water to evaporate. (You may speed up the process by carefully using the dropper pipet to siphon off some of the clear water after you allow the container to sit for at least 24 hours.) You may also set the bottle in the sun or under a desk lamp to speed up evaporation.
6. After the sediment has dried and hardened, describe its surface.
7. Carefully lay the container on its side, and cut a wide, vertical strip of plastic down the length of the bottle to expose the sediments in the container. You may find it easier if you place pieces of clay on either side of the container to stabilize it. (If the bottle is clear along its length, this step may not be required.)
8. Brush away the loose material from the sediment, and gently blow on the surface until it is clean. Examine the surface, and record your observations.

Analyze the Results

1. **Identifying Patterns** Do you see anything through the side of the bottle that could help you determine if a sedimentary rock is undisturbed? Explain your answer.
2. **Identifying Patterns** Can you observe a pattern of deposition? If so, describe the pattern of deposition of sediment that you observe from top to bottom.
3. **Explaining Events** Explain how these features might be used to identify the top of a sedimentary layer in real rock and to decide if the layer has been disturbed.
4. **Identifying Patterns** Do you see any structures through the side of the bottle that might indicate which direction is up, such as a change in particle density or size?
5. **Identifying Patterns** Use the magnifying lens to examine the boundaries between the gravel, sand, and silt. Do the size of the particles and the type of sediment change dramatically in each layer?

Draw Conclusions

6. **Making Predictions** Imagine that a layer was deposited directly above the sediment in your bottle. Describe the composition of this new layer. Will it have the same composition as the mixture in steps 1–5 in the Procedure?

Applying Your Data

With your class or with a parent, visit an outcrop of sedimentary rock. Apply the information that you have learned in this lab to see if you can determine whether the sedimentary rock layers are disturbed or undisturbed.

Chapter Review

USING KEY TERMS

1. In your own words, write a definition for the term *rock cycle*.

Complete each of the following sentences by choosing the correct term from the word bank.

stratification	foliated
extrusive igneous rock	texture

2. The ___ of a rock is determined by the sizes, shapes, and positions of the minerals the rock contains.

3. ___ metamorphic rock contains minerals that are arranged in plates or bands.

4. The most characteristic property of sedimentary rock is ___.

5. ___ forms plains called *lava plateaus*.

UNDERSTANDING KEY IDEAS

Multiple Choice

6. Sedimentary rock is classified into all of the following main categories except
   - **a.** clastic sedimentary rock.
   - **b.** chemical sedimentary rock.
   - **c.** nonfoliated sedimentary rock.
   - **d.** organic sedimentary rock.

7. An igneous rock that cools very slowly has a ___ texture.
   - **a.** foliated
   - **b.** fine-grained
   - **c.** nonfoliated
   - **d.** coarse-grained

8. Igneous rock forms when
   - **a.** minerals crystallize from a solution.
   - **b.** sand grains are cemented together.
   - **c.** magma cools and solidifies.
   - **d.** mineral grains in a rock recrystallize.

9. A ___ is a common structure found in metamorphic rock.
   - **a.** ripple mark
   - **b.** fold
   - **c.** sill
   - **d.** layer

10. The process in which sediment is removed from its source and transported is called
    - **a.** deposition.
    - **b.** erosion.
    - **c.** weathering.
    - **d.** uplift.

11. Mafic rocks are
    - **a.** light-colored rocks rich in calcium, iron, and magnesium.
    - **b.** dark-colored rocks rich in aluminum, potassium, silica, and sodium.
    - **c.** light-colored rocks rich in aluminum, potassium, silica, and sodium.
    - **d.** dark-colored rocks rich in calcium, iron, and magnesium.

Short Answer

12. Explain how composition and texture are used by scientists to classify rocks.

13. Describe two ways a rock can undergo metamorphism.

14. Explain why some minerals only occur in metamorphic rocks.

15. Describe how each type of rock changes as it moves through the rock cycle.

16 Describe two ways rocks were used by early humans and ancient civilizations.

CRITICAL THINKING

17 **Concept Mapping** Use the following terms to construct a concept map: *rocks, metamorphic, sedimentary, igneous, foliated, nonfoliated, organic, clastic, chemical, intrusive,* and *extrusive.*

18 **Making Inferences** If you were looking for fossils in the rocks around your home and the rock type that was closest to your home was metamorphic, do you think that you would find many fossils? Explain your answer.

19 **Applying Concepts** Imagine that you want to quarry, or mine, granite. You have all of the equipment, but you have two pieces of land to choose from. One area has a granite batholith underneath it. The other has a granite sill. If both intrusive bodies are at the same depth, which one would be the better choice for you to quarry? Explain your answer.

20 **Applying Concepts** The sedimentary rock coquina is made up of pieces of seashells. Which of the three kinds of sedimentary rock could coquina be? Explain your answer.

21 **Analyzing Processes** If a rock is buried deep inside the Earth, which geological processes cannot change the rock? Explain your answer.

INTERPRETING GRAPHICS

The bar graph below shows the percentage of minerals by mass that compose a sample of granite. Use the graph below to answer the questions that follow.

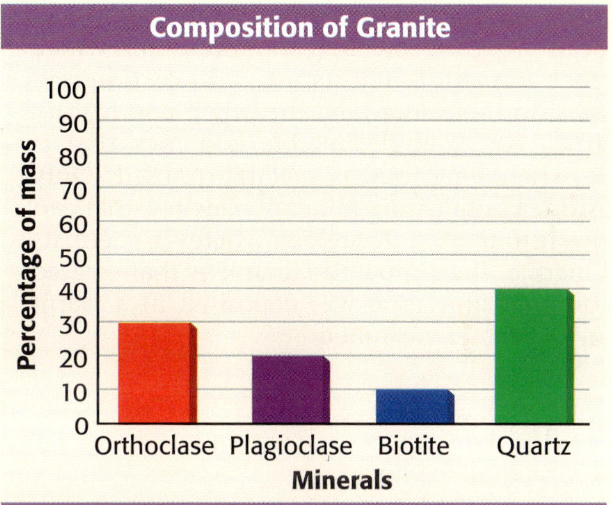

22 Your rock sample is made of four minerals. What percentage of each mineral makes up your sample?

23 Both plagioclase and orthoclase are feldspar minerals. What percentage of the minerals in your sample of granite are not feldspar minerals?

24 If your rock sample has a mass of 10 g, how many grams of quartz does it contain?

25 Use paper, a compass, and a protractor or a computer to make a pie chart. Show the percentage of each of the four minerals your sample of granite contains. (Look in the Appendix of this book for help on making a pie chart.)

Standardized Test Preparation

READING

Read each of the passages below. Then, answer the questions that follow each passage.

Passage 1 The texture and composition of a rock can provide good clues about how and where the rock formed. Scientists use both texture and composition to understand the origin and history of rocks. For example, marble is a rock that is made when limestone is metamorphosed. Only limestone contains the mineral—calcite—that can change into marble. Therefore, wherever scientists find marble, they know the sediment that created the original limestone was deposited in a warm ocean or lake environment.

1. In the passage, what does the word *origin* mean?

A size or appearance
B age
C location or surroundings
D source or formation

2. Based on the passage, what can the reader conclude?

F Marble is a sedimentary rock.
G Limestone is created by sediments deposited in warm ocean or lake environments.
H Marble is a rock that is made when sandstone has undergone metamorphism.
I In identifying a rock, the texture of a rock is more important than the composition of the rock.

3. What is the main idea of the passage?

A Scientists believe marble is the most important rock type to study.
B Scientists study the composition and texture of a rock to determine how the rock formed and what happened after it formed.
C Some sediments are deposited in warm oceans and lakes.
D When limestone undergoes metamorphism, it creates marble.

Passage 2 Fulgurites are a rare type of natural glass found in areas that have quartz-rich sediments, such as beaches and deserts. A tubular fulgurite forms when a lightning bolt strikes material such as sand and melts the quartz into a liquid. The liquid quartz cools and solidifies quickly, and a thin, glassy tube is left behind. Fulgurites usually have a rough outer surface and a smooth inner surface. Underground, a fulgurite may be shaped like the roots of a tree. The fulgurite branches out with many arms that trace the zigzag path of the lightning bolt. Some fulgurites are as short as your little finger, but others stretch 20 m into the ground.

1. In the passage, what does the word *tubular* mean?

A flat and sharp
B round and long
C funnel shaped
D pyramid shaped

2. From the information in the passage, what can the reader conclude?

F Fulgurites are formed above ground.
G Sand contains a large amount of quartz.
H Fulgurites are most often very small.
I Fulgurites are easy to find in sandy places.

3. Which of the following statements best describes a fulgurite?

A Fulgurites are frozen lightning bolts.
B Fulgurites are rootlike rocks.
C Fulgurites are glassy tubes found in deserts.
D Fulgurites are natural glass tubes formed by lightning bolts.

INTERPRETING GRAPHICS

Use the diagram below to answer the questions that follow.

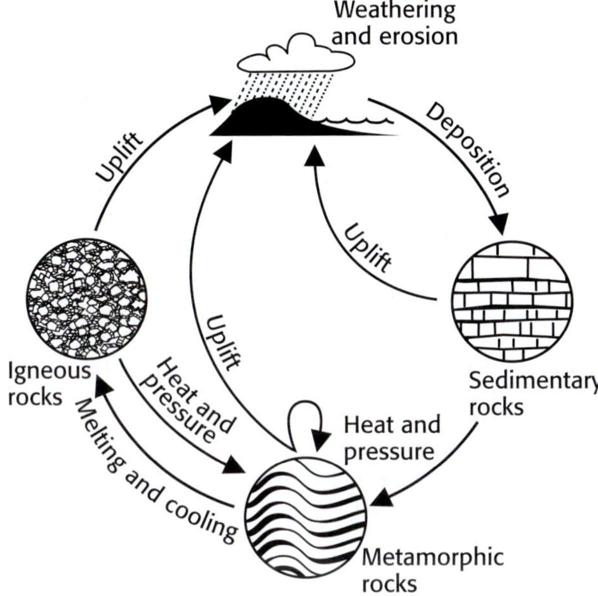

1. According to the rock cycle diagram, which of the following statements is true?
   A Only sedimentary rock gets weathered and eroded.
   B Sedimentary rocks are made from metamorphic, igneous, and sedimentary rock fragments and minerals.
   C Heat and pressure create igneous rocks.
   D Metamorphic rocks are created by melting and cooling.

2. A rock exists at the surface of the Earth. What would be the next step in the rock cycle?
   F cooling
   G weathering
   H melting
   I metamorphism

3. Which of the following processes brings rocks to Earth's surface, where they can be eroded?
   A burial
   B deposition
   C uplift
   D weathering

4. Which of the following is the best summary of the rock cycle?
   F Each type of rock gets melted. Then the magma turns into igneous, sedimentary, and metamorphic rock.
   G Magma cools to form igneous rock. Then, the igneous rock becomes sedimentary rock. Sedimentary rock is heated and forms metamorphic rock. Metamorphic rock melts to form magma.
   H All three rock types weather to create sedimentary rock. All three rock types melt to form magma. Magma forms igneous rock. All three types of rock form metamorphic rock because of heat and pressure.
   I Igneous rock is weathered to create sedimentary rock. Sedimentary rock is melted to form igneous rock. Metamorphic rock is weathered to form igneous rock.

MATH

Read each question below, and choose the best answer.

1. Eric has 25 rocks he has collected as a science project for class. Nine rocks are sedimentary, 10 are igneous, and 6 are metamorphic. If Eric chooses a rock at random, what is the probability that he will choose an igneous rock?
   A 1/2
   B 2/5
   C 3/8
   D 1/15

2. At a mineral and fossil show, Elizabeth bought two quartz crystals that cost $2.00 each and four trilobite fossils that cost $3.50 each. Which equation can be used to describe *c*, the total cost of her purchase?
   F $c = (2 \times 4) + (2.00 \times 3.50)$
   G $c = (2 \times 2.00) + (4 \times 3.50)$
   H $c = (4 \times 2.00) + (2 \times 3.50)$
   I $c = (2 + 2.00) + (4 + 3.50)$

Standardized Test Preparation

Science in Action

Science, Technology, and Society

The Moai of Easter Island

Easter island is located in the Pacific Ocean more than 3,200 km from the coast of Chile. The island is home to mysterious statues that were carved from volcanic ash. The statues, called *moai,* have human heads and large torsos. The average moai weighs 14 tons and is more than 4.5 m tall, though some are as tall as 10 m! Altogether, 887 moai have been discovered. How old are the moai? Scientists believe that the moai were built between 500 and 1,000 years ago. What purpose did moai serve for their creators? The moai may have been religious symbols or gods.

Social Studies ACTIVITY

WRITING SKILL Research another ancient society or civilization, such as the ancient Egyptians, who are believed to have used stone to construct monuments to their gods or to important people. Report your findings in a short essay.

Scientific Discoveries

Shock Metamorphism

When a large asteroid, meteoroid, or comet collides with the Earth, extremely high temperatures and pressures are created in Earth's surface rock. These high pressures and temperatures cause minerals in the surface rock to shatter and recrystallize. The new minerals that result from this recrystallization cannot be created under any other conditions. This process is called *shock metamorphism.*

When large objects from space collide with the Earth, craters are formed by the impact. However, impact craters are not always easy to find on Earth. Scientists use shock metamorphism as a clue to locate ancient impact craters.

Language Arts ACTIVITY

WRITING SKILL The impact site caused by the asteroid strike in the Yucatán 65 million years ago has been named the Chicxulub (cheeks OO loob) structure. Research the origin of the name Chicxulub, and report your findings in a short paper.

Careers

Robert L. Folk

Petrologist For Dr. Robert Folk, the study of rock takes place on the microscopic level. Dr. Folk is searching for tiny life-forms he has named nannobacteria, or dwarf bacteria, in rock. *Nannobacteria* may also be spelled *nanobacteria.* Because nannobacteria are so incredibly small, only 0.05 to 0.2 µm in diameter, Folk must use an extremely powerful 100,000× microscope, called a *scanning electron microscope,* to see the shape of the bacteria in rock. Folk's research had already led him to discover that a certain type of Italian limestone is produced by bacteria. The bacteria were consuming the minerals, and the waste of the bacteria was forming the limestone. Further research led Folk to the discovery of the tiny nannobacteria. The spherical or oval-shaped nannobacteria appeared as chains and grapelike clusters. From his research, Folk hypothesized that nannobacteria are responsible for many inorganic reactions that occur in rock. Many scientists are skeptical of Folk's nannobacteria. Some skeptics believe that the tiny size of nannobacteria makes the bacteria simply too small to contain the chemistry of life. Others believe that nannobacteria actually represent structures that do not come from living things.

If a nannobacterium is 1/10 the length, 1/10 the width, and 1/10 the height of an ordinary bacterium, how many nannobacteria can fit within an ordinary bacterium? (Hint: Draw block diagrams of both a nannobacterium and an ordinary bacterium.)

To learn more about these Science in Action topics, visit go.hrw.com and type in the keyword HZ5RCKF.

Check out Current Science® articles related to this chapter by visiting go.hrw.com. Just type in the keyword HZ5CS04.

The Rock and Fossil Record

About the 

This extremely well preserved crocodile fossil has been out of water for 49 million years. Its skeleton was collected in an abandoned mine pit in Messel, Germany.

PRE-READING ACTIVITY

FOLDNOTES **Layered Book** Before you read the chapter, create the FoldNote entitled "Layered Book" described in the **Study Skills** section of the Appendix. Label the tabs of the layered book with "Earth's history," "Relative dating," "Absolute dating," "Fossils," and "Geologic time." As you read the chapter, write information you learn about each category under the appropriate tab.

START-UP ACTIVITY

Making Fossils

How do scientists learn from fossils? In this activity, you will study "fossils" and identify the object that made each.

Procedure

1. You and three or four of your classmates will be given **several pieces** of **modeling clay** and a **paper sack** containing a few **small objects.**
2. Press each object firmly into a piece of clay. Try to leave a "fossil" imprint showing as much detail as possible.
3. After you have made an imprint of each object, exchange your model fossils with another group.
4. On a **sheet of paper,** describe the fossils you have received. List as many details as possible. What patterns and textures do you observe?
5. Work as a group to identify each fossil, and check your results. Were you right?

Analysis

1. What kinds of details were important in identifying your fossils? What kinds of details were not preserved in the imprints? For example, can you tell the materials from which the objects are made or their color?
2. Explain how scientists follow similar methods when studying fossils.

SECTION 1

Earth's Story and Those Who First Listened

READING WARM-UP

Objectives

- Compare uniformitarianism and catastrophism.
- Describe how the science of geology has changed over the past 200 years.
- Explain the role of paleontology in the study of Earth's history.

Terms to Learn

uniformitarianism
catastrophism
paleontology

READING STRATEGY

Reading Organizer As you read this section, make a table comparing uniformitarianism and catastrophism.

How do mountains form? How is new rock created? How old is the Earth? Have you ever asked these questions? Nearly 250 years ago, a Scottish farmer and scientist named James Hutton did.

Searching for answers to his questions, Hutton spent more than 30 years studying rock formations in Scotland and England. His observations led to the foundation of modern geology.

The Principle of Uniformitarianism

In 1788, James Hutton collected his notes and wrote *Theory of the Earth*. In *Theory of the Earth,* he stated that the key to understanding Earth's history was all around us. In other words, processes that we observe today—such as erosion and deposition—remain uniform, or do not change, over time. This assumption is now called uniformitarianism. **Uniformitarianism** is the idea that the same geologic processes shaping the Earth today have been at work throughout Earth's history. **Figure 1** shows how Hutton developed the idea of uniformitarianism.

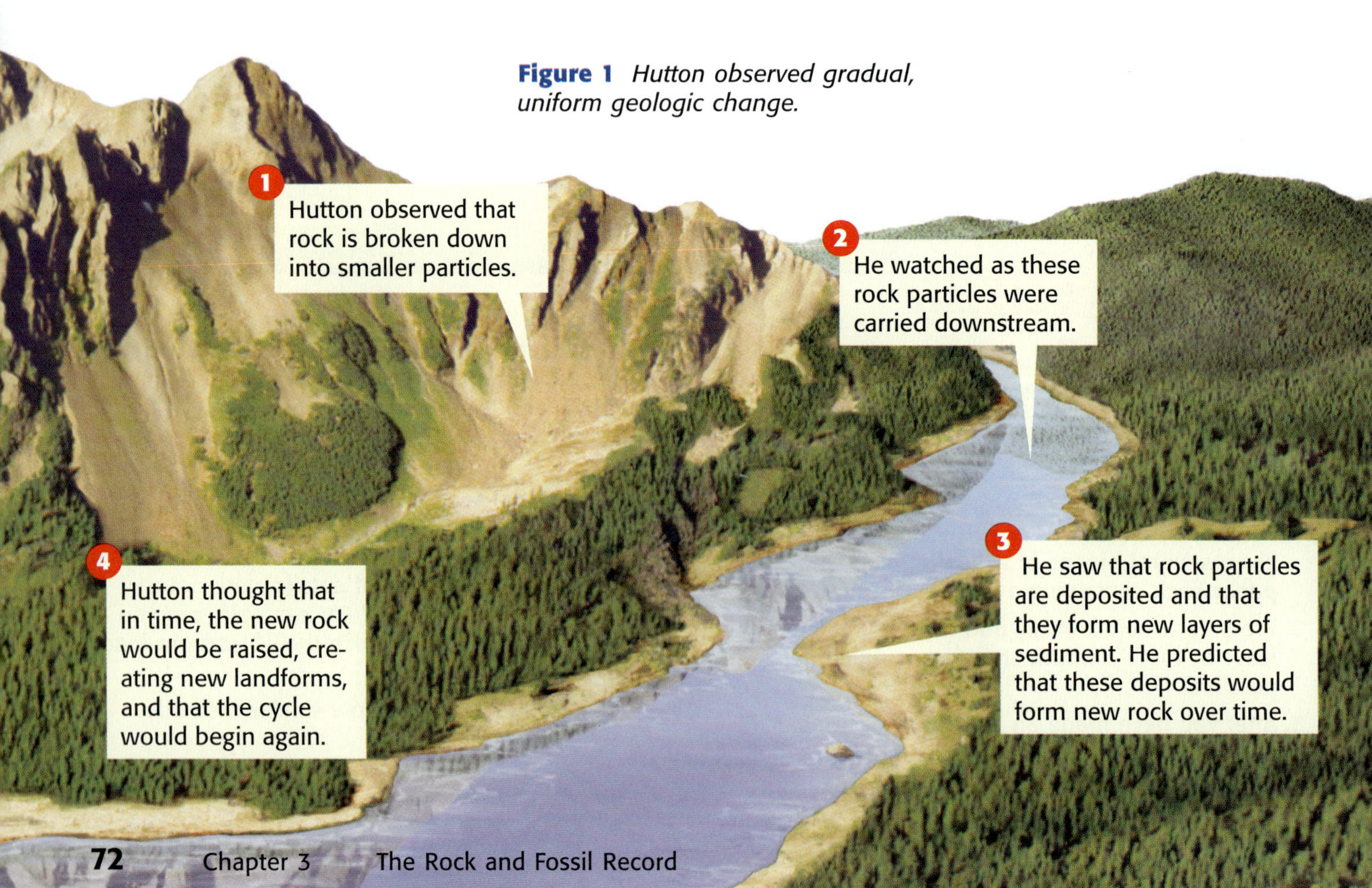

Figure 1 *Hutton observed gradual, uniform geologic change.*

Figure 2 *This photograph shows Siccar Point on the coast of Scotland. Siccar Point is one of the places where Hutton observed results of geologic processes that would lead him to form his principle of uniformitarianism.*

Uniformitarianism Versus Catastrophism

Hutton's theories sparked a scientific debate by suggesting that Earth was much older than previously thought. In Hutton's time, most people thought that Earth was only a few thousand years old. A few thousand years was not nearly enough time for the gradual geologic processes that Hutton described to have shaped our planet. The rocks that he observed at Siccar Point, shown in **Figure 2,** were deposited and folded, indicating a long geological history. To explain Earth's history, most scientists supported catastrophism. **Catastrophism** is the principle that states that all geologic change occurs suddenly. Supporters of catastrophism thought that Earth's features, such as its mountains, canyons, and seas, formed during rare, sudden events called *catastrophes*. These unpredictable events caused rapid geologic change over large areas—sometimes even globally.

uniformitarianism a principle that states that geologic processes that occurred in the past can be explained by current geologic processes

catastrophism a principle that states that geologic change occurs suddenly

Reading Check **According to catastrophists, what was the rate of geologic change?** (*See the Appendix for answers to Reading Checks.*)

A Victory for Uniformitarianism

Despite Hutton's work, catastrophism remained geology's guiding principle for decades. Only after the work of British geologist Charles Lyell did people seriously consider uniformitarianism as geology's guiding principle.

From 1830 to 1833, Lyell published three volumes, collectively titled *Principles of Geology,* in which he reintroduced uniformitarianism. Armed with Hutton's notes and new evidence of his own, Lyell successfully challenged the principle of catastrophism. Lyell saw no reason to doubt that major geologic change happened at the same rate in the past as it happens in the present—gradually.

CONNECTION TO Biology

WRITING SKILL **Darwin and Lyell** The theory of evolution was developed soon after Lyell introduced his ideas, which was no coincidence. Lyell and Charles Darwin were good friends, and their talks greatly influenced Darwin's theories. Similar to uniformitarianism, Darwin's theory of evolution proposes that changes in species occur gradually over long periods of time. Write a short essay comparing uniformitarianism and evolution.

Modern Geology—A Happy Medium

During the late 20th century, scientists such as Stephen J. Gould challenged Lyell's uniformitarianism. They believed that catastrophes do, at times, play an important role in shaping Earth's history.

Today, scientists realize that neither uniformitarianism nor catastrophism accounts for all geologic change throughout Earth's history. Although most geologic change is gradual and uniform, catastrophes that cause geologic change have occurred during Earth's long history. For example, huge craters have been found where asteroids and comets are thought to have struck Earth in the past. Some scientists think one such asteroid strike, approximately 65 million years ago, may have caused the dinosaurs to become extinct. **Figure 3** is an imaginary re-creation of the asteroid strike that is thought to have caused the extinction of the dinosaurs. The impact of this asteroid is thought to have thrown debris into the atmosphere. The debris spread around the entire planet and rained down on Earth for decades. This global debris cloud may have blocked the sun's rays, causing major changes in the global climate that doomed the dinosaurs.

Reading Check How can a catastrophe affect life on Earth?

Figure 3 *Today, scientists think that sudden events are responsible for some changes during Earth's past. An asteroid hitting Earth, for example, may have led to the extinction of the dinosaurs about 65 million years ago.*

Paleontology—The Study of Past Life

The history of the Earth would be incomplete without a knowledge of the organisms that have inhabited our planet and the conditions under which they lived. The science involved with the study of past life is called **paleontology.** Scientists who study this life are called *paleontologists*. The data paleontologists use are fossils. Fossils are the remains of organisms preserved by geologic processes. Some paleontologists specialize in the study of particular organisms. Invertebrate paleontologists study animals without backbones, whereas vertebrate paleontologists, such as the scientist in **Figure 4,** study animals with backbones. Paleobotanists study fossils of plants. Other paleontologists reconstruct past ecosystems, study the traces left behind by animals, and piece together the conditions under which fossils were formed. As you see, the study of past life is as varied and complex as Earth's history itself.

Figure 4 *Edwin Colbert was a 20th-century vertebrate paleontologist who made important contributions to the study of dinosaurs.*

paleontology the scientific study of fossils

SECTION Review

Summary

- Uniformitarianism assumes that geologic change is gradual. Catastrophism is based on the idea that geologic change is sudden.
- Modern geology is based on the idea that gradual geologic change is interrupted by catastrophes.
- Using fossils to study past life is called *paleontology*.

Using Key Terms

1. Use each of the following terms in a separate sentence: *uniformitarianism, catastrophism,* and *paleontology.*

Understanding Key Ideas

2. Which of the following words describes change according to the principle of uniformitarianism?
   a. sudden
   b. rare
   c. global
   d. gradual
3. What is the difference between uniformitarianism and catastrophism?
4. Describe how the science of geology has changed.
5. Give one example of catastrophic global change.
6. Describe the work of three types of paleontologists.

Math Skills

7. An impact crater left by an asteroid strike has a radius of 85 km. What is the area of the crater? (Hint: The area of a circle is πr^2.)

Critical Thinking

8. **Analyzing Ideas** Why is uniformitarianism considered to be the foundation of modern geology?
9. **Applying Concepts** Give an example of a type of recent catastrophe.

SECTION 2

Relative Dating: Which Came First?

READING WARM-UP

Objectives

- Explain how relative dating is used in geology.
- Explain the principle of superposition.
- Describe how the geologic column is used in relative dating.
- Identify two events and two features that disrupt rock layers.
- Explain how physical features are used to determine relative ages.

Terms to Learn

relative dating
superposition
geologic column
unconformity

READING STRATEGY

Reading Organizer As you read this section, create an outline of the section. Use the headings from the section in your outline.

Imagine that you are a detective investigating a crime scene. What is the first thing you would do?

You might begin by dusting the scene for fingerprints or by searching for witnesses. As a detective, you must figure out the sequence of events that took place before you reached the crime scene.

Geologists have a similar goal when investigating the Earth. They try to determine the order in which events have happened during Earth's history. But instead of relying on fingerprints and witnesses, geologists rely on rocks and fossils to help them in their investigation. Determining whether an object or event is older or younger than other objects or events is called **relative dating.**

The Principle of Superposition

Suppose that you have an older brother who takes a lot of photographs of your family and piles them in a box. Over the years, he keeps adding new photographs to the top of the stack. Think about the family history recorded in those photos. Where are the oldest photographs—the ones taken when you were a baby? Where are the most recent photographs—those taken last week?

Layers of sedimentary rock, such as the ones shown in **Figure 1,** are like stacked photographs. As you move from top to bottom, the layers are older. The principle that states that younger rocks lie above older rocks in undisturbed sequences is called **superposition.**

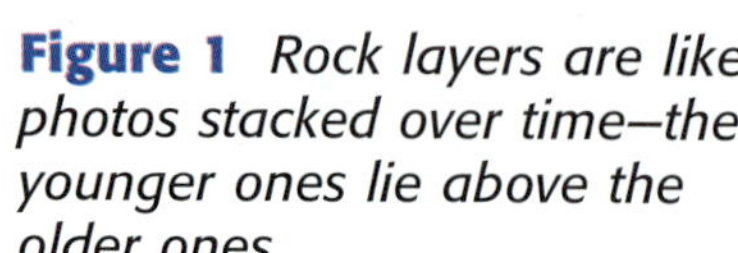

Figure 1 *Rock layers are like photos stacked over time–the younger ones lie above the older ones.*

Disturbing Forces

Not all rock sequences are arranged with the oldest layers on the bottom and the youngest layers on top. Some rock sequences are disturbed by forces within the Earth. These forces can push other rocks into a sequence, tilt or fold rock layers, and break sequences into movable parts. Sometimes, geologists even find rock sequences that are upside down! The disruptions of rock sequences pose a challenge to geologists trying to determine the relative ages of rocks. Fortunately, geologists can get help from a very valuable tool—the geologic column.

relative dating any method of determining whether an event or object is older or younger than other events or objects

superposition a principle that states that younger rocks lie above older rocks if the layers have not been disturbed

geologic column an arrangement of rock layers in which the oldest rocks are at the bottom

The Geologic Column

To make their job easier, geologists combine data from all the known undisturbed rock sequences around the world. From this information, geologists create the geologic column, as illustrated in **Figure 2.** The **geologic column** is an ideal sequence of rock layers that contains all the known fossils and rock formations on Earth, arranged from oldest to youngest.

Geologists rely on the geologic column to interpret rock sequences. Geologists also use the geologic column to identify the layers in puzzling rock sequences.

Reading Check **List two ways in which geologists use the geologic column.** *(See the Appendix for answers to Reading Checks.)*

Figure 2 Constructing the Geologic Column

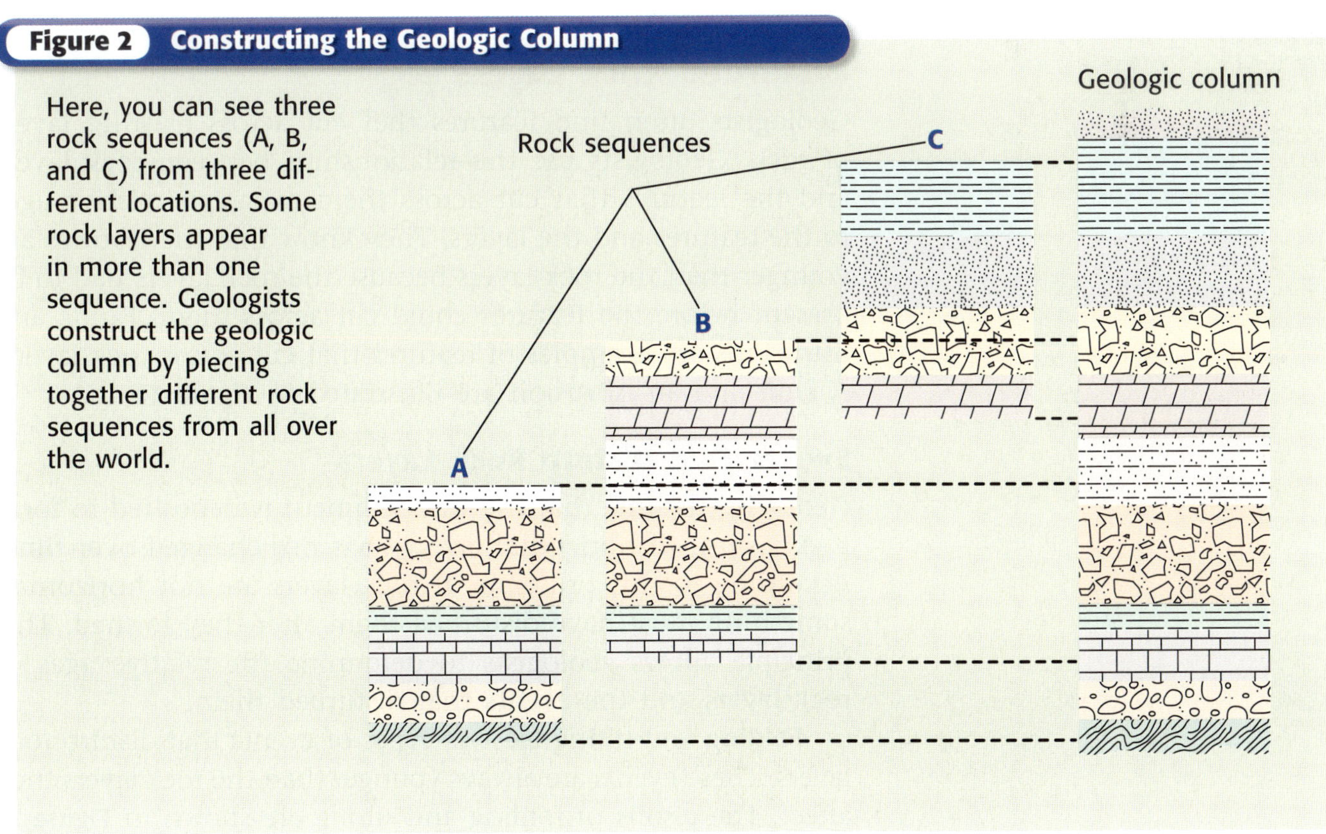

Here, you can see three rock sequences (A, B, and C) from three different locations. Some rock layers appear in more than one sequence. Geologists construct the geologic column by piecing together different rock sequences from all over the world.

Figure 3 How Rock Layers Become Disturbed

Fault A *fault* is a break in the Earth's crust along which blocks of the crust slide relative to one another.

Intrusion An *intrusion* is molten rock from the Earth's interior that squeezes into existing rock and cools.

Folding *Folding* occurs when rock layers bend and buckle from Earth's internal forces.

Tilting *Tilting* occurs when internal forces in the Earth slant rock layers.

Disturbed Rock Layers

Geologists often find features that cut across existing layers of rock. Geologists use the relationships between rock layers and the features that cut across them to assign relative ages to the features and the layers. They know that the features are younger than the rock layers because the rock layers had to be present before the features could cut across them. Faults and intrusions are examples of features that cut across rock layers. A fault and an intrusion are illustrated in **Figure 3.**

Events That Disturb Rock Layers

Geologists assume that the way sediment is deposited to form rock layers—in horizontal layers—has not changed over time. According to this principle, if rock layers are not horizontal, something must have disturbed them after they formed. This principle allows geologists to determine the relative ages of rock layers and the events that disturbed them.

Folding and tilting are two types of events that disturb rock layers. These events are always younger than the rock layers they affect. The results of folding and tilting are shown in **Figure 3.**

Gaps in the Record—Unconformities

Faults, intrusions, and the effects of folding and tilting can make dating rock layers a challenge. Sometimes, layers of rock are missing altogether, creating a gap in the geologic record. To think of this another way, let's say that you stack your newspapers every day after reading them. Now, let's suppose you want to look at a paper you read 10 days ago. You know that the paper should be 10 papers deep in the stack. But when you look, the paper is not there. What happened? Perhaps you forgot to put the paper in the stack. Now, imagine a missing rock layer instead of a missing newspaper.

Missing Evidence

Missing rock layers create breaks in rock-layer sequences called unconformities. An **unconformity** is a surface that represents a missing part of the geologic column. Unconformities also represent missing time—time that was not recorded in layers of rock. When geologists find an unconformity, they must question whether the "missing layer" was never present or whether it was somehow removed. **Figure 4** shows how *nondeposition,* or the stoppage of deposition when a supply of sediment is cut off, and *erosion* create unconformities.

unconformity a break in the geologic record created when rock layers are eroded or when sediment is not deposited for a long period of time

Reading Check Define the term unconformity.

Figure 4 **How Unconformities Are Created**

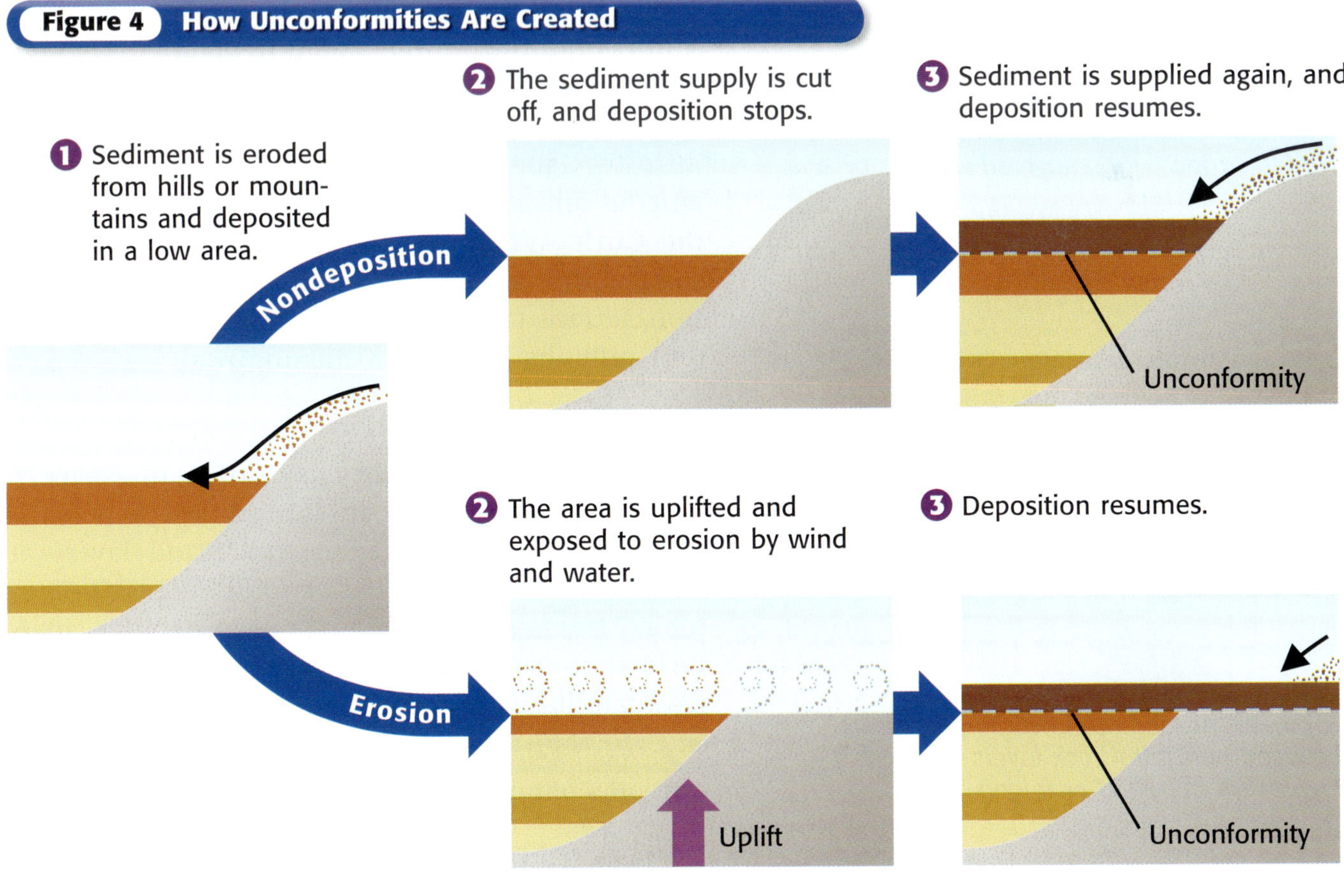

Types of Unconformities

Most unconformities form by both erosion and nondeposition. But other factors can complicate matters. To simplify the study of unconformities, geologists place them into three major categories: disconformities, nonconformities, and angular unconformities. The three diagrams at left illustrate these three categories.

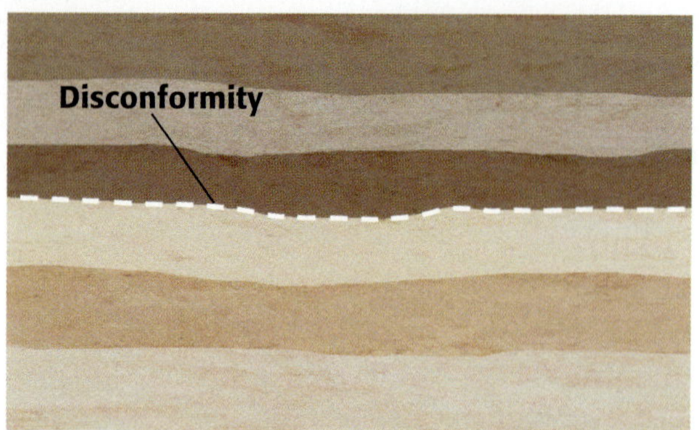

Figure 5 *A disconformity exists where part of a sequence of parallel rock layers is missing.*

Disconformities

The most common type of unconformity is a disconformity, which is illustrated in **Figure 5.** *Disconformities* are found where part of a sequence of parallel rock layers is missing. A disconformity can form in the following way. A sequence of rock layers is uplifted. Younger layers at the top of the sequence are removed by erosion, and the eroded material is deposited elsewhere. At some future time, deposition resumes, and sediment buries the old erosion surface. The disconformity that results shows where erosion has taken place and rock layers are missing. A disconformity represents thousands to many millions of years of missing time.

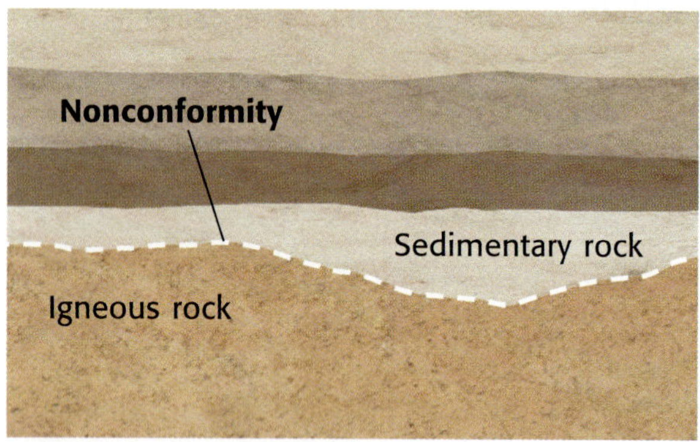

Figure 6 *A nonconformity exists where sedimentary rock layers lie on top of an eroded surface of nonlayered igneous or metamorphic rock.*

Nonconformities

A nonconformity is illustrated in **Figure 6.** *Nonconformities* are found where horizontal sedimentary rock layers lie on top of an eroded surface of older intrusive igneous or metamorphic rock. Intrusive igneous and metamorphic rocks form deep within the Earth. When these rocks are raised to Earth's surface, they are eroded. Deposition causes the erosion surface to be buried. Nonconformities represent millions of years of missing time.

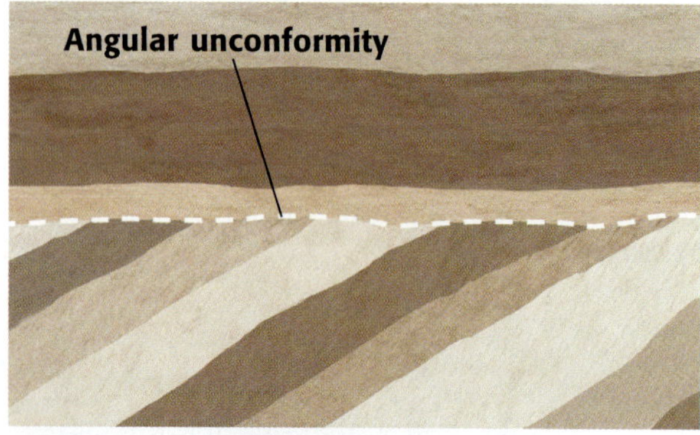

Figure 7 *An angular unconformity exists between horizontal rock layers and rock layers that are tilted or folded.*

Angular Unconformities

An angular unconformity is shown in **Figure 7.** *Angular unconformities* are found between horizontal layers of sedimentary rock and layers of rock that have been tilted or folded. The tilted or folded layers were eroded before horizontal layers formed above them. Angular unconformities represent millions of years of missing time.

Reading Check **Describe each of the three major categories of unconformities.**

Rock-Layer Puzzles

Geologists often find rock-layer sequences that have been affected by more than one of the events and features mentioned in this section. For example, as shown in **Figure 8,** intrusions may squeeze into rock layers that contain an unconformity. Determining the order of events that led to such a sequence is like piecing together a jigsaw puzzle. Geologists must use their knowledge of the events that disturb or remove rock-layer sequences to help piece together the history of Earth as told by the rock record.

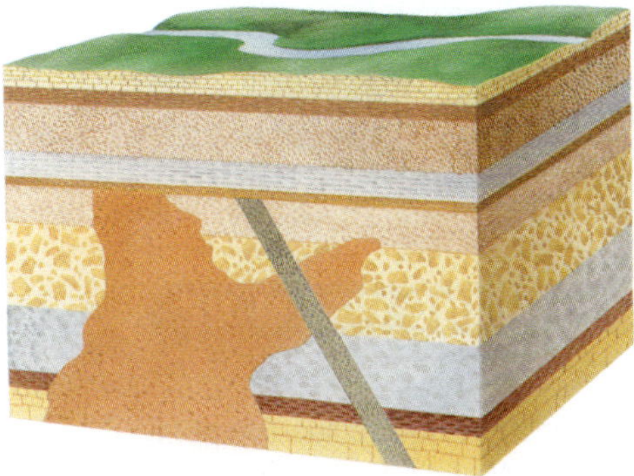

Figure 8 *Rock-layer sequences are often disturbed by more than one rock-disturbing feature.*

SECTION Review

Summary

- Geologists use relative dating to determine the order in which events happen.
- The principle of superposition states that in undisturbed rock sequences, younger layers lie above older layers.
- Folding and tilting are two events that disturb rock layers. Faults and intrusions are two features that disturb rock layers.
- The known rock and fossil record is indicated by the geologic column.
- Geologists examine the relationships between rock layers and the structures that cut across them in order to determine relative ages.

Using Key Terms

1. In your own words, write a definition for each of the following terms: *relative dating, superposition,* and *geologic column.*

Understanding Key Ideas

2. Molten rock that squeezes into existing rock and cools is called a(n)
   a. fold.
   b. fault.
   c. intrusion.
   d. unconformity.
3. List two events and two features that can disturb rock-layer sequences.
4. Explain how physical features are used to determine relative ages.

Critical Thinking

5. **Analyzing Concepts** Is there a place on Earth that has all the layers of the geologic column? Explain.
6. **Analyzing Ideas** Disconformities are hard to recognize because all of the layers are horizontal. How does a geologist know when he or she is looking at a disconformity?

Interpreting Graphics

Use the illustration below to answer the question that follows.

7. If the top rock layer were eroded and deposition later resumed, what type of unconformity would mark the boundary between older rock layers and the newly deposited rock layers?

SECTION 3

Absolute Dating: A Measure of Time

READING WARM-UP

Objectives

- Describe how radioactive decay occurs.
- Explain how radioactive decay relates to radiometric dating.
- Identify four types of radiometric dating.
- Determine the best type of radiometric dating to use to date an object.

Terms to Learn

absolute dating
isotope
radioactive decay
radiometric dating
half-life

READING STRATEGY

Reading Organizer As you read this section, make a concept map by using the terms above.

Have you ever heard the expression "turning back the clock"? With the discovery of the natural decay of uranium in 1896, French physicist Henri Becquerel provided a means of doing just that. Scientists could use radioactive elements as clocks to measure geologic time.

The process of establishing the age of an object by determining the number of years it has existed is called **absolute dating.** In this section, you will learn about radiometric dating, which is the most common method of absolute dating.

Radioactive Decay

To determine the absolute ages of fossils and rocks, scientists analyze isotopes of radioactive elements. Atoms of the same element that have the same number of protons but have different numbers of neutrons are called **isotopes.** Most isotopes are stable, meaning that they stay in their original form. But some isotopes are unstable. Scientists call unstable isotopes *radioactive.* Radioactive isotopes tend to break down into stable isotopes of the same or other elements in a process called **radioactive decay. Figure 1** shows an example of how radioactive decay occurs. Because radioactive decay occurs at a steady rate, scientists can use the relative amounts of stable and unstable isotopes present in an object to determine the object's age.

Figure 1 Radioactive Decay

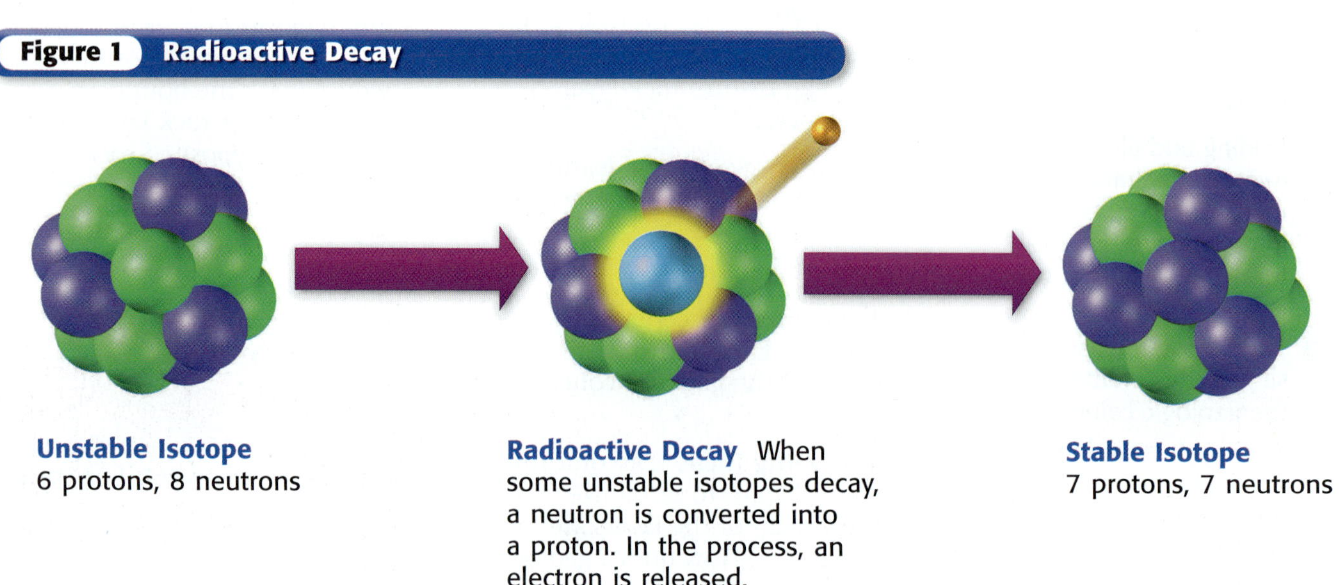

Unstable Isotope 6 protons, 8 neutrons

Radioactive Decay When some unstable isotopes decay, a neutron is converted into a proton. In the process, an electron is released.

Stable Isotope 7 protons, 7 neutrons

Dating Rocks—How Does It Work?

In the process of radioactive decay, an unstable radioactive isotope of one element breaks down into a stable isotope. The stable isotope may be of the same element or, more commonly, a different element. The unstable radioactive isotope is called the *parent isotope*. The stable isotope produced by the radioactive decay of the parent isotope is called the *daughter isotope*. The radioactive decay of a parent isotope into a stable daughter isotope can occur in a single step or a series of steps. In either case, the rate of decay is constant. Therefore, to date rock, scientists compare the amount of parent material with the amount of daughter material. The more daughter material there is, the older the rock is.

absolute dating any method of measuring the age of an event or object in years

isotope an atom that has the same number of protons (or the same atomic number) as other atoms of the same element do but that has a different number of neutrons (and thus a different atomic mass)

radioactive decay the process in which a radioactive isotope tends to break down into a stable isotope of the same element or another element

radiometric dating a method of determining the age of an object by estimating the relative percentages of a radioactive (parent) isotope and a stable (daughter) isotope

half-life the time needed for half of a sample of a radioactive substance to undergo radioactive decay

Radiometric Dating

If you know the rate of decay for a radioactive element in a rock, you can figure out the absolute age of the rock. Determining the absolute age of a sample, based on the ratio of parent material to daughter material, is called **radiometric dating.** For example, let's say that a rock sample contains an isotope with a half-life of 10,000 years. A **half-life** is the time that it takes one-half of a radioactive sample to decay. So, for this rock sample, in 10,000 years, half the parent material will have decayed and become daughter material. You analyze the sample and find equal amounts of parent material and daughter material. This means that half the original radioactive isotope has decayed and that the sample must be about 10,000 years old.

What if one-fourth of your sample is parent material and three-fourths is daughter material? You would know that it took 10,000 years for half the original sample to decay and another 10,000 years for half of what remained to decay. The age of your sample would be 2 × 10,000, or 20,000, years. **Figure 2** shows how this steady decay happens.

Reading Check **What is a half-life?** (*See the Appendix for answers to Reading Checks.*)

Figure 2 *After every half-life, the amount of parent material decreases by one-half.*

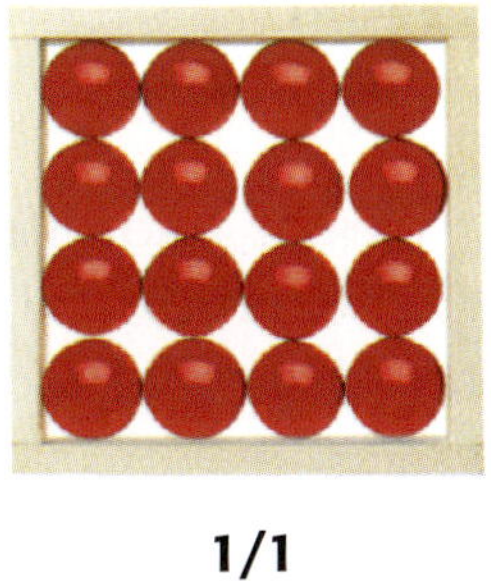

1/1
0 years

1/2
10,000 years

1/4
20,000 years

1/8
30,000 years

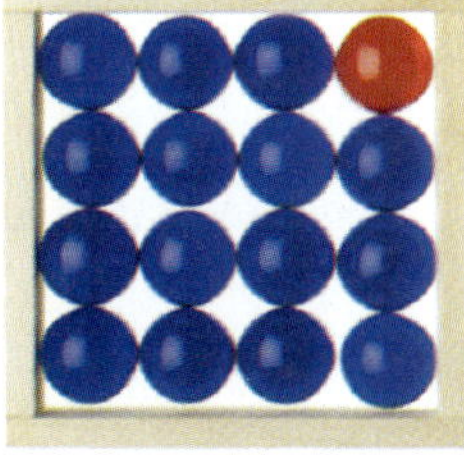

1/16
40,000 years

Figure 3 *This burial mound at Effigy Mounds resembles a snake.*

Types of Radiometric Dating

Imagine traveling back through the centuries to a time before Columbus arrived in America. You are standing along the bluffs of what will one day be called the Mississippi River. You see dozens of people building large mounds. Who are these people, and what are they building?

The people you saw in your time travel were Native Americans, and the structures they were building were burial mounds. The area you imagined is now an archaeological site called Effigy Mounds National Monument. **Figure 3** shows one of these mounds.

According to archaeologists, people lived at Effigy Mounds from 2,500 years ago to 600 years ago. How do archaeologists know these dates? They have dated bones and other objects in the mounds by using radiometric dating. Scientists use different radiometric-dating techniques based on the estimated age of an object. As you read on, think about how the half-life of an isotope relates to the age of the object being dated. Which technique would you use to date the burial mounds?

Potassium-Argon Method

One isotope that is used for radiometric dating is potassium-40. Potassium-40 has a half-life of 1.3 billion years, and it decays to argon and calcium. Geologists measure argon as the daughter material. This method is used mainly to date rocks older than 100,000 years.

Uranium-Lead Method

Uranium-238 is a radioactive isotope that decays in a series of steps to lead-206. The half-life of uranium-238 is 4.5 billion years. The older the rock is, the more daughter material (lead-206) there will be in the rock. Uranium-lead dating can be used for rocks more than 10 million years old. Younger rocks do not contain enough daughter material to be accurately measured by this method.

Rubidium-Strontium Method

Through radioactive decay, the unstable parent isotope rubidium-87 forms the stable daughter isotope strontium-87. The half-life of rubidium-87 is 49 billion years. This method is used to date rocks older than 10 million years.

Reading Check **What is the daughter isotope of rubidium-87?**

Carbon-14 Method

The element carbon is normally found in three forms, the stable isotopes carbon-12 and carbon-13 and the radioactive isotope carbon-14. These carbon isotopes combine with oxygen to form the gas carbon dioxide, which is taken in by plants during photosynthesis. As long as a plant is alive, new carbon dioxide with a constant carbon-14 to carbon-12 ratio is continually taken in. Animals that eat plants contain the same ratio of carbon isotopes.

Once a plant or an animal dies, however, no new carbon is taken in. The amount of carbon-14 begins to decrease as the plant or animal decays, and the ratio of carbon-14 to carbon-12 decreases. This decrease can be measured in a laboratory, such as the one shown in **Figure 4.** Because the half-life of carbon-14 is only 5,730 years, this dating method is used mainly for dating things that lived within the last 50,000 years.

Figure 4 *Some samples containing carbon must be cleaned and burned before their age can be determined.*

SECTION Review

Summary

- During radioactive decay, an unstable isotope decays at a constant rate and becomes a stable isotope of the same or a different element.
- Radiometric dating, based on the ratio of parent to daughter material, is used to determine the absolute age of a sample.
- Methods of radiometric dating include potassium-argon, uranium-lead, rubidium-strontium, and carbon-14 dating.

Using Key Terms

1. Use each of the following terms in a separate sentence: *absolute dating, isotope,* and *half-life.*

Understanding Key Ideas

2. Rubidium-87 has a half-life of

- **a.** 5,730 years.
- **b.** 4.5 billion years.
- **c.** 49 billion years.
- **d.** 1.3 billion years.

3. Explain how radioactive decay occurs.

4. How does radioactive decay relate to radiometric dating?

5. List four types of radiometric dating.

Math Skills

6. A radioactive isotope has a half-life of 1.3 billion years. After 3.9 billion years, how much of the parent material will be left?

Critical Thinking

7. Analyzing Methods Explain why radioactive decay must be constant in order for radiometric dating to be accurate.

8. Applying Concepts Which radiometric-dating method would be most appropriate for dating artifacts found at Effigy Mounds? Explain.

SECTION 4

Looking at Fossils

Descending from the top of a ridge in the badlands of Argentina, your expedition team suddenly stops. You look down and realize that you are walking on eggshells—dinosaur eggshells!

A paleontologist named Luis Chiappe had this experience. He had found an enormous dinosaur nesting ground.

READING WARM-UP

Objectives

- Describe five ways that different types of fossils form.
- List three types of fossils that are not part of organisms.
- Explain how fossils can be used to determine the history of changes in environments and organisms.
- Explain how index fossils can be used to date rock layers.

Terms to Learn

fossil
trace fossil
mold
cast
index fossil

READING STRATEGY

Reading Organizer As you read this section, create an outline of the section. Use the headings from this section in your outline.

Fossilized Organisms

The remains or physical evidence of an organism preserved by geologic processes is called a **fossil.** Fossils are most often preserved in sedimentary rock. But as you will see, other materials can also preserve evidence of past life.

Fossils in Rocks

When an organism dies, it either immediately begins to decay or is consumed by other organisms. Sometimes, however, organisms are quickly buried by sediment when they die. The sediment slows down decay. Hard parts of organisms, such as shells and bones, are more resistant to decay than soft tissues are. So, when sediments become rock, the hard parts of animals are much more commonly preserved than are soft tissues.

Fossils in Amber

Imagine that an insect is caught in soft, sticky tree sap. Suppose that the insect gets covered by more sap, which quickly hardens and preserves the insect inside. Hardened tree sap is called *amber.* Some of our best insect fossils are found in amber, as shown in **Figure 1.** Frogs and lizards have also been found in amber.

Reading Check **Describe how organisms are preserved in amber.** (*See the Appendix for answers to Reading Checks.*)

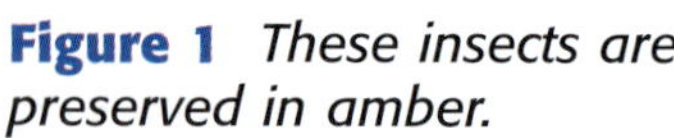

Figure 1 *These insects are preserved in amber.*

Figure 2 *Scientist Vladimir Eisner studies the upper molars of a 20,000-year-old woolly mammoth found in Siberia, Russia. The almost perfectly preserved male mammoth was excavated from a block of ice in October 1999.*

Petrifaction

Another way that organisms are preserved is by petrifaction. *Petrifaction* is a process in which minerals replace an organism's tissues. One form of petrifaction is called permineralization. *Permineralization* is a process in which the pore space in an organism's hard tissue—for example, bone or wood—is filled up with mineral. Another form of petrifaction is called *replacement,* a process in which the organism's tissues are completely replaced by minerals. For example, in some specimens of petrified wood, all of the wood has been replaced by minerals.

fossil the remains or physical evidence of an organism preserved by geological processes

Fossils in Asphalt

There are places where asphalt wells up at the Earth's surface in thick, sticky pools. The La Brea asphalt deposits in Los Angeles, California, for example, are at least 38,000 years old. These pools of thick, sticky asphalt have trapped and preserved many kinds of organisms for the past 38,000 years. From these fossils, scientists have learned about the past environment in southern California.

Frozen Fossils

In October 1999, scientists removed a 20,000-year-old woolly mammoth frozen in the Siberian tundra. The remains of this mammoth are shown in **Figure 2.** Woolly mammoths, relatives of modern elephants, became extinct approximately 10,000 years ago. Because cold temperatures slow down decay, many types of frozen fossils are preserved from the last ice age. Scientists hope to find out more about the mammoth and the environment in which it lived.

CONNECTION TO Environmental Science

WRITING SKILL **Preservation in Ice** Subfreezing climates contain almost no decomposing bacteria. The well-preserved body of John Torrington, a member of an expedition that explored the Northwest Passage in Canada in the 1840s, was uncovered in 1984. His body appeared much as it did at the time he died, more than 160 years earlier. Research another well-preserved discovery, and write a report for your class.

Figure 3 *These dinosaur tracks are located in Arizona. They leave a trace of a dinosaur that had longer legs than humans do.*

Other Types of Fossils

Besides their hard parts—and in rare cases their soft parts—do organisms leave behind any other clues about their existence? What other evidence of past life do paleontologists look for?

Trace Fossils

Any naturally preserved evidence of animal activity is called a **trace fossil.** Tracks like the ones shown in **Figure 3** are a fascinating example of a trace fossil. These fossils form when animal footprints fill with sediment and are preserved in rock. Tracks reveal a lot about the animal that made them, including how big it was and how fast it was moving. Parallel trackways showing dinosaurs moving in the same direction have led paleontologists to hypothesize that dinosaurs moved in herds.

Burrows are another trace fossil. Burrows are shelters made by animals, such as clams, that bury in sediment. Like tracks, burrows are preserved when they are filled in with sediment and buried quickly. A *coprolite* (KAHP roh LIET), a third type of trace fossil, is preserved animal dung.

trace fossil a fossilized mark that is formed in soft sediment by the movement of an animal

mold a mark or cavity made in a sedimentary surface by a shell or other body

cast a type of fossil that forms when sediments fill in the cavity left by a decomposed organism

Molds and Casts

Molds and casts are two more examples of fossils. A cavity in rock where a plant or animal was buried is called a **mold.** A **cast** is an object created when sediment fills a mold and becomes rock. A cast shows what the outside of the organism looked like. **Figure 4** shows two types of molds from the same organism—and internal mold and an external mold.

Reading Check How are a cast and a mold different?

Figure 4 *This photograph shows two molds from an ammonite. The image on the left is the internal mold of the ammonite, which formed when sediment filled the ammonite's shell, which later dissolved away. The image on the right is the external mold of the ammonite, which preserves the external features of the shell.*

Figure 5 *This scientist has found marine fossils on mountaintops in the Yoho National Park in Canada. The fossil of* Marrella, *shown above, tells the scientist that these rocks were pushed up from below sea level millions of years ago.*

Using Fossils to Interpret the Past

Think about your favorite outdoor place. Now, imagine that you are a paleontologist at the same site 65 million years from now. What types of fossils would you dig up? Based on the fossils you found, how would you reconstruct this place?

The Information in the Fossil Record

The fossil record offers only a rough sketch of the history of life on Earth. Some parts of this history are more complete than others. For example, scientists know more about organisms that had hard body parts than about organisms that had soft body parts. Scientists also know more about organisms that lived in environments that favored fossilization. The fossil record is incomplete because most organisms never became fossils. And of course, many fossils have yet to be discovered.

History of Environmental Changes

Would you expect to find marine fossils on the mountaintop shown in **Figure 5**? The presence of marine fossils means that the rocks of these mountaintops in Canada formed in a totally different environment—at the bottom of an ocean.

The fossil record reveals a history of environmental change. For example, marine fossils help scientists reconstruct ancient coastlines and the deepening and shallowing of ancient seas. Using the fossils of plants and land animals, scientists can reconstruct past climates. They can tell whether the climate in an area was cooler or wetter than it is at present.

Make a Fossil

1. Find a **common object,** such as a shell, a button, or a pencil, to use to make a mold. Keep the object hidden from your classmates.
2. To create a mold, press the items down into **modeling clay** in a **shallow pan or tray.**
3. Trade your tray with a classmate's tray, and try to identify the item that made the mold.
4. Describe how a cast could be formed from your mold.

History of Changing Organisms

By studying the relationships between fossils, scientists can interpret how life has changed over time. For example, older rock layers contain organisms that often differ from the organisms found in younger rock layers.

Only a small fraction of the organisms that have existed in Earth's history have been fossilized. Because the fossil record is incomplete, it does not provide paleontologists with a continuous record of change. Instead, they look for similarities between fossils, or between fossilized organisms and their closest living relatives, and try to fill in the blanks in the fossil record.

Fossil Hunt

Go on a fossil hunt with your family. Find out what kinds of rocks in your local area might contain fossils. Take pictures or draw sketches of your trip and any fossils that you find.

Reading Check How do paleontologists fill in missing information about changes in organisms in the fossil record?

Using Fossils to Date Rocks

Scientists have found that particular types of fossils appear only in certain layers of rock. By dating the rock layers above and below these fossils, scientists can determine the time span in which the organisms that formed the fossils lived. If a type of organism existed for only a short period of time, its fossils would show up in a limited range of rock layers. These types of fossils are called index fossils. **Index fossils** are fossils of organisms that lived during a relatively short, well-defined geologic time span.

index fossil a fossil that is found in the rock layers of only one geologic age and that is used to establish the age of the rock layers

Ammonites

To be considered an index fossil, a fossil must be found in rock layers throughout the world. One example of an index fossil is the fossil of a genus of ammonites (AM uh NIETS) called *Tropites*, shown in **Figure 6.** *Tropites* was a marine mollusk similar to a modern squid. It lived in a coiled shell. *Tropites* lived between 230 million and 208 million years ago and is an index fossil for that period of time.

Figure 6 Tropites *is a genus of coiled ammonites.* Tropites *existed for only about 20 million years, which makes this genus a good index fossil.*

Trilobites

Fossils of a genus of trilobites (TRIE loh BIETS) called *Phacops* are another example of an index fossil. Trilobites are extinct. Their closest living relative is the horseshoe crab. Through the dating of rock, paleontologists have determined that *Phacops* lived approximately 400 million years ago. So, when scientists find *Phacops* in rock layers anywhere on Earth, they assume that these rock layers are also approximately 400 million years old. An example of a *Phacops* fossil is shown in **Figure 7.**

Figure 7 *Paleontologists assume that any rock layer containing a fossil of the trilobite* Phacops *is about 400 million years old.*

Reading Check **Explain how fossils of *Phacops* can be used to establish the age of rock layers.**

SECTION Review

Summary

- Fossils are the remains or physical evidence of an organism preserved by geologic processes.
- Fossils can be preserved in rock, amber, asphalt, and ice and by petrifaction.
- Trace fossils are any naturally preserved evidence of animal activity. Tracks, burrows, and coprolites are examples of trace fossils.
- Scientists study fossils to determine how environments and organisms have changed over time.
- An index fossil is a fossil of an organism that lived during a relatively short, well-defined time span. Index fossils can be used to establish the age of rock layers.

Using Key Terms

Complete each of the following sentences by choosing the correct term from the word bank.

cast	index fossils
mold	trace fossils

1. A ___ is a cavity in rock where a plant or animal was buried.
2. ___ can be used to establish the age of rock layers.

Understanding Key Ideas

3. Fossils are most often preserved in
   a. ice.
   b. amber.
   c. asphalt.
   d. rock.
4. Describe three types of trace fossils.
5. Explain how an index fossil can be used to date rock.
6. Explain why the fossil record contains an incomplete record of the history of life on Earth.
7. Explain how fossils can be used to determine the history of changes in environments and organisms.

Math Skills

8. If a scientist finds the remains of a plant between a rock layer that contains 400 million–year-old *Phacops* fossils and a rock layer that contains 230 million–year-old *Tropites* fossils, how old could the plant fossil be?

Critical Thinking

9. **Making Inferences** If you find rock layers containing fish fossils in a desert, what can you infer about the history of the desert?
10. **Identifying Bias** Because information in the fossil record is incomplete, scientists are left with certain biases concerning fossil preservation. Explain two of these biases.

SCILINKS. NSTA Developed and maintained by the National Science Teachers Association

For a variety of links related to this chapter, go to www.scilinks.org

Topic: Looking at Fossils
SciLinks code: HSM0886

SECTION 5

Time Marches On

How old is the Earth? Well, if the Earth celebrated its birthday every million years, there would be 4,600 candles on its birthday cake! Humans have been around only long enough to light the last candle on the cake.

Try to think of the Earth's history in "fast-forward." If you could watch the Earth change from this perspective, you would see mountains rise up like wrinkles in fabric and quickly wear away. You would see life-forms appear and then go extinct. In this section, you will learn that geologists must "fast-forward" the Earth's history when they write or talk about it. You will also learn about some incredible events in the history of life on Earth.

READING WARM-UP

Objectives

- Explain how geologic time is recorded in rock layers.
- Identify important dates on the geologic time scale.
- Explain how environmental changes resulted in the extinction of some species.

Terms to Learn

geologic time scale
eon
era
period
epoch
extinction

READING STRATEGY

Brainstorming The key idea of this section is the geologic time scale. Brainstorm words and phrases related to the geologic time scale.

Geologic Time

Shown in **Figure 1** is the rock wall at the Dinosaur Quarry Visitor Center in Dinosaur National Monument, Utah. Contained within this wall are approximately 1,500 fossil bones that have been excavated by paleontologists. These are the remains of dinosaurs that inhabited the area about 150 million years ago. Granted, 150 million years seems to be an incredibly long period of time. However, in terms of the Earth's history, 150 million years is little more than 3% of the time our planet has existed. It is a little less than 4% of the time represented by the Earth's oldest known rocks.

Figure 1 *Bones of dinosaurs that lived about 150 million years ago are exposed in the quarry wall at Dinosaur National Monument in Utah.*

Figure 2 *Well-preserved plant and animal fossils are common in the Green River formation. Clockwise from the upper right are a fossil leaf, a dragonfly, a fish, and a turtle.*

The Rock Record and Geologic Time

One of the best places in North America to see the Earth's history recorded in rock layers is in Grand Canyon National Park. The Colorado River has cut the canyon nearly 2 km deep in some places. Over the course of 6 million years, the river has eroded countless layers of rock. These layers represent almost half, or nearly 2 billion years, of Earth's history.

Reading Check **How much geologic time is represented by the rock layers in the Grand Canyon?** (*See the Appendix for answers to Reading Checks.*)

The Fossil Record and Geologic Time

Figure 2 shows sedimentary rocks that belong to the Green River formation. These rocks, which are found in parts of Wyoming, Utah, and Colorado, are thousands of meters thick. These rocks were once part of a system of ancient lakes that existed for a period of millions of years. Fossils of plants and animals are common in these rocks and are very well preserved. Burial in the fine-grained lake-bed sediments preserved even the most delicate structures.

For another activity related to this chapter, go to **go.hrw.com** and type in the keyword **HZ5FOSW.**

Phanerozoic Eon

(543 million years ago to the present)
The rock and fossil record mainly represents the Phanerozoic eon, which is the eon in which we live.

Proterozoic Eon

(2.5 billion years ago to 543 million years ago)
The first organisms with well-developed cells appeared during this eon.

Archean Eon

(3.8 billion years ago to 2.5 billion years ago)
The earliest known rocks on Earth formed during this eon.

Hadean Eon

(4.6 billion years ago to 3.8 billion years ago)
The only rocks that scientists have found from this eon are meteorites and rocks from the moon.

Geologic Time Scale

Eon	Era	Period	Epoch	Millions of years ago
PHANEROZOIC EON	Cenozoic	Quaternary	Holocene	0.01
			Pleistocene	1.8
		Tertiary	Pliocene	5.3
			Miocene	23.8
			Oligocene	33.7
			Eocene	54.8
			Paleocene	65
	Mesozoic	Cretaceous		144
		Jurassic		206
		Triassic		248
	Paleozoic	Permian		290
		Pennsylvanian		323
		Mississippian		354
		Devonian		417
		Silurian		443
		Ordovician		490
		Cambrian		543
PROTEROZOIC EON				2,500
ARCHEAN EON				3,800
HADEAN EON				4,600

Figure 3 *The geologic time scale accounts for Earth's entire history. It is divided into four major parts called* eons. *Dates given for intervals on the geologic time scale are estimates.*

The Geologic Time Scale

The geologic column represents the billions of years that have passed since the first rocks formed on Earth. Altogether, geologists study 4.6 billion years of Earth's history! To make their job easier, geologists have created the geologic time scale. The **geologic time scale,** which is shown in **Figure 3,** is a scale that divides Earth's 4.6 billion–year history into distinct intervals of time.

Reading Check **Define the term *geologic time scale.***

Divisions of Time

Geologists have divided Earth's history into sections of time, as shown on the geologic time scale in **Figure 3.** The largest divisions of geologic time are **eons** (EE AHNZ). There are four eons—the Hadean eon, the Archean eon, the Proterozoic eon, and the Phanerozoic eon. The Phanerozoic eon is divided into three **eras,** which are the second-largest divisions of geologic time. The three eras are further divided into **periods,** which are the third-largest divisions of geologic time. Periods are divided into **epochs** (EP uhks), which are the fourth-largest divisions of geologic time.

The boundaries between geologic time intervals represent shorter intervals in which visible changes took place on Earth. Some changes are marked by the disappearance of index fossil species, while others are recognized only by detailed paleontological studies.

geologic time scale the standard method used to divide the Earth's long natural history into manageable parts

eon the largest division of geologic time

era a unit of geologic time that includes two or more periods

period a unit of geologic time into which eras are divided

epoch a subdivision of a geologic period

extinction the death of every member of a species

The Appearance and Disappearance of Species

At certain times during Earth's history, the number of species has increased or decreased dramatically. An increase in the number of species often comes as a result of either a relatively sudden increase or decrease in competition among species. *Hallucigenia,* shown in **Figure 4,** appeared during the Cambrian period, when the number of marine species greatly increased. On the other hand, the number of species decreases dramatically over a relatively short period of time during a mass extinction event. **Extinction** is the death of every member of a species. Gradual events, such as global climate change and changes in ocean currents, can cause mass extinctions. A combination of these events can also cause mass extinctions.

Figure 4 Hallucigenia, *named for its "bizarre and dreamlike quality," was one of numerous marine organisms to make its appearance during the early Cambrian period.*

Figure 5 *Jungles were present during the Paleozoic era, but there were no birds singing in the trees and no monkeys swinging from the branches. Birds and mammals didn't evolve until much later.*

The Paleozoic Era—Old Life

The Paleozoic era lasted from about 543 million to 248 million years ago. It is the first era well represented by fossils.

Marine life flourished at the beginning of the Paleozoic era. The oceans became home to a diversity of life. However, there were few land organisms. By the middle of the Paleozoic, all modern groups of land plants had appeared. By the end of the era, amphibians and reptiles lived on the land, and insects were abundant. **Figure 5** shows what the Earth might have looked like late in the Paleozoic era. The Paleozoic era came to an end with the largest mass extinction in Earth's history. Some scientists believe that ocean changes were a likely cause of this extinction, which killed nearly 90% of all species.

The Mesozoic Era—The Age of Reptiles

The Mesozoic era began about 248 million years ago. The Mesozoic is known as the *Age of Reptiles* because reptiles, such as the dinosaurs shown in **Figure 6,** inhabited the land.

During this time, reptiles dominated. Small mammals appeared about the same time as dinosaurs, and birds appeared late in the Mesozoic era. Many scientists think that birds evolved directly from a type of dinosaur. At the end of the Mesozoic era, about 15% to 20% of all species on Earth, including the dinosaurs, became extinct. Global climate change may have been the cause.

Reading Check **Why is the Mesozoic known as the *Age of Reptiles*?**

Figure 6 *Imagine walking in the desert and bumping into these fierce creatures! It's a good thing humans didn't evolve in the Mesozoic era, which was dominated by dinosaurs.*

The Cenozoic Era—The Age of Mammals

The Cenozoic era, as shown in **Figure 7,** began about 65 million years ago and continues to the present. This era is known as the *Age of Mammals*. During the Mesozoic era, mammals had to compete with dinosaurs and other animals for food and habitat. After the mass extinction at the end of the Mesozoic era, mammals flourished. Unique traits, such as regulating body temperature internally and bearing young that develop inside the mother, may have helped mammals survive the environmental changes that probably caused the extinction of the dinosaurs.

Figure 7 *Thousands of species of mammals evolved during the Cenozoic era. This scene shows species from the early Cenozoic era that are now extinct.*

SECTION Review

Summary

- The geologic time scale divides Earth's 4.6 billion–year history into distinct intervals of time. Divisions of geologic time include eons, eras, periods, and epochs.
- The boundaries between geologic time intervals represent visible changes that have taken place on Earth.
- The rock and fossil record represents mainly the Phanerozoic eon, which is the eon in which we live.
- At certain times in Earth's history, the number of life-forms has increased or decreased dramatically.

Using Key Terms

1. Use each of the following terms in the same sentence: *era, period,* and *epoch.*

Understanding Key Ideas

2. The unit of geologic time that began 65 million years ago and continues to the present is the
   - **a.** Holocene epoch.
   - **b.** Cenozoic era.
   - **c.** Phanerozoic eon.
   - **d.** Quaternary period.
3. What are the major time intervals represented by the geologic time scale?
4. Explain how geologic time is recorded in rock layers.
5. What kinds of environmental changes cause mass extinctions?

Critical Thinking

6. **Making Inferences** What future event might mark the end of the Cenozoic era?
7. **Identifying Relationships** How might a decrease in competition between species lead to the sudden appearance of many new species?

Interpreting Graphics

8. Look at the illustration below. On the Earth-history clock shown, 1 h equals 383 million years, and 1 min equals 6.4 million years. In millions of years, how much more time is represented by the Proterozoic eon than by the Phanerozoic eon?

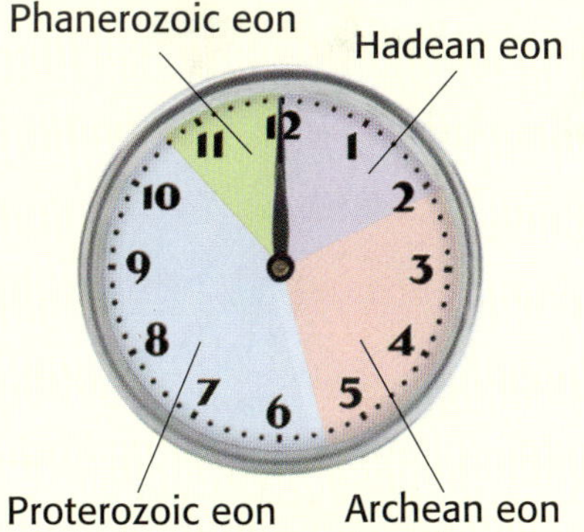

For a variety of links related to this chapter, go to www.scilinks.org

Topic: Geologic Time
SciLinks code: HSM0668

Model-Making Lab

How Do You Stack Up?

According to the principle of superposition, in undisturbed sequences of sedimentary rock, the oldest layers are on the bottom. Geologists use this principle to determine the relative age of the rocks in a small area. In this activity, you will model what geologists do by drawing sections of different rock outcrops. Then, you will create a part of the geologic column, showing the geologic history of the area that contains all of the outcrops.

OBJECTIVES

Make a model of a geologic column.

Interpret the geologic history represented by the geologic column you have made.

MATERIALS

- paper, white
- pencil
- pencils or crayons, assorted colors
- ruler, metric
- scissors
- tape, transparent

SAFETY

Procedure

1. Use a metric ruler and a pencil to draw four boxes on a blank piece of paper. Each box should be 3 cm wide and at least 6 cm tall. (You can trace the boxes shown on the next page.)
2. With colored pencils, copy the illustrations of the four outcrops on the next page. Copy one illustration in each of the four boxes. Use colors and patterns similar to those shown.
3. Pay close attention to the contact between layers—straight or wavy. Straight lines represent bedding planes, where deposition was continuous. Wavy lines represent unconformities, where rock layers may be missing. The top of each outcrop is incomplete, so it should be a jagged line. (Assume that the bottom of the lowest layer is a bedding plane.)
4. Use a black crayon or pencil to add the symbols representing fossils to the layers in your drawings. Pay attention to the shapes of the fossils and the layers that they are in.
5. Write the outcrop number on the back of each section.
6. Carefully cut the outcrops out of the paper, and lay the individual outcrops next to each other on your desk or table.
7. Find layers that have the same rocks and contain the same fossils. Move each outcrop up or down to line up similar layers next to each other.
8. If unconformities appear in any of the outcrops, there may be rock layers missing. You may need to examine other sections to find out what fits between the layers above and below the unconformities. Leave room for these layers by cutting the outcrops along the unconformities (wavy lines).

9. Eventually, you should be able to make a geologic column that represents all four of the outcrops. It will show rock types and fossils for all the known layers in the area.
10. Tape the pieces of paper together in a pattern that represents the complete geologic column.

Analyze the Results

1. **Examining Data** How many layers are in the part of the geologic column that you modeled?
2. **Examining Data** Which is the oldest layer in your column? Which rock layer is the youngest? How do you know? Describe these layers in terms of rock type or the fossils they contain.
3. **Classifying** List the fossils in your column from oldest to youngest. Label the youngest and oldest fossils.
4. **Analyzing Data** Look at the unconformity in outcrop 2. Which rock layers are partially or completely missing? How do you know?

Draw Conclusions

5. **Drawing Conclusions** Which (if any) fossils can be used as index fossils for a single layer? Why are these fossils considered index fossils? What method(s) would be required to determine the absolute age of these fossils?

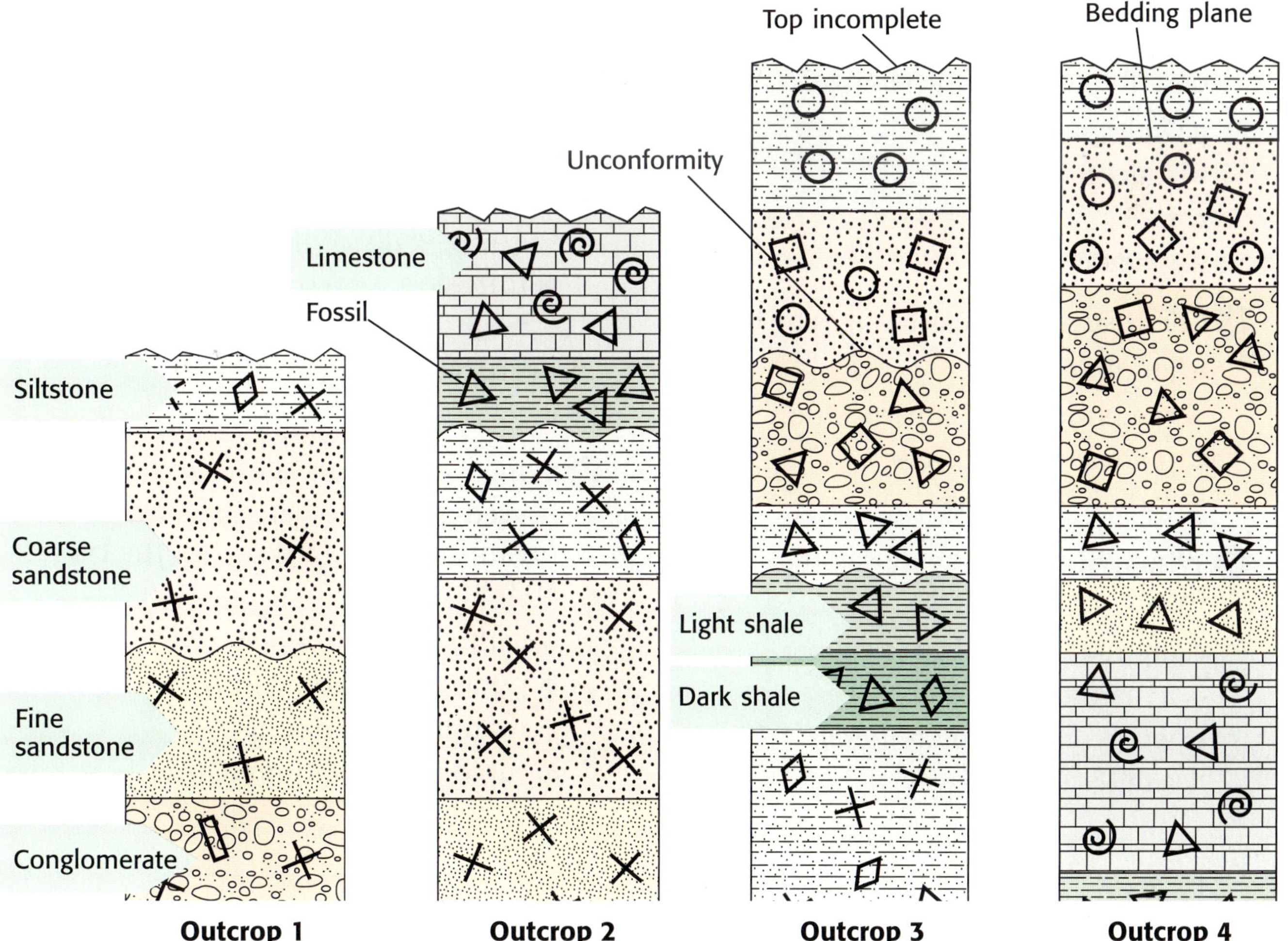

Chapter Review

USING KEY TERMS

1. In your own words, write a definition for each of the following terms: *superposition, geologic column,* and *geologic time scale.*

For each pair of terms, explain how the meanings of the terms differ.

2. *uniformitarianism* and *catastrophism*

3. *relative dating* and *absolute dating*

4. *trace fossil* and *index fossil*

UNDERSTANDING KEY IDEAS

Multiple Choice

5. Which of the following does not describe catastrophic change?
   a. widespread
   b. sudden
   c. rare
   d. gradual

6. Scientists assign relative ages by using
   a. absolute dating.
   b. the principle of superposition.
   c. radioactive half-lives.
   d. carbon-14 dating.

7. Which of the following is a trace fossil?
   a. an insect preserved in amber
   b. a mammoth frozen in ice
   c. wood replaced by minerals
   d. a dinosaur trackway

8. The largest divisions of geologic time are called
   a. periods.
   b. eras.
   c. eons.
   d. epochs.

9. Rock layers cut by a fault formed
   a. after the fault.
   b. before the fault.
   c. at the same time as the fault.
   d. There is not enough information to determine the answer.

10. Of the following isotopes, which is stable?
    a. uranium-238
    b. potassium-40
    c. carbon-12
    d. carbon-14

11. A surface that represents a missing part of the geologic column is called a(n)
    a. intrusion.
    b. fault.
    c. unconformity.
    d. fold.

12. Which method of radiometric dating is used mainly to date the remains of organisms that lived within the last 50,000 years?
    a. carbon-14 dating
    b. potassium-argon dating
    c. uranium-lead dating
    d. rubidium-strontium dating

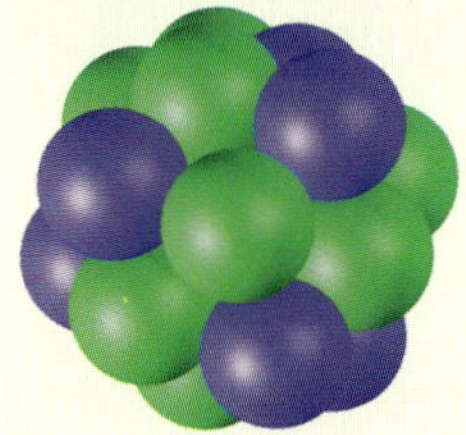

Short Answer

13. Describe three processes by which fossils form.

14. Identify the role of uniformitarianism in Earth science.

15. Explain how radioactive decay occurs.

16. Describe two ways in which scientists use fossils to determine environmental change.

17. Explain the role of paleontology in the study of Earth's history.

CRITICAL THINKING

18. **Concept Mapping** Use the following terms to create a concept map: *age, half-life, absolute dating, radioactive decay, radiometric dating, relative dating, superposition, geologic column,* and *isotopes.*

19. **Applying Concepts** Identify how changes in environmental conditions can affect the survival of a species. Give two examples.

20. **Identifying Relationships** Why do paleontologists know more about hard-bodied organisms than about soft-bodied organisms?

21. **Analyzing Processes** Why isn't a 100 million–year-old fossilized tree made of wood?

INTERPRETING GRAPHICS

Use the diagram below to answer the questions that follow.

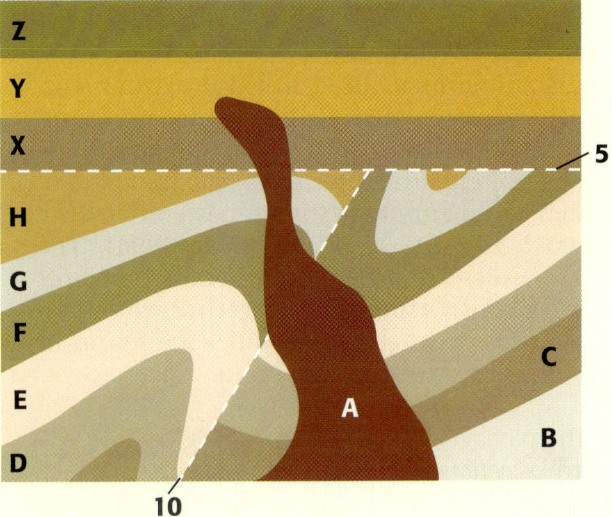

22. Is intrusion **A** younger or older than layer **X**? Explain.

23. What feature is marked by **5**?

24. Is intrusion **A** younger or older than fault **10**? Explain.

25. Other than the intrusion and faulting, what event happened in layers **B, C, D, E, F, G,** and **H**? Number this event, the intrusion, and the faulting in the order that they happened.

Standardized Test Preparation

READING

Read each of the passages below. Then, answer the questions that follow each passage.

Passage 1 Three hundred million years ago, the region that is now Illinois had a different climate than it does today. Swamps and shallow bays covered much of the area. No fewer than 500 species of plants and animals lived in this environment. Today, the remains of these organisms are found beautifully preserved within nodules. Nodules are round or oblong structures usually composed of cemented sediments that sometimes contain the fossilized hard parts of plants and animals. The Illinois nodules are exceptional because the soft parts of organisms are found together with hard parts. For this reason, these nodules are found in fossil collections around the world.

1. In the passage, what is the meaning of the word *exceptional*?

A beautiful
B extraordinary
C average
D large

2. According to the passage, which of the following statements about nodules is correct?

F Nodules are rarely round or oblong.
G Nodules are usually composed of cemented sediment.
H Nodules are not found in present-day Illinois.
I Nodules always contain fossils.

3. Which of the following is a fact in the passage?

A The Illinois nodules are not well known outside of Illinois.
B Illinois has had the same climate throughout Earth's history.
C Both the hard and soft parts of organisms are preserved in the Illinois nodules.
D Fewer than 500 species of plants and animals have been found in Illinois nodules.

Passage 2 In 1995, paleontologist Paul Sereno and his team were working in an unexplored region of Morocco when they made an astounding find—an enormous dinosaur skull! The skull measured approximately 1.6 m in length, which is about the height of a refrigerator. Given the size of the skull, Sereno concluded that the skeleton of the animal it came from must have been about 14 m long—about as big as a school bus. The dinosaur was even larger than *Tyrannosaurus rex*! The newly discovered 90 million–year-old predator most likely chased other dinosaurs by running on large, powerful hind legs, and its bladelike teeth meant certain death for its prey.

1. In the passage, what does the word *astounding* mean?

A important
B new
C incredible
D one of a kind

2. Which of the following is evidence that the dinosaur described in the passage was a predator?

F It had bladelike teeth.
G It had a large skeleton.
H It was found with the bones of a smaller animal nearby.
I It is 90 million years old.

3. What types of information do you think that fossil teeth provide about an organism?

A the color of its skin
B the types of food it ate
C the speed that it ran
D the mating habits it had

INTERPRETING GRAPHICS

Use the graph below to answer the questions that follow.

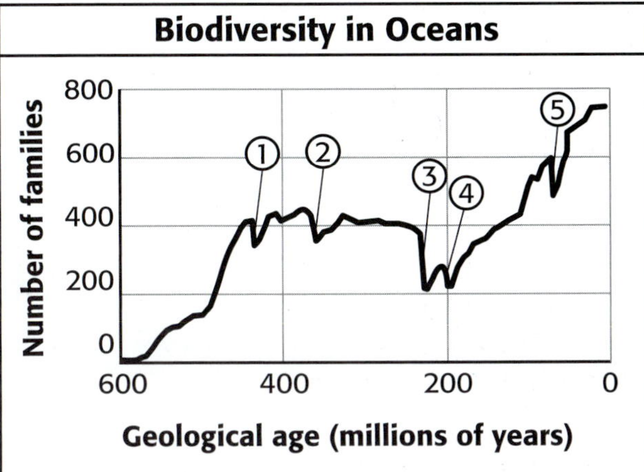

1. At which point in Earth's history did the greatest mass-extinction event take place?
   - **A** at point **1**, the Ordovician-Silurian boundary
   - **B** at point **3**, the Permian-Triassic boundary
   - **C** at point **4**, the Triassic-Jurassic boundary
   - **D** at point **5**, the Cretaceous-Tertiary boundary

2. Immediately following the Cretaceous-Tertiary extinction, represented by point **5**, approximately how many families of marine organisms remained in the Earth's oceans?
   - **F** 200 marine families
   - **G** 300 marine families
   - **H** 500 marine families
   - **I** 700 marine families

3. Approximately how many million years ago did the Ordovician-Silurian mass-extinction event, represented by point **1**, take place?
   - **A** 200 million years ago
   - **B** 250 million years ago
   - **C** 350 million years ago
   - **D** 420 million years ago

MATH

Read each question below, and choose the best answer.

1. Carbon-14 is a radioactive isotope with a half-life of 5,730 years. How much carbon-14 would remain in a sample that is 11,460 years old?
   - **A** 12.5%
   - **B** 25%
   - **C** 50%
   - **D** 100%

2. If a sample contains an isotope with a half-life of 10,000 years, how old would the sample be if 1/8 of the original isotope remained in the sample?
   - **F** 20,000 years
   - **G** 30,000 years
   - **H** 40,000 years
   - **I** 50,000 years

3. If a sample contains an isotope with a half-life of 5,000 years, how old would the sample be if 1/4 of the original isotope remained in the sample?
   - **A** 10,000 years
   - **B** 20,000 years
   - **C** 30,000 years
   - **D** 40,000 years

4. If Earth history spans 4.6 billion years and the Phanerozoic eon was 543 million years, what percentage of Earth history does the Phanerozoic eon represent?
   - **F** about 6%
   - **G** about 12%
   - **H** about 18%
   - **I** about 24%

5. Humans live in the Holocene epoch. If the Holocene epoch has lasted approximately 10,000 years, what percentage of the Quaternary period, which began 1.8 million years ago, is represented by the Holocene?
   - **A** about 0.0055%
   - **B** about 0.055%
   - **C** about 0.55%
   - **D** about 5.5%

Science in Action

Scientific Debate

Feathered Dinosaurs

One day in 1996, a Chinese farmer broke open a rock he found in the bed of an ancient dry lake. What he found inside the rock became one of the most exciting paleontological discoveries of the 20th century. Preserved inside were the remains of a dinosaur. The dinosaur had a large head; powerful jaws; sharp, jagged teeth; and, most important of all, a row of featherlike structures along the backbone. Scientists named the dinosaur *Sinosauropteryx,* or "Chinese dragon wing." *Sinosauropteryx* and the remains of other "feathered" dinosaurs recently discovered in China have led some scientists to hypothesize that feathers evolved through theropod (three-toed) dinosaurs. Other paleontologists disagree. They believe the structures along the backbone of these dinosaurs are not feathers but the remains of elongated spines, like those that run down the head and back of an iguana.

Language Arts ACTIVITY

Paleontologists often give dinosaurs names that describe something unusual about the animal's head, body, feet, or size. These names have Greek or Latin roots. Research the names of some dinosaurs, and find out what the names mean. Create a list of dinosaur names and their meanings.

Science, Technology, and Society

DNA and a Mammoth Discovery

In recent years, scientists have unearthed several mammoths that had been frozen in ice in Siberia and other remote northern locations. Bones, fur, food in the stomach, and even dung have all been found in good condition. Some scientists hoped that DNA extracted from the mammoths might lead to the cloning of this animal, which became extinct about 10,000 years ago. But the DNA might not be able to be duplicated by scientists. However, DNA samples may nevertheless help scientists understand why mammoths became extinct. One theory about why mammoths became extinct is that they were killed off by disease. Using DNA taken from fossilized mammoth bone, hair, or dung, scientists can check to see if it contains the DNA of a disease-causing pathogen that led to the extinction of the mammoths.

Math ACTIVITY

The male Siberian mammoth reached a height of about 3 m at the shoulder. Females reached a height of about 2.5 m at the shoulder. What is the ratio of the maximum height of a female Siberian mammoth to the height of a male Siberian mammoth?

People in Science

Lizzie May

Amateur Paleontologist For Lizzie May, summer vacations have meant trips into the Alaskan wilderness with her stepfather, geologist/paleontologist Kevin May. The purpose of these trips has not been for fun. Instead, Kevin and Lizzie have been exploring the Alaskan wilderness for the remains of ancient life—dinosaurs, in particular.

At age 18, Lizzie May has gained the reputation of being Alaska's most famous teenage paleontologist. It is a reputation that is well deserved. To date, Lizzie has collected hundreds of dinosaur bones and located important sites of dinosaur, bird, and mammal tracks. In her honor and as a result of her hard work in the field, scientists named the skeleton of a dinosaur discovered by the Mays "Lizzie." "Lizzie" is a duckbill dinosaur, or hadrosaur, that lived approximately 90 million years ago. "Lizzie" is the oldest dinosaur ever found in Alaska and one of the earliest known duckbill dinosaurs in North America.

The Mays have made other, equally exciting discoveries. On one summer trip, Kevin and Lizzie located six dinosaur and bird track sites that dated back 97 million to 144 million years. On another trip, the Mays found a fossil marine reptile more than 200 million years old—an ichthyosaur—that had to be removed with the help of a military helicopter. You have to wonder what other exciting adventures are in store for Lizzie and Kevin!

Social Studies ACTIVITY

WRITING SKILL Lizzie May is not the only young person to have made a mark in dinosaur paleontology. Using the Internet or another source, research people such as Bucky Derflinger, Johnny Maurice, Brad Riney, and Wendy Sloboda, who as young people made contributions to the field of dinosaur study. Write a short essay summarizing your findings.

To learn more about these Science in Action topics, visit go.hrw.com and type in the keyword HZ5FOSF.

Current Science

Check out Current Science® articles related to this chapter by visiting go.hrw.com. Just type in the keyword HZ5CS06.

Energy Resources

About the PHOTO

Would you believe that this house is made from empty soda cans and old tires? Well, it is! *The Castle,* named by its designer, architect Mike Reynolds, and located in Taos, New Mexico, not only uses recycled materials but also saves Earth's energy resources. All of the energy used to run this house comes directly from the sun, and the water used for household activities is rainwater.

PRE-READING ACTIVITY

Graphic Organizer

Comparison Table

Before you read the chapter, create the graphic organizer entitled "Comparison Table" described in the **Study Skills** section of the Appendix. Label the columns with an energy resource from the chapter. Label the rows with "Pros" and "Cons." As you read the chapter, fill in the table with details about the pros and cons of each energy resource.

START-UP ACTIVITY

What Is the Sun's Favorite Color?

Try the following activity to see which colors are better than others at absorbing the sun's energy.

Procedure

1. Obtain **at least five balloons** that are different colors but the same size and shape. One of the balloons should be white, and one should be black. Do not inflate the balloons.
2. Place **several small ice cubes** in each balloon. Each balloon should contain the same amount of ice.
3. Line up the balloons on a flat, uniformly colored surface that receives direct sunlight. Make sure that all of the balloons receive the same amount of sunlight and that the openings in the balloons are not facing directly toward the sun.
4. Record the time that it takes the ice to melt completely in each of the balloons. You can tell how much ice has melted in each balloon by pinching the balloons open and then gently squeezing the balloon.

Analysis

1. In which balloon did the ice melt first? Why?
2. What color would you paint a device used to collect solar energy? Explain your answer.

SECTION 1

Natural Resources

READING WARM-UP

Objectives

- Describe how humans use natural resources.
- Compare renewable resources with nonrenewable resources.
- Explain three ways that humans can conserve natural resources.

Terms to Learn

natural resource
renewable resource
nonrenewable resource
recycling

READING STRATEGY

Reading Organizer As you read this section, make a concept map by using the terms above.

What does the water you drink, the paper you write on, the gasoline used in the cars you ride in, and the air you breathe have in common?

Water, trees used to make paper, crude oil used to make gasoline, and air are just a few examples of Earth's resources. Can you think of other examples of Earth's resources?

Earth's Resources

The Earth provides almost everything needed for life. For example, the Earth's atmosphere provides the air you breathe, maintains air temperatures, and produces rain. The oceans and other waters of the Earth give you food and needed water. The solid part of the Earth gives nutrients, such as potassium, to the plants you eat. These resources that the Earth provides for you are called natural resources.

A **natural resource** is any natural material that is used by humans. Examples of natural resources are water, petroleum, minerals, forests, and animals. Most resources are changed and made into products that make people's lives more comfortable and convenient, as shown in **Figure 1.** The energy we get from many of these resources, such as gasoline and wind, ultimately comes from the sun's energy.

Figure 1 Natural Resources

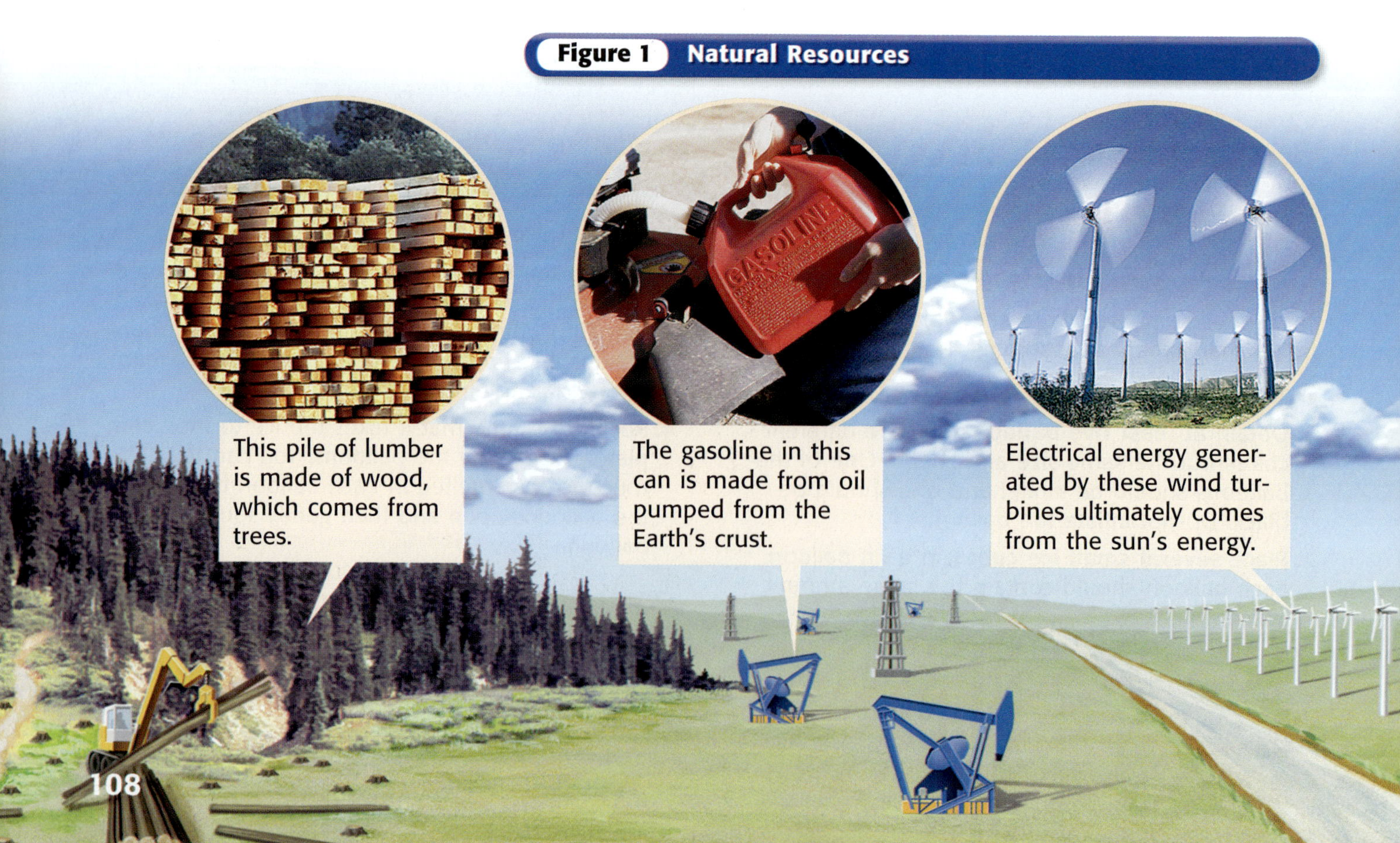

This pile of lumber is made of wood, which comes from trees.

The gasoline in this can is made from oil pumped from the Earth's crust.

Electrical energy generated by these wind turbines ultimately comes from the sun's energy.

Figure 2 *Trees and fresh water are just a few of the renewable resources available on Earth.*

Renewable Resources

Some natural resources can be renewed. A **renewable resource** is a natural resource that can be replaced at the same rate at which the resource is used. **Figure 2** shows two examples of renewable resources. Although many resources are renewable, they still can be used up before they can be renewed. Trees, for example, are renewable. However, some forests are being cut down faster than new forests can grow to replace them.

Reading Check **What is a renewable resource?** (*See the Appendix for answers to Reading Checks.*)

natural resource any natural material that is used by humans, such as water, petroleum, minerals, forests, and animals

renewable resource a natural resource that can be replaced at the same rate at which the resource is consumed

nonrenewable resource a resource that forms at a rate that is much slower than the rate at which it is consumed

Nonrenewable Resources

Not all of Earth's natural resources are renewable. A **nonrenewable resource** is a resource that forms at a rate that is much slower than the rate at which it is consumed. Coal, shown in **Figure 3,** is an example of a nonrenewable resource. It takes millions of years for coal to form. Once coal is used up, it is no longer available. Petroleum and natural gas are other examples of nonrenewable resources. When these resources become scarce, humans will have to find other resources to replace them.

Figure 3 *The coal used in the industrial process shown here is not quickly replaced by natural processes.*

Renewable?

WRITING SKILL With a parent, find five products in your home that were made from natural resources. List the resource or resources from which each product was made. Label each resource as renewable or nonrenewable. What can you do to help conserve the resources you listed? In your **science journal,** describe a personal action plan to conserve some of the resources you rely on every day.

ACTIVITY

Conserving Natural Resources

Whether the natural resources you use are renewable or nonrenewable, you should be careful how you use them. To conserve natural resources, you should try to use them only when necessary. For example, leaving the faucet on while brushing your teeth wastes clean water. Turning the faucet on only to rinse your brush saves water that you may need for other uses.

Conserving resources also means taking care of the resources even when you are not using them. For example, it is important to keep lakes, rivers, and other water resources free of pollution. Polluted lakes and rivers can affect the water you drink. Also, polluted water resources can harm the plants and animals, including humans, that depend on them to survive.

Energy Conservation

The energy we use to heat our homes, drive our cars, and run our computers comes from natural resources. The way in which we choose to use energy on a daily basis affects the availability of the natural resources. Most of the natural resources that provide us energy are nonrenewable resources. So, if we don't limit our use of energy now, the resources may not be available in the future.

As with all natural resources, conserving energy is important. You can conserve energy by being careful to use only the resources that you need. For example, turn lights off when you are not using them. And make sure the washing machine is full before you start it, as shown in **Figure 4.** You can also ride a bike, walk, or take a bus because these methods use fewer resources than a car does.

Figure 4 *Making sure the washing machine is full before running it is one way you can avoid wasting natural resources.*

Reduce, Reuse, Recycle

Another way to conserve natural resources is to recycle, as shown in **Figure 5. Recycling** is the process of reusing materials from waste or scrap. Recycling reduces the amount of natural resources that must be obtained from the Earth. For example, recycling paper reduces the number of trees that must be cut down to make new paper products. Recycling also conserves energy. Though energy is required to recycle materials, it takes less energy to recycle an aluminum can than it does to make a new one!

Newspaper, aluminum cans, most plastic containers, and cardboard boxes can be recycled. Most plastic containers have a number on them. This number informs you whether the item can be recycled. Plastic products with the numbers 1 and 2 can be recycled in most communities. Check with your community's recycling center to see what kinds of materials the center recycles.

Reading Check **What are some kinds of products that can be recycled?**

recycling the process of recovering valuable or useful materials from waste or scrap; the process of reusing some items

Figure 5 *You can recycle many household items to help conserve natural resources.*

SECTION Review

Summary

- We use natural resources such as water, petroleum, and lumber to make our lives more comfortable and convenient.
- Renewable resources can be replaced within a relatively short period of time, but nonrenewable resources may take thousands or even millions of years to form.
- Natural resources can be conserved by using only what is needed, taking care of resources, and recycling.

Using Key Terms

1. Use each of the following terms in a separate sentence: *natural resource, renewable resource, nonrenewable resource,* and *recycling.*

Understanding Key Ideas

2. How do humans use most natural resources?
3. Which of the following is a renewable resource?
   a. oil
   b. water
   c. coal
   d. natural gas
4. Describe three ways to conserve natural resources.

Math Skills

5. If a faucet dripped for 8.6 h and 3.3 L of water dripped out every hour, how many liters of water dripped out altogether?

Critical Thinking

6. **Making Inferences** How does human activity affect Earth's renewable and nonrenewable resources?
7. **Applying Concepts** List five products you regularly use that can be recycled.
8. **Making Inferences** Why is the availability of some renewable resources more of a concern now than it was 100 years ago?

SECTION 2

Fossil Fuels

How does a sunny day 200 million years ago relate to your life today?

Chances are that if you traveled to school today or used a product made of plastic, you used some of the energy from sunlight that fell on Earth several hundred million years ago. Life as you know it would be very different without the fuels or products formed from plants and animals that lived along-side the dinosaurs.

READING WARM-UP

Objectives

- Describe what energy resources are.
- Identify three different forms of fossil fuels.
- Explain how fossil fuels form.
- Describe how fossil fuels are found and obtained.
- Identify four problems with fossil fuels.

Terms to Learn

fossil fuel
petroleum
natural gas
coal
acid precipitation
smog

READING STRATEGY

Brainstorming The key idea of this section is fossil fuels. Brainstorm words and phrases related to fossil fuels.

Energy Resources

The fuels we use to run cars, ships, planes, and factories and to generate electrical energy, shown in **Figure 1,** are energy resources. *Energy resources* are natural resources that humans use to generate energy. Most of the energy we use comes from a group of natural resources called fossil fuels. A **fossil fuel** is a nonrenewable energy resource formed from the remains of plants and animals that lived long ago. Examples of fossil fuels include petroleum, coal, and natural gas.

Energy is released from fossil fuels when they are burned. For example, the energy from burning coal in a power plant is used to produce electrical energy. However, because fossil fuels are a nonrenewable resource, once they are burned, they are gone. Therefore, like other resources, fossil fuels need to be conserved. In the 21st century, societies will continue to explore alternatives to fossil fuels. But they will also focus on developing more-efficient ways to use these fuels.

fossil fuel a nonrenewable energy resource formed from the remains of organisms that lived long ago; examples include oil, coal, and natural gas

Figure 1 *Light produced from electrical energy can be seen in this satellite image taken from space.*

Figure 2 *Some refineries use a process called* distillation *to separate petroleum into various types of petroleum products.*

Types of Fossil Fuels

All living things are made up of the element carbon. Because fossil fuels are formed from the remains of plants and animals, all fossil fuels are made of carbon, too. Most of the carbon in fossil fuels exists as hydrogen-carbon compounds called *hydrocarbons*. But different fossil fuels have different forms. Fossil fuels may exist as liquids, gases, or solids.

Liquid Fossil Fuels: Petroleum

A liquid mixture of complex hydrocarbon compounds is called **petroleum.** Petroleum is also commonly known as *crude oil*. Petroleum is separated into several kinds of products in refineries, such as the one shown in **Figure 2.** Examples of fossil fuels separated from petroleum are gasoline, jet fuel, kerosene, diesel fuel, and fuel oil.

More than 40% of the world's energy comes from petroleum products. Petroleum products are the main fuel for forms of transportation, such as airplanes, trains, boats, and ships. Crude oil is so valuable that it is often called *black gold*.

petroleum a liquid mixture of complex hydrocarbon compounds; used widely as a fuel source

natural gas a mixture of gaseous hydrocarbons located under the surface of the Earth, often near petroleum deposits; used as a fuel

Gaseous Fossil Fuels: Natural Gas

A gaseous mixture of hydrocarbons is called **natural gas.** Most natural gas is used for heating, but it is also used for generating electrical energy. Your kitchen stove may be powered by natural gas. Some motor vehicles, such as the van in **Figure 3,** use natural gas as fuel. An advantage of using natural gas is that using it causes less air pollution than using oil does. However, natural gas is very flammable. Gas leaks can lead to fires or deadly explosions.

Methane, CH_4, is the main component of natural gas. But other components, such as butane and propane, can be separated from natural gas, too. Butane and propane are often used as fuel for camp stoves and outdoor grills.

Reading Check **What is natural gas most often used for?** *(See the Appendix for answers to Reading Checks.)*

Figure 3 *Vehicles powered by natural gas are becoming more common.*

Figure 4 *This coal is being gathered so that it may be burned in the power plant shown in the background.*

Solid Fossil Fuels: Coal

coal a fossil fuel that forms underground from partially decomposed plant material

The solid fossil fuel that humans use most is coal. **Coal** is a fossil fuel that is formed underground from partially decomposed plant material. Coal was once the major source of energy in the United States. People burned coal in stoves to heat their homes. They also used coal in transportation. Many trains in the 1800s and early 1900s were powered by coal-burning steam locomotives.

As cleaner energy resources became available, people reduced their use of coal. People began to use coal less because burning coal produces large amounts of air pollution. Now, people use forms of transportation that use oil instead of coal as fuel. In the United States, coal is now rarely used as a fuel for heating. However, many power plants, such as the one shown in **Figure 4,** burn coal to generate electrical energy.

INTERNET ACTIVITY

For another activity related to this chapter, go to **go.hrw.com** and type in the keyword **HZ5ENRW.**

Reading Check **In the 1800s and early 1900s, what was coal most commonly used for?**

CONNECTION TO Chemistry

Hydrocarbons Both petroleum and natural gas are made of compounds called *hydrocarbons.* A hydrocarbon is an organic compound that contains only carbon and hydrogen. A molecule of propane, C_3H_8, a gaseous fossil fuel, contains three carbons and eight hydrogens. Using a molecular model set, create a model of a propane molecule. (Hint: Each carbon atom should have four bonds, and each hydrogen atom should have one bond.)

ACTIVITY

How Do Fossil Fuels Form?

All fossil fuels form from the buried remains of ancient organisms. But different kinds of fossil fuels form in different ways and from different kinds of organisms.

Petroleum and Natural Gas Formation

Petroleum and natural gas form mainly from the remains of microscopic sea organisms. When these organisms die, their remains settle on the ocean floor. There, the remains decay, are buried, and become part of the ocean sediment. Over time, the sediment slowly becomes rock, trapping the decayed remains. Through physical and chemical changes over millions of years, the remains become petroleum and gas. Gradually, more rocks form above the rocks that contain the fossil fuels. Under the pressure of overlying rocks and sediments, the fossil fuels can move through permeable rocks. *Permeable rocks* are rocks that allow fluids, such as petroleum and gas, to move through them. As shown in **Figure 5,** these permeable rocks become reservoirs that hold petroleum and natural gas.

The formation of petroleum and natural gas is an ongoing process. Part of the remains of today's sea life will become petroleum and natural gas millions of years from now.

Rock Sponge

1. Place **samples of sandstone, limestone,** and **shale** in separate **Petri dishes.**
2. Place **five drops of light machine oil** on each rock sample.
3. Observe and record the time required for the oil to be absorbed by each of the rock samples.
4. Which rock sample absorbed the oil fastest? Why?
5. Based on your findings, describe a property that allows fossil fuels to be easily removed from reservoir rock.

Figure 5 *Petroleum and gas move through permeable rock. Eventually, these fuels are collected in reservoirs. Rocks that are folded upward are excellent fossil-fuel traps.*

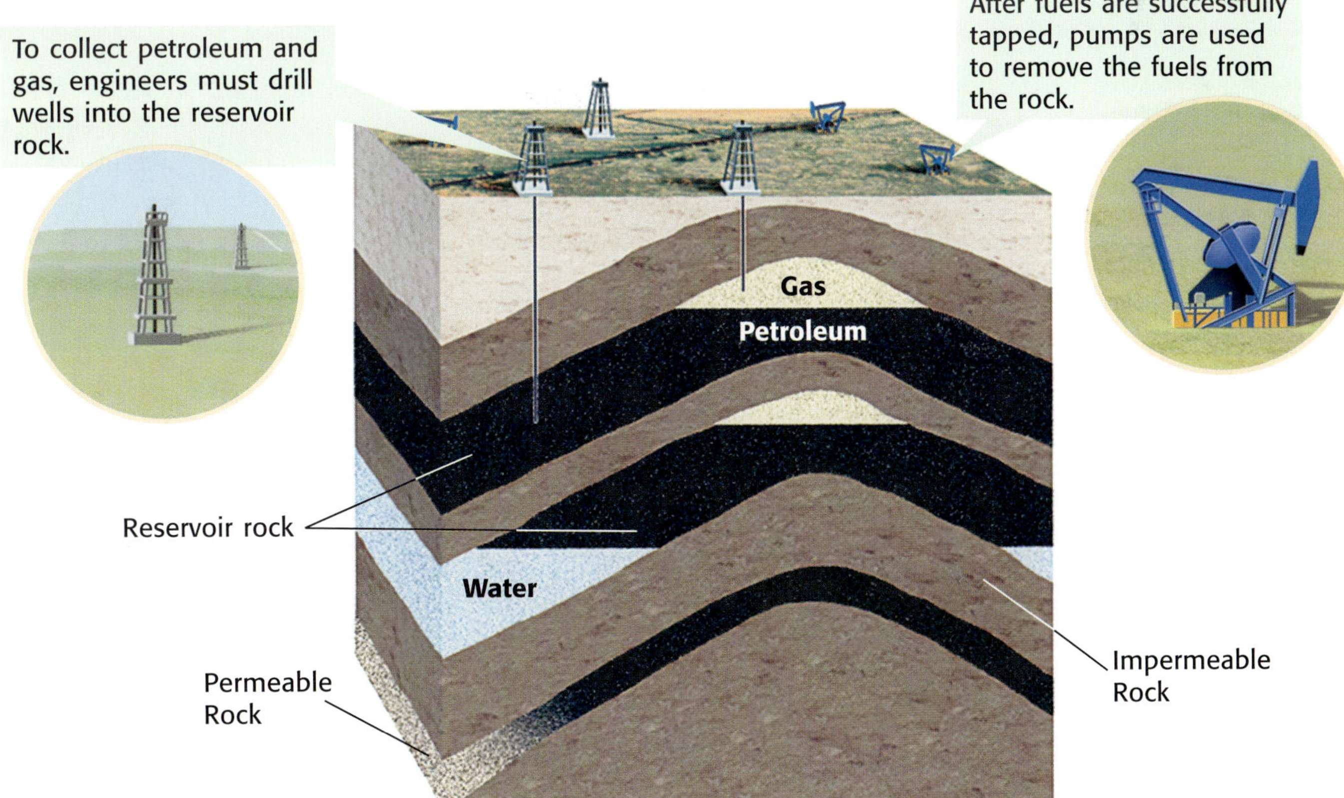

Coal Formation

Coal forms differently from the way petroleum and natural gas form. Coal forms underground from decayed swamp plants over millions of years. When the plants die, they sink to the bottom of the swamp. This begins the process of coal formation.

The four stages of coal formation, as shown in **Figure 6,** are peat, lignite, bituminous coal, and anthracite. Peat is a form of coal in which the remains of plants are only partially decomposed. Over time, the peat is buried under sediment. Water and gases are squeezed out of the peat. Pressure and high temperature then turn the peat into lignite. The process by which pressure and temperature increase due to the deposition of sediment creates different forms of coal. The percentage of carbon increases with each stage of coal formation. The higher the carbon content is, the more cleanly the material burns. Pollution controls can remove most of the pollutants produced by burning coal. However, when burned, all grades of coal pollute the air.

Figure 6 **Coal Formation**

Stage 1: Peat
Bacteria and fungi change sunken swamp plants into peat. Peat is about **60% carbon.**

Stage 2: Lignite
Sediment buries the peat, which increases the pressure and temperature. The peat slowly changes into lignite, which is about **70% carbon.**

Stage 3: Bituminous Coal
As the lignite becomes more buried, the temperature and pressure continue to increase. Eventually, lignite turns into bituminous coal, which is about **80% carbon.**

Stage 4: Anthracite
As bituminous coal becomes more buried, the temperature and pressure continue to increase. Bituminous coal turns into anthracite, which is about **90% carbon.**

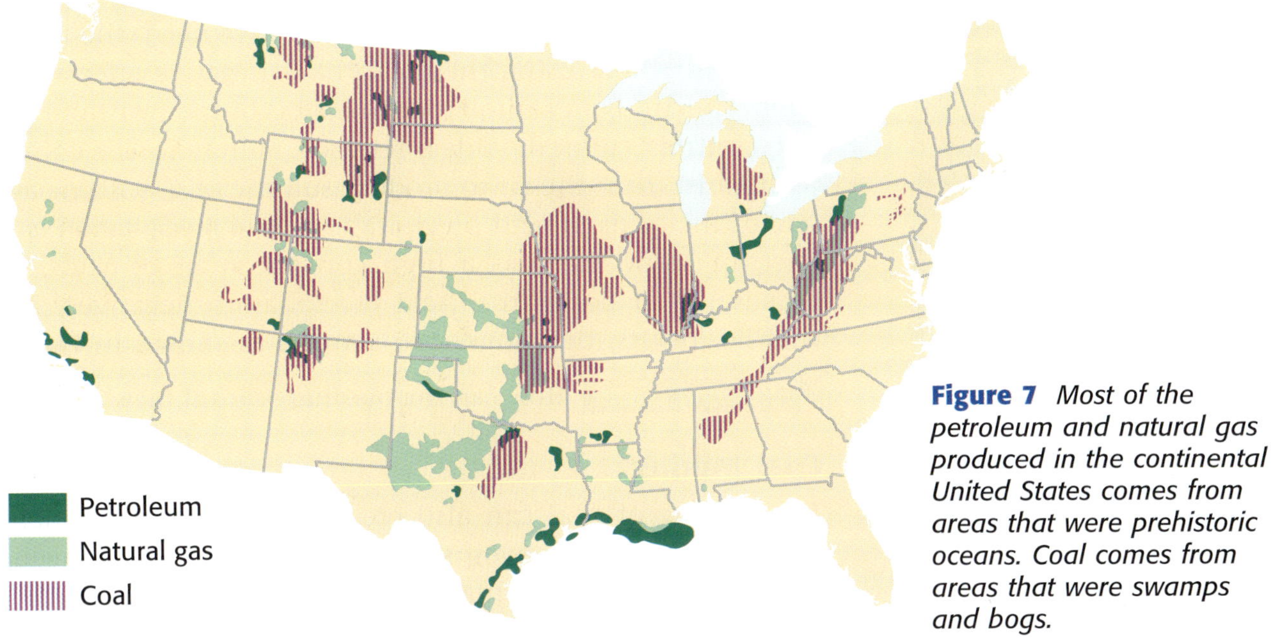

Figure 7 *Most of the petroleum and natural gas produced in the continental United States comes from areas that were prehistoric oceans. Coal comes from areas that were swamps and bogs.*

Where Are Fossil Fuels Found?

Fossil fuels are found in many parts of the world. Some fossil fuels are found on land, while other fossil fuels are found beneath the ocean. As shown in **Figure 7,** the United States has large reserves of petroleum, natural gas, and coal. Despite its large reserves of petroleum, the United States imports petroleum as well. About one-half of the petroleum used by the United States is imported from the Middle East, South America, Africa, Canada, and Mexico.

How Do We Obtain Fossil Fuels?

Humans use several methods to remove fossil fuels from the Earth's crust. The kind and location of fuel determine the method used to remove the fuel. People remove petroleum and natural gas from Earth by drilling wells into rock that contains these resources. Oil wells exist on land and in the ocean. For offshore drilling, engineers mount drills on platforms that are secured to the ocean floor or that float at the ocean's surface. **Figure 8** shows an offshore oil rig.

People obtain coal either by mining deep beneath Earth's surface or by surface mining. Surface mining, also known as *strip mining,* is the process by which soil and rock are stripped from the Earth's surface to expose the underlying coal that is to be mined.

Reading Check **How are natural gas and petroleum removed from Earth?**

Figure 8 *Large oil rigs, some of which are more than 300 m tall, operate offshore in many places, such as the Gulf of Mexico and the North Sea.*

Figure 9 *Notice how this statue looked before the effects of acid precipitation.*

Problems with Fossil Fuels

Although fossil fuels provide the energy we need, the methods of obtaining and using them can have negative effects on the environment. For example, when coal is burned without pollution controls, sulfur dioxide is released. Sulfur dioxide combines with moisture in the air to produce sulfuric acid. Sulfuric acid is one of the acids in acid precipitation. **Acid precipitation** is rain, sleet, or snow that has a high concentration of acids, often because of air pollutants. Acid precipitation negatively affects wildlife, plants, buildings, and statues, as shown in **Figure 9.**

Reading Check How can the burning of fossil fuels affect rain?

acid precipitation precipitation, such as rain, sleet, or snow, that contains a high concentration of acids, often because of the pollution of the atmosphere

smog photochemical haze that forms when sunlight acts on industrial pollutants and burning fuels

Coal Mining

The mining of coal can also create environmental problems. Surface mining removes soil, which some plants need for growth and some animals need for shelter. If land is not properly restored afterward, surface mining can destroy wildlife habitats. Coal mining can also lower water tables and pollute water supplies. The potential for underground mines to collapse endangers the lives of miners.

Petroleum Problems

Producing, transporting, and using petroleum can cause environmental problems and endanger wildlife. In June 2000, the carrier, *Treasure,* sank off the coast of South Africa and spilled more than 400 tons of oil. The toxic oil coated thousands of blackfooted penguins, as shown in **Figure 10.** The oil hindered the penguins from swimming and catching fish for food.

Smog

Burning petroleum products causes an environmental problem called smog. **Smog** is photochemical haze that forms when sunlight acts on industrial pollutants and burning fuels. Smog is particularly serious in cities such as Houston and Los Angeles as a result of millions of automobiles that burn gasoline. Also, mountains that surround Los Angeles prevent the wind from blowing pollutants away.

Figure 10 *The oil spilled from the carrier,* Treasure, *endangered the lives of many animals including the blackfooted penguins.*

SECTION Review

Summary

- Energy resources are resources that humans use to produce energy.
- Petroleum is a liquid fossil fuel that is made of hydrocarbon compounds.
- Natural gas is a gaseous fossil fuel that is made of hydrocarbon compounds.
- Coal is a solid fossil fuel that forms from decayed swamp plants.
- Petroleum and natural gas form from decayed sea life on the ocean floor.
- Fossil fuels are found all over the world. The United States imports half of the petroleum it uses from the Middle East, South America, Africa, Mexico, and Canada.
- Fossil fuels are obtained by drilling oil wells, mining below Earth's surface, and strip mining.
- Acid precipitation, smog, water pollution, and the destruction of wildlife habitat are some of the environmental problems that are created by the use of fossil fuels.

Using Key Terms

1. Use each of the following terms in a separate sentence: *energy resource, fossil fuel, petroleum, natural gas, coal, acid precipitation,* and *smog.*

Understanding Key Ideas

2. Which of the following stages of coal formation contains the highest carbon content?
   a. lignite
   b. anthracite
   c. peat
   d. bituminous coal
3. Name a solid fossil fuel, a liquid fossil fuel, and a gaseous fossil fuel.
4. Briefly describe how petroleum and natural gas form.
5. How do we obtain petroleum and natural gas?
6. Describe the advantages and disadvantages of fossil fuel use.

Critical Thinking

7. **Making Comparisons** What is the difference between the organic material from which coal forms and the organic material from which petroleum and natural gas form?
8. **Making Inferences** Why can't carpooling and using mass-transit systems eliminate the problems associated with fossil fuels?

Interpreting Graphics

Use the pie chart below to answer the questions that follow.

Oil Production by Region

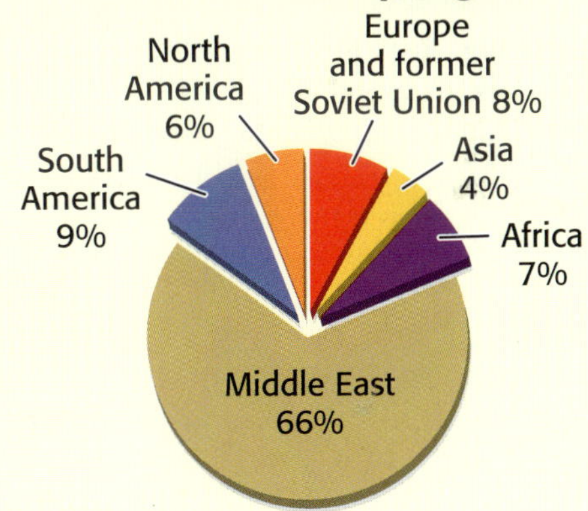

Source: International Energy Agency.

9. Which region produces the most oil?
10. If the total sales of oil in 2002 were $500 billion, what was the value of the oil produced in North America?

SECTION 3

Alternative Resources

READING WARM-UP

Objectives

- Describe alternatives to the use of fossil fuels.
- List advantages and disadvantages of using alternative energy resources.

Terms to Learn

nuclear energy
chemical energy
solar energy
wind power
hydroelectric energy
biomass
gasohol
geothermal energy

READING STRATEGY

Paired Summarizing Read this section silently. In pairs, take turns summarizing the material. Stop to discuss ideas that seem confusing.

What would your life be like if you couldn't play video games, turn on lights, microwave your dinner, take a hot shower, or take the bus to school?

Most of your energy needs and the energy needs of others are met by the use of fossil fuels. Yet, there are two main problems with fossil fuels. First, the availability of fossil fuels is limited. Fossil fuels are nonrenewable resources. Once fossil fuels are used up, new supplies won't be available for thousands—or even millions—of years.

Second, obtaining and using fossil fuels has environmental consequences. To continue to have access to energy and to overcome pollution, we must find alternative sources of energy.

Splitting the Atom: Fission

The energy released by a fission or fusion reaction is **nuclear energy.** *Fission* is a process in which the nuclei of radioactive atoms are split into two or more smaller nuclei, as shown in **Figure 1.** When fission takes place, a large amount of energy is released. This energy can be used to generate electrical energy. The SI unit for all forms of energy is the joule. However, electrical energy and nuclear energy is often measured in megawatts (MW).

Figure 1 **Fission**

A neutron from a uranium-235 atom splits the nucleus into two smaller nuclei called *fission products* and two or more neutrons.

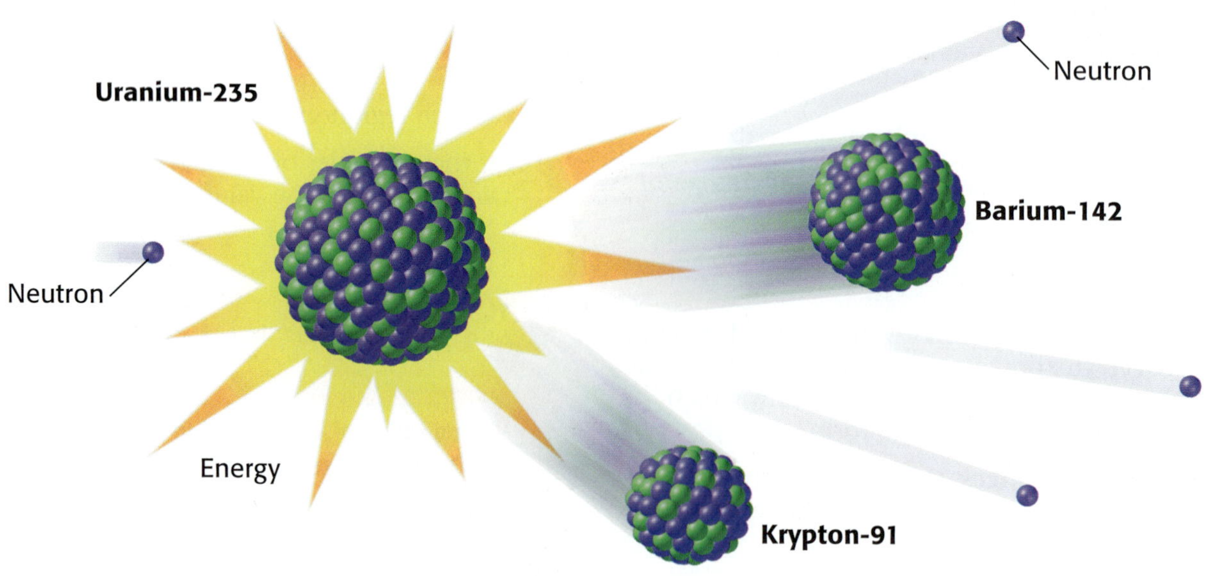

Pros and Cons of Fission

Nuclear power plants provide alternative sources of energy that do not have the problems that fossil fuels do. So, why don't we use nuclear energy more instead of using fossil fuels? Nuclear power plants produce dangerous radioactive wastes. Radioactive wastes must be removed from the plant and stored until their radioactivity decreases to a harmless level. But nuclear wastes can remain dangerously radioactive for thousands of years. These wastes must be stored in an isolated place where the radiation that they emit cannot harm anyone.

Another problem with nuclear power plants is the potential for accidental release of radiation into the environment. A release could happen if the plant overheats. If a plant's cooling system were to stop working, the plant would overheat. Then, its reactor could melt, and a large amount of radiation could escape into the environment. In addition, towers like the one shown in **Figure 2,** keep hot water from potentially disrupting the local ecosystem.

Figure 2 *Cooling towers are used to cool water leaving a nuclear power plant before the water is released into the environment.*

Combining Atoms: Fusion

Another method of getting energy from nuclei is fusion, shown in **Figure 3.** *Fusion* is the joining of two or more nuclei to form a larger nucleus. This process releases a large amount of energy and happens naturally in the sun.

nuclear energy the energy released by a fission or fusion reaction; the binding energy of the atomic nucleus

The main advantage of fusion is that it produces few dangerous wastes. The main disadvantage of fusion is that very high temperatures are required for the reaction to take place. No known material can withstand such high temperatures. Therefore, the reaction must happen within a special environment, such as a magnetic field. Controlled fusion reactions have been limited to laboratory experiments.

Reading Check **What is the advantage of producing energy through fusion?** (*See the Appendix for answers to Reading Checks.*)

Figure 3 **Fusion**

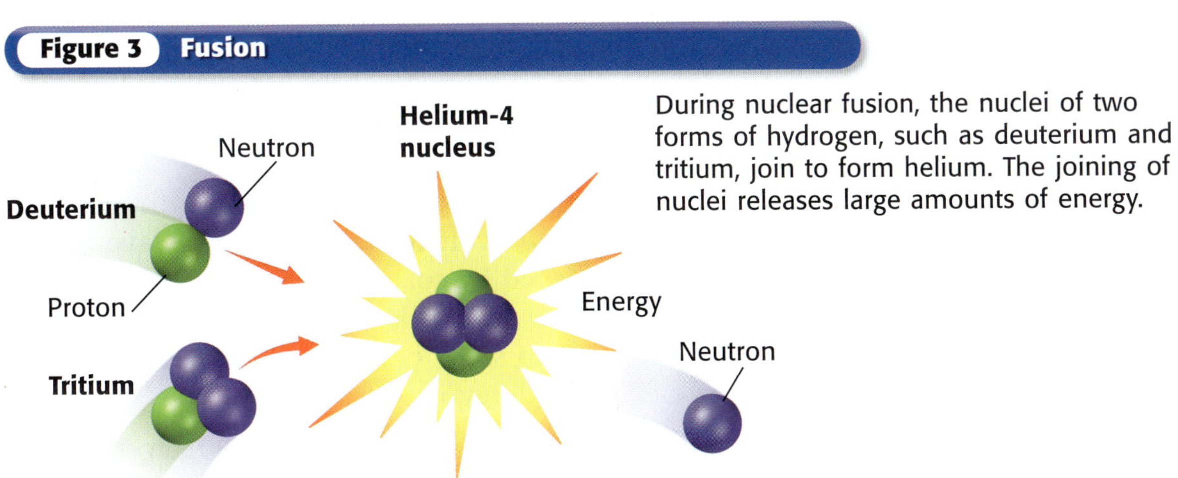

During nuclear fusion, the nuclei of two forms of hydrogen, such as deuterium and tritium, join to form helium. The joining of nuclei releases large amounts of energy.

Figure 4 *This image shows a prototype of a fuel-cell car. Power from fuel cells may be commonly used in the future.*

Chemical Energy

chemical energy the energy released when a chemical compound reacts to produce new compounds

solar energy the energy received by the Earth from the sun in the form of radiation

When you think of fuel for an automobile, you most likely think of gasoline. However, not all vehicles are fueled by gasoline. Some vehicles, such as the one shown in **Figure 4,** are powered by energy that is generated by fuel cells. Fuel cells power automobiles by converting **chemical energy** into electrical energy by reacting hydrogen and oxygen into water. One advantage of using fuel cells as energy sources is that fuel cells do not create pollution. The only byproduct of fuel cells is water. Fuel cells are also more efficient than internal combustion engines are.

The United States has been using fuel cells in space travel since the 1960s. Fuel cells have provided space crews with electrical energy and drinking water. One day, fuel-cell technology may be used to generate electrical energy in buildings, ships, and submarines, too.

Solar Energy

Almost all forms of energy, such as the energy of fossil fuels, come from the sun. The energy received by the Earth from the sun in the form of radiation is **solar energy.** The Earth receives more than enough solar energy to meet all of our energy needs. And because the Earth continuously receives solar energy, this energy is a renewable resource. Solar energy can be used directly to heat buildings and to generate electrical energy. However, we do not yet have the technology to generate the amount of electrical energy we need from solar energy.

Sunlight can be changed into electrical energy through the use of solar cells or photovoltaic cells. You may have used a calculator that is powered by solar cells. *Solar panels* are large panels made up of many solar cells wired together. Solar panels mounted on the roofs of some homes and businesses provide some of the electrical energy used in the buildings.

Reading Check Where does the energy of fossil fuels come from?

CONNECTION TO Language Arts

WRITING SKILL **Resources of the Future** In 100 years from now do you think humans will still be using fossil fuels to power their cars? Maybe humans will be using alternative energy resources. Maybe humans won't even be driving cars! Write a short, creative, science fiction story describing the energy use of humans 100 years from now.

Solar Heating

Solar energy is also used for direct heating through solar collectors. *Solar collectors* are dark-colored boxes that have glass or plastic tops. A common use of solar collectors is to heat water, as shown in **Figure 5.** More than 1 million solar water heaters have been installed in the United States. Solar water heaters are especially common in Florida and California.

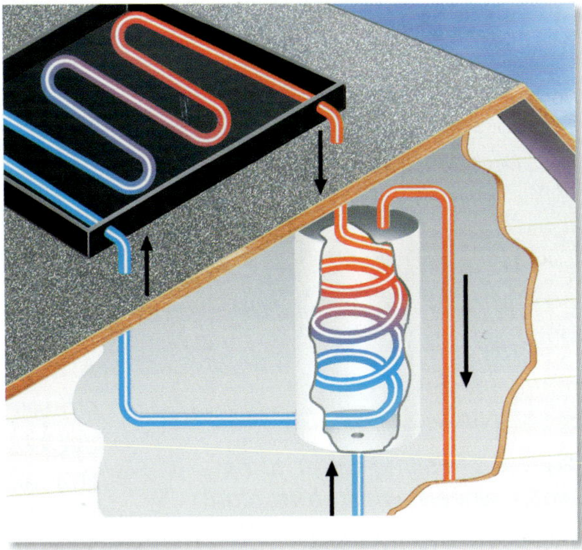

Figure 5 *The liquid in the solar collector is heated by the sun. Then, the liquid is pumped through tubes that run through a water heater, which causes the temperature of the water to increase.*

Pros and Cons of Solar Energy

One of the best things about solar energy is that it doesn't produce pollution. Also, solar energy is renewable, because it comes from the sun. However, some climates don't have enough sunny days to benefit from solar energy. Also, although solar energy is free, solar cells and solar collectors are more expensive to make than other energy systems are. The cost of installing a complete solar-power system in a house can be one-third of the total cost of the house.

Wind Power

Wind is made indirectly by solar energy through the uneven heating of air. Energy can be harnessed from wind. **Wind power** is the use of a windmill to drive an electric generator. Clusters of wind turbines, like the ones shown in **Figure 6,** can generate a significant amount of electrical energy. Wind energy is renewable, and it doesn't cause any pollution. However, in many areas, the wind isn't strong enough or frequent enough to create energy on a large scale.

wind power the use of a windmill to drive an electric generator

Figure 6 *Wind turbines take up only a small part of the ground's surface. As a result, the land on wind farms can be used for more than one purpose.*

Figure 7 *Falling water turns water wheels, which turn giant millstones used to grind grain into flour.*

Hydroelectric Energy

Humans have used the energy of falling water for thousands of years. Water wheels, such as the one shown in **Figure 7,** have been around since ancient times. In the early years of the Industrial Revolution, water wheels provided energy for many factories. Today, the energy of falling water is used to generate electrical energy. Electrical energy produced by falling water is called **hydroelectric energy.**

hydroelectric energy electrical energy produced by falling water

Pros and Cons of Hydroelectric Energy

After the dam is built, hydroelectric energy is inexpensive and causes little pollution. It is renewable because water constantly cycles from water sources to the air, to the land, and back to the water source. But like wind energy, hydroelectric energy is not available everywhere. It can be produced only where large volumes of falling water can be harnessed. Huge dams, such as the one in **Figure 8,** must be built on major rivers to capture enough water to generate significant amounts of electrical energy.

Using more hydroelectric energy could reduce the demand for fossil fuels, but there are trade-offs. Building the large dams necessary for hydroelectric power plants often destroys other resources, such as forests and wildlife habitats. For example, hydroelectric dams on the lower Snake and Columbia Rivers in Washington state disrupt the migratory paths of local populations of salmon and steelhead. Large numbers of these fish die each year because their migratory path is disrupted. Dams can also decrease water quality and create erosion problems.

Reading Check **Why is hydroelectric energy renewable?**

Figure 8 *Falling water turns turbines inside hydroelectric dams and generates electrical energy for millions of people.*

Power from Plants

Plants are similar to solar collectors. Both absorb energy from the sun and store it for later use. Leaves, wood, and other parts of plants contain the stored energy. Even the dung of plant-grazing animals is high in stored energy. These sources of energy are called biomass. **Biomass** is organic matter that can be a source of energy.

biomass organic matter that can be a source of energy

gasohol a mixture of gasoline and alcohol that is used as a fuel

Burning Biomass

Biomass energy can be released in several ways. The most common way is to burn biomass. Approximately 70% of people living in developing countries, about half the world population, burn wood or charcoal to heat their homes and cook their food. In contrast, about 5% of the people in the United States heat and cook this way. Scientists estimate that the burning of wood and animal dung accounts for approximately 14% of the world's total energy use. **Figure 9** shows a woman who is preparing cow dung that will be dried and used for fuel.

Miles per Acre

Imagine that you own a car that runs on alcohol made from corn that you grow. You drive your car about 15,000 mi per year, and you get 240 gal of alcohol from each acre of corn that you process. If your car has a gas mileage of 25 mi/gal, how many acres of corn must you process to fuel your car for a year?

Gasohol

Biomass material can also be changed into liquid fuel. Plants that contain sugar or starch can be made into alcohol. The alcohol can be burned as a fuel. Or alcohol can be mixed with gasoline to make a fuel called **gasohol.** More than 1,000 L of alcohol can be made from 1 acre of corn. But people in the United States use a large amount of fuel for their cars. And the alcohol produced from about 40% of one corn harvest in the United States would provide only 10% of the fuel used in our cars! Biomass is a renewable source of energy. However, producing biomass requires land that could be used for growing food.

Figure 9 *In many parts of the world where firewood is scarce, people burn animal dung for energy.*

Energy from Within Earth

If you have ever seen a volcanic eruption, you know how powerful the Earth can be. The energy produced by the heat within Earth is called **geothermal energy.**

geothermal energy the energy produced by heat within the Earth

Geothermal Energy

In some areas, groundwater is heated by *magma,* or melted rock. Often, the heated groundwater becomes steam. *Geysers* are natural vents that discharge this steam or water in a column into the air. The steam and hot water can also escape through wells drilled into the rock. From these wells, geothermal power plants can harness the energy from within Earth by pumping the steam and hot water, as shown in **Figure 10.** The world's largest geothermal power plant in California, called *The Geysers,* produces electrical energy for 1.7 million households.

Geothermal energy can also be used to heat buildings. In this process, hot water and steam are used to heat a fluid. Then, this fluid is pumped through a building in order to heat the building. Buildings in Iceland are heated from the country's many geothermal sites in this way.

Reading Check **How do geothermal power plants obtain geothermal energy from the Earth?**

Figure 10 **How a Geothermal Power Plant Works**

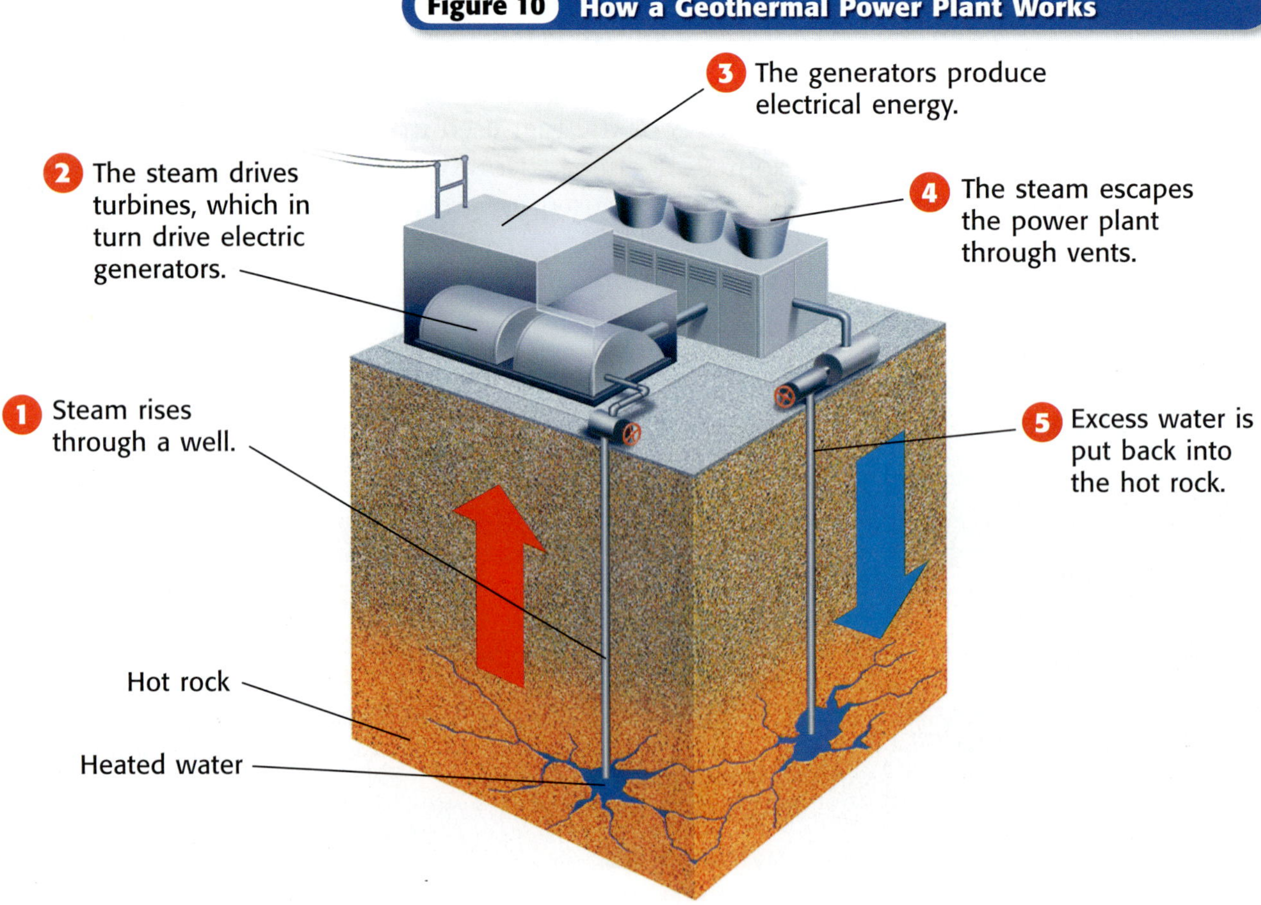

SECTION Review

Summary

- Nuclear energy can be released by fission, and or fusion. The byproduct of fission is radioactive waste.
- For fusion to take place, extremely high temperatures are required.
- Fuel cells combine hydrogen and oxygen to produce electrical energy. Fuel cells release water as a byproduct.
- Solar energy is a renewable resource that doesn't emit pollution. However, solar panels and solar collectors are expensive.
- Wind power is a renewable resource that doesn't emit pollution. However, wind energy cannot be generated in all areas.
- Hydroelectric energy is a cheap, renewable resource that causes little pollution. However, it is available only in some areas.
- Burning biomass and gasohol can release energy, but not enough to meet all of our energy needs.
- Geothermal energy comes from the Earth but is available only in certain areas.

Using Key Terms

1. In your own words, write a definition for each of the following terms: *nuclear energy, solar energy, wind power, hydroelectric energy, biomass, gasohol,* and *geothermal energy.*

Understanding Key Ideas

2. Which of the following alternative resources requires hydrogen and oxygen to produce energy?
   a. fuel cells
   b. solar energy
   c. nuclear energy
   d. geothermal energy
3. Describe two ways of using solar energy.
4. Where is the production of hydroelectric energy practical?
5. Describe two ways to release biomass energy.
6. Describe two ways to use geothermal energy.

Critical Thinking

7. **Analyzing Methods** If you were going to build a nuclear power plant, why wouldn't you build it in the middle of a desert?
8. **Predicting Consequences** If an alternative resource could successfully replace crude oil, how might the use of that resource affect the environment?

Interpreting Graphics

Use the graph below to answer the questions that follow.

How Energy Is Used in the United States

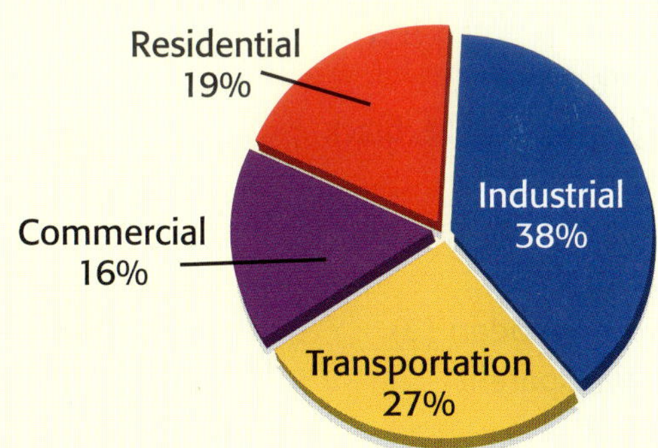

Source: International Energy Agency.

9. What is the total percentage of energy that is used for commercial and industrial purposes?
10. What is the total percentage of energy that is not used for residential purposes?

For a variety of links related to this chapter, go to www.scilinks.org

Topic: Renewable Resources
SciLinks code: HSM1291

Using Scientific Methods

Model-Making Lab

Make a Water Wheel

OBJECTIVES

Create a model of a water wheel.

Determine factors that influence the rate at which a water wheel lifts a weight.

MATERIALS

- bottle, soda, 2 L, filled with water
- card, index, 3 × 5 in.
- clay, modeling
- coin
- cork
- glue
- hole punch
- jug, milk, plastic
- marker, permanent, black
- meterstick
- safety razor (for teacher)
- scissors
- skewers, wooden (2)
- tape, transparent
- thread, 20 cm
- thumbtacks (5)
- watch or clock that indicates seconds

SAFETY

Lift Enterprises is planning to build a water wheel that will lift objects like a crane does. The president of the company has asked you to modify the basic water wheel design so that the water wheel will lift objects more quickly.

Ask a Question

1. What factors influence the rate at which a water wheel lifts a weight?

Form a Hypothesis

2. Change the question above into a statement to formulate a testable hypothesis.

Test the Hypothesis

3. Build a water wheel model. Measure and mark a 5 × 5 cm square on an index card. Cut the square out of the card. Fold the square in half to form a triangle.
4. Measure and mark a line 8 cm from the bottom of the plastic jug. Use scissors to cut along this line. (Your teacher may need to use a safety razor to start this cut for you.)
5. Use the paper triangle you made in step 3 as a template. Use a permanent marker to trace four triangles onto the flat parts of the top section of the plastic jug. Cut the triangles out of the plastic to form four fins.
6. Use a thumbtack to attach one corner of each plastic fin to the round edge of the cork, as shown below. Make sure the fins are equally spaced around the cork.

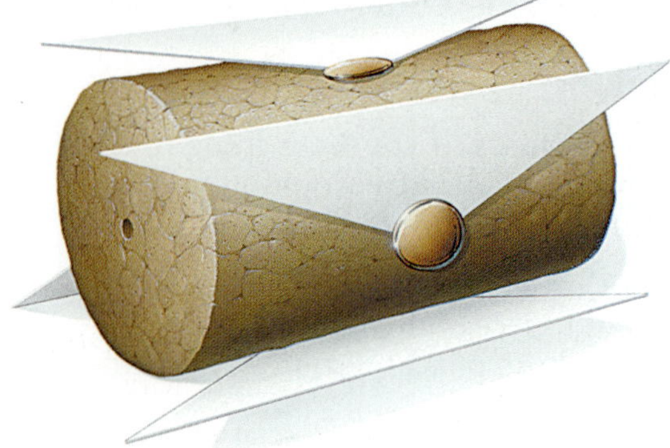

7. Press a thumbtack into one of the flat sides of the cork. Jiggle the thumbtack to widen the hole in the cork, and then remove the thumbtack. Repeat on the other side of the cork.

8. Place a drop of glue on one end of each skewer. Insert the first skewer into one of the holes in the end of the cork. Insert the second skewer into the hole in the other end of the cork.

9. Use a hole punch to carefully punch two holes in the bottom section of the plastic jug. Punch each hole 1 cm from the top edge of the jug, directly across from one another.

10. Carefully push the skewers through the holes, and suspend the cork in the center of the jug. Attach a small ball of clay to the end of each skewer. The balls should be the same size.

11. Tape one end of the thread to one skewer on the outside of the jug, next to the clay ball. Wrap the thread around the clay ball three times. (As the water wheel turns, the thread should wrap around the clay. The other ball of clay balances the weight and helps to keep the water wheel turning smoothly.)

12. Tape the free end of the thread to a coin. Wrap the thread around the coin, and tape it again.

13. Slowly pour water from the 2 L bottle onto the fins so that the water wheel spins. What happens to the coin? Record your observations.

14. Lower the coin back to the starting position. Add more clay to the skewer to increase the diameter of the wheel. Repeat step 13. Did the coin rise faster or slower this time?

15. Lower the coin back to the starting position. Modify the shape of the clay, and repeat step 13. Does the shape of the clay affect how quickly the coin rises? Explain your answer.

16. What happens if you remove two of the fins from opposite sides? What happens if you add more fins?

17. Experiment with another fin shape. How does a different fin shape affect how quickly the coin rises?

Analyze the Results

1. **Examining Data** What factors influence how quickly you can lift the coin? Explain.

Draw Conclusions

2. **Drawing Conclusions** What recommendations would you make to the president of Lift Enterprises to improve the water wheel?

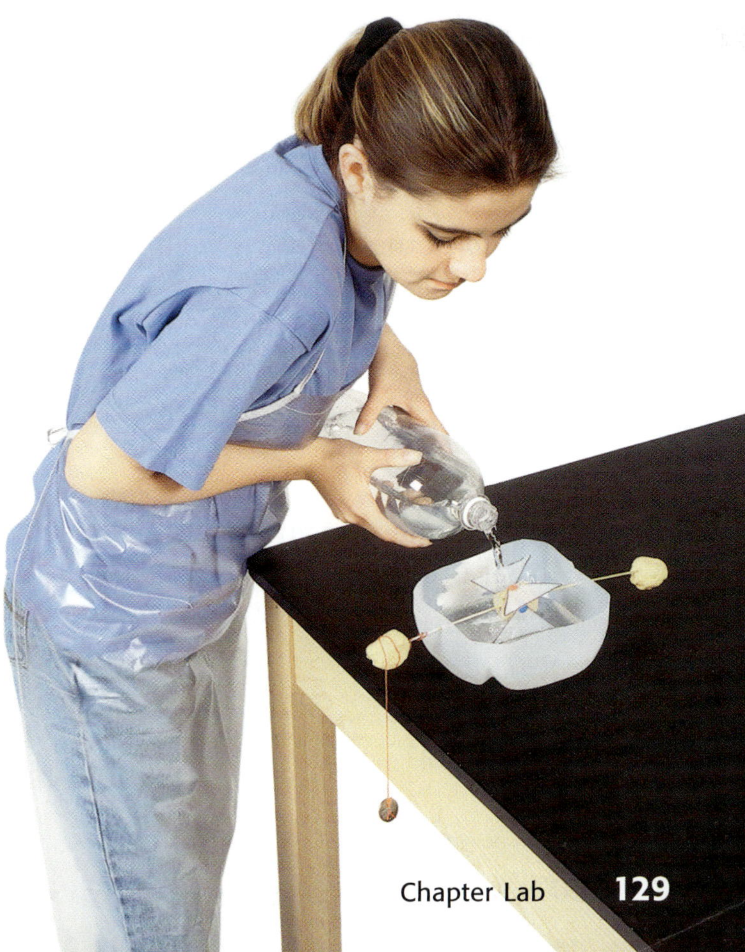

Chapter Review

USING KEY TERMS

The statements below are false. For each statement, replace the underlined term to make a true statement.

1. A liquid mixture of complex hydrocarbon compounds is called <u>natural gas</u>.

2. Energy that is released when a chemical compound reacts to produce a new compound is called <u>nuclear energy</u>.

For each pair of terms, explain how the meanings of the terms differ.

3. *solar energy* and *wind power*

4. *biomass* and *gasohol*

UNDERSTANDING KEY IDEAS

Multiple Choice

5. Which of the following resources is a renewable resource?
   - **a.** coal
   - **b.** trees
   - **c.** oil
   - **d.** natural gas

6. Which of the following fuels is NOT made from petroleum?
   - **a.** jet fuel
   - **b.** lignite
   - **c.** kerosene
   - **d.** fuel oil

7. Peat, lignite, and anthracite are all forms of
   - **a.** petroleum.
   - **b.** natural gas.
   - **c.** coal.
   - **d.** gasohol.

8. Which of the following factors contributes to smog?
   - **a.** automobiles
   - **b.** sunlight
   - **c.** mountains surrounding urban areas
   - **d.** All of the above

9. Which of the following resources is produced by fusion?
   - **a.** solar energy
   - **b.** natural gas
   - **c.** nuclear energy
   - **d.** petroleum

10. To produce energy, nuclear power plants use a process called
    - **a.** fission.
    - **b.** fusion.
    - **c.** fractionation.
    - **d.** None of the above

11. A solar-powered calculator uses
    - **a.** solar collectors.
    - **b.** solar panels.
    - **c.** solar mirrors.
    - **d.** solar cells.

Short Answer

12. How does acid precipitation form?

13. If sunlight is free, why is electrical energy from solar cells expensive?

14. Describe three ways that humans use natural resources.

15. Explain how fossil fuels are found and obtained.

CRITICAL THINKING

16 **Concept Mapping** Use the following terms to create a concept map: *fossil fuels, wind energy, energy resources, biomass, renewable resources, solar energy, nonrenewable resources, natural gas, gasohol, coal,* and *oil.*

17 **Predicting Consequences** How would your life be different if fossil fuels were less widely available?

18 **Evaluating Assumptions** Are fossil fuels nonrenewable? Explain.

19 **Evaluating Assumptions** Why do we need to conserve renewable resources even though they can be replaced?

20 **Evaluating Data** What might limit the productivity of a geothermal power plant?

21 **Identifying Relationships** Explain why the energy we get from many of our resources ultimately comes from the sun.

22 **Applying Concepts** Describe the different ways you can conserve natural resources at home.

23 **Identifying Relationships** Explain why coal usually forms in different locations from where petroleum and natural gas form.

24 **Applying Concepts** Choose an alternative energy resource that you think should be developed more. Explain the reason for your choice.

INTERPRETING GRAPHICS

Use the graph below to answer the questions that follow.

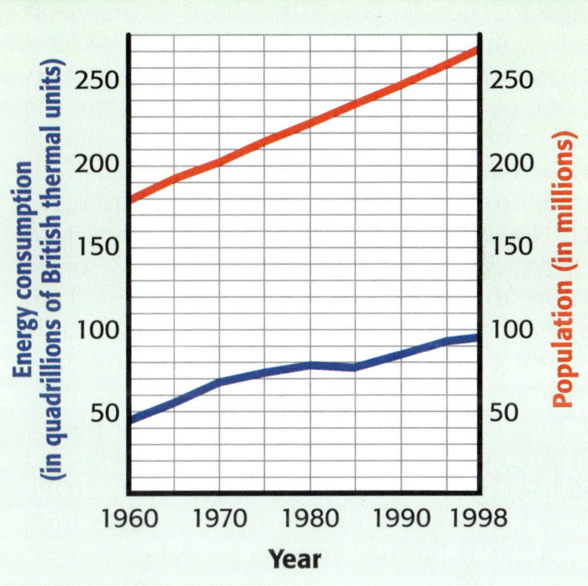

25 How many British thermal units were consumed in 1970?

26 In what year was the most energy consumed?

27 Why do you think that energy consumption has not increased at the same rate as the population has increased?

Standardized Test Preparation

READING

Read each of the passages below. Then, answer the questions that follow each passage.

Passage 1 Did you know that the average person creates about 2 kg of garbage every day? About 7% of this waste is composed of plastic products that can be recycled. Instead of adding to the landfill problem, you can recycle your plastic trash so that you can sit on it. Today, plastic is recycled into items such as park benches and highchairs. However, before plastic can be made into new products, it must be sorted. Plastic is sorted according to the resin codes that are printed on every recyclable plastic product. The resin code tells you the type of plastic that was used to make the product. The plastic most often recycled to make furniture includes the polyethylene plastic called high-density polyethylene, or HDPE, and low-density polyethylene, or LDPE. Both HDPE and LDPE are used to make items such as milk jugs, detergent bottles, and plastic bags.

1. In the passage, what does *composed* mean?
- **A** processed into
- **B** formed
- **C** crushed
- **D** melted

2. According to the passage, plastic products
- **F** can be recycled into highchairs.
- **G** can be recycled into cars.
- **H** cannot be recycled.
- **I** cause environmental problems.

3. According to the passage, which of the following statements is a fact?
- **A** The average person creates 7 kg of waste every day.
- **B** The average person weighs 7 kg.
- **C** LDPE can be used to make milk jugs, detergent bottles, plastic bags, and grocery bags.
- **D** Recycled plastics are too weak to be made into furniture.

Passage 2 You may have heard of the great California gold rush. In 1849, thousands of people moved west to California hoping to strike gold. But you may not have heard about another rush, which occurred 10 years later. What lured people to northwestern Pennsylvania in 1859? The thrill of striking oil did! However, people were using oil long before 1859. People started using oil as early as 3000 BCE. In Mesopotamia, oil was used to waterproof ships. The Egyptians and Chinese utilized oil as a medicine. It was not until the late 1700s and early 1800s that people began to use oil as a fuel for lamps to light homes and factories. Today, oil is most commonly used in transportation.

1. In the passage, what does *utilized* mean?
- **A** processed
- **B** drank
- **C** burned
- **D** used

2. According to the passage, which of the following statements is true?
- **F** Oil can be used to waterproof ships.
- **G** Oil wasn't discovered until 3000 BCE.
- **H** Oil was first used in Pennsylvania as a medicine.
- **I** Oil was used in transportation as early as 1849.

3. According to the passage, which of the following statements is a fact?
- **A** In 1849, people moved to Pennsylvania for a gold rush.
- **B** In 1649, people used oil to light homes and factories.
- **C** In 1849, people moved to California to find gold.
- **D** In 1849, people did not have any use for oil.

INTERPRETING GRAPHICS

Below is a pie chart of how various energy resources meet the world's energy needs. Use this pie chart to answer the questions that follow.

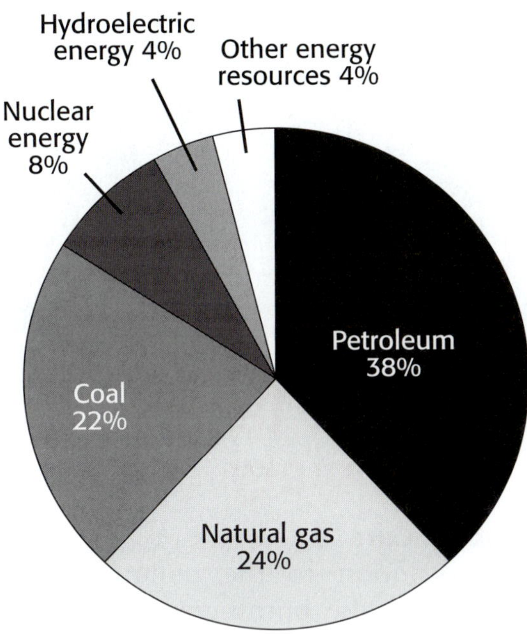

1. What percentage of the energy used in the world comes from coal?

A 22%
B 24%
C 28%
D 38%

2. What percentage of the energy used in the world comes from fossil fuels?

F 54%
G 84%
H 96%
I 100%

3. What is the total percentage of energy used for resources that do not include fossil fuels?

A 3%
B 16%
C 24%
D 64%

MATH

Read each question below, and choose the best answer.

1. The ratio of the number of kilograms of aluminum recycled to the number of kilograms of newspaper recycled by a seventh-grade class is 34 to 170. What is this ratio written as a decimal?

A 0.02
B 0.2
C 0.5
D 5

2. Peat is about 60% carbon. Approximately how many kilograms of carbon would a 130 kg sample of peat contain?

F 7.8 kg
G 52 kg
H 78 kg
I 780 kg

3. If a 24 kg sample of anthracite is examined and 21.6 kg of carbon is found in the sample, what percentage of the sample is carbon?

A 90%
B 10%
C 9%
D 2.4%

4. Cora's car runs on alcohol made from corn. Cora drives her car 12,000 km per year, and her car's gas mileage is 30 km/L. If 200 L of alcohol can be obtained from each acre of corn that is processed, about how many acres of corn would Cora have to process to fuel her car for a year?

F 1 acre
G 2 acres
H 8 acres
I 50 acres

Science in Action

Scientific Debate

The Three Gorges Dam

Dams provide hydroelectric energy, drinking water, and food for crops. Unfortunately, massive dam projects flood scenic landscapes and disrupt the environment around the dam. For example, the Three Gorges dam in China has displaced almost 2 million people living in the project area. Opponents of the project claim that the dam will also increase pollution levels in the Yangtze River. However, supporters of the dam say it will control flooding and provide millions of people with hydroelectric power. Engineers estimate that the dam's turbines will produce enough electrical energy to power a city 10 times the size of Los Angeles, California.

Language Arts ACTIVITY

WRITING SKILL Find out more about another dam project. Develop your own opinion on the project. What do you think the best outcome would be? Create a fictional story that expresses this outcome.

Science, Technology, and Society

Hybrid Cars

One solution to the pollution problem caused by the burning of fossil fuels for transportation purposes is to develop cars that depend less on fossil fuels. One such car is called a *hybrid*. Instead of using only gasoline for energy, a hybrid car uses gasoline and electricity. Because of its special batteries, the hybrid needs less gasoline to run than a car powered only by gasoline does. Some hybrids can have a gas mileage of as much as 45 mi/gal! Already, there are several models on the market to choose from. In the near future, you might see more hybrid cars on the roads.

Math ACTIVITY

Charlie's truck has a gas mileage of 17 mi/gal. Charlie drives his truck an average of 12,000 mi per year. Then, he sells the truck and buys a new hybrid car that has a gas mileage of 45 mi/gal. If gasoline costs $1.40 per gallon, how much money will Charlie save in a year by driving the hybrid car instead of his truck?

Careers

Fred Begay

Nuclear Physicist Generating energy by combining atoms is called *fusion*. This process is being developed by nuclear physicists, such as Dr. Fred Begay, at the Department of Energy's Los Alamos National Laboratory. Begay hopes to someday make fusion an alternative energy resource. Because fusion is the process that generates energy in the sun, Begay uses NASA satellites to study the sun. Begay explains that it is necessary to develop skills in abstract reasoning to study fusion. As a Navajo, Begay developed these skills while growing up at his Navajo home in Towaoc, Colorado, where his family taught him about nature. While he was growing up on a reservation, Fred Begay never dreamed that he would become a physicist. In fact, he was sent to a school that trained children to be farmers because authorities did not expect Navajo children to succeed in academics. In spite of this early adversity, Begay went on to earn his Ph.D. in physics. Today, Begay uses his skills not only to help develop a new energy resource but also to mentor Native American and minority students. In 1999, Begay won the Distinguished Scientist Award from the Society for Advancement of Chicanos and Native Americans in Science.

Social Studies ACTIVITY

Research the lifestyle of Native Americans before 1900. Then, create a poster that compares resources that Native Americans used before 1900 with resources that many people use today.

To learn more about these Science in Action topics, visit go.hrw.com and type in the keyword HZ5ENRF.

Current Science

Check out Current Science® articles related to this chapter by visiting go.hrw.com. Just type in the keyword HZ5CS05.

The Restless Earth

About the PHOTO

The San Andreas fault stretches across the California landscape like a giant wound. The fault, which is 1,000 km long, breaks the Earth's crust from Northern California to Mexico. Because the North American plate and Pacific plate are slipping past one another along the fault, many earthquakes happen.

PRE-READING ACTIVITY

FOLDNOTES **Key-Term Fold** Before you read the chapter, create the FoldNote entitled "Key-Term Fold" described in the **Study Skills** section of the Appendix. Write a key term from the chapter on each tab of the key-term fold. Under each tab, write the definition of the key term.

START-UP ACTIVITY

Continental Collisions

As you can see, continents not only move but can also crash into each other. In this activity, you will model the collision of two continents.

Procedure

1. Obtain **two stacks of paper** that are each about 1 cm thick.
2. Place the two stacks of paper on a **flat surface,** such as a desk.
3. Very slowly, push the stacks of paper together so that they collide. Continue to push the stacks until the paper in one of the stacks folds over.

Analysis

1. What happens to the stacks of paper when they collide with each other?
2. Are all of the pieces of paper pushed upward? If not, what happens to the pieces that are not pushed upward?
3. What type of landform will most likely result from this continental collision?

SECTION 1

Inside the Earth

If you tried to dig to the center of the Earth, what do you think you would find? Would the Earth be solid or hollow? Would it be made of the same material throughout?

Actually, the Earth is made of several layers. Each layer is made of different materials that have different properties. Scientists think about physical layers in two ways—by their composition and by their physical properties.

READING WARM-UP

Objectives

- Identify the layers of the Earth by their composition.
- Identify the layers of the Earth by their physical properties.
- Describe a tectonic plate.
- Explain how scientists know about the structure of Earth's interior.

Terms to Learn

crust, mantle, core, lithosphere, asthenosphere, mesosphere, tectonic plate

READING STRATEGY

Reading Organizer As you read this section, create an outline of the section. Use the headings from the section in your outline.

The Composition of the Earth

The Earth is divided into three layers—the crust, the mantle, and the core—based on the compounds that make up each layer. A *compound* is a substance composed of two or more elements. The less dense compounds make up the crust and mantle, and the densest compounds make up the core. The layers form because heavier elements are pulled toward the center of the Earth by gravity, and elements of lesser mass are found farther from the center.

The Crust

The outermost layer of the Earth is the **crust.** The crust is 5 to 100 km thick. It is the thinnest layer of the Earth.

As **Figure 1** shows, there are two types of crust—continental and oceanic. Both continental crust and oceanic crust are made mainly of the elements oxygen, silicon, and aluminum. However, the denser oceanic crust has almost twice as much iron, calcium, and magnesium, which form minerals that are denser than those in the continental crust.

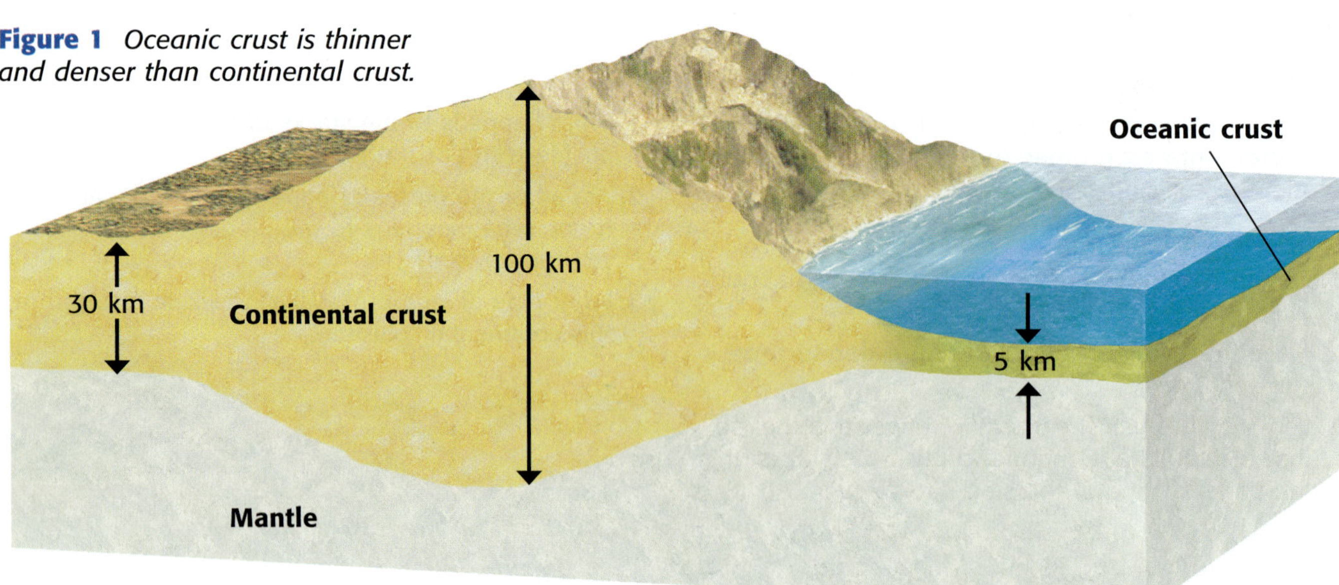

Figure 1 *Oceanic crust is thinner and denser than continental crust.*

The Mantle

The layer of the Earth between the crust and the core is the **mantle.** The mantle is much thicker than the crust and contains most of the Earth's mass.

No one has ever visited the mantle. The crust is too thick to drill through to reach the mantle. Scientists must draw conclusions about the composition and other physical properties of the mantle from observations made on the Earth's surface. In some places, mantle rock pushes to the surface, which allows scientists to study the rock directly.

As you can see in **Figure 2,** another place scientists look for clues about the mantle is the ocean floor. Magma from the mantle flows out of active volcanoes on the ocean floor. These underwater volcanoes have given scientists many clues about the composition of the mantle. Because the mantle has more magnesium and less aluminum and silicon than the crust does, the mantle is denser than the crust.

Figure 2 *Volcanic vents on the ocean floor, such as this vent off the coast of Hawaii, allow magma to rise up through the crust from the mantle.*

crust the thin and solid outermost layer of the Earth above the mantle

mantle the layer of rock between the Earth's crust and core

core the central part of the Earth below the mantle

The Core

The layer of the Earth that extends from below the mantle to the center of the Earth is the **core.** Scientists think that the Earth's core is made mostly of iron and contains smaller amounts of nickel but almost no oxygen, silicon, aluminum, or magnesium. As shown in **Figure 3,** the core makes up roughly one-third of the Earth's mass.

Reading Check **Briefly describe the layers that make up the Earth.** *(See the Appendix for answers to Reading Checks.)*

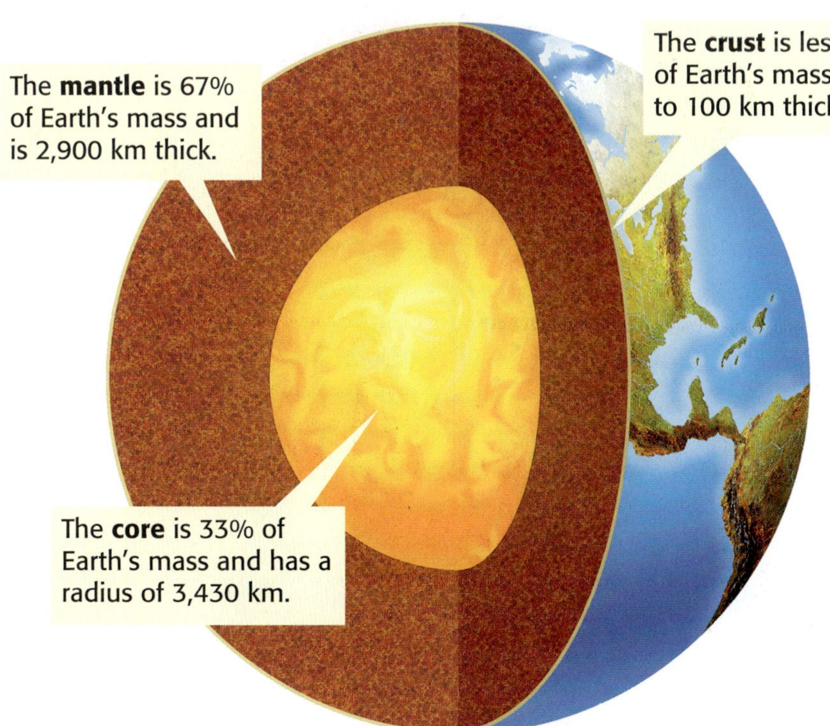

Figure 3 *The Earth is made up of three layers based on the composition of each layer.*

Using Models

Imagine that you are building a model of the Earth that will have a radius of 1 m. You find out that the average radius of the Earth is 6,380 km and that the thickness of the lithosphere is about 150 km. What percentage of the Earth's radius is the lithosphere? How thick (in centimeters) would you make the lithosphere in your model? Using these two facts and the other measurements given on this page, make both a circle chart and a bar graph showing the relative sizes of each of the Earth's layers.

The Physical Structure of the Earth

Another way to look at the Earth is to examine the physical properties of its layers. The Earth is divided into five physical layers—the lithosphere, asthenosphere, mesosphere, outer core, and inner core. As shown in the figure below, each layer has its own set of physical properties.

Reading Check What are the five physical layers of the Earth?

Lithosphere The outermost, rigid layer of the Earth is the **lithosphere.** The lithosphere is made of two parts—the crust and the rigid upper part of the mantle. The lithosphere is divided into pieces called *tectonic plates.*

Asthenosphere The **asthenosphere** is a plastic layer of the mantle on which pieces of the lithosphere move. The asthenosphere is made of solid rock that flows very slowly.

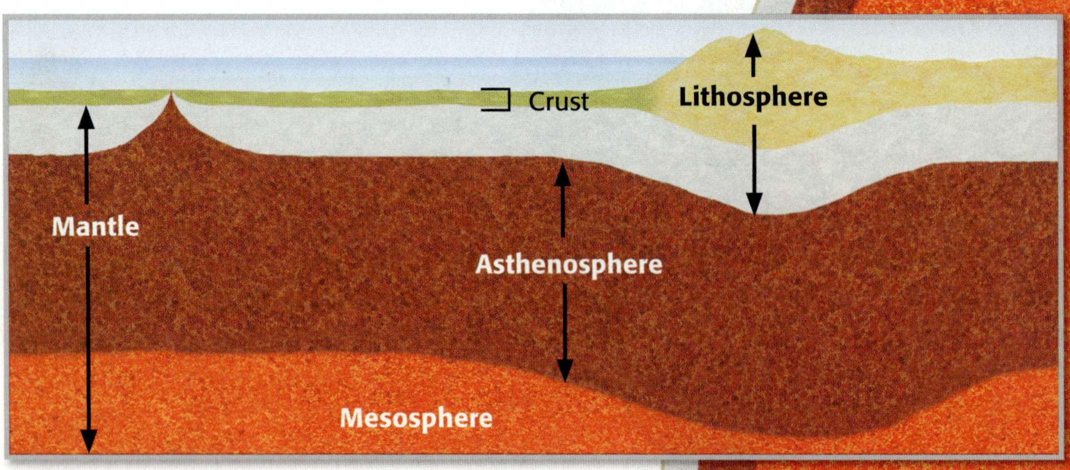

lithosphere the solid, outer layer of the Earth that consists of the crust and the rigid upper part of the mantle

asthenosphere the soft layer of the mantle on which the tectonic plates move

mesosphere the strong, lower part of the mantle between the asthenosphere and the outer core

Mesosphere Beneath the asthenosphere is the strong, lower part of the mantle called the **mesosphere.** The mesosphere extends from the bottom of the asthenosphere to the Earth's core.

Outer Core The Earth's core is divided into two parts—the outer core and the inner core. The outer core is the liquid layer of the Earth's core that lies beneath the mantle and surrounds the inner core.

Inner Core The inner core is the solid, dense center of our planet that extends from the bottom of the outer core to the center of the Earth, which is about 6,380 km beneath the surface.

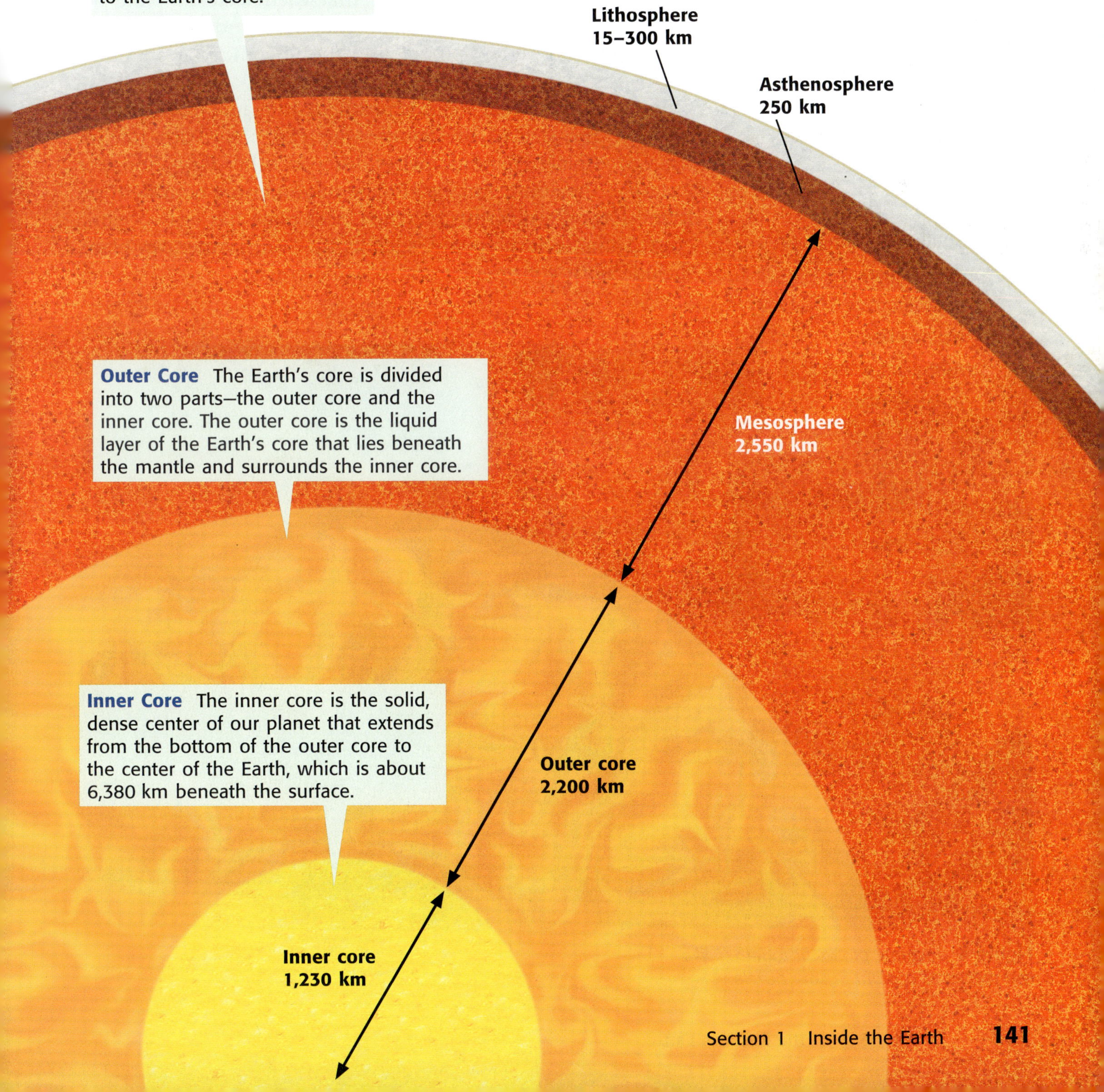

Tectonic Plates

tectonic plate a block of lithosphere that consists of the crust and the rigid, outermost part of the mantle

Pieces of the lithosphere that move around on top of the asthenosphere are called **tectonic plates**. But what exactly does a tectonic plate look like? How big are tectonic plates? How and why do they move around? To answer these questions, begin by thinking of the lithosphere as a giant jigsaw puzzle.

A Giant Jigsaw Puzzle

All of the tectonic plates have names, some of which you may already know. Some of the major tectonic plates are named on the map in **Figure 4.** Notice that each tectonic plate fits together with the tectonic plates that surround it. The lithosphere is like a jigsaw puzzle, and the tectonic plates are like the pieces of a jigsaw puzzle.

Notice that not all tectonic plates are the same. For example, compare the size of the South American plate with that of the Cocos plate. Tectonic plates differ in other ways, too. For example, the South American plate has an entire continent on it and has oceanic crust, but the Cocos plate has only oceanic crust. Some tectonic plates, such as the South American plate, include both continental and oceanic crust.

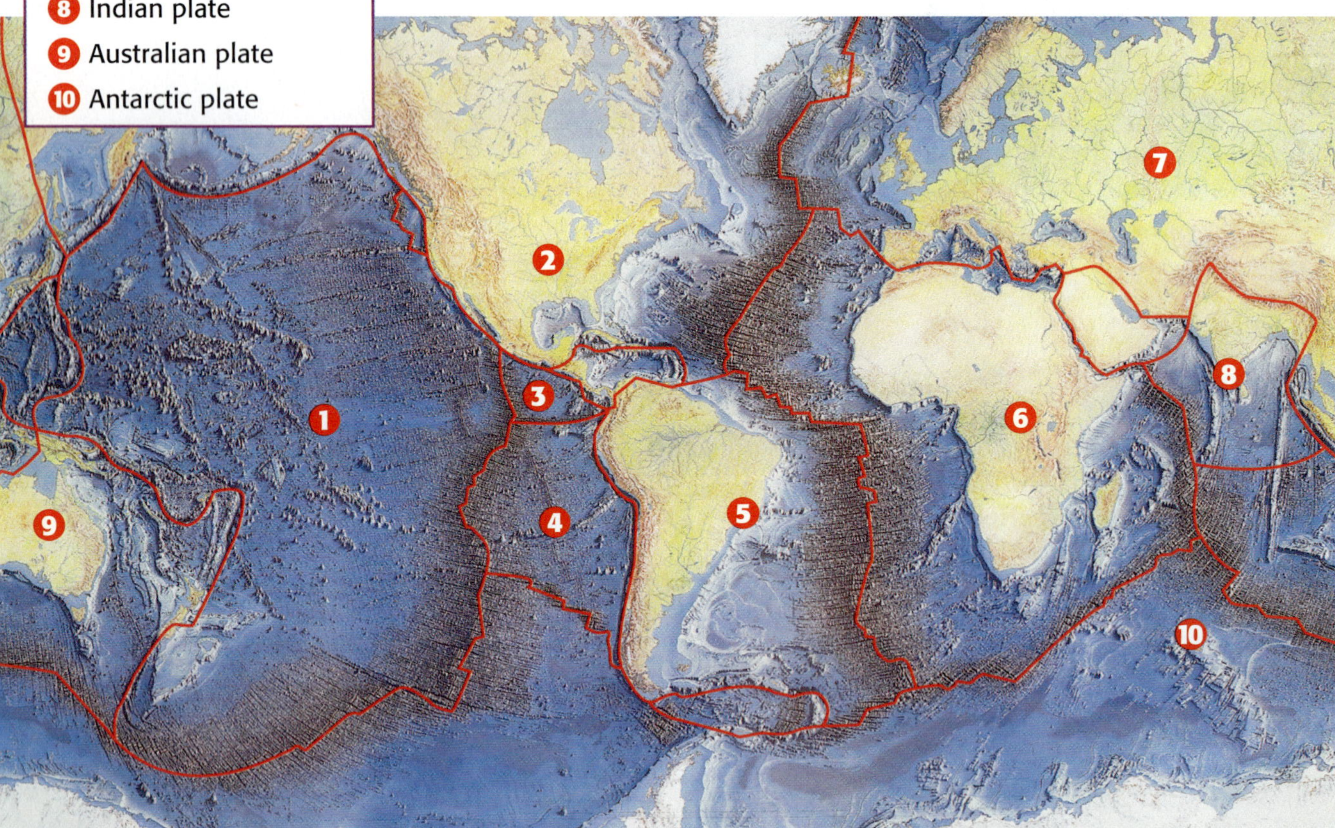

Figure 4 *Tectonic plates fit together like the pieces of a giant jigsaw puzzle.*

Figure 5 **The South American Plate**

This image shows what you might see if you could lift the South American plate out of its position between other tectonic plates.

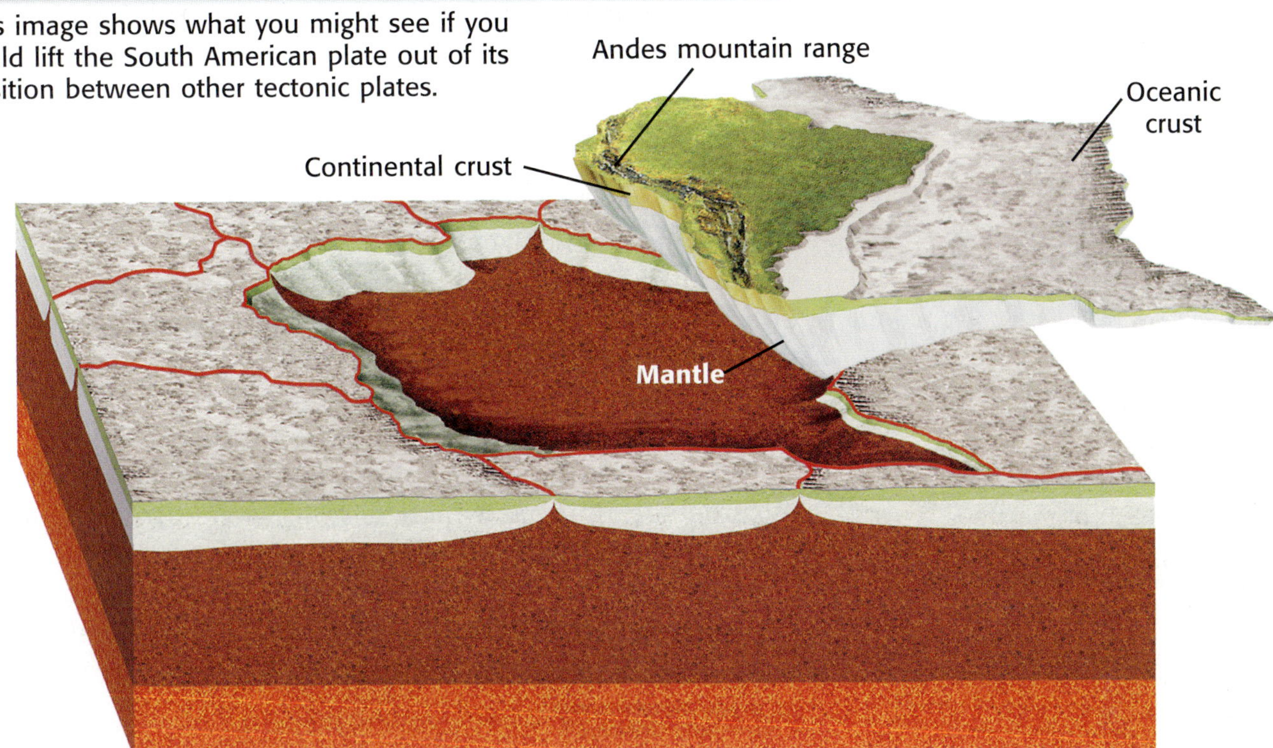

A Tectonic Plate Close-Up

What would a tectonic plate look like if you could lift it out of its place? **Figure 5** shows what the South American plate might look like if you could. Notice that this tectonic plate not only consists of the upper part of the mantle but also consists of both oceanic crust and continental crust. The thickest part of the South American plate is the continental crust. The thinnest part of this plate is in the mid-Atlantic Ocean.

Like Ice Cubes in a Bowl of Punch

Think about ice cubes floating in a bowl of punch. If there are enough cubes, they will cover the surface of the punch and bump into one another. Parts of the ice cubes are below the surface of the punch and displace the punch. Large pieces of ice displace more punch than small pieces of ice. Tectonic plates "float" on the asthenosphere in a similar way. The plates cover the surface of the asthenosphere, and they touch one another and move around. The lithosphere displaces the asthenosphere. Thick tectonic plates, such as those made of continental crust, displace more asthenosphere than do thin plates, such as those made of oceanic lithosphere.

Reading Check **Why do tectonic plates made of continental lithosphere displace more asthenosphere than tectonic plates made of oceanic lithosphere do?**

Tectonic Ice Cubes

1. Take the bottom half of a clear, **2 L soda bottle** that has been cut in half. Make sure that the label has been removed.
2. Fill the bottle with **water** to about 1 cm below the top edge of the bottle.
3. Get **three pieces of irregularly shaped ice** that are small, medium, and large.
4. Float the ice in the water, and note how much of each piece is below the surface of the water.
5. Do all pieces of ice float mostly below the surface? Which piece is mostly below the surface? Why?

Build a Seismograph

Seismographs are instruments that seismologists, scientists who study earthquakes, use to detect seismic waves. Research seismograph designs with your parent. For example, a simple seismograph can be built by using a weight suspended by a spring next to a ruler. With your parent, attempt to construct a home seismograph based on a design you have selected. Outline each of the steps used to build your seismograph, and present the written outline to your teacher.

Mapping the Earth's Interior

How do scientists know things about the deepest parts of the Earth, where no one has ever been? Scientists have never even drilled through the crust, which is only a thin skin on the surface of the Earth. So, how do we know so much about the mantle and the core?

Would you be surprised to know that some of the answers come from earthquakes? When an earthquake happens, vibrations called *seismic waves* are produced. Seismic waves travel at different speeds through the Earth. Their speed depends on the density and composition of material that they pass through. For example, a seismic wave traveling through a solid will go faster than a seismic wave traveling through a liquid.

When an earthquake happens, machines called *seismographs* measure the times at which seismic waves arrive at different distances from an earthquake. Seismologists can then use these distances and travel times to calculate the density and thickness of each physical layer of the Earth. **Figure 6** shows how seismic waves travel through the Earth.

Reading Check **What are some properties of seismic waves?**

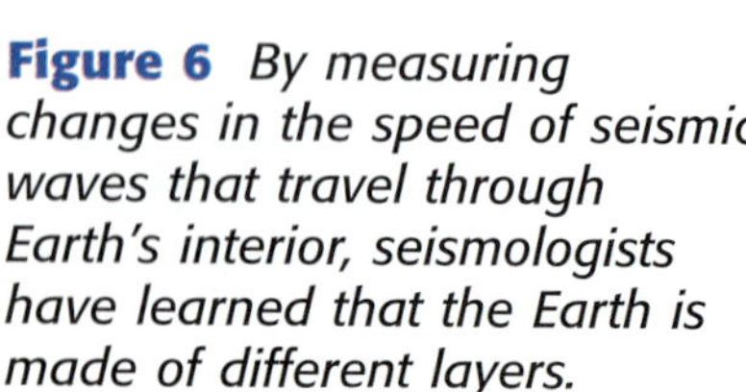

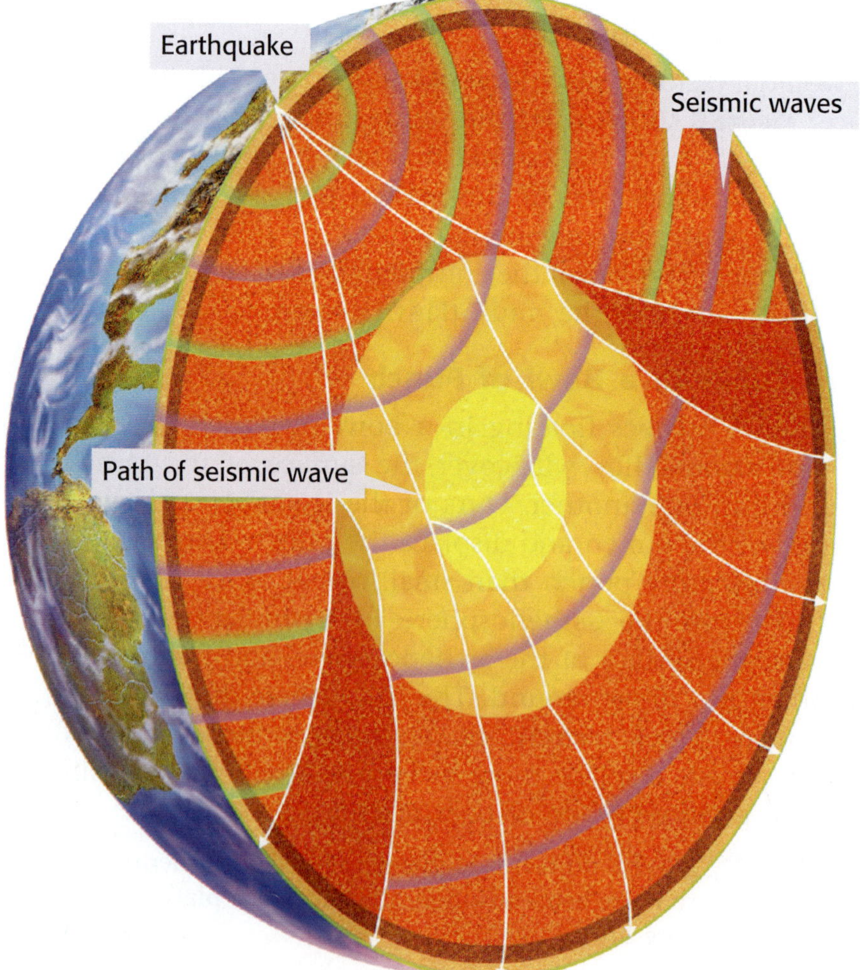

Figure 6 *By measuring changes in the speed of seismic waves that travel through Earth's interior, seismologists have learned that the Earth is made of different layers.*

SECTION Review

Summary

- The Earth is made up of three layers—the crust, the mantle, and the core—based on chemical composition. Less dense compounds make up the crust and mantle. Denser compounds make up the core.
- The Earth is made up of five main physical layers: the lithosphere, the asthenosphere, the mesosphere, the outer core, and the inner core.
- Tectonic plates are large pieces of the lithosphere that move around on the Earth's surface.
- The crust in some tectonic plates is mainly continental. Other plates have only oceanic crust. Still other plates include both continental and oceanic crust.
- Thick tectonic plates, such as those in which the crust is mainly continental, displace more asthenosphere than do thin plates, such as those in which the crust is mainly oceanic.
- Knowledge about the layers of the Earth comes from the study of seismic waves caused by earthquakes.

Using Key Terms

For each pair of terms, explain how the meanings of the terms differ.

1. *crust* and *mantle*
2. *lithosphere* and *asthenosphere*

Understanding Key Ideas

3. The part of the Earth that is molten is the
   a. crust.
   b. mantle.
   c. outer core.
   d. inner core.
4. The part of the Earth on which the tectonic plates move is the
   a. lithosphere.
   b. asthenosphere.
   c. mesosphere.
   d. crust.
5. Identify the layers of the Earth by their chemical composition.
6. Identify the layers of the Earth by their physical properties.
7. Describe a tectonic plate.
8. Explain how scientists know about the structure of the Earth's interior.

Interpreting Graphics

9. According to the wave speeds shown in the table below, which two physical layers of the Earth are densest?

Speed of Seismic Waves in Earth's Interior

Physical layer	Wave speed
Lithosphere	7 to 8 km/s
Asthenosphere	7 to 11 km/s
Mesosphere	11 to 13 km/s
Outer core	8 to 10 km/s
Inner core	11 to 12 km/s

Critical Thinking

10. **Making Comparisons** Explain the difference between the crust and the lithosphere.
11. **Analyzing Ideas** Why does a seismic wave travel faster through solid rock than through water?

SECTION 2

Restless Continents

Have you ever looked at a map of the world and noticed how the coastlines of continents on opposite sides of the oceans appear to fit together like the pieces of a puzzle? Is it just coincidence that the coastlines fit together well? Is it possible that the continents were actually together sometime in the past?

READING WARM-UP

Objectives

- Describe Wegener's hypothesis of continental drift.
- Explain how sea-floor spreading provides a way for continents to move.
- Describe how new oceanic lithosphere forms at mid-ocean ridges.
- Explain how magnetic reversals provide evidence for sea-floor spreading.

Terms to Learn

continental drift
sea-floor spreading

READING STRATEGY

Paired Summarizing Read this section silently. In pairs, take turns summarizing the material. Stop to discuss ideas that seem confusing.

Wegener's Continental Drift Hypothesis

One scientist who looked at the pieces of this puzzle was Alfred Wegener (VAY guh nuhr). In the early 1900s, he wrote about his hypothesis of *continental drift.* **Continental drift** is the hypothesis that states that the continents once formed a single landmass, broke up, and drifted to their present locations. This hypothesis seemed to explain a lot of puzzling observations, including the observation of how well continents fit together.

Continental drift also explained why fossils of the same plant and animal species are found on continents that are on different sides of the Atlantic Ocean. Many of these ancient species could not have crossed the Atlantic Ocean. As you can see in **Figure 1,** without continental drift, this pattern of fossils would be hard to explain. In addition to fossils, similar types of rock and evidence of the same ancient climatic conditions were found on several continents.

continental drift the hypothesis that states that the continents once formed a single landmass, broke up, and drifted to their present locations

Reading Check How did fossils provide evidence for Wegener's hypothesis of continental drift? (*See the Appendix for answers to Reading Checks.*)

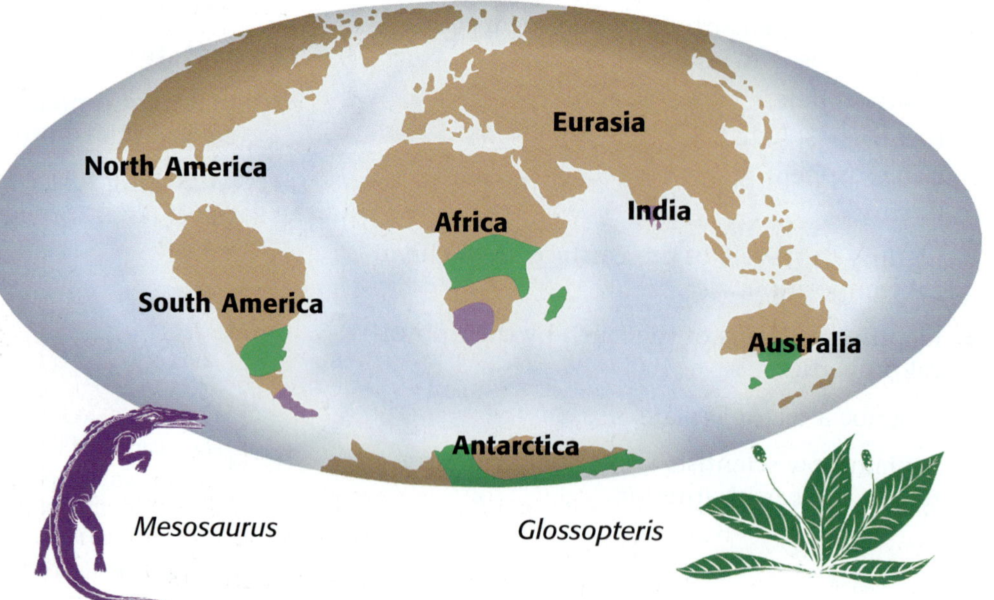

Figure 1 *Fossils of* Mesosaurus, *a small, aquatic reptile, and* Glossopteris, *an ancient plant species, have been found on several continents.*

Figure 2 **The Drifting Continents**

245 Million Years Ago
Pangaea existed when some of the earliest dinosaurs were roaming the Earth. The continent was surrounded by a sea called *Panthalassa*, which means "all sea."

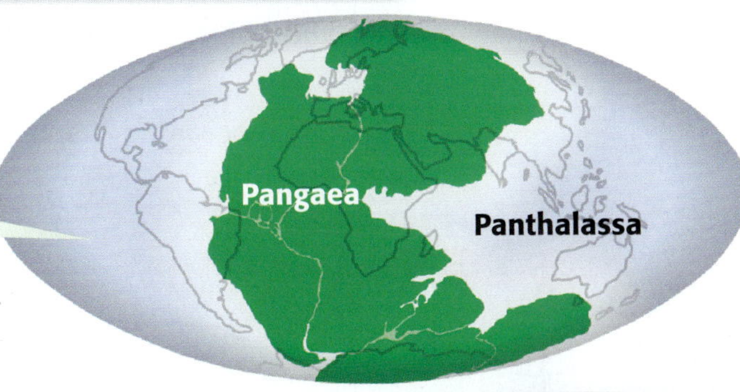

180 Million Years Ago
Gradually, Pangaea broke into two big pieces. The northern piece is called *Laurasia*. The southern piece is called *Gondwana*.

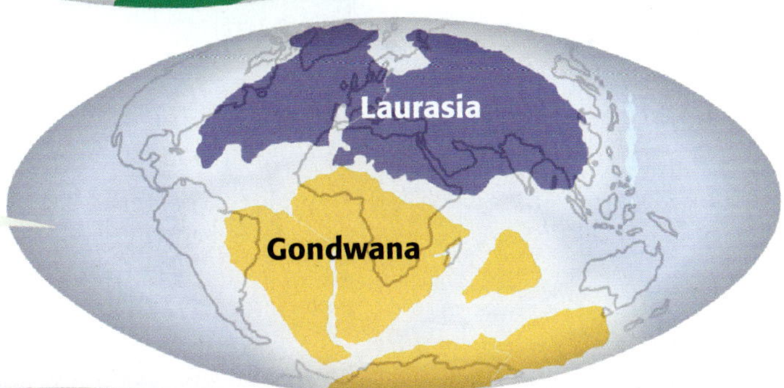

65 Million Years Ago
By the time the dinosaurs became extinct, Laurasia and Gondwana had split into smaller pieces.

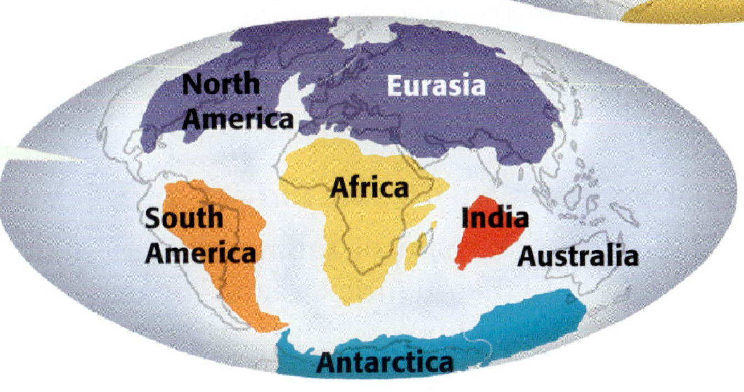

The Breakup of Pangaea

Wegener made many observations before proposing his hypothesis of continental drift. He thought that all of the present continents were once joined in a single, huge continent. Wegener called this continent *Pangaea* (pan JEE uh), which is Greek for "all earth." We now know from the hypothesis of plate tectonics that Pangaea existed about 245 million years ago. We also know that Pangaea further split into two huge continents—Laurasia and Gondwana—about 180 million years ago. As shown in **Figure 2,** these two continents split again and formed the continents we know today.

Sea-Floor Spreading

When Wegener put forth his hypothesis of continental drift, many scientists would not accept his hypothesis. From the calculated strength of the rocks, it did not seem possible for the crust to move in this way. During Wegener's life, no one knew the answer. It wasn't until many years later that evidence provided some clues to the forces that moved the continents.

Figure 3 **Sea-Floor Spreading**

Sea-floor spreading creates new oceanic lithosphere at mid-ocean ridges.

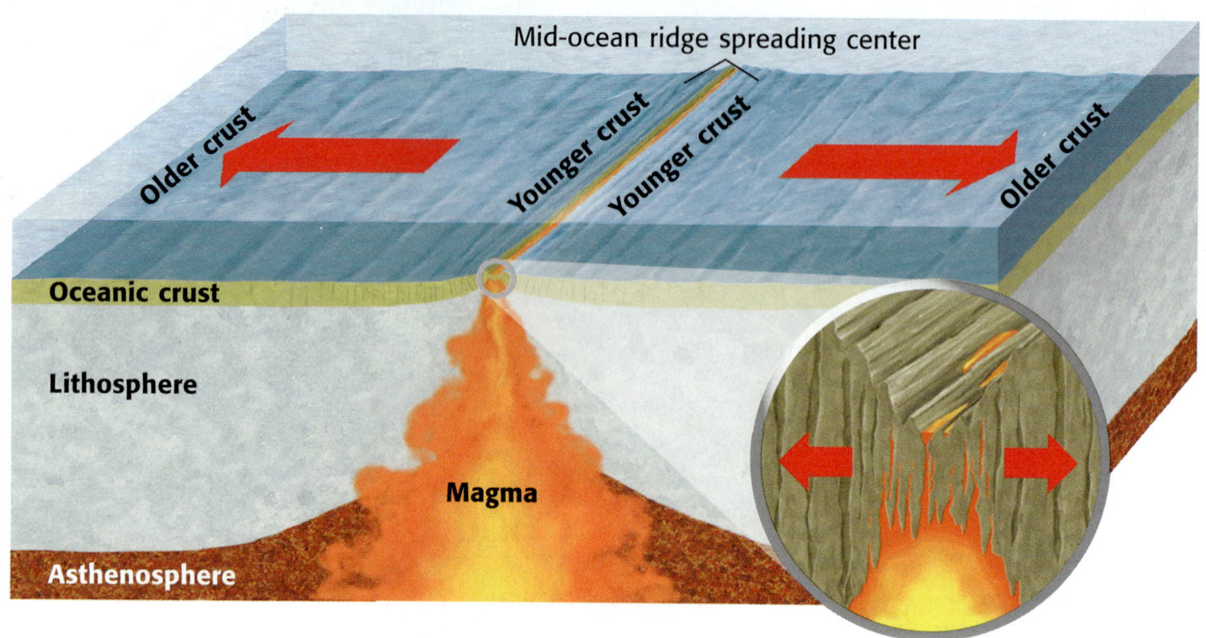

sea-floor spreading the process by which new oceanic lithosphere forms as magma rises toward the surface and solidifies

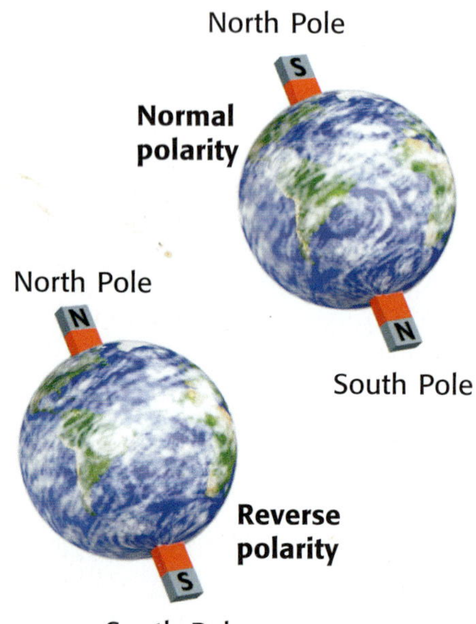

Figure 4 *The polarity of Earth's magnetic field changes over time.*

Mid-Ocean Ridges and Sea-Floor Spreading

A chain of submerged mountains runs through the center of the Atlantic Ocean. The chain is part of a worldwide system of mid-ocean ridges. Mid-ocean ridges are underwater mountain chains that run through Earth's ocean basins.

Mid-ocean ridges are places where sea-floor spreading takes place. **Sea-floor spreading** is the process by which new oceanic lithosphere forms as magma rises toward the surface and solidifies. As the tectonic plates move away from each other, the sea floor spreads apart and magma fills in the gap. As this new crust forms, the older crust gets pushed away from the mid-ocean ridge. As **Figure 3** shows, the older crust is farther away from the mid-ocean ridge than the younger crust is.

Evidence for Sea-Floor Spreading: Magnetic Reversals

Some of the most important evidence of sea-floor spreading comes from magnetic reversals recorded in the ocean floor. Throughout Earth's history, the north and south magnetic poles have changed places many times. When the poles change places, the polarity of Earth's magnetic poles changes, as shown in **Figure 4.** When Earth's magnetic poles change places, this change is called a *magnetic reversal*.

Magnetic Reversals and Sea-Floor Spreading

The molten rock at the mid-ocean ridges contains tiny grains of magnetic minerals. These mineral grains contain iron and are like compasses. They align with the magnetic field of the Earth. When the molten rock cools, the record of these tiny compasses remains in the rock. This record is then carried slowly away from the spreading center of the ridge as sea-floor spreading occurs.

As you can see in **Figure 5,** when the Earth's magnetic field reverses, the magnetic mineral grains align in the opposite direction. The new rock records the direction of the Earth's magnetic field. As the sea floor spreads away from a mid-ocean ridge, it carries with it a record of magnetic reversals. This record of magnetic reversals was the final proof that sea-floor spreading does occur.

Reading Check **How is a record of magnetic reversals recorded in molten rock at mid-ocean ridges?**

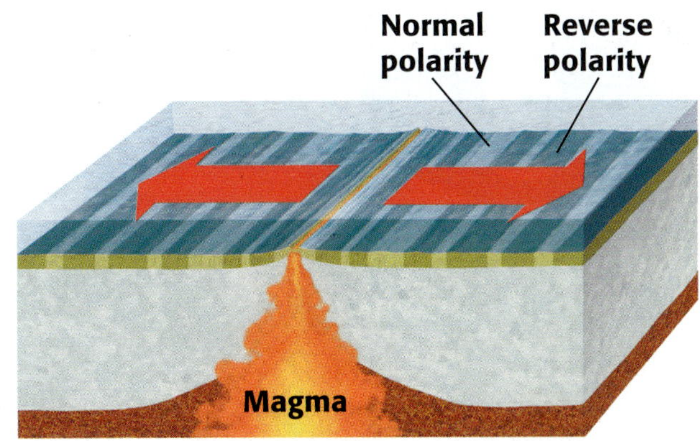

Figure 5 *Magnetic reversals in oceanic crust are shown as bands of light blue and dark blue oceanic crust. Light blue bands indicate normal polarity, and dark blue bands indicate reverse polarity.*

SECTION Review

Summary

- Wegener hypothesized that continents drift apart from one another and have done so in the past.
- The process by which new oceanic lithosphere forms at mid-ocean ridges is called sea-floor spreading.
- As tectonic plates separate, the sea floor spreads apart and magma fills in the gap.
- Magnetic reversals are recorded over time in oceanic crust.

Using Key Terms

1. In your own words, write a definition for each of the following terms: *continental drift* and *sea-floor spreading*.

Understanding Key Ideas

2. At mid-ocean ridges,
   - **a.** the crust is older.
   - **b.** sea-floor spreading occurs.
   - **c.** oceanic lithosphere is destroyed.
   - **d.** tectonic plates are colliding.
3. Explain how oceanic lithosphere forms at mid-ocean ridges.
4. What is magnetic reversal?

Math Skills

5. If a piece of sea floor has moved 50 km in 5 million years, what is the yearly rate of sea-floor motion?

Critical Thinking

6. **Identifying Relationships** Explain how magnetic reversals provide evidence for sea-floor spreading.
7. **Applying Concepts** Why do bands indicating magnetic reversals appear to be of similar width on both sides of a mid-ocean ridge?
8. **Applying Concepts** Why do you think that old rocks are rare on the ocean floor?

SECTION 3

The Theory of Plate Tectonics

READING WARM-UP

Objectives

- Describe the three types of tectonic plate boundaries.
- Describe the three forces thought to move tectonic plates.
- Explain how scientists measure the rate at which tectonic plates move.

Terms to Learn

plate tectonics
convergent boundary
divergent boundary
transform boundary

READING STRATEGY

Brainstorming The key idea of this section is plate tectonics. Brainstorm words and phrases related to plate tectonics.

It takes an incredible amount of force to move a tectonic plate! But where does this force come from?

As scientists' understanding of mid-ocean ridges and magnetic reversals grew, scientists formed a theory to explain how tectonic plates move. **Plate tectonics** is the theory that the Earth's lithosphere is divided into tectonic plates that move around on top of the asthenosphere. In this section, you will learn what causes tectonic plates to move. But first you will learn about the different types of tectonic plate boundaries.

Tectonic Plate Boundaries

A boundary is a place where tectonic plates touch. All tectonic plates share boundaries with other tectonic plates. These boundaries are divided into three types: convergent, divergent, and transform. The type of boundary depends on how the tectonic plates move relative to one another. Tectonic plates can collide, separate, or slide past each other. Earthquakes can occur at all three types of plate boundaries. The figure below shows examples of tectonic plate boundaries.

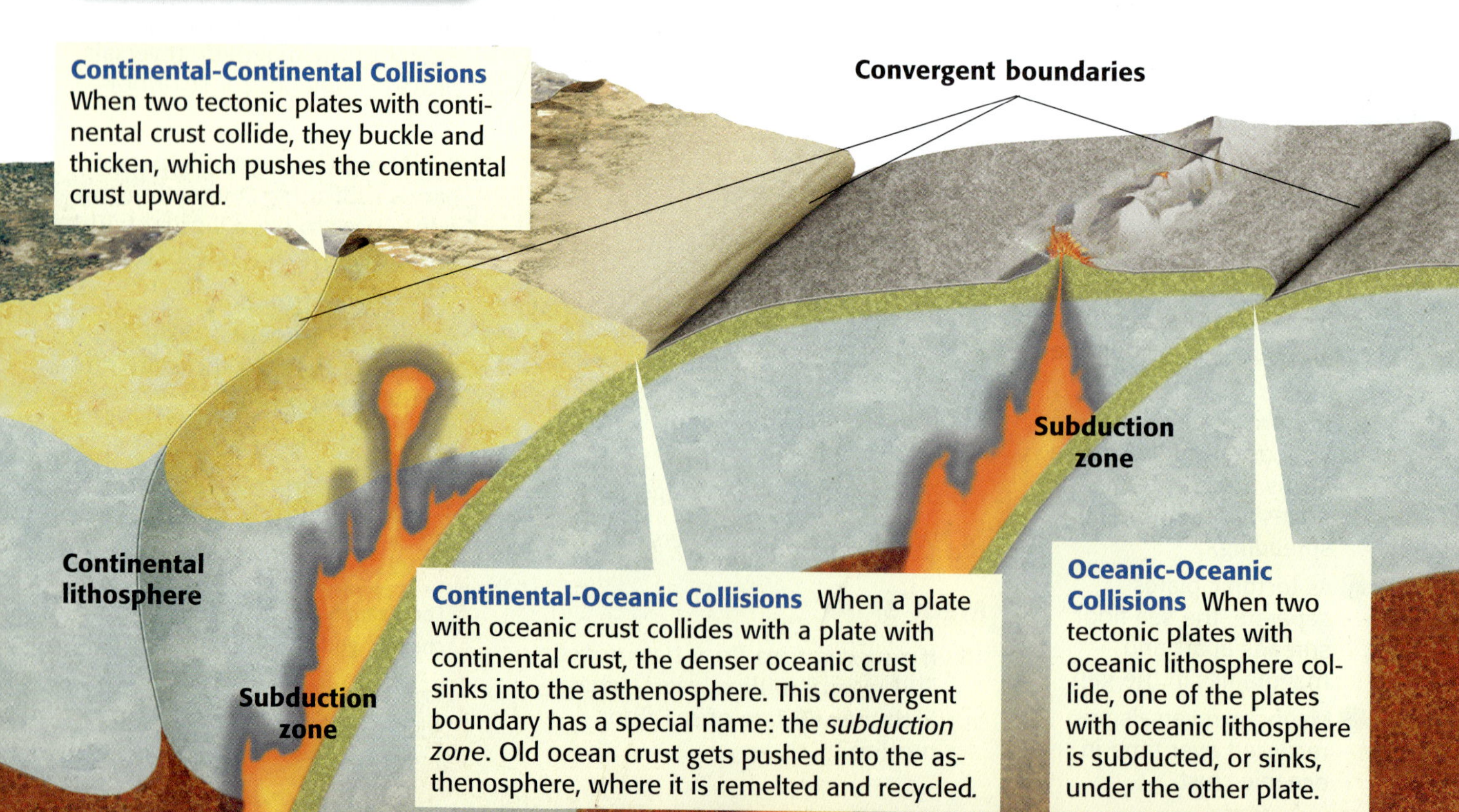

Continental-Continental Collisions When two tectonic plates with continental crust collide, they buckle and thicken, which pushes the continental crust upward.
Convergent boundaries
Subduction zone
Continental lithosphere
Subduction zone
Continental-Oceanic Collisions When a plate with oceanic crust collides with a plate with continental crust, the denser oceanic crust sinks into the asthenosphere. This convergent boundary has a special name: the *subduction zone*. Old ocean crust gets pushed into the asthenosphere, where it is remelted and recycled.
Oceanic-Oceanic Collisions When two tectonic plates with oceanic lithosphere collide, one of the plates with oceanic lithosphere is subducted, or sinks, under the other plate.

Convergent Boundaries

When two tectonic plates collide, the boundary between them is a **convergent boundary.** What happens at a convergent boundary depends on the kind of crust at the leading edge of each tectonic plate. The three types of convergent boundaries are continental-continental boundaries, continental-oceanic boundaries, and oceanic-oceanic boundaries.

Divergent Boundaries

When two tectonic plates separate, the boundary between them is called a **divergent boundary.** New sea floor forms at divergent boundaries. Mid-ocean ridges are the most common type of divergent boundary.

Transform Boundaries

When two tectonic plates slide past each other horizontally, the boundary between them is a **transform boundary.** The San Andreas Fault in California is a good example of a transform boundary. This fault marks the place where the Pacific and North American plates are sliding past each other.

Reading Check **Define the term *transform boundary*.** *(See the Appendix for answers to Reading Checks.)*

plate tectonics the theory that explains how large pieces of the Earth's outermost layer, called *tectonic plates,* move and change shape

convergent boundary the boundary formed by the collision of two lithospheric plates

divergent boundary the boundary between two tectonic plates that are moving away from each other

transform boundary the boundary between tectonic plates that are sliding past each other horizontally

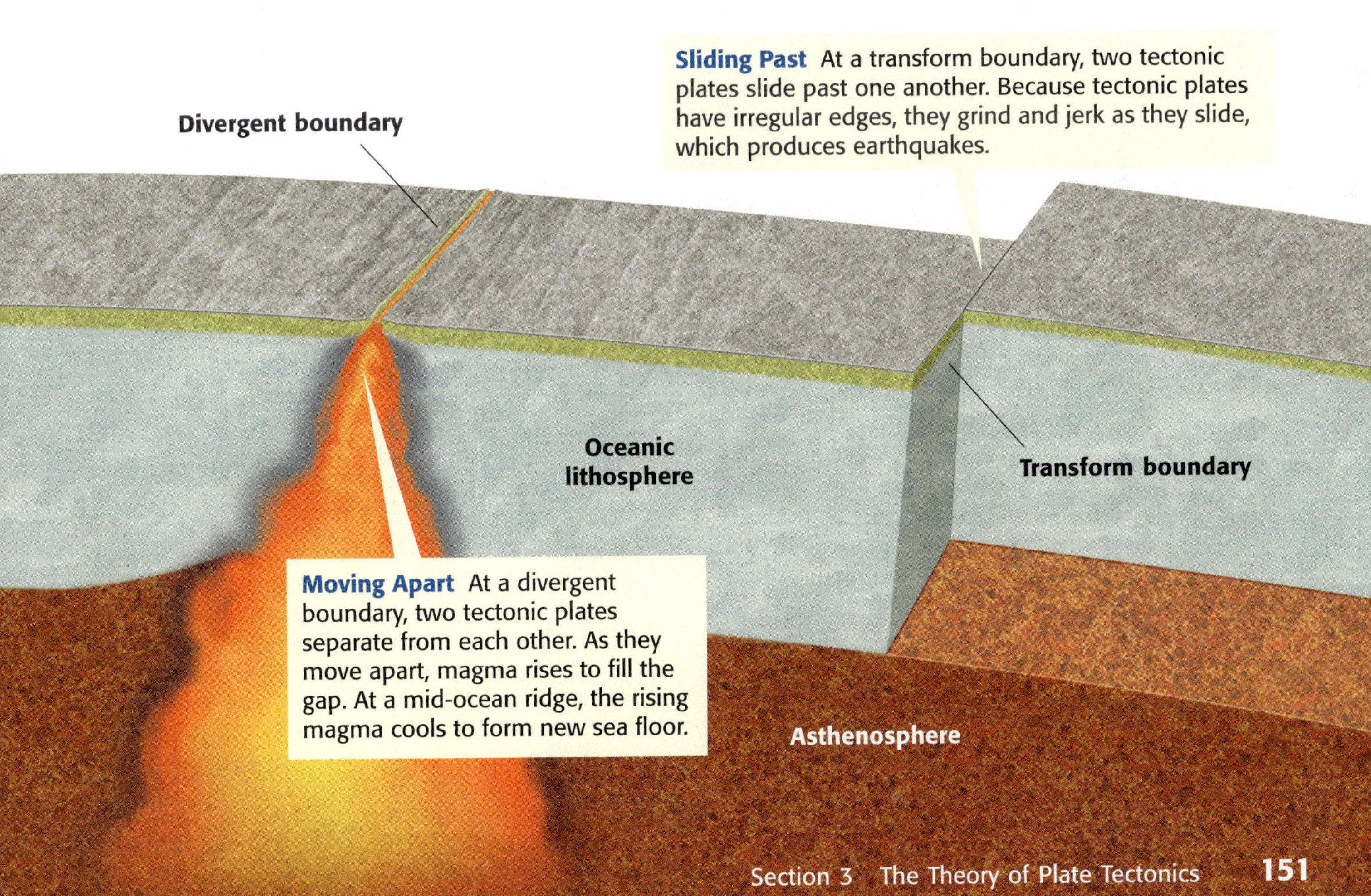

Sliding Past At a transform boundary, two tectonic plates slide past one another. Because tectonic plates have irregular edges, they grind and jerk as they slide, which produces earthquakes.

Moving Apart At a divergent boundary, two tectonic plates separate from each other. As they move apart, magma rises to fill the gap. At a mid-ocean ridge, the rising magma cools to form new sea floor.

Possible Causes of Tectonic Plate Motion

You have learned that plate tectonics is the theory that the lithosphere is divided into tectonic plates that move around on top of the asthenosphere. What causes the motion of tectonic plates? Remember that the solid rock of the asthenosphere flows very slowly. This movement occurs because of changes in density within the asthenosphere. These density changes are caused by the outward flow of thermal energy from deep within the Earth. When rock is heated, it expands, becomes less dense, and tends to rise to the surface of the Earth. As the rock gets near the surface, the rock cools, becomes more dense, and tends to sink. **Figure 1** shows three possible causes of tectonic plate motion.

Reading Check What causes changes in density in the asthenosphere?

Figure 1 Three Possible Driving Forces of Plate Tectonics

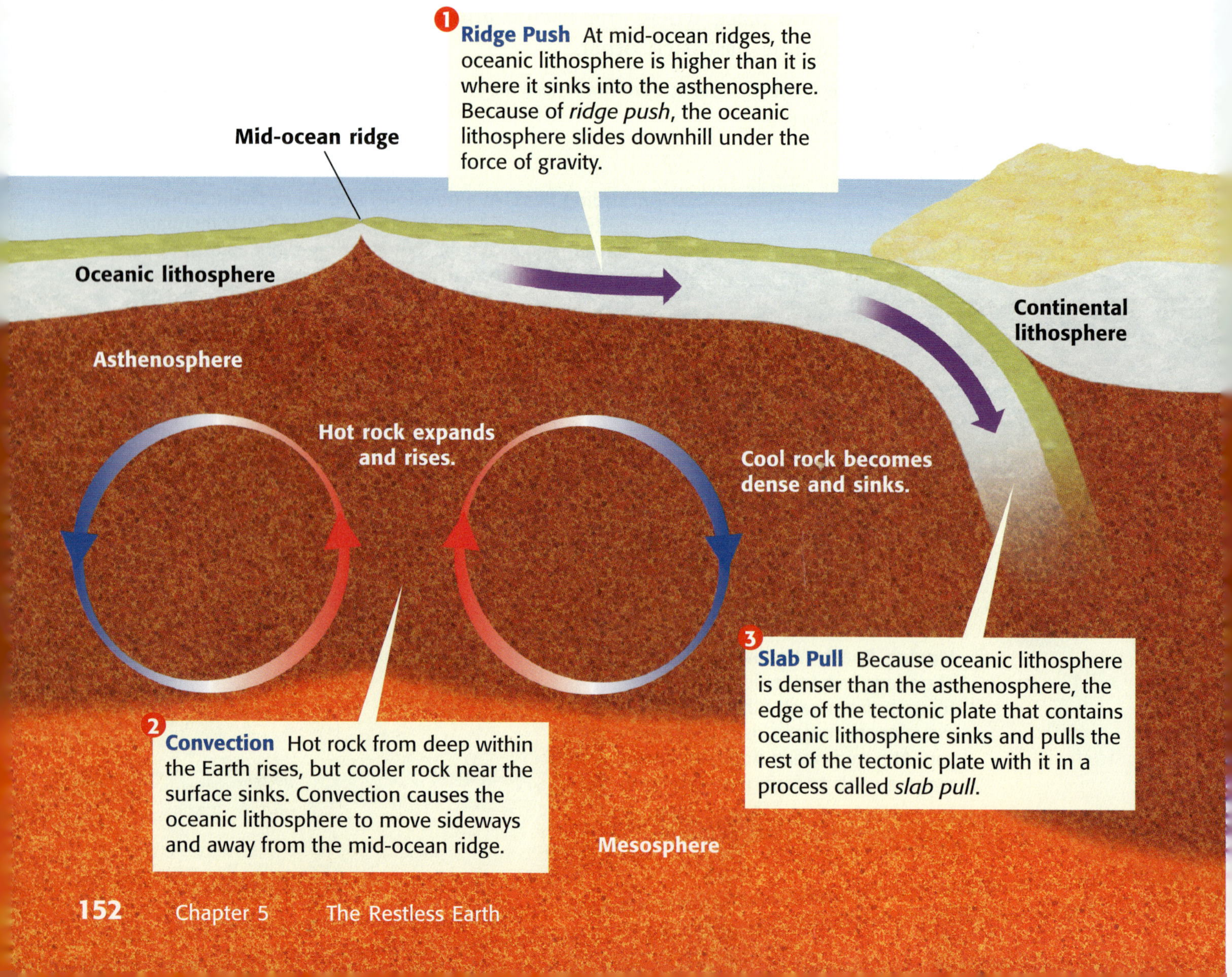

1 Ridge Push At mid-ocean ridges, the oceanic lithosphere is higher than it is where it sinks into the asthenosphere. Because of *ridge push*, the oceanic lithosphere slides downhill under the force of gravity.

2 Convection Hot rock from deep within the Earth rises, but cooler rock near the surface sinks. Convection causes the oceanic lithosphere to move sideways and away from the mid-ocean ridge.

3 Slab Pull Because oceanic lithosphere is denser than the asthenosphere, the edge of the tectonic plate that contains oceanic lithosphere sinks and pulls the rest of the tectonic plate with it in a process called *slab pull*.

Tracking Tectonic Plate Motion

How fast do tectonic plates move? The answer to this question depends on many factors, such as the type and shape of the tectonic plate and the way that the tectonic plate interacts with the tectonic plates that surround it. Tectonic plate movements are so slow and gradual that you can't see or feel them—the movement is measured in centimeters per year.

The Global Positioning System

Scientists use a system of satellites called the *global positioning system* (GPS), shown in **Figure 2,** to measure the rate of tectonic plate movement. Radio signals are continuously beamed from satellites to GPS ground stations, which record the exact distance between the satellites and the ground station. Over time, these distances change slightly. By recording the time it takes for the GPS ground stations to move a given distance, scientists can measure the speed at which each tectonic plate moves.

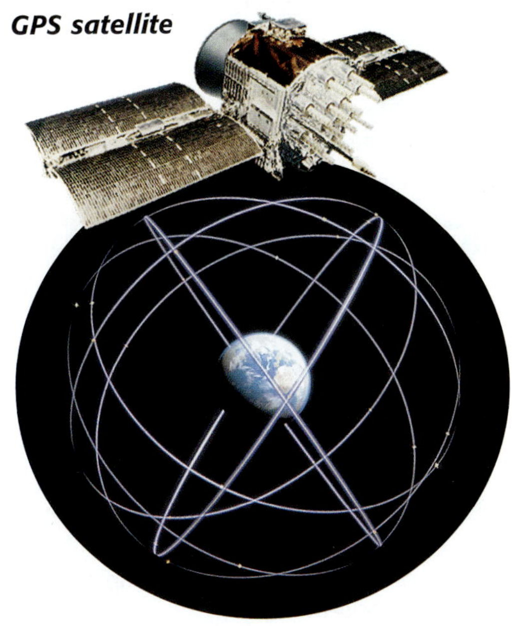

Figure 2 *The image above shows the orbits of the GPS satellites.*

SECTION Review

Summary

- Boundaries between tectonic plates are classified as convergent, divergent, or transform.
- Ridge push, convection, and slab pull are three possible driving forces of plate tectonics.
- Scientists use data from a system of satellites called the global positioning system to measure the rate of motion of tectonic plates.

Using Key Terms

1. In your own words, write a definition for the term *plate tectonics*.

Understanding Key Ideas

2. The speed a tectonic plate moves per year is best measured in
   - **a.** kilometers per year.
   - **b.** centimeters per year.
   - **c.** meters per year.
   - **d.** millimeters per year.
3. Briefly describe three possible driving forces of tectonic plate movement.
4. Explain how scientists use GPS to measure the rate of tectonic plate movement.

Math Skills

5. If an orbiting satellite has a diameter of 60 cm, what is the total surface area of the satellite? (Hint: *surface area* = $4\pi r^2$)

Critical Thinking

6. **Identifying Relationships** When convection takes place in the mantle, why does cool rock material sink and warm rock material rise?
7. **Analyzing Processes** Why does oceanic crust sink beneath continental crust at convergent boundaries?

SECTION 4

Deforming the Earth's Crust

READING WARM-UP

Objectives

- Describe two types of stress that deform rocks.
- Describe three major types of folds.
- Explain the differences between the three major types of faults.
- Identify the most common types of mountains.
- Explain the difference between uplift and subsidence.

Terms to Learn

compression
tension
folding
fault
uplift
subsidence

READING STRATEGY

Discussion Read this section silently. Write down questions that you have about this section. Discuss your questions in a small group.

Have you ever tried to bend something, only to have it break? Take long, uncooked pieces of spaghetti, and bend them very slowly but only a little. Now, bend them again, but this time, bend them much farther and faster. What happened?

How can a material bend at one time and break at another time? The answer is that the stress you put on the material was different each time. *Stress* is the amount of force per unit area on a given material. The same principle applies to the rocks in the Earth's crust. Different things happen to rock when different types of stress are applied.

Deformation

The process by which the shape of a rock changes because of stress is called *deformation*. In the example above, the spaghetti deformed in two different ways—by bending and by breaking. **Figure 1** illustrates this concept. The same thing happens in rock layers. Rock layers bend when stress is placed on them. But when enough stress is placed on rocks, they can reach their elastic limit and break.

Compression and Tension

The type of stress that occurs when an object is squeezed, such as when two tectonic plates collide, is called **compression.** When compression occurs at a convergent boundary, large mountain ranges can form.

Another form of stress is *tension*. **Tension** is stress that occurs when forces act to stretch an object. As you might guess, tension occurs at divergent plate boundaries, such as mid-ocean ridges, when two tectonic plates pull away from each other.

Reading Check **How do the forces of plate tectonics cause rock to deform?** (*See the Appendix for answers to Reading Checks.*)

Figure 1 *When a small amount of stress is placed on uncooked spaghetti, the spaghetti bends. Additional stress causes the spaghetti to break.*

Figure 2 **Folding: When Rock Layers Bend Because of Stress**

Unstressed

Horizontal stress

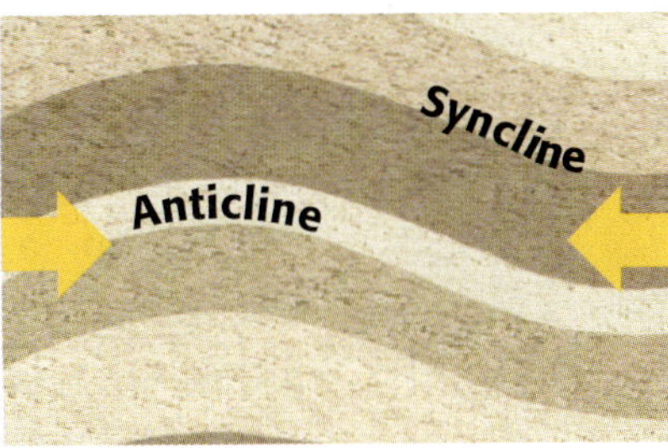

Vertical stress

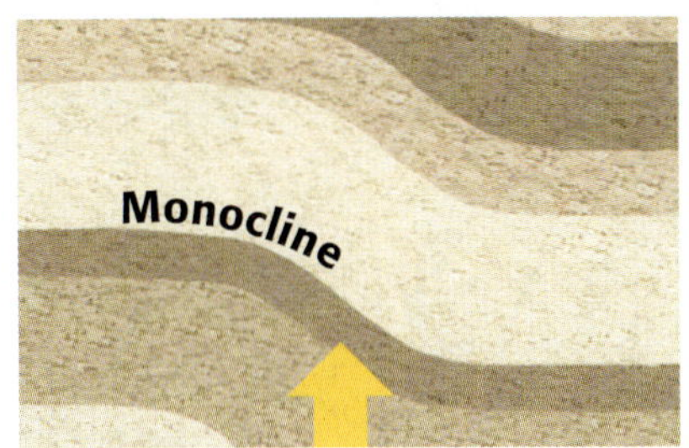

Folding

The bending of rock layers because of stress in the Earth's crust is called **folding.** Scientists assume that all rock layers started as horizontal layers. So, when scientists see a fold, they know that deformation has taken place.

compression stress that occurs when forces act to squeeze an object

tension stress that occurs when forces act to stretch an object

folding the bending of rock layers due to stress

Types of Folds

Depending on how the rock layers deform, different types of folds are made. **Figure 2** shows the two most common types of folds—*anticlines,* or upward-arching folds, and *synclines,* downward, troughlike folds. Another type of fold is a *monocline.* In a monocline, rock layers are folded so that both ends of the fold are horizontal. Imagine taking a stack of paper and laying it on a table. Think of the sheets of paper as different rock layers. Now put a book under one end of the stack. You can see that both ends of the sheets are horizontal, but all of the sheets are bent in the middle.

Folds can be large or small. The largest folds are measured in kilometers. Other folds are also obvious but are much smaller. These small folds can be measured in centimeters. **Figure 3** shows examples of large and small folds.

Figure 3 *The large photo shows mountain-sized folds in the Rocky Mountains. The small photo shows a rock that has folds smaller than a penknife.*

Faulting

Some rock layers break when stress is applied to them. The surface along which rocks break and slide past each other is called a **fault.** The blocks of crust on each side of the fault are called *fault blocks*.

When a fault is not vertical, understanding the difference between its two sides—the *hanging wall* and the *footwall*—is useful. **Figure 4** shows the difference between a hanging wall and a footwall. Two main types of faults can form. The type of fault that forms depends on how the hanging wall and footwall move in relationship to each other.

Fault
Footwall
Hanging wall

Figure 4 *The position of a fault block determines whether it is a hanging wall or a footwall.*

Normal Faults

A *normal fault* is shown in **Figure 5.** When a normal fault moves, it causes the hanging wall to move down relative to the footwall. Normal faults usually occur when tectonic forces cause tension that pulls rocks apart.

fault a break in a body of rock along which one block slides relative to another

Reverse Faults

A *reverse fault* is shown in **Figure 5.** When a reverse fault moves, it causes the hanging wall to move up relative to the footwall. This movement is the reverse of a normal fault. Reverse faults usually happen when tectonic forces cause compression that pushes rocks together.

Reading Check **How does the hanging wall in a normal fault move in relation to a reverse fault?**

Figure 5 **Normal and Reverse Faults**

Normal Fault When rocks are pulled apart because of tension, normal faults often form.

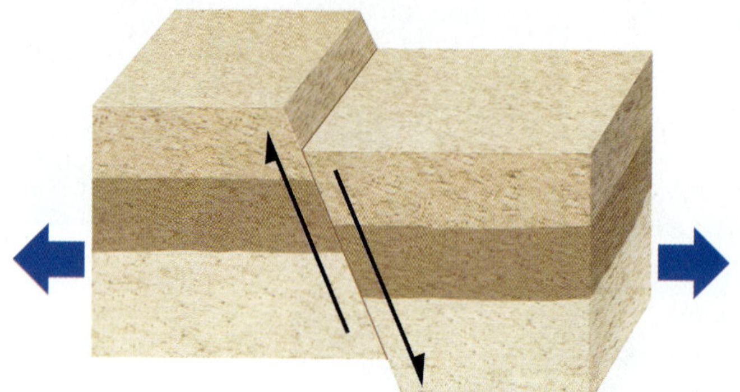

Reverse Fault When rocks are pushed together by compression, reverse faults often form.

Figure 6 *The photo at left is a normal fault. The photo at right is a reverse fault.*

Telling the Difference Between Faults

It's easy to tell the difference between a normal fault and a reverse fault in drawings with arrows. But what types of faults are shown in **Figure 6**? You can certainly see the faults, but which one is a normal fault, and which one is a reverse fault? In the top left photo in **Figure 6,** one side has obviously moved relative to the other side. You can tell this fault is a normal fault by looking at the order of sedimentary rock layers. If you compare the two dark layers near the surface, you can see that the hanging wall has moved down relative to the footwall.

Strike-Slip Faults

A third major type of fault is called a *strike-slip fault*. An illustration of a strike-slip fault is shown in **Figure 7.** *Strike-slip faults* form when opposing forces cause rock to break and move horizontally. If you were standing on one side of a strike-slip fault looking across the fault when it moved, the ground on the other side would appear to move to your left or right. The San Andreas Fault in California is a spectacular example of a strike-slip fault.

Quick Lab

Modeling Strike-Slip Faults

1. Use **modeling clay** to construct a box that is 6 in. × 6 in. × 4 in. Use different colors of clay to represent different horizontal layers.
2. Using **scissors,** cut the box down the middle. Place **two 4 in. × 6 in. index cards** inside the cut so that the two sides of the box slide freely.
3. Using gentle pressure, slide the two sides horizontally past one another.
4. How does this model illustrate the motion that occurs along a strike-slip fault?

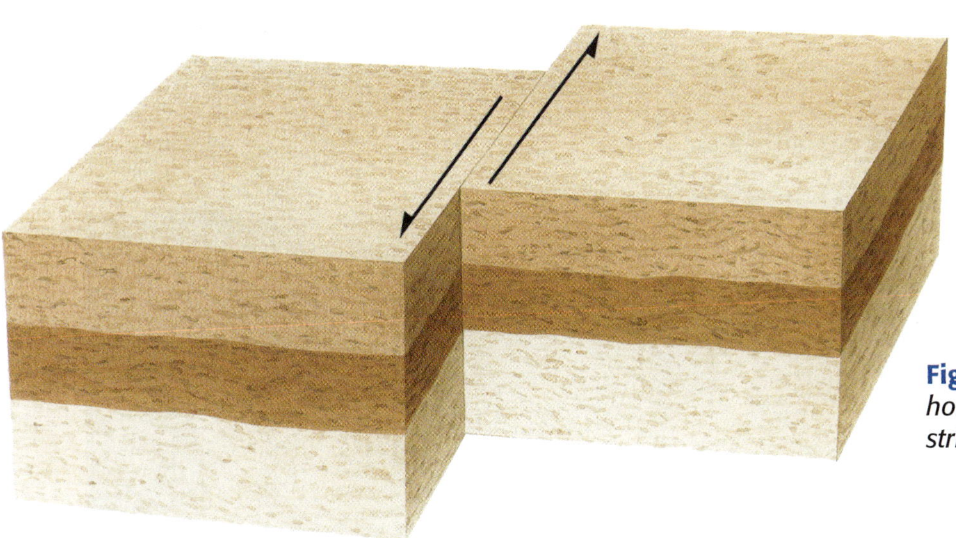

Figure 7 *When rocks are moved horizontally by opposing forces, strike-slip faults often form.*

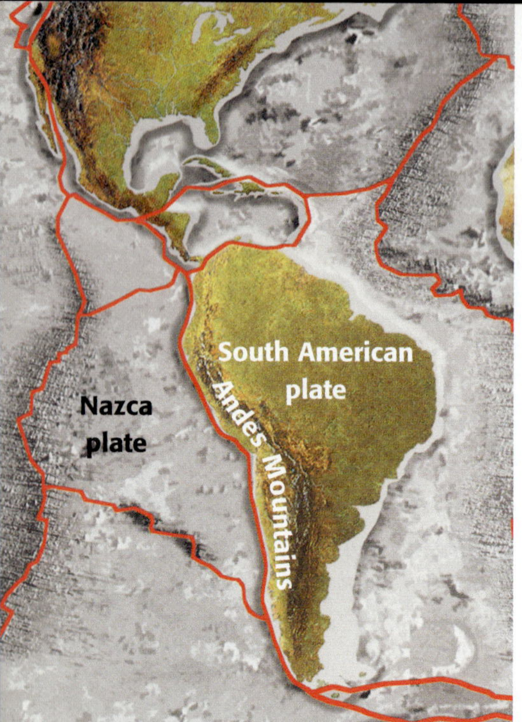

Figure 8 *The Andes Mountains formed on the edge of the South American plate where it converges with the Nazca plate.*

Plate Tectonics and Mountain Building

You have just learned about several ways the Earth's crust changes because of the forces of plate tectonics. When tectonic plates collide, land features that start as folds and faults can eventually become large mountain ranges. Mountains exist because tectonic plates are continually moving around and colliding with one another. As shown in **Figure 8,** the Andes Mountains formed above the subduction zone where two tectonic plates converge.

When tectonic plates undergo compression or tension, they can form mountains in several ways. Take a look at three of the most common types of mountains—folded mountains, fault-block mountains, and volcanic mountains.

Folded Mountains

The highest mountain ranges in the world are made up of folded mountains. These ranges form at convergent boundaries where continents have collided. *Folded mountains* form when rock layers are squeezed together and pushed upward. If you place a pile of paper on a table and push on opposite edges of the pile, you will see how folded mountains form.

An example of a folded mountain range that formed at a convergent boundary is shown in **Figure 9.** About 390 million years ago, the Appalachian Mountains formed when the landmasses that are now North America and Africa collided. Other examples of mountain ranges that consist of very large and complex folds are the Alps in central Europe, the Ural Mountains in Russia, and the Himalayas in Asia.

Reading Check **Explain how folded mountains form.**

Figure 9 *The Appalachian Mountains were once as tall as the Himalaya Mountains but have been worn down by hundreds of millions of years of weathering and erosion.*

Figure 10 *When the crust is subjected to tension, the rock can break along a series of normal faults, which creates fault-block mountains.*

Fault-Block Mountains

When tectonic forces put enough tension on the Earth's crust, a large number of normal faults can result. *Fault-block mountains* form when this tension causes large blocks of the Earth's crust to drop down relative to other blocks. **Figure 10** shows one way that fault-block mountains form.

When sedimentary rock layers are tilted up by faulting, they can produce mountains that have sharp, jagged peaks. As shown in **Figure 11,** the Tetons in western Wyoming are a spectacular example of fault-block mountains.

Volcanic Mountains

Most of the world's major volcanic mountains are located at convergent boundaries where oceanic crust sinks into the asthenosphere at subduction zones. The rock that is melted in subduction zones forms magma, which rises to the Earth's surface and erupts to form *volcanic mountains*. Volcanic mountains can also form under the sea. Sometimes these mountains can rise above the ocean surface to become islands. The majority of tectonically active volcanic mountains on the Earth have formed around the tectonically active rim of the Pacific Ocean. The rim has become known as the *Ring of Fire*.

CONNECTION TO Social Studies

WRITING SKILL **The Naming of the Appalachian Mountains** How did the Appalachian Mountains get their name? It is believed that the Appalachian Mountains were named by Spanish explorers in North America during the 16th century. It is thought that the name was taken from a Native American tribe called *Appalachee,* who lived in northern Florida. Research other geological features in the United States, including mountains and rivers, whose names are of Native American origin. Write the results of your research in a short essay.

Figure 11 *The Tetons formed as a result of tectonic forces that stretched the Earth's crust and caused it to break in a series of normal faults.*

For another activity related to this chapter, go to **go.hrw.com** and type in the keyword **HZ5TECW.**

Uplift and Subsidence

Vertical movements in the crust are divided into two types—uplift and subsidence. The rising of regions of Earth's crust to higher elevations is called **uplift.** Rocks that are uplifted may or may not be highly deformed. The sinking of regions of Earth's crust to lower elevations is known as **subsidence** (suhb SIED'ns). Unlike some uplifted rocks, rocks that subside do not undergo much deformation.

uplift the rising of regions of the Earth's crust to higher elevations

subsidence the sinking of regions of the Earth's crust to lower elevations

Uplifting of Depressed Rocks

The formation of mountains is one type of uplift. Uplift can also occur when large areas of land rise without deforming. One way areas rise without deforming is a process known as *rebound.* When the crust rebounds, it slowly springs back to its previous elevation. Uplift often happens when a weight is removed from the crust.

Subsidence of Cooler Rocks

Rocks that are hot take up more space than cooler rocks. For example, the lithosphere is relatively hot at mid-ocean ridges. The farther the lithosphere is from the ridge, the cooler and denser the lithosphere becomes. Because the oceanic lithosphere now takes up less volume, the ocean floor subsides.

Tectonic Letdown

Subsidence can also occur when the lithosphere becomes stretched in rift zones. A *rift zone* is a set of deep cracks that forms between two tectonic plates that are pulling away from each other. As tectonic plates pull apart, stress between the plates causes a series of faults to form along the rift zone. As shown in **Figure 12,** the blocks of crust in the center of the rift zone subside.

Figure 12 *The East African Rift, from Ethiopia to Kenya, is part of a divergent boundary, but you can see how the crust has subsided relative to the blocks at the edge of the rift zone.*

SECTION Review

Summary

- Compression and tension are two forces of plate tectonics that can cause rock to deform.
- Folding occurs when rock layers bend because of stress.
- Faulting occurs when rock layers break because of stress and then move on either side of the break.
- Mountains are classified as either folded, fault-block, or volcanic depending on how they form.
- Mountain building is caused by the movement of tectonic plates. Folded mountains and volcanic mountains form at convergent boundaries. Fault-block mountains form at divergent boundaries.
- Uplift and subsidence are the two types of vertical movement in the Earth's crust. Uplift occurs when regions of the crust rise to higher elevations. Subsidence occurs when regions of the crust sink to lower elevations.

Using Key Terms

For each pair of key terms, explain how the meanings of the terms differ.

1. *compression* and *tension*
2. *uplift* and *subsidence*

Understanding Key Ideas

3. The type of fault in which the hanging wall moves up relative to the footwall is called a
   a. strike-slip fault.
   b. fault-block fault.
   c. normal fault.
   d. reverse fault.
4. Describe three types of folds.
5. Describe three types of faults.
6. Identify the most common types of mountains.
7. What is rebound?
8. What are rift zones, and how do they form?

Critical Thinking

9. **Predicting Consequences** If a fault occurs in an area where rock layers have been folded, which type of fault is it likely to be? Why?
10. **Identifying Relationships** Would you expect to see a folded mountain range at a mid-ocean ridge? Explain your answer.

Interpreting Graphics

Use the diagram below to answer the questions that follow.

11. What type of fault is shown in the diagram?
12. At what kind of tectonic boundary would you most likely find this fault?

For a variety of links related to this chapter, go to www.scilinks.org

Topic: Faults; Mountain Building

SciLinks code: HSM0566; HSM0999

SECTION 5

What Are Earthquakes?

READING WARM-UP

Objectives

- Explain where earthquakes take place.
- Explain what causes earthquakes.
- Identify three different types of faults that occur at plate boundaries.
- Describe how energy from earthquakes travels through the Earth.

Terms to Learn

seismology
deformation
elastic rebound
seismic waves
P wave
S wave

READING STRATEGY

Paired Summarizing Read this section silently. In pairs, take turns summarizing the material. Stop to discuss ideas that seem confusing.

Have you ever felt the earth move under your feet? Many people have. Every day, somewhere within this planet, an earthquake is happening.

The word *earthquake* defines itself fairly well. But there is more to earthquakes than just the shaking of the ground. An entire branch of Earth science, called **seismology** (siez MAHL uh jee), is devoted to studying earthquakes. Earthquakes are complex, and they present many questions for *seismologists,* the scientists who study earthquakes.

Where Do Earthquakes Occur?

Most earthquakes take place near the edges of tectonic plates. *Tectonic plates* are giant pieces of Earth's thin, outermost layer. Tectonic plates move around on top of a layer of plastic rock. **Figure 1** shows the Earth's tectonic plates and the locations of recent major earthquakes.

Tectonic plates move in different directions and at different speeds. Two plates can push toward or pull away from each other. They can also slip slowly past each other. As a result of these movements, numerous features called faults exist in the Earth's crust. A *fault* is a break in the Earth's crust along which blocks of the crust slide relative to one another. Earthquakes occur along faults because of this sliding.

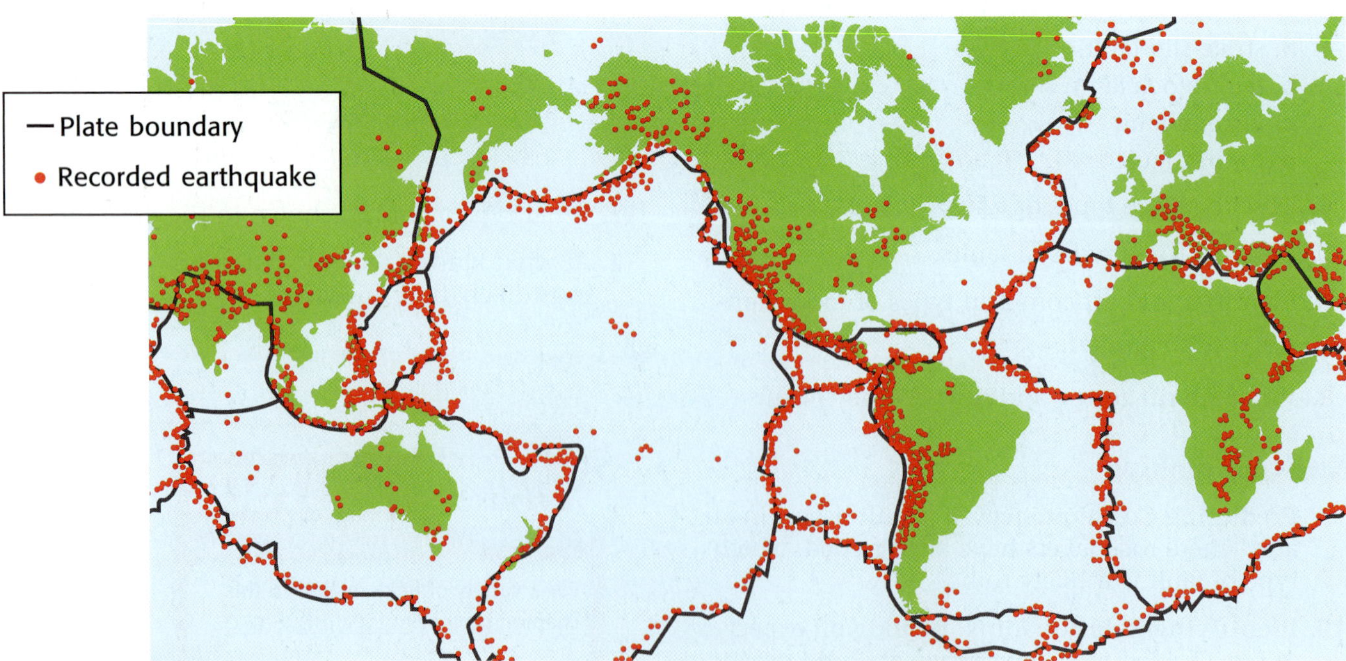

Figure 1 *The largest and most active earthquake zone lies along the plate boundaries surrounding the Pacific Ocean.*

What Causes Earthquakes?

As tectonic plates push, pull, or slip past each other, stress increases along faults near the plates' edges. In response to this stress, rock in the plates deforms. **Deformation** is the change in the shape of rock in response to stress. Rock along a fault deforms in mainly two ways. It deforms in a plastic manner, like a piece of molded clay, or in an elastic manner, like a rubber band. *Plastic deformation,* which is shown in **Figure 2,** does not lead to earthquakes.

Elastic deformation, however, does lead to earthquakes. Rock can stretch farther without breaking than steel can, but rock will break at some point. Think of elastically deformed rock as a stretched rubber band. You can stretch a rubber band only so far before it breaks. When the rubber band breaks, it releases energy. Then, the broken pieces return to their unstretched shape.

Figure 2 *This road cut is adjacent to the San Andreas Fault in southern California. The rocks in the cut have undergone deformation because of the continuous motion of the fault.*

Elastic Rebound

The sudden return of elastically deformed rock to its original shape is called **elastic rebound.** Elastic rebound is like the return of the broken rubber-band pieces to their unstretched shape. Elastic rebound occurs when more stress is applied to rock than the rock can withstand. During elastic rebound, energy is released. Some of this energy travels as seismic waves. These seismic waves cause an earthquake, as shown in **Figure 3.**

Reading Check How does elastic rebound relate to earthquakes? (*See the Appendix for answers to Reading Checks.*)

seismology the study of earthquakes

deformation the bending, tilting, and breaking of the Earth's crust; the change in the shape of rock in response to stress

elastic rebound the sudden return of elastically deformed rock to its undeformed shape

Figure 3 **Elastic Rebound and Earthquakes**

Before earthquake

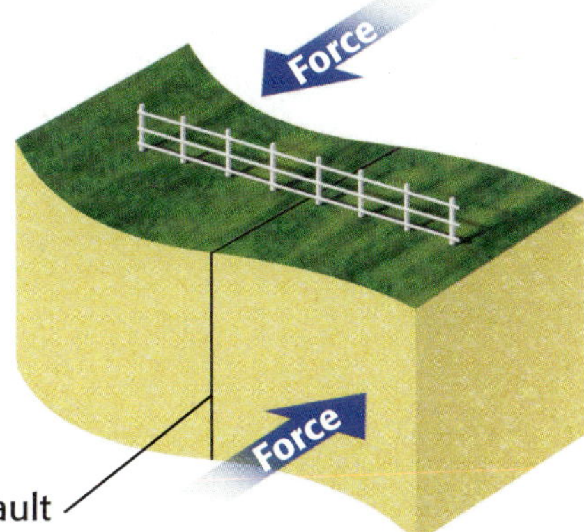

1. Tectonic forces push rock on either side of the fault in opposite directions, but the rock is locked together and does not move. The rock deforms in an elastic manner.

After earthquake

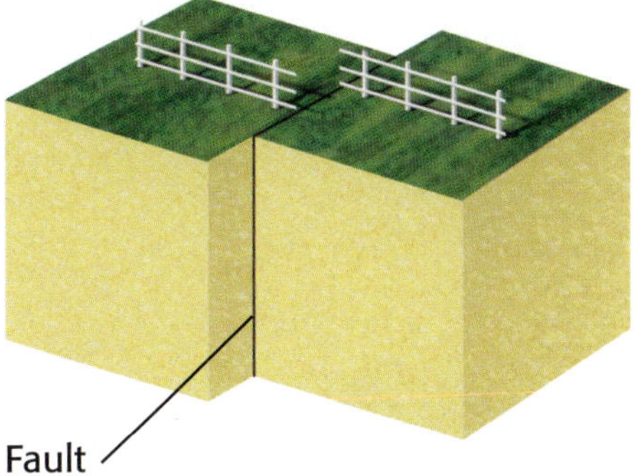

2. When enough stress is applied, the rock slips along the fault and releases energy.

Faults at Tectonic Plate Boundaries

A specific type of plate motion takes place at different tectonic plate boundaries. Each type of motion creates a particular kind of fault that can produce earthquakes. Examine **Table 1** and the diagram below to learn more about plate motion.

Table 1 Plate Motion and Fault Types	
Plate motion	**Major fault type**
Transform	strike-slip fault
Convergent	reverse fault
Divergent	normal fault

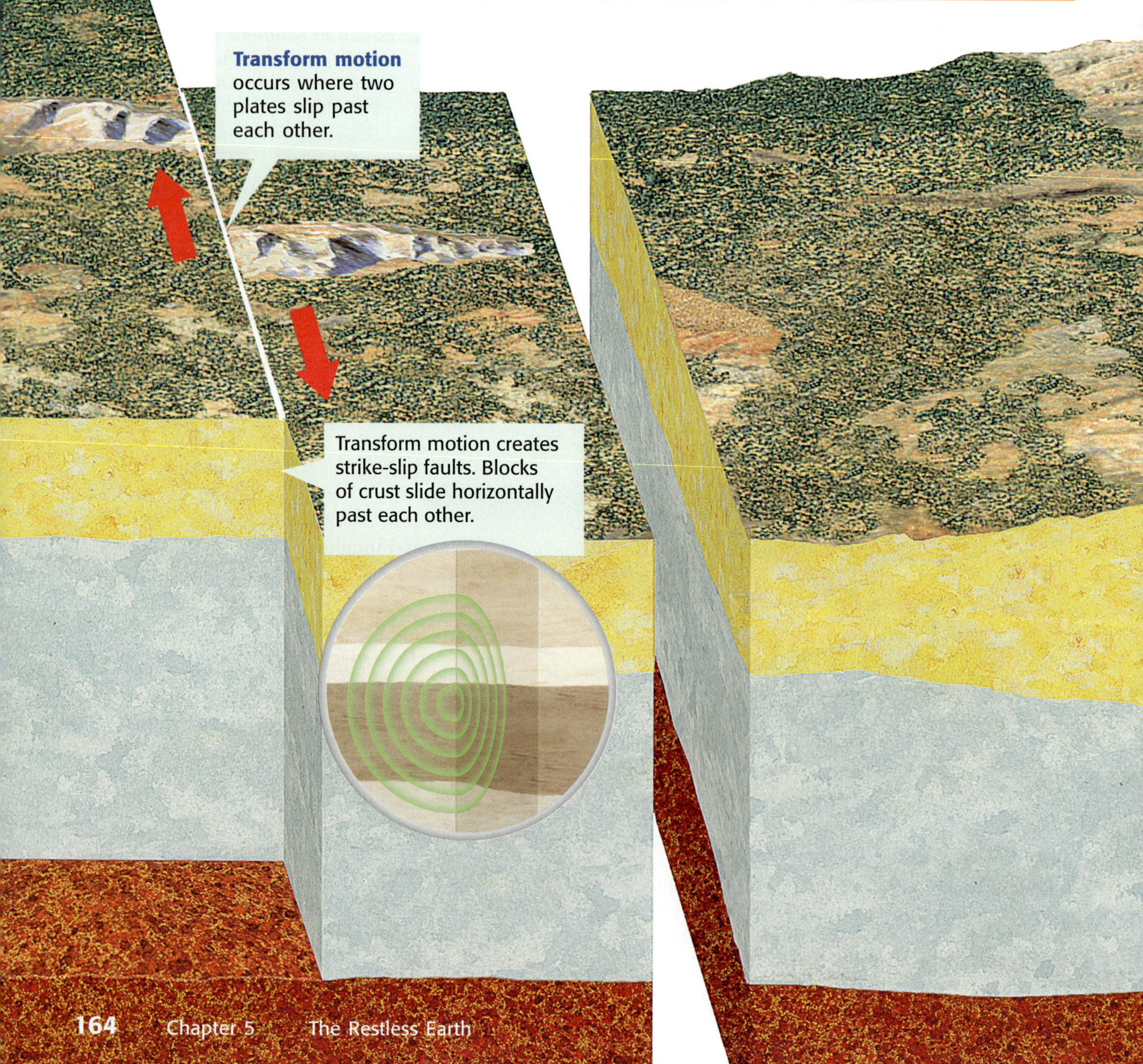

Earthquake Zones

Earthquakes can happen both near Earth's surface or far below it. Most earthquakes happen in the earthquake zones along tectonic plate boundaries. Earthquake zones are places where a large number of faults are located. The San Andreas Fault Zone in California is an example of an earthquake zone. But not all faults are located at tectonic plate boundaries. Sometimes, earthquakes happen along faults in the middle of tectonic plates.

Reading Check Where are earthquake zones located?

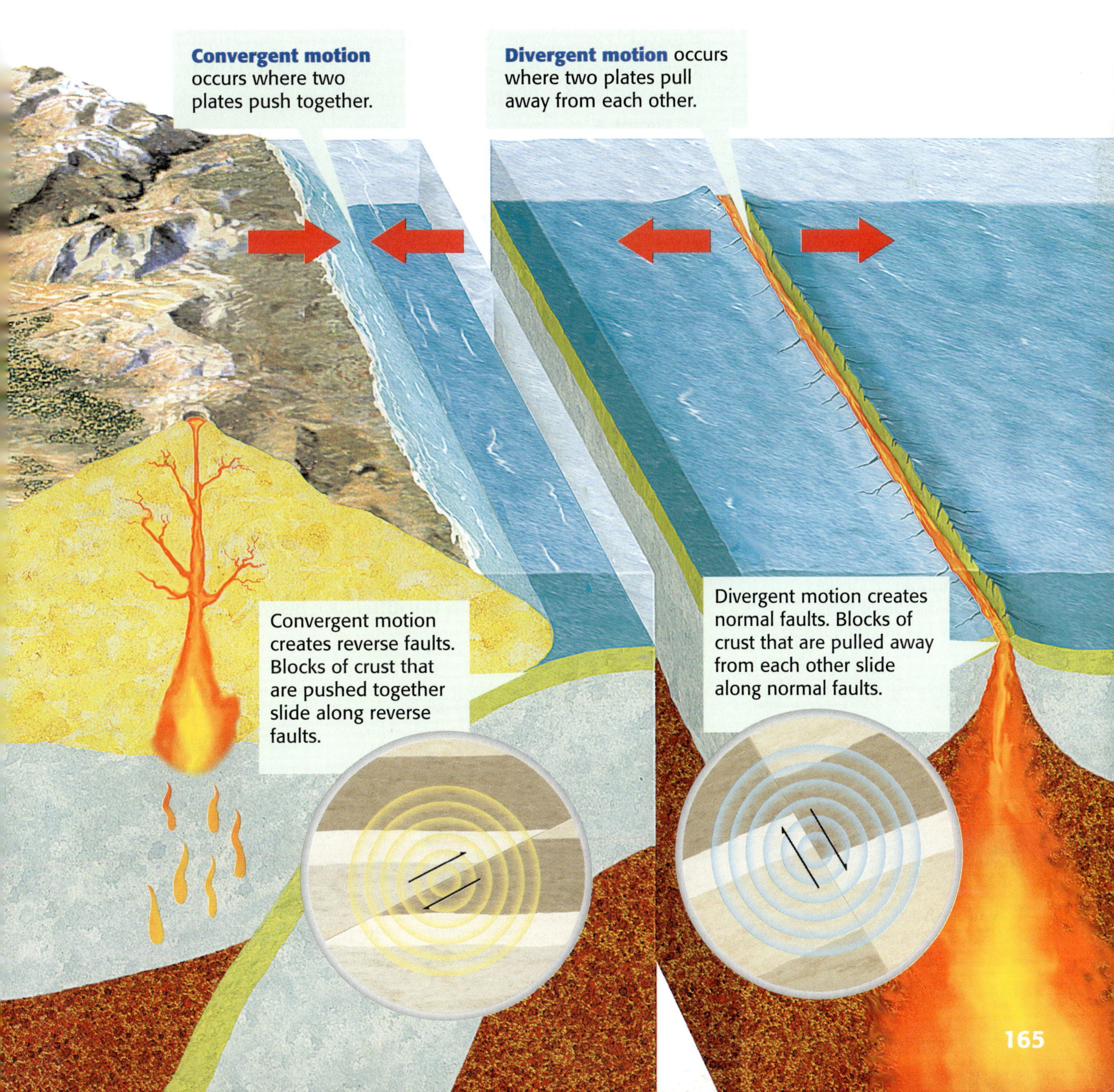

How Do Earthquake Waves Travel?

Waves of energy that travel through the Earth are called **seismic waves.** Seismic waves that travel through the Earth's interior are called *body waves.* There are two types of body waves: P waves and S waves. Seismic waves that travel along the Earth's surface are called *surface waves.* Each type of seismic wave travels through Earth's layers in a different way and at a different speed. Also, the speed of a seismic wave depends on the kind of material the wave travels through.

P Waves and S Waves

Waves that travel through solids, liquids, and gases are called **P waves** (pressure waves). P waves are the fastest seismic waves, so they are the first earthquake waves to be detected. P waves move back and forth, squeezing and stretching rock as shown in **Figure 4.**

Rock can also be deformed from side to side. After being deformed from side to side, the rock springs back to its original position and S waves are created. **S waves,** or shear waves, are the second-fastest seismic waves. S waves shear rock side to side, as shown in **Figure 4,** which means they stretch the rock sideways. Unlike P waves, S waves cannot travel through parts of the Earth that are completely liquid.

Surface Waves

Surface waves move along the Earth's surface and produce motion mostly in the upper few kilometers of Earth's crust. There are two types of surface waves. One type produces motion up, down, and around, as shown in **Figure 4.** The other type produces back-and-forth motion like the motion produced by S waves. Surface waves travel more slowly than body waves and are more destructive.

Reading Check Explain the difference between surface waves and body waves.

Quick Lab

Modeling Seismic Waves

1. Stretch a **spring toy** lengthwise on a **table.**
2. Hold one end of the spring while a partner holds the other end. Push your end toward your partner's end. Observe what happens.
3. Repeat step 2, but this time shake the spring from side to side.
4. Which type of seismic wave is represented in step 2? in step 3?

seismic wave a wave of energy that travels through the Earth, away from an earthquake in all directions

P wave a seismic wave that causes particles of rock to move in a back-and-forth direction

S wave a seismic wave that causes particles of rock to move in a side-to-side direction

Figure 4 *The largest and most active earthquake zone lies along the plate boundaries surrounding the Pacific Ocean.*

Direction of wave travel

P waves move rock back and forth, which squeezes and stretches the rock, as they travel through the rock.

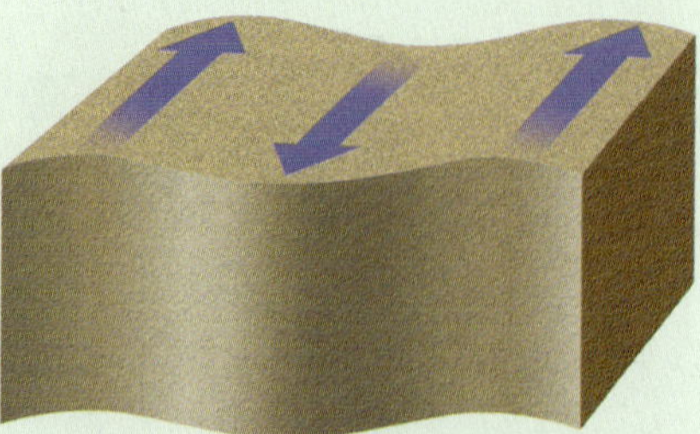

S waves shear rock side to side as they travel through the rock.

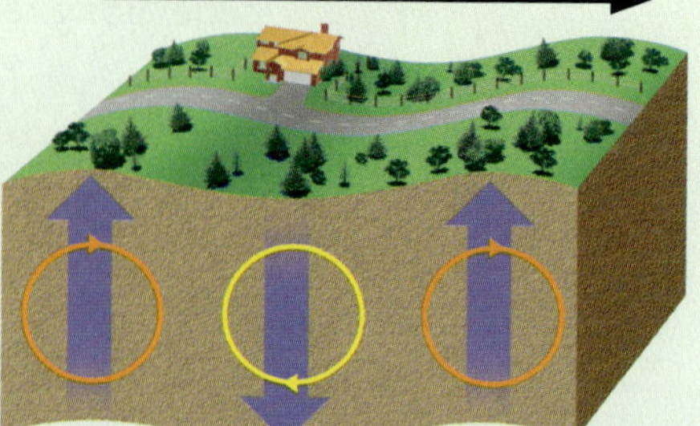

Surface waves move the ground much like ocean waves move water particles.

Earthquakes and Buildings

In many places, buildings, such as those shown in **Figure 5,** are not designed or constructed to withstand an earthquake's forces. Scientists and engineers study buildings that fail during earthquakes. As a result, they have learned a lot about making buildings more earthquake resistant.

Architects and engineers combine what they have learned with the newest technology to design and construct buildings and bridges to better withstand earthquakes. Today, older structures in California and other earthquake-prone areas are being made more earthquake resistant. The process of making older structures more earthquake resistant is called *retrofitting.*

Figure 5 *During the January 17, 1995, earthquake, the fronts of entire buildings collapsed into the streets of Kobe, Japan.*

SECTION Review

Summary

- Earthquakes occur mainly near the edges of tectonic plates.
- Elastic rebound is the direct cause of earthquakes.
- Three major types of faults occur at tectonic plate boundaries: normal faults, reverse faults, and strike-slip faults.
- Earthquake energy travels as body waves through the Earth's interior or as surface waves along the surface of the Earth.

Using Key Terms

Complete each of the following sentences by choosing the correct term from the word bank.

Deformation	P waves
Elastic rebound	S waves

1. _____ is the change in shape of rock due to stress.
2. _____ always travel ahead of other waves.

Understanding Key Ideas

3. Seismic waves that shear rock side to side are called
   a. surface waves.
   b. S waves.
   c. P waves.
   d. Both (b) and (c)
4. Where do earthquakes occur?
5. What is the direct cause of earthquakes?
6. Describe the three types of plate motion and the faults that are characteristic of each type of motion.
7. What is retrofitting?

Math Skills

8. A seismic wave is traveling through the Earth at an average rate of speed of 8 km/s. How long will it take the wave to travel 480 km?

Critical Thinking

9. **Applying Concepts** Given what you know about elastic rebound, why do you think some earthquakes are stronger than others?
10. **Identifying Relationships** Why are surface waves more destructive to buildings than P waves or S waves are?
11. **Identifying Relationships** Why do you think the majority of earthquake zones are located at tectonic plate boundaries?

For a variety of links related to this chapter, go to www.scilinks.org

Topic: What Is an Earthquake?

SciLinks code: HSM1658

Using Scientific Methods

Model-Making Lab

OBJECTIVES

Model convection currents to simulate plate tectonic movement.

Draw conclusions about the role of convection in plate tectonics.

MATERIALS

- craft sticks (2)
- food coloring
- gloves, heat-resistant
- hot plates, small (2)
- pan, aluminum, rectangular
- pencil
- ruler, metric
- thermometers (3)
- water, cold
- wooden blocks

SAFETY

Convection Connection

Some scientists think that convection currents within the Earth's mantle cause tectonic plates to move. Because these convection currents cannot be observed directly, scientists use models to simulate the process. In this activity, you will make your own model to simulate tectonic plate movement.

Ask a Question

1. How can I make a model of convection currents in the Earth's mantle?

Form a Hypothesis

2. Turn the question above into a statement in which you give your best guess about what factors will have the greatest effect on your convection model.

Test the Hypothesis

3. Place two hot plates side by side in the center of your lab table. Be sure that they are away from the edge of the table.
4. Place the pan on top of the hot plates. Slide the wooden blocks under the pan to support the ends. Make sure that the pan is level and secure.
5. Fill the pan with cold water. The water should be at least 4 cm deep. Turn on the hot plates, and put on your gloves.
6. After a minute or two, tiny bubbles will begin to rise in the water above the hot plates. Gently place two craft sticks on the water's surface.
7. Use the pencil to align the sticks parallel to the short ends of the pan. The sticks should be about 3 cm apart and near the center of the pan.
8. As soon as the sticks begin to move, place a drop of food coloring in the center of the pan. Observe what happens to the food coloring.

9. With the help of a partner, hold one thermometer bulb just under the water at the center of the pan. Hold the other two thermometers just under the water near the ends of the pan. Record the temperatures.

10. When you are finished, turn off the hot plates. After the water has cooled, carefully empty the water into a sink.

Analyze the Results

1. **Explaining Events** Based on your observations of the motion of the food coloring, how does the temperature of the water affect the direction in which the craft sticks move?

Draw Conclusions

2. **Drawing Conclusions** How does the motion of the craft sticks relate to the motion of the water?

3. **Applying Conclusions** How does this model relate to plate tectonics and the movement of the continents?

4. **Applying Conclusions** Based on your observations, what can you conclude about the role of convection in plate tectonics?

Applying Your Data

Suggest a substance other than water that might be used to model convection in the mantle. Consider using a substance that flows more slowly than water.

Chapter Review

USING KEY TERMS

1. Use the following terms in the same sentence: *crust, mantle,* and *core.*

Complete each of the following sentences by choosing the correct term from the word bank.

asthenosphere	uplift
deformation	continental drift

2. The hypothesis that continents can drift apart and have done so in the past is known as ___.

3. The ___ is the soft layer of the mantle on which the tectonic plates move.

4. ___ is the change in the shape of rock in response to stress.

5. The rising of regions of the Earth's crust to higher elevations is called ___.

UNDERSTANDING KEY IDEAS

Multiple Choice

6. The strong, lower part of the mantle is a physical layer called the
   - a. lithosphere.
   - b. mesosphere.
   - c. asthenosphere.
   - d. outer core.

7. The type of tectonic plate boundary that forms from a collision between two tectonic plates is a
   - a. divergent plate boundary.
   - b. transform plate boundary.
   - c. convergent plate boundary.
   - d. normal plate boundary.

8. The bending of rock layers due to stress in the Earth's crust is known as
   - a. uplift.
   - b. folding.
   - c. faulting.
   - d. subsidence.

9. The type of fault in which the hanging wall moves up relative to the footwall is called a
   - a. strike-slip fault.
   - b. fault-block fault.
   - c. normal fault.
   - d. reverse fault.

10. The type of mountain that forms when rock layers are squeezed together and pushed upward is the
    - a. folded mountain.
    - b. fault-block mountain.
    - c. volcanic mountain.
    - d. strike-slip mountain.

11. Scientists' knowledge of the Earth's interior has come primarily from
    - a. studying magnetic reversals in oceanic crust.
    - b. using a system of satellites called the *global positioning system.*
    - c. studying seismic waves generated by earthquakes.
    - d. studying the pattern of fossils on different continents.

Short Answer

12. Explain how scientists use seismic waves to map the Earth's interior.

13. How do magnetic reversals provide evidence of sea-floor spreading?

14 Explain how sea-floor spreading provides a way for continents to move.

15 Explain how different seismic waves affect rock as they travel through it.

16 What is the global positioning system (GPS), and how does GPS allow scientists to measure the rate of motion of tectonic plates?

CRITICAL THINKING

17 **Concept Mapping** Use the following terms to create a concept map: *sea-floor spreading, convergent boundary, divergent boundary, subduction zone, transform boundary,* and *tectonic plates.*

18 **Applying Concepts** Why does oceanic lithosphere sink at subduction zones but not at mid-ocean ridges?

19 **Identifying Relationships** New tectonic material continually forms at divergent boundaries. Tectonic plate material is also continually destroyed in subduction zones at convergent boundaries. Do you think that the total amount of lithosphere formed on the Earth is about equal to the amount destroyed? Why?

20 **Applying Concepts** Japan is located near a point where three tectonic plates converge. What would you imagine the earthquake-hazard level in Japan to be? Explain why.

INTERPRETING GRAPHICS

Imagine that you could travel to the center of the Earth. Use the diagram below to answer the questions that follow.

Composition	Structure
Crust (50 km)	Lithosphere (150 km)
Mantle (2,900 km)	Asthenosphere (250 km)
	Mesosphere (2,550 km)
Core (3,430 km)	Outer core (2,200 km)
	Inner core (1,228 km)

21 How far beneath the Earth's surface would you have to go before you were no longer passing through rock that had the composition of granite?

22 How far beneath the Earth's surface would you have to go to find liquid material in the Earth's core?

23 At what depth would you find mantle material but still be within the lithosphere?

24 How far beneath the Earth's surface would you have to go to find solid iron and nickel in the Earth's core?

Standardized Test Preparation

READING

Read each of the passages below. Then, answer the questions that follow each passage.

Passage 1 The Deep Sea Drilling Project was a program to retrieve and research rocks below the ocean to test the hypothesis of sea-floor spreading. For 15 years, scientists studying sea-floor spreading conducted research aboard the ship *Glomar Challenger*. Holes were drilled in the sea floor from the ship. Long, cylindrical lengths of rock, called *cores,* were obtained from the drill holes. By examining fossils in the cores, scientists discovered that rock closest to mid-ocean ridges was the youngest. The farther from the ridge the holes were drilled, the older the rock in the cores was. This evidence supported the idea that sea-floor spreading creates new lithosphere at mid-ocean ridges.

1. In the passage, what does *conducted* mean?
   - **A** directed
   - **B** led
   - **C** carried on
   - **D** guided

2. Why were cores drilled in the sea floor from the *Glomar Challenger*?
   - **F** to determine the depth of the crust
   - **G** to find minerals in the sea-floor rock
   - **H** to examine fossils in the sea-floor rock
   - **I** to find oil and gas in the sea-floor rock

3. Which of the following statements is a fact according to the passage?
   - **A** Rock closest to mid-ocean ridges is older than rock at a distance from mid-ocean ridges.
   - **B** One purpose of scientific research on the *Glomar Challenger* was to gather evidence for sea-floor spreading.
   - **C** Fossils examined by scientists came directly from the sea floor.
   - **D** Evidence gathered by scientists did not support sea-floor spreading.

Passage 2 The Himalayas are a range of mountains that is 2,400 km long and that arcs across Pakistan, India, Tibet, Nepal, Sikkim, and Bhutan. The Himalayas are the highest mountains on Earth. Nine mountains, including Mount Everest, the highest mountain on Earth, are more than 8,000 m tall. The formation of the Himalaya Mountains began about 80 million years ago. A tectonic plate carrying the Indian subcontinent collided with the Eurasian plate. The Indian plate was driven beneath the Eurasian plate. This collision caused the uplift of the Eurasian plate and the formation of the Himalayas. This process is continuing today.

1. In the passage, what does the word *arcs* mean?
   - **A** forms a circle
   - **B** forms a plane
   - **C** forms a curve
   - **D** forms a straight line

2. According to the passage, which geologic process formed the Himalaya Mountains?
   - **F** divergence
   - **G** subsidence
   - **H** strike-slip faulting
   - **I** convergence

3. Which of the following statements is a fact according to the passage?
   - **A** The nine tallest mountains on Earth are located in the Himalaya Mountains.
   - **B** The Himalaya Mountains are located within six countries.
   - **C** The Himalaya Mountains are the longest mountain range on Earth.
   - **D** The Himalaya Mountains formed more than 80 million years ago.

INTERPRETING GRAPHICS

The illustration below shows the relative velocities (in centimeters per year) and directions in which tectonic plates are separating and colliding. Arrows that point away from one another indicate plate separation. Arrows that point toward one another indicate plate collision. Use the illustration below to answer the questions that follow.

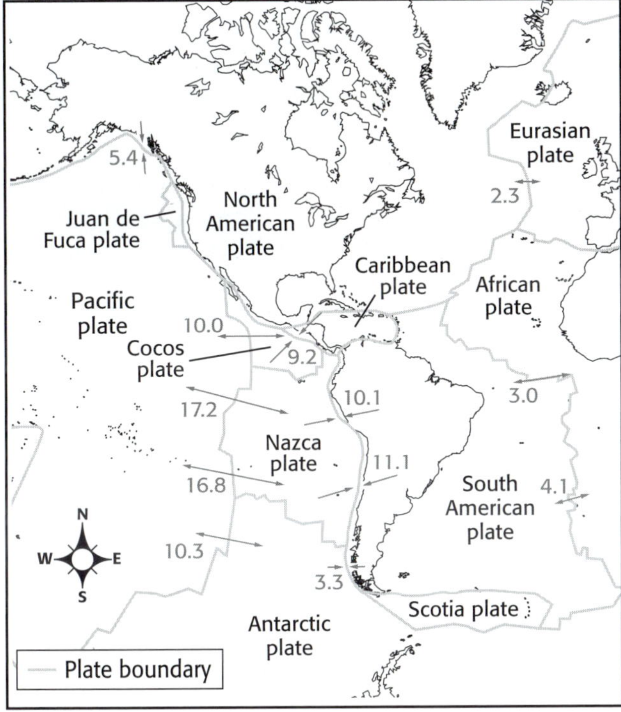

1. Between which two tectonic plates does spreading appear to be the fastest?
   - **A** the Australian plate and the Pacific plate
   - **B** the Antarctic plate and the Pacific plate
   - **C** the Nazca plate and the Pacific plate
   - **D** the Cocos plate and the Pacific plate

2. Where do you think mountain building is taking place?
   - **F** between the African plate and the South American plate
   - **G** between the Nazca plate and the South American plate
   - **H** between the North American plate and the Eurasian plate
   - **I** between the African plate and the North American plate

MATH

Read each question below, and choose the best answer.

1. The mesosphere is 2,550 km thick, and the asthenosphere is 250 km thick. If you assume that the lithosphere is 150 km thick and that the crust is 50 km thick, how thick is the mantle?
   - **A** 2,950 km
   - **B** 2,900 km
   - **C** 2,800 km
   - **D** 2,550 km

2. If a seismic wave travels through the mantle at an average velocity of 8 km/s, how many seconds will the wave take to travel through the mantle?
   - **F** 318.75 s
   - **G** 350.0 s
   - **H** 362.5 s
   - **I** 368.75 s

3. If the crust in a certain area is subsiding at the rate of 2 cm per year and has an elevation of 1,000 m, what elevation will the crust have in 10,000 years?
   - **A** 500 m
   - **B** 800 m
   - **C** 1,200 m
   - **D** 2,000 m

4. Assume that a very small oceanic plate is located between a mid-ocean ridge and a subduction zone. At the ridge, the plate is growing at a rate of 5 km every 1 million years. At the subduction zone, the plate is being destroyed at a rate of 10 km every 1 million years. If the oceanic plate is 100 km across, how long will it take the plate to disappear?
   - **F** 100 million years
   - **G** 50 million years
   - **H** 20 million years
   - **I** 5 million years

Science in Action

Science, Technology, and Society

Using Satellites to Track Plate Motion

When you think of laser beams firing, you may think of science fiction movies. However, scientists use laser beams to determine the rate and direction of motion of tectonic plates. From ground stations on Earth, laser beams are fired at several small satellites orbiting 5,900 km above Earth. From the satellites, the laser beams are reflected back to ground stations. Differences in the time it takes signals to be reflected from targets are measured over a period of time. From these differences, scientists can determine the rate and direction of plate motion.

Social Studies ACTIVITY

WRITING SKILL Research a society that lives at an active plate boundary. Find out how the people live with dangers such as volcanoes and earthquakes. Include your findings in a short report.

This scientist is using a laser to test one of the satellites that will be used to track plate motion.

Scientific Discoveries

Megaplumes

Eruptions of boiling water from the sea floor form giant, spiral disks that twist through the oceans. Do you think it's impossible? Oceanographers have discovered these disks at eight locations at mid-ocean ridges over the past 20 years. These disks, which may be tens of kilometers across, are called *megaplumes*. Megaplumes are like blenders. They mix hot water with cold water in the oceans. Megaplumes can rise hundreds of meters from the ocean floor to the upper layers of the ocean. They carry gases and minerals and provide extra energy and food to animals in the upper layers of the ocean.

Language Arts ACTIVITY

WRITING SKILL Did you ever wonder about the origin of the name *Himalaya*? Research the origin of the name *Himalaya,* and write a short report about what you find.

People in Science

Alfred Wegener

Continental Drift Alfred Wegener's greatest contribution to science was the hypothesis of continental drift. This hypothesis states that continents drift apart from one another and have done so in the past. To support his hypothesis, Wegener used geologic, fossil, and glacial evidence gathered on both sides of the Atlantic Ocean. For example, Wegener recognized similarities between rock layers in North America and Europe and between rock layers in South America and Africa. He believed that these similarities could be explained only if these geologic features were once part of the same continent.

Although continental drift explained many of his observations, Wegener could not find scientific evidence to develop a complete explanation of how continents move. Most scientists were skeptical of Wegener's hypothesis and dismissed it as foolishness. It was not until the 1950s and 1960s that the discoveries of magnetic reversals and sea-floor spreading provided evidence of continental drift.

Math Activity

The distance between South America and Africa is 7,200 km. As new crust is created at the mid-ocean ridge, South America and Africa are moving away from each other at a rate of about 3.5 cm per year. How many millions of years ago were South America and Africa joined?

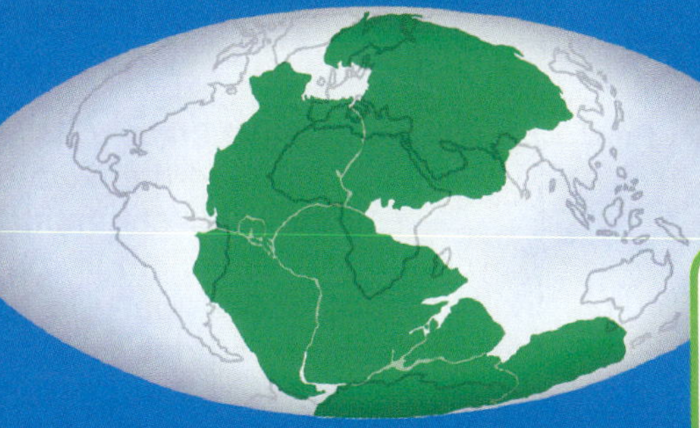

To learn more about these Science in Action topics, visit go.hrw.com and type in the keyword HZ5TECF.

Current Science

Check out Current Science® articles related to this chapter by visiting go.hrw.com. Just type in the keyword HZ5CS07.

Volcanoes

About the PHOTO

When you think of a volcanic eruption, you probably think of a cone-shaped mountain exploding and sending huge clouds of ash into the air. Some volcanic eruptions do just that! Most volcanic eruptions, such as the one shown here, which is flowing over a road in Hawaii, are slow and quiet. Volcanic eruptions happen throughout the world, and they play a major role in shaping the Earth's surface.

PRE-READING ACTIVITY

FOLDNOTES

Layered Book Before you read the chapter, create the FoldNote entitled "Layered Book" described in the **Study Skills** section of the Appendix. Label the tabs of the layered book with "Volcanic eruptions," "Effects of eruptions," and "Causes of eruptions." As you read the chapter, write information you learn about each category under the appropriate tab.

START-UP ACTIVITY

Anticipation

In this activity, you will build a simple model of a volcano and you will try to predict an eruption.

Procedure

1. Place **10 mL of baking soda** on a **sheet of tissue.** Fold the corners of the tissue over the baking soda, and place the tissue packet in a **large pan.**
2. Put **modeling clay** around the top edge of a **funnel.** Press that end of the funnel over the tissue packet to make a tight seal.
3. After you put on **safety goggles,** add **50 mL of vinegar** and **several drops of liquid dish soap** to a **200 mL beaker** and stir.
4. Predict how long it will take the volcano to erupt after the liquid is poured into the funnel. Then, carefully pour the liquid into the funnel, and use a **stopwatch** to measure how long the volcano takes to begin erupting.

Analysis

1. Based on your observations, explain what happened to cause the eruption.
2. How accurate was your prediction? By how many seconds did the class predictions vary?
3. How do the size of the funnel opening and the amount of baking soda and vinegar affect the amount of time that the volcano takes to erupt?

SECTION 1

Volcanic Eruptions

Think about the force released when the first atomic bomb exploded during World War II. Now imagine an explosion 10,000 times stronger, and you will get an idea of how powerful a volcanic eruption can be.

The explosive pressure of a volcanic eruption can turn an entire mountain into a billowing cloud of ash and rock in a matter of seconds. But eruptions are also creative forces—they help form fertile farmland. They also create some of the largest mountains on Earth. During an eruption, molten rock, or *magma,* is forced to the Earth's surface. Magma that flows onto the Earth's surface is called *lava.* **Volcanoes** are areas of Earth's surface through which magma and volcanic gases pass.

READING WARM-UP

Objectives

- Distinguish between nonexplosive and explosive volcanic eruptions.
- Identify the features of a volcano.
- Explain how the composition of magma affects the type of volcanic eruption that will occur.
- Describe four types of lava and four types of pyroclastic material.

Terms to Learn

volcano
magma chamber
vent

READING STRATEGY

Reading Organizer As you read this section, make a table comparing types of lava and pyroclastic material.

Nonexplosive Eruptions

At this moment, volcanic eruptions are occurring around the world—on the ocean floor and on land. Nonexplosive eruptions are the most common type of eruption. These eruptions produce relatively calm flows of lava, such as those shown in **Figure 1.** Nonexplosive eruptions can release huge amounts of lava. Vast areas of the Earth's surface, including much of the sea floor and the Northwest region of the United States, are covered with lava from nonexplosive eruptions.

volcano a vent or fissure in the Earth's surface through which magma and gases are expelled

Figure 1 Examples of Nonexplosive Eruptions

Sometimes, nonexplosive eruptions can spray lava into the air. Lava fountains, such as this one, pulse with the pressure of escaping gases.

The speed of a lava flow can range from a slow creep to as fast as 60 km/h.

Explosive Eruptions

Explosive eruptions, such as the one shown in **Figure 2,** are much rarer than nonexplosive eruptions. However, the effects of explosive eruptions can be incredibly destructive. During an explosive eruption, clouds of hot debris, ash, and gas rapidly shoot out from a volcano. Instead of producing lava flows, explosive eruptions cause molten rock to be blown into tiny particles that harden in the air. The dust-sized particles, called *ash,* can reach the upper atmosphere and can circle the Earth for years. Larger pieces of debris fall closer to the volcano. An explosive eruption can also blast millions of tons of lava and rock from a volcano. In a matter of seconds, an explosive eruption can demolish an entire mountainside, as shown in **Figure 3.**

Reading Check **List two differences between explosive and nonexplosive eruptions.** (*See the Appendix for answers to Reading Checks.*)

Figure 2 *In what resembles a nuclear explosion, volcanic ash rockets skyward during the 1990 eruption of Mount Redoubt in Alaska.*

Figure 3 *Within seconds, the 1980 eruption of Mount St. Helens in Washington State caused the side of the mountain to collapse. The blast scorched and flattened 600 km^2 of forest.*

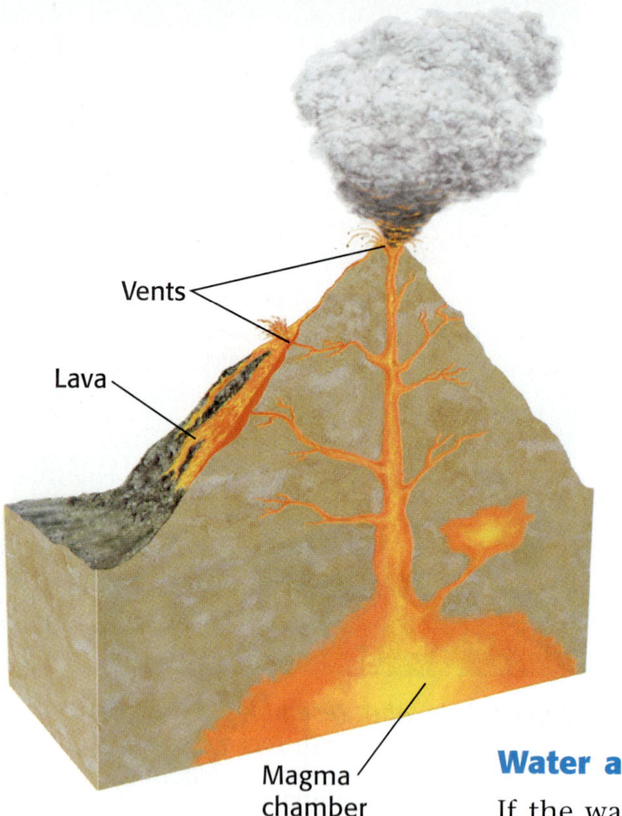

Figure 4 *Volcanoes form when lava is released from vents.*

What Is Inside a Volcano?

If you could look inside an erupting volcano, you would see the features shown in **Figure 4.** A **magma chamber** is a body of molten rock deep underground that feeds a volcano. Magma rises from the magma chamber through cracks in the Earth's crust to openings called **vents.** Magma is released from the vents during an eruption.

magma chamber the body of molten rock that feeds a volcano

vent an opening at the surface of the Earth through which volcanic material passes

What Makes Up Magma?

By comparing the composition of magma from different eruptions, scientists have made an important discovery. The composition of the magma affects how explosive a volcanic eruption is. The key to whether an eruption will be explosive lies in the silica, water, and gas content of the magma.

Water and Magma Are an Explosive Combination

If the water content of magma is high, an explosive eruption is more likely. Because magma is underground, it is under intense pressure and water stays dissolved in the magma. If the magma quickly moves to the surface, the pressure suddenly decreases and the water and other compounds, such as carbon dioxide, become gases. As the gases expand rapidly, an explosion can result. This process is similar to what happens when you shake a can of soda and open it. When a can of soda is shaken, the CO_2 dissolved in the soda is released and pressure builds up. When the can is opened, the soda shoots out, just as lava shoots out of a volcano during an explosive eruption. In fact, some lava is so frothy with gas when it reaches the surface that its solid form, called *pumice,* can float in water!

Silica-Rich Magma Traps Explosive Gases

Magma that has a high silica content also tends to cause explosive eruptions. Silica-rich magma has a stiff consistency. It flows slowly and tends to harden in a volcano's vents. As a result, it plugs the vent. As more magma pushes up from below, pressure increases. If enough pressure builds up, an explosive eruption takes place. Stiff magma also prevents water vapor and other gases from easily escaping. Gas bubbles trapped in magma can expand until they explode. When they explode, the magma shatters and ash and pumice are blasted from the vent. Magma that contains less silica has a more fluid, runnier consistency. Because gases escape this type of magma more easily, explosive eruptions are less likely to occur.

Reading Check How do silica levels affect an eruption?

What Erupts from a Volcano?

Magma erupts as either lava or pyroclastic (PIE roh KLAS tik) material. *Lava* is liquid magma that flows from a volcanic vent. *Pyroclastic material* forms when magma is blasted into the air and hardens. Nonexplosive eruptions produce mostly lava. Explosive eruptions produce mostly pyroclastic material. Over many years—or even during the same eruption—a volcano's eruptions may alternate between lava and pyroclastic eruptions.

Types of Lava

The viscosity of lava, or how lava flows, varies greatly. To understand viscosity, remember that a milkshake has high viscosity and a glass of milk has low viscosity. Lava that has high viscosity is stiff. Lava that has low viscosity is more fluid. The viscosity of lava affects the surface of a lava flow in different ways, as shown in **Figure 5.** *Blocky lava* and *pahoehoe* (puh HOY HOY) have a high viscosity and flow slowly. Other types of lava flows, such as *aa* (AH AH) and *pillow lava,* have lower viscosities and flow more quickly.

CONNECTION TO Social Studies

Fertile Farmlands Volcanic ash helps create some of the most fertile farmland in the world. Use a world map and reference materials to find the location of volcanoes that have helped create farmland in Italy, Africa, South America, and the United States. Make an illustrated map on a piece of poster board to share your findings.

ACTIVITY

Figure 5 Four Types of Lava

Aa is so named because of the painful experience of walking barefoot across its jagged surface. This lava pours out quickly and forms a brittle crust. The crust is torn into jagged pieces as molten lava continues to flow underneath.

Pahoehoe lava flows slowly, like wax dripping from a candle. Its glassy surface has rounded wrinkles.

Pillow lava forms when lava erupts underwater. As you can see here, this lava forms rounded lumps that are the shape of pillows.

Blocky lava is cool, stiff lava that does not travel far from the erupting vent. Blocky lava usually oozes from a volcano and forms jumbled heaps of sharp-edged chunks.

Figure 6 **Four Types of Pyroclastic Material**

Volcanic bombs are large blobs of magma that harden in the air. The shape of this bomb was caused by the magma spinning through the air as it cooled.

Lapilli, which means "little stones" in Italian, are pebblelike bits of magma that hardened before they hit the ground.

Volcanic ash forms when the gases in stiff magma expand rapidly and the walls of the gas bubbles explode into tiny, glasslike slivers. Ash makes up most of the pyroclastic material in an eruption.

Volcanic blocks, the largest pieces of pyroclastic material, are pieces of solid rock erupted from a volcano.

Types of Pyroclastic Material

Pyroclastic material forms when magma explodes from a volcano and solidifies in the air. This material also forms when powerful eruptions shatter existing rock. The size of pyroclastic material ranges from boulders that are the size of houses to tiny particles that can remain suspended in the atmosphere for years. **Figure 6** shows four types of pyroclastic material: volcanic bombs, volcanic blocks, lapilli (lah PIL IE), and volcanic ash.

Reading Check Describe four types of pyroclastic material.

Modeling an Explosive Eruption

1. Inflate a **large balloon,** and place it in a **cardboard box.**
2. Spread a **sheet** on the floor. Place the box in the middle of the sheet. Mound a thin layer of **sand** over the balloon to make a volcano that is taller than the edges of the box.
3. Lightly mist the volcano with **water.** Sprinkle **tempera paint** on the volcano until the volcano is completely covered.
4. Place **small objects** such as **raisins** randomly on the volcano. Draw a sketch of the volcano.
5. Put on your **safety goggles.** Pop the balloon with a **pin.**
6. Use a **metric ruler** to calculate the average distance that 10 grains of sand and 10 raisins traveled.
7. How did the relative weight of each type of material affect the average distance that the material traveled?
8. Draw a sketch of the exploded volcano.

Pyroclastic Flows

One particularly dangerous type of volcanic flow is called a *pyroclastic flow*. Pyroclastic flows are produced when enormous amounts of hot ash, dust, and gases are ejected from a volcano. This glowing cloud of pyroclastic material can race downhill at speeds of more than 200 km/h—faster than most hurricane-force winds! The temperature at the center of a pyroclastic flow can exceed 700°C. A pyroclastic flow from the eruption of Mount Pinatubo is shown in **Figure 7.** Fortunately, scientists were able to predict the eruption and a quarter of a million people were evacuated before the eruption.

Figure 7 *The 1991 eruption of Mount Pinatubo in the Philippines released terrifying pyroclastic flows.*

SECTION Review

Summary

- Volcanoes erupt both explosively and nonexplosively.
- Magma that has a high level of water, CO_2, or silica tends to erupt explosively.
- Lava can be classified by its viscosity and by the surface texture of lava flows.
- Pyroclastic material, such as ash and volcanic bombs, forms when magma solidifies as it travels through the air.

Using Key Terms

1. In your own words, write a definition for each of the following terms: *volcano, magma chamber,* and *vent.*

Understanding Key Ideas

2. Which of the following factors influences whether a volcano erupts explosively?
   a. the concentration of volcanic bombs in the magma
   b. the concentration of phosphorus in the magma
   c. the concentration of aa in the magma
   d. the concentration of water in the magma
3. How are lava and pyroclastic material classified? Describe four types of lava.
4. Which produces more pyroclastic material: an explosive eruption or a nonexplosive eruption?
5. Explain how the presence of silica and water in magma increases the chances of an explosive eruption.
6. What is a pyroclastic flow?

Math Skills

7. A sample of magma is 64% silica. Express this percentage as a simplified fraction.

Critical Thinking

8. **Analyzing Ideas** How is an explosive eruption similar to opening a can of soda that has been shaken? Be sure to describe the role of carbon dioxide.
9. **Making Inferences** Predict the silica content of aa, pillow lava, and blocky lava.
10. **Making Inferences** Explain why the names of many types of lava are Hawaiian but the names of many types of pyroclastic material are Italian and Indonesian.

For a variety of links related to this chapter, go to www.scilinks.org

Topic: Volcanic Eruptions
SciLinks code: HSM1616

SECTION 2

Effects of Volcanic Eruptions

In 1816, Chauncey Jerome, a resident of Connecticut, wrote that the clothes his wife had laid out to dry the day before had frozen during the night. This event would not have been unusual except that the date was June 10!

At that time, residents of New England did not know that the explosion of a volcanic island on the other side of the world had severely changed the global climate and was causing "The Year Without a Summer."

READING WARM-UP

Objectives

- Explain how volcanic eruptions can affect climate.
- Compare the three types of volcanoes.
- Compare craters, calderas, and lava plateaus.

Terms to Learn

crater
caldera
lava plateau

READING STRATEGY

Paired Summarizing Read this section silently. In pairs, take turns summarizing the material. Stop to discuss ideas that seem confusing.

Volcanic Eruptions and Climate Change

The explosion of Mount Tambora in 1815 blanketed most of Indonesia in darkness for three days. It is estimated that 12,000 people died directly from the explosion and 80,000 people died from the resulting hunger and disease. The global effects of the eruption were not felt until the next year, however. During large-scale eruptions, enormous amounts of volcanic ash and gases are ejected into the upper atmosphere.

As volcanic ash and gases spread throughout the atmosphere, they can block enough sunlight to cause global temperatures to drop. The Tambora eruption affected the global climate enough to cause food shortages in North America and Europe. More recently, the eruption of Mount Pinatubo, shown in **Figure 1,** caused average global temperatures to drop by as much as 0.5°C. Although this may seem insignificant, such a shift can disrupt climates all over the world.

Reading Check **How does a volcanic eruption affect climate?** *(See the Appendix for answers to Reading Checks.)*

Figure 1 *Ash from the eruption of Mount Pinatubo blocked out the sun in the Philippines for several days. The eruption also affected global climate.*

Different Types of Volcanoes

Volcanic eruptions can cause profound changes in climate. But the changes to Earth's surface caused by eruptions are probably more familiar. Perhaps the best known of all volcanic landforms are the volcanoes themselves. The three basic types of volcanoes are illustrated in **Figure 2.**

Shield Volcanoes

Shield volcanoes are built of layers of lava released from repeated nonexplosive eruptions. Because the lava is very runny, it spreads out over a wide area. Over time, the layers of lava create a volcano that has gently sloping sides. Although their sides are not very steep, shield volcanoes can be enormous. Hawaii's Mauna Kea, the shield volcano shown here, is the tallest mountain on Earth. Measured from its base on the sea floor, Mauna Kea is taller than Mount Everest.

Cinder Cone Volcanoes

Cinder cone volcanoes are made of pyroclastic material usually produced from moderately explosive eruptions. The pyroclastic material forms steep slopes, as shown in this photo of the Mexican volcano Paricutín. Cinder cones are small and usually erupt for only a short time. Paricutín appeared in a cornfield in 1943 and erupted for only nine years before stopping at a height of 400 m. Cinder cones often occur in clusters, commonly on the sides of other volcanoes. They usually erode quickly because the pyroclastic material is not cemented together.

Composite Volcanoes

Composite volcanoes, sometimes called *stratovolcanoes,* are one of the most common types of volcanoes. They form from explosive eruptions of pyroclastic material followed by quieter flows of lava. The combination of both types of eruptions forms alternating layers of pyroclastic material and lava. Composite volcanoes, such as Japan's Mount Fuji (shown here), have broad bases and sides that get steeper toward the top. Composite volcanoes in the western region of the United States include Mount Hood, Mount Rainier, Mount Shasta, and Mount St. Helens.

Figure 2 Three Types of Volcanoes

Shield volcano

Cinder cone volcano

Composite volcano

Figure 3 *A crater, such as this one in Kamchatka, Russia, forms around the central vent of a volcano.*

Other Types of Volcanic Landforms

In addition to volcanoes, other landforms are produced by volcanic activity. These landforms include craters, calderas, and lava plateaus. Read on to learn more about these landforms.

Craters

Around the central vent at the top of many volcanoes is a funnel-shaped pit called a **crater.** An example of a crater is shown in **Figure 3.** During less explosive eruptions, lava flows and pyroclastic material can pile up around the vent creating a cone with a central crater. As the eruption stops, the lava that is left in the crater often drains back underground. The vent may then collapse to form a larger crater. If the lava hardens in the crater, the next eruption may blast it away. In this way, a crater becomes larger and deeper.

crater a funnel-shaped pit near the top of the central vent of a volcano

caldera a large, semicircular depression that forms when the magma chamber below a volcano partially empties and causes the ground above to sink

Calderas

Calderas can appear similar to craters, but they are many times larger. A **caldera** is a large, semicircular depression that forms when the chamber that supplies magma to a volcano partially empties and the chamber's roof collapses. As a result, the ground above the magma chamber sinks, as shown in **Figure 4.** Much of Yellowstone Park is made up of three large calderas that formed when volcanoes collapsed between 1.9 million and 0.6 million years ago. Today, hot springs, such as Old Faithful, are heated by the thermal energy left over from those events.

Reading Check How do calderas form?

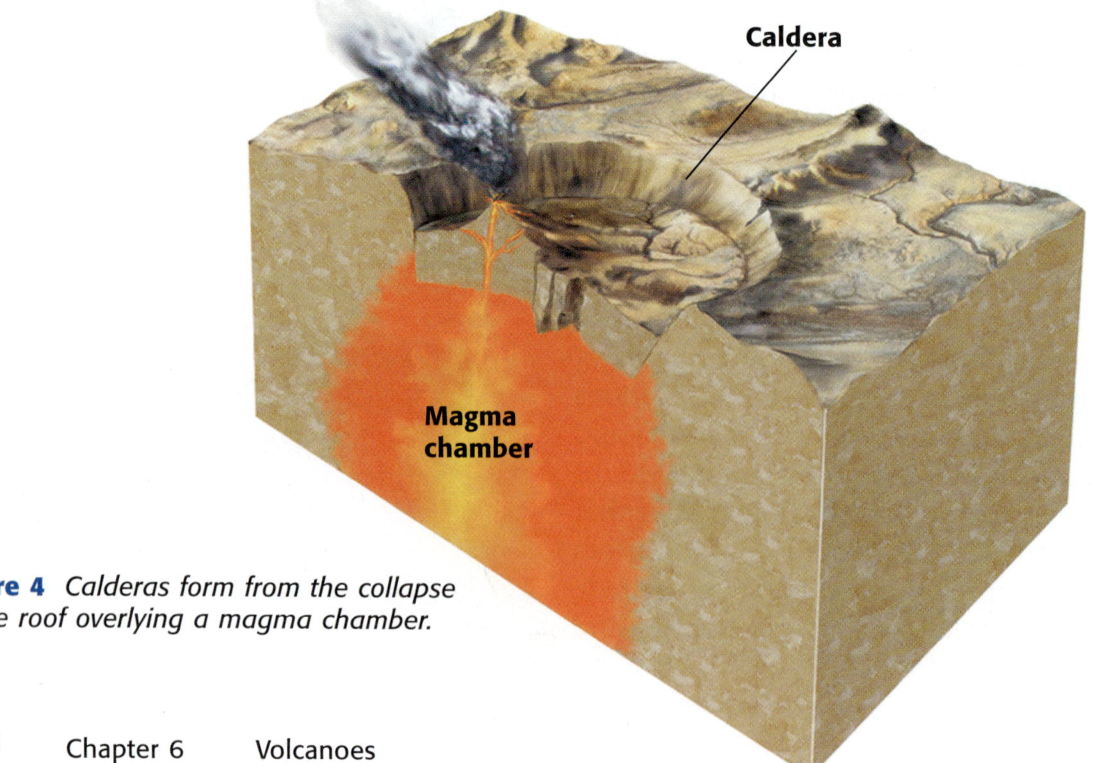

Figure 4 *Calderas form from the collapse of the roof overlying a magma chamber.*

Lava Plateaus

The most massive outpourings of lava do not come from individual volcanoes. Most of the lava on Earth's surface erupted from long cracks, or *rifts,* in the crust. In this type of eruption, runny lava can pour out for millions of years and spread over huge areas. A landform that results from repeated eruptions of lava spread over a large area is called a **lava plateau.** The Columbia River Plateau, part of which is shown in **Figure 5,** is a lava plateau that formed between 17 million and 14 million years ago in the northwestern region of the United States. In some places, the Columbia River Plateau is 3 km thick.

Figure 5 *The Columbia River Plateau formed from a massive outpouring of lava that began 17 million years ago.*

lava plateau a wide, flat landform that results from repeated nonexplosive eruptions of lava that spread over a large area

SECTION Review

Summary

- The large volumes of gas and ash released from volcanic eruptions can affect climate.
- Shield volcanoes result from many eruptions of relatively runny lava.
- Cinder cone volcanoes result from mildly explosive eruptions of pyroclastic material.
- Composite volcanoes result from alternating explosive and nonexplosive eruptions.
- Craters, calderas, and lava plateaus are volcanic landforms.

Using Key Terms

Complete each of the following sentences by choosing the correct term from the word bank.

caldera crater

1. A ___ is a funnel-shaped hole around the central vent.
2. A ___ results when a magma chamber partially empties.

Understanding Key Ideas

3. Which type of volcano results from alternating explosive and nonexplosive eruptions?
   a. composite volcano
   b. cinder cone volcano
   c. rift-zone volcano
   d. shield volcano
4. Why do cinder cone volcanoes have narrower bases and steeper sides than shield volcanoes do?
5. Why does a volcano's crater tend to get larger over time?

Math Skills

6. The fastest lava flow recorded was 60 km/h. A horse can gallop as fast as 48 mi/h. Could a galloping horse outrun the fastest lava flow? (Hint: 1 km = 0.621 mi)

Critical Thinking

7. **Making Inferences** Why did it take a year for the effects of the Tambora eruption to be experienced in New England?

For a variety of links related to this chapter, go to www.scilinks.org

Topic: Volcanic Effects
SciLinks code: HSM1615

SECTION 3

Causes of Volcanic Eruptions

More than 2,000 years ago, Pompeii was a busy Roman city near the sleeping volcano Mount Vesuvius. People did not see Vesuvius as much of a threat. Everything changed when Vesuvius suddenly erupted and buried the city in a deadly blanket of ash that was almost 20 ft thick!

Today, even more people are living on and near active volcanoes. Scientists closely monitor volcanoes to avoid this type of disaster. They study the gases coming from active volcanoes and look for slight changes in the volcano's shape that could indicate that an eruption is near. Scientists know much more about the causes of eruptions than the ancient Pompeiians did, but there is much more to be discovered.

READING WARM-UP

Objectives

- Describe the formation and movement of magma.
- Explain the relationship between volcanoes and plate tectonics.
- Summarize the methods scientists use to predict volcanic eruptions.

Terms to Learn

rift zone
hot spot

READING STRATEGY

Reading Organizer As you read this section, make a flowchart of the steps of magma formation in different tectonic environments.

The Formation of Magma

Understanding how magma forms helps explain why volcanoes erupt. Magma forms in the deeper regions of the Earth's crust and in the uppermost layers of the mantle where the temperature and pressure are very high. Changes in pressure and temperature cause magma to form.

Pressure and Temperature

Part of the upper mantle is made of very hot, puttylike rock that flows slowly. The rock of the mantle is hot enough to melt at Earth's surface, but it remains a puttylike solid because of pressure. This pressure is caused by the weight of the rock above the mantle. In other words, the rock above the mantle presses the atoms of the mantle so close together that the rock cannot melt. As **Figure 1** shows, rock melts when its temperature increases or when the pressure on the rock decreases.

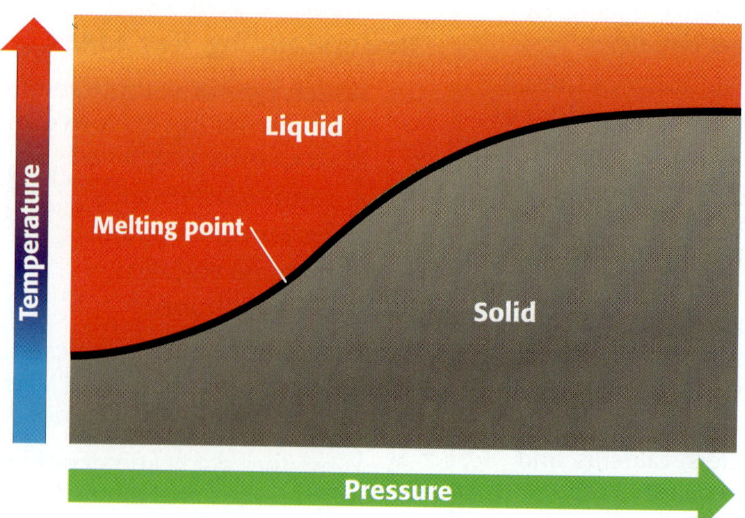

Figure 1 *The curved line indicates the melting point of a rock. As pressure decreases and temperature increases, the rock begins to melt.*

Magma Formation in the Mantle

Because the temperature of the mantle is fairly constant, a decrease in pressure is the most common cause of magma formation. Magma often forms at the boundary between separating tectonic plates, where pressure is decreased. Once formed, the magma is less dense than the surrounding rock, so the magma slowly rises toward the surface like an air bubble in a jar of honey.

Where Volcanoes Form

The locations of volcanoes give clues about how volcanoes form. The map in **Figure 2** shows the location of some of the world's major active volcanoes. The map also shows the boundaries between tectonic plates. A large number of volcanoes lie directly on tectonic plate boundaries. In fact, the plate boundaries surrounding the Pacific Ocean have so many volcanoes that the area is called the *Ring of Fire.*

Tectonic plate boundaries are areas where tectonic plates either collide, separate, or slide past one another. At these boundaries, it is possible for magma to form and travel to the surface. About 80% of active volcanoes on land form where plates collide, and about 15% form where plates separate. The remaining few occur far from tectonic plate boundaries.

Reading Check **Why are most volcanoes on plate boundaries?** *(See the Appendix for answers to Reading Checks.)*

Reaction to Stress

1. Make a pliable "rock" by pouring **60 mL of water** into a **plastic cup** and adding **150 mL of corn-starch,** 15 mL at a time. Stir well each time.
2. Pour half of the cornstarch mixture into a **clear bowl.** Carefully observe how the "rock" flows. Be patient—this process is slow!
3. Scrape the rest of the "rock" out of the cup with a **spoon.** Observe the behavior of the "rock" as you scrape.
4. What happened to the "rock" when you let it flow by itself? What happened when you put stress on the "rock"?
5. How is this pliable "rock" similar to the rock of the upper part of the mantle?

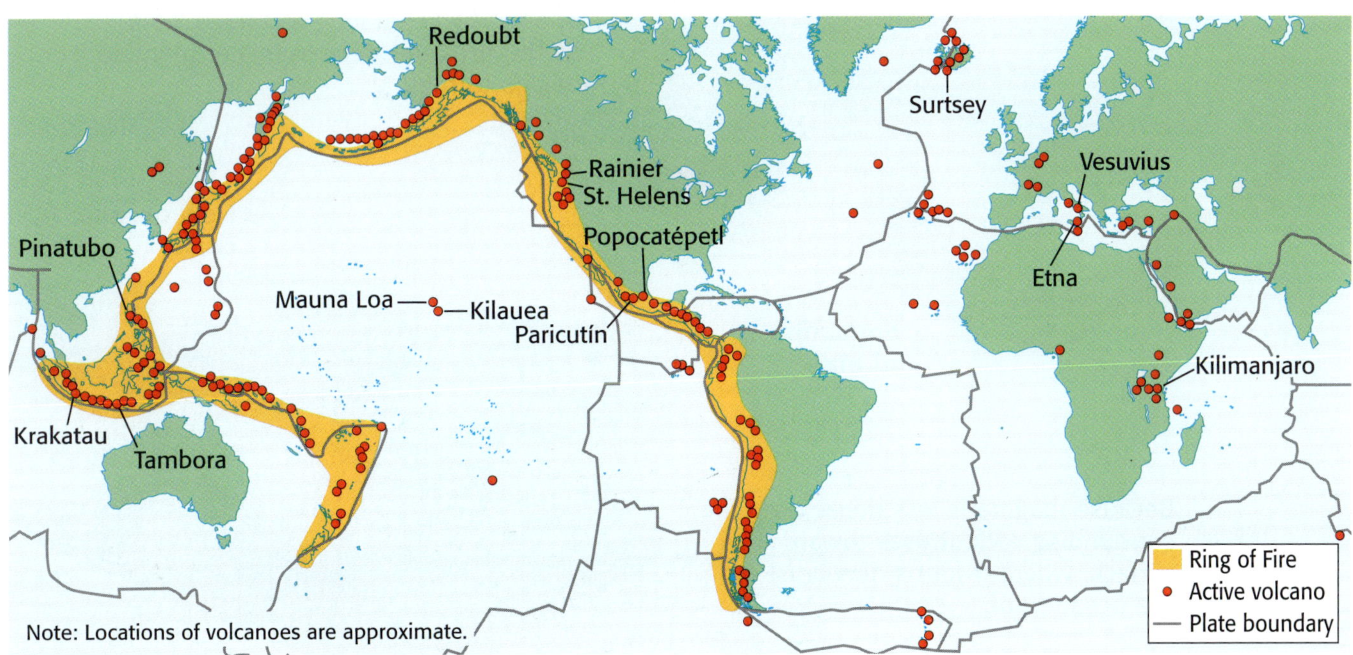

Figure 2 *Tectonic plate boundaries are likely places for volcanoes to form. The Ring of Fire contains nearly 75% of the world's active volcanoes on land.*

How Hot Is Hot?

Inside the Earth, magma can reach a burning-hot 1,400°C! You may be more familiar with Fahrenheit temperatures, so convert 1,400°C to degrees Fahrenheit by using the formula below.

$$°F = (°C \div 5 \times 9) + 32$$

What is the temperature in degrees Fahrenheit?

rift zone an area of deep cracks that forms between two tectonic plates that are pulling away from each other

When Tectonic Plates Separate

At a *divergent boundary,* tectonic plates move away from each other. As tectonic plates separate, a set of deep cracks called a **rift zone** forms between the plates. Mantle rock then rises to fill in the gap. When mantle rock gets closer to the surface, the pressure decreases. The pressure decrease causes the mantle rock to melt and form magma. Because magma is less dense than the surrounding rock, it rises through the rifts. When the magma reaches the surface, it spills out and hardens, creating new crust, as shown in **Figure 3.**

Mid-Ocean Ridges Form at Divergent Boundaries

Lava that flows from undersea rift zones produces volcanoes and mountain chains called *mid-ocean ridges.* Just as a baseball has stitches, the Earth is circled with mid-ocean ridges. At these ridges, lava flows out and creates new crust. Most volcanic activity on Earth occurs at mid-ocean ridges. While most mid-ocean ridges are underwater, Iceland, with its volcanoes and hot springs, was created by lava from the Mid-Atlantic Ridge. In 1963, enough lava poured out of the Mid-Atlantic Ridge near Iceland to form a new island called *Surtsey.* Scientists watched this new island being born!

Figure 3 How Magma Forms at a Divergent Boundary

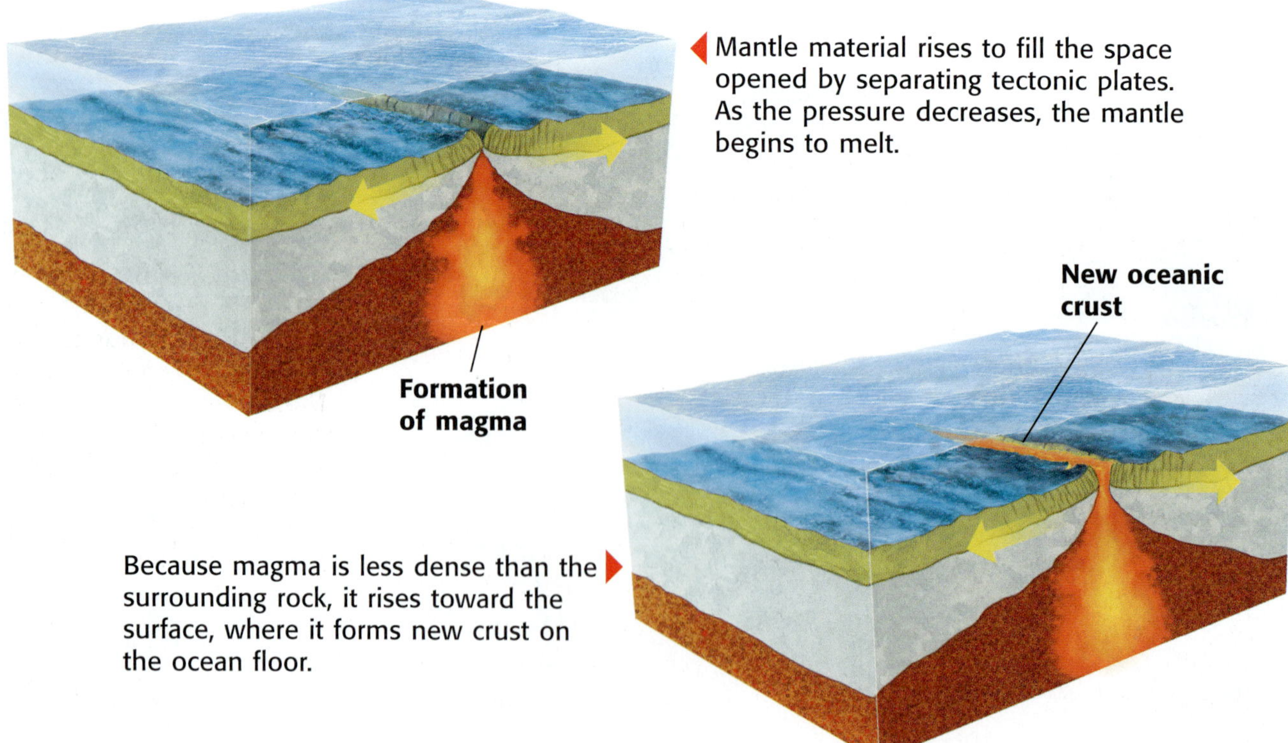

Figure 4 How Magma Forms at a Convergent Boundary

Oceanic crust

Continental crust

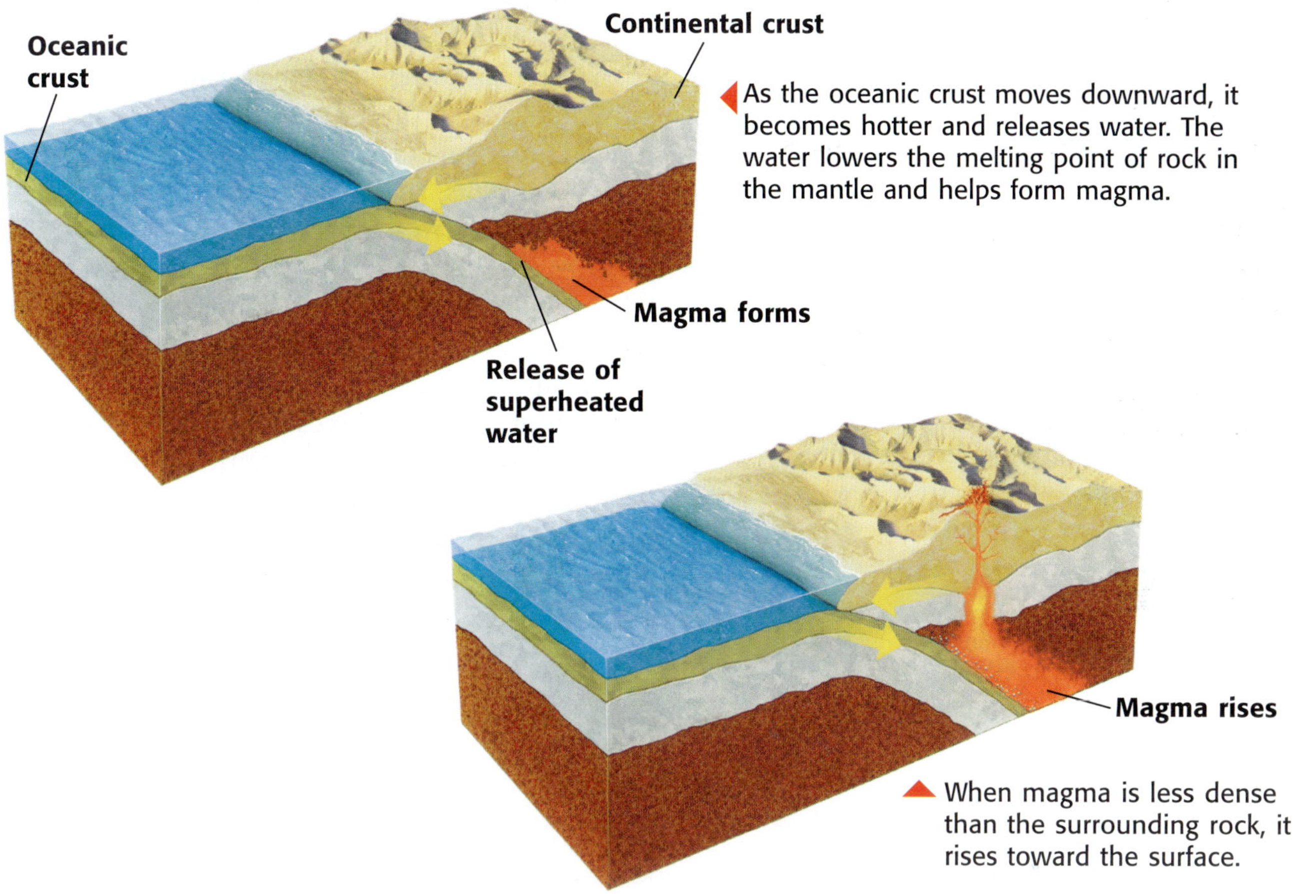

As the oceanic crust moves downward, it becomes hotter and releases water. The water lowers the melting point of rock in the mantle and helps form magma.

Magma forms

Release of superheated water

Magma rises

When magma is less dense than the surrounding rock, it rises toward the surface.

When Tectonic Plates Collide

If you slide two pieces of notebook paper into one another on a flat desktop, the papers will either buckle upward or one piece of paper will move under the other. This is similar to what happens at a convergent boundary. A *convergent boundary* is a place where tectonic plates collide. When an oceanic plate collides with a continental plate, the oceanic plate usually slides underneath the continental plate. The process of *subduction,* the movement of one tectonic plate underneath another, is shown in **Figure 4.** Oceanic crust is subducted because it is denser and thinner than continental crust.

Subduction Produces Magma

As the descending oceanic crust scrapes past the continental crust, the temperature and pressure increase. The combination of increased heat and pressure causes the water contained in the oceanic crust to be released. The water then mixes with the mantle rock, which lowers the rock's melting point, causing it to melt. This body of magma can rise to form a volcano.

Reading Check How does subduction produce magma?

Tectonic Models

Create models of convergent and divergent boundaries by using materials of your choice. Have your teacher approve your list before you start building your model at home with a parent. In class, use your model to explain how each type of boundary leads to the formation of magma.

ACTIVITY

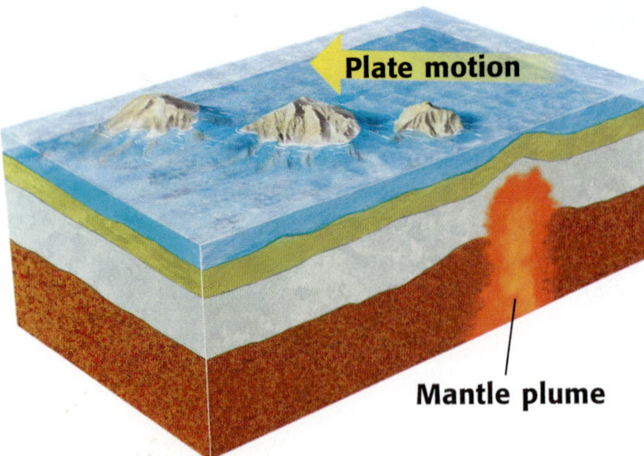

Figure 5 *According to one theory, a string of volcanic islands forms as a tectonic plate passes over a mantle plume.*

Hot Spots

Not all magma develops along tectonic plate boundaries. For example, the Hawaiian Islands, some of the most well-known volcanoes on Earth, are nowhere near a plate boundary. The volcanoes of Hawaii and several other places on Earth are known as *hot spots*. **Hot spots** are volcanically active places on the Earth's surface that are far from plate boundaries. Some scientists think that hot spots are directly above columns of rising magma, called *mantle plumes*. Other scientists think that hot spots are the result of cracks in the Earth's crust.

A hot spot often produces a long chain of volcanoes. One theory is that the mantle plume stays in the same spot while the tectonic plate moves over it, as shown in **Figure 5.** Another theory argues that hot-spot volcanoes occur in long chains because they form along the cracks in the Earth's crust. Both theories may be correct.

hot spot a volcanically active area of Earth's surface far from a tectonic plate boundary

Reading Check **Describe two theories that explain the existence of hot spots.**

Predicting Volcanic Eruptions

You now understand some of the processes that produce volcanoes, but how do scientists predict when a volcano is going to erupt? Volcanoes are classified in three categories. *Extinct volcanoes* have not erupted in recorded history and probably never will erupt again. *Dormant volcanoes* are currently not erupting, but the record of past eruptions suggests that they may erupt again. *Active volcanoes* are currently erupting or show signs of erupting in the near future. Scientists study active and dormant volcanoes for signs of a future eruption.

Measuring Small Quakes and Volcanic Gases

Most active volcanoes produce small earthquakes as the magma within them moves upward and causes the surrounding rock to shift. Just before an eruption, the number and intensity of the earthquakes increase and the occurrence of quakes may be continuous. Monitoring these quakes is one of the best ways to predict an eruption.

As **Figure 6** shows, scientists also study the volume and composition of volcanic gases. The ratio of certain gases, especially that of sulfur dioxide, SO_2, to carbon dioxide, CO_2, may be important in predicting eruptions. Changes in this ratio may indicate changes in the magma chamber below.

Figure 6 *As if being this close to an active volcano is not dangerous enough, the gases being collected are extremely poisonous.*

Measuring Slope and Temperature

As magma moves upward prior to an eruption, it can cause the Earth's surface to swell. The side of a volcano may even bulge as the magma moves upward. An instrument called a *tiltmeter* helps scientists detect small changes in the angle of a volcano's slope. Scientists also use satellite technology such as the Global Positioning System (GPS) to detect the changes in a volcano's slope that may signal an eruption.

One of the newest methods for predicting volcanic eruptions includes using satellite images. Infrared satellite images record changes in the surface temperature and gas emissions of a volcano over time. If the site is getting hotter, the magma below is probably rising!

For another activity related to this chapter, go to **go.hrw.com** and type in the keyword **HZ5VOLW.**

SECTION Review

Summary

- Temperature and pressure influence magma formation.
- Most volcanoes form at tectonic boundaries.
- As tectonic plates separate, magma rises to fill the cracks, or rifts, that develop.
- As oceanic and continental plates collide, the oceanic plate tends to subduct and cause the formation of magma.
- To predict eruptions, scientists study the frequency and type of earthquakes associated with the volcano as well as changes in slope, changes in the gases released, and changes in the volcano's surface temperature.

Using Key Terms

1. Use each of the following terms in a separate sentence: *hot spot* and *rift zone.*

Understanding Key Ideas

2. If the temperature of a rock remains constant but the pressure on the rock decreases, what tends to happen?
   - **a.** The temperature increases.
   - **b.** The rock becomes liquid.
   - **c.** The rock becomes solid.
   - **d.** The rock subducts.
3. Which of the following words is a synonym for *dormant*?
   - **a.** predictable
   - **b.** active
   - **c.** dead
   - **d.** sleeping
4. What is the Ring of Fire?
5. Explain how convergent and divergent plate boundaries cause magma formation.
6. Describe four methods that scientists use to predict volcanic eruptions.
7. Why does a oceanic plate tend to subduct when it collides with a continental plate?

Math Skills

8. If a tectonic plate moves at a rate of 2 km every 1 million years, how long would it take a hot spot to form a chain of volcanoes 100 km long?

Critical Thinking

9. **Making Inferences** New crust is constantly being created at mid-ocean ridges. So, why is the oldest oceanic crust only about 150 million years old?
10. **Identifying Relationships** If you are studying a volcanic deposit, would the youngest layers be more likely to be found on the top or on the bottom? Explain your answer.

Skills Practice Lab

OBJECTIVES

Build a working apparatus to test carbon dioxide levels.

Test the levels of carbon dioxide emitted from a model volcano.

MATERIALS

- baking soda, 15 mL
- bottle, drinking, 16 oz
- box or stand for plastic cup
- clay, modeling
- coin
- cup, clear plastic, 9 oz
- graduated cylinder
- limewater, 1 L
- straw, drinking, flexible
- tissue, bathroom (2 sheets)
- vinegar, white, 140 mL
- water, 100 mL

SAFETY

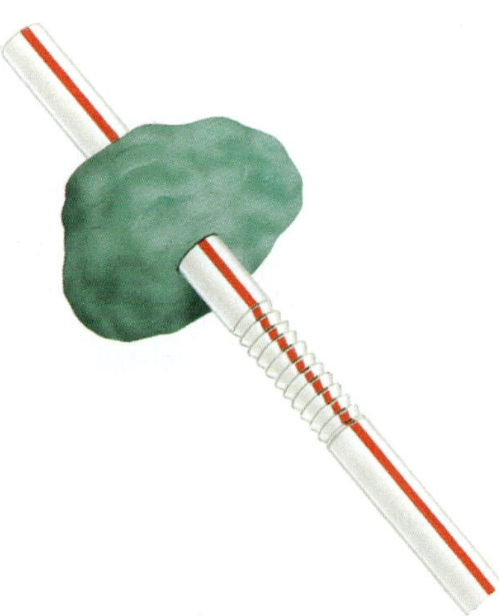

Volcano Verdict

You will need to pair up with a partner for this exploration. You and your partner will act as geologists who work in a city located near a volcano. City officials are counting on you to predict when the volcano will erupt next. You and your partner have decided to use limewater as a gas-emissions tester. You will use this tester to measure the levels of carbon dioxide emitted from a simulated volcano. The more active the volcano is, the more carbon dioxide it releases.

Procedure

1. Put on your safety goggles, and carefully pour limewater into the plastic cup until the cup is three-fourths full. You have just made your gas-emissions tester.
2. Now, build a model volcano. Begin by pouring 50 mL of water and 70 mL of vinegar into the drink bottle.
3. Form a plug of clay around the short end of the straw, as shown at left. The clay plug must be large enough to cover the opening of the bottle. Be careful not to get the clay wet.
4. Sprinkle 5 mL of baking soda along the center of a single section of bathroom tissue. Then, roll the tissue, and twist the ends so that the baking soda can't fall out.

5. Drop the tissue into the drink bottle, and immediately put the short end of the straw inside the bottle to make a seal with the clay.
6. Put the other end of the straw into the limewater, as shown at right.
7. You have just taken your first measurement of gas levels from the volcano. Record your observations.
8. Imagine that it is several days later and you need to test the volcano again to collect more data. Before you continue, toss a coin. If it lands heads up, go to step 9. If it lands tails up, go to step 10. Write down the step that you follow.
9. Repeat steps 1–7. This time, add 2 mL of baking soda to the vinegar and water. (Note: You must use fresh water, vinegar, and limewater.) Write down your observations. Go to step 11.
10. Repeat steps 1–7. This time, add 8 mL of baking soda to the vinegar and water. (Note: You must use fresh water, vinegar, and limewater.) Write down your observations. Go to step 11.
11. Return to step 8 once. Then, answer the questions below.

Analyze the Results

1. **Explaining Events** How do you explain the difference in the appearance of the limewater from one trial to the next?
2. **Recognizing Patterns** What does the data that you collected indicate about the activity in the volcano?

Draw Conclusions

3. **Evaluating Results** Based on your results, do you think it would be necessary to evacuate the city?
4. **Applying Conclusions** How would a geologist use a gas-emissions tester to predict volcanic eruptions?

Chapter Review

USING KEY TERMS

For each pair of terms, explain how the meanings of the terms differ.

1. *caldera* and *crater*
2. *lava* and *magma*
3. *lava* and *pyroclastic material*
4. *vent* and *rift*
5. *cinder cone volcano* and *shield volcano*

UNDERSTANDING KEY IDEAS

Multiple Choice

6. The type of magma that tends to cause explosive eruptions has a
   - **a.** high silica content and high viscosity.
   - **b.** high silica content and low viscosity.
   - **c.** low silica content and low viscosity.
   - **d.** low silica content and high viscosity.
7. Lava that flows slowly to form a glassy surface with rounded wrinkles is called
   - **a.** aa lava.
   - **b.** pahoehoe lava.
   - **c.** pillow lava.
   - **d.** blocky lava.
8. Magma forms within the mantle most often as a result of
   - **a.** high temperature and high pressure.
   - **b.** high temperature and low pressure.
   - **c.** low temperature and high pressure.
   - **d.** low temperature and low pressure.
9. What causes an increase in the number and intensity of small earthquakes before an eruption?
   - **a.** the movement of magma
   - **b.** the formation of pyroclastic material
   - **c.** the hardening of magma
   - **d.** the movement of tectonic plates
10. If volcanic dust and ash remain in the atmosphere for months or years, what do you predict will happen?
    - **a.** Solar reflection will decrease, and temperatures will increase.
    - **b.** Solar reflection will increase, and temperatures will increase.
    - **c.** Solar reflection will decrease, and temperatures will decrease.
    - **d.** Solar reflection will increase, and temperatures will decrease.
11. At divergent plate boundaries,
    - **a.** heat from Earth's core causes mantle plumes.
    - **b.** oceanic plates sink, which causes magma to form.
    - **c.** tectonic plates move apart.
    - **d.** hot spots cause volcanoes.
12. A theory that helps explain the causes of both earthquakes and volcanoes is the theory of
    - **a.** pyroclastics.
    - **b.** plate tectonics.
    - **c.** climatic fluctuation.
    - **d.** mantle plumes.

Short Answer

13 How does the presence of water in magma affect a volcanic eruption?

14 Describe four clues that scientists use to predict eruptions.

15 Identify the characteristics of the three types of volcanoes.

16 Describe the positive effects of volcanic eruptions.

CRITICAL THINKING

17 **Concept Mapping** Use the following terms to create a concept map: *volcanic bombs, aa, pyroclastic material, pahoehoe, lapilli, lava,* and *volcano.*

18 **Identifying Relationships** You are exploring a volcano that has been dormant for some time. You begin to keep notes on the types of volcanic debris that you see as you walk. Your first notes describe volcanic ash. Later, your notes describe lapilli. In what direction are you most likely traveling—toward the crater or away from the crater? Explain your answer.

19 **Making Inferences** Loihi is a submarine Hawaiian volcano that might grow to form a new island. The Hawaiian Islands are located on the Pacific plate, which is moving northwest. Considering how this island chain may have formed, where do you think the new volcanic island will be located? Explain your answer.

20 **Evaluating Hypotheses** What evidence could confirm the existence of mantle plumes?

INTERPRETING GRAPHICS

The graph below illustrates the average change in temperature above or below normal for a community over several years. Use the graph below to answer the questions that follow.

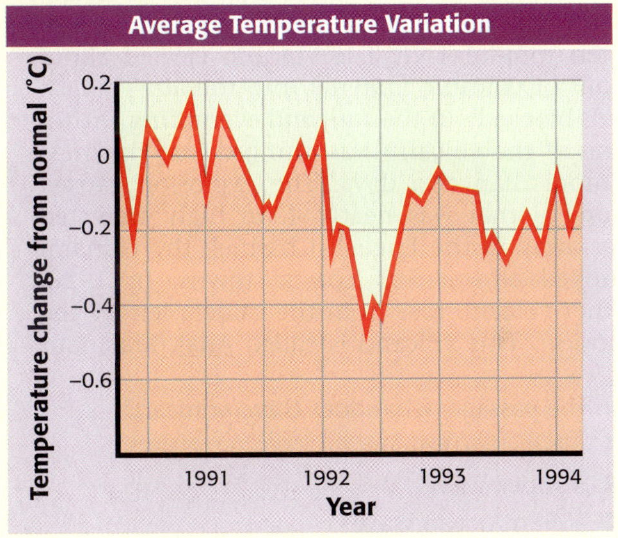

21 If the variation in temperature over the years was influenced by a major volcanic eruption, when did the eruption most likely take place? Explain.

22 If the temperature were measured only once each year (at the beginning of the year), how would your interpretation be different?

Standardized Test Preparation

READING

Read each of the passages below. Then, answer the questions that follow each passage.

Passage 1 When the volcanic island of Krakatau in Indonesia exploded in 1883, a shock wave sped around the world seven times. The explosion was probably the loudest sound in recorded human history. What caused this enormous explosion? Most likely, the walls of the volcano ruptured, and ocean water flowed into the magma chamber of the volcano. The water instantly turned into steam, and the volcano exploded with the force of 100 million tons of TNT. The volcano ejected about 18 km^3 of volcanic material into the air. The ash clouds blocked out the sun, and everything within 80 km of the volcano was plunged into darkness for more than two days. The explosion caused a <u>tsunami</u> that was nearly 40 m high. Detected as far away as the English Channel, the tsunami destroyed almost 300 coastal towns. In 1928, another volcano rose from the caldera left by the explosion. This volcano is called <u>Anak</u> Krakatau.

1. In the passage, what does *tsunami* mean?

- **A** a large earthquake
- **B** a shock wave
- **C** a giant ocean wave
- **D** a cloud of gas and dust

2. According to the passage, what was the size of the Krakatau explosion probably the result of?

- **F** pyroclastic material rapidly mixing with air
- **G** 100 million tons of TNT
- **H** an ancient caldera
- **I** the flow of water into the magma chamber

3. What does the Indonesian word *anak* probably mean?

- **A** father
- **B** child
- **C** mother
- **D** grandmother

Passage 2 Yellowstone National Park in Montana and Wyoming contains three overlapping calderas and evidence of the <u>cataclysmic</u> ash flows that erupted from them. The oldest eruption occurred 1.9 million years ago, the second eruption happened 1.3 million years ago, and the most recent eruption occurred 0.6 million years ago. Seismographs regularly detect the movement of magma beneath the caldera, and the hot springs and geysers of the park indicate that a large body of magma lies beneath the park. The geology of the area shows that major eruptions occurred about once every 0.6 or 0.7 million years. Thus, a devastating eruption is long overdue. People living near the park should be evacuated immediately.

1. In the passage, what does *cataclysmic* mean?

- **A** nonexplosive
- **B** ancient
- **C** destructive
- **D** characterized by ash flows

2. Which of the following clues are evidence of an active magma body beneath the park?

- **F** cataclysmic ash flows
- **G** the discovery of seismoclasts
- **H** minor eruptions
- **I** seismograph readings

3. Which of the following contradicts the author's conclusion that an eruption is "long overdue"?

- **A** Magma has been detected beneath the park.
- **B** With a variation of 0.1 million years, an eruption may occur in the next 100,000 years.
- **C** The composition of gases emitted indicates that an eruption is near.
- **D** Seismographs have detected the movement of magma.

INTERPRETING GRAPHICS

The map below shows some of the Earth's major volcanoes and the tectonic plate boundaries. Use the map below to answer the questions that follow.

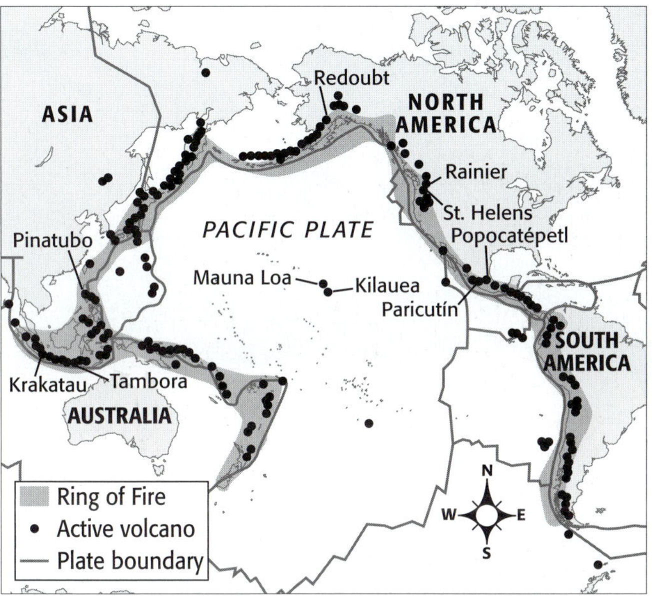

1. If ash from Popocatépetl landed on the west coast of the United States, what direction did the ash travel?
   - **A** northeast
   - **B** northwest
   - **C** southeast
   - **D** southwest

2. Why aren't there any active volcanoes in Australia?
   - **F** Australia is not located on a plate boundary.
   - **G** Australia is close to Krakatau and Tambora.
   - **H** Australia is near a plate boundary.
   - **I** Australia is near a rift zone.

3. If a scientist traveled along the Ring of Fire from Mt. Redoubt to Krakatau, which of the following most accurately describes the directions in which she traveled?
   - **A** west, southeast, east
   - **B** west, southeast, west
   - **C** west, southwest, east
   - **D** west, southwest, west

MATH

Read each question below, and choose the best answer.

1. Midway Island is 1,935 km northwest of Hawaii. If the Pacific plate is moving to the northwest at a rate of 9 cm per year, how long ago was Midway Island over the hot spot that formed the island?
   - **A** 215,000 years
   - **B** 2,150,000 years
   - **C** 21,500,000 years
   - **D** 215,000,000 years

2. In the first year that the Mexican volcano Paricutín appeared in a cornfield, it grew 360 m. The volcano stopped growing at about 400 m. What percentage of the volcano's total growth occurred in the first year?
   - **F** 67%
   - **G** 82%
   - **H** 90%
   - **I** 92%

3. A pyroclastic flow is moving down a hill at 120 km/h. If you lived in a town 5 km away, how much time would you have before the flow reached your town?
   - **A** 2 min and 30 s
   - **B** 1 min and 21 s
   - **C** 3 min and 12 s
   - **D** 8 min and 3 s

4. The Columbia River plateau is a lava plateau that contains 350,000 km^3 of solidified lava. The plateau took 3 million years to form. What was the average rate of lava deposition each century?
   - **F** 0.116 km^3
   - **G** 11.6 km^3
   - **H** 116 km^3
   - **I** 11,600 km^3

Science in Action

Weird Science

Pele's Hair

It is hard to believe that the fragile specimen shown below is a volcanic rock. This strange type of lava, called *Pele's hair,* forms when volcanic gases spray molten rock high into the air. When conditions are right, the lava can harden into strands of volcanic glass as thin as a human hair. This type of lava is named after Pele, the Hawaiian goddess of volcanoes. Several other types of lava are named in Pele's honor. Pele's tears are tear-shaped globs of volcanic glass often found at the end of strands of Pele's hair. Pele's diamonds are green, gemlike stones found in hardened lava flows.

Language Arts ACTIVITY

Volcanic terms come from many languages. Research some volcanic terms on the Internet, and create an illustrated volcanic glossary to share with your class.

Science, Technology, and Society

Fighting Lava with Fire Hoses

What would you do if a 60 ft wall of lava was advancing toward your home? Most people would head for safety. But when an eruption threatened to engulf the Icelandic fishing village of Heimaey in 1973, some villagers held their ground and fought back. Working 14-hour days in conditions so hot that their boots would catch on fire, villagers used fire-hoses to spray sea water on the lava flow. For several weeks, the lava advanced toward the town, and it seemed as if there was no hope. But the water eventually cooled the lava fast enough to divert the flow and save the village. It took 5 months and about 1.5 billion gallons of water to fight the lava flow. When the eruption stopped, villagers found that the island had grown by 20%!

Social Studies ACTIVITY

WRITING SKILL To try to protect the city of Hilo, Hawaii, from an eruption in 1935, planes dropped bombs on the lava. Find out if this mission was successful, and write a report about other attempts to stop lava flows.

Careers

Tina Neal

Volcanologist Would you like to study volcanoes for a living? Tina Neal is a volcanologist at the Alaska Volcano Observatory in Anchorage, Alaska. Her job is to monitor and study some of Alaska's 41 active volcanoes. Much of her work focuses on studying volcanoes in order to protect the public. According to Neal, being near a volcano when it is erupting is a wonderful adventure for the senses. "Sometimes you can get so close to an erupting volcano that you can feel the heat, hear the activity, and smell the lava. It's amazing! In Alaska, erupting volcanoes are too dangerous to get very close to, but they create a stunning visual display even from a distance."

Neal also enjoys the science of volcanoes. "It's fascinating to be near an active volcano and become aware of all the chemical and physical processes taking place. When I'm watching a volcano, I think about everything we understand and don't understand about what is happening. It's mind-boggling!" Neal says that if you are interested in becoming a volcanologist, it is important to be well rounded as a scientist. So, you would have to study math, geology, chemistry, and physics. Having a good understanding of computer tools is also important because volcanologists use computers to manage a lot of data and to create models. Neal also suggests learning a second language, such as Spanish. In her spare time, Neal is learning Russian so that she can better communicate with research partners in Kamchatka, Siberia.

Math ACTIVITY

The 1912 eruption of Mt. Katmai in Alaska could be heard 5,620 km away in Atlanta, Georgia. If the average speed of sound in the atmosphere is 342 m/s, how many hours after the eruption did the citizens of Atlanta hear the explosion?

To learn more about these Science in Action topics, visit go.hrw.com and type in the keyword HZ5VOLF.

Current Science

Check out Current Science® articles related to this chapter by visiting go.hrw.com. Just type in the keyword HZ5CS09.

Agents of Erosion and Deposition

About the PHOTO

The results of erosion can often be dramatic. For example, this sinkhole formed in a parking lot in Atlanta, Georgia, when water running underground eventually caused the surface of the land to collapse.

PRE-READING ACTIVITY

FOLDNOTES **Layered Book** Before you read the chapter, create the FoldNote entitled "The Layered Book" described in the **Study Skills** section of the Appendix. Label the tabs of the layered book with "Shoreline erosion and deposition," "Wind erosion and deposition," and "Erosion and deposition by ice." As you read the chapter, write information you learn about each category under the appropriate tab.

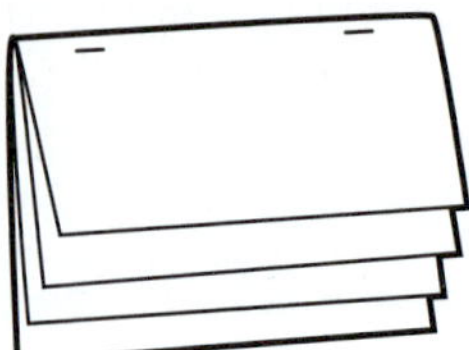

START-UP ACTIVITY

Making Waves

Above ground or below, water plays an important role in the erosion and deposition of rock and soil. A shoreline is a good example of how water shapes the Earth's surface by erosion and deposition. Did you know that shorelines are shaped by crashing waves? Build a model shoreline, and see for yourself!

Procedure

1. Make a shoreline by adding **sand** to one end of a **washtub**. Fill the washtub with **water** to a depth of 5 cm. Sketch the shoreline profile (side view), and label it "A."
2. Place a **block** at the end of the washtub opposite the beach.
3. Move the block up and down very slowly to create small waves for 2 min. Sketch the new shoreline profile, and label it "B."
4. Now, move the block up and down more rapidly to create large waves for 2 min. Sketch the new shoreline profile, and label it "C."

Analysis

1. Compare the three shoreline profiles. What is happening to the shoreline?
2. How do small waves and large waves erode the shoreline differently?

SECTION 1

Shoreline Erosion and Deposition

READING WARM-UP

Objectives

- Explain how energy from waves affects a shoreline.
- Identify six shoreline features created by wave erosion.
- Explain how wave deposits form beaches.
- Describe how sand moves along a beach.

Terms to Learn

shoreline
beach

READING STRATEGY

Reading Organizer As you read this section, create an outline of the section. Use the headings from the section in your outline.

Think about the last time you were at a beach. Where did all of the sand come from?

Two basic ingredients are necessary to make sand: rock and energy. The rock is usually available on the shore. The energy is provided by waves that travel through water. When waves crash into rocks over long periods of time, the rocks are broken down into smaller and smaller pieces until they become sand.

As you read on, you will learn how wave erosion and deposition shape the shoreline. A **shoreline** is simply the place where land and a body of water meet. Waves usually play a major role in building up and breaking down the shoreline.

Wave Energy

As the wind moves across the ocean surface, it produces ripples called *waves*. The size of a wave depends on how hard the wind is blowing and how long the wind blows. The harder and longer the wind blows, the bigger the wave.

The wind that results from summer hurricanes and severe winter storms produces large waves that cause dramatic shoreline erosion. Waves may travel hundreds or even thousands of kilometers from a storm before reaching the shoreline. Some of the largest waves to reach the California coast are produced by storms as far away as Australia. So, the California surfer in **Figure 1** can ride a wave that formed on the other side of the Pacific Ocean!

Figure 1 *Waves produced by storms on the other side of the Pacific Ocean propel this surfer toward a California shore.*

Wave Trains

When you drop a pebble into a pond, is there just one ripple? Of course not. Waves, like ripples, don't move alone. As shown in **Figure 2,** waves travel in groups called *wave trains*. As wave trains move away from their source, they travel through the ocean water uninterrupted. But when waves reach shallow water, the bottom of the wave drags against the sea floor, slowing the wave down. The upper part of the wave moves more rapidly and grows taller. When the top of the wave becomes so tall that it cannot support itself, it begins to curl and break. These breaking waves are known as *surf*. Now you know how surfers got their name. The *wave period* is the time interval between breaking waves. Wave periods are usually 10 to 20 s long.

Figure 2 *Because waves travel in wave trains, they break at regular intervals.*

The Pounding Surf

Look at **Figure 3,** and you will get an idea of how sand is made. A tremendous amount of energy is released when waves break. A crashing wave can break solid rock and throw broken rocks back against the shore. As the rushing water in breaking waves enters cracks in rock, it helps break off large boulders and wash away fine grains of sand. The loose sand picked up by waves wears down and polishes coastal rocks. As a result of these actions, rock is broken down into smaller and smaller pieces that eventually become sand.

shoreline the boundary between land and a body of water

Reading Check **How do waves help break down rock into sand?** (*See the Appendix for answers to Reading Checks.*)

Counting Waves

If the wave period is 10 s, approximately how many waves reach a shoreline in a day? (Hint: Calculate how many waves occur in an hour, and multiply that number by the number of hours in a day.)

Figure 3 *Breaking waves crash against the rocky shore, releasing their energy.*

Wave Erosion

Wave erosion produces a variety of features along a shoreline. *Sea cliffs* are formed when waves erode and undercut rock to produce steep slopes. Waves strike the base of the cliff, which wears away the soil and rock and makes the cliff steeper. The rate at which the sea cliffs erode depends on the hardness of the rock and the energy of the waves. Sea cliffs made of hard rock, such as granite, erode very slowly. Sea cliffs made of soft rock, such as shale, erode more rapidly, especially during storms.

Figure 4 **Coastal Landforms Created by Wave Erosion**

Shaping a Shoreline

Much of the erosion responsible for landforms you might see along the shoreline takes place during storms. Large waves generated by storms release far more energy than normal waves do. This energy is so powerful that it is capable of removing huge chunks of rock. **Figure 4** shows some of the major landscape features that result from wave erosion.

Reading Check **Why are large waves more capable of removing large chunks of rock from a shoreline than normal waves are?**

Figure 5 *Beaches are made of different types of material deposited by waves.*

Wave Deposits

Waves carry a variety of materials, including sand, rock fragments, dead coral, and shells. Often, this material is deposited on a shoreline, where it forms a beach.

Beaches

You would probably recognize a beach if you saw one. However, scientifically speaking, a **beach** is any area of the shoreline made up of material deposited by waves. Some beach material is also deposited by rivers.

Compare the beaches shown in **Figure 5.** Notice that the colors and textures vary. They vary because the type of material found on a beach depends on its source. Light-colored sand is the most common beach material. Much of this sand comes from the mineral quartz. But not all beaches are made of light-colored sand. For example, on many tropical islands, such as the Virgin Islands, beaches are made of fine, white coral material. Some Florida beaches are made of tiny pieces of broken seashells. Black sand beaches in Hawaii are made of eroded volcanic lava. In areas where stormy seas are common, beaches are made of pebbles and boulders.

Reading Check Where does beach material come from?

beach an area of the shoreline made up of material deposited by waves

Wave Angle and Sand Movement

The movement of sand along a beach depends on the angle at which the waves strike the shore. Most waves approach the beach at a slight angle and retreat in a direction more perpendicular to the shore. This movement of water is called a longshore current. A *longshore current* is a water current that moves the sand in a zigzag pattern along the beach, as you can see in **Figure 6.**

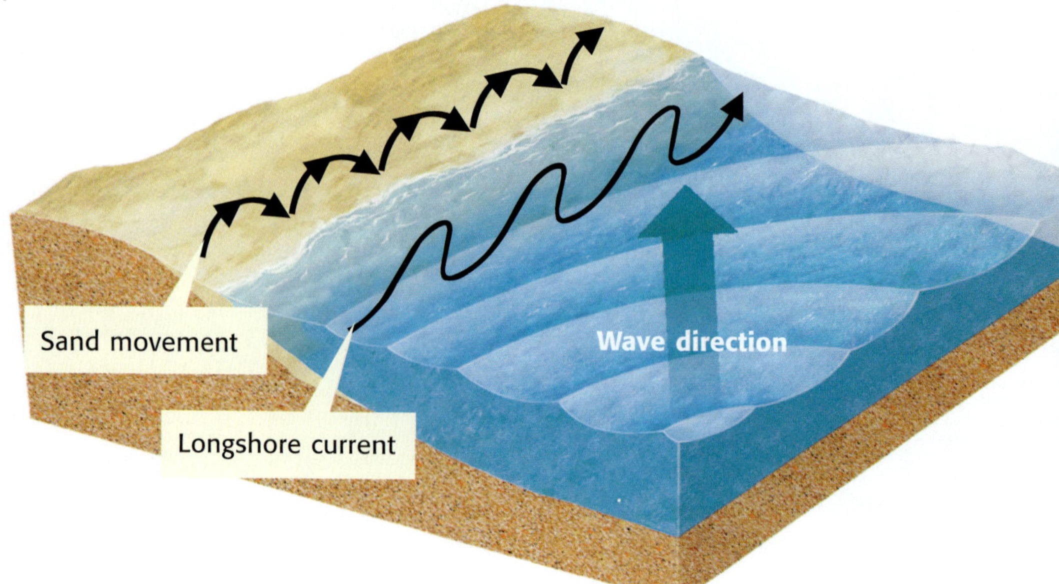

Figure 6 *When waves strike the shoreline at an angle, sand migrates along the beach in a zigzag path.*

Offshore Deposits

Waves moving at an angle to the shoreline push water along the shore and create longshore currents. When waves erode material from the shoreline, longshore currents can transport and deposit this material offshore, which creates landforms in open water. A *sandbar* is an underwater or exposed ridge of sand, gravel, or shell material. A *barrier spit* is an exposed sandbar that is connected to the shoreline. Cape Cod, Massachusetts, shown in **Figure 7,** is an example of a barrier spit. A barrier island is a long, narrow island usually made of sand that forms offshore parallel to the shoreline.

Figure 7 *A barrier spit, such as Cape Cod, Massachusetts, occurs when an exposed sandbar is connected to the shoreline.*

SECTION Review

Summary

- As waves break against a shoreline, rock is broken down into sand.
- Six shoreline features created by wave erosion include sea cliffs, sea stacks, sea caves, sea arches, headlands, and wave-cut terraces.
- Beaches are made from material deposited by waves.
- Longshore currents cause sand to move in a zigzag pattern along the shore.

Using Key Terms

Complete each of the following sentences by choosing the correct term from the word bank.

shoreline beach

1. A ___ is an area made up of material deposited by waves.
2. An area in which land and a body of water meet is a ___.

Understanding Key Ideas

3. Which of the following is a result of wave deposition?
   a. sea arch
   b. sea cave
   c. barrier spit
   d. headland
4. How do wave deposits affect a shoreline?
5. Describe how sand moves along a beach.
6. What are six shoreline features created by wave erosion?
7. How can the energy of waves traveling through water affect a shoreline?
8. Would a small wave or a large wave have more energy? Explain your answer.

Math Skills

9. Imagine that there is a large boulder on the edge of a shoreline. If the wave period is 15 s long, how many times is the boulder hit in a year?

Critical Thinking

10. **Applying Concepts** Not all beaches are made from light-colored sand. Explain why this statement is true.
11. **Making Inferences** How can severe storms over the ocean affect shoreline erosion and deposition?
12. **Making Predictions** How could a headland change in 250 years? Describe some of the features that may form.

SECTION 2

Wind Erosion and Deposition

Have you ever been working outside and had a gusty wind blow an important stack of papers all over the place?

READING WARM-UP

Objectives

- Explain why some areas are more affected by wind erosion than other areas are.
- Describe the process of saltation.
- Identify three landforms that result from wind erosion and deposition.
- Explain how dunes move.

Terms to Learn

saltation
deflation
abrasion
loess
dune

READING STRATEGY

Reading Organizer As you read this section, make a table comparing deflation and abrasion.

Do you remember how fast and far the papers traveled and how long it took to pick them up? Every time you caught up with them, they were on the move again. If this has happened to you, then you have seen how wind erosion works. As an agent of erosion, the wind removes soil, sand, and rock particles and transports them from one place to another.

Certain locations are more vulnerable to wind erosion than others. An area with little plant cover can be severely affected by wind erosion because plant roots anchor sand and soil in place. Deserts and coastlines that are made of fine, loose rock material and have little plant cover are shaped most dramatically by the wind.

The Process of Wind Erosion

Wind moves material in different ways. In areas where strong winds occur, material is moved by saltation. **Saltation** is the skipping and bouncing movement of sand-sized particles in the direction the wind is blowing. As you can see in **Figure 1,** the wind causes the particles to bounce. When moving sand grains knock into one another, some grains bounce up in the air, fall forward, and strike other sand grains. These impacts cause other grains to roll and bounce forward.

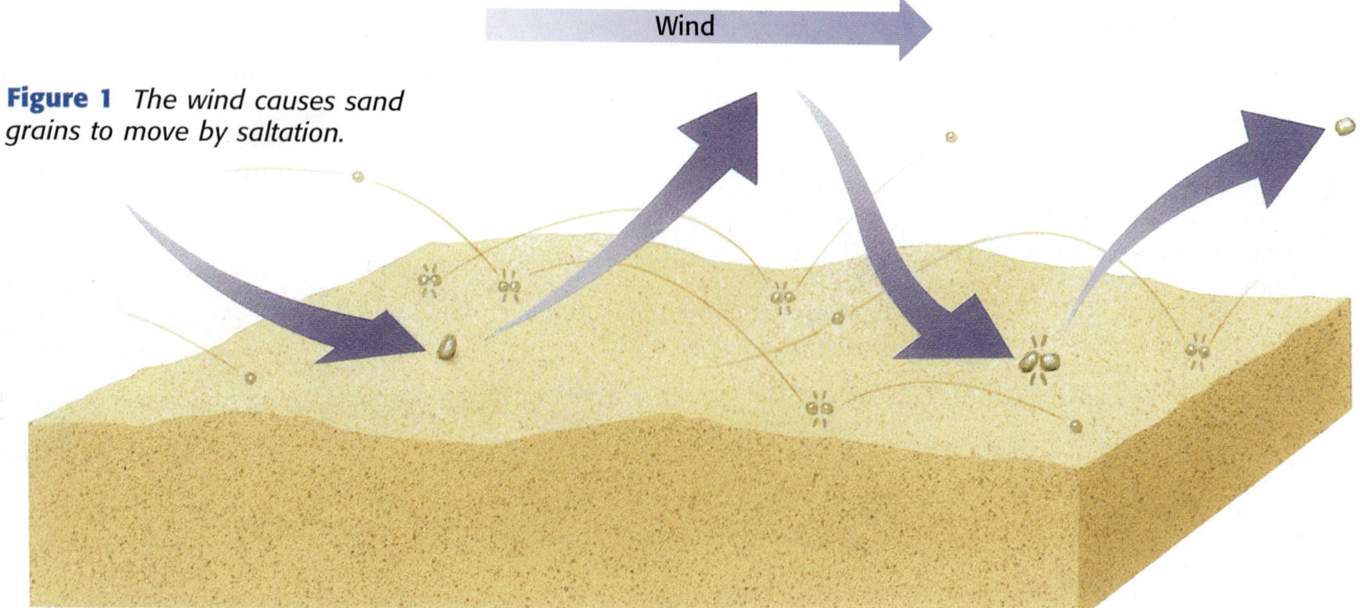

Figure 1 *The wind causes sand grains to move by saltation.*

Figure 2 *Desert pavement, such as that found in the Painted Desert in Arizona, forms when wind removes all the fine materials.*

Deflation

The removal of fine sediment by wind is called **deflation.** During deflation, wind removes the top layer of fine sediment or soil and leaves behind rock fragments that are too heavy to be lifted by the wind. Deflation may cause *desert pavement,* which is a surface consisting of pebbles and small broken rocks. An example of desert pavement is shown in **Figure 2.**

Have you ever blown on a layer of dust while cleaning off a dresser? If you have, you may have noticed that in addition to your face getting dirty, a little scooped-out depression formed in the dust. Similarly, in areas where there is little vegetation, the wind may scoop out depressions in the landscape. These depressions are called *deflation hollows.*

Reading Check **Where do deflation hollows form?** (*See the Appendix for answers to Reading Checks.*)

saltation the movement of sand or other sediments by short jumps and bounces that is caused by wind or water

deflation a form of wind erosion in which fine, dry soil particles are blown away

abrasion the grinding and wearing away of rock surfaces through the mechanical action of other rock or sand particles

Abrasion

The grinding and wearing down of rock surfaces by other rock or sand particles is called **abrasion.** Abrasion commonly happens in areas where there are strong winds, loose sand, and soft rocks. The blowing of millions of sharp sand grains creates a sandblasting effect. This effect helps to erode, smooth, and polish rocks.

Making Desert Pavement

1. Spread a mixture of **dust, sand,** and **gravel** on an **outdoor table.**
2. Place an **electric fan** at one end of the table.
3. Put on **safety goggles** and a **filter mask.** Aim the fan across the sediment. Start the fan on its lowest speed. Record your observations.
4. Turn the fan to a medium speed. Record your observations.
5. Finally, turn the fan to a high speed to imitate a desert windstorm. Record your observations.
6. What is the relationship between the wind speed and the size of the sediment that is moved?
7. Does the remaining sediment fit the definition of desert pavement?

Wind-Deposited Materials

Much like rivers, the wind also carries sediment. And just as rivers deposit their loads, the wind eventually drops all the material it carries. The amount and the size of particles the wind can carry depend on the wind speed. The faster the wind blows, the more material and the heavier the particles it can carry. As wind speed slows, heavier particles are deposited first.

CONNECTION TO Language Arts

WRITING SKILL **The Dust Bowl** During the 1930s, a severe drought occurred in a section of the Great Plains that became known as the *Dust Bowl*. The wind carried so much dust that some cities left street lights on during the day. Research the Dust Bowl, and describe it in a series of three journal entries written from the perspective of a farmer.

Loess

Wind can deposit extremely fine material. Thick deposits of this windblown, fine-grained sediment are known as **loess** (LOH ES). Loess feels like the talcum powder a person may use after a shower.

Because wind carries fine-grained material much higher and farther than it carries sand, loess deposits are sometimes found far away from their source. Many loess deposits came from glacial sources during the last Ice Age. In the United States, loess is present in the Midwest, along the eastern edge of the Mississippi Valley, and in eastern Oregon and Washington.

loess very fine sediments deposited by the wind

dune a mound of wind-deposited sand that keeps its shape even though it moves

Dunes

When the wind hits an obstacle, such as a plant or a rock, the wind slows down. As it slows, the wind deposits, or drops, the heavier material. The material collects, which creates an additional obstacle. This obstacle causes even more material to be deposited, forming a mound. Eventually, the original obstacle becomes buried. The mounds of wind-deposited sand are called **dunes.** Dunes are common in sandy deserts and along the sandy shores of lakes and oceans. **Figure 3** shows a large dune in a desert area.

Figure 3 *Dunes migrate in the direction of the wind.*

The Movement of Dunes

Dunes tend to move in the direction of strong winds. Different wind conditions produce dunes in various shapes and sizes. A dune usually has a gently sloped side and a steeply sloped side, or *slip face,* as shown in **Figure 4.** In most cases, the gently sloped side faces the wind. The wind is constantly transporting material up this side of the dune. As sand moves over the crest, or peak, of the dune, it slides down the slip face, creating a steep slope.

Reading Check In what direction do dunes move?

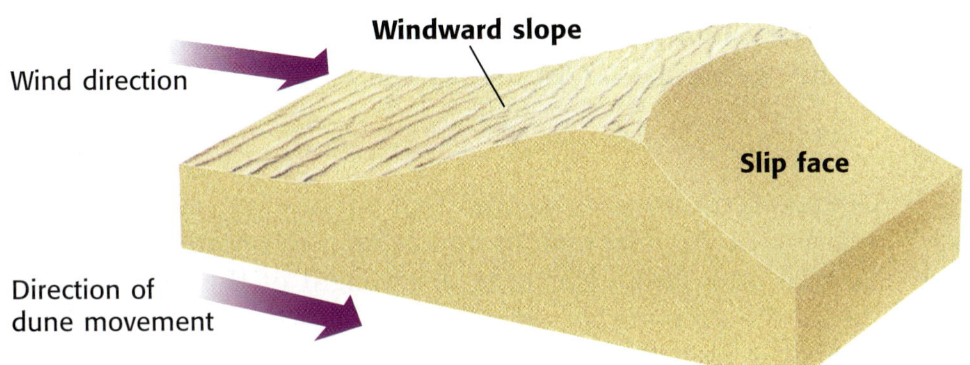

Figure 4 *Dunes are formed from material deposited by wind.*

SECTION Review

Summary

- Areas with little plant cover and desert areas covered with fine rock material are more vulnerable than other areas to wind erosion.
- Saltation is the process in which sand-sized particles move in the direction of the wind.
- Three landforms that are created by wind erosion and deposition are desert pavement, deflation hollows, and dunes.
- Dunes move in the direction of the wind.

Using Key Terms

In each of the following sentences, replace the incorrect term with the correct term from the word bank.

dune	saltation
deflation	abrasion

1. Deflation hollows are mounds of wind-deposited sand.
2. The removal of fine sediment by wind is called abrasion.

Understanding Key Ideas

3. Which of the following landforms is the result of wind deposition?
   a. deflation hollow
   b. desert pavement
   c. dune
   d. abrasion
4. Describe how material is moved in areas where strong winds blow.
5. Explain the process of abrasion.

Math Skills

6. If a dune moves 40 m per year, how far does it move in 1 day?

Critical Thinking

7. **Identifying Relationships** Explain the relationship between plant cover and wind erosion.
8. **Applying Concepts** If you climbed up the steep side of a sand dune, is it likely that you traveled in the direction the wind was blowing?

SECTION 3

Erosion and Deposition by Ice

READING WARM-UP

Objectives

- Explain the difference between alpine glaciers and continental glaciers.
- Describe two ways in which glaciers move.
- Identify five landscape features formed by alpine glaciers.
- Identify four types of moraines.

Terms to Learn

glacier
glacial drift
till
stratified drift

READING STRATEGY

Discussion Read this section silently. Write down questions that you have about this section. Discuss your questions in a small group.

Can you imagine an ice cube that is the size of a football stadium? Well, glaciers can be even bigger than that.

A **glacier** is an enormous mass of moving ice. Because glaciers are very heavy and have the ability to move across the Earth's surface, they are capable of eroding, moving, and depositing large amounts of rock materials. And while you will never see a glacier chilling a punch bowl, you might one day visit some of the spectacular landscapes carved by glacial activity!

Glaciers—Rivers of Ice

Glaciers form in areas so cold that snow stays on the ground year-round. In polar regions and at high elevations, snow piles up year after year. Over time, the weight of the snow on top causes the deep-packed snow to become ice crystals. These ice crystals eventually form a giant ice mass. Because glaciers are so massive, the pull of gravity causes them to flow slowly, like "rivers of ice." In this section, you will learn about two main types of glaciers, alpine and continental.

Alpine Glaciers

Alpine glaciers form in mountainous areas. One common type of alpine glacier is a valley glacier. Valley glaciers form in valleys originally created by stream erosion. As these glaciers slowly flow downhill, they widen and straighten the valleys into broad U shapes as shown in **Figure 1.**

glacier a large mass of moving ice

Reading Check **Where do alpine glaciers form?** (*See the Appendix for answers to Reading Checks.*)

Figure 1 *Alpine glaciers start as snowfields in mountainous areas.*

Figure 2 *Eleven U.S. states were covered by ice during the last glacial ice period. Because much of the Earth's water was frozen in glaciers, sea levels fell. Blue lines show the coastline at that time.*

Continental Glaciers

Not all glaciers are true "rivers of ice." In fact, some glaciers spread across entire continents. These glaciers, called *continental glaciers*, are huge, continuous masses of ice. The largest continental glacier in the world covers almost all of Antarctica. This ice sheet is approximately one and a half times the size of the United States. It is so thick—more than 4,000 m in places—that it buries everything but the highest mountain peaks.

Glaciers on the Move

When enough ice builds up on a slope, the ice begins to move downhill. Thick glaciers move faster than thin glaciers, and the steeper the slope is, the faster the glaciers will move. Glaciers move in two ways: by sliding and by flowing. A glacier slides when its weight causes the ice at the bottom of the glacier to melt. As the water from a melting ice cube causes the ice cube to travel across a table, the water from the melting ice causes a glacier to move forward. A glacier also flows slowly as ice crystals within the glacier slip over each other. Think of placing a deck of cards on a table and then tilting the table. The top cards will slide farther than the lower cards. Similarly, the upper part of the glacier flows faster than the base.

Glacier movement is affected by climate. As the Earth cools, glaciers grow. About 10,000 years ago, a continental glacier covered most of North America, as shown in **Figure 2.** In some places, the ice sheet was several kilometers thick!

SCHOOL to HOME

The *Titanic*

WRITING SKILL An area where an ice sheet is resting on open water is called an *ice shelf.* When pieces of the ice shelf break off, they are called *icebergs.* How far do you think the iceberg that struck the *Titanic* drifted before the two met that fateful night in 1912? Together with a parent, plot on a map of the North Atlantic Ocean the route of the *Titanic* from Southampton, England, to New York. Then, plot a possible route of the drifting iceberg from Greenland to where the ship sank, just south of the Canadian island province of Newfoundland. Describe your findings in your **science journal.**

ACTIVITY

Speed of a Glacier

An alpine glacier is estimated to be moving forward at 5 m per day. Calculate how long the ice will take to reach a road and campground located 0.5 km from the front of the advancing glacier. (Hint: 1 km = 1,000 m)

Landforms Carved by Glaciers

Continental glaciers and alpine glaciers produce landscapes that are very different from one another. Continental glaciers smooth the landscape by scraping and eroding features that existed before the ice appeared. Alpine glaciers carve out rugged features in the mountain rocks through which they flow. **Figure 3** shows the very different landscapes that each type of glacier produces.

Alpine glaciers, such as those in the Rocky Mountains and the Alps, carve out large amounts of rock material and create spectacular landforms. **Figure 4** shows the kinds of landscape features that are sculpted by alpine glaciers.

Figure 3 Landscapes Created by Glaciers

Continental glaciers smooth and flatten the landscape.

Alpine glaciers carved out this rugged landscape.

Figure 4 Landscape Features Carved by Alpine Glaciers

Types of Glacial Deposits

glacial drift the rock material carried and deposited by glaciers

till unsorted rock material that is deposited directly by a melting glacier

stratified drift a glacial deposit that has been sorted and layered by the action of streams or meltwater

As a glacier melts, it drops all the material it is carrying. **Glacial drift** is the general term used to describe all material carried and deposited by glaciers. Glacial drift is divided into two main types, *till* and *stratified drift*.

Till Deposits

Unsorted rock material that is deposited directly by the ice when it melts is called **till.** *Unsorted* means that the till is made up of rock material of different sizes—from large boulders to fine sediment. When the glacier melts, the unsorted material is deposited on the surface of the ground.

The most common till deposits are *moraines*. Moraines generally form ridges along the edges of glaciers. Moraines are produced when glaciers carry material to the front of and along the sides of the ice. As the ice melts, the sediment and rock it is carrying are dropped, which forms different types of moraines. The various types of moraines are shown in **Figure 5.**

Figure 5 **Types of Moraines**

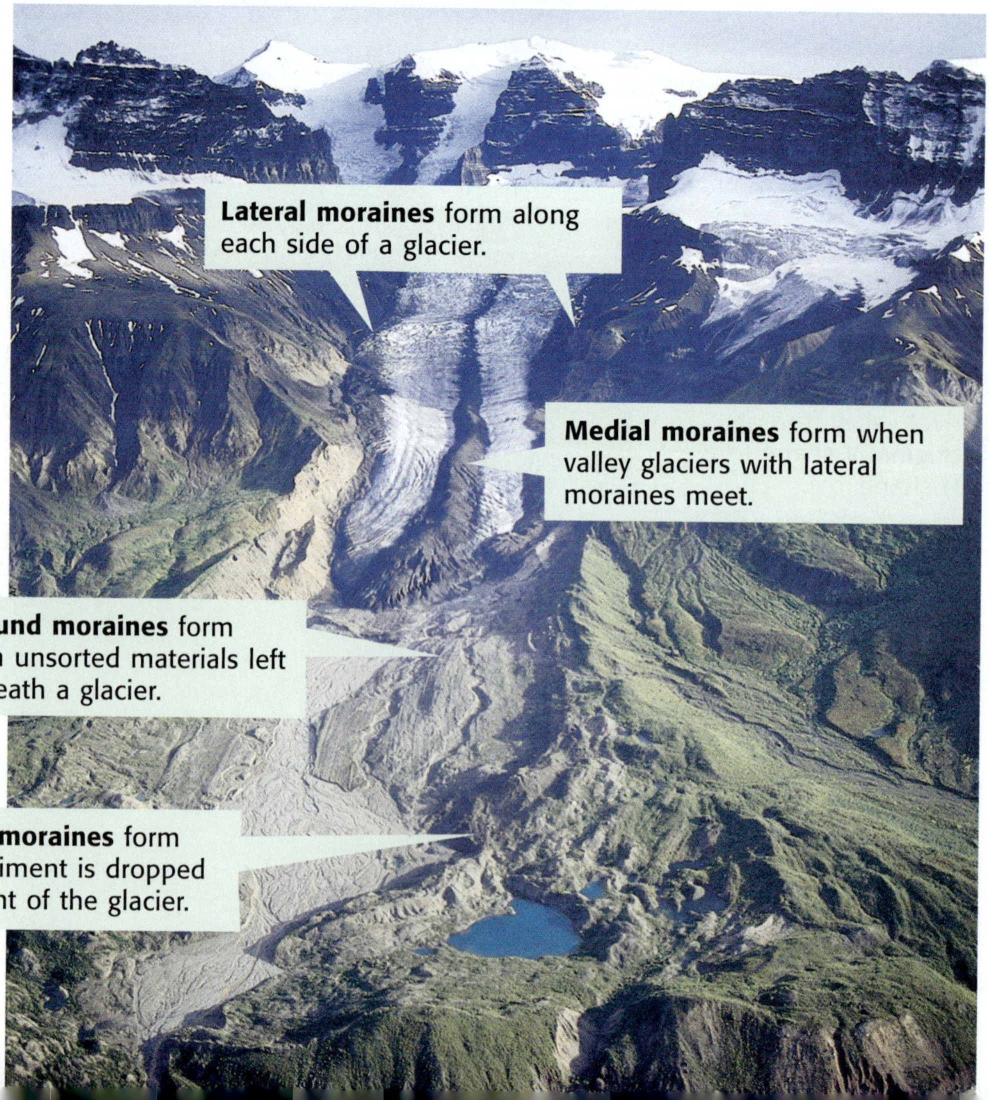

Stratified Drift

When a glacier melts, streams form that carry rock material away from the shrinking glacier. A glacial deposit that is sorted into layers based on the size of the rock material is called **stratified drift.** Streams carry sorted material and deposit it in front of the glacier in a broad area called an *outwash plain*. Sometimes, a block of ice is left in the outwash plain when a glacier retreats. As the ice melts, sediment builds up around the block of ice, and a depression called a *kettle* forms. Kettles commonly fill with water to form lakes or ponds, as **Figure 6** shows.

Reading Check **Explain the difference between a till deposit and stratified drift.**

Figure 6 *Kettle lakes form in outwash plains and are common in states such as Minnesota.*

SECTION Review

Summary

- Alpine glaciers form in mountainous areas. Continental glaciers spread across entire continents.
- Glaciers can move by sliding or by flowing.
- Alpine glaciers can carve cirques, arêtes, horns, U-shaped valleys, and hanging valleys.
- Two types of glacial drift are till and stratified drift.
- Four types of moraines are lateral, medial, ground, and terminal moraines.

Using Key Terms

Complete each of the following sentences by choosing the correct term from the word bank.

glacial drift glacier
stratified drift till

1. A glacial deposit that is sorted into layers based on the size of the rock material is called ___.
2. ___ is all of the material carried and deposited by glaciers.
3. Unsorted rock material that is deposited directly by the ice when it melts is ___.
4. A ___ is an enormous mass of moving ice.

Understanding Key Ideas

5. Which of the following is not a type of moraine?
   a. lateral
   b. horn
   c. ground
   d. medial
6. Explain the difference between alpine and continental glaciers.
7. Name five landscape features formed by alpine glaciers.
8. Describe two ways in which glaciers move.

Math Skills

9. A recent study shows that a glacier in Alaska is melting at a rate of 23 ft per year. At what rate is the glacier melting in meters? (Hint: 1 ft = 0.3 m)

Critical Thinking

10. **Analyzing Ideas** Explain why continental glaciers smooth the landscape and alpine glaciers create a rugged landscape.
11. **Applying Concepts** How can a glacier deposit both sorted and unsorted material?
12. **Applying Concepts** Why are glaciers such effective agents of erosion and deposition?

SECTION 4

The Effect of Gravity on Erosion and Deposition

READING WARM-UP

Objectives

- Explain the role of gravity as an agent of erosion and deposition.
- Explain how angle of repose is related to mass movement.
- Describe four types of rapid mass movement.
- Describe three factors that affect creep.

Terms to Learn

mass movement
rock fall
landslide
mudflow
creep

READING STRATEGY

Prediction Guide Before reading this section, write the title of each heading in this section. Next, under each heading, write what you think you will learn.

Did you know that the Appalachian Mountains may have once been almost five times as tall as they are now? Why are they shorter now? Part of the answer lies in the effect that gravity has on all objects on Earth.

Although you can't see it, the force of gravity is also an agent of erosion and deposition. Gravity not only influences the movement of water and ice but also causes rocks and soil to move downslope. **Mass movement** is the movement of any material, such as rock, soil, or snow, downslope. Whether mass movement happens rapidly or slowly, it plays a major role in shaping the Earth's surface.

mass movement a movement of a section of land down a slope

Angle of Repose

If dry sand is piled up, it will move downhill until the slope becomes stable. The *angle of repose* is the steepest angle, or slope, at which loose material will not slide downslope. This is demonstrated in **Figure 1.** The angle of repose is different for each type of surface material. Characteristics of the surface material, such as its size, weight, shape, and moisture level, determine at what angle the material will move downslope.

Figure 1 *If the slope on which material rests is less than the angle of repose, the material will stay in place. If the slope is greater than the angle of repose, the material will move downslope.*

Rapid Mass Movement

The most destructive mass movements happen suddenly and rapidly. Rapid mass movement can be very dangerous and can destroy everything in its path.

Rock Falls

While driving along a mountain road, you may have noticed signs along the road that warn of falling rocks. A **rock fall** happens when loose rocks fall down a steep slope. Steep slopes are sometimes created to make room for a road in mountainous areas. Loosened and exposed rocks above the road tend to fall as a result of gravity. The rocks in a rock fall can range in size from small fragments to large boulders.

Landslides

Another type of rapid mass movement is a landslide. A **landslide** is the sudden and rapid movement of a large amount of material downslope. A *slump,* shown in **Figure 2,** is the most common type of landslide. Slumping occurs when a block of material moves downslope over a curved surface. Heavy rains, deforestation, construction on unstable slopes, and earthquakes increase the chances that a landslide will happen. **Figure 3** shows a landslide in India.

Reading Check What is a slump? (*See the Appendix for answers to Reading Checks.*)

Figure 2 *A slump is a type of landslide that occurs when a block of land becomes detached and slides downhill.*

rock fall a group of loose rocks that fall down a steep slope

landslide the sudden movement of rock and soil down a slope

Figure 3 *This landslide in Bombay, India, happened after heavy monsoon rains.*

Figure 4 *This photo shows one of the many mudflows that have occurred in California during rainy winters.*

Mudflows

A rapid movement of a large mass of mud is a **mudflow.** Mudflows happen when a large amount of water mixes with soil and rock. The water causes the slippery mass of mud to flow rapidly downslope. Mudflows commonly happen in mountainous regions when a long dry season is followed by heavy rains. Deforestation and the removal of ground cover can often result in devastating mudflows. As you can see in **Figure 4,** a mudflow can carry trees, houses, cars, and other objects that lie in its path.

mudflow the flow of a mass of mud or rock and soil mixed with a large amount of water

Lahars

Volcanic eruptions or heavy rains on volcanic ash can produce some of the most dangerous mudflows. Mudflows of volcanic origin are called *lahars*. Lahars can travel at speeds greater than 80 km/h and can be as thick as cement. On volcanoes with snowy peaks, an eruption can suddenly melt a great amount of ice. The water from the ice liquefies the soil and volcanic ash to produce a hot mudflow that rushes downslope. **Figure 5** shows the effects of a massive lahar in Japan.

Reading Check Explain how a lahar occurs.

Figure 5 *This lahar overtook the city of Kyushu in Japan.*

Slow Mass Movement

Sometimes, you don't even notice mass movement happening. Although rapid mass movements are visible and dramatic, slow mass movements happen a little at a time. However, because slow mass movements occur more frequently, more material is moved collectively over time.

Figure 6 *Bent tree trunks are evidence that creep is happening.*

Creep

Even though most slopes appear to be stable, they are actually undergoing slow mass movement, as shown in **Figure 6.** The extremely slow movement of material downslope is called **creep.** Many factors contribute to creep. Water loosens soil and allows it to move freely. In addition, plant roots act as a wedge that forces rocks and soil particles apart. Burrowing animals, such as gophers and groundhogs, also loosen rock and soil particles. In fact, rock and soil on every slope travels slowly downhill.

creep the slow downhill movement of weathered rock material

SECTION Review

Summary

- Gravity causes rocks and soil to move downslope.
- If the slope on which material rests is greater than the angle of repose, mass movement will occur.
- Four types of rapid mass movement are rock falls, landslides, mudflows, and lahars.
- Water, plant roots, and burrowing animals can cause creep.

Using Key Terms

Complete each of the following sentences by choosing the correct term from the word bank.

creep	mass movement
mudflow	rock fall

1. A ___ occurs when a large amount of water mixes with soil and rock.
2. The extremely slow movement of material downslope is called ___.

Understanding Key Ideas

3. Which of the following is a factor that affects creep?
   a. water
   b. burrowing animals
   c. plant roots
   d. All of the above.
4. How is the angle of repose related to mass movement?

Math Skills

5. If a lahar is traveling at 80 km/h, how long will it take the lahar to travel 20 km?

Critical Thinking

6. **Identifying Relationships** Which types of mass movement are most dangerous to humans? Explain your answer.
7. **Making Inferences** How does deforestation increase the likelihood of mudflows?

For a variety of links related to this chapter, go to www.scilinks.org

Topic: Mass Movements
SciLinks code: HSM0917

Model-Making Lab

Gliding Glaciers

OBJECTIVES

Build a model of a glacier.

Demonstrate the effects of glacial erosion by various materials.

Observe the effect of pressure on the melting rate of a glacier.

MATERIALS

- brick (3)
- clay, modeling (2 lb)
- container, empty large margarine (3)
- freezer
- graduated cylinder, 50 mL
- gravel (1 lb)
- pan, aluminum rectangular (3)
- rolling pin, wood
- ruler, metric
- sand (1 lb)
- stopwatch
- towel, small hand
- water

A glacier is a large, moving mass of ice. Glaciers are responsible for shaping many of Earth's natural features. Glaciers are set in motion by the pull of gravity and by the gradual melting of the glacier. As a glacier moves, it changes the landscape by eroding the surface over which it passes.

Part A: Getting in the Groove

Procedure

The material that is carried by a glacier erodes Earth's surface by gouging out grooves called *striations*. Different materials have varying effects on the landscape. In this activity, you will create a model glacier with which to demonstrate the effects of glacial erosion by various materials.

1. Fill one margarine container with sand to a depth of 1 cm. Fill another margarine container with gravel to a depth of 1 cm. Leave the third container empty. Fill the containers with water.
2. Put the three containers in a freezer, and leave them there overnight.
3. Retrieve the containers from the freezer, and remove the three ice blocks from the containers.
4. Use a rolling pin to flatten the modeling clay.
5. Hold the ice block from the third container firmly with a towel, and press as you move the ice along the length of the clay. Do this three times. In a notebook, sketch the pattern that the ice block makes in the clay.

6. Repeat steps 4 and 5 using the ice block that contains sand.
7. Repeat steps 4 and 5 using the ice block that contains gravel.

Analyze the Results

1. **Describing Events** Did any material from the clay become mixed with the material in the ice blocks? Explain.
2. **Describing Events** Was any material from the ice blocks deposited on the clay surface? Explain.
3. **Examining Data** What glacial features are represented in your clay model?

Draw Conclusions

4. **Evaluating Data** Compare the patterns formed by the three model glaciers. Do the patterns look like features carved by alpine glaciers or by continental glaciers? Explain.

Part B: Melting Away

Procedure

As the layers of ice build up and a glacier gets larger, a glacier will eventually begin to melt. The water from the melted ice allows a glacier to move forward. In this activity, you'll explore the effect of pressure on the melting rate of a glacier.

1. If possible, make three identical ice blocks without any sand or gravel in them. If that is not possible, use the ice blocks from Part A. Place one ice block upside down in each pan.
2. Place one brick on top of one of the ice blocks. Place two bricks on top of another ice block. Do not put any bricks on the third ice block.
3. After 15 min, remove the bricks from the ice blocks.
4. Using the graduated cylinder, measure the amount of water that has melted from each ice block.
5. Observe and record your findings.

Analyze the Results

1. **Analyzing Data** Which ice block produced the most water?
2. **Explaining Events** What did the bricks represent?
3. **Analyzing Results** What part of the ice blocks melted first? Explain.

Draw Conclusions

4. **Interpreting Information** How could you relate this investigation to the melting rate of glaciers? Explain.

Applying Your Data

Replace the clay with different materials, such as soft wood or sand. How does each ice block affect the different surface materials? What types of surfaces do the different materials represent?

Chapter Review

USING KEY TERMS

For each pair of terms, explain how the meanings of the terms differ.

1. *shoreline* and *longshore current*
2. *beaches* and *dunes*
3. *deflation* and *saltation*
4. *continental glacier* and *alpine glacier*
5. *stratified drift* and *till*
6. *mudflow* and *creep*

UNDERSTANDING KEY IDEAS

Multiple Choice

7. *Surf* refers to
   - a. large storm waves in the open ocean.
   - b. giant waves produced by hurricanes.
   - c. breaking waves near the shoreline.
   - d. small waves on a calm sea.
8. When waves cut completely through a headland, a ___ is formed.
   - a. sea cave
   - b. sea arch
   - c. wave-cut terrace
   - d. sandbar
9. A narrow strip of sand that is formed by wave deposition and is connected to the shore is called a
   - a. barrier spit.
   - b. sandbar.
   - c. wave-cut terrace.
   - d. headland.
10. A wind-eroded depression is called a
    - a. deflation hollow.
    - b. desert pavement.
    - c. dune.
    - d. dust bowl.
11. What term describes all types of glacial deposits?
    - a. glacial drift
    - b. dune
    - c. till
    - d. outwash
12. Which of the following is NOT a landform created by an alpine glacier?
    - a. cirque
    - b. deflation hollow
    - c. horn
    - d. arête
13. What is the term for a mass movement that is of volcanic origin?
    - a. lahar
    - b. slump
    - c. creep
    - d. rock fall
14. Which of the following is a slow mass movement?
    - a. mudflow
    - b. landslide
    - c. creep
    - d. rock fall

Short Answer

15. Why do waves break when they near the shore?
16. Why are some areas more affected by wind erosion than other areas are?

17 What kind of mass movement happens continuously, day after day?

18 In what direction do sand dunes move?

19 Describe the different types of glacial moraines.

CRITICAL THINKING

20 **Concept Mapping** Use the following terms to create a concept map: *deflation, strong winds, saltation, dune,* and *desert pavement.*

21 **Making Inferences** How do humans increase the likelihood that wind erosion will occur?

22 **Identifying Relationships** If the large ice sheet covering Antarctica were to melt completely, what type of landscape would you expect Antarctica to have?

23 **Applying Concepts** You are a geologist who is studying rock to determine the direction of flow of an ancient glacier. What clues might help you determine the glacier's direction of flow?

24 **Applying Concepts** You are interested in purchasing a home that overlooks the ocean. The home that you want to buy sits atop a steep sea cliff. Given what you have learned about shoreline erosion, what factors would you take into consideration when deciding whether to buy the home?

INTERPRETING GRAPHICS

The graph below illustrates coastal erosion and deposition at an imaginary beach over a period of 8 years. Use the graph below to answer the questions that follow.

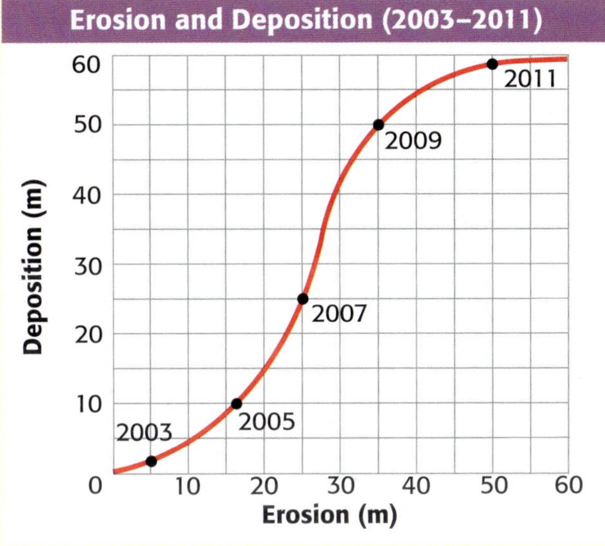

25 What is happening to the beach over time?

26 In what year does the amount of erosion equal the amount of deposition?

27 Based on the erosion and deposition data for 2005, what might happen to the beach in the years that follow 2005?

Standardized Test Preparation

READING

Read each of the passages below. Then, answer the questions that follow each passage.

Passage 1 When you drop a pebble into a pond, is there just one ripple? Of course not. Waves, like ripples, don't move alone. Waves travel in groups called wave trains. As wave trains move away from their sources, they travel through the ocean water uninterrupted. But when waves reach shallow water, they change form because the ocean floor crowds the lower part of the wave. As a result, the waves get closer together and taller.

1. In this passage, what does the word *uninterrupted* mean?

A not continuous
B not broken
C broken again
D not interpreted

2. In this passage, what does the word *train* mean?

F to teach someone a skill
G the part of a gown that trails behind the person who is wearing the gown
H a series of moving things
I a series of railroad cars

3. According to the passage, what is the cause of taller waves?

A shallow water
B deep ocean water
C rippling
D wave trains

4. If certain waves are short and far apart, which of the following can be concluded?

F The waves are approaching the shore.
G The waves are moving toward their source.
H The waves were interrupted.
I The waves are in deep ocean water.

Passage 2 Winter storms create powerful waves that crash into cliffs and break off pieces of rock that fall into the ocean. On February 8, 1998, unusually large waves crashed against the cliffs along Broad Beach Road in Malibu, California. Eventually, the ocean-eroded cliffs buckled, which caused a landslide. One house collapsed into the ocean, and two more houses dangled on the edge of the cliff's newly eroded face. Powerful waves, buckled cliffs, and landslides are part of the ongoing natural process of coastal erosion that is taking place along the California shoreline and along similar shorelines throughout the world.

1. In this passage, what does *buckled* mean?

A tightened
B collapsed
C formed
D heated up

2. Which of the following describes how this coastal area was damaged?

F The area was damaged by collapsing houses.
G The area was damaged an earthquake.
H The area was damaged by ocean currents.
I The area was damaged by unusually large waves produced by a winter storm.

3. Which of the following can be concluded from this passage?

A This area may have landslides in the future.
B This area is safe from future landslides.
C This type of landslide is common only to the California coastline.
D Erosion in this area happens very rarely.

INTERPRETING GRAPHICS

Use each figure below to answer the questions that follow each figure.

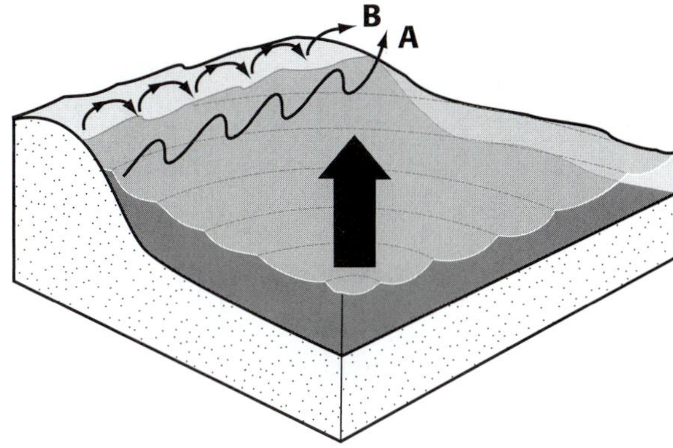

1. In the illustration, what does A label?
   A wave direction
   B wave amplitude
   C wavelength
   D a longshore current

2. In the illustration, what does B label?
   F wave direction
   G wave period
   H the movement of sand
   I a longshore current

3. What process created the landform in the illustration above?
   A erosion by waves
   B saltation
   C abrasion
   D deposition by waves

MATH

Read each question below, and choose the best answer.

1. Wind erosion caused a deflation hollow that was circular in shape. The hollow is 100 m wide. What is the circumference of this deflation hollow?
   A 31.4 m
   B 62.8 m
   C 314 m
   D 628 m

2. A homeowner needs to buy and plant 28 trees to prevent wind erosion. Each tree costs $29.99. What is a reasonable estimate for the total cost of these trees before tax?
   F a little more than $200
   G a little less than $600
   H a little less than $900
   I a little more than $1,000

Use the equation below to answer the questions that follow.

$$\frac{\textit{number of waves}}{\textit{per minute}} = \frac{60\text{ s}}{\textit{wave period}\text{ (s)}}$$

3. If the wave period is 15 s, how many waves occur in 1 min?
   A 4
   B 60
   C 75
   D 240

4. If the wave period is 30 s, how many waves occur in 1 min?
   F 1
   G 2
   H 3
   I 5

5. If 480 waves broke in 40 min, what is the wave period?
   A 5 s
   B 12 s
   C 15 s
   D 20 s

Science in Action

Weird Science

Long-Runout Landslides

At 4:10 A.M. on April 29, 1903, the town of Frank, Canada, was changed forever when disaster struck without warning. An enormous chunk of limestone fell suddenly from the top of nearby Turtle Mountain. In less than two minutes, the huge mass of rock buried most of the town! Landslides such as the Frank landslide are now known as *long-runout landslides*. Most landslides travel a horizontal distance that is less than twice the vertical distance that they have fallen. But long-runout landslides carry enormous amounts of rock and thus can travel many times farther than they fall. The physics of long-runout landslides are still a mystery to scientists.

Math ACTIVITY

The Frank landslide traveled 4 km in 100 s. Calculate this speed in meters per second.

Scientific Discoveries

The Lost Squadron

During World War II, an American squadron of eight planes crash-landed on the ice of Greenland. The crew was rescued, but the planes were lost. After the war, several people tried to find the "Lost Squadron." Finally, in 1988, a team of adventurers found the planes by using radar. The planes were buried by 40 years of snowfall and had become part of the Greenland ice sheet! When the planes were found, they were buried under 80 m of glacial ice. Incredibly, the team tunneled down through the ice and recovered a plane. The plane is now named Glacier Girl, and it still flies today!

Language Arts ACTIVITY

WRITING SKILL The crew of the Lost Squadron had to wait 10 days to be rescued by dog sled. Imagine that you were part of the crew—what would you have done to survive? Write a short story describing your adventure on the ice sheet of Greenland.

Careers

Johan Reinhard

High-Altitude Anthropologist Imagine discovering the mummified body of a girl from 500 years ago! In 1995, while climbing Mount Ampato, one of the tallest mountains in the Andes, Johan Reinhard made an incredible discovery—the well-preserved mummy of a young Inca girl. The recent eruption of a nearby volcano had caused the snow on Mount Ampato to melt and uncover the mummy. The discovery of the "Inca Ice Maiden" gave scientists a wealth of new information about Incan culture. Today, Reinhard considers the discovery of the Inca Ice Maiden his most exciting moment in the field.

Johan Reinhard is an anthropologist. Anthropologists study the physical and cultural characteristics of human populations. Reinhard studied anthropology at the University of Arizona and at the University of Vienna, in Austria. Early in his career, Reinhard worked on underwater archeology projects in Austria and Italy and on projects in the mountains of Nepal and Tibet. He soon made mountains and mountain peoples the focus of his career as an anthropologist. Reinhard spent 10 years in the highest mountains on Earth, the Himalayas. There, he studied the role of sacred mountains in Tibetan religions. Now, Reinhard studies the culture of the ancient Inca in the Andes of South America.

Social Studies ACTIVITY

Find out more about the Inca Ice Maiden or about Ötzi, a mummy that is more than 5,000 years old that was found in a glacier in Italy. Create a poster that summarizes what scientists have learned from these discoveries.

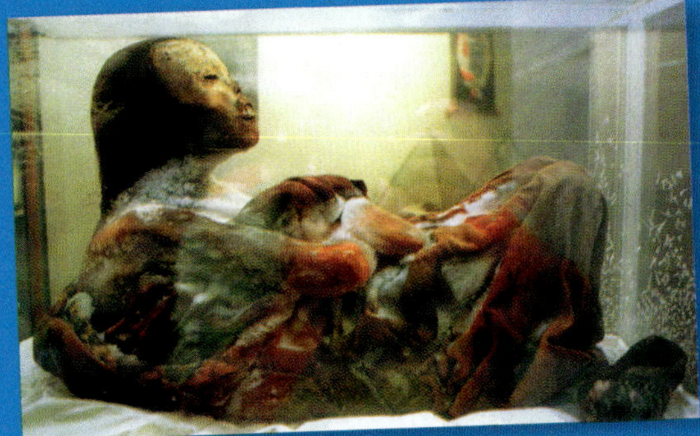

The Inca Ice Maiden was buried under ice and snow for more than 500 years.

To learn more about these Science in Action topics, visit go.hrw.com and type in the keyword **HZ5ICEF.**

Current Science

Check out Current Science® articles related to this chapter by visiting go.hrw.com. Just type in the keyword HZ5CS12.

Climate

About the PHOTO

Would you like to hang out on this ice with the penguins? You probably would not. You would be shivering, and your teeth would be chattering. However, these penguins feel comfortable. They have thick feathers and lots of body fat to keep them warm. Like other animals, penguins have adapted to their climate, which allows them to live comfortably in that climate. So, you will never see one of these penguins living comfortably on a hot, sunny beach in Florida!

PRE-READING ACTIVITY

FOLDNOTES

Pyramid Before you read the chapter, create the FoldNote entitled "Pyramid" described in the **Study Skills** section of the Appendix. Label the sides of the pyramid with "Tropical climate," "Temperate climate," and "Polar climate." As you read the chapter, define each climate zone, and write characteristics of each climate zone on the appropriate pyramid side.

START-UP ACTIVITY

What's Your Angle?

Try this activity to see how the angle of the sun's solar rays influences temperatures on Earth.

Procedure

1. Place a **lamp** 30 cm from a **globe.**
2. Point the lamp so that the light shines directly on the globe's equator.
3. Using **adhesive putty,** attach a **thermometer** to the globe's equator in a vertical position. Attach **another thermometer** to the globe's North Pole so that the tip points toward the lamp.
4. Record the temperature reading of each thermometer.
5. Turn on the lamp, and let the light shine on the globe for 3 minutes.
6. After 3 minutes, turn off the lamp and record the temperature reading of each thermometer again.

Analysis

1. Was there a difference between the final temperature at the globe's North Pole and the final temperature at the globe's equator? If so, what was it?
2. Explain why the temperature readings at the North Pole and the equator may be different.

SECTION 1

What Is Climate?

Suppose you receive a call from a friend who is coming to visit you tomorrow. To decide what clothing to bring, he asks about the current weather in your area.

You step outside to see if rain clouds are in the sky and to check the temperature. But what would you do if your friend asked you about the climate in your area? What is the difference between weather and climate?

READING WARM-UP

Objectives

- Explain the difference between weather and climate.
- Identify five factors that determine climates.
- Identify the three climate zones of the world.

Terms to Learn

weather	elevation
climate	surface current
latitude	biome
prevailing winds	

READING STRATEGY

Discussion Read this section silently. Write down questions that you have about this section. Discuss your questions in a small group.

Climate Vs. Weather

The main difference between weather and climate is the length of time over which both are measured. **Weather** is the condition of the atmosphere at a particular time. Weather conditions vary from day to day and include temperature, humidity, precipitation, wind, and visibility. **Climate,** on the other hand, is the average weather condition in an area over a long period of time. Climate is mostly determined by two factors—temperature and precipitation. Different parts of the world can have different climates, as shown in **Figure 1.** But why are climates so different? The answer is complicated. It includes factors in addition to temperature and precipitation, such as latitude, wind patterns, mountains, large bodies of water, and ocean currents.

Reading Check How is climate different from weather? *(See the Appendix for answers to Reading Checks.)*

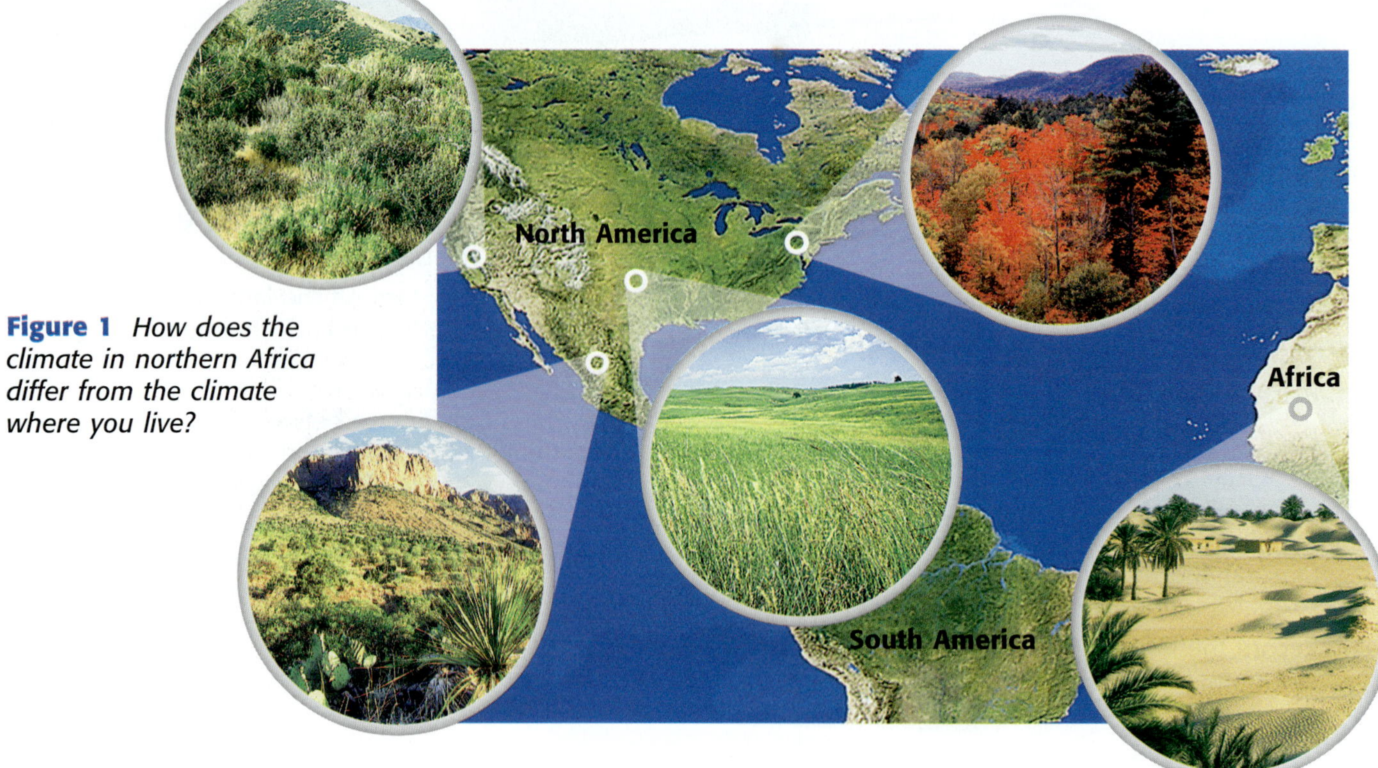

Figure 1 *How does the climate in northern Africa differ from the climate where you live?*

Latitude

Think of the last time you looked at a globe. Do you recall the thin, horizontal lines that circle the globe? Those lines are called lines of latitude. **Latitude** is the distance north or south, measured in degrees, from the equator. In general, the temperature of an area depends on its latitude. The higher the latitude is, the colder the climate tends to be. One of the coldest places on Earth, the North Pole, is 90° north of the equator. However, the equator, at latitude 0°, is usually hot.

As shown in **Figure 2,** if you were to take a trip to different latitudes in the United States, you would experience different climates. For example, the climate in Washington, D.C., which is at a higher latitude, is different from the climate in Texas.

Figure 2 *Winter in south Texas (top) is different from winter in Washington D.C. (bottom).*

Solar Energy and Latitude

Solar energy, which is energy from the sun, heats the Earth. The amount of direct solar energy a particular area receives is determined by latitude. **Figure 3** shows how the curve of the Earth affects the amount of direct solar energy at different latitudes. Notice that the sun's rays hit the equator directly, at almost a 90° angle. At this angle, a small area of the Earth's surface receives more direct solar energy than at a lesser angle. As a result, that area has high temperatures. However, the sun's rays strike the poles at a lesser angle than they do the equator. At this angle, the same amount of direct solar energy that hits the area at the equator is spread over a larger area at the poles. The result is lower temperatures at the poles.

weather the short-term state of the atmosphere, including temperature, humidity, precipitation, wind, and visibility

climate the average weather condition in an area over a long period of time

latitude the distance north or south from the equator; expressed in degrees

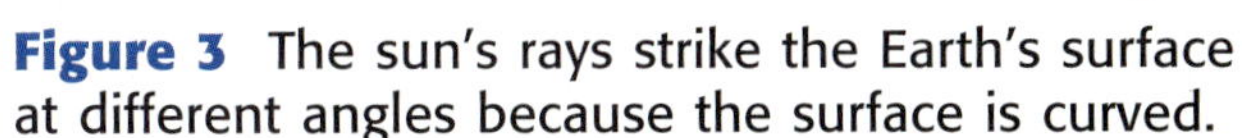

Figure 3 The sun's rays strike the Earth's surface at different angles because the surface is curved.

Seasons and Latitude

In most places in the United States, the year consists of four seasons. But there are places in the world that do not have such seasonal changes. For example, areas near the equator have approximately the same temperatures and same amount of daylight year-round. Seasons happen because the Earth is tilted on its axis at a 23.5° angle. This tilt affects how much solar energy an area receives as Earth moves around the sun. **Figure 4** shows how latitude and the tilt of the Earth determine the seasons and the length of the day in a particular area.

Reading Check Why is there less seasonal change near the equator?

Figure 4 **The Seasons**

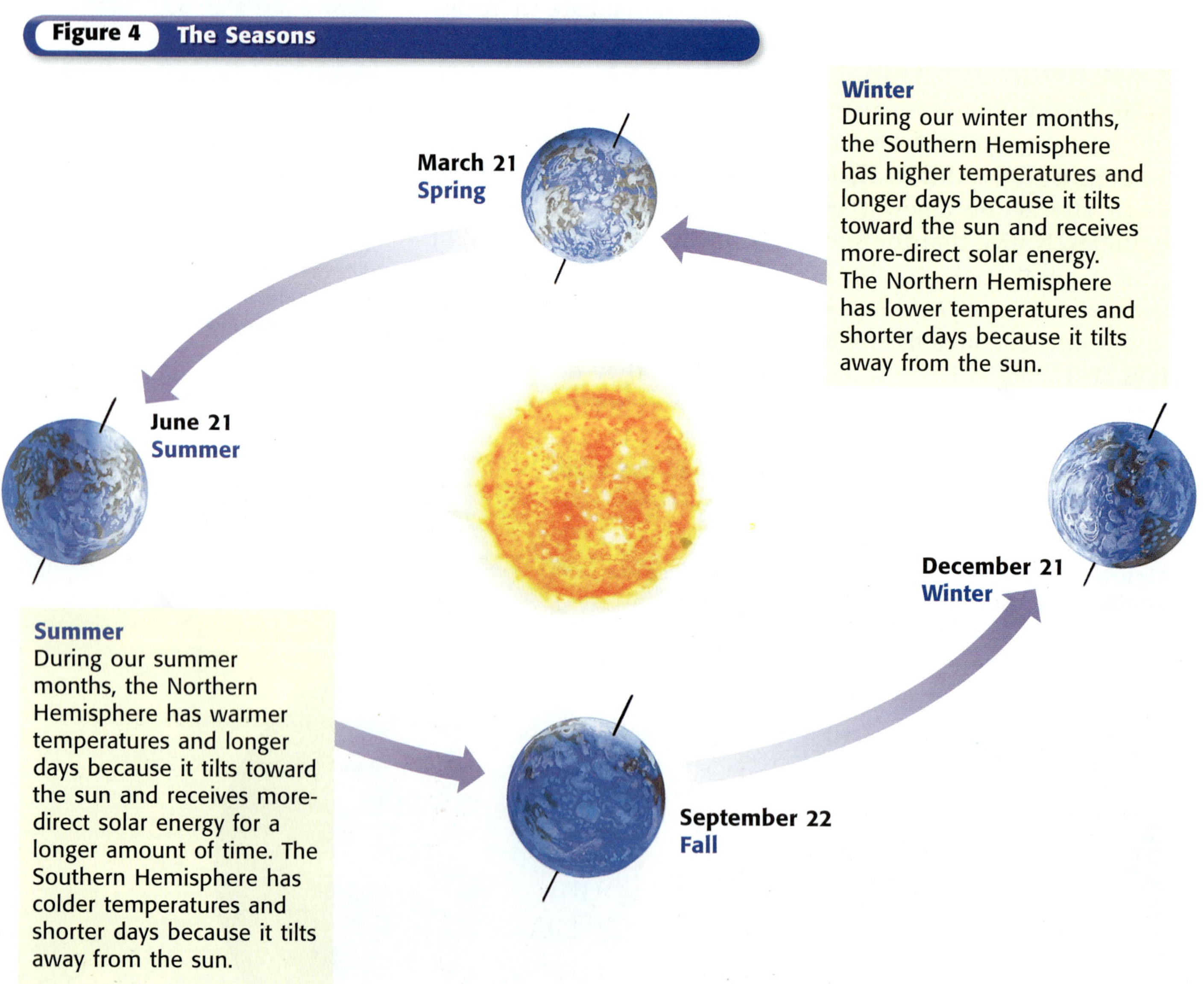

Winter
During our winter months, the Southern Hemisphere has higher temperatures and longer days because it tilts toward the sun and receives more-direct solar energy. The Northern Hemisphere has lower temperatures and shorter days because it tilts away from the sun.

Summer
During our summer months, the Northern Hemisphere has warmer temperatures and longer days because it tilts toward the sun and receives more-direct solar energy for a longer amount of time. The Southern Hemisphere has colder temperatures and shorter days because it tilts away from the sun.

Figure 5 The Circulation of Warm Air and Cold Air

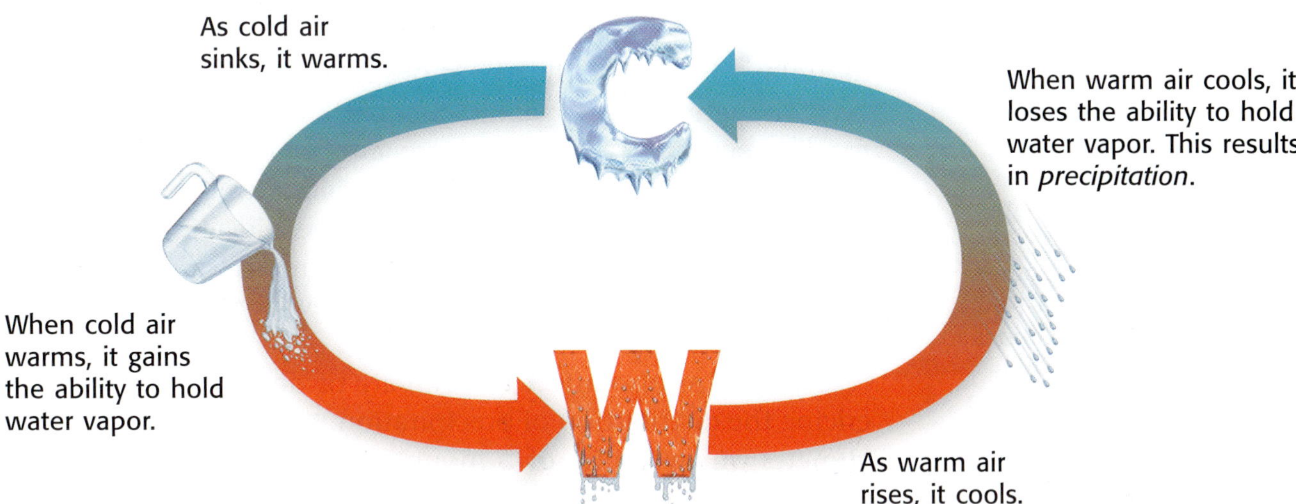

Prevailing Winds

Winds that blow mainly from one direction are **prevailing winds**. Before you learn how the prevailing winds affect climate, take a look at **Figure 5** to learn about some of the basic properties of air.

prevailing winds winds that blow mainly from one direction during a given period

Prevailing winds affect the amount of precipitation that a region receives. If the prevailing winds form from warm air, they may carry moisture. If the prevailing winds form from cold air, they will probably be dry.

The amount of moisture in prevailing winds is also affected by whether the winds blow across land or across a large body of water. Winds that travel across large bodies of water absorb moisture. Winds that travel across land tend to be dry. Even if a region borders the ocean, the area might be dry. **Figure 6** shows an example of how dry prevailing winds can cause the land to be dry though the land is near an ocean.

A Cool Breeze

1. Hold a **thermometer** next to the top edge of a **cup** of **water** containing two **ice cubes.** Record the temperature next to the cup.
2. Have your lab partner fan the surface of the cup with a **paper fan.** Record the temperature again. Has the temperature changed? Why or why not?

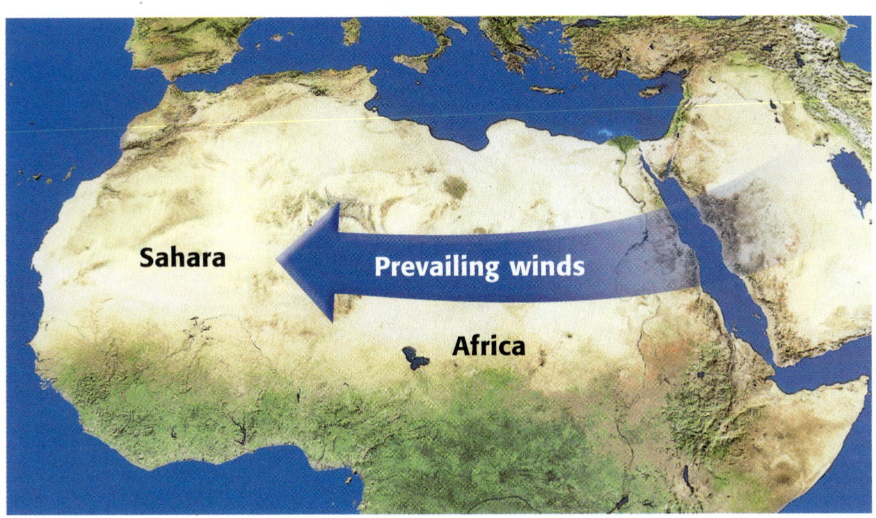

Figure 6 *The Sahara Desert, in northern Africa, is extremely dry because of the dry prevailing winds that blow across the continent.*

Using a Map

With your parent, use a physical map to locate the mountain ranges in the United States. Does climate vary from one side of a mountain range to the other? If so, what does this tell you about the climatic conditions on either side of the mountain? From what direction are the prevailing winds blowing?

Mountains

Mountains can influence an area's climate by affecting both temperature and precipitation. Kilimanjaro is the tallest mountain in Africa. It has snow-covered peaks year-round, even though it is only about 3° (320 km) south of the equator. Temperatures on Kilimanjaro and in other mountainous areas are affected by elevation. **Elevation** is the height of surface landforms above sea level. As the elevation increases, the ability of air to transfer energy from the ground to the atmosphere decreases. Therefore, as elevation increases, temperature decreases.

Mountains also affect the climate of nearby areas by influencing the distribution of precipitation. **Figure 7** shows how the climates on two sides of a mountain can be very different.

Reading Check **Why does the atmosphere become cooler at higher elevations?**

Figure 7 *Mountains block the prevailing winds and affect the climate on the other side.*

The Wet Side
Mountains force air to rise. The air cools as it rises, releasing moisture as snow or rain. The land on the windward side of the mountain is usually green and lush because the wind releases its moisture.

The Dry Side
After dry air crosses the mountain, the air begins to sink. As the air sinks, it is warmed and absorbs moisture. The dry conditions created by the sinking, warm air usually produce a desert. This side of the mountain is in a *rain shadow*.

Large Bodies of Water

Large bodies of water can influence an area's climate. Water absorbs and releases heat slower than land does. Because of this quality, water helps to moderate the temperatures of the land around it. So, sudden or extreme temperature changes rarely take place on land near large bodies of water. For example, the state of Michigan, which is surrounded by the Great Lakes, has more-moderate temperatures than other places at the same latitude. The lakes also increase the moisture content of the air, which leads to heavy snowfall in the winter. This "lake effect" can cause 350 inches of snow to drop in one year!

Ocean Currents

The circulation of ocean surface currents has a large effect on an area's climate. **Surface currents** are streamlike movements of water that occur at or near the surface of the ocean. **Figure 8** shows the pattern of the major ocean surface currents.

As surface currents move, they carry warm or cool water to different locations. The surface temperature of the water affects the temperature of the air above it. Warm currents heat the surrounding air and cause warmer temperatures. Cool currents cool the surrounding air and cause cooler temperatures. The Gulf Stream current carries warm water northward off the east coast of North America and past Iceland. Iceland is an island country located just below the Arctic Circle. The warm water from the Gulf Stream heats the surrounding air and creates warmer temperatures in southern Iceland. Iceland experiences milder temperatures than Greenland, its neighboring island. Greenland's climate is cooler because Greenland is not influenced by the Gulf Stream.

Reading Check **Why does Iceland experience milder temperatures than Greenland?**

elevation the height of an object above sea level

surface current a horizontal movement of ocean water that is caused by wind and that occurs at or near the ocean's surface

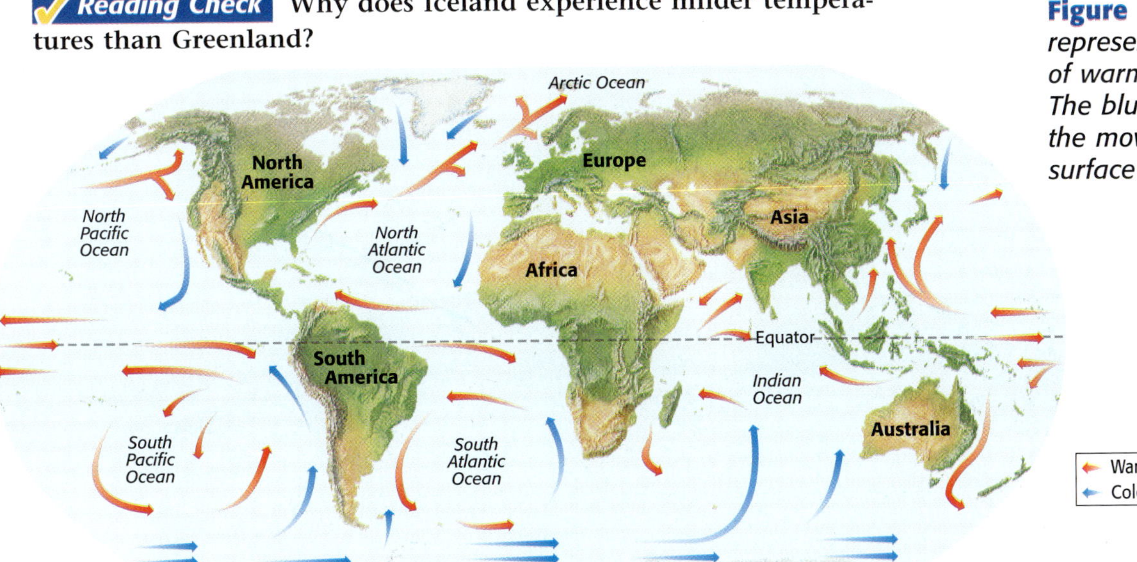

Figure 8 *The red arrows represent the movement of warm surface currents. The blue arrows represent the movement of cold surface currents.*

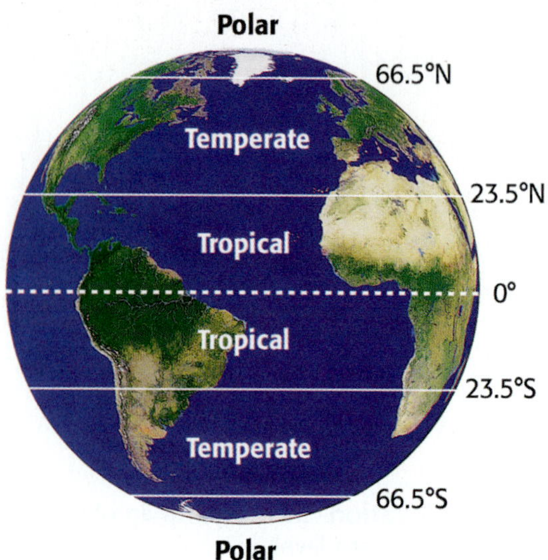

Figure 9 *The three major climate zones are determined by latitude.*

Climates of the World

Have you seen any polar bears in your neighborhood lately? You probably have not. That's because polar bears live only in very cold arctic regions. Why are the animals in one part of the world so different from the animals in other parts? One of the differences has to do with climate. Plants and animals that have adapted to one climate may not be able to live in another climate. For example, frogs would not be able to survive at the North Pole.

Climate Zones

The Earth's three major climate zones—tropical, temperate, and polar—are shown in **Figure 9.** Each zone has a temperature range that relates to its latitude. However, in each of these zones, there are several types of climates because of differences in the geography and the amount of precipitation. Because of the various climates in each zone, there are different biomes in each zone. A **biome** is a large region characterized by a specific type of climate and certain types of plant and animal communities. **Figure 10** shows the distribution of the Earth's land biomes. In which biome do you live?

biome a large region characterized by a specific type of climate and certain types of plant and animal communities

Reading Check What factors distinguish one biome from another biome?

Figure 10 **The Earth's Land Biomes**

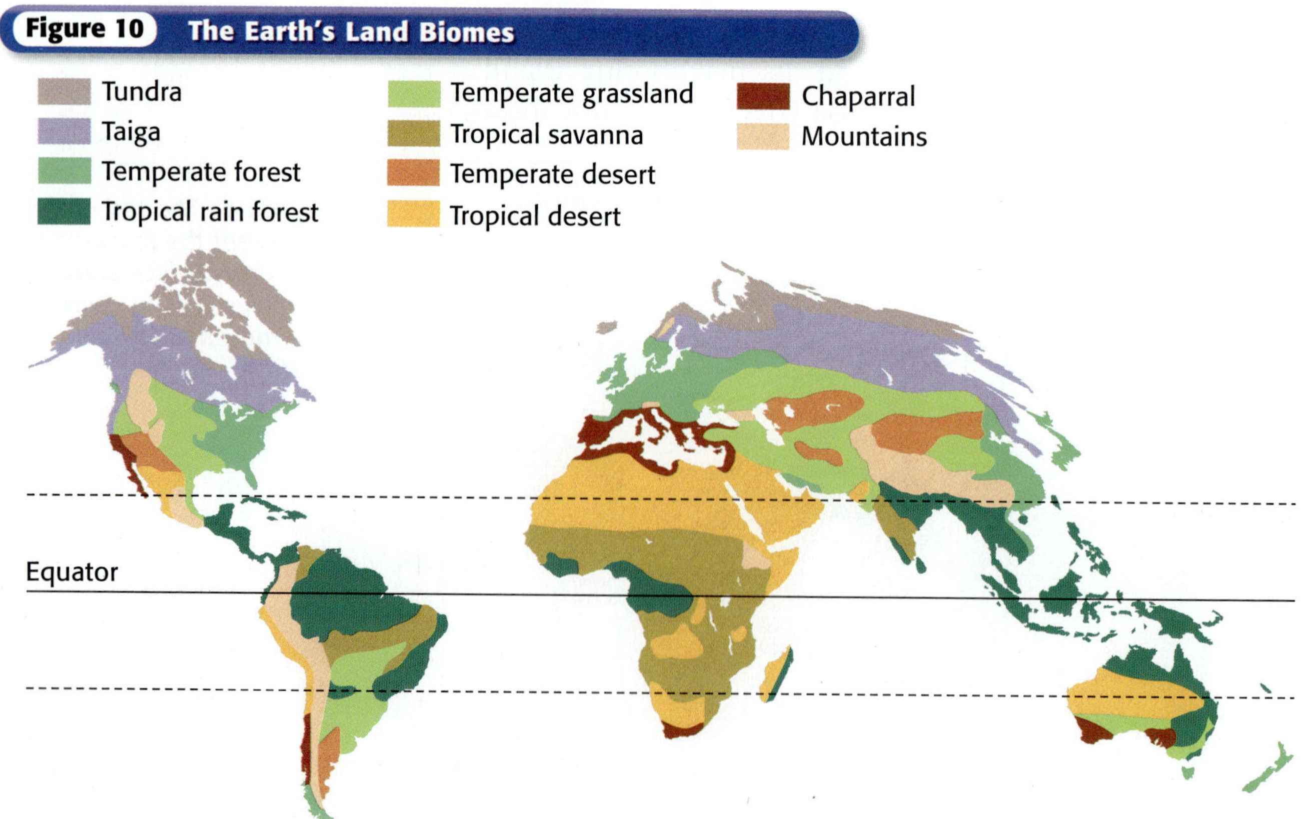

SECTION Review

Summary

- Weather is the condition of the atmosphere at a particular time. This condition includes temperature, humidity, precipitation, wind, and visibility.
- Climate is the average weather condition in an area over a long period of time.
- The higher the latitude, the cooler the climate.
- Prevailing winds affect the climate of an area by the amount of moisture they carry.
- Mountains influence an area's climate by affecting both temperature and precipitation.
- Large bodies of water and ocean currents influence the climate of an area by affecting the temperature of the air over the water.
- The three climate zones of the world are the tropical zone, the temperate zone, and the polar zone.

Using Key Terms

1. In your own words, write a definition for each of the following terms: *weather, climate, latitude, prevailing winds, elevation, surface currents,* and *biome.*

Understanding Key Ideas

2. Which of the following affects climate by causing the air to rise?
   a. mountains
   b. ocean currents
   c. large bodies of water
   d. latitude
3. What is the difference between weather and climate?
4. List five factors that determine climates.
5. Explain why there is a difference in climate between areas at 0° latitude and areas at 45° latitude.
6. List the three climate zones of the world.

Critical Thinking

7. **Analyzing Relationships** How would seasons be different if the Earth did not tilt on its axis?
8. **Applying Concepts** During what months does Australia have summer? Explain.

Interpreting Graphics

Use the map below to answer the questions that follow.

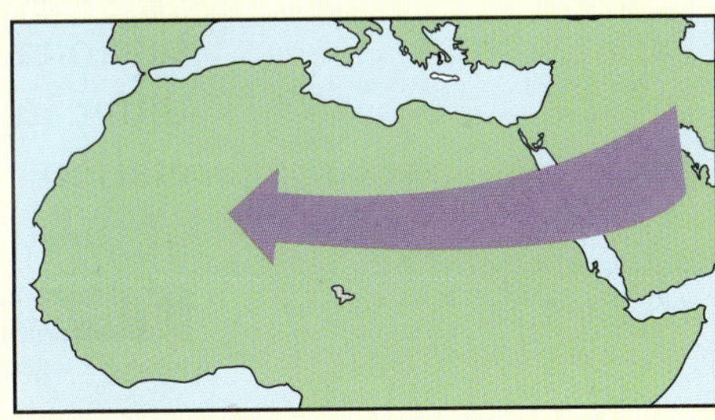

9. Would you expect the area that the arrow points to to be moist or dry? Explain your answer.
10. Describe how the climate of the same area would change if the prevailing winds traveled from the opposite direction. Explain how you came to this conclusion.

For a variety of links related to this chapter, go to www.scilinks.org

Topic: What Is Climate?
SciLinks code: HSM1659

SECTION 2

The Tropics

Where in the world do you think you could find a flying dragon gliding above you from one treetop to the next?

Don't worry. This flying dragon, or tree lizard, is only about 20 cm long, and it eats only insects. With winglike skin flaps, the flying dragon can glide from one treetop to the next. But, you won't find this kind of animal in the United States. These flying dragons live in Southeast Asia, which is in the tropical zone.

READING WARM-UP

Objectives

- Locate and describe the tropical zone.
- Describe the biomes found in the tropical zone.

Terms to Learn

tropical zone

READING STRATEGY

Reading Organizer As you read this section, make a table comparing *tropical rain forests, tropical savannas,* and *tropical deserts.*

The Tropical Zone

The region that surrounds the equator and that extends from about 23.5° north latitude to 23.5° south latitude is called the **tropical zone.** The tropical zone is also known as the Tropics. Latitudes in the tropical zone receive the most solar radiation. Temperatures are therefore usually hot, except at high elevations.

Within the tropical zone, there are three major types of biomes—tropical rain forest, tropical desert, and tropical savanna. These three biomes have high temperatures. But they differ in the amount of precipitation, soil characteristics, vegetation, and kinds of animals. **Figure 1** shows the distribution of these biomes.

Reading Check At what latitudes would you find the tropical zone? (*See the Appendix for answers to Reading Checks.*)

Figure 1 Biomes of the Tropical Zone

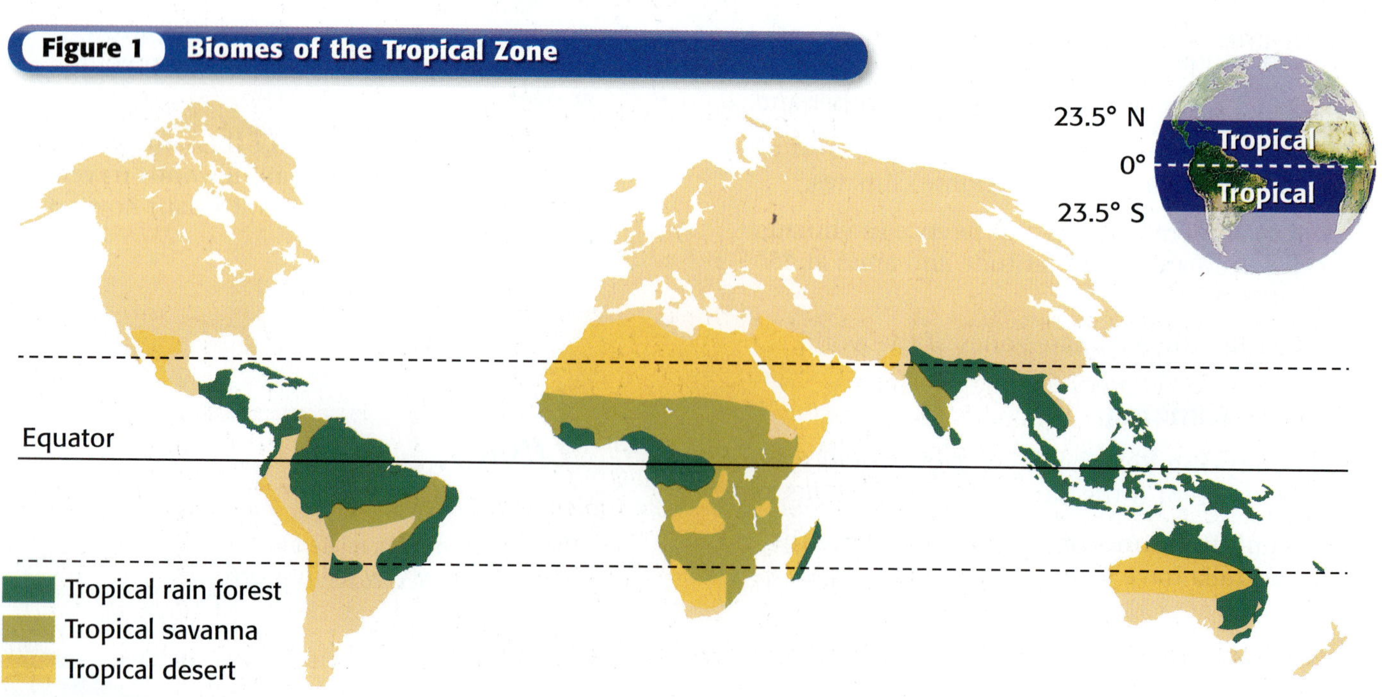

Tropical Rain Forest

- **Average Temperature Range** 25°C to 28°C (77°F to 82°F)
- **Average Yearly Precipitation** 200 cm or more
- **Soil Characteristics** thin and nutrient poor

Figure 2 *In tropical rain forests, many of the trees form aboveground roots that provide extra support for the trees in the thin, nutrient-poor soil.*

Tropical Rain Forests

Tropical rain forests are always warm and wet. Because they are located near the equator, they receive strong sunlight year-round. So, there is little difference between seasons in tropical rain forests.

Tropical rain forests contain the greatest number of animal and plant species of any biome. Animals found in tropical rain forests include monkeys, parrots, tree frogs, tigers, and leopards. Plants found in tropical rain forests include mahogany, vines, ferns, and bamboo. But in spite of the lush vegetation, shown in **Figure 2,** the soil in rain forests is poor. The rapid decay of plants and animals returns nutrients to the soil. But these nutrients are quickly absorbed and used by the plants. The nutrients that are not immediately used by the plants are washed away by the heavy rains. The soil is left thin and nutrient poor.

tropical zone the region that surrounds the equator and that extends from about 23.5° north latitude to 23.5° south latitude

CONNECTION TO Social Studies

WRITING SKILL **Living in the Tropics** The tropical climate is very hot and humid. People who live in the Tropics have had to adapt to feel comfortable in that climate. For example, in the country of Samoa, some people live in homes that have no walls, which are called *fales*. Fales have only a roof, which provides shade. The openness of the home allows cool breezes to flow through the home. Research other countries in the Tropics. See how the climate influences the way the people live in those countries. Then, in your **science journal,** describe how the people's lifestyle helps them adapt to the climate.

Tropical Savanna

- **Average Temperature Range** 27°C to 32°C (80°F to 90°F)
- **Average Yearly Precipitation** 100 cm
- **Soil Characteristics** generally nutrient poor

Figure 3 *The grass of a tropical savanna can be as tall as 5 m.*

Tropical Savannas

Tropical savannas, or grasslands, are composed of tall grasses and a few scattered trees. The climate is usually very warm. Tropical savannas have a dry season that lasts four to eight months and that is followed by short periods of rain. Savanna soils are generally nutrient poor. However, grass fires, which are common during the dry season, leave the soils nutrient enriched. An African savanna is shown in **Figure 3.**

Many plants have adapted to fire and use it to promote development. For example, some species need fire to break open their seeds' outer skin. Only after this skin is broken can each seed grow. For other species, heat from the fire triggers the plants to drop their seeds into the newly enriched soil.

Animals that live in tropical savannas include giraffes, lions, crocodiles, and elephants. Plants include tall grasses, trees, and thorny shrubs.

WRITING SKILL **Animal and Plant Adaptations** Animals and plants adapt to the climate in which they live. These adaptations cause certain animals and plants to be unique to particular biomes. For example, the camel, which is unique to the desert, has adapted to going for long periods of time without water. Research other animals or plants that live in the Tropics. Then, in your **science journal,** describe the characteristics that help them survive in the Tropics.

Tropical Deserts

A desert is an area that receives less than 25 cm of rainfall per year. Because of this low yearly rainfall, deserts are the driest places on Earth. Desert plants, such as those shown in **Figure 4,** are adapted to survive in places that have little water. Animals such as rats, lizards, snakes, and scorpions have also adapted to survive in these deserts.

There are two kinds of deserts—hot deserts and cold deserts. Hot deserts are caused by cool, sinking air masses. Many hot deserts, such as the Sahara, in Africa, are tropical deserts. Daily temperatures in tropical deserts often vary from very hot daytime temperatures (50°C) to cool nighttime temperatures (20°C). Because of the dryness of deserts, the soil is poor in organic matter, which is needed for plants to grow.

Reading Check What animals would you find in a tropical desert?

Tropical Desert

- Average Temperature Range 16°C to 50°C (61°F to 120°F)
- Average Yearly Precipitation 0–25 cm
- Soil Characteristics poor in organic matter

Figure 4 *Plants such as succulents have fleshy stems and leaves to store water.*

SECTION Review

Summary

- The tropical zone is located around the equator, between 23.5° north and 23.5° south latitude.
- Temperatures are usually hot in the tropical zone.
- Tropical rain forests are warm and wet. They have the greatest number of plant and animal species of any biome.
- Tropical savannas are grasslands that have a dry season.
- Tropical deserts are hot and receive little rain.

Using Key Terms

1. In your own words, write a definition for the term *tropical zone.*

Understanding Key Ideas

2. Which of the following tropical biomes has less than 50 cm of precipitation a year?
   a. rain forest
   b. desert
   c. grassland
   d. savanna
3. What are the soil characteristics of a tropical rain forest?
4. In what ways have savanna vegetation adapted to fire?

Math Skills

5. Suppose that in a tropical savanna, the temperature was recorded every hour for 4 h. The recorded temperatures were 27°C, 28°C, 29°C, and 29°C. Calculate the average temperature for this 4 h period.

Critical Thinking

6. **Analyzing Relationships** How do the tropical biomes differ?
7. **Making Inferences** How would you expect the adaptations of a plant in a tropical rain forest to differ from the adaptations of a tropical desert plant? Explain.
8. **Analyzing Data** An area has a temperature range of 30°C to 40°C and received 10 cm of rain this year. What biome is this area in?

SECTION 3

Temperate and Polar Zones

Which season is your favorite? Do you like the change of colors in the fall, the flowers in the spring, or do you prefer the hot days of summer?

READING WARM-UP

Objectives

- Locate and describe the temperate zone and the polar zone.
- Describe the different biomes found in the temperate zone and the polar zone.
- Explain what a microclimate is.

Terms to Learn

temperate zone
polar zone
microclimate

READING STRATEGY

Reading Organizer As you read this section, create an outline of the section. Use the headings from the section in your outline.

If you live in the continental United States, chances are you live in a biome that experiences seasonal change. Seasonal change is one characteristic of the temperate zone. Most of the continental United States is in the temperate zone, which is the climate zone between the Tropics and the polar zone.

The Temperate Zone

The climate zone between the Tropics and the polar zone is the **temperate zone.** Latitudes in the temperate zone receive less solar energy than latitudes in the Tropics do. Because of this, temperatures in the temperate zone tend to be lower than in the Tropics. Some biomes in the temperate zone have a mild change of seasons. Other biomes in the country can experience freezing temperatures in the winter and very hot temperatures in the summer. The temperate zone consists of the following four biomes—temperate forest, temperate grassland, chaparral, and temperate desert. Although these biomes have four distinct seasons, the biomes differ in temperature and precipitation and have different plants and animals. **Figure 1** shows the distribution of thc biomes found in the temperate zone.

Reading Check **Where is the temperate zone?** *(See the Appendix for answers to Reading Checks.)*

Figure 1 **Biomes of the Temperate Zone**

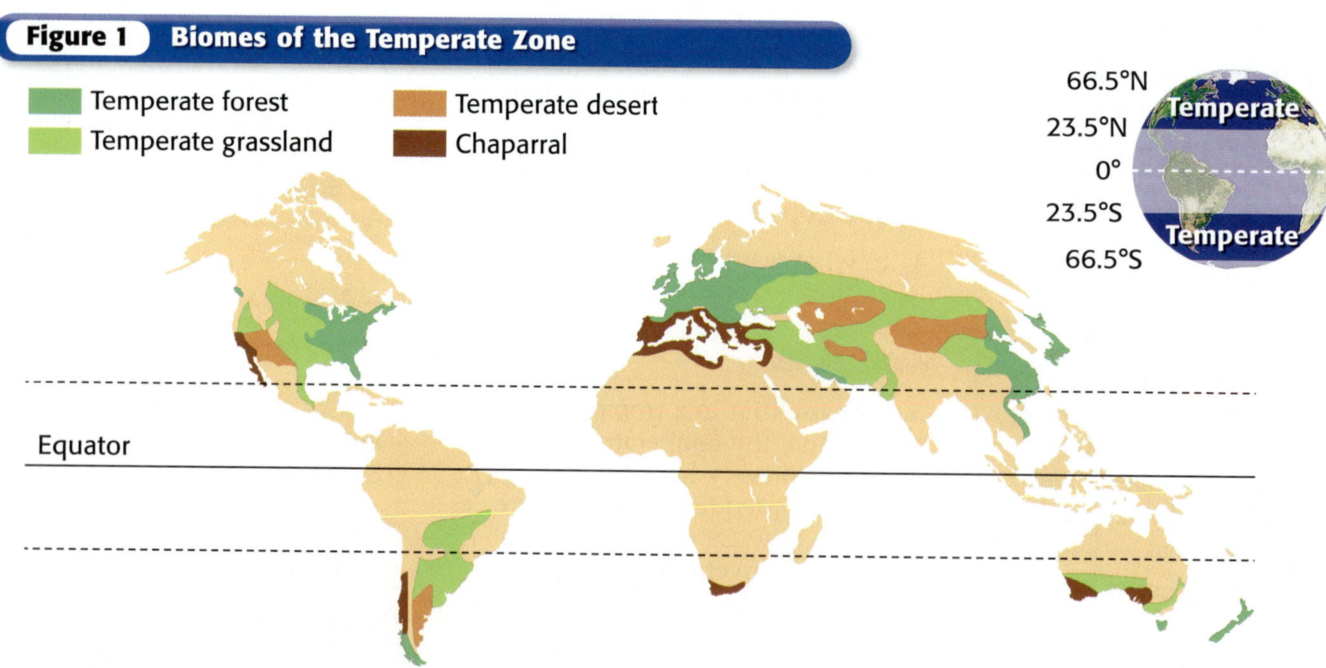

Figure 2 *Deciduous trees have leaves that change color and drop when temperatures become cold.*

Temperate Forests

The temperate forest biomes tend to have high amounts of rainfall and seasonal temperature differences. Summers are often warm, and winters are often cold. Animals such as deer, bears, and foxes live in temperate forests. **Figure 2** shows deciduous trees in a temperate forest. *Deciduous* describes trees that lose their leaves at the end of the growing season. The soils in deciduous forests are usually fertile because of the high organic content from decaying leaves that drop every winter. Another type of tree found in the temperate forest is the evergreen. *Evergreens* are trees that keep their leaves year-round.

temperate zone the climate zone between the Tropics and the polar zone

Temperate Grasslands

Temperate grasslands, such as those shown in **Figure 3,** are regions that receive too little rainfall for trees to grow. This biome has warm summers and cold winters. Examples of animals that are found in temperate grasslands include bison in North America and kangaroo in Australia. Grasses are the most common kind of plant found in this biome. Because grasslands have the most-fertile soils of all biomes, much of the grassland has been plowed to make room for croplands.

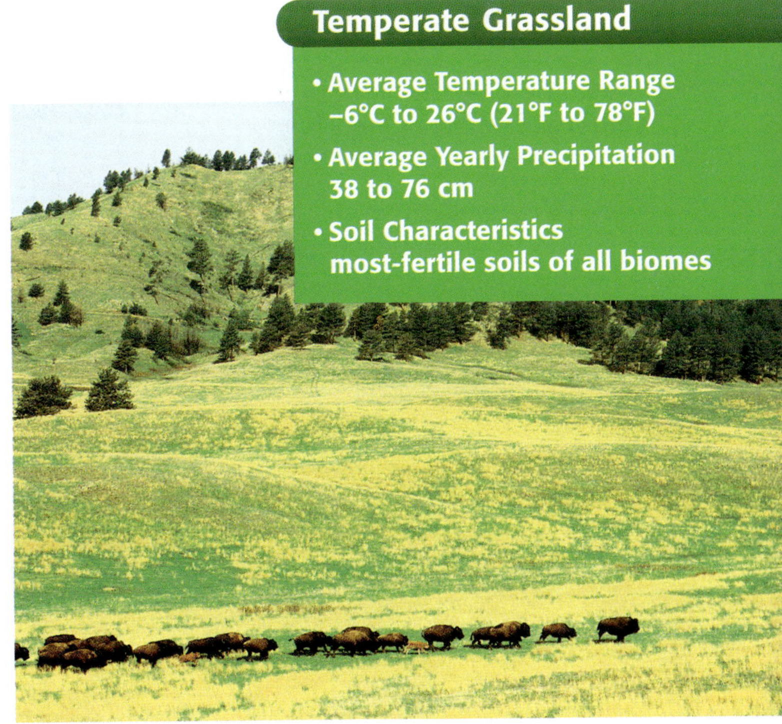

Figure 3 *At one time, the world's grasslands covered about 42% of Earth's total land surface. Today, they occupy only about 12% of the Earth's total land surface.*

Chaparral

- **Average Temperature Range** 11°C to 26°C (51°F to 78°F)
- **Average Yearly Precipitation** 48 to 56 cm
- **Soil Characteristics** rocky, nutrient-poor soils

Figure 4 *Some plant species found in chaparral require fire to reproduce.*

Chaparrals

Chaparral regions, as shown in **Figure 4,** have cool, wet winters and hot, dry summers. Animals, such as coyotes and mountain lions live in chaparrals. The vegetation is mainly evergreen shrubs. These shrubs are short, woody plants with thick, waxy leaves. The waxy leaves are adaptations that help prevent water loss in dry conditions. These shrubs grow in rocky, nutrient-poor soil. Like tropical-savanna vegetation, chaparral vegetation has adapted to fire. In fact, some plants, such as chamise, can grow back from their roots after a fire.

Temperate Deserts

The temperate desert biomes, like the one shown in **Figure 5,** tend to be cold deserts. Like all deserts, cold deserts receive less than 25 cm of precipitation yearly. Examples of animals that live in temperate deserts are lizards, snakes, bats, and toads. And the types of plants found in temperate deserts include cacti, shrubs, and thorny trees.

Temperate deserts can be very hot in the daytime. But, unlike hot deserts, they are often very cold at night. This large change in temperature between day and night is caused by low humidity and cloudless skies. These conditions allow for a large amount of energy to heat the Earth's surface during the day. However, these same characteristics allow the energy to escape at night. This causes temperatures to drop. You probably rarely think of snow and deserts together. But temperate deserts often receive light snow during the winter.

Reading Check Why are temperate deserts cold at night?

Figure 5 *The Great Basin Desert is in the rain shadow of the Sierra Nevada.*

Figure 6 Biomes of the Polar Zone

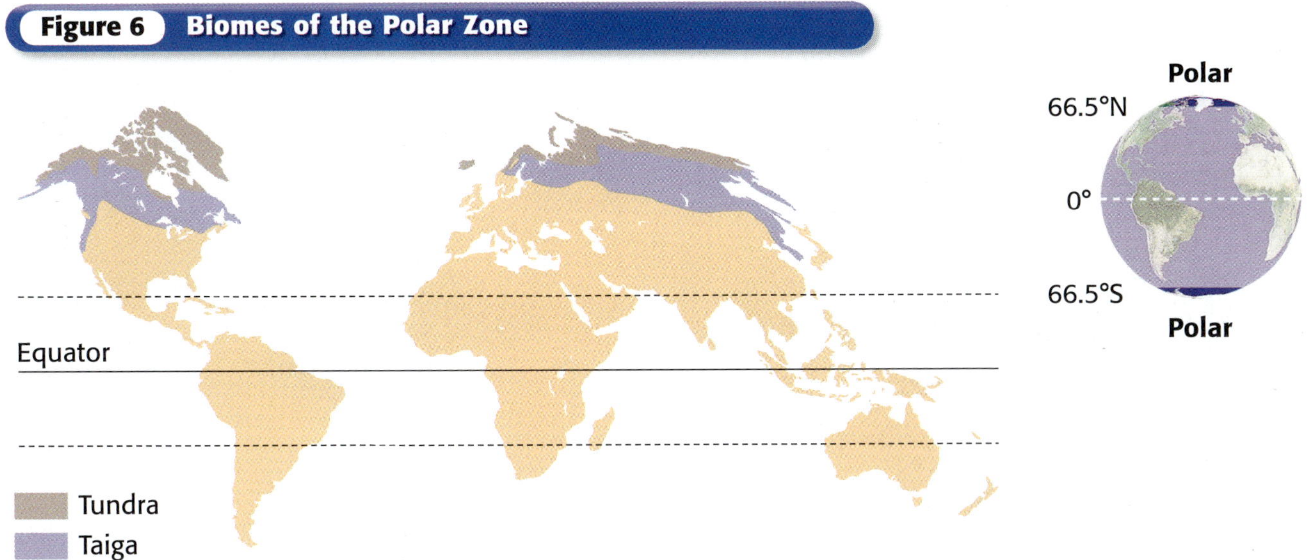

The Polar Zone

The climate zone located at the North or South Pole and its surrounding area is called the **polar zone.** Polar climates have the coldest average temperatures of all the climate zones. Temperatures in the winter stay below freezing. The temperatures during the summer remain cool. **Figure 6** shows the distribution of the biomes found in the polar zone.

polar zone the North or South Pole and its surrounding area

Tundra

The tundra biome, as shown in **Figure 7,** has long, cold winters with almost 24 hours of night. It also has short, cool summers with almost 24 hours of daylight. In the summer, only the top meter of soil thaws. Underneath the thawed soil lies a permanently frozen layer of soil, called *permafrost.* This frozen layer prevents the water in the thawed soil from draining. Because of the poor drainage, the upper soil layer is muddy. This muddy layer of soil makes a great breeding ground for insects, such as mosquitoes. Many birds migrate to the tundra during the summer to feed on the insects. Other animals that live in the tundra are caribou, reindeer, and polar bears. Plants in this biome include mosses and lichens.

Tundra

- **Average Temperature Range** –27°C to 5°C (–17°F to 41°F)
- **Average Yearly Precipitation** 0 to 25 cm
- **Soil Characteristics** frozen

Figure 7 *In the tundra, mosses and lichens cover rocks.*

Taiga

- **Average Temperature Range** −10°C to 15°C (14°F to 59°F)
- **Average Yearly Precipitation** 40 to 61 cm
- **Soil Characteristics** acidic

Figure 8 *The taiga, such as this one in Washington, have mostly evergreens for trees.*

Taiga (Northern Coniferous Forest)

Just south of the tundra lies the taiga biome. The taiga, as shown in **Figure 8,** has long, cold winters and short, warm summers. Animals commonly found here are moose, bears, and rabbits. The majority of the trees are evergreen needle-leaved trees called *conifers,* such as pine, spruce, and fir trees. The needles and flexible branches allow these trees to shed heavy snow before they can be damaged. Conifer needles are made of acidic substances. When the needles die and fall to the soil, they make the soil acidic. Most plants cannot grow in acidic soil. Because of the acidic soil, the forest floor is bare except for some mosses and lichens.

microclimate the climate of a small area

Microclimates

The climate and the biome of a particular place can also be influenced by local conditions. **Microclimate** is the climate of a small area. The alpine biome is a cold biome found on mountains all around the world. The alpine biome can even be found on mountains in the Tropics! How is this possible? The high elevation affects the area's climate and therefore its biome. As the elevation increases, the air's ability to transfer heat from the ground to the atmosphere by conduction decreases, which causes temperatures to decrease. In winter, the temperatures are below freezing. In summer, average temperatures range from 10°C to 15°C. Plants and animals have had to develop special adaptations to live in this severe climate.

WRITING SKILL **Your Biome** With your parents, explore the biome in the area where you live. What kinds of animals and plants live in your area? Write a one-page paper that describes the biome and why the biome of your area has its particular climate.

Cities

Cities are also microclimates. In a city, temperatures can be 1°C to 2°C warmer than the surrounding rural areas. Have you ever walked barefoot on a black asphalt street on a hot summer day? Doing so burns your feet because buildings and pavement made of dark materials absorb solar radiation instead of reflecting it. There is also less vegetation in a city to take in the sun's rays. This absorption and re-radiation of heat by buildings and pavement heats the surrounding air. In turn, the temperatures rise.

Reading Check **Why do cities have higher temperatures than the surrounding rural areas?**

CONNECTION TO Physics

Hot Roofs! Scientists studied roofs on a sunny day when the air temperature was 13°C. They recorded roof temperatures ranging from 18°C to 61°C depending on color and material of the roof. Place thermometers on outside objects that are made of different types of materials and that are different colors. Please stay off the roof! Is there a difference in temperatures?

ACTIVITY

SECTION Review

Summary

- The temperate zone is located between the Tropics and the polar zone. It has moderate temperatures.
- Temperate forests, temperate grasslands, and temperate deserts are biomes in the temperate zone.
- The polar zone includes the North or South Pole and its surrounding area. The polar zone has the coldest temperatures.
- The tundra and the taiga are biomes within the polar zone.

Using Key Terms

1. In your own words, write a definition for the term *microclimate*.

Complete each of the following sentences by choosing the correct term from the word bank.

temperate zone polar zone microclimate

2. The coldest temperatures are found in the ___.
3. The ___ has moderate temperatures.

Understanding Key Ideas

4. Which of the following biomes has the driest climate?
   - **a.** temperate forests
   - **b.** temperate grasslands
   - **c.** chaparrals
   - **d.** temperate deserts
5. Explain why the temperate zone has lower temperatures than the Tropics.
6. Describe how the latitude of the polar zone affects the climate in that area.
7. Explain why the tundra can sometimes experience 24 hours of daylight or 24 hours of night.
8. How do conifers make the soil they grow in too acidic for other plants to grow?

Math Skills

9. Texas has an area of about 700,000 square kilometers. Grasslands compose about 20% of this area. About how many square kilometers of grassland are there in Texas?

Critical Thinking

10. **Identifying Relationships** Which biome would be more suitable for growing crops, temperate forest or taiga? Explain.
11. **Making Inferences** Describe the types of animals and vegetation you might find in the Alpine biome.

SECTION 4

Changes in Climate

As you have probably noticed, the weather changes from day to day. Sometimes, the weather can change several times in one day! But have you ever noticed the climate change?

READING WARM-UP

Objectives

- Describe how the Earth's climate has changed over time.
- Summarize four different theories that attempt to explain why the Earth's climate has changed.
- Explain the greenhouse effect and its role in global warming.

Terms to Learn

ice age
global warming
greenhouse effect

READING STRATEGY

Paired Summarizing Read this section silently. In pairs, take turns summarizing the material. Stop to discuss ideas that seem confusing.

On Saturday, your morning baseball game was canceled because of rain, but by that afternoon the sun was shining. Now, think about the climate where you live. You probably haven't noticed a change in climate, because climates change slowly. What causes climatic change? Studies indicate that human activity may cause climatic change. However, natural factors also can influence changes in the climate.

Ice Ages

The geologic record indicates that the Earth's climate has been much colder than it is today. In fact, much of the Earth was covered by sheets of ice during certain periods. An **ice age** is a period during which ice collects in high latitudes and moves toward lower latitudes. Scientists have found evidence of many major ice ages throughout the Earth's geologic history. The most recent ice age began about 2 million years ago.

ice age a long period of climate cooling during which ice sheets cover large areas of Earth's surface; also known as a glacial period

Glacial Periods

During an ice age, there are periods of cold and periods of warmth. These periods are called glacial and interglacial periods. During *glacial periods,* the enormous sheets of ice advance. As they advance, they get bigger and cover a larger area, as shown in **Figure 1.** Because a large amount of water is frozen during glacial periods, the sea level drops.

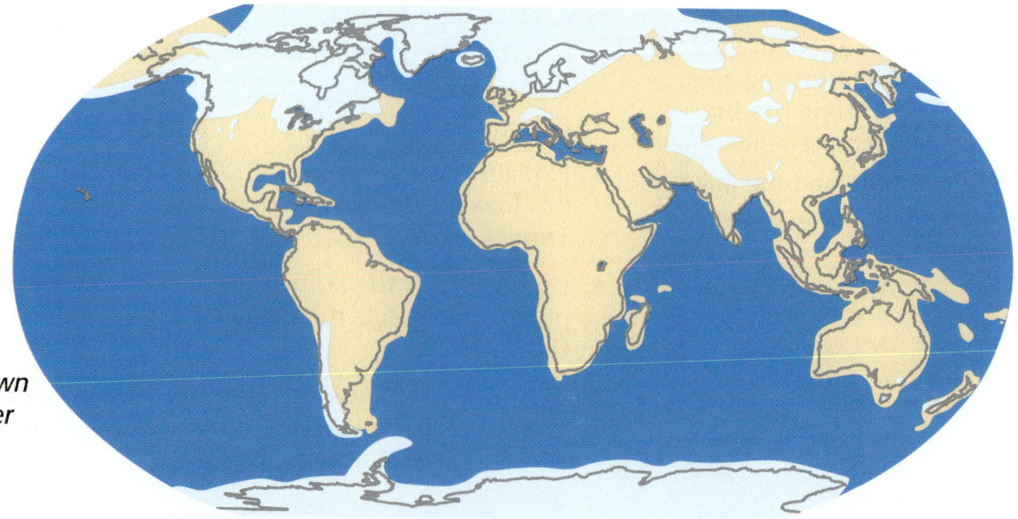

Figure 1 *During glacial periods, ice sheets (as shown in light blue), cover a larger portion of the Earth.*

Interglacial Periods

Warmer times that happen between glacial periods are called *interglacial periods*. During an interglacial period, the ice begins to melt and the sea level rises again. The last interglacial period began 10,000 years ago and is still happening. Why do these periods occur? Will the Earth have another glacial period in the future? These questions have been debated by scientists for the past 200 years.

INTERNET ACTIVITY

For another activity related to this chapter, go to **go.hrw.com** and type in the keyword **HZ5CLMW.**

Motions of the Earth

There are many theories about the causes of ice ages. Each theory tries to explain the gradual cooling that begins an ice age. This cooling leads to the development of large ice sheets that periodically cover large areas of the Earth's surface.

The *Milankovitch theory* explains why an ice age isn't just one long cold spell. Instead, the ice age alternates between cold and warm periods. Milutin Milankovitch, a Yugoslavian scientist, proposed that changes in the Earth's orbit and in the tilt of the Earth's axis cause ice ages. His theory is shown in **Figure 2.** In a 100,000 year period, the Earth's orbit changes from elliptical to circular. This changes the Earth's distance from the sun. In turn, it changes the temperature on Earth. Changes in the tilt of the Earth also influence the climate. The more the Earth is tilted, the closer the poles are to the sun.

Reading Check **What are the two things Milankovitch says causes ice ages?** (*See the Appendix for answers to Reading Checks.*)

Figure 2 **The Milankovitch Theory**

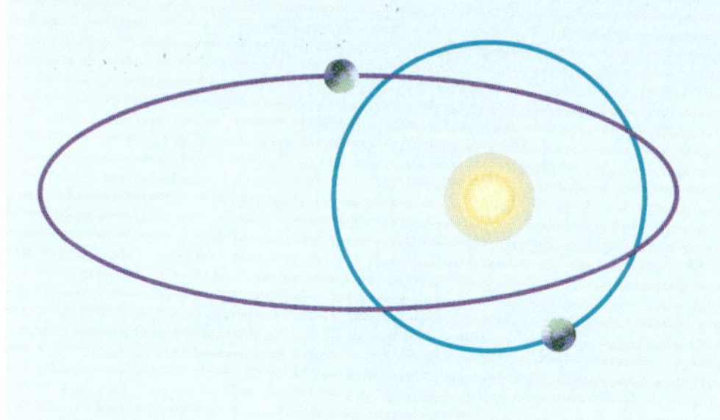

1 Over a period of 100,000 years, the Earth's orbit slowly changes from a more circular shape to a more elliptical shape and back again. When Earth's orbit is elliptical, Earth receives more energy from the sun. When its orbit is more circular, Earth receives less energy from the sun.

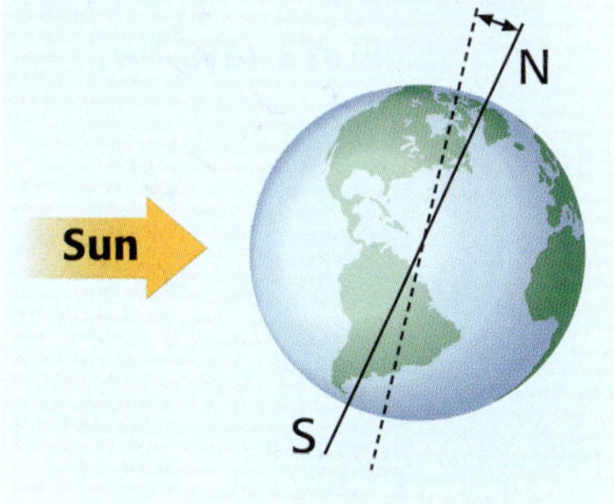

2 Over a period of 41,000 years, the tilt of the Earth's axis varies between 22.2° and 24.5°. When the tilt is at 24.5°, the poles receive more solar energy.

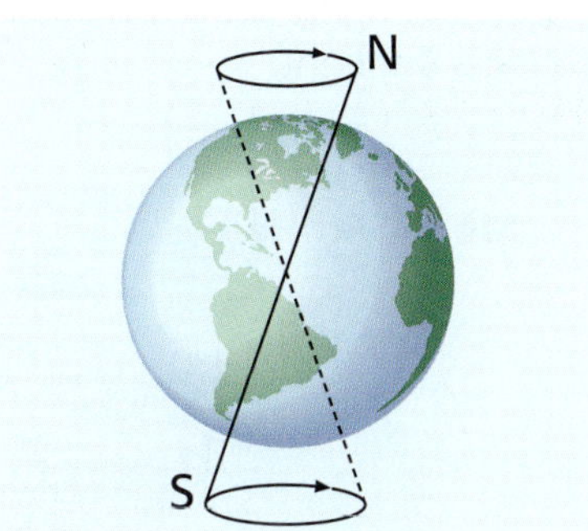

3 The Earth's axis traces a complete circle every 26,000 years. The circular motion of the Earth's axis determines the time of year that the Earth is closest to the sun.

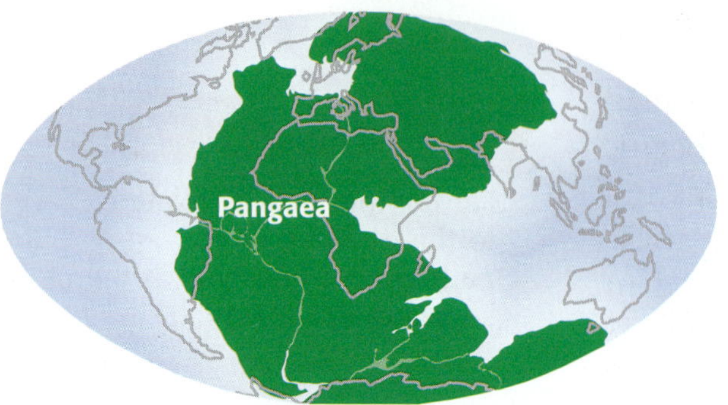

Figure 3 *Much of Pangaea—the part that is now Africa, South America, India, Antarctica, Australia, and Saudi Arabia—was covered by continental ice sheets.*

Plate Tectonics

The Earth's climate is further influenced by plate tectonics and continental drift. One theory proposes that ice ages happen when the continents are positioned closer to the polar regions. About 250 million years ago, all the continents were connected near the South Pole in one giant landmass called *Pangaea,* as shown in **Figure 3.** During this time, ice covered a large area of the Earth's surface. As Pangaea broke apart, the continents moved toward the equator, and the ice age ended. During the last ice age, many large landmasses were positioned in the polar zones. Antarctica, northern North America, Europe, and Asia were covered by large sheets of ice.

Volcanic Eruptions

Many natural factors can affect global climate. Catastrophic events, such as volcanic eruptions, can influence climate. Volcanic eruptions send large amounts of dust, ash, and smoke into the atmosphere. Once in the atmosphere, the dust, smoke, and ash particles act as a shield. This shield blocks the sun's rays, which causes the Earth to cool. **Figure 4** shows how dust particles from a volcanic eruption block the sun.

Reading Check **How can volcanoes change the climate?**

Figure 4 Volcanic Dust in the Atmosphere

Volcanic eruptions, such as the 1980 eruption of Mount St. Helens, as shown at right, produce dust that reflects sunlight.

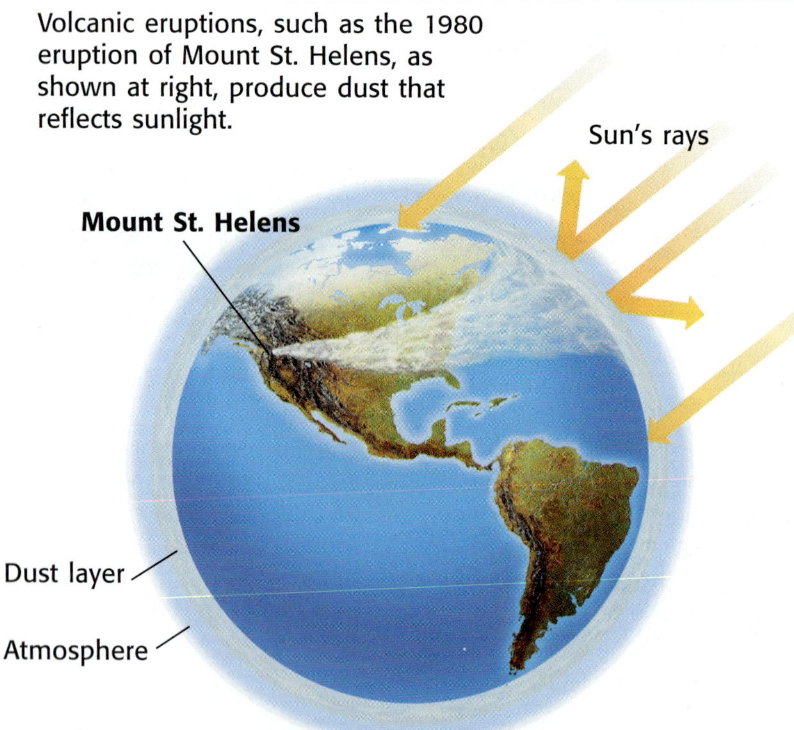

Figure 5 *Some scientists believe that a 10 km chunk of rock smashed into the Earth 65 million years ago, which caused the climatic change that resulted in the extinction of dinosaurs.*

Asteroid Impact

Imagine a rock the size of a car flying in from outer space and crashing in your neighborhood. This rock, like the one shown in **Figure 5,** is called an asteroid. An *asteroid* is a small, rocky object that orbits the sun. Sometimes, asteroids enter our atmosphere and crash into the Earth. What would happen if an asteroid 1 km wide, which is more than half a mile long, hit the Earth? Scientists believe that if an asteroid this big hit the Earth, it could change the climate of the entire world.

When a large piece of rock slams into the Earth, it causes debris to shoot into the atmosphere. *Debris* is dust and smaller rocks. This debris can block some of the sunlight and thermal energy. This would lower average temperatures, which would change the climate. Plants wouldn't get the sunlight they needed to grow, and animals would find surviving difficult. Scientists believe such an event is what caused dinosaurs to become extinct 65 million years ago when a 10 km asteroid slammed into the Earth and changed the Earth's climate.

The Sun's Cycle

Some changes in the climate can be linked to changes in the sun. You might think that the sun always stays the same. However, the sun follows an 11-year cycle. During this cycle, the sun changes from a solar maximum to a solar minimum. During a solar minimum, the sun produces a low percentage of high-energy radiation. But when the sun is at its solar maximum, it produces a large percentage of high-energy radiation. This increase in high-energy radiation warms the winds in the atmosphere. This change in turn affects climate patterns around the world.

CONNECTION TO Astronomy

Sunspots Sunspots are dark areas on the sun's surface. The number of sunspots changes with the sun's cycle. When the cycle is at a solar maximum, there are many sunspots. When the cycle is at a solar minimum, there are fewer sunspots. If the number of sunspots was low in 1997, in what year will the next low point in the cycle happen?

The Ride to School

1. The round-trip distance from your home to school is 20 km.
2. You traveled from home to school and from school to home 23 times in a month.
3. The vehicle in which you took your trips travels 30 km/gal.
4. If burning 1 gal of gasoline produces 9 kg of carbon dioxide, how much carbon dioxide did the vehicle release during the month?

Global Warming

A gradual increase in the average global temperature that is due to a higher concentration of gases, such as carbon dioxide in the atmosphere, is called **global warming.** To understand how global warming works, you must first learn about the greenhouse effect.

Greenhouse Effect

The Earth's natural heating process, in which gases in the atmosphere trap thermal energy, is called the **greenhouse effect.** The car in **Figure 6** shows how the greenhouse effect works. The car's windows stop most of the thermal energy from escaping, and the inside of the car gets hot. On Earth, instead of glass stopping the thermal energy, atmospheric gases absorb the thermal energy. When this happens, the thermal energy stays in the atmosphere and keeps the Earth warm. Many scientists believe that the rise in global temperatures is due to an increase of carbon dioxide, an atmospheric gas. Most evidence shows that the increase in carbon dioxide is caused by the burning of fossil fuels.

Another factor that may add to global warming is the clearing of forests. In many countries, forests are being burned to clear land for farming. Burning of the forests releases more carbon dioxide. Because plants use carbon dioxide to make food, destroying the trees decreases a natural way of removing carbon dioxide from the atmosphere.

global warming a gradual increase in the average global temperature

greenhouse effect the warming of the surface and lower atmosphere of Earth that occurs when carbon dioxide, water vapor, and other gases in the air absorb and trap thermal energy

Figure 6 *Sunlight streams into the car through the clear, glass windows. The seats absorb the radiant energy and change it into thermal energy. The energy is then trapped in the car.*

Consequences of Global Warming

Many scientists think that if the global temperature continues to rise, the ice caps will melt and cause flooding. Melted icecaps would raise the sea level and flood low-lying areas, such as the coasts.

Areas that receive little rainfall, such as deserts, might receive even less because of increased evaporation. Desert animals and plants would find surviving harder. Warmer and drier climates could harm crops in the Midwest of the United States. But farther north, such as in Canada, weather conditions for farming could improve.

Reading Check How would warmer temperatures affect deserts?

Reducing Pollution
Your city just received a warning from the Environmental Protection Agency for exceeding the automobile fuel emissions standards. Discuss with your parent ways that the city can reduce the amount of automobile emissions.

SECTION Review

Summary

- The Earth's climate experiences glacial and interglacial periods.
- The Milankovitch theory states that the Earth's climate changes as its orbit and the tilt of its axis change.
- Climate changes can be caused by volcanic eruptions, asteroid impact, the sun's cycle, and by global warming.
- Excess carbon dioxide is believed to contribute to global warming.

Using Key Terms

1. Use the following term in a sentence: *ice age.*
2. In your own words, write a definition for each of the following terms: *global warming* and *greenhouse effect.*

Understanding Key Ideas

3. Describe the possible causes of an ice age.
4. Which of the following can cause a change in the climate due to dust particles?
   a. volcanic eruptions
   b. plate tectonics
   c. solar cycles
   d. ice ages
5. How has the Earth's climate changed over time?
6. What might have caused the Earth's climate to change?
7. Which period of an ice age are we in currently? Explain.
8. Explain how the greenhouse effect warms the Earth.

Math Skills

9. After a volcanic eruption, the average temperature in a region dropped from 30° to 18°C. By how many degrees Celsius did the temperature drop?

Critical Thinking

10. **Analyzing Relationships** How will the warming of the Earth affect agriculture in different parts of the world? Explain.
11. **Predicting Consequences** How would deforestation (the cutting of trees) affect global warming?

Skills Practice Lab

Biome Business

OBJECTIVES

Interpret data in a climatograph.

Identify the biome for each climatograph.

You have just been hired as an assistant to a world-famous botanist. You have been provided with climatographs for three biomes. A *climatograph* is a graph that shows the monthly temperature and precipitation of an area in a year.

You can use the information provided in the three graphs to determine what type of climate each biome has. Next to the climatograph for each biome is an unlabeled map of the biome. Using the maps and the information provided in the graphs, you must figure out what the environment is like in each biome. You can find the exact location of each biome by tracing the map of the biome and matching it to the map at the bottom of the page.

Procedure

1. Look at each climatograph. The shaded areas show the average precipitation for the biome. The red line shows the average temperature.
2. Use the climatographs to determine the climate patterns for each biome. Compare the map of each biome with the map below to find the exact location of each biome.

Tundra
Taiga
Temperate forest
Tropical rain forest
Temperate grassland
Tropical savanna
Temperate desert
Tropical desert
Chaparral
Mountains

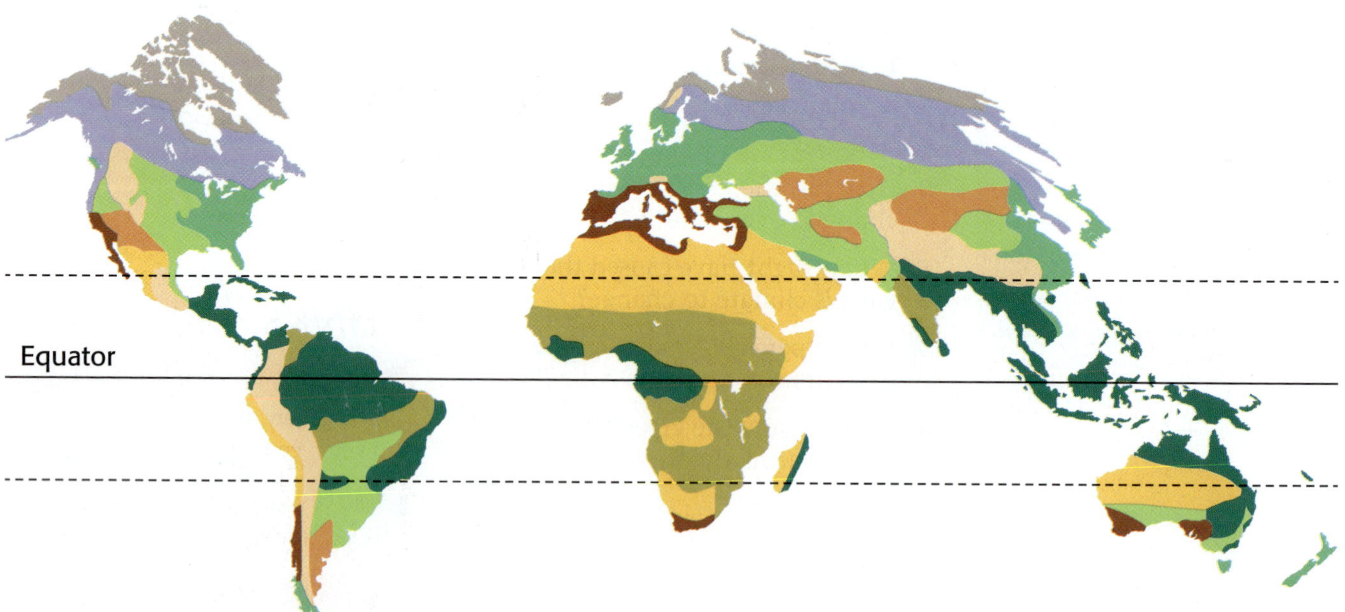

Analyze Results

1. **Analyzing Data** Describe the precipitation patterns of each biome by answering the following questions:
   a. In which month does the biome receive the most precipitation?
   b. Do you think that the biome is dry, or do you think that it is wet from frequent rains?

2. **Analyzing Data** Describe the temperature patterns of each biome by answering the following questions:
   a. In the biome, which months are warmest?
   b. Does the biome seem to have temperature cycles, like seasons, or is the temperature almost always the same?
   c. Do you think that the biome is warm or cold? Explain.

Draw Conclusions

3. **Drawing Conclusions** Name each biome.

4. **Applying Conclusions** Where is each biome located?

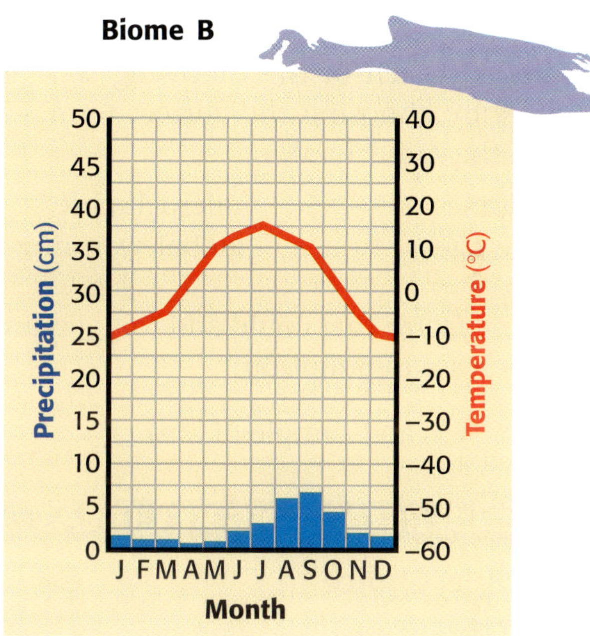

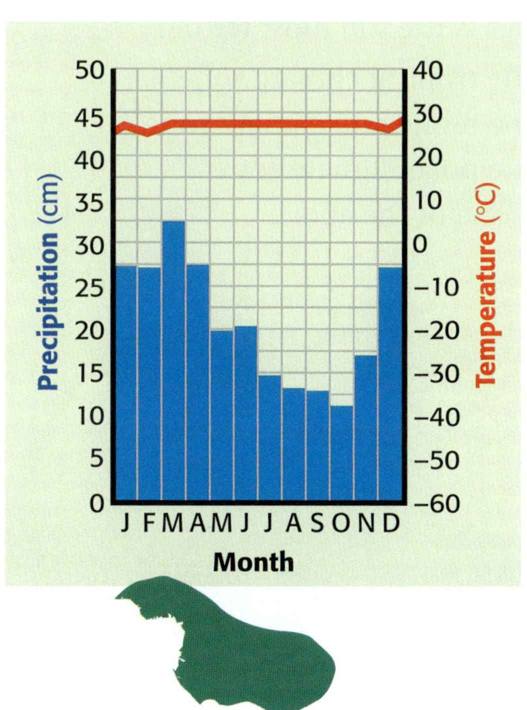

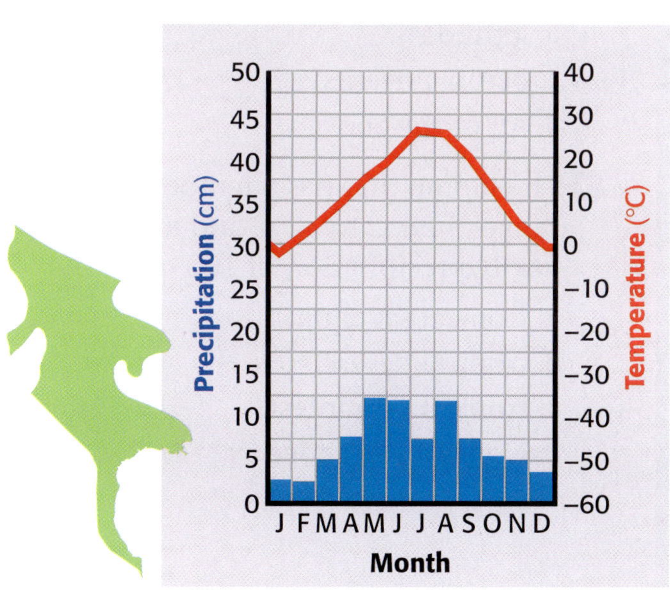

Chapter Review

USING KEY TERMS

For each pair of terms, explain how the meanings of the terms differ.

1. *biome* and *tropical zone*
2. *weather* and *climate*
3. *temperate zone* and *polar zone*

Complete each of the following sentences by choosing the correct term from the word bank.

biome	microclimate
ice age	global warming

4. One factor that could add to ___ is an increase in pollution.
5. A city is an example of a(n) ___.

UNDERSTANDING KEY IDEAS

Multiple Choice

6. Which of the following is a factor that affects climate?
   - **a.** prevailing winds
   - **b.** latitude
   - **c.** ocean currents
   - **d.** All of the above
7. The biome that has a temperature range of 28°C to 32°C and an average yearly precipitation of 100 cm is the
   - **a.** tropical savanna.
   - **b.** tropical desert.
   - **c.** tropical rain forest.
   - **d.** None of the above
8. Which of the following biomes is NOT found in the temperate zone?
   - **a.** temperate forest
   - **b.** taiga
   - **c.** chaparral
   - **d.** temperate grassland
9. In which of the following is the tilt of the Earth's axis considered to have an effect on climate?
   - **a.** global warming
   - **b.** the sun's cycle
   - **c.** the Milankovitch theory
   - **d.** asteroid impact
10. Which of the following substances contributes to the greenhouse effect?
    - **a.** smoke
    - **b.** smog
    - **c.** carbon dioxide
    - **d.** All of the above
11. In which of the following climate zones is the soil most fertile?
    - **a.** the tropical climate zone
    - **b.** the temperate climate zone
    - **c.** the polar climate zone
    - **d.** None of the above

Short Answer

12. Why do higher latitudes receive less solar radiation than lower latitudes do?

13. How does wind influence precipitation patterns?

14. Give an example of a microclimate. What causes the unique temperature and precipitation characteristics of this area?

15. How are tundras and deserts similar?

16. How does deforestation influence global warming?

CRITICAL THINKING

17. **Concept Mapping** Use the following terms to create a concept map: *global warming, deforestation, changes in climate, greenhouse effect, ice ages,* and *the Milankovitch theory.*

18. **Analyzing Processes** Explain how ocean surface currents cause milder climates.

19. **Identifying Relationships** Describe how the tilt of the Earth's axis affects seasonal changes in different latitudes.

20. **Evaluating Conclusions** Explain why the climate on the eastern side of the Rocky Mountains differs drastically from the climate on the western side.

21. **Applying Concepts** What are some steps you and your family can take to reduce the amount of carbon dioxide that is released into the atmosphere?

22. **Applying Concepts** If you wanted to live in a warm, dry area, which biome would you choose to live in?

23. **Evaluating Data** Explain why the vegetation in areas that have a tundra climate is sparse even though these areas receive precipitation that is adequate to support life.

INTERPRETING GRAPHICS

Use the diagram below to answer the questions that follow.

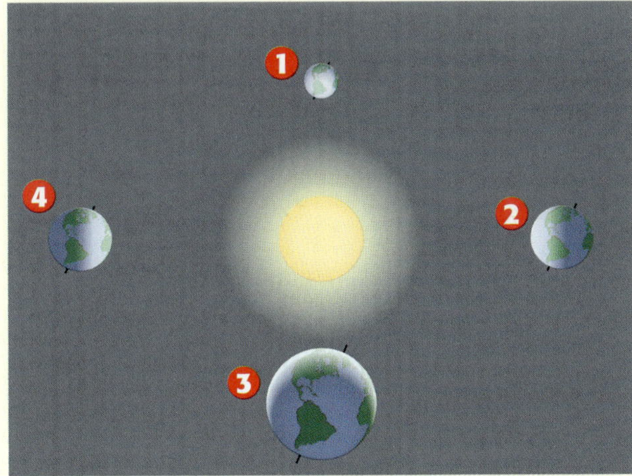

24. At what position—1, 2, 3, or 4—is it spring in the Southern Hemisphere?

25. At what position does the South Pole receive almost 24 hours of daylight?

26. Explain what is happening in each climate zone in both the Northern and Southern Hemispheres at position 4.

Standardized Test Preparation

READING

Read each of the passages below. Then, answer the questions that follow each passage.

Passage 1 Earth's climate has gone through many changes. For example, 6,000 years ago today's desert in North Africa was grassland and shallow lakes. Hippopotamuses, crocodiles, and early Stone Age people shared the shallow lakes that covered the area. For many years, scientists have known that Earth's climate has changed. What they didn't know was why it changed. Today, scientists can use supercomputers and complex computer programs to help them find the answer. Now, scientists may be able to decipher why North Africa's lakes and grasslands became a desert. And that information may be useful for predicting future heat waves and ice ages.

1. In this passage, what does *decipher* mean?

A to question
B to cover up
C to explain
D to calculate

2. According to the passage, which of the following statements is true?

F Scientists did not know that Earth's climate has changed.
G Scientists have known that Earth's climate has changed.
H Scientists have known why Earth's climate has changed.
I Scientists know that North Africa was always desert.

3. Which of the following is a fact in the passage?

A North African desert areas never had lakes.
B North American desert areas never had lakes.
C North African desert areas had shallow lakes.
D North Africa is covered with shallow lakes.

Passage 2 El Niño, which is Spanish for "the child," is the name of a weather event that occurs in the Pacific Ocean. Every 2 to 12 years, the interaction between the ocean surface and atmospheric winds creates El Niño. This event influences weather patterns in many regions of the world. For example, in Indonesia and Malaysia, El Niño meant drought and forest fires in 1998. Thousands of people in these countries suffered respiratory ailments caused by breathing the smoke from these fires. Heavy rains in San Francisco created extremely high mold-spore counts. These spores caused problems for people who have allergies. In San Francisco, the spore count in February is usually between 0 and 100. In 1998, the count was often higher than 8,000.

1. In this passage, what does *drought* mean?

A windy weather
B stormy weather
C long period of dry weather
D rainy weather

2. What can you infer about mold spores from reading the passage?

F Some people in San Francisco are allergic to mold spores.
G Mold spores are only in San Francisco.
H A higher mold-spore count helps people with allergies.
I The mold-spore count was low in 1998.

3. According to the passage, which of the following statements is true?

A El Niño causes droughts in Indonesia and Malaysia.
B El Niño occurs every year.
C El Niño causes fires in San Francisco.
D El Niño last occurred in 1998.

INTERPRETING GRAPHICS

The chart below shows types of organisms in an unknown biome. Use the chart below to answer the questions that follow.

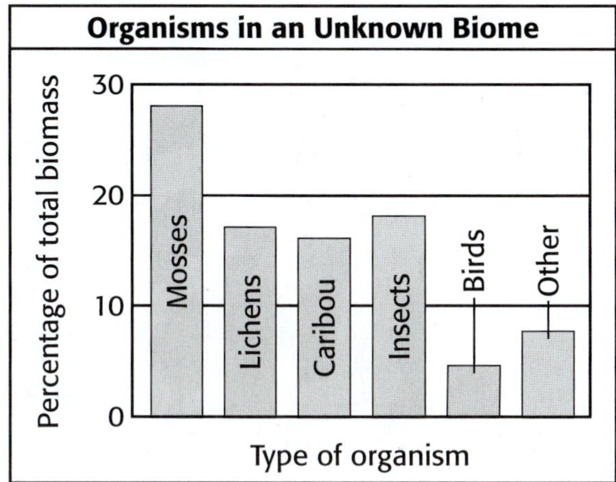

1. *Biomass* is a term that means "the total mass of all living things in a certain area." The graph above shows the relative percentages of the total biomass for different plants and animals in a given area. What type of biome does the graph represent?

A rain forest
B chaparral
C tundra
D taiga

2. Approximately what percentage of biomass is made up of caribou?

F 28%
G 25%
H 16%
I 5%

3. Approximately what percentage of biomass is made up of lichens and mosses?

A 45%
B 35%
C 25%
D 16%

MATH

Read each question below, and choose the best answer.

1. In a certain area of the savanna that is 12 km long and 5 km wide, there are 180 giraffes. How many giraffes are there per square kilometer in this area?

A 12
B 6
C 4
D 3

2. If the air temperature near the shore of a lake measures 24°C and the temperature increases by 0.055°C every 10 m traveled away from the lake, what would the air temperature 1 km from the lake be?

F 5°C
G 25°C
H 29.5°C
I 35°C

3. In a temperate desert, the temperature dropped from 50°C at noon to 37°C by nightfall. By how many degrees Celsius did the noon temperature drop?

A 13°C
B 20°C
C 26°C
D 50°C

4. Earth is tilted on its axis at a 23.5° angle. What is the measure of the angle that is complementary to a 23.5° angle?

F 66.5°
G 67.5°
H 156.5°
I 336.5°

5. After a volcanic eruption, the average temperature in a region dropped from 30°C to 18°C. By what percentage did the temperature drop?

A 30%
B 25%
C 40%
D 15%

Science in Action

Scientific Debate

Global Warming

Many scientists believe that pollution from burning fossil fuels is causing temperatures on Earth to rise. Higher average temperatures can cause significant changes in climate. These changes may make survival difficult for animals and plants that have adapted to a biome.

However, other scientists believe that there isn't enough evidence to prove that global warming exists. They argue that any increase in temperatures around the world can be caused by a number of factors other than pollution, such as the sun's cycle.

Language Arts ACTIVITY

WRITING SKILL Read articles that present a variety of viewpoints on global warming. Then, write your own article supporting your viewpoint on global warming.

Science, Technology, and Society

Ice Cores

How do scientists know what Earth's climate was like thousands of years ago? Scientists learn about Earth's past climates by studying ice cores. An ice core is collected by drilling a tube of ice from glaciers and polar ice sheets. Layers in the ice core contain substances that landed in the snow during a particular year or season, such as dust from desert storms, ash from volcanic eruptions, and carbon dioxide from pollution. By studying the layers of the ice cores, scientists can learn what factors influenced the past climates.

Math ACTIVITY

An area has an average yearly rainfall of 20 cm. In 1,000 years, if the average yearly rainfall decreases by 6%, what would the new average yearly rainfall be?

People in Science

Mercedes Pascual

Climate Change and Disease Mercedes Pascual is a theoretical ecologist at the University of Michigan. Pascual has been able to help the people of Bangladesh save lives by using information about climate changes to predict outbreaks of the disease cholera. Cholera can be a deadly disease that people usually contract by drinking contaminated water. Pascual knew that in Bangladesh, outbreaks of cholera peak every 3.7 years. She noticed that this period matches the frequency of the El Niño Southern Oscillations, which is a weather event that occurs in the Pacific Ocean. El Niño affects weather patterns in many regions of the world, including Bangladesh. El Niño increases the temperatures of the sea off the coast of Bangladesh. Pascual found that increased sea temperatures lead to higher numbers of the bacteria that cause cholera. In turn, more people contract cholera. But because of the research conducted by Pascual and other scientists, the people of Bangladesh can better predict and prepare for outbreaks of cholera.

Social Studies ACTIVITY

WRITING SKILL Research the effects of El Niño. Write a report describing El Niño and its affect on a country other than Bangladesh.

To learn more about these Science in Action topics, visit go.hrw.com and type in the keyword HZ5CLMF.

Current Science

Check out Current Science® articles related to this chapter by visiting go.hrw.com. Just type in the keyword HZ5CS17.

Stars, Galaxies, and the Universe

About the PHOTO

This image was taken by the *Hubble Space Telescope* and shows the IC 2163 galaxy (right) swinging past the NGC 2207 galaxy (left). Strong forces from NGC 2207 have caused stars and gas to fling out of IC 2163 into long streamers.

PRE-READING ACTIVITY

FOLDNOTES

Three-Panel Flip Chart

Before you read the chapter, create the FoldNote entitled "Three-Panel Flip Chart" described in the **Study Skills** section of the Appendix. Label the flaps of the three-panel flip chart with "Stars," "Galaxies," and "The universe." As you read the chapter, write information you learn about each category under the appropriate flap.

START-UP ACTIVITY

Exploring the Movement of Galaxies in the Universe

Not all galaxies are the same. Galaxies can differ by size, shape, and how they move in space. In this activity, you will explore how the galaxies in the photo move in space.

Procedure

1. Fill a **one-quart glass jar** three-fourths of the way with **water.**
2. Take a pinch of **glitter,** and sprinkle it on the surface of the water.
3. Quickly stir the water with a **wooden spoon.** Be sure to stir the water in a circular pattern.
4. After you stop stirring, look at the water from the sides of the jar and from the top of the jar.

Analysis

1. What kind of motion did the water make after you stopped stirring the water?
2. How is the motion similar to the galaxies in the photo?
3. Make up a name that describes the galaxies in the photo.

SECTION 1

Stars

Do you remember the children's song "Twinkle, Twinkle Little Star"? In the song, you sing "How I wonder what you are!" Well, what are stars? And what are they made of?

Most stars look like faint dots of light in the night sky. But stars are actually huge, hot, bright balls of gas that are trillions of kilometers away from Earth. How do astronomers learn about stars when the stars are too far away to visit? Astronomers study starlight!

READING WARM-UP

Objectives

- Describe how color indicates the temperature of a star.
- Explain how a scientist can identify a of star's composition.
- Describe how scientists classify stars.
- Compare absolute magnitude with apparent magnitude.
- Identify how astronomers measure distances from Earth to stars.
- Describe the difference between the apparent motion and the actual motion of stars.

Terms to Learn

spectrum
apparent magnitude
absolute magnitude
light-year
parallax

READING STRATEGY

Prediction Guide Before reading this section, write the title of each heading in this section. Next, under each heading, write what you think you will learn.

Color of Stars

Look at the flames on the candle and the Bunsen burner shown in **Figure 1.** Which flame is hottest? How can you tell? Although red and yellow may be thought of as "warm" colors and blue may be thought of as a "cool" color, scientists consider red and yellow to be cool colors and blue to be a warm color. For example, the blue flame of the Bunsen burner is much hotter than the yellow flame of the candle.

If you look carefully at the night sky, you might notice the different colors of some stars. Betelgeuse (BET uhl JOOZ), which is red, and Rigel (RIE juhl), which is blue, are the stars that form two corners of the constellation Orion, shown in **Figure 1.** Because these two stars are different colors, we can conclude that they have different temperatures.

Reading Check Which star is hotter, Betelgeuse or Rigel? Explain your answer. (*See the Appendix for answers to Reading Checks.*)

Figure 1 *In the same way that we know the blue flame of the Bunsen burner is hotter than the yellow flame of the candle, astronomers know that Rigel is hotter than Betelgeuse.*

Composition of Stars

A star is made up of different elements in the form of gases. The inner layers of a star are very dense and hot. But the outer layers of a star, or a star's atmosphere, are made up of cool gases. Elements in a star's atmosphere absorb some of the light that radiates from the star. Because different elements absorb different wavelengths of light, astronomers can tell what elements a star is made of from the light they observe from the star.

The Colors of Light

When you look at white light through a glass prism, you see a rainbow of colors called a **spectrum.** The spectrum consists of millions of colors, including red, orange, yellow, green, blue, indigo, and violet. A hot, solid object, such as the glowing wire inside a light bulb, gives off a *continuous spectrum*—a spectrum that shows all the colors. However, the spectrum of a star is different. Astronomers use an instrument called a *spectrograph* to break a star's light into a spectrum. The spectrum gives astronomers information about the composition and temperature of a star. To understand how to read a star's spectrum, think about something more familiar—a neon sign.

Making an ID

Many restaurants use neon signs to attract customers. The gas in a neon sign glows when an electric current flows through the gas. If you were to look at the sign with a spectrograph, you would not see a continuous spectrum. Instead, you would see *emission lines*. Emission lines are lines that are made when certain wavelengths of light, or colors, are given off by hot gases. When an element emits light, only some colors in the spectrum show up, while all the other colors are missing. Each element has a unique set of bright emission lines. Emission lines are like fingerprints for the elements. You can see emission lines for four elements in **Figure 2.**

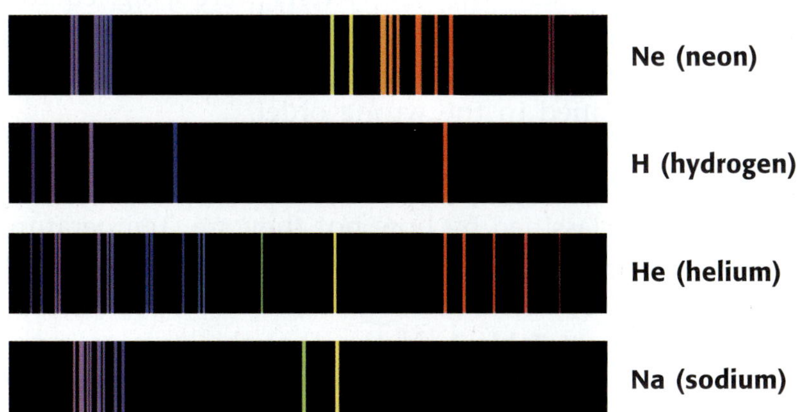

Figure 2 *Neon gas produces a unique set of emission lines, as do the elements hydrogen, helium, and sodium.*

CONNECTION TO Physics

WRITING SKILL **Fingerprinting Cars** Police use spectrographs to "fingerprint" cars. Car makers put trace elements in the paint of cars. Each make of car has a special paint and thus its own combination of trace elements. When a car is in a hit-and-run accident, police officers can identify the make of the car by the paint left behind. Using a spectrograph to identify a car is one of many scientific methods that police use to solve crimes. Solving crimes by using scientific equipment and methods is part of a science called *forensic science*. Research the topic of forensic science. In your **science journal,** write a short paragraph about the other scientific methods that forensic scientists use to solve crimes.

spectrum the band of color produced when white light passes through a prism

Trapping the Light—Cosmic Detective Work

Like an element that is charged by an electric current, a star also produces a spectrum. However, while the spectrum of an electrically charged element is made of bright emission lines, a star's spectrum is made of dark emission lines. A star's atmosphere absorbs certain colors of light in the spectrum, which causes black lines to appear.

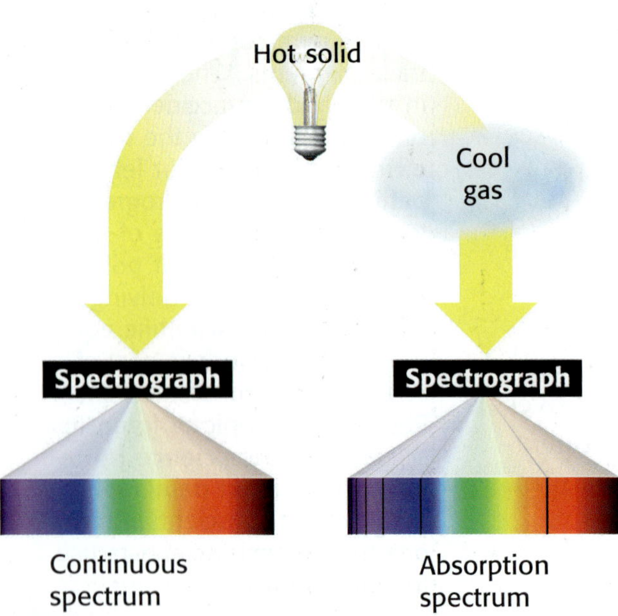

Figure 3 *A continuous spectrum (left) shows all colors while an absorption spectrum (right) absorbs some colors. Black lines appear in the spectrum where colors are absorbed.*

Identifying Elements Using Dark Lines

Because a star's atmosphere absorbs colors of light instead of emitting them, the spectrum of a star is called an *absorption spectrum*. An absorption spectrum is produced when light from a hot solid or dense gas passes through a cooler gas. Therefore, a star gives off an absorption spectrum because a star's atmosphere is cooler than the inner layers of the star. The black lines of a star's spectrum represent places where less light gets through. **Figure 3** compares a continuous spectrum and an absorption spectrum. What do you notice about the absorption spectrum that is different?

The pattern of lines in a star's absorption spectrum shows some of the elements that are in the star's atmosphere. If a star were made of one element, we could easily identify the element from the star's absorption spectrum. But a star is a mixture of elements and all the different sets of lines for a star's elements appear together in its spectrum. Sorting the patterns is often a puzzle.

Reading Check **What does a star's absorption spectrum show?**

Rods, Cones, and Stars WRITING SKILL Have you ever wondered why it's hard to see the different colors of stars? Our eyes are not sensitive to colors when light levels are low. There are two types of light-sensitive cells in the eye: rods and cones. Research the functions of rods and cones. In your **science journal,** write a paragraph that explains why we can't see colors well in low light.

Classifying Stars

In the 1800s, astronomers started to collect and classify the spectra of many stars. At first, letters were assigned to each type of spectra. Stars were classified according to the elements of which they were made. Later, scientists realized that the stars were classified in the wrong order.

Differences in Temperature

Stars are now classified by how hot they are. Temperature differences between stars result in color differences that you can see. For example, the original class O stars are blue—the hottest stars. Look at **Table 1.** Notice that the stars are arranged in order from highest temperature to lowest temperature.

Table 1 Types of Stars

Class	Color	Surface temperature (°C)	Elements detected	Examples of stars
O	blue	above 30,000	helium	10 Lacertae
B	blue-white	10,000–30,000	helium and hydrogen	Rigel, Spica
A	blue-white	7,500–10,000	hydrogen	Vega, Sirius
F	yellow-white	6,000–7,500	hydrogen and heavier elements	Canopus, Procyon
G	yellow	5,000–6,000	calcium and other metals	the sun, Capella
K	orange	3,500–5,000	calcium and molecules	Arcturus, Aldebaran
M	red	less than 3,500	molecules	Betelgeuse, Antares

Differences in Brightness

With only their eyes to aid them, early astronomers created a system to classify stars based on their brightness. They called the brightest stars in the sky *first-magnitude* stars and the dimmest stars *sixth-magnitude* stars. But when they began to use telescopes, astronomers were able to see many stars that had been too dim to see before. Rather than replace the old system of magnitudes, they added to it. Positive numbers represent dimmer stars, and negative numbers represent brighter stars. For example, by using large telescopes, astronomers can see stars as dim as 29th magnitude. And the brightest star in the night sky, Sirius, has a magnitude of -1.4. The Big Dipper, shown in **Figure 4,** contains both bright stars and dim stars.

Stargazing

WRITING SKILL Someone looking at the night sky in a city would not see as many stars as someone looking at the sky in the country. With a parent, research why this is true. Try to find a place near your home that would be ideal for stargazing. If you find one, schedule a night to stargaze. Write down what you see in the night sky.

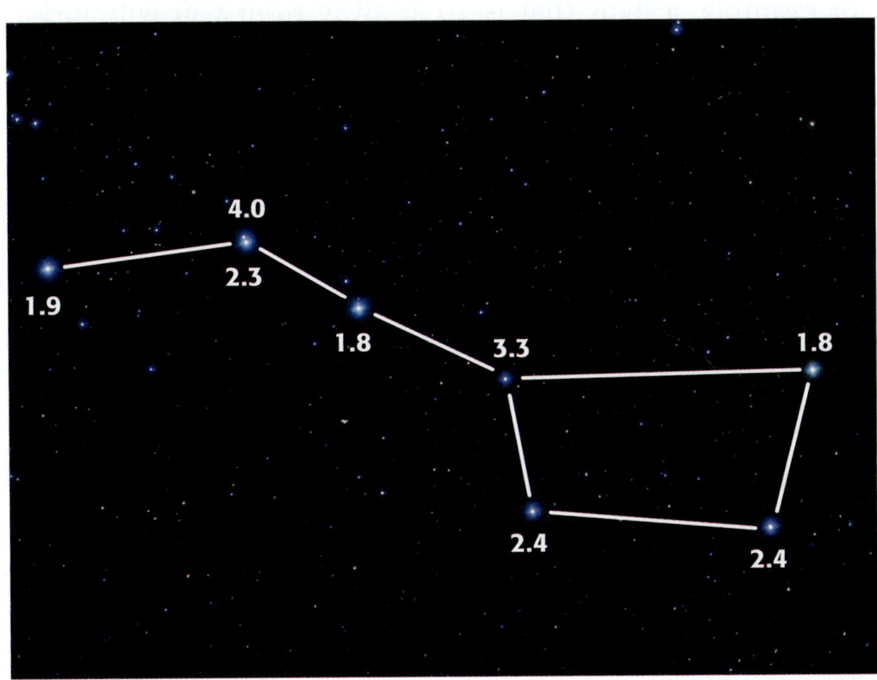

Figure 4 *The Big Dipper contains both bright stars and dim stars. What is the magnitude of the brightest star in the Big Dipper?*

Figure 5 *You can estimate how far away each street light is by looking at its apparent brightness. Does this process work when estimating the distance of stars from Earth?*

apparent magnitude the brightness of a star as seen from the Earth

absolute magnitude the brightness that a star would have at a distance of 32.6 light-years from Earth

How Bright Is That Star?

If you look at a row of street lights, such as those shown in **Figure 5,** do they all look the same? Of course not! The nearest ones look bright, and the farthest ones look dim.

Apparent Magnitude

The brightness of a light or star is called **apparent magnitude.** If you measure the brightness of a street light with a light meter, you will find that the light's brightness depends on the square of the ratio between the light and the light meter. For example, a light that is 10 m away from you will appear 4 (2×2, or 2^2) times brighter than a light that is 20 m away from you. The same light will appear 9 (3×3, or 3^2) times brighter than a light that is 30 m away. But unlike street lights, some stars are brighter than other stars because of their size or energy output, not because of their distance from Earth. So, how can you tell how bright a star is and why?

Reading Check What is apparent magnitude?

Absolute Magnitude

Astronomers use a star's apparent magnitude and its distance from Earth to calculate its absolute magnitude. **Absolute magnitude** is the actual brightness of a star. If all stars were the same distance away, their absolute magnitudes would be the same as their apparent magnitudes. The sun, for example, has an absolute magnitude of +4.8, which is ordinary for a star. But because the sun is so close to Earth, the sun's apparent magnitude is −26.8, which makes it the brightest object in the sky.

Starlight, Star Bright

Magnitude is used to show how bright one object is compared with another object. Every five magnitudes is equal to a factor of 100 times in brightness. The brightest blue stars, for example, have an absolute magnitude of −10. The sun has an absolute magnitude of about +5. How much brighter is a blue star than the sun? Because each five magnitudes is a factor of 100 and the blue star is 15 magnitudes greater than the sun, the blue star must be 100 × 100 × 100, or 1,000,000 (1 million), times brighter than the sun!

Figure 6 Measuring a Star's Parallax

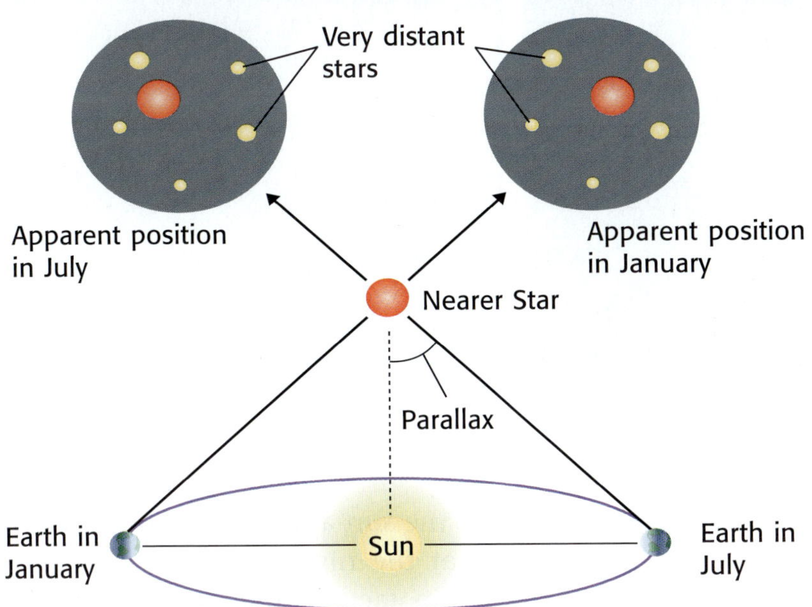

light-year the distance that light travels in one year; about 9.5 trillion kilometers

parallax an apparent shift in the position of an object when viewed from different locations

Distance to the Stars

Because stars are so far away, astronomers use light-years to measure the distances from Earth to the stars. A **light-year** is the distance that light travels in one year. Obviously, it would be easier to give the distance to the North Star as 431 light-years than as 4,080,000,000,000,000 km. But how do astronomers measure a star's distance from Earth?

Stars near the Earth seem to move, while more-distant stars seem to stay in one place as Earth revolves around the sun, as shown in **Figure 6.** A star's apparent shift in position is called **parallax.** Notice that the location of the nearer star in **Figure 6** seems to shift in relation to the pattern of more-distant stars. This shift can be seen only through telescopes. Astronomers use parallax and simple trigonometry (a type of math) to find the actual distance to stars that are close to Earth.

Reading Check What is a light-year?

Motions of Stars

As you know, daytime and nighttime are caused by the Earth's rotation. The Earth's tilt and revolution around the sun cause the seasons. During each season, the Earth faces a different part of the sky at night. Look again at **Figure 6.** In January, the Earth's night side faces a different part of the sky than it faces in July. This is why you see a different set of constellations at different times of the year.

Not All Thumbs!

1. Hold your thumb in front of your face at arm's length.
2. Close one eye, and focus on an **object** some distance behind your thumb.
3. Slowly turn your head side to side a small amount. Notice how your thumb seems to be moving compared with the background you are looking at.
4. Now, move your thumb in close to your face, and move your head the same amount. Does your thumb seem to move more?

Figure 7 *As Earth rotates on its axis, the stars appear to rotate around Polaris.*

The Apparent Motion of Stars

Because of Earth's rotation, the sun appears to move across the sky. Likewise, if you look at the night sky long enough, the stars also appear to move. In fact, at night you can observe that the whole sky is rotating above us. Look at **Figure 7.** All the stars you see appear to rotate around Polaris, the North Star, which is almost directly above Earth's North Pole. Because of Earth's rotation, all of the stars in the sky appear to make one complete circle around Polaris every 24 h.

The Actual Motion of Stars

You now know that the apparent motion of the sun and stars in our sky is due to Earth's rotation. But each star is also moving in space. Because stars are so distant, however, their actual motion is hard to see. If you could put thousands of years into one hour, a star's movement would be obvious. **Figure 8** shows how familiar star patterns slowly change their shapes.

Reading Check **Why is the actual motion of stars hard to see?**

Figure 8 *Over time, the shapes of star patterns, such as the Big Dipper and other groups, change.*

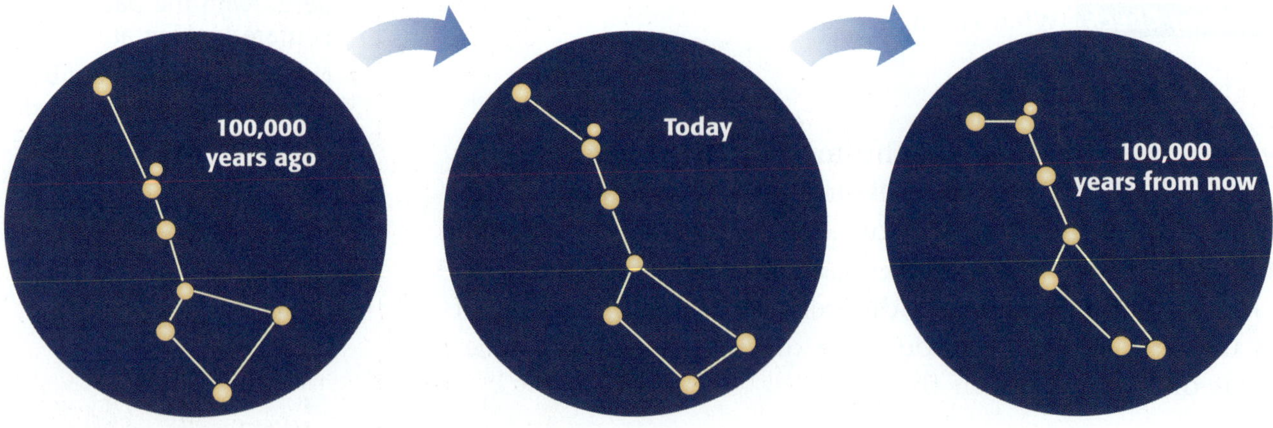

SECTION Review

Summary

- The color of a star depends on its temperature. Hot stars are blue. Cool stars are red.
- The spectrum of a star shows the composition of a star.
- Scientists classify stars by temperature and brightness.
- Apparent magnitude is the brightness of a star as seen from Earth.
- Absolute magnitude is the measured brightness of a star at a distance of 32.6 light-years.
- Astronomers use parallax and trigonometry to measure distances from Earth to stars.
- Stars appear to move because of Earth's rotation. However, the actual motion of stars is very hard to see because stars are so distant.

Using Key Terms

1. Use the following terms in the same sentence: *apparent magnitude* and *absolute magnitude*.
2. Use each of the following terms in a separate sentence: *spectrum, light-year,* and *parallax*.

Understanding Key Ideas

3. When you look at white light through a glass prism, you see a rainbow of colors called a
   a. spectograph.
   b. spectrum.
   c. parallax.
   d. light-year.
4. Class F stars are
   a. blue.
   b. yellow.
   c. yellow-white.
   d. red.
5. Describe how scientists classify stars.
6. Explain how color indicates the temperature of a star.

Critical Thinking

7. **Applying Concepts** If a certain star displayed a large parallax, what could you say about the star's distance from Earth?
8. **Making Comparisons** Compare a continuous spectrum with an absorption spectrum. Then, explain how an absorption spectrum can identify a star's composition.
9. **Making Comparisons** Compare apparent motion with actual motion.

Interpreting Graphics

10. Look at the two figures below. How many hours passed between the first image and the second image? Explain your answer.

SECTION 2

The Sun: Our Very Own Star

Some stars are so far away from Earth that it takes years for their light to reach us! Because the sun is our closest star, it only takes minutes for light from the sun to reach Earth.

Energy from the sun lights and heats Earth's surface. Energy from the sun even drives the weather. Making up more than 99% of the solar system's mass, the sun is the dominant member of our solar system. The sun is basically a large ball of gas made mostly of hydrogen and helium held together by gravity. But what does the inside of the sun look like?

READING WARM-UP

Objectives

- Describe the basic structure and composition of the sun.
- Explain how the sun generates energy.
- Describe the surface activity of the sun, and identify how this activity affects Earth.

Terms to Learn

nuclear fusion
sunspot

READING STRATEGY

Reading Organizer As you read this section, create an outline of the section. Use the headings from the section in your outline.

The Structure of the Sun

Although the sun may appear to have a solid surface, it does not. When you see a picture of the sun, you are really seeing through the sun's outer atmosphere. The visible surface of the sun starts at the point where the gas becomes so thick that you cannot see through it. As **Figure 1** shows, the sun is made of several layers.

Figure 1 The Structure and Atmosphere of the Sun

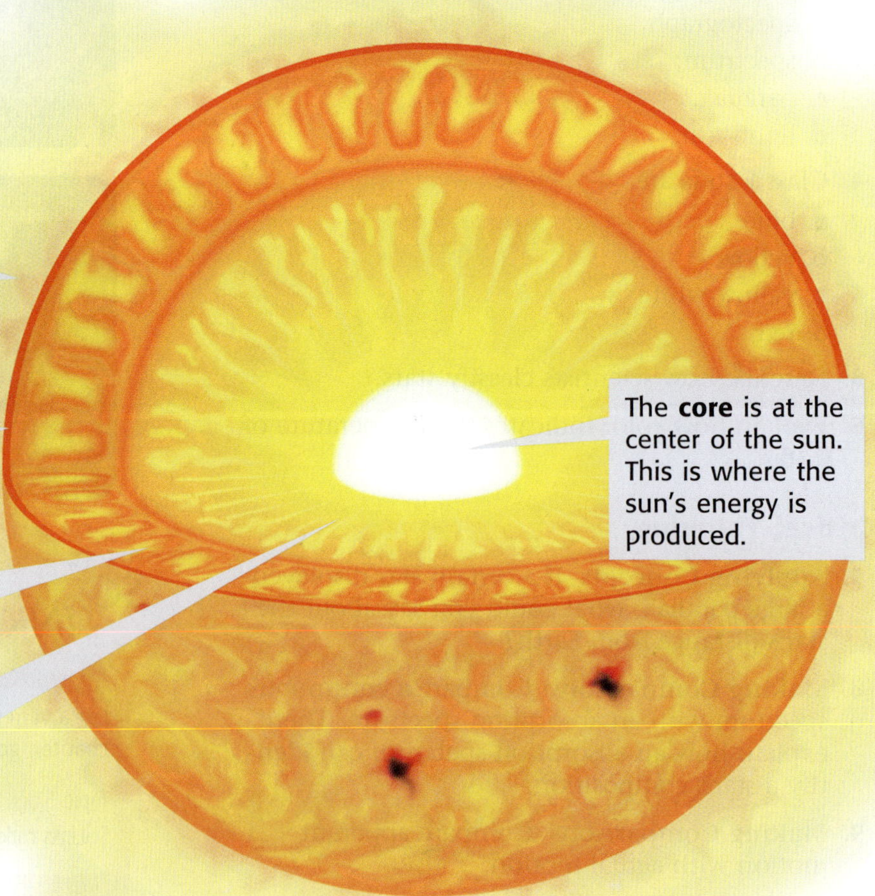

Figure 2 *Ideas about the source of the sun's energy have changed over time.*

Energy Production in the Sun

The sun has been shining on Earth for about 4.6 billion years. How can the sun stay hot for so long? And what makes it shine? **Figure 2** shows two theories that were proposed to answer these questions. Many scientists thought that the sun burned fuel to generate its energy. But the amount of energy that is released by burning would not be enough to power the sun. If the sun were simply burning, it would last for only 10,000 years.

Burning or Shrinking?

It eventually became clear to scientists that burning wouldn't last long enough to keep the sun shining. Then, scientists began to think that gravity was causing the sun to slowly shrink. They thought that perhaps gravity would release enough energy to heat the sun. While the release of gravitational energy is more powerful than burning, it is not enough to power the sun. If all of the sun's gravitational energy were released, the sun would last for only 45 million years. However, fossils that have been discovered prove that dinosaurs roamed the Earth more than 65 million years ago, so this couldn't be the case. Therefore, something even more powerful than gravity was needed.

✓ Reading Check **Why isn't energy from gravity enough to power the sun?** (*See the Appendix for answers to Reading Checks.*)

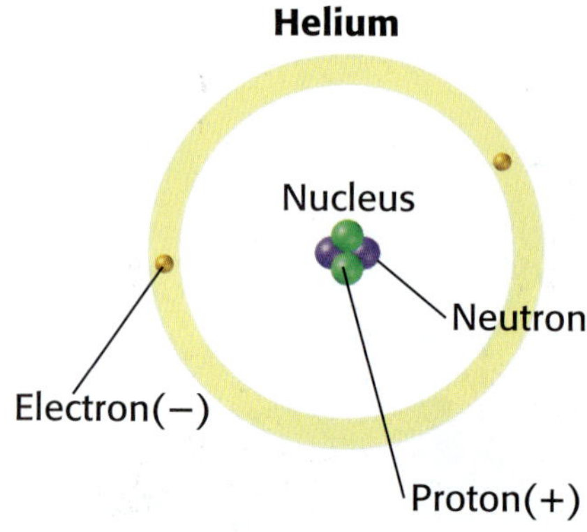

CONNECTION TO Chemistry

Atoms An atom consists of a nucleus surrounded by one or more electrons. Electrons have a negative charge. In most elements, the atom's nucleus is made up of two types of particles: *protons,* which have a positive charge, and *neutrons,* which have no charge. The protons in the nucleus are usually balanced by an equal number of electrons. The number of protons and electrons gives the atom its chemical identity. A helium atom, shown at left, has two protons, two neutrons, and two electrons. Use a Periodic Table to find the chemical identity of the following atoms: nitrogen, oxygen, and carbon.

Nuclear Fusion

At the beginning of the 20th century, Albert Einstein showed that matter and energy are interchangeable. Matter can change into energy according to his famous formula: $E = mc^2$. (E is energy, m is mass, and c is the speed of light.) Because c is such a large number, tiny amounts of matter can produce a huge amount of energy. With this idea, scientists began to understand a very powerful source of energy.

nuclear fusion the combination of the nuclei of small atoms to form a larger nucleus; releases energy

Nuclear fusion is the process by which two or more low-mass nuclei join together, or fuse, to form another nucleus. In this way, four hydrogen nuclei can fuse to form a single nucleus of helium. During the process, energy is produced. Scientists now know that the sun gets its energy from nuclear fusion. Einstein's equation, shown in **Figure 3,** changed ideas about the sun's energy source by equating mass and energy.

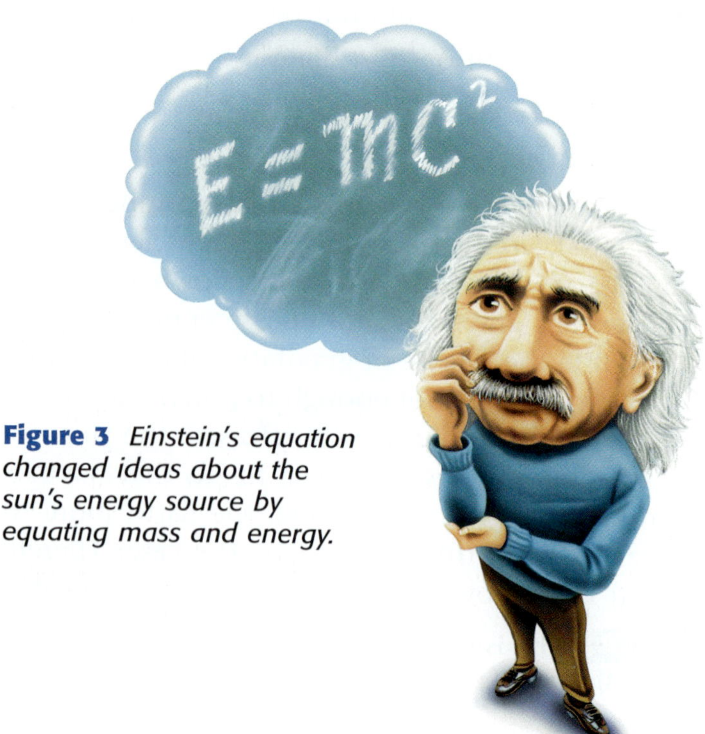

Figure 3 *Einstein's equation changed ideas about the sun's energy source by equating mass and energy.*

Fusion in the Sun

Under normal conditions, the nuclei of hydrogen atoms never get close enough to combine. The reason is that they are positively charged. Like charges repel each other, as shown in **Figure 4.** In the center of the sun, however, the temperature and pressure are very high. As a result, the hydrogen nuclei have enough energy to overcome the repulsive force, and hydrogen fuses into helium, as shown in **Figure 5.**

The energy produced in the center, or core of the sun takes millions of years to reach the sun's surface. The energy passes from the core through a very dense region called the *radiative zone*. The matter in the radiative zone is so crowded that the light and energy are blocked and sent in different directions. Eventually, the energy reaches the *convective zone*. Gases circulate in the convective zone. Hot gases in the convective zone carry the energy up to the *photosphere,* the visible surface of the sun. From there, the energy leaves the sun as light, which takes only 8.3 min to reach Earth. Only a small fraction of the sun's light reaches the surface of Earth.

Reading Check **What causes the nuclei of hydrogen atoms to repel each other?**

Figure 4 *Like charges repel just as similar poles on a pair of magnets do.*

Figure 5 Fusion of Hydrogen in the Sun

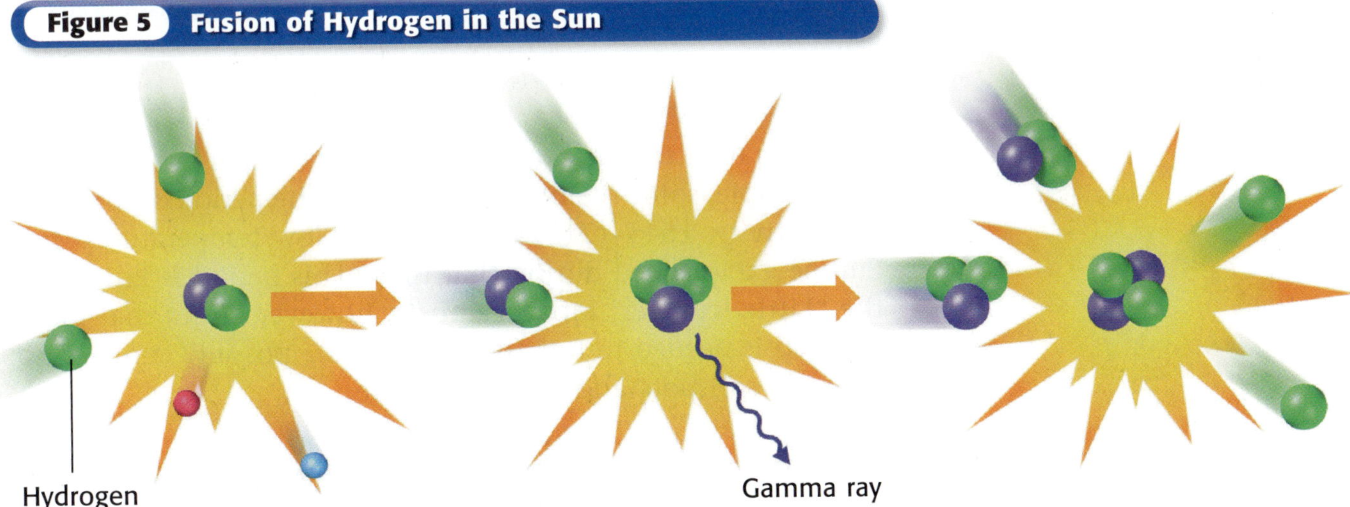

1 Deuterium Two hydrogen nuclei (protons) collide. One proton emits particles and energy and then becomes a neutron. The proton and neutron combine to produce a heavy form of hydrogen called *deuterium*.

2 Helium-3 Deuterium combines with another hydrogen nucleus to form a variety of helium called *helium-3*. More energy, as well as gamma rays, is released.

3 Helium-4 Two helium-3 atoms then combine to form ordinary helium-4, which releases more energy and a pair of hydrogen nuclei.

Solar Activity

The photosphere is an ever-changing place. Thermal energy moves from the sun's interior by the circulation of gases in the convective zone. This movement of energy causes the gas in the photosphere to boil and churn. This circulation, combined with the sun's rotation, creates magnetic fields that reach far out into space.

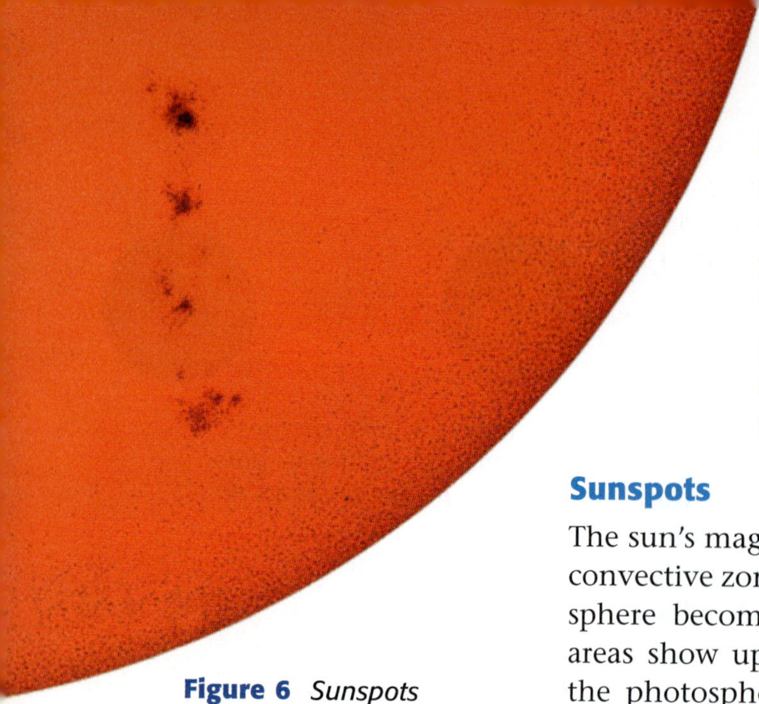

Sunspots

The sun's magnetic fields tend to slow down the activity in the convective zone. When activity slows down, areas of the photosphere become cooler than surrounding areas. These cooler areas show up as sunspots. **Sunspots** are cooler, dark spots of the photosphere of the sun, as shown in **Figure 6.** Sunspots can vary in shape and size. Some sunspots can be as large as 50,000 miles in diameter.

Figure 6 *Sunspots mark cooler areas on the sun's surface. They are related to changes in the magnetic properties of the sun.*

The numbers and locations of sunspots on the sun change in a regular cycle. Scientists have found that the sunspot cycle lasts about 11 years. Every 11 years, the amount of sunspot activity in the sun reaches a peak intensity and then decreases. **Figure 7** shows the sunspot cycle since 1610, excluding the years 1645–1715, which was a period of unusually low sunspot activity.

sunspot a dark area of the photosphere of the sun that is cooler than the surrounding areas and that has a strong magnetic field

Reading Check **What are sunspots? What causes sunspots to occur?**

Climate Confusion

Scientists have found that sunspot activity can affect the Earth. For example, some scientists have linked the period of low sunspot activity, 1645–1715, with the very low temperatures that Europe experienced during that time. This period is known as the "Little Ice Age." Most scientists, however, think that more research is needed to fully understand the possible connection between sunspots and Earth's climate.

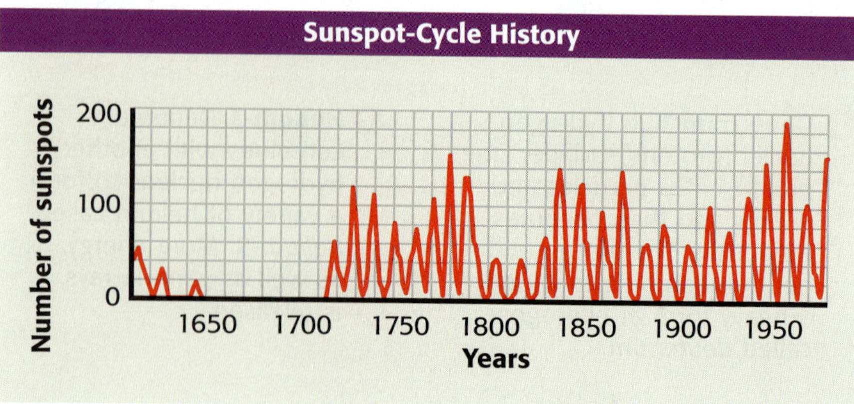

Figure 7 *This graph shows the number of sunspots that have occurred each year since Galileo's first observation in 1610.*

Solar Flares

The magnetic fields that cause sunspots also cause solar flares. *Solar flares,* as shown in **Figure 8,** are regions of extremely high temperature and brightness that develop on the sun's surface. When a solar flare erupts, it sends huge streams of electrically charged particles into the solar system. Solar flares can extend upward several thousand kilometers within minutes. Solar flares are usually associated with sunspots and can interrupt radio communications on Earth and in orbit. Scientists are trying to find ways to give advance warning of solar flares.

Figure 8 *Solar flares are giant eruptions on the sun's surface.*

SECTION Review

Summary

- The sun is a large ball of gas made mostly of hydrogen and helium. The sun consists of many layers.
- The sun's energy comes from nuclear fusion that takes place in the center of the sun.
- The visible surface of the sun, or the photosphere, is very active.
- Sunspots and solar flares are the result of the sun's magnetic fields that reach space.
- Sunspot activity may affect Earth's climate, and solar flares can interact with Earth's atmosphere.

Using Key Terms

1. In your own words, write a definition for each of the following terms: *sunspot* and *nuclear fusion.*

Understanding Key Ideas

2. Which of the following statements describes how energy is produced in the sun?
   - **a.** The sun burns fuels to generate energy.
   - **b.** As hydrogen changes into helium deep inside the sun, a great deal of energy is made.
   - **c.** Energy is released as the sun shrinks because of gravity.
   - **d.** None of the above
3. Describe the composition of the sun.
4. Name and describe the layers of the sun.
5. In which area of the sun do sunspots appear?
6. Explain how sunspots form.
7. Describe how sunspots can affect the Earth.
8. What are solar flares, and how do they form?

Math Skills

9. If the equatorial diameter of the sun is 1.39 million kilometers, how many kilometers is the sun's radius?

Critical Thinking

10. **Applying Concepts** If nuclear fusion in the sun's core suddenly stopped today, would the sky be dark in the daytime tomorrow? Explain.
11. **Making Comparisons** Compare the theories that scientists proposed about the source of the sun's energy with the process of nuclear fusion in the sun.

SECTION 3

The Life Cycle of Stars

Some stars exist for billions of years. But how are they born? And what happens when a star dies?

READING WARM-UP

Objectives

- Describe different types of stars.
- Describe the quantities that are plotted in the H-R diagram.
- Explain how stars at different stages in their life cycle appear on the H-R diagram.

Terms to Learn

red giant	supernova
white dwarf	neutron star
H-R diagram	pulsar
main sequence	black hole

READING STRATEGY

Paired Summarizing Read this section silently. In pairs, take turns summarizing the material. Stop to discuss ideas that seem confusing.

Because stars exist for billions of years, scientists cannot observe a star throughout its entire life. Therefore, scientists have developed theories about the life cycle of stars by studying them in different stages of development.

The Beginning and End of Stars

A star enters the first stage of its life cycle as a ball of gas and dust. Gravity pulls the gas and dust together into a sphere. As the sphere becomes denser, it gets hotter and the hydrogen changes to helium in a process called *nuclear fusion.*

As stars get older, they lose some of their material. Stars usually lose material slowly, but sometimes they can lose material in a big explosion. Either way, when a star dies, much of its material returns to space. In space, some of the material combines with more gas and dust to form new stars.

Different Types of Stars

Stars can be classified by their size, mass, brightness, color, temperature, spectrum, and age. Some types of stars include *main-sequence stars, giants, supergiants,* and *white dwarf stars.* A star can be classified as one type of star early in its life cycle and then can be classified as another star when it gets older. For example, the star shown in **Figure 1** has reached the final stage in its life cycle. It has run out of fuel, which has caused the central parts of the star to collapse inward.

Figure 1 *This star (center) has entered the last stage of its life cycle.*

Main-Sequence Stars

After a star forms, it enters the second and longest stage of its life cycle known as the main sequence. During this stage, energy is generated in the core of the star as hydrogen atoms fuse into helium atoms. This process releases an enormous amount of energy. The size of a main-sequence star will change very little as long as the star has a continuous supply of hydrogen atoms to fuse into helium atoms.

Giants and Supergiants

After the main-sequence stage, a star can enter the third stage of its life cycle. In this third stage, a star can become a red giant. A **red giant** is a star that expands and cools once it uses all of its hydrogen. Eventually, the loss of hydrogen causes the center of the star to shrink. As the center of the star shrinks, the atmosphere of the star grows very large and cools to form a red giant or a red supergiant, as shown in **Figure 2.** Red giants can be 10 or more times bigger than the sun. Supergiants are at least 100 times bigger than the sun.

Figure 2 *The red supergiant star Antares is shown above. Antares is located in the constellation of Scorpius.*

Reading Check **What is the difference between a red giant star and a red supergiant star?** *(See the Appendix for answers to Reading Checks.)*

White Dwarfs

In the final stages of a star's life cycle, a star that has the same mass as the sun or smaller can be classified as a white dwarf. A **white dwarf** is a small hot star that is the leftover center of an older star. A white dwarf has no hydrogen left and can no longer generate energy by fusing hydrogen atoms into helium atoms. White dwarfs can shine for billions of years before they cool completely.

red giant a large, reddish star late in its life cycle

white dwarf a small, hot, dim star that is the leftover center of an old star

CONNECTION TO Astronomy

WRITING SKILL **Long Live the Sun** Our sun probably took about 10 million years to become a main-sequence star. It has been shining for about 5 billion years. In another 5 billion years, our sun will burn up most of its hydrogen and expand to become a red giant. When this change happens, the sun's diameter will increase. How will this change affect Earth and our solar system? Use the Internet or library resources to find out what might happen as the sun gets older and how the changes in the sun might affect our solar system. Gather your findings, and write a report on what you find out about the life cycle the sun.

A Tool for Studying Stars

H-R diagram **H**ertzsprung-**R**ussell diagram, a graph that shows the relationship between a star's surface temperature and absolute magnitude

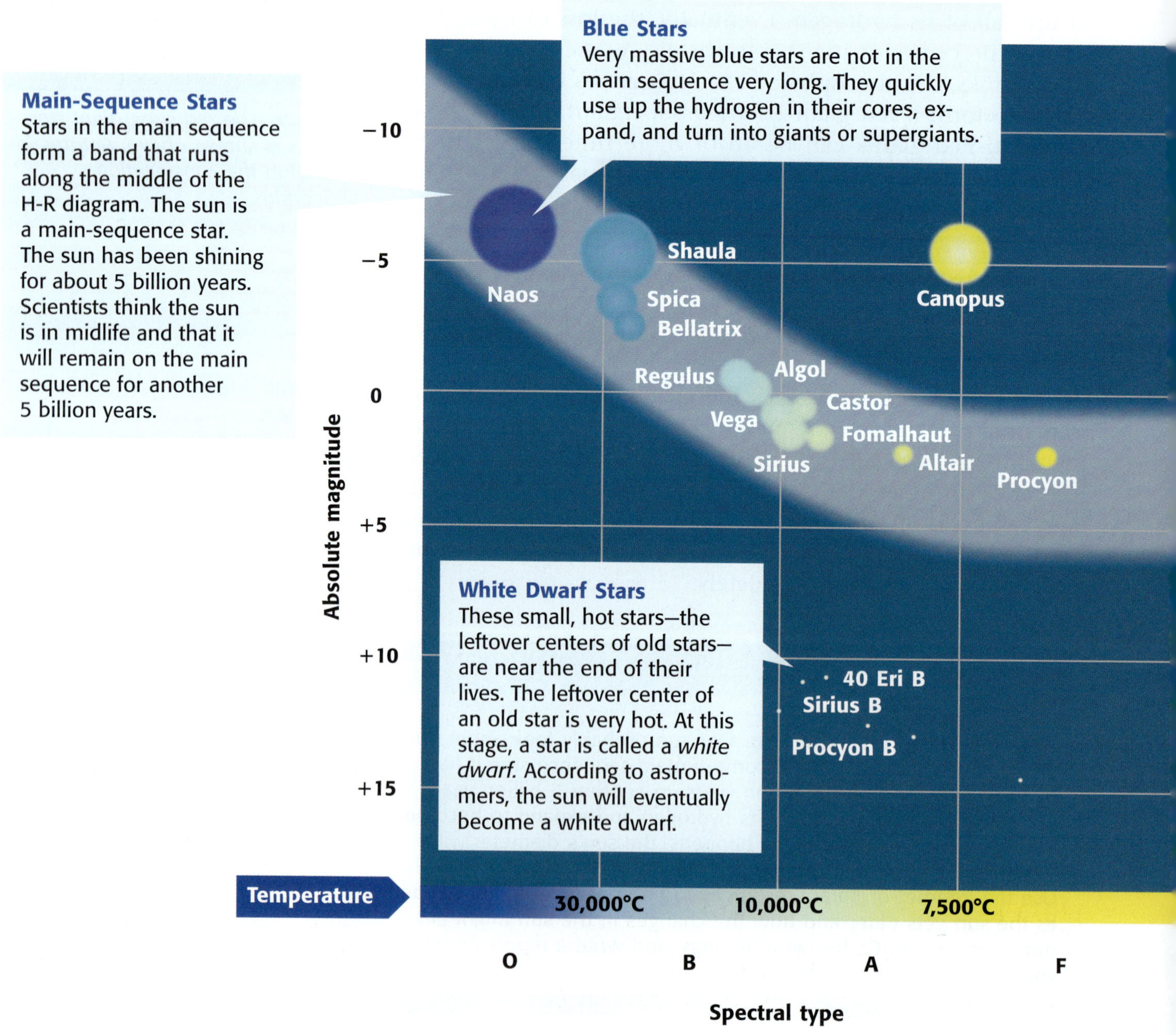

In 1911, a Danish astronomer named Ejnar Hertzsprung (IE nawr HUHRTS sproong) compared the brightness and temperature of stars on a graph. Two years later, American astronomer Henry Norris Russell made some similar graphs. Although these astronomers used different data, they had similar results. The combination of their ideas is now called the Hertzsprung-Russell diagram, or H-R diagram. The **H-R diagram** is a graph that shows the relationship between a star's surface temperature and its absolute magnitude. Over the years, the H-R diagram has become a tool for studying the lives of stars. It shows not only how stars are classified by brightness and temperature but also how stars change over time.

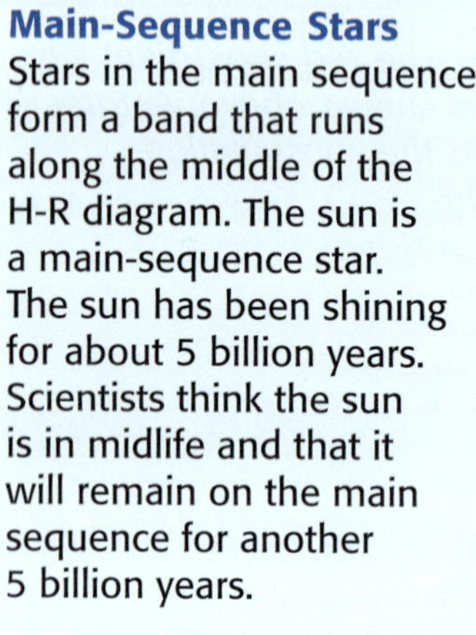

Reading the H-R Diagram

The modern H-R diagram is shown below. Temperature is given along the bottom of the diagram and absolute magnitude, or brightness, is given along the left side. Hot (blue) stars are located on the left, and cool (red) stars are on the right. Bright stars are at the top, and dim stars are at the bottom. The brightest stars are 1 million times brighter than the sun. The dimmest stars are 1/10,000 as bright as the sun. The diagonal pattern on the H-R diagram where most stars lie, is called the **main sequence.** A star spends most of its lifetime in the main sequence. As main-sequence stars age, they move up and to the right on the H-R diagram to become giants or supergiants and then down and to the left to become white dwarfs.

main sequence the location on the H-R diagram where most stars lie

Giants and Supergiants
When a star runs out of hydrogen in its core, the center of the star shrinks inward and the outer parts expand outward. For a star the size of our sun, the star's atmosphere will grow very large and become cool. When this change happens, the star becomes a *red giant.* If the star is very massive, it becomes a supergiant.

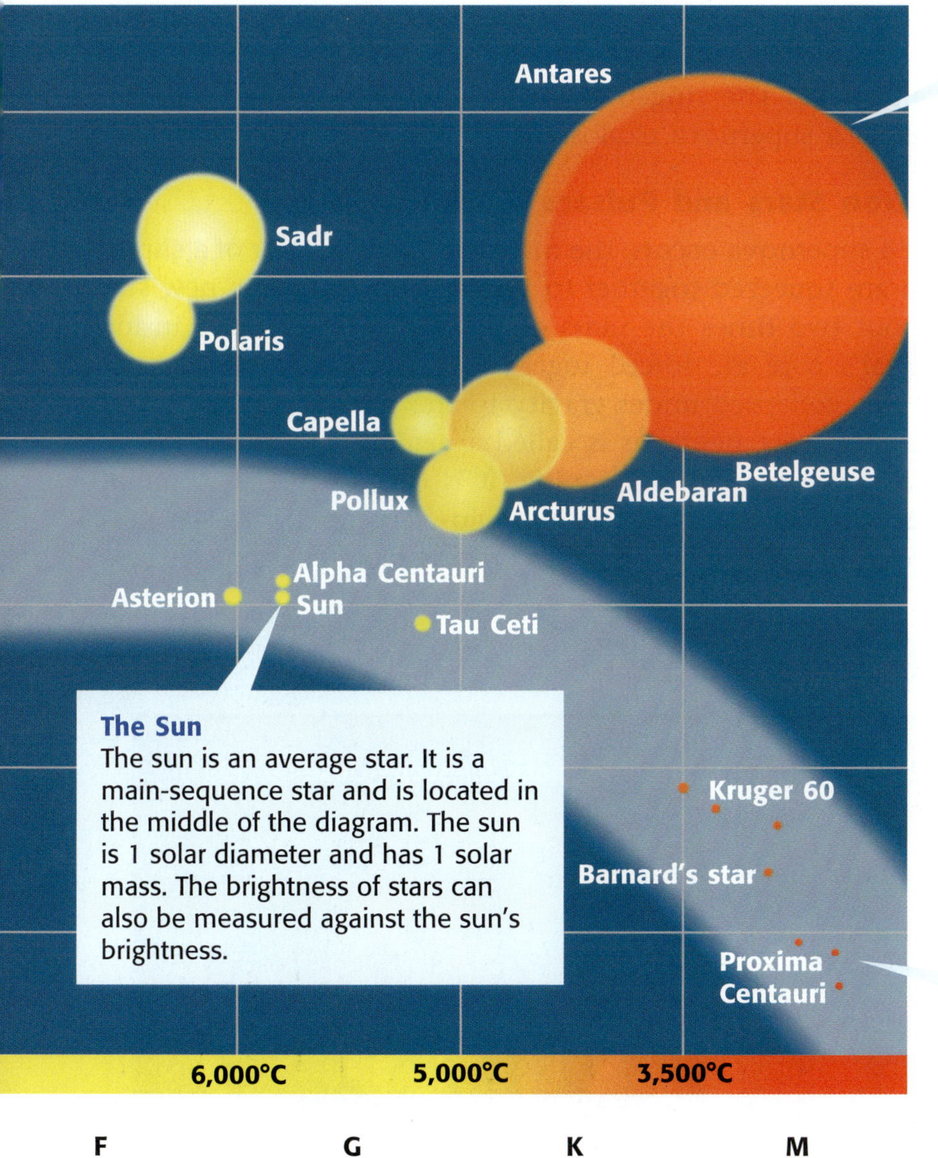

Red Dwarf Stars
At the lower end of the main sequence are the red dwarf stars, which are low-mass stars. Low-mass stars remain on the main sequence a long time. The stars that have the lowest mass are among the oldest stars in the universe.

When Stars Get Old

Although stars may stay on the main sequence for a long time, they don't stay there forever. Average stars, such as the sun, become red giants and then white dwarfs. However, stars that are more massive than the sun may explode with such intensity that they become a variety of strange objects such as supernovas, neutron stars, pulsars, and black holes.

Supernovas

Massive blue stars use their hydrogen much faster than stars like the sun do. Therefore, blue stars generate more energy than stars like the sun do, which makes blue stars very hot and blue! And compared with other stars, blue stars don't have long lives. At the end of its life, a blue star may explode in a large, bright flash called a *supernova*. A **supernova** is a gigantic explosion in which a massive star collapses. The explosion is so powerful that it can be brighter than an entire galaxy for several days. The ringed structure shown in **Figure 3** is the result of a supernova explosion.

supernova a gigantic explosion in which a massive star collapses and throws its outer layers into space

neutron star a star that has collapsed under gravity to the point that the electrons and protons have smashed together to form neutrons

pulsar a rapidly spinning neutron star that emits rapid pulses of radio and optical energy

Neutron Stars and Pulsars

After a supernova occurs, the materials in the center of a supernova are squeezed together to form a new star. This new star is about two times the mass of the sun. The particles inside the star's core are forced together to form neutrons. A star that has collapsed under gravity to the point at which all of its particles are neutrons is called a **neutron star.**

If a neutron star is spinning, it is called a **pulsar.** A pulsar sends out a beam of radiation that spins very rapidly. The beam is detected on Earth by radio telescopes as rapid clicks, or pulses.

Figure 3 **Explosion of a Supernova**

Supernova 1987A was the first supernova visible to the unaided eye in 400 years. The first image shows what the original star must have looked like only a few hours before the explosion. Today, the star's remains form a double ring of gas and dust, as shown at right.

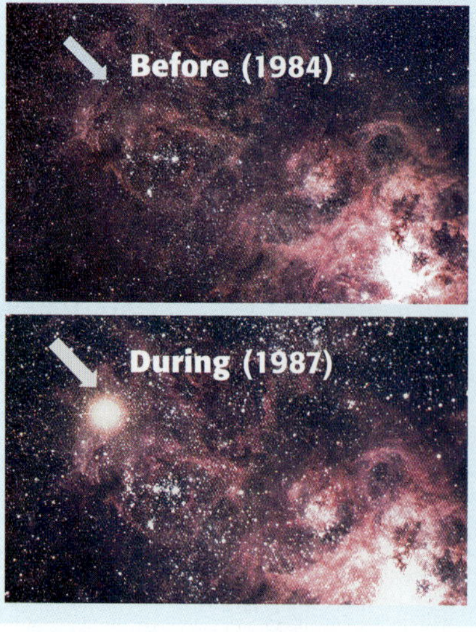

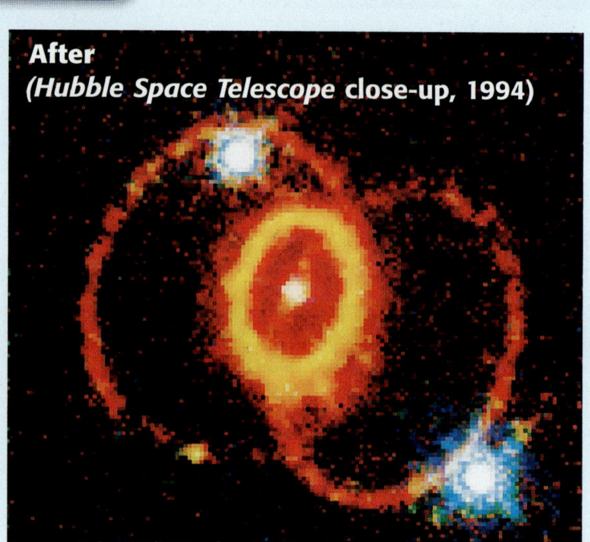

Black Holes

Sometimes the leftovers of a supernova are so massive that they collapse to form a black hole. A **black hole** is an object that is so massive that even light cannot escape its gravity. So, it is called a *black hole*. A black hole doesn't gobble up other stars like some movies show. Because black holes do not give off light, locating them is difficult. If a star is nearby, some gas or dust from the star will spiral into the black hole and give off X rays. These X rays allow astronomers to detect the existence of black holes.

black hole an object so massive and dense that even light cannot escape its gravity

Reading Check **What is a black hole? How do astronomers detect the presence of black holes?**

SECTION Review

Summary

- New stars form from the material of old stars that have gone through their lives.
- Types of stars include main-sequence stars, giants and supergiants, and white dwarf stars.
- The H-R diagram shows the brightness of a star in relation to the temperature of a star. It also shows the life cycle of stars.
- Most stars are main-sequence stars.
- Massive stars become supernovas. Their cores can change into neutron stars or black holes.

Using Key Terms

For each pair of terms, explain how the meanings of the terms differ.

1. *white dwarf* and *red giant*
2. *supernova* and *neutron star*
3. *pulsar* and *black hole*

Understanding Key Ideas

4. The sun is a
   a. white dwarf.
   b. main-sequence star.
   c. red giant.
   d. red dwarf.
5. A star begins as a ball of gas and dust pulled together by
   a. black holes.
   b. electrons and protons.
   c. heavy metals.
   d. gravity.
6. Are blue stars young or old? How can you tell?
7. In main-sequence stars, what is the relationship between brightness and temperature?
8. Arrange the following stages in order of their appearance in the life cycle of a star: white dwarf, red giant, and main-sequence star. Explain your answer.

Math Skills

9. The sun's present radius is 700,000 km. If the sun's radius increased by 150 times, what would its radius be?

Critical Thinking

10. **Applying Concepts** Given that there are more low-mass stars than high-mass stars in the universe, do you think there are more white dwarfs or more black holes in the universe? Explain.
11. **Analyzing Processes** Describe what might happen to a star after it becomes a supernova.
12. **Evaluating Data** How does the H-R diagram explain the life cycle of a star?

SECTION 4

Galaxies

Your complete address is part of a much larger system than your street, city, state, country, and even the planet Earth. You also live in the Milky Way galaxy.

READING WARM-UP

Objectives

- Identify three types of galaxies.
- Describe the contents and characteristics of galaxies.
- Explain why looking at distant galaxies reveals what young galaxies looked like.

Terms to Learn

galaxy
nebula
globular cluster
open cluster
quasar

READING STRATEGY

Reading Organizer As you read this section, make a table comparing the different types of galaxies.

Large groups of stars, dust, and gas are called **galaxies.** Galaxies come in a variety of sizes and shapes. The largest galaxies contain more than a trillion stars. Astronomers don't count the stars, of course. They estimate how many sun-sized stars the galaxy might have by studying the size and brightness of the galaxy.

galaxy a collection of stars, dust, and gas bound together by gravity

Types of Galaxies

There are many different types of galaxies. Edwin Hubble, the astronomer for whom the *Hubble Space Telescope* is named, began to classify galaxies, mostly by their shapes, in the 1920s. Astronomers still use the galaxy classification that Hubble developed.

Spiral Galaxies

When someone says the word *galaxy,* most people probably think of a spiral galaxy. *Spiral galaxies,* such as the one shown in **Figure 1,** have a bulge at the center and spiral arms. The spiral arms are made up of gas, dust, and new stars that have formed in these denser regions of gas and dust.

Reading Check What are two characteristics of spiral galaxies? What makes up the arms of a spiral galaxy? (*See the Appendix for answers to Reading Checks.*)

Figure 1 Types of Galaxies

Spiral Galaxy
The Andromeda galaxy is a spiral galaxy that looks similar to what our galaxy, the Milky Way, is thought to look like.

The Milky Way

It is hard to tell what type of galaxy we live in because the gas, dust, and stars keep astronomers from having a good view of our galaxy. Observing other galaxies and making measurements inside our galaxy, the Milky Way, has led astronomers to think that our solar system is in a disk-shaped, spiral galaxy. The sun is a medium-sized star located near the edge of the Milky Way. There are hundreds of billions of stars in our galaxy!

Elliptical Galaxies

About one-third of all galaxies are simply massive blobs of stars. Many look like spheres, and others are more stretched out. Because we don't know how they are oriented, some of these galaxies could be cucumber shaped, with the round end facing our galaxy. These galaxies are called *elliptical galaxies.* Elliptical galaxies usually have very bright centers and very little dust and gas. Elliptical galaxies contain mostly old stars. Some elliptical galaxies, such as M87, shown in **Figure 1,** are huge and are called *giant elliptical galaxies.* Other elliptical galaxies are much smaller and are called *dwarf elliptical galaxies.*

Irregular Galaxies

When Hubble first classified galaxies, he had a group of leftovers. He named the leftovers "irregulars." *Irregular galaxies* are galaxies that don't fit into any other class. As their name suggests, their shape is irregular. Many of these galaxies, such as the Large Magellanic Cloud, shown in **Figure 1,** are close companions of large spiral galaxies. The large spiral galaxies may be distorting the shape of these irregular galaxies.

CONNECTION TO Language Arts

WRITING SKILL **Alien Observer** As you read earlier, it's hard to tell what type of galaxy we live in because the gas, dust, and stars keep us from having a good view. But our galaxy might look different to an alien observer. Write a short story describing how our galaxy would look to an alien observer in another galaxy.

Elliptical Galaxy
Unlike the Milky Way, the galaxy known as M87 has no spiral arms.

Irregular Galaxy
The Large Magellanic Cloud, an irregular galaxy, is located within our galactic neighborhood.

Contents of Galaxies

Galaxies are composed of billions of stars and some planetary systems, too. Some of these stars form large features, such as gas clouds and star clusters, as shown in **Figure 2.**

Gas Clouds

nebula a large cloud of dust and gas in interstellar space; a region in space where stars are born or where stars explode at the end of their lives

globular cluster a tight group of stars that looks like a ball and contains up to 1 million stars

open cluster a group of stars that are close together relative to surrounding stars

The Latin word for "cloud" is *nebula.* In space, **nebulas** (or nebulae) are large clouds of gas and dust. Some types of nebulas glow, while others absorb light and hide stars. Still, other nebulas reflect starlight and produce some amazing images. Some nebulas are regions in which new stars form. **Figure 2** shows part of the Eagle nebula. Spiral galaxies usually contain nebulas, but elliptical galaxies contain very few.

Star Clusters

Globular clusters are groups of older stars. A **globular cluster** is a group of stars that looks like a ball, as shown in **Figure 2.** There may be up to one million stars in a globular cluster. Globular clusters are located in a spherical *halo* that surrounds spiral galaxies such as the Milky Way. Globular clusters are also common near giant elliptical galaxies.

Open clusters are groups of closely grouped stars that are usually located along the spiral disk of a galaxy. Newly formed open clusters have many bright blue stars, as shown in **Figure 2.** There may be a few hundred to a few thousand stars in an open cluster.

Reading Check **What is the difference between a globular cluster and an open cluster?**

Figure 2 **Gas Clouds and Star Clusters**

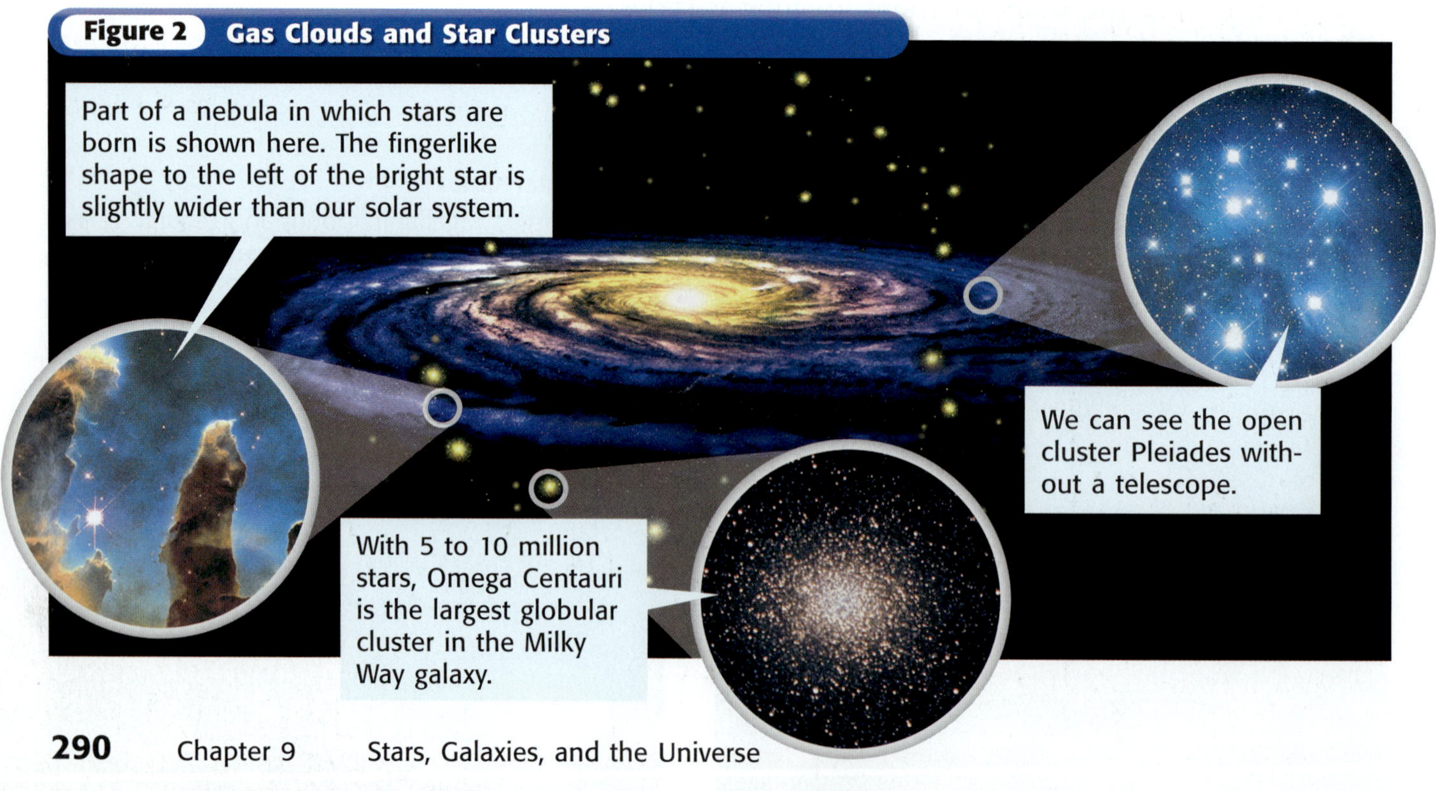

Part of a nebula in which stars are born is shown here. The fingerlike shape to the left of the bright star is slightly wider than our solar system.

We can see the open cluster Pleiades without a telescope.

With 5 to 10 million stars, Omega Centauri is the largest globular cluster in the Milky Way galaxy.

Origin of Galaxies

Scientists investigate the early universe by observing objects that are extremely far away in space. Because it takes time for light to travel through space, looking through a telescope is like looking back in time. Looking at distant galaxies reveals what early galaxies looked like. This information gives scientists an idea of how galaxies change over time and may give them insight about what caused the galaxies to form.

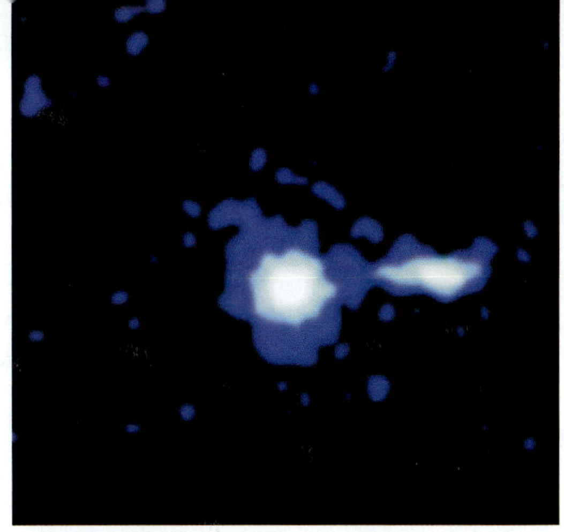

Figure 3 *The quasar known as PKS 0637-752 is as massive as 10 billion suns.*

Quasars

Among the most distant objects are quasars. **Quasars** are starlike sources of light that are extremely far away. They are among the most powerful energy sources in the universe. Some scientists think that quasars may be caused by massive black holes in the cores of some galaxies. **Figure 3** shows a quasar that is 6 billion light-years away.

quasar a very luminous, starlike object that generates energy at a high rate; quasars are thought to be the most distant objects in the universe

Reading Check **What are quasars? What do some scientists think quasars might be?**

SECTION Review

Summary

- Edwin Hubble classified galaxies according to their shape including spiral, elliptical, and irregular galaxies.
- Some galaxies consist of nebulas and star clusters.
- Nebulas are large clouds of gas and dust. Globular clusters are tightly grouped stars. Open clusters are closely grouped stars.
- Scientists look at distant galaxies to learn what early galaxies looked like.

Using Key Terms

1. Use the following terms in the same sentence: *nebula, globular cluster,* and *open cluster.*

Understanding Key Ideas

2. Arrange the following galaxies in order of decreasing size: spiral, giant elliptical, dwarf elliptical, and irregular.
3. All of the following are shapes used to classify galaxies EXCEPT
   a. elliptical.
   b. irregular.
   c. spiral.
   d. triangular.

Critical Thinking

4. **Making Comparisons** Describe the difference between an elliptical galaxy and a globular cluster.
5. **Identifying Relationships** Explain how looking through a telescope is like looking back in time.

Math Skills

6. The quasar known as PKS 0637-752 is 6 billion light-years away from Earth. The North Star is 431 light-years away from Earth. What is the ratio of the distances in kilometers these two celestial objects are from Earth? (Hint: One light-year is equal to 9.46 trillion km.)

SECTION 5

Formation of the Universe

Imagine explosions, bright lights, and intense energy. Does that scene sound like an action movie? This scene could also describe a theory about the formation of the universe.

READING WARM-UP

Objectives

- Describe the big bang theory.
- Explain evidence used to support the big bang theory.
- Describe the structure of the universe.
- Describe two ways scientists calculate the age of the universe.
- Explain what will happen if the universe expands forever.

Terms to Learn

cosmology
big bang theory

READING STRATEGY

Prediction Guide Before reading this section, write the title of each heading in this section. Next, under each heading, write what you think you will learn.

The study of the origin, structure, and future of the universe is called **cosmology.** Like other scientific theories, theories about the beginning and end of the universe must be tested by observations or experiments.

cosmology the study of the origin, properties, processes, and evolution of the universe

Universal Expansion

To understand how the universe formed, scientists study the movement of galaxies. Careful measurements have shown that most galaxies are moving apart.

A Raisin-Bread Model

To understand how the galaxies are moving, imagine a loaf of raisin bread before it is baked. Inside the dough, each raisin is a certain distance from every other raisin. As the dough gets warm and rises, it expands and all of the raisins begin to move apart. No matter which raisin you observe, the other raisins are moving farther away from it. The universe, like the rising bread dough, is expanding. Think of the raisins as galaxies. As the universe expands, the galaxies move farther apart.

The Big Bang Theory

With the discovery that the universe is expanding, scientists began to wonder what it would be like to watch the formation of the universe in reverse. The universe would appear to be contracting, not expanding. All matter would eventually come together at a single point. Thinking about what would happen if all of the matter in the universe were squeezed into such a small space led scientists to the big bang theory.

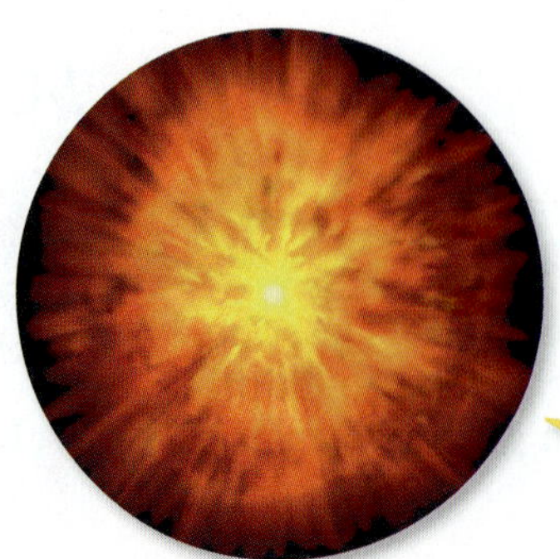

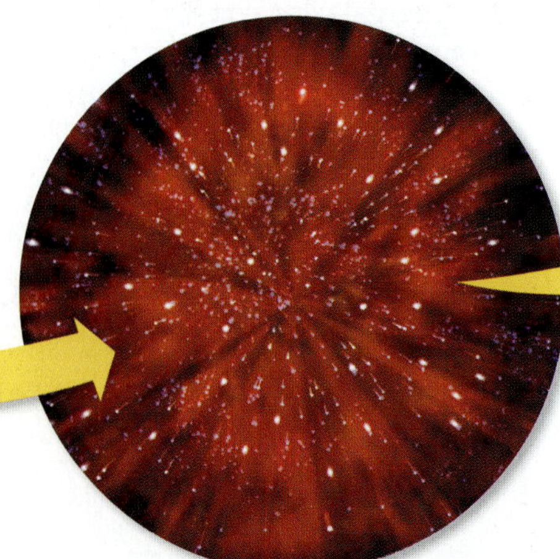

Figure 1 *Some astronomers think the big bang caused the universe to expand in all directions.*

A Tremendous Explosion

The theory that the universe began with a tremendous explosion is called the **big bang theory.** According to the theory, 13.7 billion years ago all the contents of the universe was compressed under extreme pressure, temperature, and density in a very tiny spot. Then, the universe rapidly expanded, and matter began to come together and form galaxies. **Figure 1** illustrates what the big bang might have looked like.

big bang theory the theory that states the universe began with a tremendous explosion 13.7 billion years ago

Cosmic Background Radiation

In 1964, two scientists using a huge antenna accidentally found radiation coming from all directions in space. One explanation for this radiation is that it is *cosmic background radiation* left over from the big bang. To understand the connection between the big bang theory and cosmic background radiation, think about a kitchen oven. When an oven door is left open after the oven has been used, thermal energy is transferred throughout the kitchen and the oven cools. Eventually, the room and the oven are the same temperature. According to the big bang theory, the thermal energy from the original explosion was distributed in every direction as the universe expanded. This cosmic background radiation now fills all of space.

Reading Check **Explain the relationship between cosmic background radiation and the big bang theory.** (*See the Appendix for answers to Reading Checks.*)

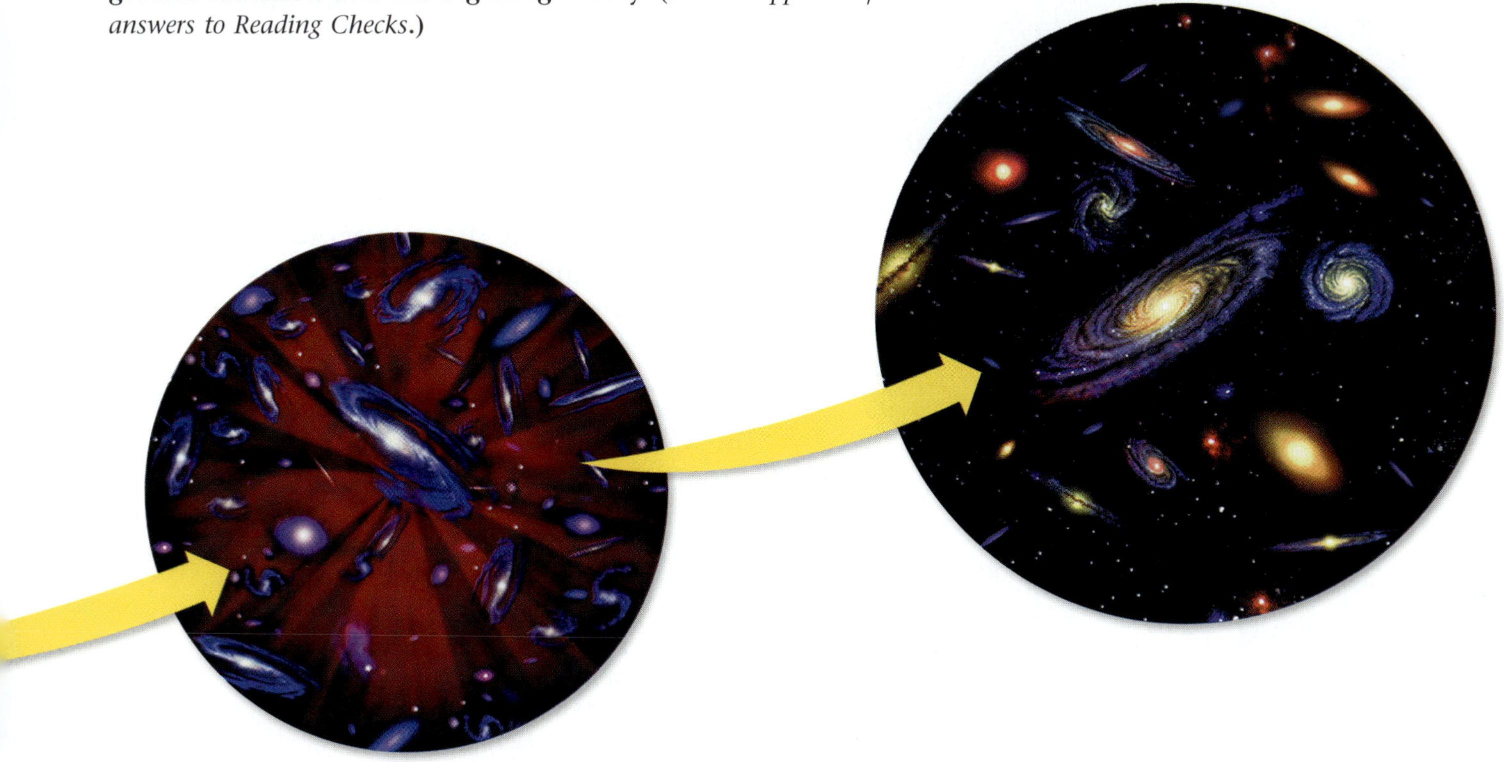

Structure of the Universe

From our home on Earth, the universe stretches out farther than astronomers can see with their most advanced instruments. The universe contains a variety of objects. But these objects in the universe are not simply scattered through the universe in a random pattern. The universe has a structure that is loosely repeated over and over again.

A Cosmic Repetition

Every object in the universe is part of a larger system. As illustrated in **Figure 2,** a cluster or group of galaxies can be made up of smaller star clusters and galaxies. Galaxies, such as the Milky Way, can include planetary systems, such as our solar system. Earth is part of our solar system. Although our solar system is the planetary system that we are most familiar with, other planets have been detected in orbit around other stars. Scientists think that planetary systems are common in the universe.

How Old Is the Universe?

One way scientists can calculate the age of the universe is to measure the distance from Earth to various galaxies. By using these distances, scientists can estimate the age of the universe and predict its rate of expansion.

Another way to estimate the age of the universe is to calculate the ages of old, nearby stars. Because the universe must be at least as old as the oldest stars it contains, the ages of the stars provide a clue to the age of the universe.

Reading Check **What is one way that scientists calculate the age of the universe?**

Figure 2 *Every object in the universe is part of a larger system. Earth is part of our solar system, which is in turn part of the Milky Way galaxy.*

A Forever Expanding Universe

What will happen to the universe? As the galaxies move farther apart, they get older and stop forming stars. The farther galaxies move apart from each other, the less visible to us they will become. The expansion of the universe depends on how much matter the universe contains. Scientists predict that if there is enough matter, gravity could eventually stop the expansion of the universe. If the universe stops expanding, it could start collapsing to its original state. This process would be a reverse of what might have happened during the big bang.

However, scientists now think that there may not be enough matter in the universe, so the universe will continue to expand forever. Therefore, stars will age and die, and the universe will probably become cold and dark after many billions of years. Even after the universe becomes cold and dark, it will continue to expand forever.

Reading Check **If the universe expanded to the point at which gravity stopped the expansion, what would happen? What will happen if the expansion of the universe continues forever?**

CONNECTION TO Physics

WRITING SKILL **Origin of the Universe** The big bang theory is one scientific theory about the origin of the universe. Use library resources to research these other scientific theories. In your **science journal,** describe in your own words the different theories of the origin of the universe. Use charts or tables to examine and evaluate these differences.

SECTION Review

Summary

- Observations show that the universe is expanding.
- The big bang theory states that the universe began with an explosion about 13.7 billion years ago.
- Cosmic background radiation helps support the big bang theory.
- Scientists use different ways to calculate the age of the universe.
- Scientists think that the universe may expand forever.

Using Key Terms

1. In your own words, write a definition for the following terms: *cosmology* and *big bang theory*.

Understanding Key Ideas

2. Describe two ways scientists calculate the age of the universe.
3. The expansion of the universe can be compared to
   a. cosmology.
   b. raisin bread baking in an oven.
   c. thermal energy leaving an oven as the oven cools.
   d. bread pudding.
4. How does cosmic background radiation support the big bang theory?
5. What do scientists think will eventually happen to the universe?

Math Skills

6. The North Star is 4.08×10^{12} km from Earth. What is this number written in its long form?

Critical Thinking

7. **Applying Concepts** Explain how every object in the universe is part of a larger system.
8. **Analyzing Ideas** Why do scientists think that the universe will expand forever?

Using Scientific Methods

Skills Practice Lab

Red Hot, or Not?

OBJECTIVES

Discover what the color of a glowing object reveals about the temperature of the object.

Describe how the color and temperature of a star are related.

MATERIALS

- battery, D cell (2)
- battery, D cell, weak
- flashlight bulb
- tape, electrical
- wire, insulated copper, with ends stripped, 20 cm long (2)

SAFETY

When you look at the night sky, some stars are brighter than others. Some are even different colors. For example, Betelgeuse, a bright star in the constellation Orion, glows red. Sirius, one of the brightest stars in the sky, glows bluish white. Astronomers use color to estimate the temperature of stars. In this activity, you will experiment with a light bulb and some batteries to discover what the color of a glowing object reveals about the temperature of the object.

Ask a Question

1. How are the color and temperature of a star related?

Form a Hypothesis

2. On a sheet of paper, change the question above into a statement that gives your best guess about the relationship between a star's color and temperature.

Test the Hypothesis

3. Tape one end of an insulated copper wire to the positive pole of the weak D cell. Tape one end of the second wire to the negative pole.
4. Touch the free end of each wire to the light bulb. Hold one of the wires against the bottom tip of the light bulb. Hold the second wire against the side of the metal portion of the bulb. The bulb should light.
5. Record the color of the filament in the light bulb. Carefully touch your hand to the bulb. Observe the temperature of the bulb. Record your observations.
6. Repeat steps 3–5 with one of the two fresh D cells.
7. Use the electrical tape to connect two fresh D cells so that the positive pole of the first cell is connected to the negative pole of the second cell.
8. Repeat steps 3–5 using the fresh D cells that are taped together.

Analyze the Results

1. **Describing Events** What was the color of the filament in each of the three trials? For each trial, compare the bulb temperature to the temperature of the bulb in the other two trials.
2. **Analyzing Results** What information does the color of a star tell you about the star?
3. **Classifying** What color are stars that have relatively high surface temperatures? What color are stars that have relatively low surface temperatures?

Draw Conclusions

4. **Applying Conclusions** Arrange the following stars in order from highest to lowest surface temperature: Sirius, which is bluish white; Aldebaran, which is orange; Procyon, which is yellow-white; Capella, which is yellow; and Betelgeuse, which is red.

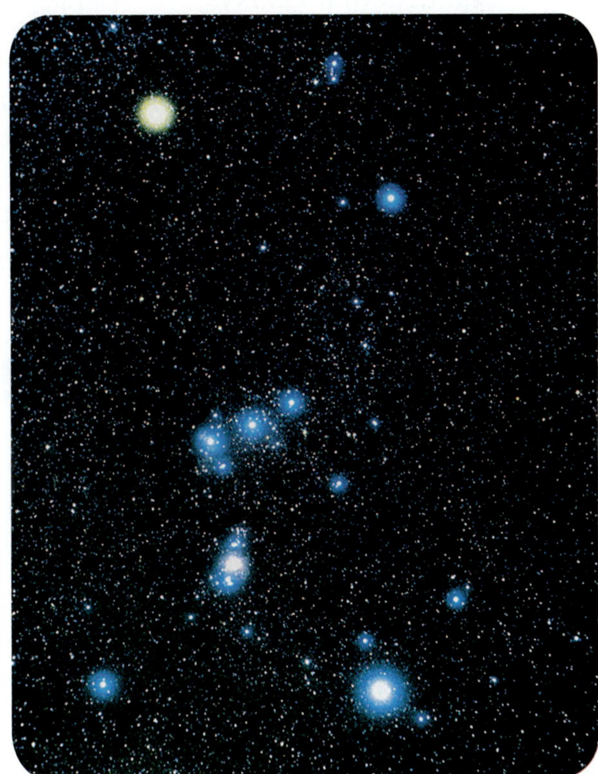

Chapter Review

USING KEY TERMS

The statements below are false. For each statement, replace the underlined term to make a true statement.

1. The distance that light travels in space in 1 year is called apparent magnitude.
2. Globular clusters are groups of stars that are usually located along the spiral disk of a galaxy.
3. Galaxies that have very bright centers and very little dust and gas are called spiral galaxies.
4. When you look at white light through a glass prism, you see a rainbow of colors called a supernova.

UNDERSTANDING KEY IDEAS

Multiple Choice

5. A scientist can identify a star's composition by looking at
   a. the star's prism.
   b. the star's continuous spectrum.
   c. the star's absorption spectrum.
   d. the star's color.
6. If the universe expands forever,
   a. the universe will collapse.
   b. the universe will repeat itself.
   c. the universe will remain just as it is today.
   d. stars will age and die and the universe will become cold and dark.
7. The majority of stars in our galaxy are
   a. blue stars.
   b. white dwarfs.
   c. main-sequence stars.
   d. red giants.
8. Which of the following is used to measure the distance between objects in space?
   a. parallax
   b. magnitude
   c. zenith
   d. altitude
9. Which of the following stars would be seen as the brightest star?
   a. Alcyone, which has an apparent magnitude of 3
   b. Alpheratz, which has an apparent magnitude of 2
   c. Deneb, which has an apparent magnitude of 1
   d. Rigel, which has an apparent magnitude of 0

Short Answer

10. Compare a sunspot with a solar flare.
11. Describe the structure of the universe.
12. Explain how stars at different stages in their life cycle appear on the H-R diagram.
13. Define nuclear fusion in your own words. Describe how nuclear fusion generates the sun's energy.
14. Describe how color indicates the temperature of a star.
15. Describe two ways that scientists calculate the age of the universe.

CRITICAL THINKING

16 **Concept Mapping** Use the following terms to create a concept map: *main-sequence star, nebula, red giant, white dwarf, neutron star,* and *black hole.*

17 **Evaluating Conclusions** While looking through a telescope, you see a galaxy that doesn't appear to contain any blue stars. What kind of galaxy is it most likely to be? Explain your answer.

18 **Making Comparisons** Explain the differences between main-sequence stars, giant stars, supergiant stars, and white dwarfs.

19 **Evaluating Data** Why do astronomers use absolute magnitudes to plot stars? Why don't astronomers use apparent magnitudes to plot stars?

20 **Evaluating Sources** According to the big bang theory, how did the universe begin? What evidence supports this theory?

21 **Evaluating Data** If a certain star displayed a large parallax, what could you say about the star's distance from Earth?

INTERPRETING GRAPHICS

The graph below shows Hubble's law, which relates how far galaxies are from Earth and how fast they are moving away from Earth. Use the graph below to answer the questions that follow.

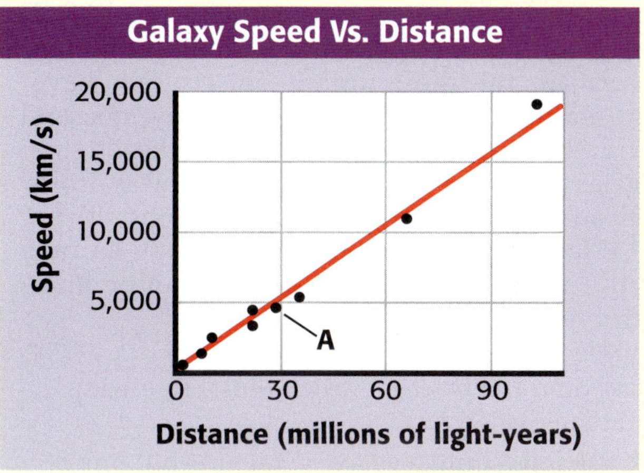

22 Look at the point that represents galaxy A in the graph. How far is galaxy A from Earth, and how fast is it moving away from Earth?

23 If a galaxy is moving away from Earth at 15,000 km/s, how far is the galaxy from Earth?

24 If a galaxy is 90,000,000 light-years from Earth, how fast is it moving away from Earth?

Standardized Test Preparation

READING

Read each of the passages below. Then, answer the questions that follow each passage.

Passage 1 Quasars are some of the most puzzling objects in the sky. If viewed through an optical telescope, a quasar appears as a small, dim star. Quasars are the most distant objects that have been observed from Earth. But many quasars are hundreds of times brighter than the brightest galaxy. Because quasars are so far away from Earth and yet are very bright, they most likely emit a large amount of energy. Scientists do not yet understand exactly how quasars can emit so much energy.

1. Based on the passage, which of the following statements is a fact?

A Quasars, unlike galaxies, include billions of bright objects.
B Galaxies are brighter than quasars.
C Quasars are hundreds of times brighter than the brightest galaxy.
D Galaxies are the most distant objects observed from Earth.

2. Based on the information in the passage, what can the reader conclude?

F Quasars are the same as galaxies.
G Quasars appear as small, dim stars, but they emit a large amount of energy.
H Quasars can be viewed only by using an optical telescope.
I Quasars will never be understood.

3. Why do scientists think that quasars emit a large amount of energy?

A because quasars are the brightest stars in the universe
B because quasars can be viewed only through an optical telescope
C because quasars are very far away and are still bright
D because quasars are larger than galaxies

Passage 2 If you live away from bright outdoor lights, you may be able to see a faint, narrow band of light and dark patches across the sky. This band is called the Milky Way. Our galaxy, the Milky Way, consists of stars, gases, and dust. Between the stars of the Milky Way are clouds of gas and dust called interstellar matter. These clouds provide materials that form new stars.

Every star that you can see in the night sky is a part of the Milky Way, because our solar system is inside the Milky Way. Because we are inside the galaxy, we cannot see the entire galaxy. But scientists can use astronomical data to create a picture of the Milky Way.

1. In the passage, what does the term *interstellar matter* mean?

A stars in the Milky Way
B the Milky Way
C a narrow band of light and dark patches across the sky
D the clouds of gas and dust between the stars in the Milky Way

2. Based on the information in the passage, what can the reader conclude?

F The Milky Way can be seen in the night sky near a large city.
G The entire Milky Way can be seen all at once.
H Every star that is seen in the night sky is a part of the Milky Way.
I Scientists have no idea what the entire Milky Way looks like.

INTERPRETING GRAPHICS

The graph below shows the relationship between a star's age and mass. Use the graph below to answer the questions that follow.

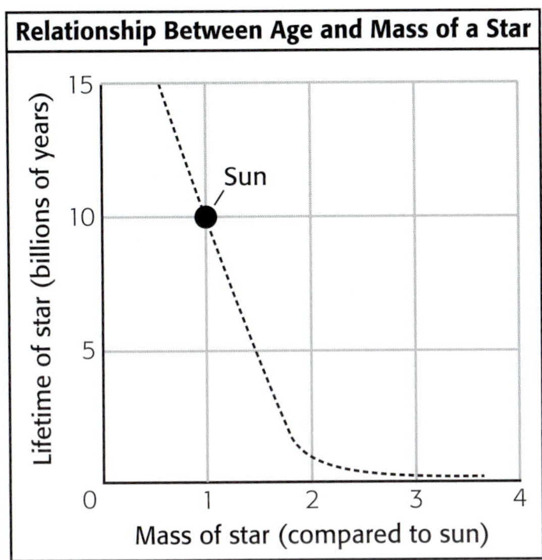

1. How long does a star that has 1.2 times the mass of the sun live?
   - **A** 10 billion years
   - **B** 8 billion years
   - **C** 6 billion years
   - **D** 5 billion years

2. How long does a star that has 2 times the mass of the sun live?
   - **F** 4 billion years
   - **G** 1 billion years
   - **H** 10 billion years
   - **I** 5 billion years

3. If the sun's mass was reduced by half, how long would the sun live?
   - **A** 2 billion years
   - **B** 8 billion years
   - **C** 10 billion years
   - **D** more than 15 billion years

4. According to the graph, how long is the sun predicted to live?
   - **F** 15 billion years
   - **G** 10 billion years
   - **H** 5 billion years
   - **I** 2 billion years

MATH

Read each question below, and choose the best answer.

1. How many kilometers away from Earth is an object that is 8 light-years away from Earth? (Hint: One light-year is equal to 9.46 trillion kilometers.)
   - **A** 77 trillion kilometers
   - **B** 76 trillion kilometers
   - **C** 7.66 trillion kilometers
   - **D** 7.6 trillion kilometers

2. An astronomer observes two stars of about the same temperature and size. Alpha Centauri B is about 4 light-years away from Earth, and Sigma 2 Eridani A is about 16 light-years away from Earth. How many times as bright as Sigma 2 Eridani A does Alpha Centauri B appear? (Hint: One light-year is equal to 9.46 trillion kilometers.)
   - **F** 2 times as bright
   - **G** 4 times as bright
   - **H** 16 times as bright
   - **I** 32 times as bright

3. Star A is 5 million kilometers from Star B. What is this distance expressed in meters?
   - **A** 0.5 m
   - **B** 5,000 m
   - **C** 5×10^6 m
   - **D** 5×10^9 m

4. In the vacuum of space, light travels 3×10^8 m/s. How far does light travel in 1 h in space?
   - **F** 3,600 m
   - **G** 1.80×10^{10} m
   - **H** 1.08×10^{12} m
   - **I** 1.08×10^{16} m

5. The mass of the known universe is about 10^{23} solar masses, which is 10^{50} metric tons. How many metric tons is one solar mass?
   - **A** 10^{27} solar masses
   - **B** 10^{27} metric tons
   - **C** 10^{73} solar masses
   - **D** 10^{73} metric tons

Science in Action

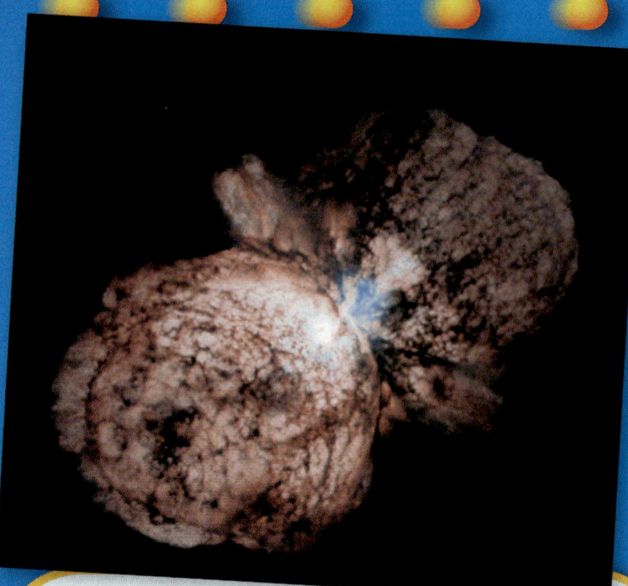

Weird Science

Holes Where Stars Once Were

An invisible phantom lurks in space, ready to swallow everything that comes near it. Once trapped in its grasp, matter is stretched, torn, and crushed into oblivion. Does this tale sound like a horror story? Guess again! Scientists call this phantom a *black hole*. As a star runs out of fuel, it cools and eventually collapses under the force of its own gravity. If the collapsing star is massive enough, it may shrink to become a black hole. The resulting gravitational attraction is so strong that even light cannot escape! Many astronomers think that black holes lie at the heart of many galaxies. Some scientists suggest that there is a giant black hole at the center of our own Milky Way.

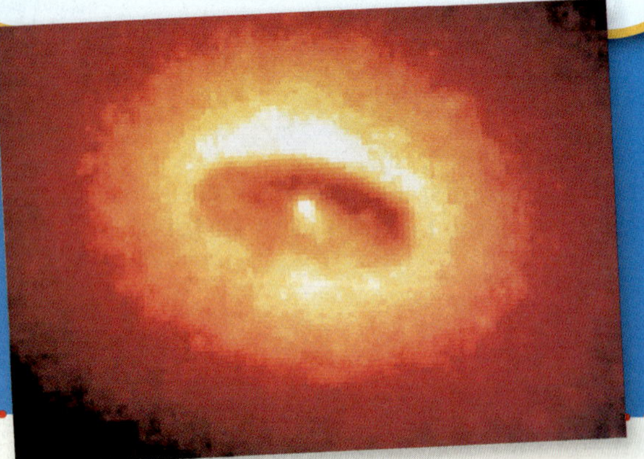

Language Arts ACTIVITY

WRITING SKILL Can you imagine traveling through a black hole? Write a short story that describes what you would see if you led a space mission to a black hole.

Scientific Discoveries

Eta Carinae: The Biggest Star Ever Discovered

In 1841, Eta Carinae was the second-brightest star in the night sky. Why is this observation a part of history? Eta Carinae's brightness is historic because before 1837, Eta Carinae wasn't even visible to the naked eye! Strangely, a few years later Eta Carinae faded again and disappeared from the night sky. Something unusual was happening to Eta Carinae, and scientists wanted to know what it was. As soon as scientists had telescopes with which they could see far into space, they took a closer look at Eta Carinae. Scientists discovered that this star is highly unstable and prone to violent outbursts. These outbursts, the last of which was seen in 1841, can be seen on Earth. Scientists also discovered that Eta Carinae is 150 times as big as our sun and about 4 million times as bright. Eta Carinae is the biggest and brightest star ever found!

Math ACTIVITY

If Eta Carinae is 8,000 light-years from our solar system, how many kilometers is Eta Carinae from our solar system? (Hint: One light-year is equal to 9.46 trillion kilometers.)

Careers

Jocelyn Bell-Burnell

Astrophysicist Imagine getting a signal from far out in space and not knowing what or whom it's coming from. That's what happened to astrophysicist Jocelyn Bell-Burnell. Bell-Burnell is known for discovering pulsars, objects in space that emit radio waves at short, regular intervals. But before she and her advisor discovered that the signals came from pulsars, they thought that the signals may have come from aliens!

Born in 1943 in Belfast, Northern Ireland, Jocelyn Bell-Burnell became interested in astronomy at an early age. At Cambridge University in 1967, Bell-Burnell, who was a graduate student, and her advisor, Anthony Hewish, completed work on a huge radio telescope designed to pick up signals from quasars. Bell-Burnell's job was to operate the telescope and analyze its chart paper recordings on a graph. Each day, the telescope recordings used 29.2 m of chart paper! After a month, Bell-Burnell noticed that the recordings showed a few "bits of scruff"—very short, pulsating radio signals—that she could not explain. Bell-Burnell and Hewish struggled to find the source of the mysterious signal. They checked the equipment and began eliminating possible sources of the signal, such as satellites, television, and radar. Shortly after finding the first signal, Bell-Burnell discovered a second. The second signal was similar to the first but came from a different position in the sky. By January 1968, Bell-Burnell had discovered two more pulsating signals. In March of 1968, her findings that the signals were from a new kind of star were published and amazed the scientific community. The scientific press named the newly discovered stars *pulsars*.

Today, Bell-Burnell is a leading expert in the field of astrophysics and the study of stars. She is currently head of the physics department at the Open University, in Milton Keynes, England.

Social Studies ACTIVITY

Use the Internet or library resources to research historical events that occurred during 1967 and 1968. Find out if the prediction that the signals from pulsars were coming from aliens affected historical events during this time.

To learn more about these Science in Action topics, visit go.hrw.com and type in the keyword HZ5UNVF.

Current Science

Check out Current Science® articles related to this chapter by visiting go.hrw.com. Just type in the keyword HZ5CS19.

TIMELINE

Physical Science

It's hard to imagine a world where nothing ever moves. Without motion or forces to cause motion, life would be very dull! The relationship between force and motion is one subject you will learn about in this unit. You will also study the interactions through which matter can change its identity. In addition, you will learn how sound and light energy travel in waves. This timeline shows some events and discoveries that have shaped the history of physical science.

Around 250 BCE

Archimedes, a Greek mathematician, develops the principle that bears his name. The principle relates the buoyant force on an object in a fluid to the amount of fluid displaced by the object.

1764

In London, Wolfgang Amadeus Mozart composes his first symphony—at the age of 8.

1846

After determining that the orbit of Uranus is different from what is predicted from the law of universal gravitation, scientists discover Neptune whose gravitational force is causing Uranus's unusual orbit.

1947

While flying a Bell X-1 rocket-powered airplane, American pilot Chuck Yeager becomes the first human to travel faster than the speed of sound.

PHILOSOPHIÆ
NATURALIS
PRINCIPIA
MATHEMATICA.

Autore J. S. NEWTON, Trin. Coll. Cantab. Soc. Matheseos Professore Lucasiano, & Societatis Regalis Sodali.

Around 240 BCE

Chinese astronomers are the first to record a sighting of Halley's Comet.

1519

Portuguese explorer Ferdinand Magellan begins the first voyage around the world.

1687

Sir Isaac Newton, a British mathematician and scientist, publishes *Principia*, a book describing his laws of motion and the law of universal gravitation.

1905

While employed as a patent clerk, German physicist Albert Einstein publishes his special theory of relativity. The theory states that the speed of light is constant no matter what the reference frame is.

1921

Bessie Coleman becomes the first African American woman licensed to fly an airplane.

1971

American astronaut Alan Shepard takes a break from gathering lunar data to play golf on the moon during the *Apollo 14* mission.

1990

The *Magellan* spacecraft begins orbiting Venus for a four-year mission to map the planet. By using the sun's gravitational forces, it propels itself to Venus without burning much fuel.

2003

NASA launches *Spirit* and *Opportunity*, two Mars Exploration Rovers, to study Mars.

Matter in Motion

About the PHOTO

Speed skaters are fast. In fact, some skaters can skate at a rate of 12 m/s! That's equal to a speed of 27 mi/h. To reach such a speed, skaters must exert large forces. They must also use friction to turn corners on the slippery surface of the ice.

PRE-READING ACTIVITY

FOLDNOTES

Four-Corner Fold

Before you read the chapter, create the FoldNote entitled "Four-Corner Fold" described in the **Study Skills** section of the Appendix. Label the flaps of the four-corner fold with "Motion," "Forces," "Friction," and "Gravity." Write what you know about each topic under the appropriate flap. As you read the chapter, add other information that you learn.

START-UP ACTIVITY

The Domino Derby

Speed is the distance traveled by an object in a certain amount of time. In this activity, you will observe one factor that affects the speed of falling dominoes.

Procedure

1. Set up **25 dominoes** in a straight line. Try to keep equal spacing between the dominoes.
2. Use a **meterstick** to measure the total length of your row of dominoes, and record the length.
3. Use a **stopwatch** to time how long it takes for the dominoes to fall. Record this measurement.
4. Predict what would happen to that amount of time if you changed the distance between the dominoes. Write your predictions.
5. Repeat steps 2 and 3 several times using distances between the dominoes that are smaller and larger than the distance used in your first setup. Use the same number of dominoes in each trial.

Analysis

1. Calculate the average speed for each trial by dividing the total distance (the length of the domino row) by the time the dominoes take to fall.
2. How did the spacing between dominoes affect the average speed? Is this result what you expected? If not, explain.

SECTION 1

Measuring Motion

Look around you—you are likely to see something in motion. Your teacher may be walking across the room, or perhaps your friend is writing with a pencil.

Even if you don't see anything moving, motion is still occurring all around you. Air particles are moving, the Earth is circling the sun, and blood is traveling through your blood vessels!

READING WARM-UP

Objectives

- Describe the motion of an object by the position of the object in relation to a reference point.
- Identify the two factors that determine speed.
- Explain the difference between speed and velocity.
- Analyze the relationship between velocity and acceleration.
- Demonstrate that changes in motion can be measured and represented on a graph.

Terms to Learn

motion
speed
velocity
acceleration

READING STRATEGY

Discussion Read this section silently. Write down questions that you have about this section. Discuss your questions in a small group.

Observing Motion by Using a Reference Point

You might think that the motion of an object is easy to detect—you just watch the object. But you are actually watching the object in relation to another object that appears to stay in place. The object that appears to stay in place is a *reference point.* When an object changes position over time relative to a reference point, the object is in **motion.** You can describe the direction of the object's motion with a reference direction, such as north, south, east, west, up, or down.

Reading Check What is a reference point? (*See the Appendix for answers to Reading Checks.*)

Common Reference Points

The Earth's surface is a common reference point for determining motion, as shown in **Figure 1.** Nonmoving objects, such as trees and buildings, are also useful reference points.

A moving object can also be used as a reference point. For example, if you were on the hot-air balloon shown in **Figure 1,** you could watch a bird fly by and see that the bird was changing position in relation to your moving balloon.

Figure 1 *During the interval between the times that these pictures were taken, the hot-air balloon changed position relative to a reference point–the mountain.*

Speed Depends on Distance and Time

Speed is the distance traveled by an object divided by the time taken to travel that distance. Look again at **Figure 1.** Suppose the time interval between the pictures was 10 s and that the balloon traveled 50 m in that time. The speed of the balloon is (50 m)/(10 s), or 5 m/s.

The SI unit for speed is meters per second (m/s). Kilometers per hour (km/h), feet per second (ft/s), and miles per hour (mi/h) are other units commonly used to express speed.

motion an object's change in position relative to a reference point

speed the distance traveled divided by the time interval during which the motion occurred

Determining Average Speed

Most of the time, objects do not travel at a constant speed. For example, you probably do not walk at a constant speed from one class to the next. So, it is very useful to calculate *average speed* using the following equation:

$$\textit{average speed} = \frac{\textit{total distance}}{\textit{total time}}$$

Recognizing Speed on a Graph

Suppose a person drives from one city to another. The blue line in the graph in **Figure 2** shows the total distance traveled during a 4 h period. Notice that the distance traveled during each hour is different. The distance varies because the speed is not constant. The driver may change speed because of weather, traffic, or varying speed limits. The average speed for the entire trip can be calculated as follows:

$$\textit{average speed} = \frac{360 \text{ km}}{4 \text{ h}} = 90 \text{ km/h}$$

The red line on the graph shows how far the driver must travel each hour to reach the same city if he or she moved at a constant speed. The slope of this line is the average speed.

What's Your Speed?

Measure a distance of 5 m or a distance of 25 ft inside or outside. Ask an adult at home to use a stopwatch or a watch with a second hand to time you as you travel the distance you measured. Then, find your average speed. Find the average speed of other members of your family in the same way.

ACTIVITY

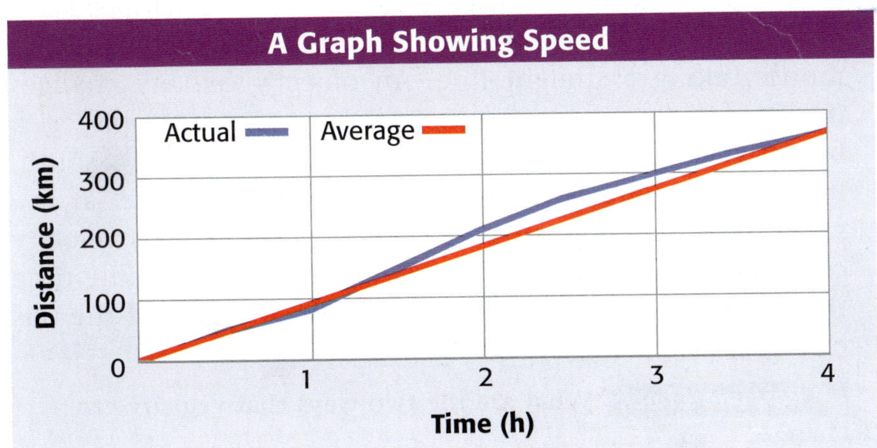

Figure 2 *Speed can be shown on a graph of distance versus time.*

Calculating Average Speed An athlete swims a distance from one end of a 50 m pool to the other end in a time of 25 s. What is the athlete's average speed?

Step 1: Write the equation for average speed.

$$\textit{average speed} = \frac{\textit{total distance}}{\textit{total time}}$$

Step 2: Replace the total distance and total time with the values given, and solve.

$$\textit{average speed} = \frac{50 \text{ m}}{25 \text{ s}} = 2 \text{ m/s}$$

Now It's Your Turn

1. Kira jogs to a store 72 m away in a time of 36 s. What is Kira's average speed?
2. If you travel 7.5 km and walk for 1.5 h, what is your average speed?
3. An airplane traveling from San Francisco to Chicago travels 1,260 km in 3.5 h. What is the airplane's average speed?

Velocity: Direction Matters

Imagine that two birds leave the same tree at the same time. They both fly at 10 km/h for 5 min, 12 km/h for 8 min, and 5 km/h for 10 min. Why don't they end up at the same place?

Have you figured out the answer? The birds went in different directions. Their speeds were the same, but they had different velocities. **Velocity** (vuh LAHS uh tee) is the speed of an object in a particular direction.

Be careful not to confuse the terms *speed* and *velocity.* They do not have the same meaning. Velocity must include a reference direction. If you say that an airplane's velocity is 600 km/h, you would not be correct. But you could say the plane's velocity is 600 km/h south. **Figure 3** shows an example of the difference between speed and velocity.

Figure 3 *The speeds of these cars may be similar, but the velocities of the cars differ because the cars are going in different directions.*

Changing Velocity

You can think of velocity as the rate of change of an object's position. An object's velocity is constant only if its speed and direction don't change. Therefore, constant velocity is always motion along a straight line. An object's velocity changes if either its speed or direction changes. For example, as a bus traveling at 15 m/s south speeds up to 20 m/s south, its velocity changes. If the bus continues to travel at the same speed but changes direction to travel east, its velocity changes again. And if the bus slows down at the same time that it swerves north to avoid a cat, the velocity of the bus changes, too.

Reading Check **What are the two ways that velocity can change?**

Figure 4 Finding Resultant Velocity

When you combine two velocities that are **in the same direction,** add them together to find the resultant velocity.

Person's resultant velocity
15 m/s east + 1 m/s east = 16 m/s east

When you combine two velocities that are **in opposite directions,** subtract the smaller velocity from the larger velocity to find the resultant velocity. The resultant velocity is in the direction of the larger velocity.

Person's resultant velocity
15 m/s east − 1 m/s west = 14 m/s east

Combining Velocities

Imagine that you are riding in a bus that is traveling east at 15 m/s. You and the other passengers are also traveling at a velocity of 15 m/s east. But suppose you stand up and walk down the bus's aisle while the bus is moving. Are you still moving at the same velocity as the bus? No! **Figure 4** shows how you can combine velocities to find the *resultant velocity.*

velocity the speed of an object in a particular direction

acceleration the rate at which velocity changes over time; an object accelerates if its speed, direction, or both change

Acceleration

Although the word *accelerate* is commonly used to mean "speed up," the word means something else in science. **Acceleration** (ak SEL uhr AY shuhn) is the rate at which velocity changes. Velocity changes if speed changes, if direction changes, or if both change. So, an object accelerates if its speed, its direction, or both change.

An increase in velocity is commonly called *positive acceleration*. A decrease in velocity is commonly called *negative acceleration,* or *deceleration*. Keep in mind that acceleration is not only how much velocity changes but also how fast velocity changes. The faster the velocity changes, the greater the acceleration is.

Figure 5 *This cyclist is accelerating at 1 m/s² south.*

Calculating Average Acceleration

You can find average acceleration by using the equation:

$$\textit{average acceleration} = \frac{\textit{final velocity} - \textit{starting velocity}}{\textit{time it takes to change velocity}}$$

Velocity is expressed in meters per second (m/s), and time is expressed in seconds (s). So acceleration is expressed in meters per second per second, or (m/s)/s, which equals m/s^2. For example, look at **Figure 5.** Every second, the cyclist's southward velocity increases by 1 m/s. His average acceleration can be calculated as follows:

$$\textit{average acceleration} = \frac{5 \text{ m/s} - 1 \text{ m/s}}{4 \text{ s}} = 1 \text{ m/s}^2 \text{ south}$$

Reading Check What are the units of acceleration?

MATH PRACTICE

Calculating Acceleration

Use the equation for average acceleration to do the following problem.

A plane passes over point A at a velocity of 240 m/s north. Forty seconds later, it passes over point B at a velocity of 260 m/s north. What is the plane's average acceleration?

Recognizing Acceleration on a Graph

Suppose that you are riding a roller coaster. The roller-coaster car moves up a hill until it stops at the top. Then, you are off! The graph in **Figure 6** shows your acceleration for the next 10 s. During the first 8 s, you move down the hill. You can tell from the graph that your acceleration is positive for the first 8 s because your velocity increases as time passes. During the last 2 s, your car starts climbing the next hill. Your acceleration is negative because your velocity decreases as time passes.

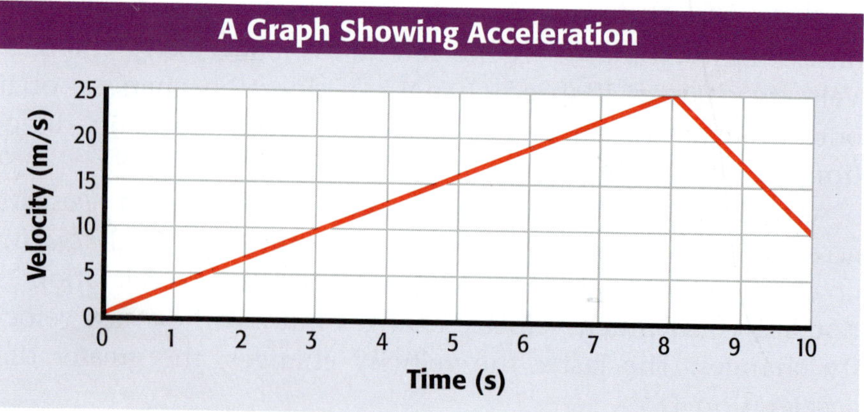

Figure 6 *Acceleration can be shown on a graph of velocity versus time.*

Circular Motion: Continuous Acceleration

You may be surprised to know that even when you are completely still, you are experiencing acceleration. You may not seem to be changing speed or direction, but you are! You are traveling in a circle as the Earth rotates. An object traveling in a circular motion is always changing its direction. Therefore, its velocity is always changing, so it is accelerating. The acceleration that occurs in circular motion is known as *centripetal acceleration* (sen TRIP uht uhl ak SEL uhr AY shuhn). Centripetal acceleration occurs on a Ferris wheel at an amusement park or as the moon orbits Earth. Another example of centripetal acceleration is shown in **Figure 7.**

Figure 7 *The blades of these windmills are constantly changing direction. Thus, centripetal acceleration is occurring.*

SECTION Review

Summary

- An object is in motion if it changes position over time in relation to a reference point.
- Speed is the distance traveled by an object divided by the time the object takes to travel that distance.
- Velocity is speed in a given direction.
- Acceleration is the rate at which velocity changes.
- An object can accelerate by changing speed, direction, or both.
- Speed can be represented on a graph of distance versus time.
- Acceleration can be represented by graphing velocity versus time.

Using Key Terms

1. In your own words, write definitions for each of the following terms: *motion* and *acceleration*.
2. Use each of the following terms in a separate sentence: *speed* and *velocity*.

Understanding Key Ideas

3. Which of the following is NOT an example of acceleration?
   a. a person jogging at 3 m/s along a winding path
   b. a car stopping at a stop sign
   c. a cheetah running 27 m/s east
   d. a plane taking off
4. Which of the following would be a good reference point to describe the motion of a dog?
   a. the ground
   b. another dog running
   c. a tree
   d. All of the above
5. Explain the difference between speed and velocity.
6. What two things must you know to determine speed?
7. How are velocity and acceleration related?

Math Skills

8. Find the average speed of a person who swims 105 m in 70 s.
9. What is the average acceleration of a subway train that speeds up from 9.6 m/s to 12 m/s in 0.8 s on a straight section of track?

Critical Thinking

10. **Applying Concepts** Why is it more helpful to know a tornado's velocity rather than its speed?
11. **Evaluating Data** A wolf is chasing a rabbit. Graph the wolf's motion using the following data: 15 m/s at 0 s, 10 m/s at 1 s, 5 m/s at 2 s, 2.5 m/s at 3 s, 1 m/s at 4 s, and 0 m/s at 5 s. What does the graph tell you?

SECTION 2

What Is a Force?

You have probably heard the word force *in everyday conversation. People say things such as "That storm had a lot of force" or "Our football team is a force to be reckoned with." But what, exactly, is a force?*

READING WARM-UP

Objectives

- Describe forces, and explain how forces act on objects.
- Determine the net force when more than one force is acting on an object.
- Compare balanced and unbalanced forces.
- Describe ways that unbalanced forces cause changes in motion.

Terms to Learn

force
newton
net force

READING STRATEGY

Reading Organizer As you read this section, make a table comparing balanced forces and unbalanced forces.

In science, a **force** is simply a push or a pull. All forces have both size and direction. A force can change the acceleration of an object. This acceleration can be a change in the speed or direction of the object. In fact, any time you see a change in an object's motion, you can be sure that the change in motion was created by a force. Scientists express force using a unit called the **newton** (N).

Forces Acting on Objects

All forces act on objects. For any push to occur, something has to receive the push. You can't push nothing! The same is true for any pull. When doing schoolwork, you use your fingers to pull open books or to push the buttons on a computer keyboard. In these examples, your fingers are exerting forces on the books and the keys. So, the forces act on the books and keys. Another example of a force acting on an object is shown in **Figure 1.**

However, just because a force acts on an object doesn't mean that motion will occur. For example, you are probably sitting on a chair. But the force you are exerting on the chair does not cause the chair to move. The chair doesn't move because the floor is also exerting a force on the chair.

Figure 1 *The bulldozer is exerting a force on the pile of soil. But the pile of soil also exerts a force by just sitting on the ground!*

Unseen Sources and Receivers of Forces

It is not always easy to tell what is exerting a force or what is receiving a force, as shown in **Figure 2.** You cannot see what exerts the force that pulls magnets to refrigerators. And you cannot see that the air around you is held near Earth's surface by a force called *gravity.*

Figure 2 *Something that you cannot see exerts a force that makes this cat's fur stand up.*

Determining Net Force

Usually, more than one force is acting on an object. The **net force** is the combination all of the forces acting on an object. So, how do you determine the net force? The answer depends on the directions of the forces.

Forces in the Same Direction

Suppose the music teacher asks you and a friend to move a piano. You pull on one end and your friend pushes on the other end, as shown in **Figure 3.** The forces you and your friend exert on the piano act in the same direction. The two forces are added to determine the net force because the forces act in the same direction. In this case, the net force is 45 N. This net force is large enough to move the piano—if it is on wheels, that is!

Reading Check **How do you determine the net force on an object if all forces act in the same direction?** (*See the Appendix for answers to Reading Checks.*)

force a push or a pull exerted on an object in order to change the motion of the object; force has size and direction

newton the SI unit for force (symbol, N)

net force the combination of all of the forces acting on an object

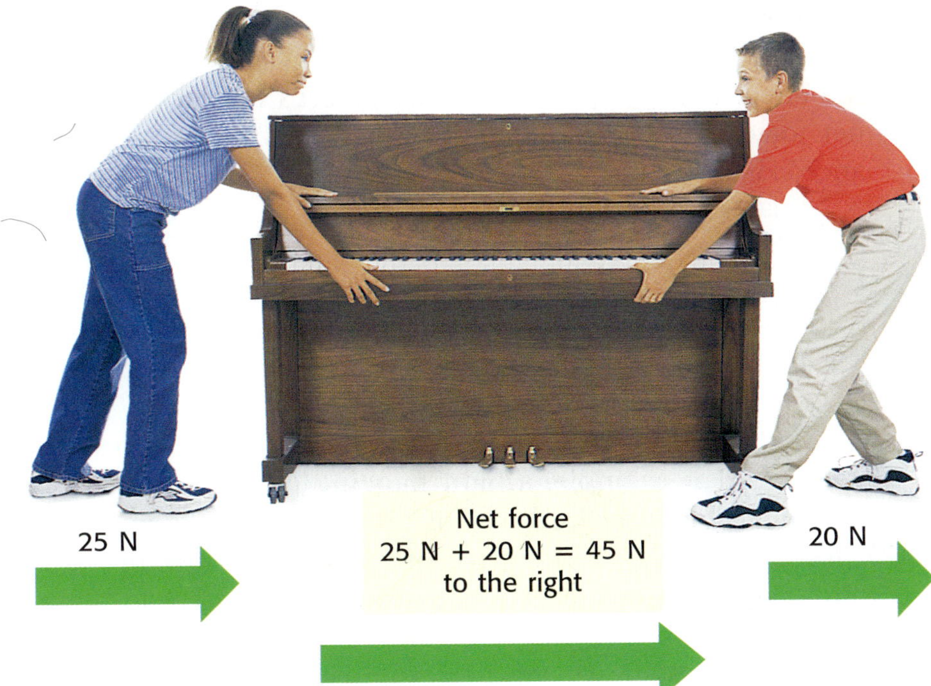

Figure 3 *When forces act in the same direction, you add the forces to determine the net force. The net force will be in the same direction as the individual forces.*

Figure 4 *When two forces act in opposite directions, you subtract the smaller force from the larger force to determine the net force. The net force will be in the same direction as the larger force.*

Forces in Different Directions

Look at the two dogs playing tug of war in **Figure 4.** Each dog is exerting a force on the rope. But the forces are in opposite directions. Which dog will win the tug of war?

Because the forces are in opposite directions, the net force on the rope is found by subtracting the smaller force from the larger one. In this case, the net force is 2 N in the direction of the dog on the right. Give that dog a dog biscuit!

Reading Check **What is the net force on an object when you combine a force of 7 N north with a force of 5 N south?**

Balanced and Unbalanced Forces

If you know the net force on an object, you can determine the effect of the net force on the object's motion. Why? The net force tells you whether the forces on the object are balanced or unbalanced.

Balanced Forces

When the forces on an object produce a net force of 0 N, the forces are *balanced*. Balanced forces will not cause a change in the motion of a moving object. And balanced forces do not cause a nonmoving object to start moving.

Many objects around you have only balanced forces acting on them. For example, a light hanging from the ceiling does not move because the force of gravity pulling down on the light is balanced by the force of the cord pulling upward. A bird's nest in a tree and a hat resting on your head are also examples of objects that have only balanced forces acting on them. **Figure 5** shows another example of balanced forces.

Figure 5 *Because all the forces on this house of cards are balanced, none of the cards move.*

Unbalanced Forces

When the net force on an object is not 0 N, the forces on the object are *unbalanced*. Unbalanced forces produce a change in motion, such as a change in speed or a change in direction. Unbalanced forces are necessary to cause a nonmoving object to start moving.

Unbalanced forces are also necessary to change the motion of moving objects. For example, consider the soccer game shown in **Figure 6.** The soccer ball is already moving when it is passed from one player to another. When the ball reaches another player, that player exerts an unbalanced force—a kick—on the ball. After the kick, the ball moves in a new direction and has a new speed.

An object can continue to move when the unbalanced forces are removed. For example, when it is kicked, a soccer ball receives an unbalanced force. The ball continues to roll on the ground long after the force of the kick has ended.

Figure 6 *The soccer ball moves because the players exert an unbalanced force on the ball each time they kick it.*

SECTION Review

Summary

- A force is a push or a pull. Forces have size and direction and are expressed in newtons.
- Force is always exerted by one object on another object.
- Net force is determined by combining forces. Forces in the same direction are added. Forces in opposite directions are subtracted.
- Balanced forces produce no change in motion. Unbalanced forces produce a change in motion.

Using Key Terms

1. In your own words, write a definition for each of the following terms: *force* and *net force*.

Understanding Key Ideas

2. Which of the following may happen when an object receives unbalanced forces?
   a. The object changes direction.
   b. The object changes speed.
   c. The object starts to move.
   d. All of the above
3. Explain the difference between balanced and unbalanced forces.
4. Give an example of an unbalanced force causing a change in motion.
5. Give an example of an object that has balanced forces acting on it.
6. Explain the meaning of the phrase "Forces act on objects."

Math Skills

7. A boy pulls a wagon with a force of 6 N east as another boy pushes it with a force of 4 N east. What is the net force?

Critical Thinking

8. **Making Inferences** When finding net force, why must you know the directions of the forces acting on an object?
9. **Applying Concepts** List three forces that you exert when riding a bicycle.

SECTION 3

Friction: A Force That Opposes Motion

READING WARM-UP

Objectives

- Explain why friction occurs.
- List the two types of friction, and give examples of each type.
- Explain how friction can be both harmful and helpful.

Terms to Learn

friction

READING STRATEGY

Brainstorming The key idea of this section is friction. Brainstorm words and phrases related to friction.

While playing ball, your friend throws the ball out of your reach. Rather than running for the ball, you walk after it. You know that the ball will stop. But do you know why?

You know that the ball is slowing down. An unbalanced force is needed to change the speed of a moving object. So, what force is stopping the ball? The force is called friction. **Friction** is a force that opposes motion between two surfaces that are in contact. Friction can cause a moving object, such as a ball, to slow down and eventually stop.

friction a force that opposes motion between two surfaces that are in contact

The Source of Friction

Friction occurs because the surface of any object is rough. Even surfaces that feel smooth are covered with microscopic hills and valleys. When two surfaces are in contact, the hills and valleys of one surface stick to the hills and valleys of the other surface, as shown in **Figure 1.** This contact causes friction.

The amount of friction between two surfaces depends on many factors. Two factors include the force pushing the surfaces together and the roughness of the surfaces.

The Effect of Force on Friction

The amount of friction depends on the force pushing the surfaces together. If this force increases, the hills and valleys of the surfaces can come into closer contact. The close contact increases the friction between the surfaces. Objects that weigh less exert less downward force than objects that weigh more do, as shown in **Figure 2.** But changing how much of the surfaces come in contact does not change the amount of friction.

Figure 1 *When the hills and valleys of one surface stick to the hills and valleys of another surface, friction is created.*

Figure 2 Force and Friction

a There is more friction between the book with more weight and the table than there is between the book with less weight and the table. A harder push is needed to move the heavier book.

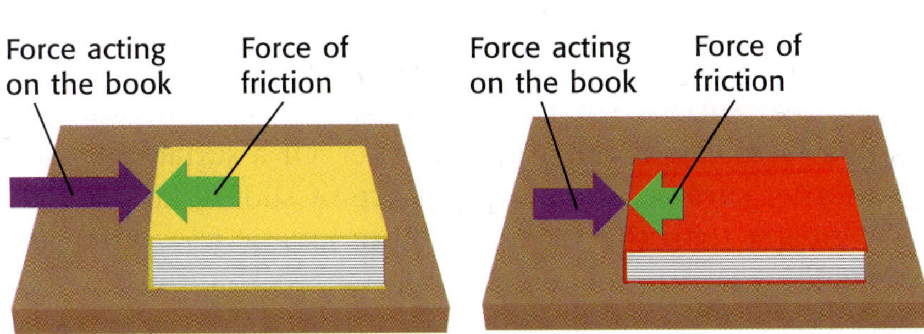

b Turning a book on its edge does not change the amount of friction between the table and the book.

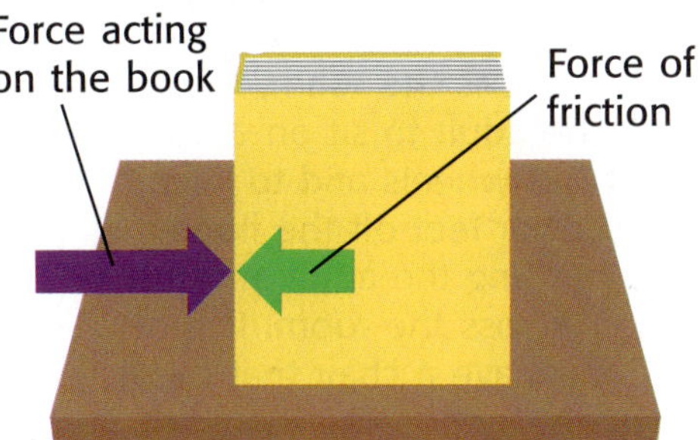

The Effect of Rougher Surfaces on Friction

Rough surfaces have more microscopic hills and valleys than smooth surfaces do. So, the rougher the surface is, the greater the friction is. For example, a ball rolling on the ground slows down because of the friction between the ball and the ground. A large amount of friction is produced because the ground has a rough surface. But imagine that you were playing ice hockey. If the puck passed out of your reach, it would slide across the ice for a long while before stopping. The reason the puck would continue to slide is that the ice is a smooth surface that has very little friction.

✓ *Reading Check* **Why is friction greater between surfaces that are rough?** (*See the Appendix for answers to Reading Checks.*)

The Friction 500

1. Make a short ramp out of **a piece of cardboard** and **one or two books** on a table.
2. Put a **toy car** at the top of the ramp, and let go of the car. If necessary, adjust the ramp height so that your car does not roll off the table.
3. Put the car at the top of the ramp again, and let go of the car. Record the distance the car travels after leaving the ramp.
4. Repeat step 3 two more times, and calculate the average for your results.
5. Change the surface of the table by covering the table with **sandpaper.** Repeat steps 3 and 4.
6. Change the surface of the table one more time by covering the table with **cloth.** Repeat steps 3 and 4 again.
7. Which surface had the most friction? Why? What do you predict would happen if the car were heavier?

Types of Friction

There are two types of friction. The friction you observe when sliding books across a tabletop is called *kinetic friction*. The other type of friction is *static friction*. You observe static friction when you push on a piece of furniture and it does not move.

Comparing Friction

Ask an adult at home to sit on the floor. Try to push the adult across the room. Next, ask the adult to sit on a chair that has wheels and to keep his or her feet off the floor. Try pushing the adult and the chair across the room. If you do not have a chair that has wheels, try pushing the adult on different kinds of flooring. Explain why there was a difference between the two trials in your **science journal.**

Kinetic Friction

The word *kinetic* means "moving." So, kinetic friction is friction between moving surfaces. The amount of kinetic friction between two surfaces depends in part on how the surfaces move. Surfaces can slide past each other. Or a surface can roll over another surface. Usually, the force of sliding kinetic friction is greater than the force of rolling kinetic friction. Thus, it is usually easier to move objects on wheels than to slide the objects along the floor, as shown in **Figure 3.**

Kinetic friction is very useful in everyday life. You use sliding kinetic friction when you apply the brakes on a bicycle and when you write with a pencil or a piece of chalk. You also use sliding kinetic friction when you scratch a part of your body that is itchy!

Rolling kinetic friction is an important part of almost all means of transportation. Anything that has wheels—bicycles, in-line skates, cars, trains, and planes—uses rolling kinetic friction.

Figure 3 **Comparing Kinetic Friction**

a Moving a heavy piece of furniture in your room can be hard work because **the force of sliding kinetic friction is large.**

b Moving a heavy piece of furniture is easier if you put it on wheels. **The force of rolling kinetic friction is smaller** and easier to overcome.

Figure 4 Static Friction

Block

Table

a There is no friction between the block and the table when no force is applied to the block.

Force applied

Static friction

b If a small force (purple arrow) is exerted on the block, the block does not move. The force of static friction (green arrow) balances the force applied.

Force applied

Kinetic friction

c When the force exerted on the block is greater than the force of static friction, the block starts moving. When the block starts moving, all static friction is gone, and only kinetic friction (green arrow) opposes the force applied.

Static Friction

When a force is applied to an object but does not cause the object to move, *static friction* occurs. The word *static* means "not moving." The object does not move because the force of static friction balances the force applied. Static friction can be overcome by applying a large enough force. Static friction disappears as soon as an object starts moving, and then kinetic friction immediately occurs. Look at **Figure 4** to understand under what conditions static friction affects an object.

Reading Check What does the word *static* mean?

For another activity related to this chapter, go to **go.hrw.com** and type in the keyword **HP5MOTW.**

Friction: Harmful and Helpful

Think about how friction affects a car. Without friction, the tires could not push against the ground to move the car forward, and the brakes could not stop the car. Without friction, a car is useless. However, friction can also cause problems in a car. Friction between moving engine parts increases their temperature and causes the parts to wear down. A liquid coolant is added to the engine to keep the engine from overheating. And engine parts need to be changed as they wear out.

Friction is both harmful and helpful to you and the world around you. Friction can cause holes in your socks and in the knees of your jeans. Friction by wind and water can cause erosion of the topsoil that nourishes plants. On the other hand, friction between your pencil and your paper is necessary to allow the pencil to leave a mark. Without friction, you would just slip and fall when you tried to walk. Because friction can be both harmful and helpful, it is sometimes necessary to decrease or increase friction.

WRITING SKILL **Invention of the Wheel** Archeologists have found evidence that the first vehicles with wheels were used in ancient Mesopotamia sometime between 3500 and 3000 BCE. Before wheels were invented, people used planks or sleds to carry loads. In your **science journal,** write a paragraph about how your life would be different if wheels did not exist.

Quick Lab

Reducing Friction

1. Stack **two or three heavy books** on a table. Use one finger to push the books across the table.
2. Place **five round pens or pencils** under the books, and push the books again.
3. Compare the force used in step 1 with the force used in step 2. Explain.
4. Open a **jar** with your hands, and close it again.
5. Spread a small amount of **liquid soap** on your hands.
6. Try to open the jar again. Was the jar easier or harder to open with the soap? Explain your observations.
7. In which situation was friction helpful? In which situation was friction harmful?

Some Ways to Reduce Friction

One way to reduce friction is to use lubricants (LOO bri kuhnts). *Lubricants* are substances that are applied to surfaces to reduce the friction between the surfaces. Some examples of common lubricants are motor oil, wax, and grease. Lubricants are usually liquids, but they can be solids or gases. An example of a gas lubricant is the air that comes out of the tiny holes of an air-hockey table. **Figure 5** shows one use of a lubricant.

Friction can also be reduced by switching from sliding kinetic friction to rolling kinetic friction. Ball bearings placed between the wheels and axles of in-line skates and bicycles make it easier for the wheels to turn by reducing friction.

Another way to reduce friction is to make surfaces that rub against each other smoother. For example, rough wood on a park bench is painful to slide across because there is a large amount of friction between your leg and the bench. Rubbing the bench with sandpaper makes the bench smoother and more comfortable to sit on. The reason the bench is more comfortable is that the friction between your leg and the bench is reduced.

Reading Check List three common lubricants.

Figure 5 *When you work on a bicycle, watch out for the chain! You might get dirty from the grease or oil that keeps the chain moving freely. Without this lubricant, friction between the sections of the chain would quickly wear the chain out.*

Some Ways to Increase Friction

One way to increase friction is to make surfaces rougher. For example, sand scattered on icy roads keeps cars from skidding. Baseball players sometimes wear textured batting gloves to increase the friction between their hands and the bat so that the bat does not fly out of their hands.

Another way to increase friction is to increase the force pushing the surfaces together. For example, if you are sanding a piece of wood, you can sand the wood faster by pressing harder on the sandpaper. Pressing harder increases the force pushing the sandpaper and wood together. So, the friction between the sandpaper and wood increases. **Figure 6** shows another example of friction increased by pushing on an object.

Figure 6 *No one likes cleaning dirty pans. To get this chore done quickly, press down with the scrubber to increase friction.*

SECTION Review

Summary

- Friction is a force that opposes motion.
- Friction is caused by hills and valleys on the surfaces of two objects touching each other.
- The amount of friction depends on factors such as the roughness of the surfaces and the force pushing the surfaces together.
- Two kinds of friction are kinetic friction and static friction.
- Friction can be helpful or harmful.

Using Key Terms

1. In your own words, write a definition for the term *friction*.

Understanding Key Ideas

2. Why is it easy to slip when there is water on the floor?
   a. The water is a lubricant and reduces the friction between your feet and the floor.
   b. The friction between your feet and the floor changes from kinetic to static friction.
   c. The water increases the friction between your feet and the floor.
   d. The friction between your feet and the floor changes from sliding kinetic friction to rolling kinetic friction.
3. Explain why friction occurs.
4. How does the roughness of surfaces that are touching affect the friction between the surfaces?
5. Describe how the amount of force pushing two surfaces together affects friction.
6. Name two ways in which friction can be increased.
7. List the two types of friction, and give an example of each.

Interpreting Graphics

8. Why do you think the sponge shown below has a layer of plastic bristles attached to it?

Critical Thinking

9. **Applying Concepts** Name two ways that friction is harmful and two ways that friction is helpful to you when riding a bicycle.
10. **Making Inferences** Describe a situation in which static friction is useful.

SECTION 4

Gravity: A Force of Attraction

READING WARM-UP

Objectives

- Describe gravity and its effect on matter.
- Explain the law of universal gravitation.
- Describe the difference between mass and weight.

Terms to Learn

gravity
weight
mass

READING STRATEGY

Paired Summarizing Read this section silently. In pairs, take turns summarizing the material. Stop to discuss ideas that seem confusing.

Have you ever seen a video of astronauts on the moon? They bounce around like beach balls even though they wear big, bulky spacesuits. Why is leaping on the moon easier than leaping on Earth?

The answer is gravity. **Gravity** is a force of attraction between objects that is due to their masses. The force of gravity can change the motion of an object by changing its speed, direction, or both. In this section, you will learn about gravity and its effects on objects, such as the astronaut in **Figure 1.**

gravity a force of attraction between objects that is due to their masses

The Effects of Gravity on Matter

All matter has mass. Gravity is a result of mass. Therefore, all matter is affected by gravity. That is, all objects experience an attraction toward all other objects. This gravitational force pulls objects toward each other. Right now, because of gravity, you are being pulled toward this book, your pencil, and every other object around you.

These objects are also being pulled toward you and toward each other because of gravity. So why don't you see the effects of this attraction? In other words, why don't you notice objects moving toward each other? The reason is that the mass of most objects is too small to cause a force large enough to move objects toward each other. However, you are familiar with one object that is massive enough to cause a noticeable attraction—the Earth.

Figure 1 *Because the moon has less gravity than the Earth does, walking on the moon's surface was a very bouncy experience for the Apollo astronauts.*

The Size of Earth's Gravitational Force

Compared with all objects around you, Earth has a huge mass. Therefore, Earth's gravitational force is very large. You must apply forces to overcome Earth's gravitational force any time you lift objects or even parts of your body.

Earth's gravitational force pulls everything toward the center of Earth. Because of this force, the books, tables, and chairs in the room stay in place, and dropped objects fall to Earth rather than moving together or toward you.

Reading Check **Why must you exert a force to pick up an object?** (*See the Appendix for answers to Reading Checks.*)

CONNECTION TO Biology

Seeds and Gravity Seeds respond to gravity. The ability to respond to gravity causes seeds to send roots down and the green shoot up. But scientists do not understand how seeds can sense gravity. Plan an experiment to study how seedlings respond to gravity. After getting your teacher's approval, do your experiment and report your observations in a poster.

ACTIVITY

Newton and the Study of Gravity

For thousands of years, people asked two very puzzling questions: Why do objects fall toward Earth, and what keeps the planets moving in the sky? The two questions were treated separately until 1665 when a British scientist named Sir Isaac Newton realized that they were two parts of the same question.

The Core of an Idea

The legend is that Newton made the connection between the two questions when he watched a falling apple, as shown in **Figure 2.** He knew that unbalanced forces are needed to change the motion of objects. He concluded that an unbalanced force on the apple made the apple fall. And he reasoned that an unbalanced force on the moon kept the moon moving circularly around Earth. He proposed that these two forces are actually the same force—a force of attraction called *gravity*.

The Birth of a Law

Newton summarized his ideas about gravity in a law now known as the *law of universal gravitation*. This law describes the relationships between gravitational force, mass, and distance. The law is called *universal* because it applies to all objects in the universe.

Figure 2 *Sir Isaac Newton realized that the same unbalanced force affected the motions of the apple and the moon.*

The Law of Universal Gravitation

The law of universal gravitation is the following: All objects in the universe attract each other through gravitational force. The size of the force depends on the masses of the objects and the distance between the objects. Understanding the law is easier if you consider it in two parts.

CONNECTION TO Astronomy

WRITING SKILL **Black Holes** Black holes are 4 times to 1 billion times as massive as our sun. So, the gravitational effects around a black hole are very large. The gravitational force of a black hole is so large that objects that enter a black hole can never get out. Even light cannot escape from a black hole. Because black holes do not emit light, they cannot be seen. Research how astronomers can detect black holes without seeing them. Write a one-page paper that details the results of your research.

Part 1: Gravitational Force Increases as Mass Increases

Imagine an elephant and a cat. Because an elephant has a larger mass than a cat does, the amount of gravity between an elephant and Earth is greater than the amount of gravity between a cat and Earth. So, a cat is much easier to pick up than an elephant! There is also gravity between the cat and the elephant, but that force is very small because the cat's mass and the elephant's mass are so much smaller than Earth's mass. **Figure 3** shows the relationship between mass and gravitational force.

This part of the law of universal gravitation also explains why the astronauts on the moon bounce when they walk. The moon has less mass than Earth does. Therefore, the moon's gravitational force is less than Earth's. The astronauts bounced around on the moon because they were not being pulled down with as much force as they would have been on Earth.

Reading Check How does mass affect gravitational force?

Figure 3 **How Mass Affects Gravitational Force**

The gravitational force between objects increases as the masses of the objects increase. The arrows indicate the gravitational force between two objects. The length of the arrows indicates the strength of the force.

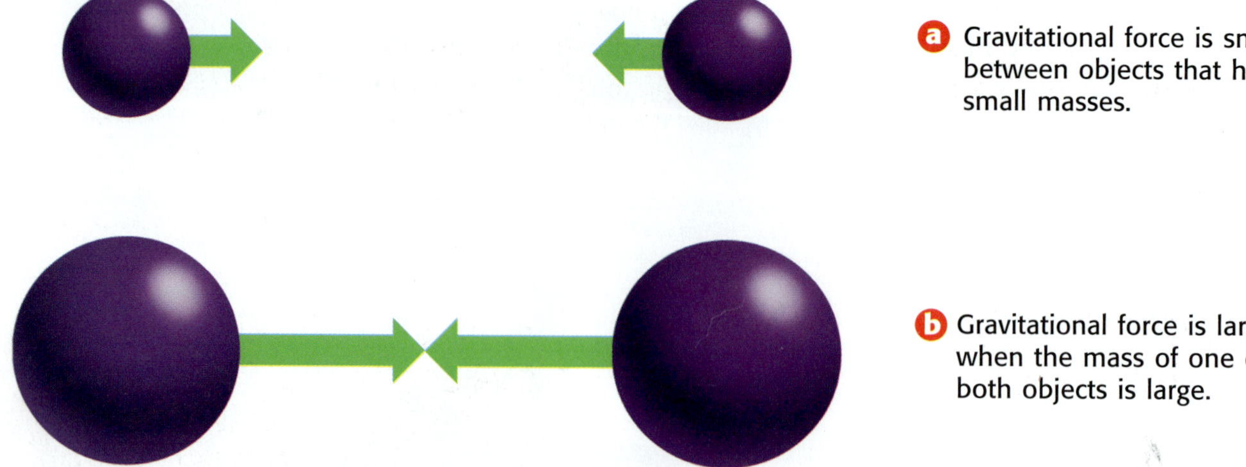

a Gravitational force is small between objects that have small masses.

b Gravitational force is large when the mass of one or both objects is large.

Part 2: Gravitational Force Decreases as Distance Increases

The gravitational force between you and Earth is large. Whenever you jump up, you are pulled back down by Earth's gravitational force. On the other hand, the sun is more than 300,000 times more massive than Earth. So why doesn't the sun's gravitational force affect you more than Earth's does? The reason is that the sun is so far away.

You are about 150 million kilometers (93 million miles) away from the sun. At this distance, the gravitational force between you and the sun is very small. If there were some way you could stand on the sun, you would find it impossible to move. The gravitational force acting on you would be so great that you could not move any part of your body!

Although the sun's gravitational force on your body is very small, the force is very large on Earth and the other planets, as shown in **Figure 4.** The gravity between the sun and the planets is large because the objects have large masses. If the sun's gravitational force did not have such an effect on the planets, the planets would not stay in orbit around the sun. **Figure 5** will help you understand the relationship between gravitational force and distance.

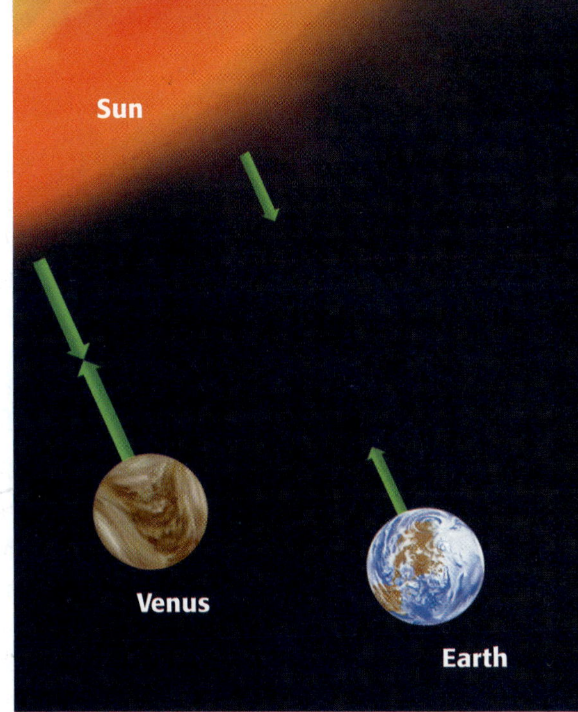

Figure 4 *Venus and Earth have approximately the same mass. But because Venus is closer to the sun, the gravitational force between Venus and the sun is greater than the gravitational force between Earth and the sun.*

Figure 5 How Distance Affects Gravitational Force

The gravitational force between objects decreases as the distance between the objects increases. The length of the arrows indicates the strength of the gravitational force between two objects.

a Gravitational force is strong when the distance between two objects is small.

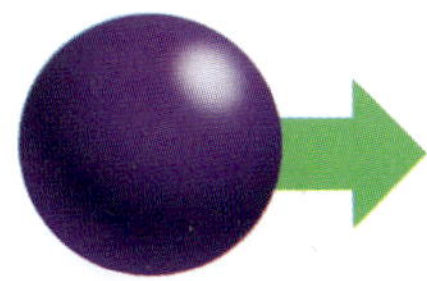

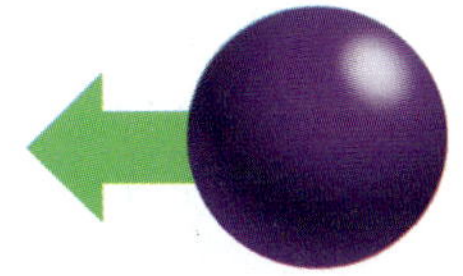

b If the distance between two objects increases, the gravitational force pulling them together decreases rapidly.

Weight as a Measure of Gravitational Force

Gravity is a force of attraction between objects. **Weight** is a measure of the gravitational force on an object. When you see or hear the word *weight,* it usually refers to Earth's gravitational force on an object. But weight can also be a measure of the gravitational force exerted on objects by the moon or other planets.

CONNECTION TO Language Arts

WRITING SKILL **Gravity Story** Suppose you had a device that could increase or decrease the gravitational force of Earth. In your **science journal,** write a short story describing what you might do with the device, what you would expect to see, and what effect the device would have on the weight of objects.

The Differences Between Weight and Mass

Weight is related to mass, but they are not the same. Weight changes when gravitational force changes. **Mass** is the amount of matter in an object. An object's mass does not change. Imagine that an object is moved to a place that has a greater gravitational force—such as the planet Jupiter. The object's weight will increase, but its mass will remain the same. **Figure 6** shows the weight and mass of an astronaut on Earth and on the moon. The moon's gravitational force is about one-sixth of Earth's gravitational force.

Gravitational force is about the same everywhere on Earth. So, the weight of any object is about the same everywhere. Because mass and weight are constant on Earth, the terms *weight* and *mass* are often used to mean the same thing. This can be confusing. Be sure you understand the difference!

weight a measure of the gravitational force exerted on an object; its value can change with the location of the object in the universe

mass a measure of the amount of matter in an object

Reading Check How is gravitational force related to the weight of an object?

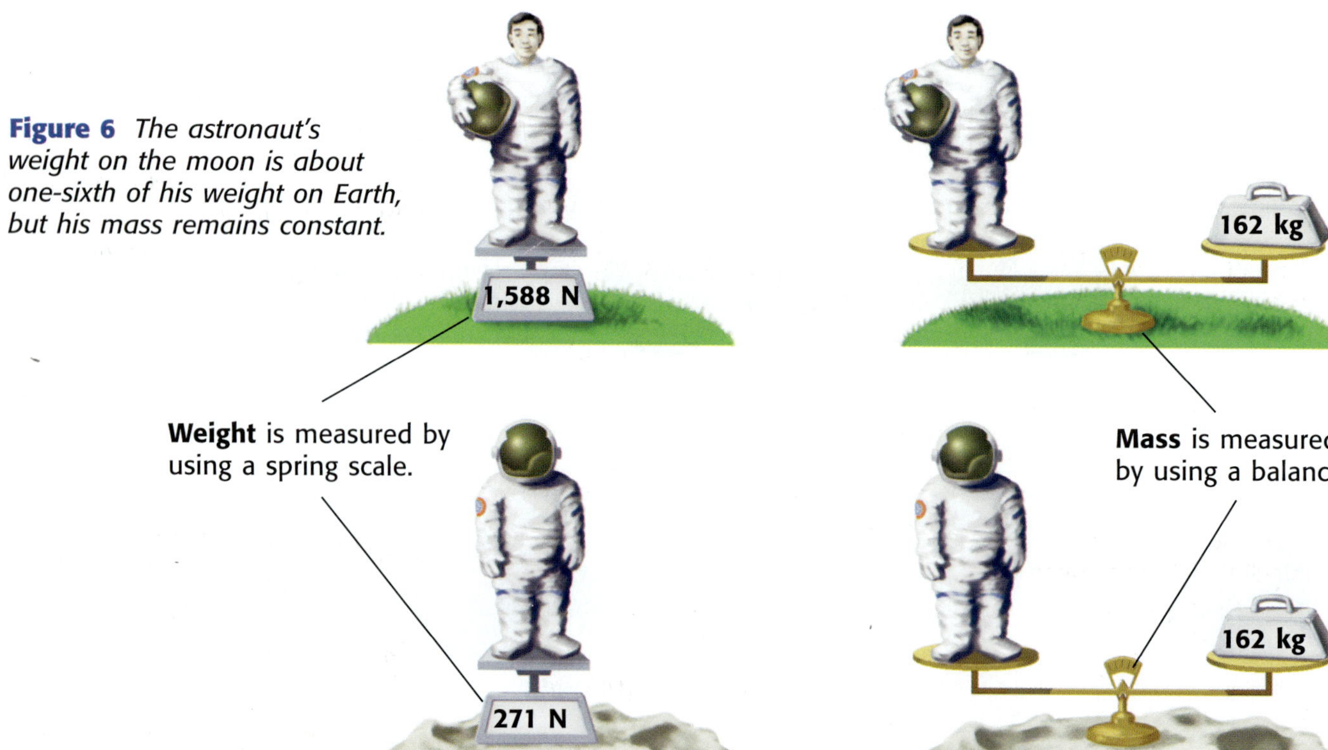

Figure 6 *The astronaut's weight on the moon is about one-sixth of his weight on Earth, but his mass remains constant.*

Units of Weight and Mass

You have learned that the SI unit of force is a newton (N). Gravity is a force, and weight is a measure of gravity. So, weight is also measured in newtons. The SI unit of mass is the kilogram (kg). Mass is often measured in grams (g) and milligrams (mg) as well. On Earth, a 100 g object, such as the apple shown in **Figure 7,** weighs about 1 N.

When you use a bathroom scale, you are measuring the gravitational force between your body and Earth. So, you are measuring your weight, which should be given in newtons. However, many bathroom scales have units of pounds and kilograms instead of newtons. Thus, people sometimes mistakenly think that the kilogram (like the pound) is a unit of weight.

Figure 7 *A small apple weighs approximately 1 N.*

SECTION Review

Summary

- Gravity is a force of attraction between objects that is due to their masses.
- The law of universal gravitation states that all objects in the universe attract each other through gravitational force.
- Gravitational force increases as mass increases.
- Gravitational force decreases as distance increases.
- Weight and mass are not the same. Mass is the amount of matter in an object. Weight is a measure of the gravitational force on an object.

Using Key Terms

1. In your own words, write a definition for the term *gravity*.
2. Use each of the following terms in a separate sentence: *mass* and *weight*.

Understanding Key Ideas

3. If Earth's mass doubled without changing its size, your weight would
   - **a.** increase because gravitational force increases.
   - **b.** decrease because gravitational force increases.
   - **c.** increase because gravitational force decreases.
   - **d.** not change because you are still on Earth.
4. What is the law of universal gravitation?
5. How does the mass of an object relate to the gravitational force that the object exerts on other objects?
6. How does the distance between objects affect the gravitational force between them?
7. Why are mass and weight often confused?

Math Skills

8. The gravitational force on Jupiter is approximately 2.3 times the gravitational force on Earth. If an object has a mass of 70 kg and a weight of 686 N on Earth, what would the object's mass and weight on Jupiter be?

Critical Thinking

9. **Applying Concepts** Your friend thinks that there is no gravity in space. How could you explain to your friend that there must be gravity in space?
10. **Making Comparisons** Explain why it is your weight and not your mass that would change if you landed on Mars.

Skills Practice Lab

OBJECTIVES

Build an accelerometer.

Explain how an accelerometer works.

MATERIALS

- container, 1 L, with watertight lid
- cork or plastic-foam ball, small
- modeling clay
- pushpin
- scissors
- string
- water

SAFETY

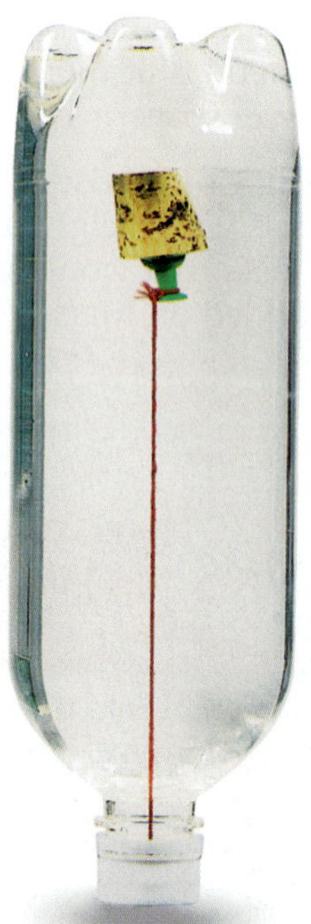

Detecting Acceleration

Have you ever noticed that you can "feel" acceleration? In a car or in an elevator, you may notice changes in speed or direction—even with your eyes closed! You are able to sense these changes because of tiny hair cells in your ears. These cells detect the movement of fluid in your inner ear. The fluid accelerates when you do, and the hair cells send a message about the acceleration to your brain. This message allows you to sense the acceleration. In this activity, you will build a device that detects acceleration. This device is called an *accelerometer* (ak SEL uhr AHM uht uhr).

Procedure

1. Cut a piece of string that reaches three-quarters of the way into the container.
2. Use a pushpin to attach one end of the string to the cork or plastic-foam ball.
3. Use modeling clay to attach the other end of the string to the center of the inside of the container lid. The cork or ball should hang no farther than three-quarters of the way into the container.
4. Fill the container with water.
5. Put the lid tightly on the container. The string and cork or ball should be inside the container.
6. Turn the container upside down. The cork should float about three-quarters of the way up inside the container, as shown at right. You are now ready to detect acceleration by using your accelerometer and completing the following steps.
7. Put the accelerometer on a tabletop. The container lid should touch the tabletop. Notice that the cork floats straight up in the water.
8. Now, gently push the accelerometer across the table at a constant speed. Notice that the cork quickly moves in the direction you are pushing and then swings backward. If you did not see this motion, repeat this step until you are sure you can see the first movement of the cork.

9 After you are familiar with how to use your accelerometer, try the following changes in motion. For each change, record your observations of the cork's first motion.

a. As you move the accelerometer across the table, gradually increase its speed.

b. As you move the accelerometer across the table, gradually decrease its speed.

c. While moving the accelerometer across the table, change the direction in which you are pushing.

d. Make any other changes in motion you can think of. You should make only one change to the motion for each trial.

Analyze the Results

1 **Analyzing Results** When you move the bottle at a constant speed, why does the cork quickly swing backward after it moves in the direction of acceleration?

2 **Explaining Events** The cork moves forward (in the direction you were moving the bottle) when you speed up but moves backward when you slow down. Explain why the cork moves this way. (Hint: Think about the direction of acceleration.)

Draw Conclusions

3 **Making Predictions** Imagine you are standing on a corner and watching a car that is waiting at a stoplight. A passenger inside the car is holding some helium balloons. Based on what you observed with your accelerometer, what do you think will happen to the balloons when the car begins moving?

Applying Your Data

If you move the bottle in a circle at a constant speed, what do you predict the cork will do? Try it, and check your answer.

Chapter Review

USING KEY TERMS

Complete each of the following sentences by choosing the correct term from the word bank.

mass	gravity
friction	weight
speed	velocity
net force	newton

1. ___ opposes motion between surfaces that are touching.
2. The ___ is the unit of force.
3. ___ is determined by combining forces.
4. Acceleration is the rate at which ___ changes.
5. ___ is a measure of the gravitational force on an object.

UNDERSTANDING KEY IDEAS

Multiple Choice

6. If a student rides her bicycle on a straight road and does not speed up or slow down, she is traveling with a
   - a. constant acceleration.
   - b. constant velocity.
   - c. positive acceleration.
   - d. negative acceleration.
7. A force
   - a. is expressed in newtons.
   - b. can cause an object to speed up, slow down, or change direction.
   - c. is a push or a pull.
   - d. All of the above
8. If you are in a spacecraft that has been launched into space, your weight would
   - a. increase because gravitational force is increasing.
   - b. increase because gravitational force is decreasing.
   - c. decrease because gravitational force is decreasing.
   - d. decrease because gravitational force is increasing.
9. The gravitational force between 1 kg of lead and Earth is ___ the gravitational force between 1 kg of marshmallows and Earth.
   - a. greater than
   - b. less than
   - c. the same as
   - d. None of the above
10. Which of the following is a measurement of velocity?
    - a. 16 m east
    - b. 25 m/s^2
    - c. 55 m/h south
    - d. 60 km/h

Short Answer

11. Describe the relationship between motion and a reference point.
12. How is it possible to be accelerating and traveling at a constant speed?
13. Explain the difference between mass and weight.

Math Skills

14 A kangaroo hops 60 m to the east in 5 s. Use this information to answer the following questions.

a. What is the kangaroo's average speed?

b. What is the kangaroo's average velocity?

c. The kangaroo stops at a lake for a drink of water and then starts hopping again to the south. Each second, the kangaroo's velocity increases 2.5 m/s. What is the kangaroo's acceleration after 5 s?

CRITICAL THINKING

15 **Concept Mapping** Use the following terms to create a concept map: *speed, velocity, acceleration, force, direction,* and *motion.*

16 **Applying Concepts** Your family is moving, and you are asked to help move some boxes. One box is so heavy that you must push it across the room rather than lift it. What are some ways you could reduce friction to make moving the box easier?

17 **Analyzing Ideas** Considering the scientific meaning of the word *acceleration,* how could using the term *accelerator* when talking about a car's gas pedal lead to confusion?

18 **Identifying Relationships** Explain why it is important for airplane pilots to know wind velocity and not just wind speed during a flight.

INTERPRETING GRAPHICS

Use the figures below to answer the questions that follow.

19 Is the graph below showing positive acceleration or negative acceleration? How can you tell?

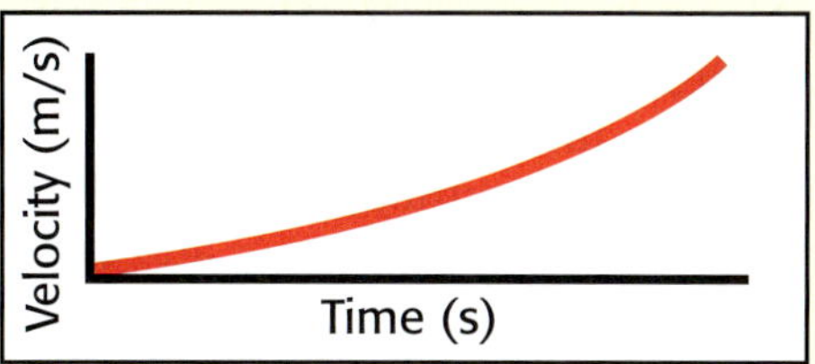

20 You know how to combine two forces that act in one or two directions. The same method can be used to combine several forces acting in several directions. Look at the diagrams, and calculate the net force in each diagram. Predict the direction each object will move.

a. 3 N (up), 6 N (right), 9 N (left), 3 N (down)

b. 5 N (right), 5 N (left), 5 N (up)

c. 4 N (up), 8 N (right), 4 N (down)

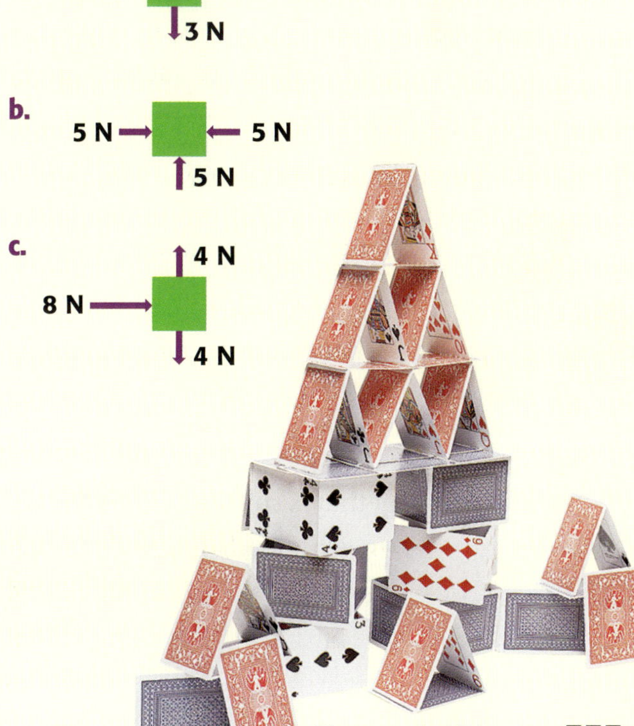

Standardized Test Preparation

READING

Read each of the passages below. Then, answer the questions that follow each passage.

Passage 1 If you look closely at the surface of a golf ball, you'll see dozens of tiny dimples. When air flows past these dimples, the air is stirred up and stays near the surface of the ball. By keeping air moving near the surface of the ball, the dimples help the golf ball move faster and farther through the air. Jeff DiTullio, a teacher at MIT in Cambridge, Massachusetts, decided to apply this principle to a baseball bat. When DiTullio tested his dimpled bat in a wind tunnel, he found that the bat could be swung 3% to 5% faster than a bat without dimples. That increase may not seem like much, but the dimpled bat could add about 5 m of distance to a fly ball!

1. Who is Jeff DiTullio?

- **A** the inventor of the dimpled golf ball
- **B** a teacher at Cambridge University
- **C** the inventor of the dimpled bat
- **D** a professional baseball player

2. Which of the following ideas is NOT stated in the passage?

- **F** Dimples make DiTullio's bat move faster.
- **G** MIT is in Cambridge, Massachusetts.
- **H** Air that is stirred up near the surface of DiTullio's bat makes it easier to swing the bat faster.
- **I** DiTullio will make a lot of money from his invention.

3. In the passage, what does *wind tunnel* mean?

- **A** a place to practice batting
- **B** a place to test the speed of objects in the air
- **C** a baseball stadium
- **D** a passageway that is shielded from the wind

Passage 2 The Golden Gate Bridge in San Francisco, California, is one of the most famous landmarks in the world. Approximately 9 million people from around the world visit the bridge each year.

The Golden Gate Bridge is a suspension bridge. A suspension bridge is one in which the roadway is hung, or suspended, from huge cables that extend from one end of the bridge to the other. The main cables on the Golden Gate Bridge are 2.33 km long. Many forces act on the main cables. For example, smaller cables pull down on the main cables to connect the roadway to the main cables. And two towers that are 227 m tall push up on the main cables. The forces on the main cable must be balanced, or the bridge will collapse.

1. In this passage, what does *landmarks* mean?

- **A** large areas of land
- **B** well-known places
- **C** street signs
- **D** places where people meet

2. Which of the following statements is a fact from the passage?

- **F** The roadway of the Golden Gate Bridge is suspended from huge cables.
- **G** The towers of the Golden Gate Bridge are 2.33 km tall.
- **H** The main cables connect the roadway to the towers.
- **I** The forces on the cables are not balanced.

3. According to the passage, why do people from around the world visit the Golden Gate Bridge?

- **A** It is the longest bridge in the world.
- **B** It is a suspension bridge.
- **C** It is the only bridge that is painted orange.
- **D** It is a famous landmark.

INTERPRETING GRAPHICS

The graph below shows the data collected by a student as she watched a squirrel running on the ground. Use the graph below to answer the questions that follow.

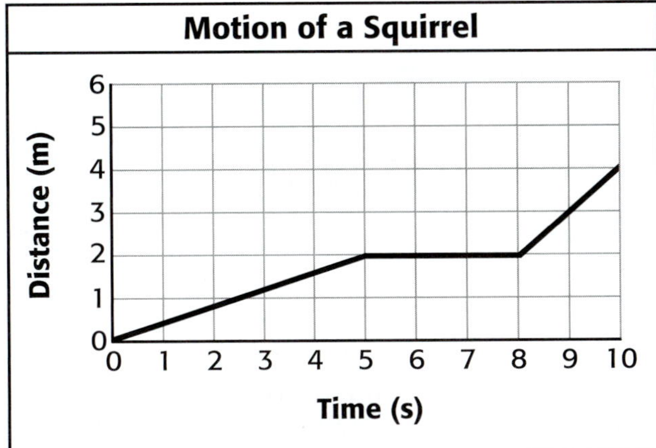

1. Which of the following best describes the motion of the squirrel between 5 s and 8 s?

A The squirrel's speed increased.
B The squirrel's speed decreased.
C The squirrel's speed did not change.
D The squirrel moved backward.

2. Which of the following statements about the motion of the squirrel is true?

F The squirrel moved with the greatest speed between 0 s and 5 s.
G The squirrel moved with the greatest speed between 8 s and 10 s.
H The squirrel moved with a constant speed between 0 s and 8 s.
I The squirrel moved with a constant speed between 5 s and 10 s.

3. What is the average speed of the squirrel between 8 s and 10 s?

A 0.4 m/s
B 1 m/s
C 2 m/s
D 4 m/s

MATH

Read each question below, and choose the best answer.

1. The distance between Cedar Rapids, Iowa, and Sioux Falls, South Dakota, is about 660 km. How long will it take a car traveling with an average speed of 95 km/h to drive from Cedar Rapids to Sioux Falls?

A less than 1 h
B about 3 h
C about 7 h
D about 10 h

2. Martha counted the number of people in each group that walked into her school's cafeteria. In the first 10 groups, she counted the following numbers of people: 6, 4, 9, 6, 4, 10, 9, 5, 9, and 8. What is the mode of this set of data?

F 6
G 7
H 9
I 10

3. Which of the following terms describes the angle marked in the triangle below.

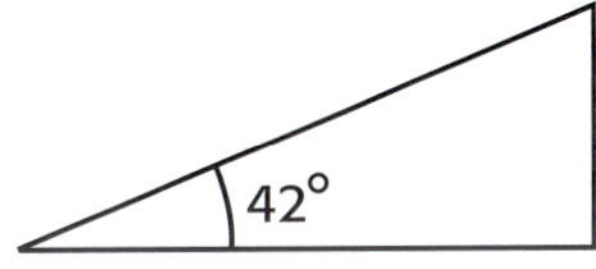

A acute
B obtuse
C right
D None of the above

4. Donnell collected money for a charity fundraiser. After one hour, he counted the money and found that he had raised $10.00 in bills and $3.74 in coins. Which of the following represents the number of coins he collected?

F 4 pennies, 9 nickels, 18 dimes, and 6 quarters
G 9 pennies, 7 nickels, 18 dimes, and 6 quarters
H 6 pennies, 7 nickels, 15 dimes, and 8 quarters
I 9 pennies, 8 nickels, 12 dimes, and 3 quarters

Science in Action

Science, Technology, and Society

GPS Watch System

Some athletes are concerned about knowing their speed during training. To calculate speed, they need to know distance and time. Finding time by using a watch is easy to do. But determining distance is more difficult. However, a GPS watch system is now available to help with this problem. *GPS* stands for *global positioning system*. A GPS unit, which is worn on an athlete's upper arm, monitors the athlete's position by using signals from satellites. As the athlete moves, the GPS unit calculates the distance traveled. The GPS unit sends a signal to the watch, which keeps the athlete's time, and the watch displays the athlete's speed.

Math ACTIVITY

Suppose an athlete wishes to finish a 5 K race in under 25 min. The distance of a 5 K is 5 km. (Remember that 1 km = 1,000 m.) If the athlete runs the race at a constant speed of 3.4 m/s, will she meet her goal?

Weird Science

The Segway™ Human Transporter

In November 2002, a new people-moving machine was introduced, and people have been fascinated by the odd-looking device ever since. The device is called the *Segway Human Transporter*. The Segway is a two-wheeled device that is powered by a rechargeable battery. To move forward, the rider simply leans forward. Sensors detect this motion and send signals to the onboard computer. The computer, in turn, tells the motor to start going. To slow down, the rider leans backward, and to stop, the rider stands straight up. The Segway has a top speed of 20 km/h (about 12.5 mi/h) and can travel up to 28 km (about 17.4 mi) on a single battery charge.

Language Arts ACTIVITY

WRITING SKILL The inventor of the Segway thinks that the machine will make a good alternative to walking and bicycle riding. Write a one-page essay explaining whether you think using a Segway is better or worse than riding a bicycle.

People in Science

Victor Petrenko

Snowboard and Ski Brakes Have you ever wished for emergency brakes on your snowboard or skis? Thanks to Victor Petrenko and the Ice Research Lab of Dartmouth College, snowboards and skis that have braking systems may soon be available.

Not many people know more about the properties of ice and ice-related technologies than Victor Petrenko does. He has spent most of his career researching the electrical and mechanical properties of ice. Through his research, Petrenko learned that ice can hold an electric charge. He used this property to design a braking system for snowboards. The system is a form of electric friction control.

The power source for the brakes is a battery. The battery is connected to a network of wires embedded on the bottom surface of a snowboard. When the battery is activated, the bottom of the snowboard gains a negative charge. This negative charge creates a positive charge on the surface of the snow. Because opposite charges attract, the snowboard and the snow are pulled together. The force that pulls the surfaces together increases friction, and the snowboard slows down.

Social Studies ACTIVITY

Research the history of skiing. Make a poster that includes a timeline of significant dates in the history of skiing. Illustrate your poster with photos or drawings.

To learn more about these Science in Action topics, visit **go.hrw.com** and type in the keyword **HP5MOTF.**

Current Science

Check out Current Science® articles related to this chapter by visiting **go.hrw.com.** Just type in the keyword **HP5CS05.**

Forces and Motion

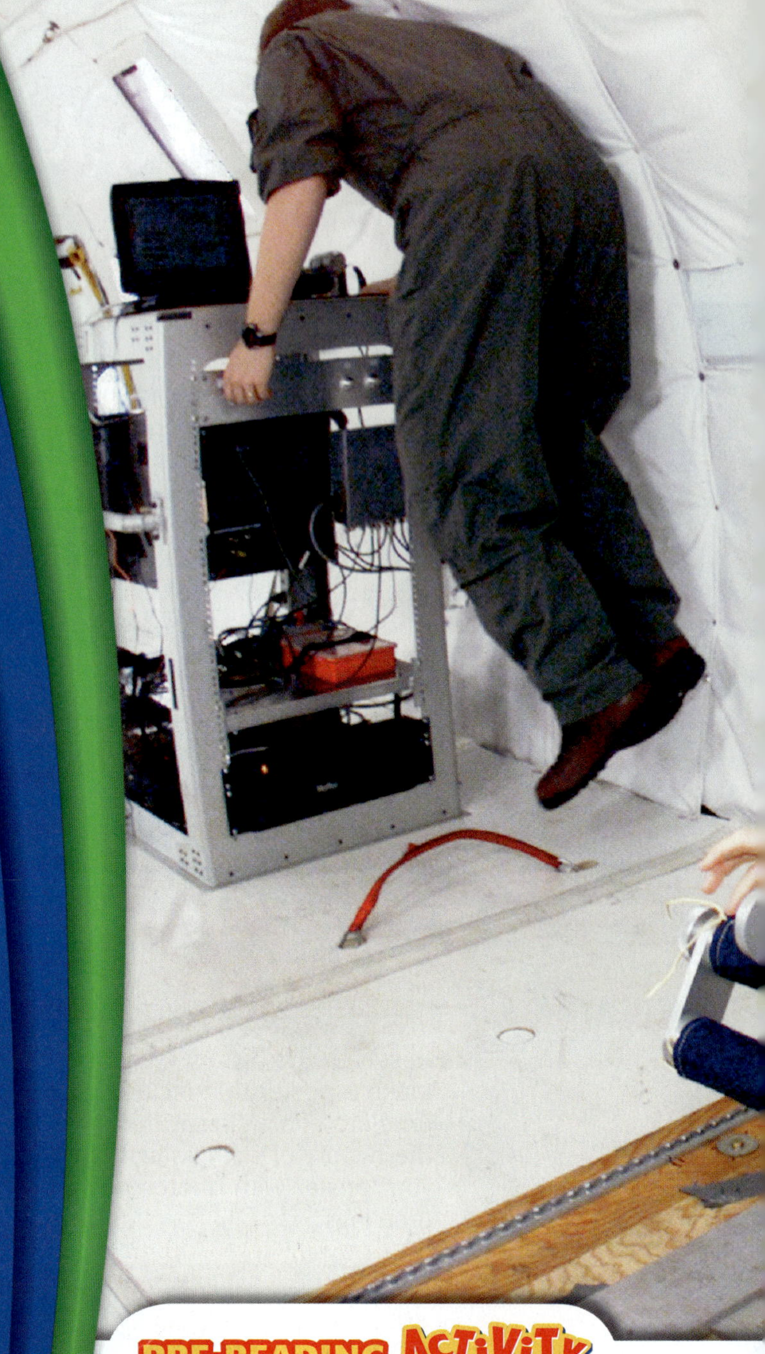

About the PHOTO

To train for space flight, astronauts fly in a modified KC-135 cargo airplane. The airplane first flies upward at a steep angle. Then, it flies downward at a 45° angle, which causes the feeling of reduced gravity inside. Under these conditions, the astronauts in the plane can float and can practice carrying out tasks that they will need to perform when they are in orbit. Because the floating makes people queasy, this KC-135 is nicknamed the "Vomit Comet."

PRE-READING ACTIVITY

Graphic Organizer

Spider Map Before you read the chapter, create the graphic organizer entitled "Spider Map" described in the **Study Skills** section of the Appendix. Label the circle "Motion." Create a leg for each law of motion, a leg for gravity, and a leg for momentum. As you read the chapter, fill in the map with details about how motion is related to the laws of motion, gravity, and momentum.

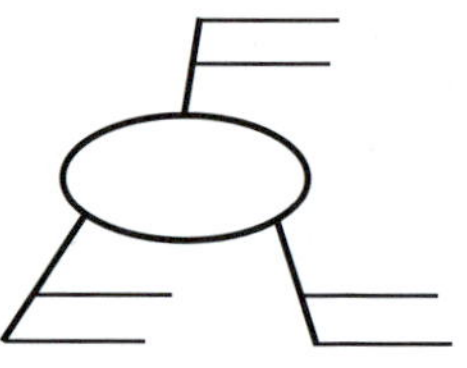

START-UP ACTIVITY

Falling Water

Gravity is one of the most important forces in your life. In this activity, you will observe the effect of gravity on a falling object.

Procedure

1. Place a **wide plastic tub** on the floor. Punch a small hole in the side of a **paper cup,** near the bottom.
2. Hold your finger over the hole, and fill the cup with **water.** Keep your finger over the hole, and hold the cup waist-high above the tub.
3. Uncover the hole. Record your observations as Trial 1.
4. Predict what will happen to the water if you drop the cup at the same time you uncover the hole.
5. Cover the hole, and refill the cup with water.
6. Uncover the hole, and drop the cup at the same time. Record your observations as Trial 2.
7. Clean up any spilled water with **paper towels.**

Analysis

1. What differences did you observe in the behavior of the water during the two trials?
2. In Trial 2, how fast did the cup fall compared with how fast the water fell?
3. How did the results of Trial 2 compare with your prediction?

SECTION 1

Gravity and Motion

Suppose you dropped a baseball and a marble at the same time from the top of a tall building. Which do you think would land on the ground first?

READING WARM-UP

Objectives

- Explain the effect of gravity and air resistance on falling objects.
- Explain why objects in orbit are in free fall and appear to be weightless.
- Describe how projectile motion is affected by gravity.

Terms to Learn

terminal velocity
free fall
projectile motion

READING STRATEGY

Reading Organizer As you read this section, create an outline of the section. Use the headings from the section in your outline.

In ancient Greece around 400 BCE, a philosopher named Aristotle (AR is TAWT uhl) thought that the rate at which an object falls depended on the object's mass. If you asked Aristotle whether the baseball or the marble would land first, he would have said the baseball. But Aristotle never tried dropping objects with different masses to test his idea about falling objects.

Gravity and Falling Objects

In the late 1500s, a young Italian scientist named Galileo Galilei (GAL uh LAY oh GAL uh LAY) questioned Aristotle's idea about falling objects. Galileo argued that the mass of an object does not affect the time the object takes to fall to the ground. According to one story, Galileo proved his argument by dropping two cannonballs of different masses from the top of the Leaning Tower of Pisa in Italy. The people watching from the ground below were amazed to see the two cannonballs land at the same time. Whether or not this story is true, Galileo's work changed people's understanding of gravity and falling objects.

Gravity and Acceleration

Objects fall to the ground at the same rate because the acceleration due to gravity is the same for all objects. Why is this true? Acceleration depends on both force and mass. A heavier object experiences a greater gravitational force than a lighter object does. But a heavier object is also harder to accelerate because it has more mass. The extra mass of the heavy object exactly balances the additional gravitational force. **Figure 1** shows objects that have different masses falling with the same acceleration.

Figure 1 *This stop-action photo shows that a table-tennis ball and a golf ball fall at the same rate even though they have different masses.*

Acceleration Due to Gravity

Acceleration is the rate at which velocity changes over time. So, the acceleration of an object is the object's change in velocity divided by the amount of time during which the change occurs. All objects accelerate toward Earth at a rate of 9.8 meters per second per second. This rate is written as 9.8 m/s/s, or 9.8 m/s^2. So, for every second that an object falls, the object's downward velocity increases by 9.8 m/s, as shown in **Figure 2.**

Reading Check **What is the acceleration due to gravity?** (*See the Appendix for answers to Reading Checks.*)

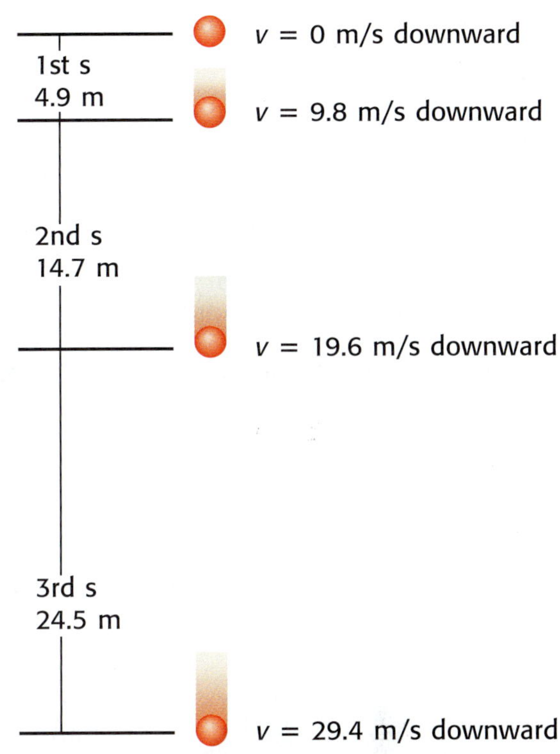

Figure 2 *A falling object accelerates at a constant rate. The object falls faster and farther each second than it did the second before.*

Velocity of Falling Objects

You can calculate the change in velocity (Δv) of a falling object by using the following equation:

$$\Delta v = g \times t$$

In this equation, g is the acceleration due to gravity on Earth (9.8 m/s^2), and t is the time the object takes to fall (in seconds). The change in velocity is the difference between the final velocity and the starting velocity. If the object starts at rest, this equation yields the velocity of the object after a certain time period.

MATH FOCUS

Calculating the Velocity of Falling Objects A stone at rest is dropped from a cliff, and the stone hits the ground after a time of 3 s. What is the stone's velocity when it hits the ground?

Step 1: Write the equation for change in velocity.

$$\Delta v = g \times t$$

Step 2: Replace g with its value and t with the time given in the problem, and solve.

$$\Delta v = 9.8 \frac{\text{m/s}}{\cancel{\text{s}}} \times 3 \cancel{\text{s}}$$
$$= 29.4 \text{ m/s}$$

To rearrange the equation to find time, divide by the acceleration due to gravity:

$$t = \frac{\Delta v}{g}$$

Now It's Your Turn

1. A penny at rest is dropped from the top of a tall stairwell. What is the penny's velocity after it has fallen for 2 s?
2. The same penny hits the ground in 4.5 s. What is the penny's velocity as it hits the ground?
3. A marble at rest is dropped from a tall building. The marble hits the ground with a velocity of 98 m/s. How long was the marble in the air?
4. An acorn at rest falls from an oak tree. The acorn hits the ground with a velocity of 14.7 m/s. How long did it take the acorn to land?

Figure 3 Effect of Air Resistance on a Falling Object

(a) The **force of gravity** is pulling down on the apple. If gravity were the only force acting on the apple, the apple would accelerate at a rate of 9.8 m/s^2.

(b) The **force of air resistance** is pushing up on the apple. This force is subtracted from the force of gravity to yield the net force.

(c) The **net force** on the apple is equal to the force of air resistance subtracted from the force of gravity. Because the net force is not 0 N, the apple accelerates downward. But the apple does not accelerate as fast as it would without air resistance.

Air Resistance and Falling Objects

Try dropping two sheets of paper—one crumpled in a tight ball and the other kept flat. What happened? Does this simple experiment seem to contradict what you just learned about falling objects? The flat paper falls more slowly than the crumpled paper because of *air resistance.* Air resistance is the force that opposes the motion of objects through air.

The amount of air resistance acting on an object depends on the size, shape, and speed of the object. Air resistance affects the flat sheet of paper more than the crumpled one. The larger surface area of the flat sheet causes the flat sheet to fall slower than the crumpled one. **Figure 3** shows the effect of air resistance on the downward acceleration of a falling object.

Figure 4 *The parachute increases the air resistance of this sky diver and slows him to a safe terminal velocity.*

Reading Check **Will air resistance have more effect on the acceleration of a falling leaf or the acceleration of a falling acorn?**

Acceleration Stops at the Terminal Velocity

As the speed of a falling object increases, air resistance increases. The upward force of air resistance continues to increase until it is equal to the downward force of gravity. At this point, the net force is 0 N and the object stops accelerating. The object then falls at a constant velocity called the **terminal velocity.**

Terminal velocity can be a good thing. Every year, cars, buildings, and vegetation are severely damaged in hailstorms. The terminal velocity of hailstones is between 5 and 40 m/s, depending on their size. If there were no air resistance, hailstones would hit the Earth at velocities near 350 m/s! **Figure 4** shows another situation in which terminal velocity is helpful.

terminal velocity the constant velocity of a falling object when the force of air resistance is equal in magnitude and opposite in direction to the force of gravity

Free Fall Occurs When There Is No Air Resistance

Sky divers are often described as being in free fall before they open their parachutes. However, that is an incorrect description, because air resistance is always acting on the sky diver.

An object is in **free fall** only if gravity is pulling it down and no other forces are acting on it. Because air resistance is a force, free fall can occur only where there is no air. Two places that have no air are in space and in a vacuum. A vacuum is a place in which there is no matter. **Figure 5** shows objects falling in a vacuum. Because there is no air resistance in a vacuum, the two objects are in free fall.

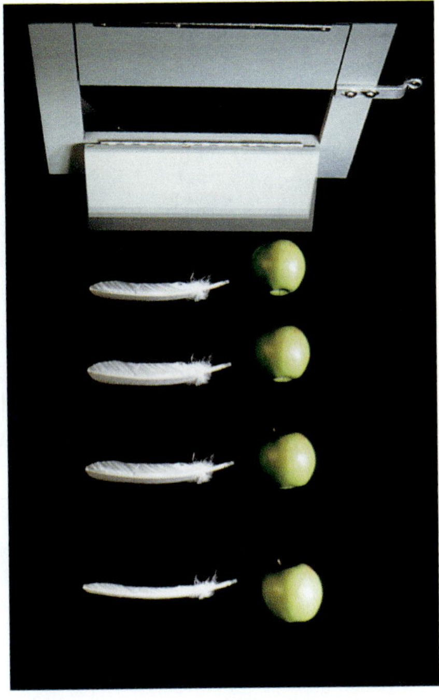

Figure 5 *Air resistance usually causes a feather to fall more slowly than an apple falls. But in a vacuum, a feather and an apple fall with the same acceleration because both are in free fall.*

Orbiting Objects Are in Free Fall

Look at the astronaut in **Figure 6.** Why is the astronaut floating inside the space shuttle? You may be tempted to say that she is weightless in space. However, it is impossible for any object to be weightless anywhere in the universe.

Weight is a measure of gravitational force. The size of the force depends on the masses of objects and the distances between them. Suppose you traveled in space far away from all the stars and planets. The gravitational force acting on you would be very small because the distance between you and other objects would be very large. But you and all the other objects in the universe would still have mass. Therefore, gravity would attract you to other objects—even if just slightly—so you would still have weight.

Astronauts float in orbiting spacecrafts because of free fall. To better understand why astronauts float, you need to know what *orbiting* means.

free fall the motion of a body when only the force of gravity is acting on the body

Figure 6 *Astronauts appear to be weightless while they are floating inside the space shuttle—but they are not weightless!*

Figure 7 **How an Orbit Is Formed**

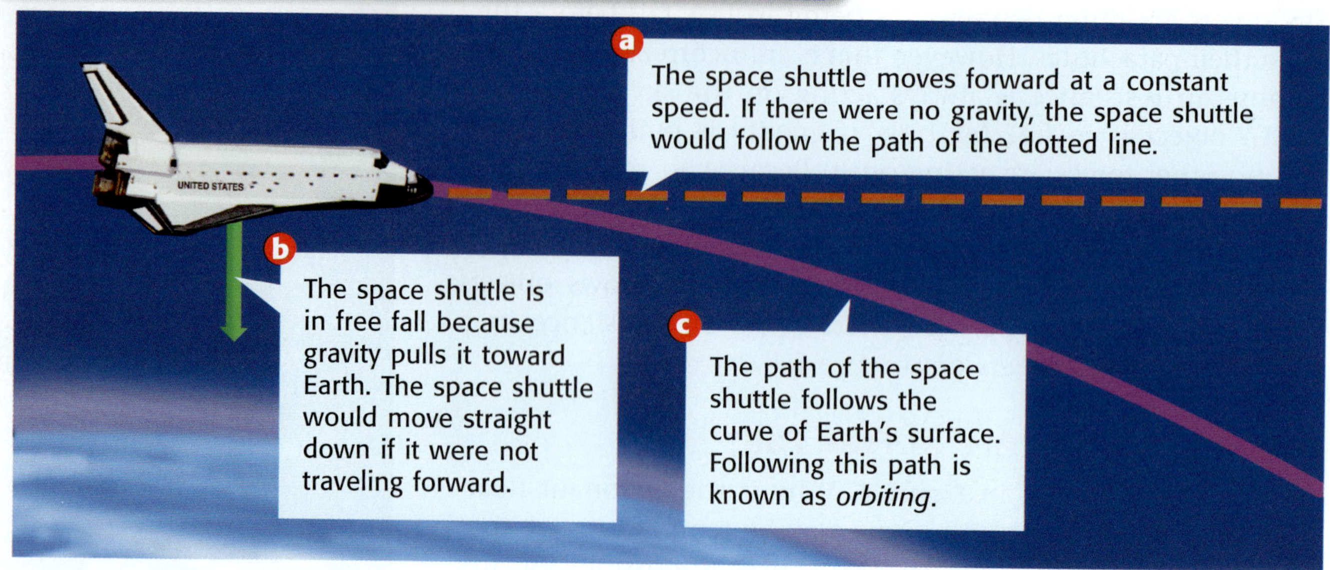

Two Motions Combine to Cause Orbiting

An object is orbiting when it is traveling around another object in space. When a spacecraft orbits Earth, it is moving forward. But the spacecraft is also in free fall toward Earth. **Figure 7** shows how these two motions combine to cause orbiting.

As you can see in **Figure 7,** the space shuttle is always falling while it is in orbit. So why don't astronauts hit their heads on the ceiling of the falling shuttle? Because they are also in free fall—they are always falling, too. Because astronauts are in free fall, they float.

Orbiting and Centripetal Force

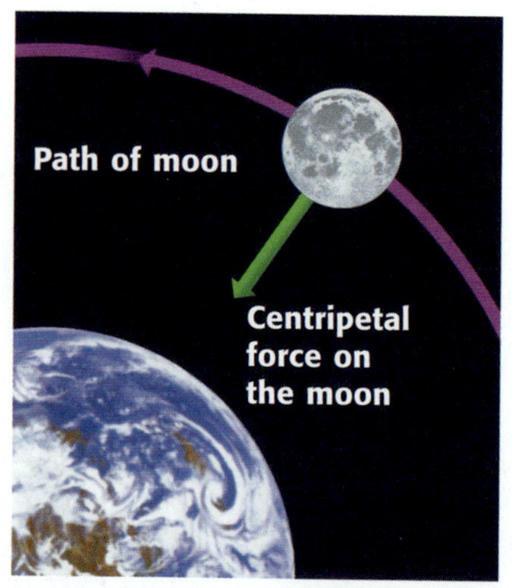

Figure 8 *The moon stays in orbit around Earth because Earth's gravitational force provides a centripetal force on the moon.*

Besides spacecrafts and satellites, many other objects in the universe are in orbit. The moon orbits the Earth. Earth and the other planets orbit the sun. In addition, many stars orbit large masses in the center of galaxies. Many of these objects are traveling in a circular or nearly circular path. Any object in circular motion is constantly changing direction. Because an unbalanced force is necessary to change the motion of any object, there must be an unbalanced force working on any object in circular motion.

The unbalanced force that causes objects to move in a circular path is called a *centripetal force* (sen TRIP uht uhl FOHRS). Gravity provides the centripetal force that keeps objects in orbit. The word *centripetal* means "toward the center." As you can see in **Figure 8,** the centripetal force on the moon points toward the center of the moon's circular orbit.

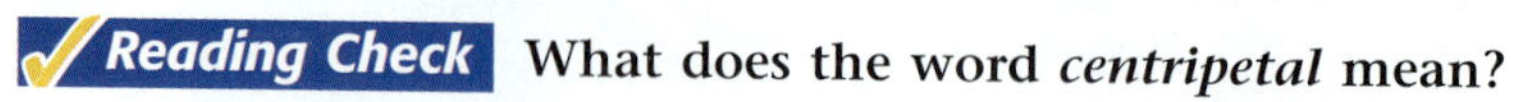

What does the word *centripetal* mean?

Projectile Motion and Gravity

The motion of a hopping grasshopper is an example of projectile motion (proh JEK tuhl MOH shuhn). **Projectile motion** is the curved path an object follows when it is thrown or propelled near the surface of the Earth. Projectile motion has two components—horizontal motion and vertical motion. The two components are independent, so they have no effect on each other. When the two motions are combined, they form a curved path, as shown in **Figure 9.** Some examples of projectile motion include the following:

- a frog leaping
- water sprayed by a sprinkler
- a swimmer diving into water
- balls being juggled
- an arrow shot by an archer

projectile motion the curved path that an object follows when thrown, launched, or otherwise projected near the surface of Earth

Horizontal Motion

When you throw a ball, your hand exerts a force on the ball that makes the ball move forward. This force gives the ball its horizontal motion, which is motion parallel to the ground.

After you release the ball, no horizontal forces are acting on the ball (if you ignore air resistance). Even gravity does not affect the horizontal component of projectile motion. So, there are no forces to change the ball's horizontal motion. Thus, the horizontal velocity of the ball is constant after the ball leaves your hand, as shown in **Figure 9.**

Figure 9 Projectile Motion

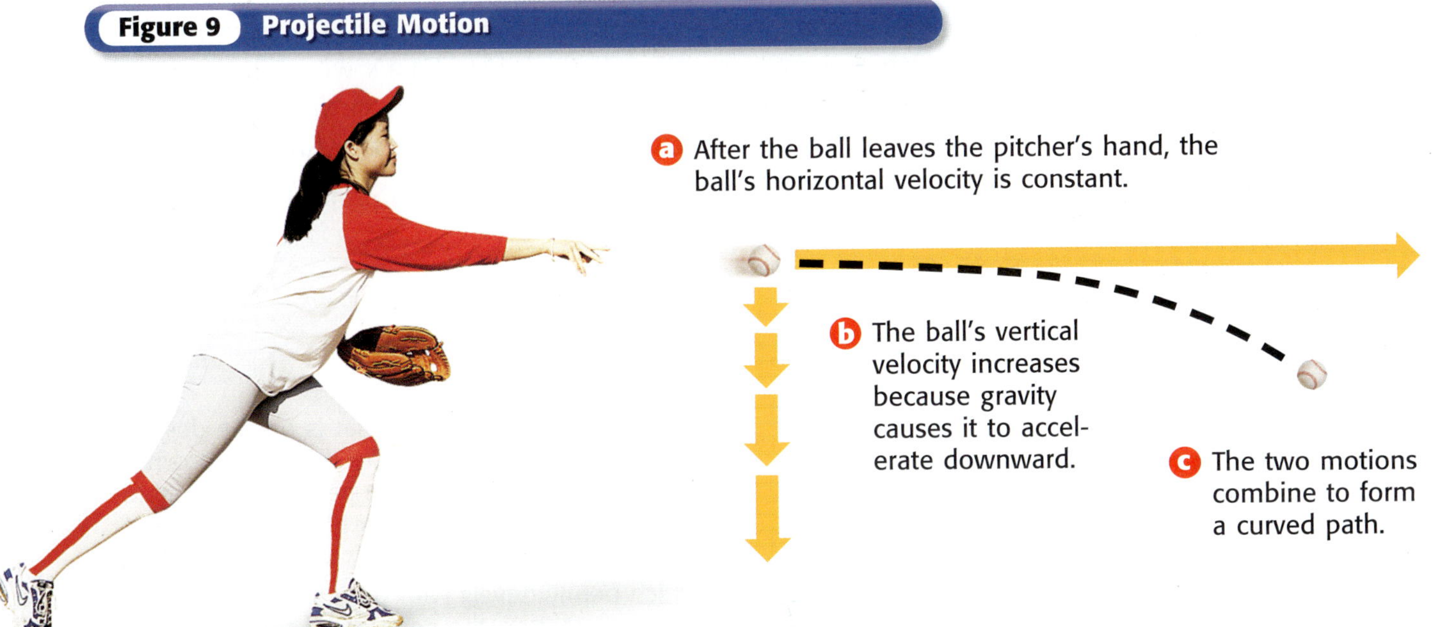

INTERNET ACTIVITY

For another activity related to this chapter, go to **go.hrw.com** and type in the keyword **HP5FORW.**

Figure 10 **Projectile Motion and Acceleration Due to Gravity**

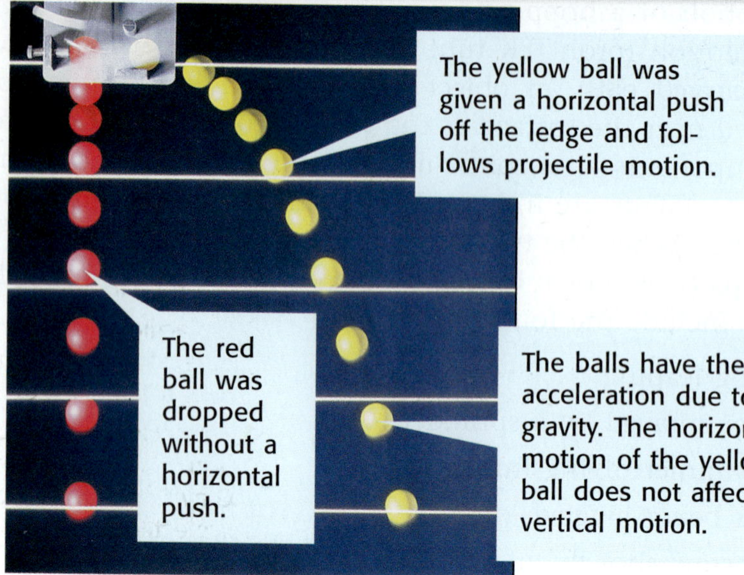

Vertical Motion

Gravity pulls everything on Earth downward toward the center of Earth. A ball in your hand is prevented from falling by your hand. After you throw the ball, gravity pulls it downward and gives the ball vertical motion. Vertical motion is motion that is perpendicular to the ground. Gravity pulls objects in projectile motion down at an acceleration of 9.8 m/s^2 (if air resistance is ignored). This rate is the same for all falling objects. **Figure 10** shows that the downward acceleration of a thrown object and a falling object are the same.

Because objects in projectile motion accelerate downward, you always have to aim above a target if you want to hit it with a thrown or propelled object. That's why when you aim an arrow directly at a bull's-eye, your arrow strikes the bottom of the target rather than the middle of the target.

Reading Check **What gives an object in projectile motion its vertical motion?**

Penny Projectile Motion

1. Position a **flat ruler** and **two pennies** on a **desk or table** as shown below.

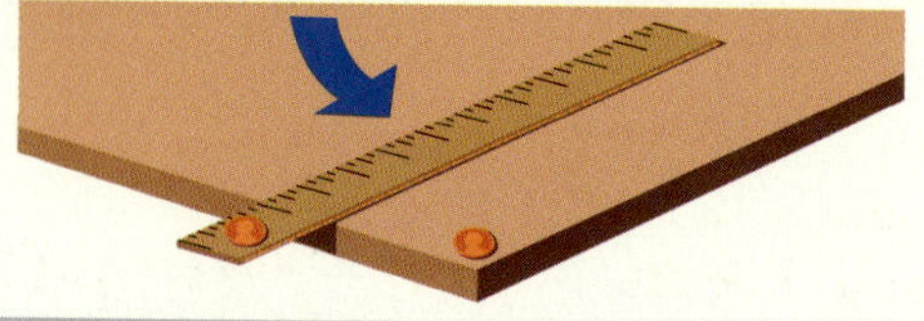

2. Hold the ruler by the end that is on the desk. Move the ruler quickly in the direction shown so that the ruler knocks the penny off the table and so that the other penny also drops. Repeat this step several times.
3. Which penny travels with projectile motion? In what order do the pennies hit the ground? Record and explain your answers.

SECTION Review

Summary

- Gravity causes all objects to accelerate toward Earth at a rate of 9.8 m/s^2.
- Air resistance slows the acceleration of falling objects. An object falls at its terminal velocity when the upward force of air resistance equals the downward force of gravity.
- An object is in free fall if gravity is the only force acting on it.
- Objects in orbit appear to be weightless because they are in free fall.
- A centripetal force is needed to keep objects in circular motion. Gravity acts as a centripetal force to keep objects in orbit.
- Projectile motion is the curved path an object follows when thrown or propelled near the surface of Earth.
- Projectile motion has two components—horizontal motion and vertical motion. Gravity affects only the vertical motion of projectile motion.

Using Key Terms

1. Use each of the following terms in a separate sentence: *terminal velocity* and *free fall.*

Understanding Key Ideas

2. Which of the following is in projectile motion?
   - **a.** a feather falling in a vacuum
   - **b.** a cat leaping on a toy
   - **c.** a car driving up a hill
   - **d.** a book laying on a desk
3. How does air resistance affect the acceleration of falling objects?
4. How does gravity affect the two components of projectile motion?
5. How is the acceleration of falling objects affected by gravity?
6. Why is the acceleration due to gravity the same for all objects?

Math Skills

7. A rock at rest falls off a tall cliff and hits the valley below after 3.5 s. What is the rock's velocity as it hits the ground?

Critical Thinking

8. **Applying Concepts** Think about a sport that uses a ball. Identify four examples from that sport in which an object is in projectile motion.
9. **Making Inferences** The moon has no atmosphere. Predict what would happen if an astronaut on the moon dropped a hammer and a feather at the same time from the same height.

Interpreting Graphics

10. Whenever Jon delivers a newspaper to the Zapanta house, the newspaper lands in the bushes, as shown below. What should Jon do to make sure the newspaper lands on the porch?

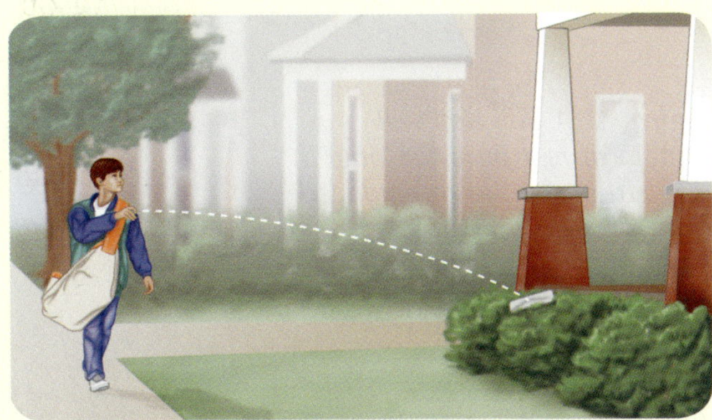

SECTION 2

Newton's Laws of Motion

Imagine that you are playing baseball. The pitch comes in, and—crack—you hit the ball hard! But instead of flying off the bat, the ball just drops to the ground. Is that normal?

You would probably say no. You know that force and motion are related. When you exert a force on a baseball by hitting it with a bat, the baseball should move. In 1686, Sir Isaac Newton explained this relationship between force and the motion of an object with his three laws of motion.

READING WARM-UP

Objectives

- Describe Newton's first law of motion, and explain how it relates to objects at rest and objects in motion.
- State Newton's second law of motion, and explain the relationship between force, mass, and acceleration.
- State Newton's third law of motion, and give examples of force pairs.

Terms to Learn

inertia

READING STRATEGY

Paired Summarizing Read this section silently. In pairs, take turns summarizing the material. Stop to discuss ideas that seem confusing.

Newton's First Law of Motion

An object at rest remains at rest, and an object in motion remains in motion at constant speed and in a straight line unless acted on by an unbalanced force.

Newton's first law of motion describes the motion of an object that has a net force of 0 N acting on it. This law may seem complicated when you first read it. But, it is easy to understand when you consider its two parts separately.

Part 1: Objects at Rest

An object that is not moving is said to be at rest. A chair on the floor and a golf ball balanced on a tee are examples of objects at rest. Newton's first law says that objects at rest will stay at rest unless they are acted on by an unbalanced force. For example, objects will not start moving until a push or a pull is exerted on them. So, a chair won't slide across the room unless you push the chair. And, a golf ball won't move off the tee unless the ball is struck by a golf club, as shown in **Figure 1.**

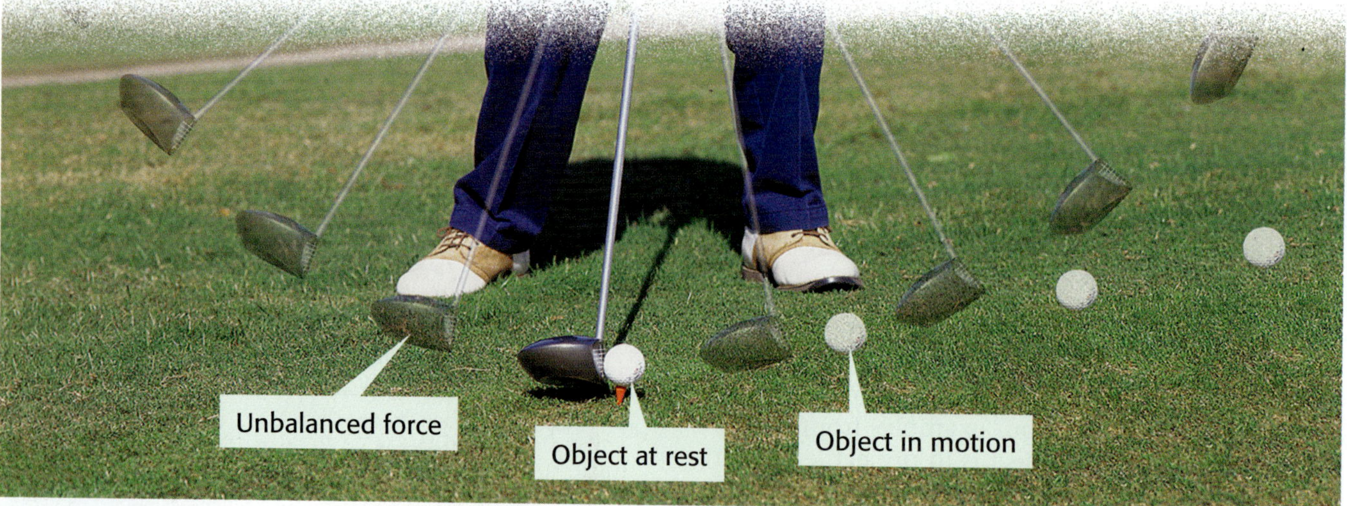

Figure 1 *A golf ball will remain at rest on a tee until it is acted on by the unbalanced force of a moving club.*

Part 2: Objects in Motion

The second part of Newton's first law is about objects moving with a certain velocity. Such objects will continue to move forever with the same velocity unless an unbalanced force acts on them.

Think about driving a bumper car at an amusement park. Your ride is pleasant as long as you are driving in an open space. But the name of the game is bumper cars! Sooner or later you are likely to run into another car, as shown in **Figure 2.** Your bumper car stops when it hits another car. But, you continue to move forward until the force from your seat belt stops you.

Figure 2 *Bumper cars let you have fun with Newton's first law.*

Friction and Newton's First Law

An object in motion will stay in motion forever unless it is acted on by an unbalanced force. So, you should be able to give your desk a push and send it sliding across the floor. If you push your desk, the desk quickly stops. Why?

There must be an unbalanced force that acts on the desk to stop its motion. That unbalanced force is friction. The friction between the desk and the floor works against the motion of the desk. Because of friction, observing the effects of Newton's first law is often difficult. For example, friction will cause a rolling ball to slow down and stop. Friction will also make a car slow down if the driver lets up on the gas pedal. Because of friction, the motion of objects changes.

Reading Check **When you ride a bus, why do you fall forward when the bus stops moving?** (*See the Appendix for answers to Reading Checks.*)

First Law Skateboard

1. Place an **empty soda can** on top of a **skateboard.**
2. Ask a friend to catch the skateboard after you push it. Now, give the skateboard a quick, firm push. What happened to the soda can?
3. Put the can on the skateboard again. Push the skateboard gently so that the skateboard moves quickly but so that the can does not fall.
4. Ask your friend to stop the skateboard after he or she allows it to travel a short distance. What happened to the can?
5. Explain how Newton's first law applies to what happened.

Inertia and Newton's First Law

Newton's first law of motion is sometimes called the *law of inertia*. **Inertia** (in UHR shuh) is the tendency of all objects to resist any change in motion. Because of inertia, an object at rest will remain at rest until a force makes it move. Likewise, inertia is the reason a moving object stays in motion with the same velocity unless a force changes its speed or direction. For example, because of inertia, you slide toward the side of a car when the driver turns a corner. Inertia is also why it is impossible for a plane, car, or bicycle to stop immediately.

inertia the tendency of an object to resist being moved or, if the object is moving, to resist a change in speed or direction until an outside force acts on the object

Mass and Inertia

Mass is a measure of inertia. An object that has a small mass has less inertia than an object that has a large mass. So, changing the motion of an object that has a small mass is easier than changing the motion of an object that has a large mass. For example, a softball has less mass and therefore less inertia than a bowling ball. Because the softball has a small amount of inertia, it is easy to pitch a softball and to change its motion by hitting it with a bat. Imagine how difficult it would be to play softball with a bowling ball! **Figure 3** further shows the relationship between mass and inertia.

Quick Lab

First-Law Magic

1. On a **table or desk**, place a **large, empty plastic cup** on top of a **paper towel.**
2. Without touching the cup or tipping it over, remove the paper towel from under the cup. How did you accomplish this? Repeat this step.
3. Fill the cup half full with **water,** and place the cup on the paper towel.
4. Once again, remove the paper towel from under the cup. Was it easier or harder to do this time?
5. Explain your observations in terms of mass, inertia, and Newton's first law of motion.

Figure 3 *Inertia makes it harder to accelerate a car than to accelerate a bicycle. Inertia also makes it easier to stop a moving bicycle than a car moving at the same speed.*

Newton's Second Law of Motion

> *The acceleration of an object depends on the mass of the object and the amount of force applied.*

Newton's second law describes the motion of an object when an unbalanced force acts on the object. As with Newton's first law, you should consider the second law in two parts.

Part 1: Acceleration Depends on Mass

Suppose you are pushing an empty cart. You have to exert only a small force on the cart to accelerate it. But, the same amount of force will not accelerate the full cart as much as the empty cart. Look at the first two photos in **Figure 4.** They show that the acceleration of an object decreases as its mass increases and that its acceleration increases as its mass decreases.

Part 2: Acceleration Depends on Force

Suppose you give the cart a hard push, as shown in the third photo in **Figure 4.** The cart will start moving faster than if you gave it only a soft push. So, an object's acceleration increases as the force on the object increases. On the other hand, an object's acceleration decreases as the force on the object decreases.

The acceleration of an object is always in the same direction as the force applied. The cart in **Figure 4** moved forward because the push was in the forward direction.

Reading Check What is the relationship between the force on an object and the object's acceleration?

CONNECTION TO Environmental Science

Car Sizes and Pollution

On average, newer cars pollute the air less than older cars do. One reason for this is that newer cars have less mass than older cars have. An object that has less mass requires less force to achieve the same acceleration as an object that has more mass. So, a small car can have a small engine and still have good acceleration. Because small engines use less fuel than large engines use, small engines create less pollution. Research three models of cars from the same year, and make a chart to compare the mass of the cars with the amount of fuel they use.

Figure 4 **Mass, Force, and Acceleration**

Acceleration

Acceleration

Acceleration

If the force applied to the carts is the same, the acceleration of the empty cart is greater than the acceleration of the loaded cart.

Acceleration will increase when a larger force is exerted.

Figure 5 Newton's Second Law and Acceleration Due to Gravity

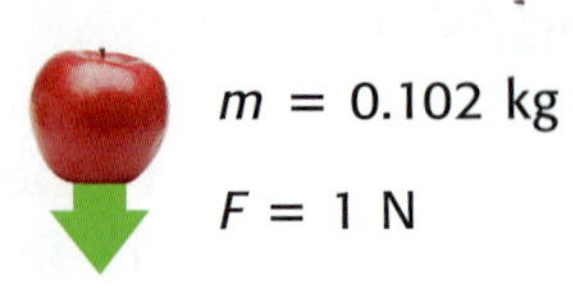

1 N = 1 kg•m/s²

$$a = \frac{1 \text{ kg•m/s}^2}{0.102 \text{ kg}} = 9.8 \text{ m/s}^2$$

10 N = 10 kg•m/s²

$$a = \frac{10 \text{ kg•m/s}^2}{1.02 \text{ kg}} = 9.8 \text{ m/s}^2$$

The apple has less mass than the watermelon does. So, less force is needed to give the apple the same acceleration that the watermelon has.

Expressing Newton's Second Law Mathematically

The relationship of acceleration (a) to mass (m) and force (F) can be expressed mathematically with the following equation:

$$a = \frac{F}{m}, \text{ or } F = m \times a$$

Notice that the equation can be rearranged to find the force applied. Both forms of the equation can be used to solve problems.

Newton's second law explains why objects fall to Earth with the same acceleration. In **Figure 5,** you can see how the large force of gravity on the watermelon is offset by its large mass. Thus, you find that the accelerations of the watermelon and the apple are the same when you solve for acceleration.

MATH FOCUS

Second-Law Problems What is the acceleration of a 3 kg mass if a force of 14.4 N is used to move the mass? (Note: 1 N is equal to 1 kg•m/s²)

Step 1: Write the equation for acceleration.

$$a = \frac{F}{m}$$

Step 2: Replace F and m with the values given in the problem, and solve.

$$a = \frac{14.4 \text{ kg•m/s}^2}{3 \text{ kg}} = 4.8 \text{ m/s}^2$$

Now It's Your Turn

1. What is the acceleration of a 7 kg mass if a force of 68.6 N is used to move it toward Earth?
2. What force is necessary to accelerate a 1,250 kg car at a rate of 40 m/s²?
3. Zookeepers carry a stretcher that holds a sleeping lion. The total mass of the lion and the stretcher is 175 kg. The lion's forward acceleration is 2 m/s². What is the force necessary to produce this acceleration?

Newton's Third Law of Motion

> ***Whenever one object exerts a force on a second object, the second object exerts an equal and opposite force on the first.***

Newton's third law can be simply stated as follows: All forces act in pairs. If a force is exerted, another force occurs that is equal in size and opposite in direction. The law itself addresses only forces. But the way that force pairs interact affects the motion of objects.

How do forces act in pairs? Study **Figure 6** to learn how one force pair helps propel a swimmer through water. Action and reaction force pairs are present even when there is no motion. For example, you exert a force on a chair when you sit on it. Your weight pushing down on the chair is the action force. The reaction force is the force exerted by the chair that pushes up on your body. The force is equal to your weight.

Reading Check How are the forces in each force pair related?

Force Pairs Do Not Act on the Same Object

A force is always exerted by one object on another object. This rule is true for all forces, including action and reaction forces. However, action and reaction forces in a pair do not act on the same object. If they did, the net force would always be 0 N and nothing would ever move! To understand how action and reaction forces act on objects, look at **Figure 6** again. The action force was exerted on the water by the swimmer's hands. But the reaction force was exerted on the swimmer's hands by the water. The forces did not act on the same object.

Newton Ball

Play catch with an adult. As you play, discuss how Newton's laws of motion are involved in the game. After you finish your game, make a list in your **science journal** of what you discussed.

Figure 6 *The action force and reaction force are a pair. The two forces are equal in size but opposite in direction.*

Figure 7 Examples of Action and Reaction Force Pairs

The space shuttle's thrusters push the exhaust gases downward as the gases push the shuttle upward with an equal force.

The rabbit's legs exert a force on Earth. Earth exerts an equal force on the rabbit's legs and causes the rabbit to accelerate upward.

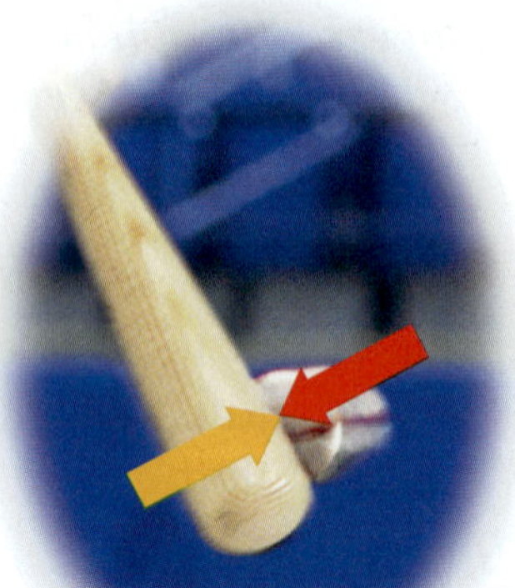

The bat exerts a force on the ball and sends the ball flying. The ball exerts an equal force on the bat, but the bat does not move backward because the batter is exerting another force on the bat.

All Forces Act in Pairs—Action and Reaction

Newton's third law says that all forces act in pairs. When a force is exerted, there is always a reaction force. A force never acts by itself. **Figure 7** shows some examples of action and reaction force pairs. In each example, the action force is shown in yellow and the reaction force is shown in red.

The Effect of a Reaction Can Be Difficult to See

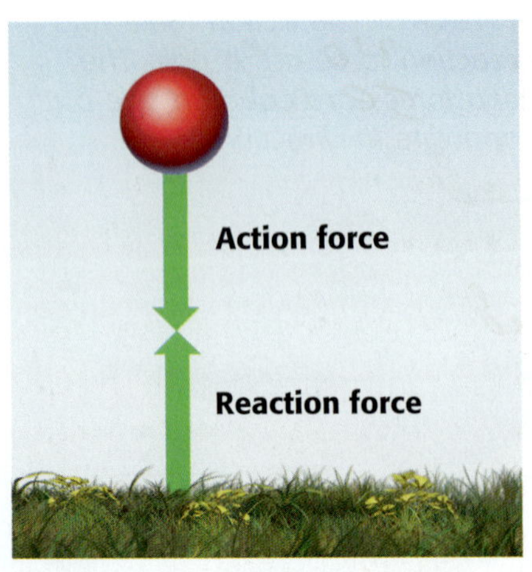

Figure 8 *The force of gravity between Earth and a falling object is a force pair.*

Another example of a force pair is shown in **Figure 8.** Gravity is a force of attraction between objects that is due to their masses. If you drop a ball, gravity pulls the ball toward Earth. This force is the action force exerted by Earth on the ball. But gravity also pulls Earth toward the ball. The force is the reaction force exerted by the ball on Earth.

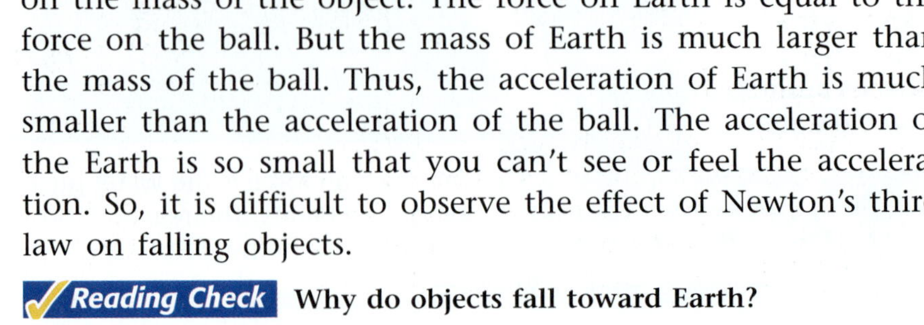

It's easy to see the effect of the action force—the ball falls to Earth. Why don't you notice the effect of the reaction force—Earth being pulled upward? To find the answer to this question, think about Newton's second law. It states that the acceleration of an object depends on the force applied to it and on the mass of the object. The force on Earth is equal to the force on the ball. But the mass of Earth is much larger than the mass of the ball. Thus, the acceleration of Earth is much smaller than the acceleration of the ball. The acceleration of the Earth is so small that you can't see or feel the acceleration. So, it is difficult to observe the effect of Newton's third law on falling objects.

✓ *Reading Check* **Why do objects fall toward Earth?**

SECTION Review

Summary

- Newton's first law of motion states that the motion of an object will not change if no unbalanced forces act on it.
- Objects at rest will not move unless acted upon by an unbalanced force.
- Objects in motion will continue to move at a constant speed and in a straight line unless acted upon by an unbalanced force.
- Inertia is the tendency of matter to resist a change in motion. Mass is a measure of inertia.
- Newton's second law of motion states that the acceleration of an object depends on its mass and on the force exerted on it.
- Newton's second law is represented by the following equation: $F = m \times a$.
- Newton's third law of motion states that whenever one object exerts a force on a second object, the second object exerts an equal and opposite force on the first object.

Using Key Terms

1. In your own words, write a definition for the term *inertia*.

Understanding Key Ideas

2. Which of the following will increase the acceleration of an object that is pushed by a force?
   a. decreasing the mass of the object
   b. increasing the mass of the object
   c. increasing the force pushing the object
   d. Both (a) and (c)
3. Give three examples of force pairs that occur when you do your homework.
4. What does Newton's first law of motion say about objects at rest and objects in motion?
5. Use Newton's second law to describe the relationship between force, mass, and acceleration.

Math Skills

6. What force is necessary to accelerate a 70 kg object at a rate of 4.2 m/s^2?

Critical Thinking

7. **Applying Concepts** When a truck pulls a trailer, the trailer and truck accelerate forward even though the action and reaction forces are the same size but are in opposite directions. Why don't these forces balance each other?
8. **Making Inferences** Use Newton's first law of motion to explain why airbags in cars are important during head-on collisions.

Interpreting Graphics

9. Imagine you accidentally bumped your hand against a table, as shown in the photo below. Your hand hurts after it happens. Use Newton's third law of motion to explain what caused your hand to hurt.

For a variety of links related to this chapter, go to www.scilinks.org

Topic: Newton's Laws of Motion
SciLinks code: HSM1028

SECTION 3

Momentum

Imagine a compact car and a large truck traveling with the same velocity. The drivers of both vehicles put on the brakes at the same time. Which vehicle will stop first?

You would probably say that the compact car will stop first. You know that smaller objects are easier to stop than larger objects. But why? The answer is momentum (moh MEN tuhm).

READING WARM-UP

Objectives

- Calculate the momentum of moving objects.
- Explain the law of conservation of momentum.

Terms to Learn

momentum

READING STRATEGY

Prediction Guide Before reading this section, write the title of each heading in this section. Next, under each heading, write what you think you will learn.

Momentum, Mass, and Velocity

The **momentum** of an object depends on the object's mass and velocity. The more momentum an object has, the harder it is to stop the object or change its direction. In the example above, the truck has more mass and more momentum than the car has. So, a larger force is needed to stop the truck. Similarly, a fast-moving car has a greater velocity and thus more momentum than a slow-moving car of the same mass. So, a fast-moving car is harder to stop than a slow-moving car. **Figure 1** shows another example of an object that has momentum.

momentum a quantity defined as the product of the mass and velocity of an object

Calculating Momentum

Momentum (p) can be calculated with the equation below:

$$p = m \times v$$

In this equation, m is the mass of an object in kilograms and v is the object's velocity in meters per second. The units of momentum are kilograms multiplied by meters per second, or kg•m/s. Like velocity, momentum has a direction. Its direction is always the same as the direction of the object's velocity.

Figure 1 *The teen on the right has less mass than the teen on the left. But, the teen on the right can have a large momentum by moving quickly when she kicks.*

Momentum Calculations What is the momentum of an ostrich with a mass of 120 kg that runs with a velocity of 16 m/s north?

Step 1: Write the equation for momentum.

$$p = m \times v$$

Step 2: Replace m and v with the values given in the problem, and solve.

$$p = 120 \text{ kg} \times 16 \text{ m/s north}$$

$$p = 19{,}200 \text{ kg} \bullet \text{m/s north}$$

Now It's Your Turn

1. What is the momentum of a 6 kg bowling ball that is moving at 10 m/s down the alley toward the pins?
2. An 85 kg man is jogging with a velocity of 2.6 m/s to the north. Nearby, a 65 kg person is skateboarding and is traveling with a velocity of 3 m/s north. Which person has greater momentum? Show your calculations.

The Law of Conservation of Momentum

When a moving object hits another object, some or all of the momentum of the first object is transferred to the object that is hit. If only some of the momentum is transferred, the rest of the momentum stays with the first object.

Imagine that a cue ball hits a billiard ball so that the billiard ball starts moving and the cue ball stops, as shown in **Figure 2.** The white cue ball had a certain amount of momentum before the collision. During the collision, all of the cue ball's momentum was transferred to the red billiard ball. After the collision, the billiard ball moved away with the same amount of momentum the cue ball had. This example shows the *law of conservation of momentum.* The law of conservation of momentum states that any time objects collide, the total amount of momentum stays the same. The law of conservation of momentum is true for any collision if no other forces act on the colliding objects. This law applies whether the objects stick together or bounce off each other after they collide.

Reading Check **What can happen to momentum when two objects collide?** (*See the Appendix for answers to Reading Checks.*)

Figure 2 *The momentum before a collision is equal to the momentum after the collision.*

Objects Sticking Together

Sometimes, objects stick together after a collision. The football players shown in **Figure 3** are an example of such a collision. A dog leaping and catching a ball and a teen jumping on a skateboard are also examples. After two objects stick together, they move as one object. The mass of the combined objects is equal to the masses of the two objects added together. In a head-on collision, the combined objects move in the direction of the object that had the greater momentum before the collision. But together, the objects have a velocity that differs from the velocity of either object before the collision. The objects have a different velocity because momentum is conserved and depends on mass and velocity. So, when mass changes, the velocity must change, too.

CONNECTION TO Language Arts

WRITING SKILL Momentum and Language The word *momentum* is often used in everyday language. For example, a sports announcer may say that the momentum of a game has changed. Or you may read that an idea is gaining momentum. In your **science journal,** write a paragraph that explains how the everyday use of the word *momentum* differs from momentum in science.

Objects Bouncing Off Each Other

In some collisions, the objects bounce off each other. The bowling ball and bowling pins shown in **Figure 3** are examples of objects that bounce off each other after they collide. Billiard balls and bumper cars are other examples. During these types of collisions, momentum is usually transferred from one object to another object. The transfer of momentum causes the objects to move in different directions at different speeds. However, the total momentum of all the objects will remain the same before and after the collision.

Reading Check What are two ways that objects may interact after a collision?

Figure 3 **Examples of Conservation of Momentum**

When football players tackle another player, they stick together. The velocity of each player changes after the collision because of conservation of momentum.

Although the bowling ball and bowling pins bounce off each other and move in different directions after a collision, momentum is neither gained nor lost.

Conservation of Momentum and Newton's Third Law

Conservation of momentum can be explained by Newton's third law of motion. In the example of the billiard ball, the cue ball hit the billiard ball with a certain amount of force. This force was the action force. The reaction force was the equal but opposite force exerted by the billiard ball on the cue ball. The action force made the billiard ball start moving, and the reaction force made the cue ball stop moving, as shown in **Figure 4.** Because the action and reaction forces are equal and opposite, momentum is neither gained nor lost.

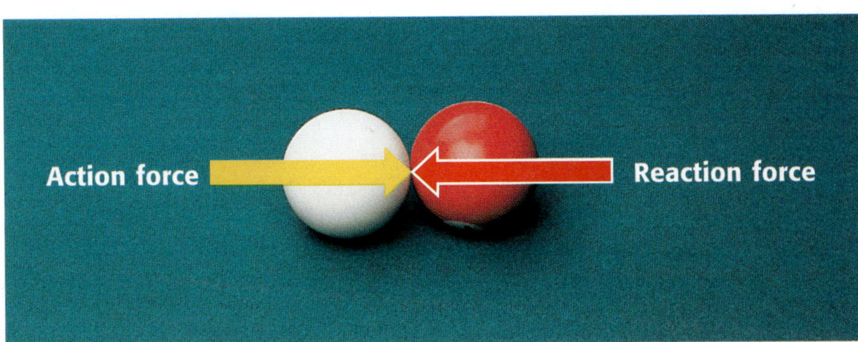

Figure 4 *The action force makes the billiard ball begin moving, and the reaction force stops the cue ball's motion.*

SECTION Review

Summary

- Momentum is a property of moving objects.
- Momentum is calculated by multiplying the mass of an object by the object's velocity.
- When two or more objects collide, momentum may be transferred, but the total amount of momentum does not change. This is the law of conservation of momentum.

Using Key Terms

1. Use the following term in a sentence: *momentum.*

Understanding Key Ideas

2. Which of the following has the smallest amount of momentum?
   a. a loaded truck driven at highway speeds
   b. a track athlete running a race
   c. a baby crawling on the floor
   d. a jet airplane being towed toward an airport
3. Explain the law of conservation of momentum.
4. How is Newton's third law of motion related to the law of conservation of momentum?

Math Skills

5. Calculate the momentum of a 2.5 kg puppy that is running with a velocity of 4.8 m/s south.

Critical Thinking

6. **Applying Concepts** A car and a train are traveling with the same velocity. Do the two objects have the same momentum? Explain your answer.
7. **Analyzing Ideas** When you catch a softball, your hand and glove move in the same direction that the ball is moving. Analyze the motion of your hand and glove in terms of momentum.

Skills Practice Lab

OBJECTIVES

Observe several effects of inertia.

Describe the motion of objects in terms of inertia.

MATERIALS

Station 1
- egg, hard-boiled
- egg, raw

Station 2
- card, index
- coin
- cup

Station 3
- mass, hanging, 1 kg
- meterstick
- scissors
- thread, spool

SAFETY

Inertia-Rama!

Inertia is a property of all matter, from small particles of dust to enormous planets and stars. In this lab, you will investigate the inertia of various shapes and kinds of matter. Keep in mind that each investigation requires you to either overcome or use the object's inertia.

Station 1: Magic Eggs

Procedure

1. There are two eggs at this station—one is hard-boiled (solid all the way through) and the other is raw (liquid inside). The masses of the two eggs are about the same. The eggs are not marked. You should not be able to tell them apart by their appearance. Without breaking them open, how can you tell which egg is raw and which egg is hard-boiled?
2. Before you do anything to either egg, make some predictions. Will there be any difference in the way the two eggs spin? Which egg will be the easier to stop?
3. First, spin one egg. Then, place your finger on it gently to make it stop spinning. Record your observations.
4. Repeat step 3 with the second egg.
5. Compare your predictions with your observations. (Repeat steps 3 and 4 if necessary.)
6. Which egg is hard-boiled and which one is raw? Explain.

Analyze the Results

1. **Explaining Events** Explain why the eggs behave differently when you spin them even though they should have the same inertia. (Hint: Think about what happens to the liquid inside the raw egg.)

Draw Conclusions

2. **Drawing Conclusions** Explain why the eggs react differently when you try to stop them.

Station 2: Coin in a Cup

Procedure

1. At this station, you will find a coin, an index card, and a cup. Place the card over the cup. Then, place the coin on the card over the center of the cup, as shown below.

2. Write down a method for getting the coin into the cup without touching the coin and without lifting the card.
3. Try your method. If it doesn't work, try again until you find a method that does work.

Analyze the Results

1. **Describing Events** Use Newton's first law of motion to explain why the coin falls into the cup if you remove the card quickly.

Draw Conclusions

2. **Defending Conclusions** Explain why pulling on the card slowly will not work even though the coin has inertia. (Hint: Friction is a force.)

Station 3: The Magic Thread

Procedure

1. At this station, you will find a spool of thread and a mass hanging from a strong string. Cut a piece of thread about 40 cm long. Tie the thread around the bottom of the mass, as shown at right.

2. Pull gently on the end of the thread. Observe what happens, and record your observations.
3. Stop the mass from moving. Now hold the end of the thread so that there is a lot of slack between your fingers and the mass.
4. Give the thread a quick, hard pull. You should observe a very different event. Record your observations. Throw away the thread.

Analyze the Results

1. **Analyzing Results** Use Newton's first law of motion to explain why the result of a gentle pull is different from the result of a hard pull.

Draw Conclusions

2. **Applying Conclusions** Both moving and nonmoving objects have inertia. Explain why throwing a bowling ball and catching a thrown bowling ball are hard.
3. **Drawing Conclusions** Why is it harder to run with a backpack full of books than to run with an empty backpack?

Chapter Review

USING KEY TERMS

Complete each of the following sentences by choosing the correct term from the word bank.

free fall
inertia
momentum
projectile motion
terminal velocity

1. An object in motion has ___, so it tends to stay in motion.

2. An object is falling at its ___ if it falls at a constant velocity.

3. ___ is the path that a thrown object follows.

4. ___ is a property of moving objects that depends on mass and velocity.

5. ___ occurs only when air resistance does not affect the motion of a falling object.

UNDERSTANDING KEY IDEAS

Multiple Choice

6. When a soccer ball is kicked, the action and reaction forces do not cancel each other out because
   - **a.** the forces are not equal in size.
   - **b.** the forces act on different objects.
   - **c.** the forces act at different times.
   - **d.** All of the above

7. An object is in projectile motion if it
   - **a.** is thrown with a horizontal push.
   - **b.** is accelerated downward by gravity.
   - **c.** does not accelerate horizontally.
   - **d.** All of the above

8. Newton's first law of motion applies to
   - **a.** moving objects.
   - **b.** objects that are not moving.
   - **c.** objects that are accelerating.
   - **d.** Both (a) and (b)

9. To accelerate two objects at the same rate, the force used to push the object that has more mass should be
   - **a.** smaller than the force used to push the object that has less mass.
   - **b.** larger than the force used to push the object that has less mass.
   - **c.** the same as the force used to push the object that has less mass.
   - **d.** equal to the object's weight.

10. A golf ball and a bowling ball are moving at the same velocity. Which of the two has more momentum?
    - **a.** The golf ball has more momentum because it has less mass.
    - **b.** The bowling ball has more momentum because it has more mass.
    - **c.** They have the same momentum because they have the same velocity.
    - **d.** There is not enough information to determine the answer.

Short Answer

11 Give an example of an object that is in free fall.

12 Describe how gravity and air resistance are related to an object's terminal velocity.

13 Why can friction make observing Newton's first law of motion difficult?

Math Skills

14 A 12 kg rock falls from rest off a cliff and hits the ground in 1.5 s.

a. Without considering air resistance, what is the rock's velocity just before it hits the ground?

b. What is the rock's momentum just before it hits the ground?

CRITICAL THINKING

15 **Concept Mapping** Use the following terms to create a concept map: *gravity, free fall, terminal velocity, projectile motion,* and *air resistance.*

16 **Identifying Relationships** During a space shuttle launch, about 830,000 kg of fuel is burned in 8 min. The fuel provides the shuttle with a constant thrust, or forward force. How does Newton's second law of motion explain why the shuttle's acceleration increases as the fuel is burned?

17 **Analyzing Processes** When using a hammer to drive a nail into wood, you have to swing the hammer through the air with a certain velocity. Because the hammer has both mass and velocity, it has momentum. Describe what happens to the hammer's momentum after the hammer hits the nail.

18 **Applying Concepts** Suppose you are standing on a skateboard or on in-line skates and you toss a backpack full of heavy books toward your friend. What do you think will happen to you? Explain your answer in terms of Newton's third law of motion.

INTERPRETING GRAPHICS

19 The picture below shows a common desk toy. If you pull one ball up and release it, it hits the balls at the bottom and comes to a stop. In the same instant, the ball on the other side swings up and repeats the cycle. How does conservation of momentum explain how this toy works?

Standardized Test Preparation

READING

Read each of the passages below. Then, answer the questions that follow each passage.

Passage 1 How do astronauts prepare for trips in the space shuttle? One method is to use simulations on Earth that mimic the conditions in space. For example, underwater training lets astronauts experience reduced gravity. They can also ride on NASA's modified KC-135 airplane. NASA's KC-135 simulates how it feels to be in a space shuttle. How does this airplane work? It flies upward at a steep angle and then flies downward at a 45° angle. When the airplane flies downward, the effect of reduced gravity is produced. As the plane falls, the astronauts inside the plane can float like astronauts in the space shuttle do!

1. What is the purpose of this passage?

- **A** to explain how astronauts prepare for missions in space
- **B** to convince people to become astronauts
- **C** to show that space is similar to Earth
- **D** to describe what it feels like to float in space

2. What can you conclude about NASA's KC-135 from the passage?

- **F** NASA's KC-135 is just like other airplanes.
- **G** All astronauts train in NASA's KC-135.
- **H** NASA's KC-135 simulates the space shuttle by reducing the effects of gravity.
- **I** Being in NASA's KC-135 is not very much like being in the space shuttle.

3. Based on the passage, which of the following statements is a fact?

- **A** Astronauts always have to train underwater.
- **B** Flying in airplanes is similar to riding in the space shuttle.
- **C** People in NASA's KC-135 float at all times.
- **D** Astronauts use simulations to learn what reduced gravity is like.

Passage 2 There once was a game that could be played by as few as 5 or as many as 1,000 players. The game could be played on a small field for a few hours or on a huge tract of land for several days. The game was not just for fun—in fact, it was often used as a substitute for war. One of the few rules was that the players couldn't touch the ball with their hands—they had to use a special stick with webbing on one end. Would you believe that this game is the same as the game of lacrosse that is played today?

Lacrosse is a game that was originally played by Native Americans. They called the game *baggataway*, which means "little brother of war." Although lacrosse has changed and is now played all over the world, it still requires special, webbed sticks.

1. What is the purpose of this passage?

- **A** to explain the importance of rules in lacrosse
- **B** to explain why sticks are used in lacrosse
- **C** to describe the history of lacrosse
- **D** to describe the rules of lacrosse

2. Based on the passage, what does the word *substitute* mean?

- **F** something that occurs before war
- **G** something that is needed to play lacrosse
- **H** something that is of Native American origin
- **I** something that takes the place of something else

INTERPRETING GRAPHICS

Read each question below, and choose the best answer.

1. Which of the following images shows an object with no momentum that is about to be set in motion by an unbalanced force?

A

B

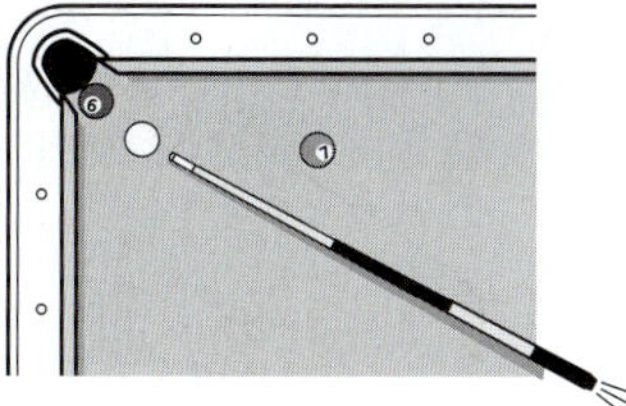

C

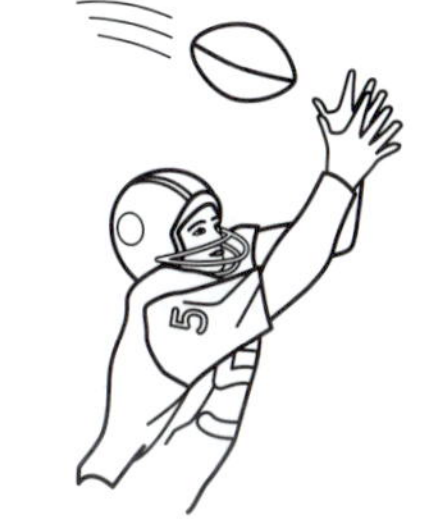

D

2. During a laboratory experiment, liquid was collected in a graduated cylinder. What is the volume of the liquid?

F 30 mL
G 35 mL
H 40 mL
I 45 mL

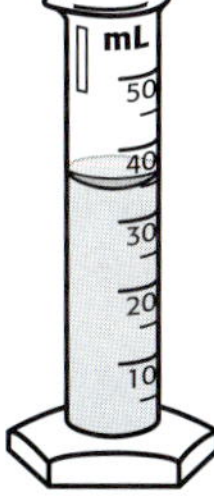

MATH

Read each question below, and choose the best answer.

1. The table below shows the accelerations produced by different forces for a 5 kg mass. Assuming that the pattern continues, use this data to predict what acceleration would be produced by a 100 N force.

Force	Acceleration
25 N	5 m/s^2
50 N	10 m/s^2
75 N	15 m/s^2

A 10 m/s^2
B 20 m/s^2
C 30 m/s^2
D 100 m/s^2

2. The average radius of the moon is 1.74×10^6 m. What is another way to express the radius of the moon?

F 0.00000174 m
G 0.000174 m
H 174,000 m
I 1,740,000 m

3. The half price bookstore is selling 4 paperback books for a total of $5.75. What would the price of 20 paperback books be?

A $23.00
B $24.75
C $28.75
D $51.75

4. A 75 kg speed skater is moving with a velocity of 16 m/s east. What is the speed skater's momentum? (Momentum is calculated with the equation: *momentum* = *mass* × *velocity*.)

F 91 kg•m/s
G 91 kg•m/s east
H 1,200 kg•m/s east
I 1,200 kg•m/s^2 east

Science in Action

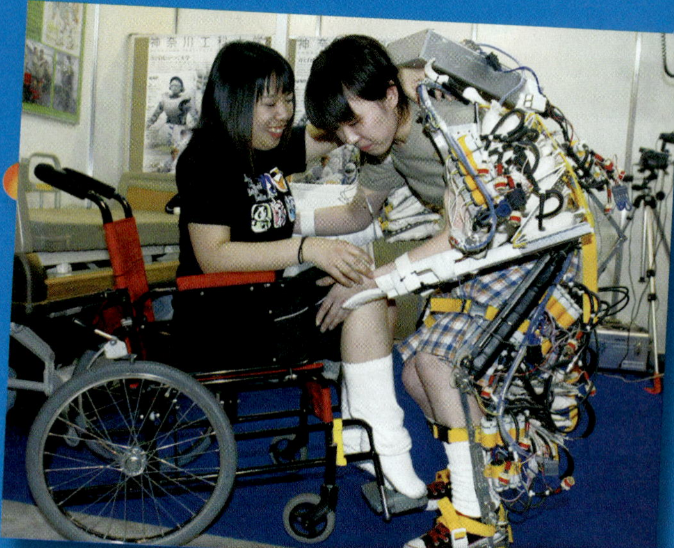

Scientific Discoveries

The Millennium Bridge

You may have heard the children's song, "London Bridge is falling down . . .". London Bridge never fell. But some people who walked on the Millennium Bridge thought that it might fall instead! The Millennium Bridge is a pedestrian bridge in London, England. The bridge opened on June 10, 2000, and more than 80,000 people crossed it that day. Immediately, people noticed something wrong—the bridge was swaying! The bridge was closed after two days so that engineers could determine what was wrong. After much research, the engineers learned that the force of the footsteps of the people crossing the bridge caused the bridge to sway.

Language Arts ACTIVITY

WRITING SKILL Imagine that you were in London on June 10, 2000 and walked across the Millennium Bridge. Write a one-page story about what you think it was like on the bridge that day.

Science, Technology, and Society

Power Suit for Lifting Patients

Imagine visiting a hospital and seeing someone who looked half human and half robot. No, it isn't a scene from a science fiction movie—it is a new invention that may some day help nurses lift patients easily. The invention, called a power suit, is a metal framework that a nurse would wear on his or her back. The suit calculates how much force a nurse needs to lift a patient, and then the robotic joints on the suit help the nurse exert the right amount of force. The suit will also help nurses avoid injuring their backs.

Math ACTIVITY

The pound (symbol £) is the currency in England. The inventor of the suit thinks that it will be sold for £1200. How much will the suit cost in dollars if $1 is equal to £0.60?

Steve Okamoto

Roller Coaster Designer Roller coasters have fascinated Steve Okamoto ever since his first ride on one. "I remember going to Disneyland as a kid. My mother was always upset with me because I kept looking over the sides of the rides, trying to figure out how they worked," he says. To satisfy his curiosity, Okamoto became a mechanical engineer. Today he uses his scientific knowledge to design and build machines, systems, and buildings. But his specialty is roller coasters.

Roller coasters really do coast along the track. A motor pulls the cars up a high hill to start the ride. After that, the cars are powered by only gravity. Designing a successful roller coaster is not a simple task. Okamoto has to calculate the cars' speed and acceleration on each part of the track. He must also consider the safety of the ride and the strength of the structure that supports the track.

Social Studies ACTIVITY

Research the history of roller coasters to learn how roller coaster design has changed over time. Make a poster to summarize your research.

To learn more about these Science in Action topics, visit go.hrw.com and type in the keyword HP5FORF.

Current Science

Check out Current Science® articles related to this chapter by visiting go.hrw.com. Just type in the keyword HP5CS06.

Chemical Reactions

About the 

Dazzling fireworks and the Statue of Liberty are great examples of chemical reactions. Chemical reactions cause fireworks to soar, explode, and light up the sky. And the Statue of Liberty has its distinctive green color because of the reaction between the statue's copper and chemicals in the air.

PRE-READING ACTIVITY

FOLDNOTES

Four-Corner Fold

Before you read the chapter, create the FoldNote entitled "Four-Corner Fold" described in the **Study Skills** section of the Appendix. Label the flaps of the four-corner fold with "Chemical formulas," "Chemical equations," "Types of chemical reactions," and "Rates of chemical reactions." Write what you know about each topic under the appropriate flap. As you read the chapter, add other information that you learn.

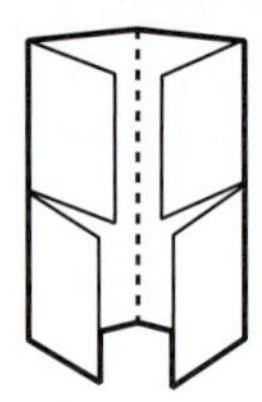

START-UP ACTIVITY

A Model Formula

Chemicals react in very precise ways. In this activity, you will model a chemical reaction and will predict how chemicals react.

Procedure

1. You will receive **several marshmallow models.** The models are marshmallows attached by **toothpicks.** Each of these models is a Model A.
2. Your teacher will show you an example of Model B and Model C. Take apart one or more Model As to make copies of Model B and Model C.
3. If you have marshmallows left over, use them to make more Model Bs and Model Cs. If you need more parts to complete a Model B or Model C, take apart another Model A.
4. Repeat step 3 until you have no parts left over.

Analysis

1. How many Model As did you use to make copies of Model B and Model C?
2. How many Model Bs did you make? How many Model Cs did you make?
3. Suppose you needed to make six Model Bs. How many Model As would you need? How many Model Cs could you make with the leftover marshmallows?

SECTION 1

Forming New Substances

Each fall, a beautiful change takes place when leaves turn colors. You see bright oranges and yellows that had been hidden by green all summer. What causes this change?

READING WARM-UP

Objectives

- Describe how chemical reactions produce new substances that have different chemical and physical properties.
- Identify four signs that indicate that a chemical reaction might be taking place.
- Explain what happens to chemical bonds during a chemical reaction.

Terms to Learn

chemical reaction
precipitate

READING STRATEGY

Reading Organizer As you read this section, create an outline of the section. Use the headings from the section in your outline.

To answer this question, you need to know what causes leaves to be green. Leaves are green because they contain a green substance, or *pigment*. This pigment is called *chlorophyll* (KLAWR uh FIL). During the spring and summer, the leaves have a large amount of chlorophyll in them. But in the fall, when temperatures drop and there are fewer hours of sunlight, chlorophyll breaks down to form new substances that have no color. The green chlorophyll is no longer present to hide the other pigments. You can now see the orange and yellow colors that were present all along.

Chemical Reactions

A chemical change takes place when chlorophyll breaks down into new substances. This change is an example of a chemical reaction. A **chemical reaction** is a process in which one or more substances change to make one or more new substances. The chemical and physical properties of the new substances differ from those of the original substances. Some results of chemical reactions are shown in **Figure 1.**

chemical reaction the process by which one or more substances change to produce one or more different substances

Figure 1 Results of Chemical Reactions

When you mix water with baking powder, substances in the baking powder react to form bubbles of carbon dioxide gas. These bubbles give the muffin its sponge-like texture.

The change of color in the fall is a result of chemical changes in the leaves.

Signs of Chemical Reactions

The more signs that you see, such as those in **Figure 2,** the more likely it is that a chemical reaction is taking place. But even if the changes seem to have stopped, reactions may still be happening. The reaction might be in equilibrium, or a balance, with the reverse reaction. The system remains that way unless the surroundings change. For example, some reactions in solution form a solid **precipitate** (pree SIP uh TAYT). When the reaction that makes the solid balances the reaction that breaks it apart, no more solid forms. The system remains that way unless a factor such as concentration or temperature changes.

precipitate a solid that is produced as a result of a chemical reaction in solution

Reading Check What is a precipitate? *(See the Appendix for answers to Reading Checks.)*

Figure 2 Some Signs of Chemical Reactions

Gas Formation
The chemical reaction in the beaker has formed a brown gas, nitrogen dioxide. This gas is formed when a strip of copper is placed into nitric acid.

Solid Formation
Here you see potassium chromate solution being added to a silver nitrate solution. The dark red solid is a precipitate of silver chromate.

Energy Change
Energy is released during some chemical reactions. The fire in this photo gives off light energy and thermal energy. During some other chemical reactions, energy is taken in.

Color Change
Don't spill chlorine bleach on your jeans! The bleach reacts with the blue dye on the fabric and causes the color of the material to change.

A Change of Properties

Even though the signs we look for to see if a reaction is taking place are good signals of chemical reactions, they do not guarantee that a reaction is happening. For example, gas can be given off when a liquid boils. But this example is a physical change, not a chemical reaction.

So, how can you be sure that a chemical reaction is occurring? The most important sign is the formation of new substances that have different properties. Look at **Figure 3.** The starting materials in this reaction are sugar and sulfuric acid. Several things tell you that a chemical reaction is taking place. Bubbles form, a gas is given off, and the beaker becomes very hot. But most important, new substances form. And the properties of these substances are very different from those of the starting substances.

Figure 3 *The top photo shows the starting substances: table sugar and sulfuric acid, a clear liquid. The substances formed in this chemical reaction are very different from the starting substances.*

Bonds: Holding Molecules Together

A *chemical bond* is a force that holds two atoms together in a molecule. For a chemical reaction to take place, the original bonds must break and new bonds must form.

Breaking and Making Bonds

How do new substances form in a chemical reaction? First, chemical bonds in the starting substances must break. Molecules are always moving. If the molecules bump into each other with enough energy, the chemical bonds in the molecules break. The atoms then rearrange, and new bonds form to make the new substances. **Figure 4** shows how bonds break and form in the reaction between hydrogen and chlorine.

Reading Check What happens to the bonds of substances during a chemical reaction?

Figure 4 Reaction of Hydrogen and Chlorine

hydrogen + chlorine → hydrogen chloride

Breaking Bonds Hydrogen and chlorine are diatomic. Diatomic molecules are two atoms bonded together. The bonds joining these atoms must first break before the atoms can react with each other.

Making Bonds A new substance, hydrogen chloride, forms as new bonds are made between hydrogen atoms and chlorine atoms.

New Bonds, New Substances

What happens when hydrogen and chlorine are combined? A chlorine gas molecule is a diatomic (DIE uh TAHM ik) molecule. That is, a chlorine molecule is made of two atoms of chlorine. Chlorine gas has a greenish yellow color. Hydrogen gas is also a diatomic molecule. Hydrogen gas is a flammable, colorless gas. When chlorine gas and hydrogen gas react, the bond between the hydrogen atoms breaks. And the bond between the chlorine atoms also breaks. A new bond forms between each hydrogen and chlorine atom. A new substance, hydrogen chloride, is formed. Hydrogen chloride is a nonflammable, colorless gas. Its properties differ from the properties of both of the starting substances.

Let's look at another example. Sodium is a metal that reacts violently in water. Chlorine gas is poisonous. When chlorine gas and sodium react, the result is a familiar compound—table salt. Sodium chloride, or table salt, is a harmless substance that almost everyone uses. The salt's properties are very different from sodium's or chlorine's. Salt is a new substance.

Reaction Ready

1. Place a **piece of chalk** in a **plastic cup.**
2. Add **5 mL of vinegar** to the cup. Record your observations.
3. What evidence of a chemical reaction do you see?
4. What type of new substance was formed?

SECTION Review

Summary

- A chemical reaction is a process by which substances change to produce new substances with new chemical and physical properties.
- Signs that indicate a chemical reaction has taken place are a color change, formation of a gas or a solid, and release of energy.
- During a reaction, bonds are broken, atoms are rearranged, and new bonds are formed.

Using Key Terms

1. Use the following terms in the same sentence: *chemical reaction* and *precipitate.*

Understanding Key Ideas

2. Most chemical reactions
   a. have starting substances that collide with each other.
   b. do not break bonds.
   c. do not rearrange atoms.
   d. cannot be seen.
3. If the chemical properties of a substance have not changed, has a chemical reaction occurred?

Critical Thinking

4. **Analyzing Processes** Steam is escaping from a teapot. Is this a chemical reaction? Explain.
5. **Applying Concepts** Explain why charcoal burning in the grill is a chemical change.

Interpreting Graphics

Use the photo below to answer the questions that follow.

6. What evidence of a chemical reaction is shown in the photo?
7. What is happening to the bonds of the starting substances?

For a variety of links related to this chapter, go to www.scilinks.org

Topic: Chemical Reactions
SciLinks code: HSM0274

SECTION 2

Chemical Formulas and Equations

READING WARM-UP

Objectives

- Interpret and write simple chemical formulas.
- Write and balance simple chemical equations.
- Explain how a balanced equation shows the law of conservation of mass.

Terms to Learn

chemical formula
chemical equation
reactant
product
law of conservation of mass

READING STRATEGY

Discussion Read this section silently. Write down questions that you have about this section. Discuss your questions in a small group.

How many words can you make using the 26 letters of the alphabet? Many thousands? Now, think of how many sentences you can make with all of those words.

Letters are used to form words. In the same way, chemical symbols are put together to make chemical formulas that describe substances. Chemical formulas can be placed together to describe a chemical reaction, just like words can be put together to make a sentence.

Chemical Formulas

All substances are formed from about 100 elements. Each element has its own chemical symbol. A **chemical formula** is a shorthand way to use chemical symbols and numbers to represent a substance. A chemical formula shows how many atoms of each kind are present in a molecule.

chemical formula a combination of chemical symbols and numbers to represent a substance

As shown in **Figure 1,** the chemical formula for water is H_2O. This formula tells you that one water molecule is made of two atoms of hydrogen and one atom of oxygen. The small 2 in the formula is a subscript. A *subscript* is a number written below and to the right of a chemical symbol in a formula. Sometimes, a symbol, such as O for oxygen in water's formula, has no subscript. If there is no subscript, only one atom of that element is present. Look at **Figure 1** for more examples of chemical formulas.

Figure 1 Chemical Formulas of Different Substances

Water	Oxygen	Glucose
H_2O	O_2	$C_6H_{12}O_6$
Water molecules are made up of 3 atoms—2 atoms of hydrogen bonded to 1 atom of oxygen.	**Oxygen** is a diatomic molecule. Each molecule has 2 atoms of oxygen bonded together.	**Glucose** molecules have 6 atoms of carbon, 12 atoms of hydrogen, and 6 atoms of oxygen.

Carbon dioxide

CO_2

The *absence of a prefix* indicates one carbon atom.

The prefix *di-* indicates two oxygen atoms.

Dinitrogen monoxide

N_2O

The prefix *di-* indicates two nitrogen atoms.

The prefix *mono-* indicates one oxygen atom.

Figure 2 *The formulas of these covalent compounds can be written by using the prefixes in the names of the compounds.*

Writing Formulas for Covalent Compounds

If you know the name of the covalent compound, you can often write the chemical formula for that compound. Covalent compounds are usually composed of two nonmetals. The names of many covalent compounds use prefixes. Each prefix represents a number, as shown in **Table 1.** The prefixes tell you how many atoms of each element are in a formula. **Figure 2** shows you how to write a chemical formula from the name of a covalent compound.

Table 1 Prefixes Used in Chemical Names

mono-	1	*hexa-*	6
di-	2	*hepta-*	7
tri-	3	*octa-*	8
tetra-	4	*nona-*	9
penta-	5	*deca-*	10

Writing Formulas for Ionic Compounds

If the name of a compound contains the name of a metal and the name of a nonmetal, the compound is ionic. To write the formula for an ionic compound, make sure the compound's charge is 0. In other words, the formula must have subscripts that cause the charges of the ions to cancel out. **Figure 3** shows you how to write a chemical formula from the name of an ionic compound.

Reading Check **What kinds of elements make up an ionic compound?** (*See the Appendix for answers to Reading Checks.*)

Sodium chloride

NaCl

A sodium ion has a **1+ charge.**

A chloride ion has a **1− charge.**

One sodium ion and one chloride ion have an overall **charge of (1+) + (1−) = 0.**

Magnesium chloride

$MgCl_2$

A magnesium ion has a **2+ charge.**

A chloride ion has a **1− charge.**

One magnesium ion and two chloride ions have an overall **charge of (2+) + 2(1−) = 0.**

Figure 3 *The formula of an ionic compound is written by using enough of each ion so that the overall charge is 0.*

Figure 4 *Like chemical symbols, the symbols on this musical score are understood around the world!*

Chemical Equations

Think about a piece of music, such as the one in **Figure 4.** Someone writing music must tell the musician what notes to play, how long to play each note, and how each note should be played. Words aren't used to describe the musical piece. Instead, musical symbols are used. The symbols can be understood by anyone who can read music.

Describing Reactions by Using Equations

In the same way that composers use musical symbols, chemists around the world use chemical symbols and chemical formulas. Instead of changing words and sentences into other languages to describe reactions, chemists use chemical equations. A **chemical equation** uses chemical symbols and formulas as a shortcut to describe a chemical reaction. A chemical equation is short and is understood by anyone who understands chemical formulas.

chemical equation a representation of a chemical reaction that uses symbols to show the relationship between the reactants and the products

reactant a substance or molecule that participates in a chemical reaction

product the substance that forms in a chemical reaction

From Reactants to Products

When carbon burns, it reacts with oxygen to form carbon dioxide. **Figure 5** shows how a chemist would use an equation to describe this reaction. The starting materials in a chemical reaction are **reactants** (ree AK tuhnts). The substances formed from a reaction are **products.** In this example, carbon and oxygen are reactants. Carbon dioxide is the product.

Reading Check **What is the difference between reactants and products in a chemical reaction?**

Figure 5 The Parts of a Chemical Equation

Charcoal is used to cook food on a barbecue grill. When carbon in charcoal reacts with oxygen in the air, the primary product is carbon dioxide, as shown by the chemical equation.

The formulas of the **reactants** are written before the arrow.

The formulas of the **products** are written after the arrow.

$$C + O_2 \rightarrow CO_2$$

A **plus sign** separates the formulas of two or more reactants or products from one another.

The **arrow,** also called the *yields sign,* separates the formulas of the reactants from the formulas of the products.

Figure 6 Examples of Similar Symbols and Formulas

CO_2

The chemical formula for the compound **carbon dioxide** is CO_2. Carbon dioxide is a colorless, odorless gas that you exhale.

CO

The chemical formula for the compound **carbon monoxide** is CO. Carbon monoxide is a colorless, odorless, and poisonous gas.

Co

The chemical symbol for the element **cobalt** is Co. Cobalt is a hard, bluish gray metal.

The Importance of Accuracy

The symbol or formula for each substance in the equation must be written correctly. For a compound, use the correct chemical formula. For an element, use the proper chemical symbol. An equation that has the wrong chemical symbol or formula will not correctly describe the reaction. In fact, even a simple mistake can make a huge difference. **Figure 6** shows how formulas and symbols can be mistaken.

The Reason Equations Must Be Balanced

Atoms are never lost or gained in a chemical reaction. They are just rearranged. Every atom in the reactants becomes part of the products. When writing a chemical equation, make sure the number of atoms of each element in the reactants equals the number of atoms of those elements in the products. This is called balancing the equation.

Balancing equations comes from the work of a French chemist, Antoine Lavoisier (lah vwah ZYAY). In the 1700s, Lavoisier found that the total mass of the reactants was always the same as the total mass of the products. Lavoisier's work led to the **law of conservation of mass.** This law states that mass is neither created nor destroyed in ordinary chemical and physical changes. This law means that a chemical equation must show the same numbers and kinds of atoms on both sides of the arrow.

Counting Atoms

Some chemical formulas contain parentheses. When counting atoms, multiply everything inside the parentheses by the subscript. For example, $Ca(NO_3)_2$ has one calcium atom, two (2 × 1) nitrogen atoms, and six (2 × 3) oxygen atoms. Find the number of atoms of each element in the formulas $Mg(OH)_2$ and $Al_2(SO_4)_3$.

law of conservation of mass the law that states that mass cannot be created or destroyed in ordinary chemical and physical changes

How to Balance an Equation

To balance an equation, you must use coefficients (KOH uh FISH uhnts). A *coefficient* is a number that is placed in front of a chemical symbol or formula. For example, 2CO represents two carbon monoxide molecules. The number *2* is the coefficient.

For an equation to be balanced, all atoms must be counted. So, you must multiply the subscript of each element in a formula by the formula's coefficient. For example, $2H_2O$ contains a total of four hydrogen atoms and two oxygen atoms. Only coefficients—not subscripts—are changed when balancing equations. Changing the subscripts in the formula of a compound would change the compound. **Figure 7** shows you how to use coefficients to balance an equation.

Reading Check **What is the coefficient if $4O_2$ is in an equation?**

CONNECTION TO Language Arts

WRITING SKILL **Diatomic Molecules** Seven of the chemical elements exist as diatomic molecules. Do research to find out which seven elements these are. Write a short report that describes each diatomic molecule. Be sure to include the formula for each molecule.

Figure 7 **Balancing a Chemical Equation**

Follow these steps to write a balanced equation for $H_2 + O_2 \longrightarrow H_2O$.

1. **Count the atoms** of each element in the reactants and in the products. You can see that there are fewer oxygen atoms in the product than in the reactants.

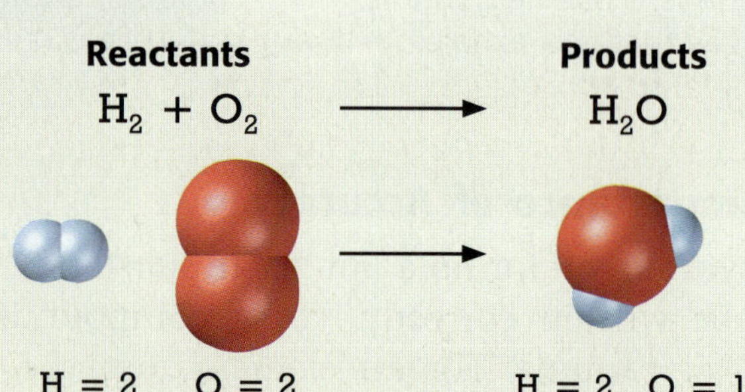

2. **To balance the oxygen atoms,** place the coefficient 2 in front of H_2O. Doing so gives you two oxygen atoms in both the reactants and the products. But now there are too few hydrogen atoms in the reactants.

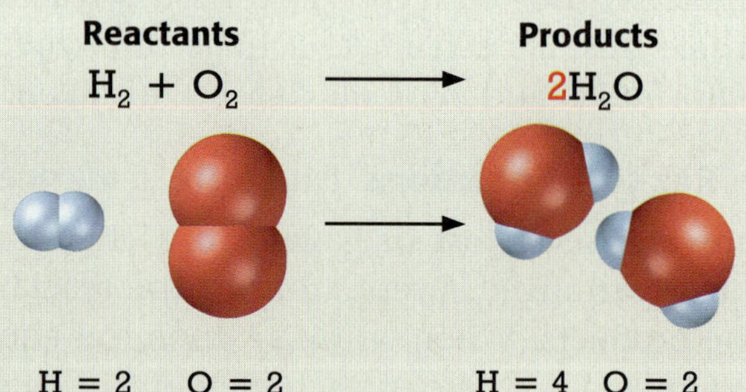

3. **To balance the hydrogen atoms,** place the coefficient 2 in front of H_2. But to be sure that your answer is correct, always double-check your work!

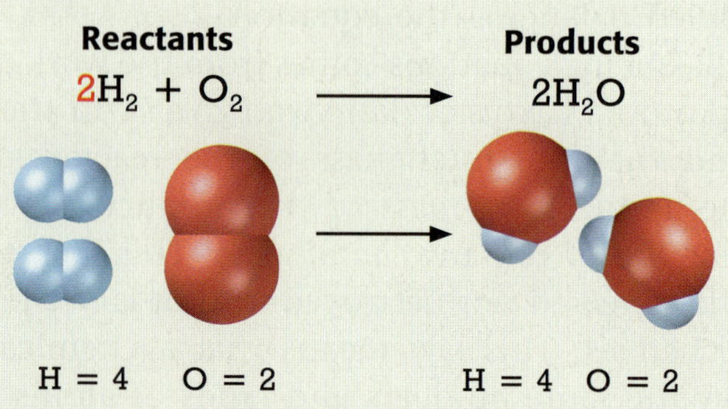

Conservation of Mass

1. Place **5 g of baking soda** into a **sealable plastic bag.**
2. Place **5 mL of vinegar** into a **plastic film canister.** Put the lid on the canister.
3. Place the canister into the bag. Squeeze the air out of the bag. Seal the bag tightly.
4. Use a **balance** to measure the mass of the bag and its contents. Record the mass.
5. Keeping the bag closed, open the canister in the bag. Mix the vinegar with the baking soda. Record your observations.
6. When the reaction has stopped, measure the mass of the bag and its contents. Record the mass.
7. Compare the mass of the materials before the reaction and the mass of the materials after the reaction. Explain your observations.

SECTION Review

Summary

- A chemical formula uses symbols and subscripts to describe the makeup of a compound.
- Chemical formulas can often be written from the names of covalent and ionic compounds.
- A chemical equation uses chemical formulas, chemical symbols, and coefficients to describe a reaction.
- Balancing an equation requires that the same numbers and kinds of atoms be on each side of the equation.
- A balanced equation illustrates the law of conservation of mass: mass is neither created nor destroyed during ordinary physical and chemical changes.

Using Key Terms

The statements below are false. For each statement, replace the underlined word to make a true statement.

1. A chemical <u>formula</u> describes a chemical reaction.
2. The substances formed from a chemical reaction are <u>reactants</u>.

Understanding Key Ideas

3. The correct chemical formula for carbon tetrachloride is
   a. CCl_3.
   b. C_3Cl.
   c. CCl.
   d. CCl_4.
4. Calcium oxide is used to make soil less acidic. Its formula is
   a. Ca_2O_2.
   b. CaO.
   c. CaO_2.
   d. Ca_2O.
5. Balance the following equations by adding the correct coefficients.
   a. $Na + Cl_2 \longrightarrow NaCl$
   b. $Mg + N_2 \longrightarrow Mg_3N_2$
6. How does a balanced chemical equation illustrate that mass is never lost or gained in a chemical reaction?
7. What is the difference between a subscript and a coefficient?

Math Skills

8. Calculate the number of atoms of each element represented in each of the following: $2Na_3PO_4$, $4Al_2(SO_4)_3$, and $6PCl_5$.

Critical Thinking

9. **Analyzing Methods** Describe how to write a formula for a covalent compound. Give an example of a covalent compound.
10. **Applying Concepts** Explain why the subscript in a formula of a chemical compound cannot be changed when balancing an equation.

SECTION 3

Types of Chemical Reactions

READING WARM-UP

Objectives

- Describe four types of chemical reactions.
- Classify a chemical equation as one of four types of chemical reactions.

Terms to Learn

synthesis reaction
decomposition reaction
single-displacement reaction
double-displacement reaction

READING STRATEGY

Mnemonics As you read this section, create a mnemonic device to help you remember the four types of chemical reactions.

There are thousands of known chemical reactions. Can you imagine having to memorize even 50 of them?

Remembering all of them would be impossible! But fortunately, there is help. In the same way that the elements are divided into groups based on their properties, reactions can be classified based on what occurs during the reaction.

Most reactions can be placed into one of four categories: synthesis (SIN thuh sis), decomposition, single-displacement, and double-displacement. Each type of reaction has a pattern that shows how reactants become products. One way to remember what happens in each type of reaction is to imagine people at a dance. As you learn about each type of reaction, study the models of students at a dance. The models will help you recognize each type of reaction.

Synthesis Reactions

A **synthesis reaction** is a reaction in which two or more substances combine to form one new compound. For example, a synthesis reaction takes place when sodium reacts with chlorine. This synthesis reaction produces sodium chloride, which you know as table salt. A synthesis reaction would be modeled by two people pairing up to form a dancing couple, as shown in **Figure 1.**

synthesis reaction a reaction in which two or more substances combine to form a new compound

Reading Check What is a synthesis reaction? (*See the Appendix for answers to Reading Checks.*)

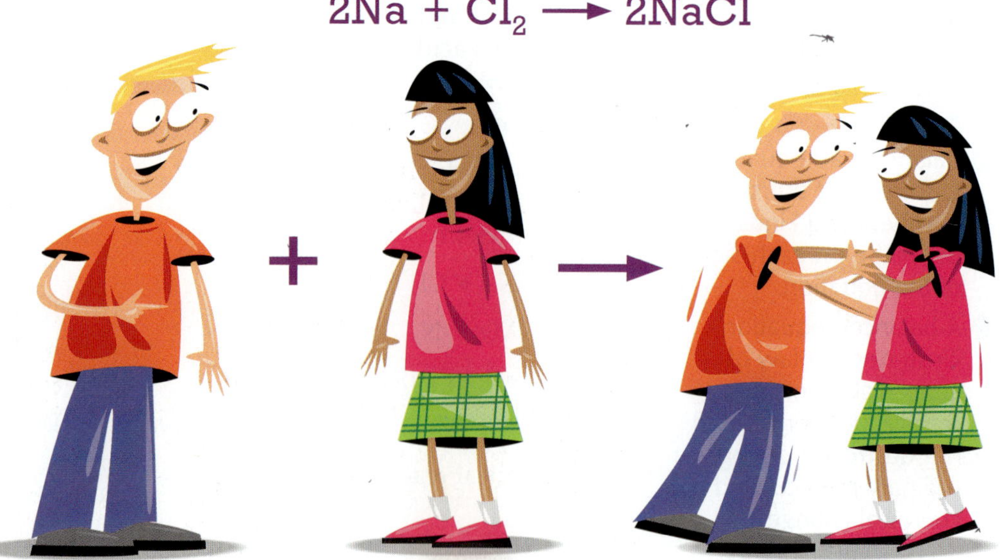

Figure 1 *Sodium reacts with chlorine to form sodium chloride in this synthesis reaction.*

Figure 2 *In this decomposition reaction, carbonic acid, H_2CO_3, decomposes to form water and carbon dioxide.*

Decomposition Reactions

A **decomposition reaction** is a reaction in which a single compound breaks down to form two or more simpler substances. Decomposition is the reverse of synthesis. The dance model for a decomposition reaction would be a couple that finishes a dance and separates, as shown in **Figure 2.**

Reading Check How is a decomposition reaction different from a synthesis reaction?

decomposition reaction a reaction in which a single compound breaks down to form two or more simpler substances

single-displacement reaction a reaction in which one element takes the place of another element in a compound

Single-Displacement Reactions

Sometimes, an element replaces another element that is a part of a compound. This type of reaction is called a **single-displacement reaction.** The products of single-displacement reactions are a new compound and a different element. The dance model for a single-displacement reaction would show a person cutting in on a couple who is dancing. A new couple is formed. And a different person is left alone, as shown in **Figure 3.**

Figure 3 *Zinc replaces the hydrogen in hydrochloric acid to form zinc chloride and hydrogen gas in this single-displacement reaction.*

Figure 4 **Reactivity of Elements**

$Cu + 2AgNO_3 \rightarrow 2Ag + Cu(NO_3)_2$
Copper is more reactive than silver.

$Ag + Cu(NO_3)_2 \rightarrow$ no reaction
Silver is less reactive than copper.

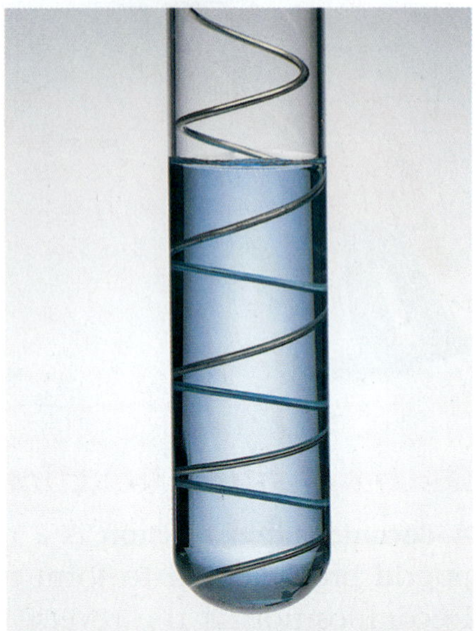

Reactivity of Elements

In a single-displacement reaction, a more reactive element can displace a less reactive element in a compound. For example, **Figure 4** shows that copper is more reactive than silver. Copper (Cu) can replace the silver (Ag) ion in the compound silver nitrate. But the opposite reaction does not occur, because silver is less reactive than copper.

The elements in Group 1 of the periodic table are the most reactive metals. Very few nonmetals are involved in single-displacement reactions. In fact, only Group 17 nonmetals participate in single-displacement reactions.

Reading Check **Why can one element sometimes replace another element in a single-displacement reaction?**

For another activity related to this chapter, go to **go.hrw.com** and type in the keyword **HP5REAW.**

QUICK LAB

Identifying Reactions

1. Study each of the following equations:
   $4Na + O_2 \rightarrow 2Na_2O$ $\quad$ $P_4 + 5O_2 \rightarrow 2P_2O_5$
   $2Ag_3N \rightarrow 6Ag + N_2$ $\quad$ $Zn + 2HCl \rightarrow ZnCl_2 + H_2$
2. Build models of each of these reactions using **colored clay.** Choose a different color of clay to represent each kind of atom.
3. Identify each type of reaction as a synthesis, decomposition, or single-displacement reaction.

Double-Displacement Reactions

A **double-displacement reaction** is a reaction in which ions from two compounds exchange places. One of the products of this type of reaction is often a gas or a precipitate. A dance model of a double-displacement reaction would be two couples dancing and then trading partners, as shown in **Figure 5.**

double-displacement reaction a reaction in which a gas, a solid precipitate, or a molecular compound forms from the exchange of ions between two compounds

Figure 5 *A double-displacement reaction occurs when sodium chloride reacts with silver fluoride to form sodium fluoride and silver chloride (a precipitate).*

SECTION Review

Summary

- A synthesis reaction is a reaction in which two or more substances combine to form a compound.
- A decomposition reaction is a reaction in which a compound breaks down to form two or more simpler substances.
- A single-displacement reaction is a reaction in which an element takes the place of another element that is part of a compound.
- A double-displacement reaction is a reaction in which ions in two compounds exchange places.

Using Key Terms

1. In your own words, write a definition for each of the following terms: *synthesis reaction* and *decomposition reaction.*

Understanding Key Ideas

2. What type of reaction does the following equation represent?

   $FeS + 2HCl \rightarrow FeCl_2 + H_2S$

   a. synthesis reaction
   b. double-displacement reaction
   c. single-displacement reaction
   d. decomposition reaction

3. Describe the difference between single- and double-displacement reactions.

Math Skills

4. Write the balanced equation in which potassium iodide, KI, reacts with chlorine to form potassium chloride, KCl, and iodine.

Critical Thinking

5. **Analyzing Processes** The first reaction below is a single-displacement reaction that could occur in a laboratory. Explain why the second single-displacement reaction could not occur.

   $CuCl_2 + Fe \rightarrow FeCl_2 + Cu$

   $CaS + Al \rightarrow$ no reaction

6. **Making Inferences** When two white compounds are mixed in a solution, a yellow solid forms. What kind of reaction has taken place? Explain your answer.

SECTION 4

Energy and Rates of Chemical Reactions

READING WARM-UP

Objectives

- Compare exothermic and endothermic reactions.
- Explain activation energy.
- Interpret an energy diagram.
- Describe five factors that affect the rate of a reaction.

Terms to Learn

exothermic reaction
endothermic reaction
law of conservation of energy
activation energy
inhibitor
catalyst

READING STRATEGY

Paired Summarizing Read this section silently. In pairs, take turns summarizing the material. Stop to discuss ideas that seem confusing.

What is the difference between eating a meal and running a mile? You could say that a meal gives you energy, while running "uses up" energy.

Chemical reactions can be described in the same way. Some reactions release energy, and other reactions absorb energy.

Reactions and Energy

Chemical energy is part of all chemical reactions. Energy is needed to break chemical bonds in the reactants. As new bonds form in the products, energy is released. By comparing the chemical energy of the reactants with the chemical energy of the products, you can decide if energy is released or absorbed in the overall reaction.

Exothermic Reactions

A chemical reaction in which energy is released is called an **exothermic reaction.** *Exo* means "go out" or "exit." *Thermic* means "heat" or "energy." Exothermic reactions can give off energy in several forms, as shown in **Figure 1.** The energy released in an exothermic reaction is often written as a product in a chemical equation, as in this equation:

$$2Na + Cl_2 \rightarrow 2NaCl + \textit{energy}$$

Figure 1 Types of Energy Released in Exothermic Reactions

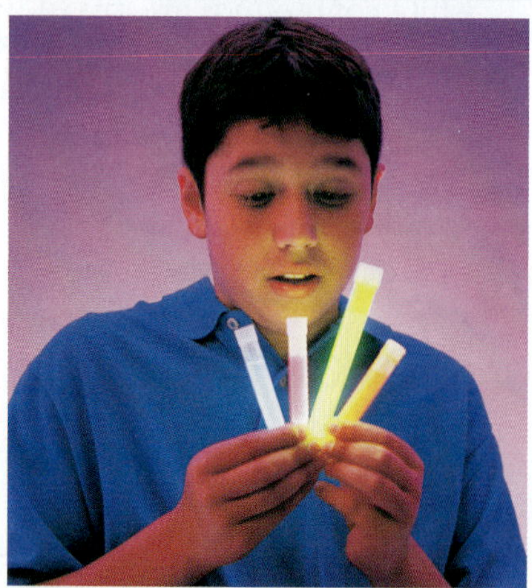

Light energy is released in the exothermic reaction that is taking place in these light sticks.

Electrical energy is released in the exothermic reaction that will take place in this battery.

Light and thermal energy are released in the exothermic reaction taking place in this campfire.

Endothermic Reactions

A chemical reaction in which energy is taken in is called an **endothermic reaction.** *Endo* means "go in." The energy that is taken in during an endothermic reaction is often written as a reactant in a chemical equation. Energy as a reactant is shown in the following equation:

$$2H_2O + \textit{energy} \rightarrow 2H_2 + O_2$$

An example of an endothermic process is photosynthesis. In photosynthesis, plants use light energy from the sun to produce glucose. Glucose is a simple sugar that is used for nutrition. The equation that describes photosynthesis is the following:

$$6CO_2 + 6H_2O + \textit{energy} \rightarrow C_6H_{12}O_6 + 6O_2$$

exothermic reaction a chemical reaction in which heat is released to the surroundings

endothermic reaction a chemical reaction that requires heat

law of conservation of energy the law that states that energy cannot be created or destroyed but can be changed from one form to another

Reading Check **What is an endothermic reaction?** (*See the Appendix for answers to Reading Checks.*)

The Law of Conservation of Energy

Neither mass nor energy can be created or destroyed in chemical reactions. The **law of conservation of energy** states that energy cannot be created or destroyed. However, energy can change forms within a system. And like a baton in a relay race, shown in **Figure 2,** energy can be transferred from one object to another in a system.

The energy released in exothermic reactions was first stored in the chemical bonds in the reactants. And the energy taken in during endothermic reactions is stored in the products. If you could measure all the energy in a reaction, you would find that the total amount of energy (of all types) is the same before and after the reaction.

Figure 2 *Energy can be transferred from one object to another object in the same way that a baton is transferred from one runner to another runner in a relay race.*

Endo Alert

1. Fill a **plastic cup** half full with **calcium chloride solution.**
2. Measure the temperature of the solution by using a **thermometer.**
3. Carefully add **1 tsp of baking soda.**
4. Record your observations.
5. When the reaction has stopped, record the temperature of the solution.
6. What evidence that an endothermic reaction took place did you observe?

Figure 3 *Chemical reactions need energy to get started in the same way that a bowling ball needs a push to get rolling.*

Rates of Reactions

A reaction takes place only if the particles of reactants collide. But there must be enough energy to break the bonds that hold particles together in a molecule. The speed at which new particles form is called the *rate of a reaction.*

activation energy the minimum amount of energy required to start a chemical reaction

Activation Energy

Before the bowling ball in **Figure 3** can roll down the alley, the bowler must first put in some energy to start the ball rolling. A chemical reaction must also get a boost of energy before the reaction can start. This boost of energy is called activation energy. **Activation energy** is the smallest amount of energy that molecules need to react.

Another example of activation energy is striking a match. Before a match can be used to light a campfire, the match has to be lit! A strike-anywhere match has all the reactants it needs to burn. The chemicals on a match react and burn. But, the chemicals will not light by themselves. You must strike the match against a surface. The heat produced by this friction provides the activation energy needed to start the reaction.

Reading Check What is activation energy?

Apples and Acids Investigate how acidity influences the rate of reactions that cause food to spoil. With a parent, dip an apple slice in water. Dip a second slice in lemon juice. Observe the slices for 15 minutes, and discuss how acidity affects the reaction rate. Describe the effect of the substances in the lemon juice by comparing the reaction rate in lemon juice with the reaction rate in water.

ACTIVITY

Sources of Activation Energy

Friction is one source of activation energy. In the match example, friction provides the energy needed to break the bonds in the reactants and allow new bonds to form. An electric spark in a car's engine is another source of activation energy. This spark begins the burning of gasoline. Light can also be a source of activation energy for a reaction. **Figure 4** shows how activation energy relates to exothermic reactions and endothermic reactions.

Figure 4 **Energy Diagrams**

Exothermic Reaction Once an exothermic reaction starts, it can continue. The energy given off as the product forms continues to supply the activation energy needed for the substances to react.

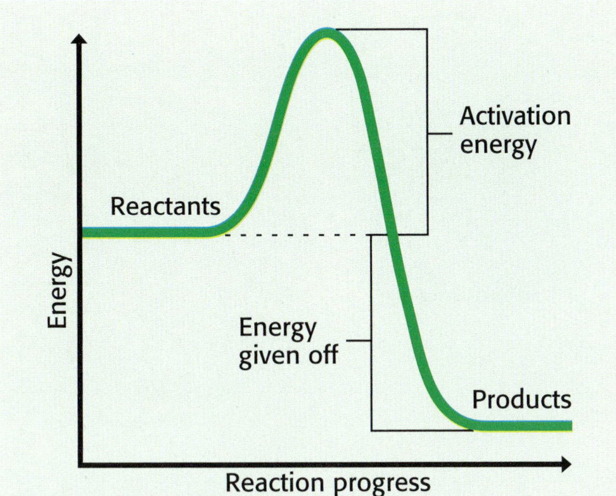

Endothermic Reaction An endothermic reaction continues to absorb energy. Energy must be used to provide the activation energy needed for the substances to react.

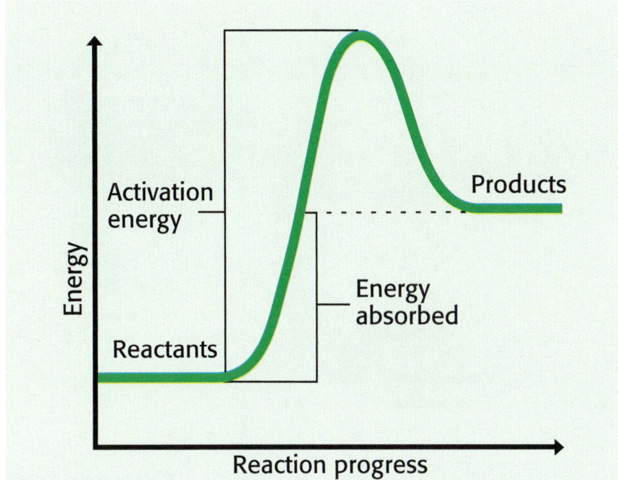

Factors Affecting Rates of Reactions

The rate of a reaction is a measure of how fast the reaction takes place. Recall that the rate of a reaction depends on how fast new particles form. There are four factors that affect the rate of a reaction. These factors are temperature, concentration, surface area, and the presence of an inhibitor or catalyst.

Temperature

A higher temperature causes a faster rate of reaction, as shown in **Figure 5.** At high temperatures, particles of reactants move quickly and collide hard and often. So, reactants quickly change into products. At low temperatures, particles move slowly and collide less often. By keeping food cold, you slow down the reactions that cause food to spoil.

Figure 5 *The light stick on the right glows brighter than the one on the left because the one on the right is warmer. The higher temperature causes the rate of the reaction to increase.*

Which Is Quicker?

1. Fill a **clear plastic cup** with **250 mL of warm water.** Fill a **second clear plastic cup** with **250 mL of cold water.**
2. Place **one-quarter of an effervescent tablet** in each of the two cups of water at the same time. Using a **stopwatch,** time each reaction.
3. Observe each reaction, and record your observations.
4. In which cup did the reaction occur at a faster rate?

Figure 6 **Concentration of Solutions**

When the amount of copper sulfate crystals dissolved in water is **small,** the concentration of the copper sulfate solution is **low.**

When the amount of copper sulfate crystals dissolved in water is **large,** the concentration of the copper sulfate solution is **high.**

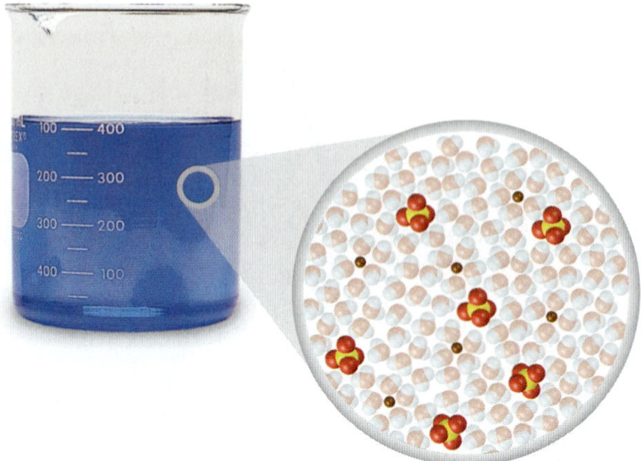

Concentration

In general, a high concentration of reactants causes a fast rate of a reaction. *Concentration* is a measure of the amount of one substance dissolved in another substance, as shown in **Figure 6.** When the concentration is high, there are many reactant particles in a given volume. So, there is a small distance between particles. The particles collide more often and react faster.

Reading Check **How does a high concentration of reactants increase the rate of a reaction?**

CONNECTION TO Biology

Feedback Often, reaction rates are controlled through feedback. The output from one reaction of a system is the input to other reactions of the system. Research blood-glucose feedback control. Make a poster or a model that explains how products of reactions affect other parts of the system and how feedback serves to control what goes on in the system as a whole.

Surface Area

Surface area is the amount of exposed surface of a substance. Increasing the surface area of solid reactants increases the rate of a reaction. Grinding a solid into a powder makes a larger surface area. Greater surface area exposes more particles of the reactant to other reactant particles. The particles collide more often. So, the rate of the reaction increases.

Inhibitors

Many substances dissolve in water. The presence or absence of these substances often affects the rates of reactions that take place in water. For example, an **inhibitor** is a substance that slows down or stops a chemical reaction. Preservatives are added to foods to inhibit, or slow down, the growth of bacteria and fungi. They prevent bacteria and fungi from producing substances that can spoil food. Some antibiotics are also examples of inhibitors. For example, penicillin prevents certain kinds of bacteria from making a cell wall. So, the bacteria die.

inhibitor a substance that slows down or stops a chemical reaction

catalyst a substance that changes the rate of a chemical reaction without being used up or changed very much

Catalysts

Some chemical reactions would be too slow to be useful without a catalyst (KAT uh LIST). A **catalyst** is a substance that speeds up a reaction without being permanently changed. Because it is not changed, a catalyst is not a reactant. A catalyst lowers the activation energy of a reaction, which allows the reaction to happen more quickly. Catalysts called *enzymes* speed up most reactions in your body. Catalysts are even found in cars, as seen in **Figure 7.** The catalytic converter decreases air pollution. It does this by increasing the rate of reactions that involve the harmful products given off by cars.

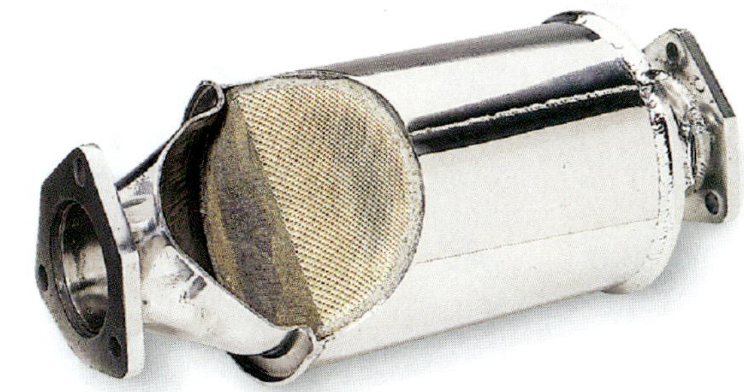

Figure 7 *This catalytic converter contains platinum and palladium. These two catalysts increase the rate of reactions that make the car's exhaust less harmful.*

SECTION Review

Summary

- Energy is given off in exothermic reactions.
- Energy is absorbed in an endothermic reaction.
- The law of conservation of energy states that energy is neither created nor destroyed.
- Activation energy is the energy needed for a reaction to occur.
- The rate of a chemical reaction is affected by temperature, concentration, surface area, and the presence of an inhibitor or catalyst.

Using Key Terms

The statements below are false. For each statement, replace the underlined term to make a true statement.

1. An <u>exothermic</u> reaction absorbs energy.
2. The rate of a reaction can be <u>increased</u> by adding an inhibitor.

Understanding Key Ideas

3. Which of the following will not increase the rate of a reaction?
   a. adding a catalyst
   b. increasing the temperature of the reaction
   c. decreasing the concentration of reactants
   d. grinding a solid into powder
4. How does the concentration of a solution affect the rate of reaction?

Critical Thinking

5. **Making Comparisons** Compare exothermic and endothermic reactions.
6. **Applying Concepts** Explain how chewing your food thoroughly can help your body digest food.

Interpreting Graphics

Use the diagram below to answer the questions that follow.

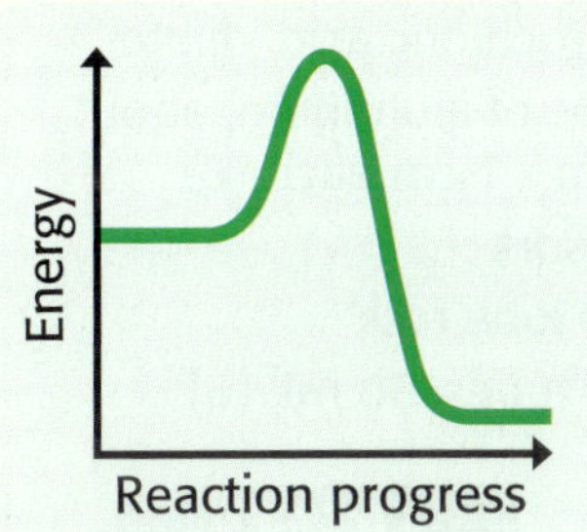

7. Does this energy diagram show an exothermic or an endothermic reaction? How can you tell?
8. A catalyst lowers the amount of activation energy needed to get a reaction started. What do you think the diagram would look like if a catalyst were added?

For a variety of links related to this chapter, go to www.scilinks.org
Topic: Exothermic and Endothermic Reactions
SciLinks code: HSM0555

Using Scientific Methods

Skills Practice Lab

Speed Control

The reaction rate (how fast a chemical reaction happens) is an important factor to control. Sometimes, you want a reaction to take place rapidly, such as when you are removing tarnish from a metal surface. Other times, you want a reaction to happen very slowly, such as when you are depending on a battery as a source of electrical energy.

In this lab, you will discover how changing the surface area and concentration of the reactants affects reaction rate. In this lab, you can estimate the rate of reaction by observing how fast bubbles form.

OBJECTIVES

Describe how the surface area of a solid affects the rate of a reaction.

Explain how concentration of reactants will speed up or slow down a reaction.

MATERIALS

- funnels (2)
- graduated cylinders, 10 mL (2)
- hydrochloric acid, concentrated
- hydrochloric acid, dilute
- strips of aluminum, about 5 cm x 1 cm each (6)
- scissors
- test-tube rack
- test tubes, 30 mL (6)

SAFETY

Part A: Surface Area

Ask a Question

1. How does changing the surface area of a metal affect reaction rate?

Form a Hypothesis

2. Write a statement that answers the question above. Explain your reasoning.

Test the Hypothesis

3. Use three identical strips of aluminum. Put one strip into a test tube. Place the test tube in the test-tube rack. **Caution:** The strips of metal may have sharp edges.

4. Carefully fold a second strip in half and then in half again. Use a textbook or other large object to flatten the folded strip as much as possible. Place the strip in a second test tube in the test-tube rack.

5. Use scissors to cut a third strip of aluminum into the smallest possible pieces. Place all of the pieces into a third test tube, and place the test tube in the test-tube rack.

6. Use a funnel and a graduated cylinder to pour 10 mL of concentrated hydrochloric acid into each of the three test tubes. **Caution:** Hydrochloric acid is corrosive. If any acid should spill on you, immediately flush the area with water and notify your teacher.

7. Observe the rate of bubble formation in each test tube. Record your observations.

Analyze the Results

1. **Organizing Data** Which form of aluminum had the greatest surface area? the smallest surface area?

2. **Analyzing Data** The amount of aluminum and the amount of acid were the same in all three test tubes. Which form of the aluminum seemed to react the fastest? Which form reacted the slowest? Explain your answers.

3. **Analyzing Results** Do your results support the hypothesis you made? Explain.

Draw Conclusions

4. **Making Predictions** Would powdered aluminum react faster or slower than the forms of aluminum you used? Explain your answer.

Part B: Concentration

Ask a Question

1. How does changing the concentration of acid affect the reaction rate?

Form a Hypothesis

2. Write a statement that answers the question above. Explain your reasoning.

Test the Hypothesis

3. Place one of the three remaining aluminum strips in each of the three clean test tubes. (Note: Do not alter the strips.) Place the test tubes in the test-tube rack.

4. Using the second funnel and graduated cylinder, pour 10 mL of water into one of the test tubes. Pour 10 mL of dilute acid into the second test tube. Pour 10 mL of concentrated acid into the third test tube.

5. Observe the rate of bubble formation in the three test tubes. Record your observations.

Analyze the Results

1. **Explaining Events** In this set of test tubes, the strips of aluminum were the same, but the concentration of the acid was different. Was there a difference between the test tube that contained water and the test tubes that contained acid? Which test tube formed bubbles the fastest? Explain.

2. **Analyzing Results** Do your results support the hypothesis you made? Explain.

Draw Conclusions

3. **Applying Conclusions** Why should spilled hydrochloric acid be diluted with water before it is wiped up?

Chapter Review

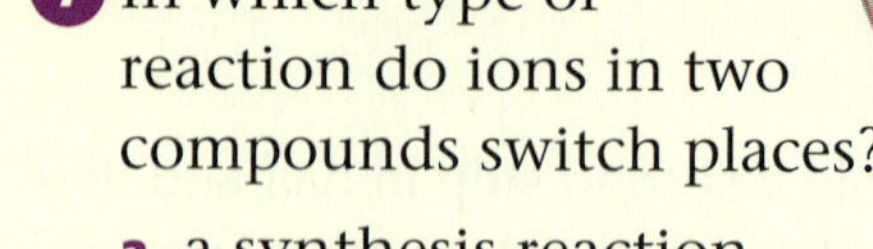

USING KEY TERMS

Complete each of the following sentences by choosing the correct term from the word bank.

subscript	exothermic reaction
inhibitor	synthesis reaction
coefficient	reactant

1. Adding a(n) ___ will slow down a chemical reaction.

2. A chemical reaction that gives off heat is called a(n) ___.

3. A chemical reaction that forms one compound from two or more substances is called a(n) ___.

4. The 2 in the formula Ag_2S is a (an) ___.

UNDERSTANDING KEY IDEAS

Multiple Choice

5. Balancing a chemical equation so that the same number of atoms of each element is found in both the reactants and the products is an example of
   - **a.** activation energy.
   - **b.** the law of conservation of energy.
   - **c.** the law of conservation of mass.
   - **d.** a double-displacement reaction.

6. Which of the following is the correct chemical formula for dinitrogen tetroxide?
   - **a.** N_4O_2
   - **b.** NO_2
   - **c.** N_2O_5
   - **d.** N_2O_4

7. In which type of reaction do ions in two compounds switch places?
   - **a.** a synthesis reaction
   - **b.** a decomposition reaction
   - **c.** a single-displacement reaction
   - **d.** a double-displacement reaction

8. Which of the following actions is an example of the use of activation energy?
   - **a.** plugging in an iron
   - **b.** playing basketball
   - **c.** holding a lit match to paper
   - **d.** eating

9. Enzymes in your body act as catalysts. Thus, the role of enzymes is
   - **a.** to increase the rate of chemical reactions.
   - **b.** to decrease the rate of chemical reactions.
   - **c.** to help you breathe.
   - **d.** to inhibit chemical reactions.

Short Answer

10. Name the type of reaction that each of the following equations represents.
    - **a.** $2Cu + O_2 \rightarrow 2CuO$
    - **b.** $2Na + MgSO_4 \rightarrow Na_2SO_4 + Mg$
    - **c.** $Ba(CN)_2 + H_2SO_4 \rightarrow BaSO_4 + 2HCN$

11. Describe what happens to chemical bonds during a chemical reaction.

12. Name four ways that you can change the rate of a chemical reaction.

13. Describe four clues that signal that a chemical reaction is taking place.

Math Skills

14 Write balanced equations for the following:

a. $Fe + O_2 \rightarrow Fe_2O_3$
b. $Al + CuSO_4 \rightarrow Al_2(SO_4)_3 + Cu$
c. $Mg(OH)_2 + HCl \rightarrow MgCl_2 + H_2O$

15 Calculate the number of atoms of each element shown in the formulas below:

a. $CaSO_4$
b. $4NaOCl$
c. $Fe(NO_3)_2$
d. $2Al_2(CO_3)_3$

CRITICAL THINKING

16 **Concept Mapping** Use the following terms to create a concept map: *products, chemical reaction, chemical equation, chemical formulas, reactants, coefficients,* and *subscripts.*

17 **Evaluating Assumptions** Your friend is very worried by rumors that he has heard about a substance called *dihydrogen monoxide* in the city's water system. What could you say to your friend to calm his fears? (Hint: Write the formula of the substance.)

18 **Analyzing Ideas** As long as proper safety precautions have been taken, why can explosives be transported long distances without exploding?

19 **Applying Concepts** You measured the mass of a steel pipe before leaving it outdoors. One month later, the pipe had rusted, and its mass had increased. Does this change violate the law of conservation of mass? Explain your answer.

20 **Applying Concepts** Acetic acid, a compound found in vinegar, reacts with baking soda to produce carbon dioxide, water, and sodium acetate. Without writing an equation, identify the reactants and the products of this reaction.

INTERPRETING GRAPHICS

Use the photo below to answer the questions that follow.

21 What evidence in the photo supports the claim that a chemical reaction is taking place?

22 Is this reaction an exothermic or endothermic reaction? Explain your answer.

23 Draw and label an energy diagram of both an exothermic and endothermic reaction. Identify the diagram that describes the reaction shown in the photo above.

Standardized Test Preparation

READING

Read each of the passages below. Then, answer the questions that follow each passage.

Passage 1 The key to an air bag's success during a crash is the speed at which it inflates. Inside the bag is a gas generator that contains the compounds sodium azide, potassium nitrate, and silicon dioxide. At the moment of a crash, an electronic sensor in the car detects the sudden change in speed. The sensor sends a small electric current to the gas generator. This electric current provides the activation energy for the chemicals in the gas generator. The rate at which the reaction happens is very fast. In 1/25 of a second, the gas formed in the reaction inflates the bag. The air bag fills upward and outward. By filling the space between a person and the car's dashboard, the air bag protects him or her from getting hurt.

1. Which of the following events happens first?

- **A** The sensor sends an electric current to the gas generator.
- **B** The air bag inflates.
- **C** The air bag fills the space between the person and the dashboard.
- **D** The sensor detects a change in speed.

2. What provides the activation energy for the reaction to occur?

- **F** the speed of the car
- **G** the inflation of the air bag
- **H** the hot engine
- **I** the electric current from the sensor

3. What is the purpose of this passage?

- **A** to convince the reader to wear a seat belt
- **B** to describe the series of events that inflate an air bag
- **C** to explain why air bags are an important safety feature in cars
- **D** to show how chemical reactions protect pedestrians

Passage 2 An important tool in fighting forest fires is a slimy, red goop. This mixture of powder and water is a very powerful fire retardant. The burning of trees, grass, and brush is an exothermic reaction. The fire retardant slows or stops this self-feeding reaction by increasing the activation energy for the materials to which it sticks. A plane can carry between 4,500 and 11,000 L of the goop. The plane then drops it all in front of the raging flames of a forest fire when the pilot presses the button. Firefighters on the ground can gain valuable time when a fire is slowed with a fire retardant. This extra time allows the ground team to create a fire line that will finally stop the fire.

1. Which of the following sentences best summarizes the passage?

- **A** The burning of forests and other brush is an exothermic reaction.
- **B** Dropping fire retardants ahead of a flame can help firefighters on the ground stop a fire.
- **C** Firefighters on the ground create a fire line that will help stop the fire from spreading.
- **D** The slimy, red goop used as a fire retardant is made of a mixture of powder and water.

2. Based on the passage, which of the following statements is a fact?

- **F** Fire retardants are always successful in putting out fires.
- **G** No more than 4,500 L of red goop are loaded onto a plane.
- **H** A fire retardant works by increasing the activation energy for the materials that it sticks on.
- **I** The burning of trees is an endothermic reaction.

INTERPRETING GRAPHICS

Use the energy diagram below to answer the questions that follow.

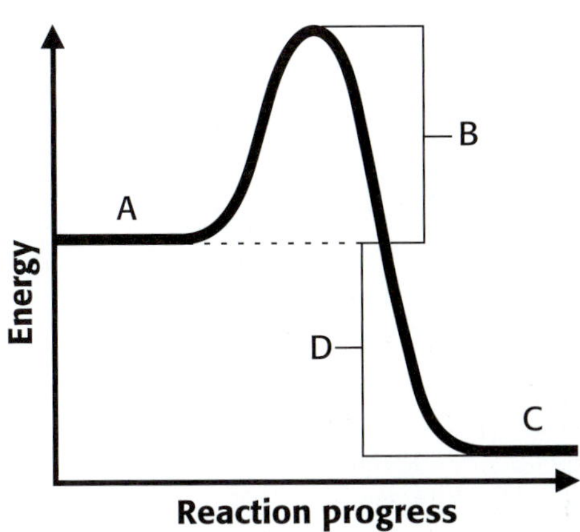

1. Which letter represents the energy of the products?
   A A
   B B
   C C
   D D

2. Which letter represents the activation energy of the reaction?
   F A
   G B
   H C
   I D

3. Which of the following statements best describes the reaction represented by the graph?
   A The reaction is endothermic because the energy of the products is greater than the energy of the reactants.
   B The reaction is endothermic because the energy of the reactants is greater than the energy of the products.
   C The reaction is exothermic because the energy of the products is greater than the energy of the reactants.
   D The reaction is exothermic because the energy of the reactants is greater than the energy of the products.

MATH

Read each question below, and choose the best answer.

1. Nina has 15 pens in her backpack. She has 3 red pens, 10 black pens, and 2 blue pens. If Ben selects a pen to borrow at random, what is the probability that the pen selected is red?
   A 2/15
   B 1/5
   C 1/3
   D 2/3

2. How many atoms of nitrogen, N, are in the formula for calcium nitrate, $Ca(NO_3)_2$?
   F 3
   G 2
   H 6
   I 1

3. Which letter best represents the number 2 3/5 on the number line?

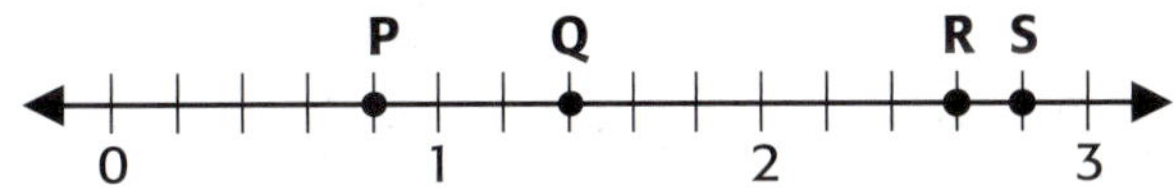

   A P
   B Q
   C R
   D S

4. According to the following chemical equation, how many reactants are needed to form water and carbon dioxide?

$$H_2CO_3 \rightarrow H_2O + CO_2$$

   F one
   G two
   H three
   I four

Standardized Test Preparation

Science in Action

Weird Science

Light Sticks

Have you ever seen light sticks at a concert? Your family may even keep them in the car for emergencies. But how do light sticks work? To activate the light stick, you have to bend it. Most light sticks are made of a plastic tube that contains a mixture of two chemicals. Also inside the tube is a thin glass vial, which contains hydrogen peroxide. As long as the glass vial is unbroken, the two chemicals are kept separate. But bending the ends of the tube breaks the glass vial. This action releases the hydrogen peroxide into the other chemicals and a chemical reaction occurs, which makes the light stick glow.

Social Studies ACTIVITY

Who invented light sticks? What was their original purpose? Research the answers to these questions. Make a poster that shows what you have learned.

Science, Technology, and Society

Bringing Down the House!

Have you ever watched a building being demolished? It takes only minutes to demolish it, but a lot of time was spent planning the demolition. And it takes time to remove hazardous chemicals from the building. For example, asbestos, which is found in insulation, can cause lung cancer. Mercury found in thermostats can cause brain damage, birth defects, and death. It is important to remove these substances because most of the rubble is sent to a landfill. If hazardous chemicals are not removed, they could leak into the groundwater and enter the water supply.

Math ACTIVITY

A city sends 4,000,000 tons of waste to landfills in 1 year. Of this waste, 82% is solid waste, and 38% of the solid waste comes from building construction and demolition. Build a circle chart and a bar graph of the categories of waste by weight. Incorporate the charts as evidence in a written argument to support a claim that the city should start a recycling program for building materials.

Larry McKee

Arson Investigator Once a fire dies down, you might see an arson investigator like Lt. Larry McKee on the scene. "After the fire is out, I can investigate the fire scene to determine where the fire started and how it started," says McKee, who questions witnesses and firefighters about what they have seen. He knows that the color of the smoke can indicate certain chemicals. He also has help detecting chemicals from an accelerant-sniffing dog, Nikki. Nikki has been trained to detect about 11 different chemicals. If Nikki finds one of these chemicals, she begins to dig. McKee takes a sample of the suspicious material to the laboratory. He treats the sample so that any chemicals present will dissolve in a liquid. A sample of this liquid is placed into an instrument called a *gas chromatograph* and tested. The results of this test are printed out in a graph, from which the suspicious chemical is identified. Next, McKee begins to search for suspects. By combining detective work with scientific evidence, fire investigators can help find clues that can lead to the conviction of the arsonist.

Language Arts 

WRITING SKILL Write a one-page story about an arson investigator. Begin the story at the scene of a fire. Take the story through the different steps that you think an investigator would have to go through to solve the crime.

To learn more about these Science in Action topics, visit go.hrw.com and type in the keyword HP5REAF.

Current Science

Check out Current Science® articles related to this chapter by visiting go.hrw.com. Just type in the keyword HP5CS14.

Chemical Compounds

About the PHOTO

The bean weevil feeds on bean seeds, which are rich in chemical compounds such as proteins, carbohydrates, and lipids. The bean weevil begins life as a tiny grub that lives in the seed where it eats starch and protein. The adult then cuts holes in the seed coat and crawls out, as you can see in this photo.

PRE-READING ACTIVITY

Layered Book Before you read the chapter, create the FoldNote entitled "Layered Book" described in the **Study Skills** section of the Appendix. Label the tabs of the layered book with "Ionic and covalent compounds," "Acids and bases," "Solutions of acids and bases," and "Organic compounds." As you read the chapter, write information you learn about each category under the appropriate tab.

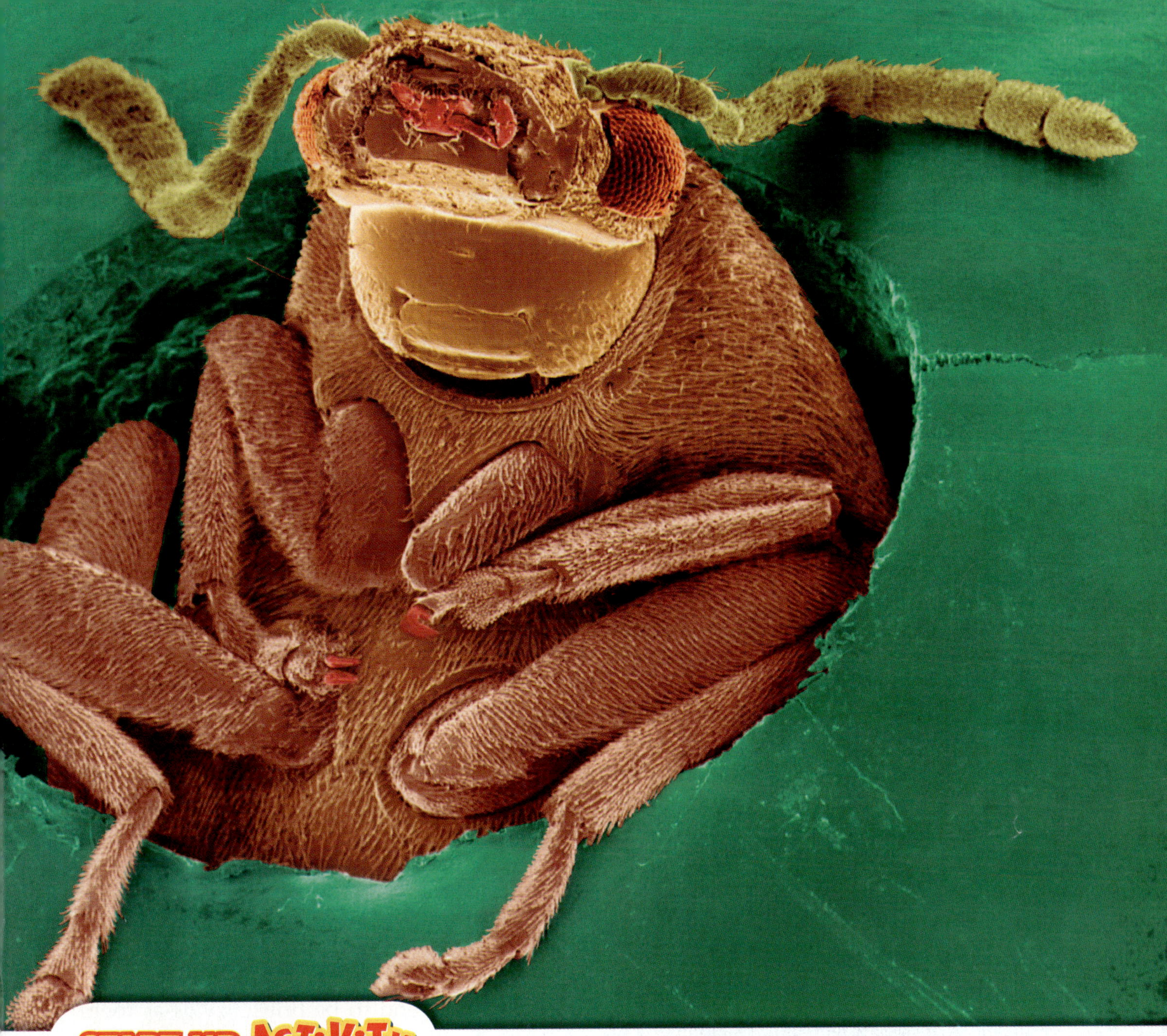

START-UP ACTIVITY

Sticking Together

In this activity, you will demonstrate the force that keeps particles together in some compounds.

Procedure

1. Rub **two balloons** with a **wool cloth.** Move the balloons near each other. Describe what you see.
2. Put one balloon against a wall. Record your observations.

Analysis

1. The balloons are charged by rubbing them with the wool cloth. Like charges repel each other. Opposite charges attract each other. Do the balloons have like or opposite charges? Explain.
2. If the balloon that was placed against the wall has a negative charge, what is the charge on the wall? Explain your answer.
3. The particles that make up compounds are attracted to each other in the same way that the balloon is attracted to the wall. What can you infer about the particles that make up such compounds?

SECTION 1

Ionic and Covalent Compounds

When ions or molecules combine, they form compounds. Because there are millions of compounds, it is helpful to organize them into groups. But how can scientists tell the difference between compounds?

READING WARM-UP

Objectives

- Describe the properties of ionic and covalent compounds.
- Classify compounds as ionic or covalent based on their properties.

Terms to Learn

chemical bond
ionic compound
covalent compound

READING STRATEGY

Reading Organizer As you read this section, create an outline of the section. Use the headings from the section in your outline.

One way to group compounds is by the kind of chemical bond they have. A **chemical bond** is the combining of atoms to form molecules or compounds. Bonding can occur between valence electrons of different atoms. *Valence electrons* are electrons in the outermost energy level of an atom. The behavior of valence electrons determines if an ionic compound or a covalent compound is formed.

chemical bond the combining of atoms to form molecules or compounds

ionic compound a compound made of oppositely charged ions

Ionic Compounds and Their Properties

The properties of ionic compounds are a result of strong attractive forces called ionic bonds. An *ionic bond* is an attraction between oppositely charged ions. Compounds that contain ionic bonds are called **ionic compounds.** Ionic compounds can be formed by the reaction of a metal with a nonmetal. Metal atoms become positively charged ions when electrons are transferred from the metal atoms to the nonmetal atoms. This transfer of electrons also causes the nonmetal atom to become a negatively charged ion. Sodium chloride, commonly known as *table salt,* is an ionic compound.

Brittleness

Ionic compounds tend to be brittle solids at room temperature. So, they usually break apart when hit. This property is due to the arrangement of ions in a repeating three-dimensional pattern called a *crystal lattice,* shown in **Figure 1.** Each ion in a lattice is surrounded by ions of the opposite charge. And each ion is bonded to the ions around it. When an ionic compound is hit, the pattern of ions shifts. Ions that have the same charge line up and repel one another, which causes the crystal to break.

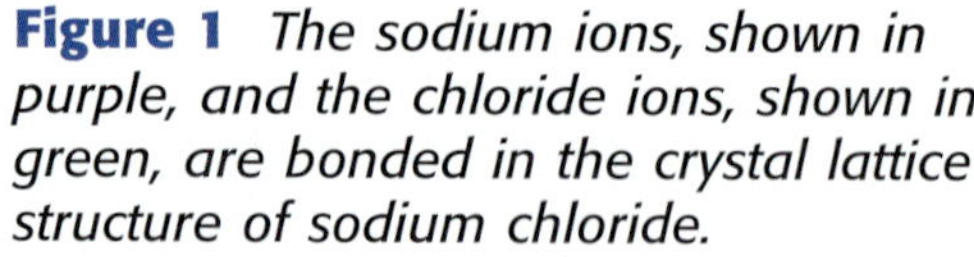

Figure 1 *The sodium ions, shown in purple, and the chloride ions, shown in green, are bonded in the crystal lattice structure of sodium chloride.*

Figure 2 **Melting Points of Some Ionic Compounds**

Potassium dichromate
Melting point: 398°C

Magnesium oxide
Melting point: 2,800°C

Nickel(II) oxide
Melting point: 1,984°C

High Melting Points

Because of the strong ionic bonds that hold ions together, ionic compounds have high melting points. These high melting points are the reason that most ionic compounds are solids at room temperature. For example, solid sodium chloride must be heated to 801°C before it will melt. The melting points of three other ionic compounds are given in **Figure 2.**

Solubility and Electrical Conductivity

Many ionic compounds are highly soluble. So, they dissolve easily in water. Water molecules attract each of the ions of an ionic compound and pull the ions away from one another. The solution that forms when an ionic compound dissolves in water can conduct an electric current, as shown in **Figure 3.** The solution can conduct an electric current because the ions are charged and are able to move freely past one another. However, an undissolved crystal of an ionic compound does not conduct an electric current.

Reading Check **Why do ionic solutions conduct an electric current?** (*See the Appendix for answers to Reading Checks.*)

Figure 3 *The pure water does not conduct an electric current. However, the solution of salt water conducts an electric current, so the bulb lights up.*

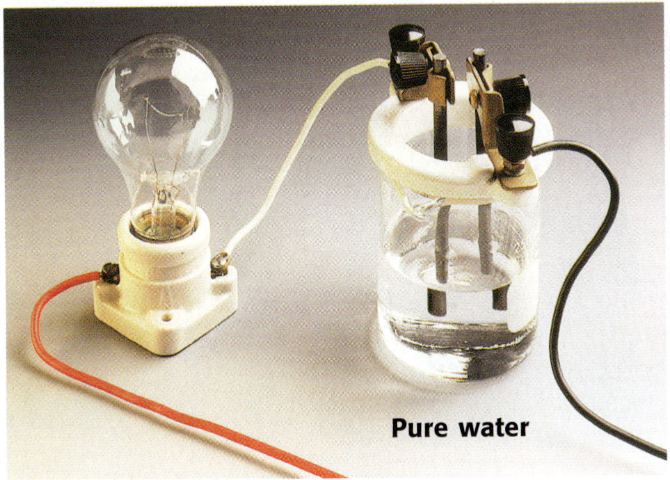

covalent compound a chemical compound formed by the sharing of electrons

Covalent Compounds and Their Properties

Most compounds are covalent compounds. **Covalent compounds** are compounds that form when a group of atoms shares electrons. This sharing of electrons forms a covalent bond. A *covalent bond* is a weaker attractive force than an ionic bond is. The group of atoms that make up a covalent compound is called a molecule. A *molecule* is the smallest particle into which a covalently bonded compound can be divided and still be the same compound. Properties of covalent compounds are very different from the properties of ionic compounds.

Low Solubility

Many covalent compounds are not soluble in water, which means that they do not dissolve well in water. You may have noticed this if you have ever left off the top of a soda bottle. The carbon dioxide gas that gives the soda its fizz eventually escapes, and your soda pop goes "flat." The attraction between water molecules is much stronger than their attraction to the molecules of most other covalent compounds. So, water molecules stay together instead of mixing with the covalent compounds. If you have ever made salad dressing, you probably know that oil and water don't mix. Oils, such as the oil in the salad dressing in **Figure 4,** are made of covalent compounds.

Reading Check Why won't most covalent compounds dissolve in water?

CONNECTION TO Language Arts

WRITING SKILL **Electrolyte Solutions** Ionic compounds that conduct electricity when they are dissolved in water are called *electrolytes*. Some electrolytes play important roles in the functioning of living cells. Electrolytes can be lost by the body during intense physical activity or illness and must be replenished for cells to work properly. Research two electrolytes that your body cells need and the function that they serve. Present your findings in a one-page research paper.

Figure 4 *Olive oil, which is used in salad dressings, is made of very large covalent molecules that do not mix with water.*

Low Melting Points

The forces of attraction between molecules of covalent compounds are much weaker than the bonds holding ionic solids together. Less heat is needed to separate the molecules of covalent compounds, so these compounds have much lower melting and boiling points than ionic compounds do.

Electrical Conductivity

Although most covalent compounds don't dissolve in water, some do. Most of the covalent compounds that dissolve in water form solutions that have uncharged molecules. Sugar is a covalent compound that dissolves in water and that does not form ions. So, a solution of sugar and water does not conduct an electric current, as shown in **Figure 5.** However, some covalent compounds do form ions when they dissolve in water. Many acids, for example, form ions in water. These solutions, like ionic solutions, conduct an electric current.

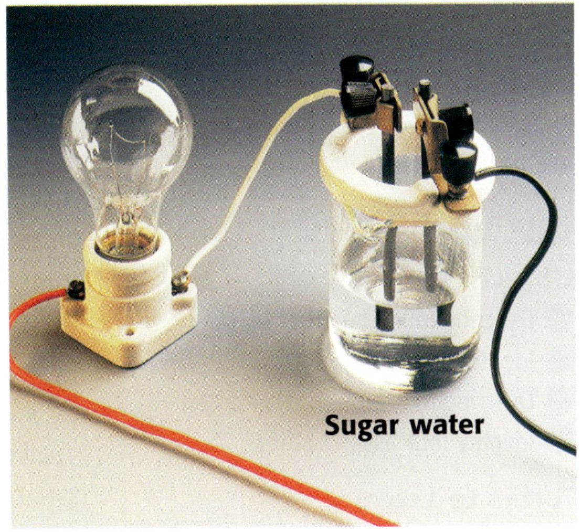

Figure 5 *This solution of sugar, a covalent compound, and water does not conduct an electric current because the molecules of sugar are not charged.*

SECTION Review

Summary

- Ionic compounds have ionic bonds between ions of opposite charges.
- Ionic compounds are usually brittle, have high melting points, dissolve in water, and often conduct an electric current.
- Covalent compounds have covalent bonds and consist of particles called molecules.
- Covalent compounds have low melting points, don't dissolve easily in water, and do not conduct an electric current.

Using Key Terms

1. Use each of the following terms in a separate sentence: *ionic compound, covalent compound,* and *chemical bond.*

Understanding Key Ideas

2. Which of the following describes an ionic compound?
   - **a.** It has a low melting point.
   - **b.** It consists of shared electrons.
   - **c.** It conducts electric current in water solutions.
   - **d.** It consists of two nonmetals.
3. List two properties of covalent compounds.

Math Skills

4. A compound contains 39.37% chromium, 38.10% oxygen, and potassium. What percentage of the compound is potassium?

Critical Thinking

5. **Making Inferences** Solid crystals of ionic compounds do not conduct an electric current. But when the crystals dissolve in water, the solution conducts an electric current. Explain.
6. **Applying Concepts** Some white solid crystals are dissolved in water. If the solution does not conduct an electric current, is the solid an ionic compound or a covalent compound? Explain.

For a variety of links related to this chapter, go to www.scilinks.org

Topic: Ionic Compounds; Covalent Compounds

SciLinks code: HSM0817; HSM0365

SECTION 2

Acids and Bases

Would you like a nice, refreshing glass of acid? This is just what you get when you have a glass of lemonade.

Lemons contain a substance called an *acid*. One property of acids is a sour taste. In this section, you will learn about the properties of acids and bases.

READING WARM-UP

Objectives

- Describe four properties of acids.
- Identify four uses of acids.
- Describe four properties of bases.
- Identify four uses of bases.

Terms to Learn

acid
indicator
base

READING STRATEGY

Reading Organizer As you read this section, make a table comparing acids and bases.

Acids and Their Properties

A sour taste is not the only property of an acid. Have you noticed that when you squeeze lemon juice into tea, the color of the tea becomes lighter? This change happens because acids cause some substances to change color. An **acid** is any compound that increases the number of hydronium ions, H_3O^+, when dissolved in water. Hydronium ions form when a hydrogen ion, H^+, separates from the acid and bonds with a water molecule, H_2O, to form a hydronium ion, H_3O^+.

Reading Check How is a hydronium ion formed? (*See the Appendix for answers to Reading Checks.*)

acid any compound that increases the number of hydronium ions when dissolved in water

Acids Have a Sour Flavor

Have you ever taken a bite of a lemon or lime? If so, like the boy in **Figure 1,** you know the sour taste of an acid. The taste of lemons, limes, and other citrus fruits is a result of citric acid. However, taste, touch, or smell should NEVER be used to identify an unknown chemical. Many acids are *corrosive,* which means that they destroy body tissue, clothing, and many other things. Most acids are also poisonous.

Figure 1 *Foods that have a sour taste usually contain acids.*

Figure 2 Detecting Acids with Indicators

The indicator, bromthymol blue, is pale blue in water.

When acid is added, the color changes to yellow because of the presence of the indicator.

indicator a compound that can reversibly change color depending on conditions such as pH

Acids Change Colors in Indicators

A substance that changes color in the presence of an acid or base is an **indicator.** Look at **Figure 2.** The flask on the left contains water and an indicator called *bromthymol blue* (BROHM THIE MAWL BLOO). Acid has been added to the flask on the right. The color changes from pale blue to yellow because the indicator detects the presence of an acid.

Another indicator commonly used in the lab is litmus. Paper strips containing litmus are available in both blue and red. When an acid is added to blue litmus paper, the color of the litmus changes to red.

Acids React with Metals

Acids react with some metals to produce hydrogen gas. For example, hydrochloric acid reacts with zinc metal to produce hydrogen gas, as shown in **Figure 3.** The equation for the reaction is the following:

$$2HCl + Zn \longrightarrow H_2 + ZnCl_2$$

In this reaction, zinc displaces hydrogen in the compound, hydrochloric acid. This displacement happens because zinc is an active metal. But if the element silver were put into hydrochloric acid, nothing would happen. Silver is not an active metal, so no reaction would take place.

Figure 3 *Bubbles of hydrogen gas form when zinc metal reacts with hydrochloric acid.*

Acids Conduct Electric Current

When acids are dissolved in water, they break apart and form ions in the solution. The ions make it possible for the solution to conduct an electric current. A car battery is one example of how an acid can be used to produce an electric current. The sulfuric acid in the battery conducts electricity to help start the car's engine.

CONNECTION TO Biology

Acids Can Curl Your Hair! Permanents contain acids. Acids make hair curly by denaturing a certain amino acid in hair proteins. Research how acids are used in products that either curl or straighten hair. Then, make a poster that demonstrates this process. Present your poster to your classmates.

ACTIVITY

Uses of Acids

Acids are used in many areas of industry and in homes. Sulfuric acid is the most widely made industrial chemical in the world. It is used to make many products, including paper, paint, detergents, and fertilizers. Nitric acid is used to make fertilizers, rubber, and plastics. Hydrochloric acid is used to make metals from their ores by separating the metals from the materials with which they are combined. It is also used in swimming pools to help keep them free of algae. Hydrochloric acid is even found in your stomach, where it aids in digestion. Hydrofluoric acid is used to etch glass, as shown in **Figure 4.** Citric acid and ascorbic acid (Vitamin C) are found in orange juice. And carbonic acid and phosphoric acid help give a sharp taste to soft drinks.

Reading Check What are three uses of acids?

Figure 4 *The image of the swan was etched into the glass through the use of hydrofluoric acid.*

Figure 5 Examples of Bases

Soaps are made from sodium hydroxide, which is a base. Soaps remove dirt and oils from skin and feel slippery when you touch them.

Bleach and detergents contain bases and are used for removing stains from clothing. Detergents feel slippery like soap.

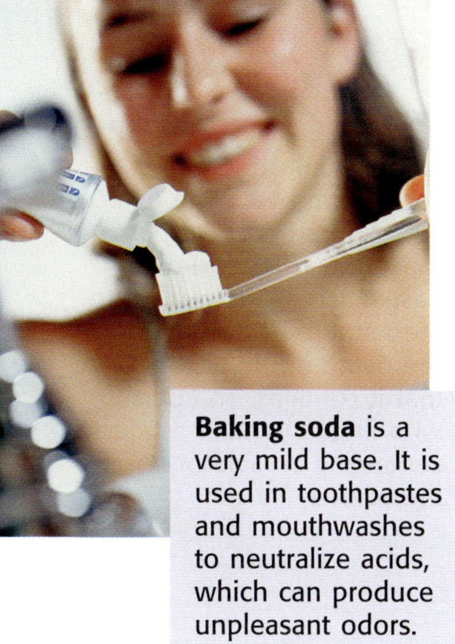

Baking soda is a very mild base. It is used in toothpastes and mouthwashes to neutralize acids, which can produce unpleasant odors.

Bases and Their Properties

A **base** is any compound that increases the number of hydroxide ions, OH^-, when dissolved in water. For example, sodium hydroxide breaks apart to form sodium ions and hydroxide ions as shown below.

base any compound that increases the number of hydroxide ions when dissolved in water

$$NaOH \longrightarrow Na^+ + OH^-$$

Hydroxide ions give bases their properties. **Figure 5** shows examples of bases that you are probably familiar with.

Bases Have a Bitter Flavor and a Slippery Feel

The properties of a base solution include a bitter taste and a slippery feel. If you have ever accidentally tasted soap, you know the bitter taste of a base. Soap will also have the slippery feel of a base. However, taste, touch or smell should NEVER be used to identify an unknown chemical. Like acids, many bases are corrosive. If your fingers feel slippery when you are using a base in an experiment, you may have gotten the base on your hands. You should immediately rinse your hands with large amounts of water and tell your teacher.

NEVER touch or taste a concentrated solution of a strong base.

Acids and Bases at Home

Ask an adult to join in a contest with you. The object is to find products at home that contain an acid or a base. Each person will write the name of the product and the name of the acid or base that it contains. The person who finds the most products containing an acid or base is the winner.

ACTIVITY

Figure 6 Detecting Bases with Indicators

The indicator, bromthymol blue, is pale blue in water.

When a base is added to the indicator, the indicator turns dark blue.

Bases Change Color in Indicators

Like acids, bases change the color of an indicator. Most indicators turn a different color in the presence of bases than they do in the presence of acids. For example, bases change the color of red litmus paper to blue. And the indicator, bromthymol blue, turns blue when a base is added to it, as shown in **Figure 6.**

Bases Conduct Electric Current

Solutions of bases conduct an electric current because bases increase the number of hydroxide ions, OH^-, in a solution. A hydroxide ion is actually a hydrogen atom and an oxygen atom bonded together. The extra electron gives the hydroxide ion a negative charge.

Blue to Red—Acid!

1. Pour about 5 mL of **test solution** into a **spot plate.** Test the solution using **red litmus paper** and **blue litmus paper** by dipping a **stirring rod** into it and then touching the rod to a piece of litmus paper.
2. Record any color changes. Clean the stirring rod.
3. Repeat the above steps with each solution. Use new pieces of litmus paper as needed.
4. Identify each solution as acidic or basic.

Uses of Bases

Like acids, bases have many uses. Sodium hydroxide is a base used to make soap and paper. It is also used in oven cleaners and in products that unclog drains. Calcium hydroxide, $Ca(OH)_2$, is used to make cement and plaster. Ammonia is found in many household cleaners and is used to make fertilizers. And magnesium hydroxide and aluminum hydroxide are used in antacids to treat heartburn. **Figure 7** shows some of the many products that contain bases. Carefully follow the safety instructions when using these products. Remember that bases can harm your skin.

✓ Reading Check What three ways can bases be used at home?

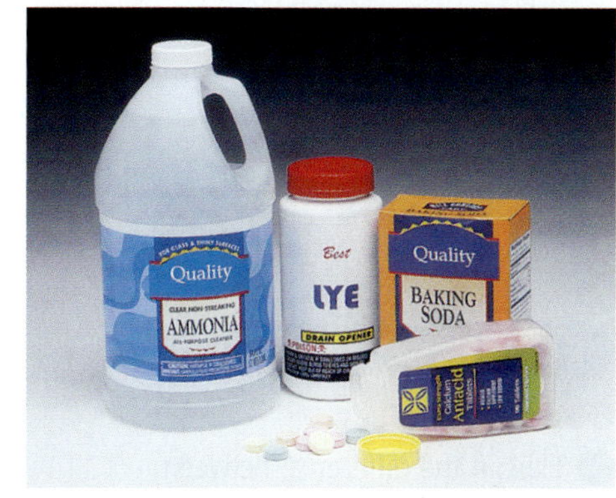

Figure 7 *Bases are common around the house. They are useful as cleaning agents, as cooking aids, and as medicines.*

SECTION Review

Summary

- An acid is a compound that increases the number of hydronium ions in solution.
- Acids taste sour, turn blue litmus paper red, react with metals to produce hydrogen gas, and may conduct an electric current when in solution.
- Acids are used for industrial purposes and in household products.
- A base is a compound that increases the number of hydroxide ions in solution.
- Bases taste bitter, feel slippery, and turn red litmus paper blue. Most solutions of bases conduct an electric current.
- Bases are used in cleaning products and acid neutralizers.

Using Key Terms

1. In your own words, write a definition for each of the following terms: *acid, base,* and *indicator.*

Understanding Key Ideas

2. A base is a substance that
   a. feels slippery.
   b. tastes sour.
   c. reacts with metals to produce hydrogen gas.
   d. turns blue litmus paper red.
3. Acids are important in
   a. making antacids.
   b. preparing detergents.
   c. keeping algae out of swimming pools.
   d. manufacturing cement.
4. What happens to red litmus paper when when it touches a base?

Math Skills

5. A cake recipe calls for 472 mL of milk. You don't have a metric measuring cup at home, so you need to convert milliliters to cups. You know that 1 L equals 1.06 quarts and that there are 4 cups in 1 quart. How many cups of milk will you need to use?

Critical Thinking

6. **Making Comparisons** Compare the properties of acids and bases.
7. **Applying Concepts** Why would it be useful for a gardener or a vegetable farmer to use litmus paper to test soil samples?
8. **Analyzing Processes** Suppose that your teacher gives you a solution of an unknown chemical. The chemical is either an acid or a base. You know that touching or tasting acids and bases is not safe. What two tests could you perform on the chemical to determine whether it is an acid or a base? What results would help you decide if the chemical was an acid or a base?

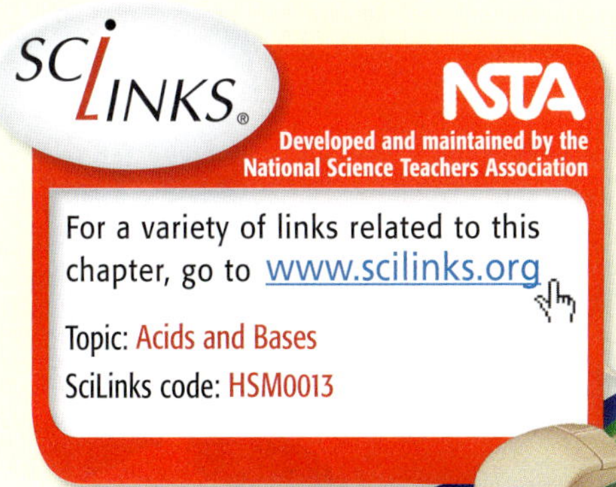

SECTION 3

Solutions of Acids and Bases

READING WARM-UP

Objectives

- Explain the difference between strong acids and bases and weak acids and bases.
- Identify acids and bases by using the pH scale.
- Describe the formation and uses of salts.

Terms to Learn

neutralization reaction
pH
salt

READING STRATEGY

Discussion Read this section silently. Write down questions that you have about this section. Discuss your questions in a small group.

Suppose that at your friend's party, you ate several large pieces of pepperoni pizza followed by cake and ice cream. Now, you have a terrible case of indigestion.

If you have ever had an upset stomach, you may have felt very much like the boy in **Figure 1.** And you may have taken an antacid. But do you know how antacids work? An antacid is a weak base that neutralizes a strong acid in your stomach. In this section, you will learn about the strengths of acids and bases. You will also learn about reactions between acids and bases.

Strengths of Acids and Bases

Acids and bases can be strong or weak. The strength of an acid or a base is not the same as the concentration of an acid or a base. The concentration of an acid or a base is the amount of acid or base dissolved in water. But the strength of an acid or a base depends on the number of molecules that break apart when the acid or base is dissolved in water.

Figure 1 *Antacids may help relieve your stomachache by reacting with the acid in your stomach.*

Strong Versus Weak Acids

As an acid dissolves in water, the acid's molecules break apart and produce hydrogen ions, H^+. If all of the molecules of an acid break apart, the acid is called a *strong acid*. Strong acids include sulfuric acid, nitric acid, and hydrochloric acid. If only a few molecules of an acid break apart, the acid is a weak acid. Weak acids include acetic (uh SEET ik) acid, citric acid, and carbonic acid.

Reading Check **What is the difference between a strong acid and a weak acid?** (*See the Appendix for answers to Reading Checks.*)

Strong Versus Weak Bases

When all molecules of a base break apart in water to produce hydroxide ions, OH^-, the base is a strong base. Strong bases include sodium hydroxide, calcium hydroxide, and potassium hydroxide. When only a few molecules of a base break apart, the base is a weak base, such as ammonium hydroxide and aluminum hydroxide.

Acids, Bases, and Neutralization

When the base in an antacid meets stomach acid, a reaction occurs. The reaction between acids and bases is a **neutralization reaction** (NOO truhl i ZA shuhn ree AK shuhn). Acids and bases neutralize one another because the hydrogen ions (H^+), which are present in an acid, and the hydroxide ions (OH^-), which are present in a base, react to form water, H_2O, which is neutral. Other ions from the acid and base dissolve in the water. If the water evaporates, these ions join to form a compound called a *salt*.

The pH Scale

An *indicator*, such as litmus, can identify whether a solution contains an acid or base. To describe how acidic or basic a solution is, the pH scale is used. The **pH** of a solution is a measure of the hydronium ion concentration in the solution. A solution that has a pH of 7 is neutral, which means that the solution is neither acidic nor basic. Pure water has a pH of 7. Basic solutions have a pH greater than 7. Acidic solutions have a pH less than 7. **Figure 2** shows the pH values for many common materials.

pHast Relief!

1. Pour **vinegar** into a **small plastic cup** until the cup is half full. Test the vinegar with **red and blue litmus paper.** Record your results.
2. Crush one **antacid tablet,** and mix it with the vinegar. Test the mixture with litmus paper. Record your results.
3. Compare the acidity of the solution before the antacid was added with the acidity of the solution after it was added.

neutralization reaction the reaction of an acid and a base to form a neutral solution of water and a salt

pH a value that is used to express the acidity or basicity (alkalinity) of a system

Figure 2 pH Values of Common Materials

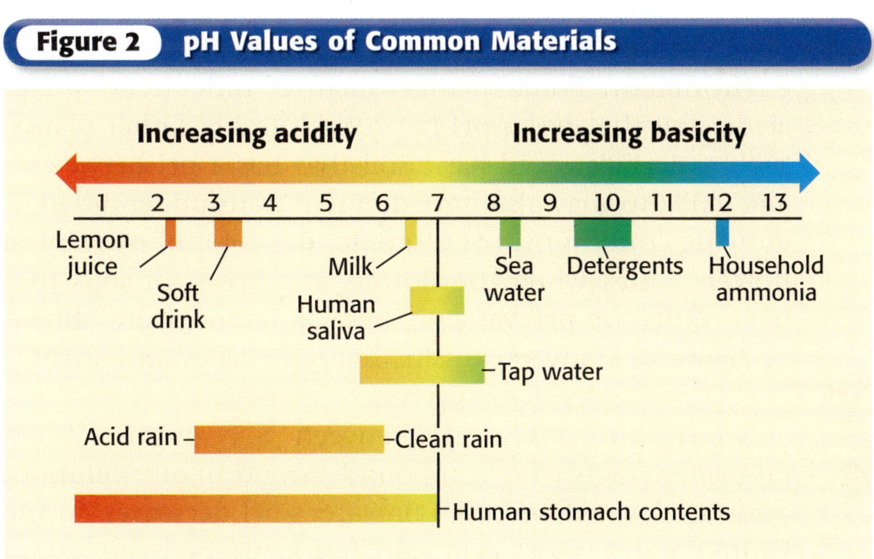

Figure 3 **Using Indicators to Find pH**

pH Indicator Scale

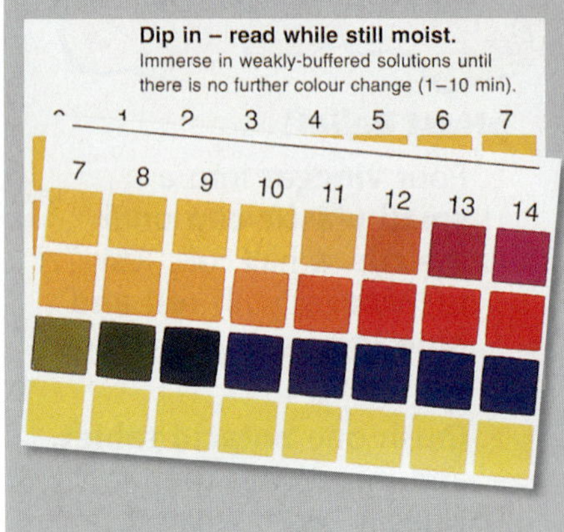

pH 4

pH 10

CONNECTION TO Biology

WRITING SKILL **Blood and pH** Human blood has a pH between 7.38 and 7.42. If the blood pH is lower or higher, the body cannot function properly. Research what can cause the pH of blood to rise above or fall below normal ranges. Write a one-page paper that details your findings.

Using Indicators to Determine pH

A combination of indicators can be used to find out how basic or how acidic a solution is. This can be done if the colors of the indicators are known at different pH values. **Figure 3** shows strips of pH paper, which contains several different indicators. These strips were dipped into two different solutions. The pH of each solution is found by comparing the colors on each strip with the colors on the indicator scale provided. This kind of indicator is often used to test the pH of water in pools and aquariums. Another way to find the pH of a solution is to use a pH meter. These meters can detect and measure hydronium ion concentration electronically.

Reading Check How can indicators determine pH?

pH and the Environment

Living things depend on having a steady pH in their environment. Some plants, such as pine trees, prefer acidic soil that has a pH between 4 and 6. Other plants, such as lettuce, need basic soil that has a pH between 8 and 9. Plants may also have different traits under different growing conditions. For example, the color of hydrangea flowers varies when the flowers are grown in soils that have different pH values. These differences are shown in **Figure 4.** Many organisms living in lakes and streams need a neutral pH to survive.

Most rain has a pH between 5.5 and 6. When rainwater reacts with compounds found in air pollution, acids are formed and the rainwater's pH decreases. In the United States, most acid rain has a pH between 4 and 4.5, but some precipitation has a pH as low as 3.

Figure 4 *To grow blue flowers, plant hydrangeas in soil that has a low pH. To grow pink flowers, use soil that has a high pH.*

Salts

When an acid neutralizes a base, a salt and water are produced. A **salt** is an ionic compound formed from the positive ion of a base and the negative ion of an acid. When you hear the word *salt,* you probably think of the table salt you use to season your food. But the sodium chloride found in your salt shaker is only one example of a large group of compounds called *salts.*

Figure 5 *Salts help keep roads free of ice by decreasing the freezing point of water.*

salt an ionic compound that forms when a metal atom replaces the hydrogen of an acid

Uses of Salts

Salts have many uses in industry and in homes. You already know that sodium chloride is used to season foods. It is also used to make other compounds, including lye (sodium hydroxide) and baking soda. Sodium nitrate is a salt that is used to preserve food. And calcium sulfate is used to make wallboard, which is used in construction. Another use of salt is shown in **Figure 5.**

SECTION Review

Summary

- Every molecule of a strong acid or base breaks apart to form ions. Few molecules of weak acids and bases break apart to form ions.
- An acid and a base can neutralize one another to make salt and water.
- pH is a measure of hydronium ion concentration in a solution.
- A salt is an ionic compound formed in a neutralization reaction. Salts have many industrial and household uses.

Using Key Terms

1. Use the following terms in the same sentence: *neutralization reaction* and *salt.*

Understanding Key Ideas

2. A neutralization reaction
   a. includes an acid and a base.
   b. produces a salt.
   c. forms water.
   d. All of the above
3. Explain the difference between a strong acid and a weak acid.

Math Skills

4. For each point lower on the pH scale, the hydrogen ions in solution increase tenfold. For example, a solution of pH 3 is not twice as acidic as a solution of pH 6 but is 1,000 times as acidic. How many times more acidic is a solution of pH 2 than a solution of pH 4?

Critical Thinking

5. **Analyzing Processes** Predict what will happen to the hydrogen ion concentration and the pH of water if hydrochloric acid is added to the water.
6. **Analyzing Relationships** Would fish be healthy in a lake that has a low pH? Explain.
7. **Applying Concepts** Soap is made from a strong base and oil. Would you expect the pH of soap to be 4 or 9? Explain.

For a variety of links related to this chapter, go to www.scilinks.org

Topic: pH scale; Salts
SciLinks code: HSM1130; HSM1347

SECTION 4

Organic Compounds

Can you believe that more than 90% of all compounds are members of a single group of compounds? It's true!

READING WARM-UP

Objectives

- Explain why there are so many organic compounds.
- Identify and describe saturated, unsaturated, and aromatic hydrocarbons.
- Describe the characteristics of carbohydrates, lipids, proteins, and nucleic acids and their functions in the body.

Terms to Learn

organic compound
hydrocarbon
carbohydrate
lipid
protein
nucleic acid

READING STRATEGY

Paired Summarizing Read this section silently. In pairs, take turns summarizing the material. Stop to discuss ideas that seem confusing.

Most compounds are members of a group called organic compounds. **Organic compounds** are covalent compounds composed of carbon-based molecules. Fuel, rubbing alcohol, and sugar are organic compounds. Even cotton, paper, and plastic belong to this group. Why are there so many kinds of organic compounds? Learning about the carbon atom can help you understand why.

The Four Bonds of a Carbon Atom

All organic compounds contain carbon. Each carbon atom has four valence electrons. So, each carbon atom can make four bonds with four other atoms.

Carbon Backbones

The models in **Figure 1** are called *structural formulas.* They are used to show how atoms in a molecule are connected. Each line represents a pair of electrons that form a covalent bond. Many organic compounds are based on the types of carbon backbones shown in **Figure 1.** Some compounds have hundreds or thousands of carbon atoms as part of their backbone! Organic compounds may also contain hydrogen, oxygen, sulfur, nitrogen, and phosphorus.

Reading Check What is the purpose of structural formulas? (*See the Appendix for answers to Reading Checks.*)

Figure 1 Three Models of Carbon Backbones

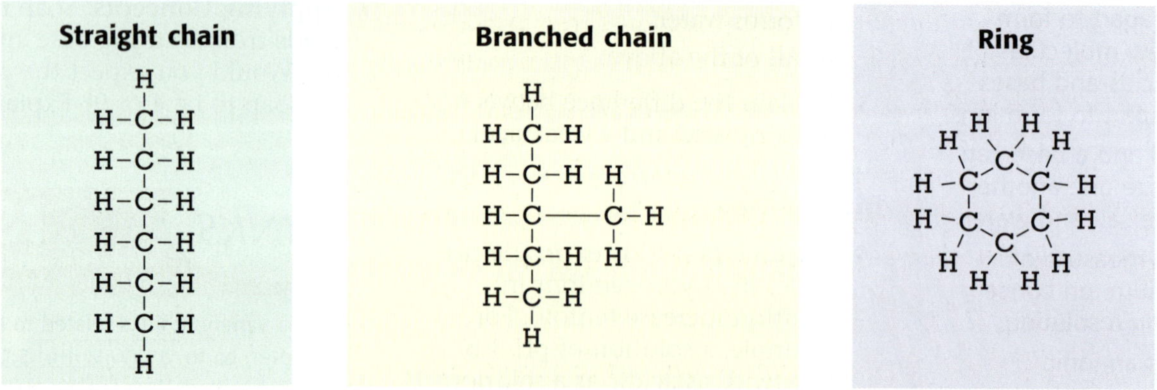

Straight chain

All carbon atoms are connected in a straight line.

Branched chain

The chain of carbon atoms branches into different directions when a carbon atom is bonded to more than one other carbon atom.

Ring

The chain of carbon atoms forms a ring.

Figure 2 Three Types of Hydrocarbons

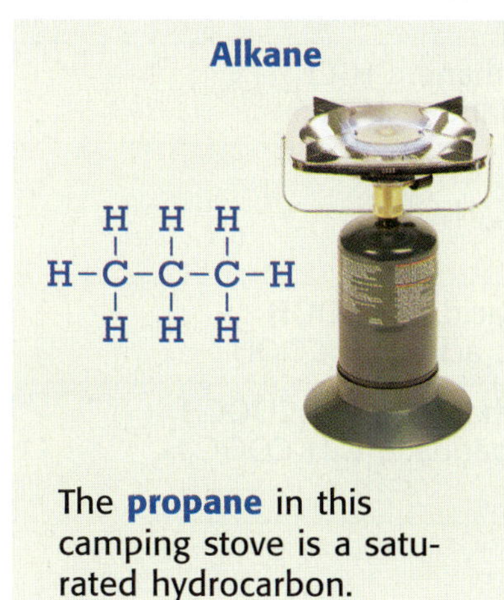

The **propane** in this camping stove is a saturated hydrocarbon.

Fruits make **ethene,** which is a compound that helps ripen the fruit.

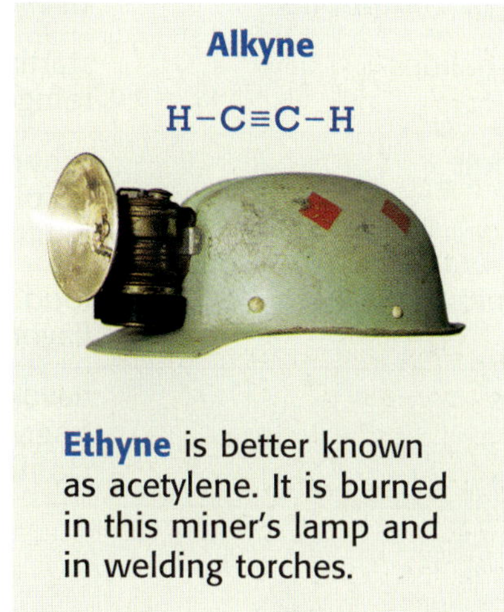

Ethyne is better known as acetylene. It is burned in this miner's lamp and in welding torches.

Hydrocarbons and Other Organic Compounds

Although many organic compounds contain several kinds of atoms, some contain only two. Organic compounds that contain only carbon and hydrogen are called **hydrocarbons.**

organic compound a covalently bonded compound that contains carbon

hydrocarbon an organic compound composed only of carbon and hydrogen

Saturated Hydrocarbons

The propane shown in **Figure 2** is a saturated hydrocarbon. A *saturated hydrocarbon,* or *alkane,* is a hydrocarbon in which each carbon atom in the molecule shares a single bond with each of four other atoms. A single bond is a covalent bond made up of one pair of shared electrons.

Unsaturated Hydrocarbons

An *unsaturated hydrocarbon,* such as ethene or ethyne shown in **Figure 2,** is a hydrocarbon in which at least one pair of carbon atoms shares a double bond or a triple bond. A double bond is a covalent bond made up of two pairs of shared electrons. A triple bond is a covalent bond made up of three pairs of shared electrons. Hydrocarbons that contain double or triple bonds are unsaturated because these bonds can be broken and more atoms can be added to the molecules.

Compounds that contain two carbon atoms connected by a double bond are called *alkenes.* Hydrocarbons that contain two carbon atoms connected by a triple bond are called *alkynes.*

Aromatic Hydrocarbons

Most aromatic (AR uh MAT ik) compounds are based on benzene. As shown in **Figure 3,** benzene has a ring of six carbons that have alternating double and single bonds. Aromatic hydrocarbons often have strong odors.

Figure 3 *Benzene is the starting material for manufacturing many products, including medicines.*

Table 1 Types and Uses of Organic Compounds

Type of compound	Uses	Examples
Alkyl halides	starting material for Teflon™ refrigerant (Freon™)	chloromethane, CH_3Cl bromoethane, C_2H_5Br
Alcohols	rubbing alcohol gasoline additive antifreeze	methanol, CH_3OH ethanol, C_2H_5OH
Organic acids	food preservatives flavorings	ethanoic acid, CH_3COOH propanoic acid, C_2H_5COOH
Esters	flavorings fragrances clothing (polyester)	methyl ethanoate, CH_3COOCH_3 ethyl propanoate, $C_2H_5COOC_2H_5$

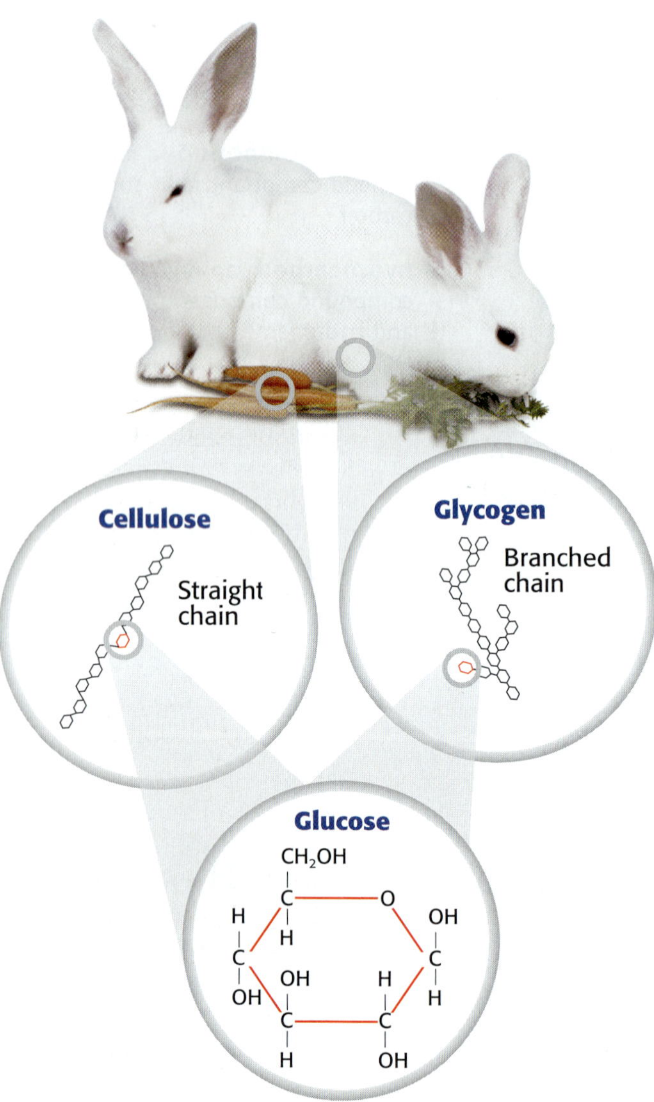

Figure 4 *Glucose molecules, represented by hexagons, can bond to form complex carbohydrates, such as cellulose and glycogen.*

Other Organic Compounds

There are many other kinds of organic compounds. Some have atoms of halogens, oxygen, sulfur, and phosphorus in their molecules. A few of these compounds and their uses are listed in **Table 1.**

Biochemicals: The Compounds of Life

Organic compounds that are made by living things are called *biochemicals*. Biochemicals are divided into four categories: carbohydrates, lipids, proteins, and nucleic acids (noo KLEE ik AS idz).

Carbohydrates

Carbohydrates are biochemicals that are composed of one or more simple sugar molecules bonded together. Carbohydrates are used as a source of energy. There are two kinds of carbohydrates: simple carbohydrates and complex carbohydrates.

Simple carbohydrates include simple sugars, such as glucose. **Figure 4** shows how glucose molecules can bond to form different complex carbohydrates. Complex carbohydrates may be made of hundreds or thousands of sugar molecules bonded together. *Cellulose* gives plant cell walls their rigid structure, and *glycogen* supplies energy to muscle cells.

Lipids

Lipids are biochemicals that do not dissolve in water. Fats, oils, and waxes are kinds of lipids. Lipids have many functions, including storing energy and making up cell membranes. Although too much fat in your diet can be unhealthy, some fat is important to good health. The foods in **Figure 5** are sources of lipids.

Lipids store excess energy in the body. Animals tend to store lipids as fats, while plants store lipids as oils. When an organism has used up most of its carbohydrates, it can obtain energy by breaking down lipids. Lipids are also used to store some vitamins.

carbohydrate a class of energy-giving nutrients that includes sugars, starches, and fiber; composed of one or more simple sugars bonded together

lipid a type of biochemical that does not dissolve in water; fats and steroids are lipids

protein a molecule that is made up of amino acids and that is needed to build and repair body structures and to regulate processes in the body

Proteins

Most of the biochemicals found in living things are proteins. In fact, after water, proteins are the most common molecules in your cells. **Proteins** are biochemicals that are composed of "building blocks" called *amino acids*.

Amino acids are small molecules made up of carbon, hydrogen, oxygen, and nitrogen atoms. Some amino acids also include sulfur atoms. Amino acids bond to form proteins of many shapes and sizes. The shape of a protein determines the function of the protein. If even a single amino acid is missing or out of place, the protein may not function correctly or at all. Proteins have many functions. They regulate chemical activities, transport and store materials, and provide structural support.

Reading Check What are proteins made of?

Food Facts

1. Select **four empty food packages.**
2. Without reading the Nutrition Facts labels, rank the items from most carbohydrate content to least carbohydrate content.
3. Rank the items from most fat content to least fat content.
4. Read the Nutrition Facts labels, and compare your rankings with the real rankings.
5. Why do you think your rankings were right, or why were they wrong? Explain your answers.

Figure 5 *Vegetable oil, meat, cheese, nuts, eggs, and milk are sources of lipids in your diet.*

Figure 6 *Spider webs are made up of proteins that are shaped like long fibers.*

Examples of Proteins

Proteins have many roles in your body and in living things. Enzymes (EN ZIEMZ) are proteins that are catalysts. *Catalysts* regulate chemical reactions in the body by increasing the rate at which the reactions occur. Some hormones are proteins. For example, insulin is a protein hormone that helps regulate your blood-sugar level. Another kind of protein, called *hemoglobin,* is found in red blood cells and delivers oxygen throughout the body. There are also large proteins that extend through cell membranes. These proteins help control the transport of materials into and out of cells. Some proteins, such as those in your hair, provide structural support. The structural proteins of silk fibers make the spider web shown in **Figure 6** strong and lightweight.

Nucleic Acids

nucleic acid a molecule made up of subunits called *nucleotides*

The largest molecules made by living organisms are nucleic acids. **Nucleic acids** are biochemicals made up of *nucleotides* (NOO klee oh TIEDZ). Nucleotides are molecules made of carbon, hydrogen, oxygen, nitrogen, and phosphorus atoms. There are only five kinds of nucleotides. But nucleic acids may have millions of nucleotides bonded together. The only reason living things differ from each other is that each living thing has a different order of nucleotides.

Nucleic acids have several functions. One function of nucleic acids is to store genetic information. They also help build proteins and other nucleic acids. Nucleic acids are sometimes called *the blueprints of life,* because they contain all the information needed for a cell to make all of its proteins.

Reading Check What are two functions of nucleic acids?

CONNECTION TO Social Studies

DNA "Fingerprinting" and Crime-Scene Investigation The chemical structure of all human DNA is the same. The only difference between one person's DNA and another's is the order, or sequence, of the building blocks in the DNA. The number of ways these building blocks can be sequenced are countless.

DNA fingerprinting is new process. However, it has changed the way that criminal investigations are carried out. Research DNA fingerprinting. Find out when DNA fingerprinting was first used, who developed the process, and how DNA fingerprinting is used in crime-scene investigations. Present your findings in an oral presentation to your class. Include a model or a poster to help explain the process to your classmates.

DNA and RNA

There are two kinds of nucleic acids: DNA and RNA. A model of DNA (**d**eoxyribo**n**ucleic **a**cid) is shown in **Figure 7.** DNA is the genetic material of the cell. DNA molecules can store a huge amount of information because of their length. The DNA molecules in a single human cell have a length of about 2 m—which is more than 6 ft long! When a cell needs to make a certain protein, it copies a certain part of the DNA. The information copied from the DNA directs the order in which amino acids are bonded to make that protein. DNA also contains information used to build the second type of nucleic acid, RNA (**r**ibo**n**ucleic **a**cid). RNA is involved in the actual building of proteins.

Figure 7 *Two strands of DNA are twisted in a spiral shape. Four different nucleotides make up the rungs of the DNA ladder.*

SECTION Review

Summary

- Organic compounds contain carbon, which can form four bonds.
- Hydrocarbons are composed of only carbon and hydrogen.
- Hydrocarbons may be saturated, unsaturated, or aromatic hydrocarbons.
- Carbohydrates are made of simple sugars.
- Lipids store energy and make up cell membranes.
- Proteins are composed of amino acids.
- Nucleic acids store genetic information and help cells make proteins.

Using Key Terms

1. Use the following terms in the same sentence: *organic compound, hydrocarbon,* and *biochemical.*
2. In your own words, write a definition for each of the following terms: *carbohydrate, lipid, protein,* and *nucleic acid.*

Understanding Key Ideas

3. A saturated hydrocarbon has
   a. only single bonds.
   b. double bonds.
   c. triple bonds.
   d. double and triple bonds.
4. List two functions of proteins.
5. What is an aromatic hydrocarbon?

Critical Thinking

6. **Identifying Relationships** Hemoglobin is a protein that is in blood and that transports oxygen to the tissues of the body. Information stored in nucleic acids tells a cell how to make proteins. What might happen if there is a mistake in the information needed to make hemoglobin?
7. **Making Comparisons** Compare saturated hydrocarbons with unsaturated hydrocarbons.

Interpreting Graphics

Use the structural formula of this organic compound to answer the questions that follow.

```
    H   H   H
    |   |   |
H – C – C – C – H
    |   |   |
    H   H   H
```

8. What type of bonds are present in this molecule?
9. Can you determine the shape of the molecule from this structural formula? Explain your answer.

Skills Practice Lab

Cabbage Patch Indicators

Indicators are weak acids or bases that change color due to the pH of the substance to which they are added. Red cabbage contains a natural indicator. It turns specific colors at specific pHs. In this lab you will extract the indicator from red cabbage. Then, you will use it to determine the pH of several liquids.

OBJECTIVES

Make a natural acid-base indicator solution.

Determine the pH of various common substances.

MATERIALS

- beaker, 250 mL
- beaker tongs
- eyedropper
- hot plate
- litmus paper
- pot holder
- red cabbage leaf
- sample liquids provided by teacher
- tape, masking
- test tubes
- test-tube rack
- water, distilled

SAFETY

Procedure

1. Copy the table below. Be sure to include one line for each sample liquid.

Data Collection Table

Liquid	Color with indicator	pH	Effect on litmus paper
Control			

2. Put on protective gloves. Place 100 mL of distilled water in the beaker. Tear the cabbage leaf into small pieces. Place the pieces in the beaker.
3. Use the hot plate to heat the cabbage and water to boiling. Continue boiling until the water is deep blue. **Caution:** Use extreme care when working near a hot plate.
4. Use tongs to remove the beaker from the hot plate. Turn the hot plate off. Allow the solution to cool on a pot holder for 5 to 10 minutes.
5. While the solution is cooling, use masking tape and a pen to label the test tubes for each sample liquid. Label one test tube as the control. Place the tubes in the rack.
6. Use the eyedropper to place a small amount (about 5 mL) of the indicator (cabbage juice) in the test tube labeled as the control.
7. Pour a small amount (about 5 mL) of each sample liquid into the appropriate test tube.
8. Using the eyedropper, place several drops of the indicator into each test tube. Swirl gently. Record the color of each liquid in the table.
9. Use the chart below to the find the pH of each sample. Record the pH values in the table.
10. Litmus paper has an indicator that turns red in an acid and blue in a base. Test each liquid with a strip of litmus paper. Record the results.

Analyze the Results

1. **Analyzing Data** What purpose does the control serve? What is the pH of the control?
2. **Examining Data** What colors in your samples indicate the presence of an acid? What colors indicate the presence of a base?
3. **Analyzing Results** Why is red cabbage juice considered a good indicator?

Draw Conclusions

4. **Interpreting Information** Which do you think would be more useful to help identify an unknown liquid—litmus paper or red cabbage juice? Why?

Applying Your Data

Unlike distilled water, rainwater has some carbon dioxide dissolved in it. Is rainwater acidic, basic, or neutral? To find out, place a small amount of the cabbage juice indicator (which is water-based) in a clean test tube. Use a straw to gently blow bubbles in the indicator. Continue blowing bubbles until you see a color change. What can you conclude about the pH of your "rainwater?" What is the purpose of blowing bubbles in the cabbage juice?

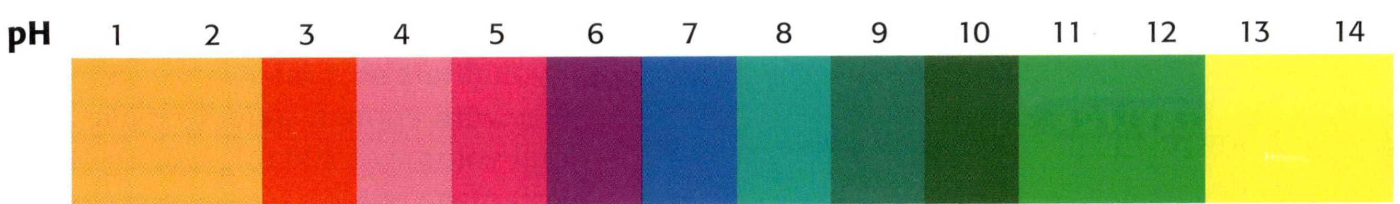

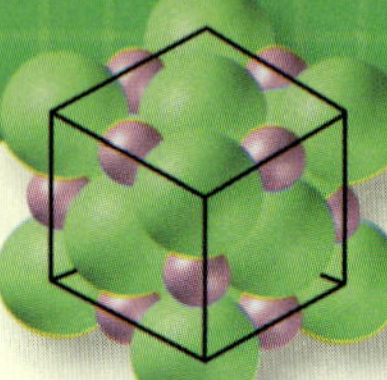

Chapter Review

USING KEY TERMS

For each pair of terms, explain how the meanings of the terms differ.

1. *ionic compound* and *covalent compound*
2. *acid* and *base*
3. *pH* and *indicator*
4. *hydrocarbon* and *organic compound*
5. *carbohydrate* and *lipid*
6. *protein* and *nucleic acid*

UNDERSTANDING KEY IDEAS

Multiple Choice

7. Which of the following statements describes lipids?
   - **a.** Lipids are used to store energy.
   - **b.** Lipids do not dissolve in water.
   - **c.** Lipids make up part of the cell membrane.
   - **d.** All of the above
8. Ionic compounds
   - **a.** have a low melting point.
   - **b.** are often brittle.
   - **c.** do not conduct electric current in water.
   - **d.** do not dissolve easily in water.
9. An increase in the concentration of hydronium ions in solution
   - **a.** raises the pH.
   - **b.** lowers the pH.
   - **c.** does not affect the pH.
   - **d.** doubles the pH.
10. The compounds that store information for building proteins are
    - **a.** lipids.
    - **b.** hydrocarbons.
    - **c.** nucleic acids.
    - **d.** carbohydrates.

Short Answer

11. What type of compound would you use to neutralize a solution of potassium hydroxide?
12. Explain why the reaction of an acid with a base is called *neutralization.*
13. What characteristic of carbon atoms helps to explain the wide variety of organic compounds?
14. What kind of ions are produced when an acid is dissolved in water and when a base is dissolved in water?

Math Skills

15. Most of the vinegar used to make pickles is 5% acetic acid. So, in 100 mL of vinegar, 5 mL is acid diluted with 95 mL of water. If you bought a 473 mL bottle of 5% vinegar, how many milliliters of acetic acid would be in the bottle? How many milliliters of water were used to dilute the acetic acid?
16. If you dilute a 75 mL can of orange juice with enough water to make a total volume of 300 mL, what is the percentage of juice in the mixture?

CRITICAL THINKING

17 **Concept Mapping** Use the following terms to create a concept map: *acid, base, salt, neutral,* and *pH.*

18 **Applying Concepts** Fish give off the base, ammonia, NH_3, as waste. How does the release of ammonia affect the pH of the water in the aquarium? What can be done to correct the pH of the water?

19 **Analyzing Methods** Many insects, such as fire ants, inject formic acid, a weak acid, when they bite or sting. Describe the type of compound that should be used to treat the bite.

20 **Making Comparisons** Organic compounds are also covalent compounds. What properties would you expect organic compounds to have as a result?

21 **Applying Concepts** Farmers have been known to taste their soil to determine whether the soil has the correct acidity for their plants. How would taste help the farmer determine the acidity of the soil?

22 **Analyzing Ideas** A diet that includes a high level of lipids is unhealthy. Why is a diet containing no lipids also unhealthy?

INTERPRETING GRAPHICS

Use the structural formulas below to answer the questions that follow.

a.
```
        H
        |
      H-C-H
        |
   H    C    H
    \ // \ /
     C    C
     |    ||
     C    C
    / \\ / \
   H    C    H
        |
        H
```

b.
```
     H
     |
   H-C-H
     |
   H-C-H  H
     |    |
   H-C----C-H
     |    |
   H-C-H  H
     |
   H-C-H
     |
     H
```

c.
```
     H
     |
   H-C-H  H
     |    |
   H-C----C-H
     |    |
   H-C-H  H
     |
   H-C-H
     |
     H
```

d.
```
        H     H
         \   /
           C
           ||
     H     C
      \   / \
   H   C     H
    \ / \
     C   H
    / \
   H   H
```

23 A saturated hydrocarbon is represented by which structural formula(s)?

24 An unsaturated hydrocarbon is represented by which structural formula(s)?

25 An aromatic hydrocarbon is represented by which structural formula(s)?

Standardized Test Preparation

READING

Read each of the passages below. Then, answer the questions that follow each passage.

Passage 1 Spider webs often resemble a bicycle wheel. The "spokes" of the web are made of a silk thread called *dragline silk*. The sticky, stretchy part of the web is called *capture silk* because the spiders use this silk to capture their prey. Spider silk is made of proteins, and proteins are made of amino acids. There are 20 naturally occurring amino acids, but spider silk has only 7 of them. Scientists used a technique called *nuclear magnetic resonance* (NMR) to see the structure of dragline silk. The silk fiber is made of two tough strands of alanine-rich protein embedded in a glycine-rich substance. This protein resembles tangled spaghetti. Scientists believe that this tangled part makes the silk springy and that a repeating sequence of 5 amino acids makes the protein stretchy.

1. According to the passage, how many types of amino acids does spider silk contain?

- **A** 20 amino acids
- **B** 5 amino acids
- **C** 7 amino acids
- **D** all naturally occurring amino acids

2. Based on the passage, which of the following statements is a fact?

- **F** Capture silk makes up the "spokes" of the web.
- **G** The silk fiber is made of two strands of glycine-rich protein.
- **H** Proteins are made of amino acids.
- **I** Spider webs are strong because of a repeating sequence of 5 amino acids.

3. In this passage, what does *capture* mean?

- **A** to kill
- **B** to eat
- **C** to free
- **D** to trap

Passage 2 The earliest evidence of soapmaking dates back to 2,800 BCE. A soaplike material was found in clay cylinders in ancient Babylon. According to Roman legend, soap was named after Mount Sapo, where animals were sacrificed. A soaplike substance was made when rain washed the melted animal fat and wood ashes into the clay soil along the Tiber River. In 1791, a major step toward large scale commercial soapmaking began when Nicholas Leblanc, a French chemist, patented a process for making soda ash from salt. About 20 years later, the science of modern soapmaking was born. At that time, Michel Chevreul, another French chemist, discovered how fats, glycerin, and fatty acids interact. This interaction is the basis of saponification, or soap chemistry, today.

1. In this passage, what does *commercial* mean?

- **A** for advertising purposes
- **B** for public sale
- **C** from French manufacturers
- **D** for a limited time period

2. Based on the passage, which of the following statements is a fact?

- **F** Saponification is a process used to make soap.
- **G** The word *soap* probably originated from the French.
- **H** Modern soapmaking began around 1791.
- **I** Soapmaking began 2,000 years ago.

3. Which of the following statements is the best summary for the passage?

- **A** The process of soapmaking has a history of at least 4,000 years.
- **B** Most of the scientists responsible for soapmaking were from France.
- **C** Commercial soapmaking began in 1791.
- **D** Soap chemistry is called *saponification*.

INTERPRETING GRAPHICS

The diagram below shows a model of a water molecule (H_2O). Use the diagram to answer the questions that follow.

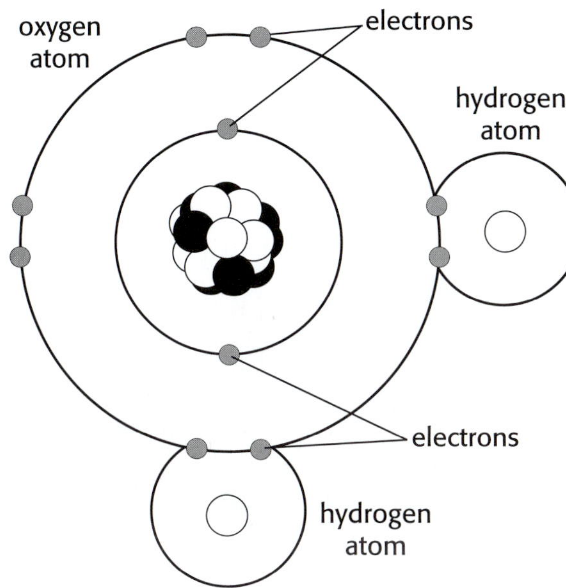

1. Which statement best describes the oxygen atom?
   A The oxygen atom has four valence electrons.
   B The oxygen atom is sharing two electrons with two other atoms.
   C The oxygen atom lost two electrons.
   D The oxygen atom has two valence electrons.

2. Which of the following features cannot be determined by looking at the model?
   F the number of atoms in the molecule
   G the number of electrons in each atom
   H the type of bonds joining the atoms
   I the physical state of the substance

3. Which statement best describes each of the smaller atoms in the molecule?
   A Each atom has eight total electrons.
   B Each atom has two protons.
   C Each atom has two valence electrons.
   D Each atom has lost two electrons.

MATH

Read each question below, and choose the best answer.

1. Marty is making 8 gal of lemonade. How much sugar does he need if 1 1/2 cups of sugar are needed for every 2 gal of lemonade?
   A 4 cups
   B 6 cups
   C 3 cups
   D 8 cups

2. The Jimenez family went to the science museum. They bought four tickets at $12.95 each, four snacks at $3 each, and two souvenirs at $7.95 each. What is the best estimate of the total cost of tickets, snacks, and souvenirs?
   F $24
   G $50
   H $63
   I $80

3. Each whole number on the pH scale represents a tenfold change in the concentration of hydronium ions. An acid that has a pH of 2 has how many times more hydronium ions than an acid that has a pH of 5?
   A 30 times more hydronium ions
   B 100 times more hydronium ions
   C 1,000 times more hydronium ions
   D 10,000 times more hydronium ions

Science in Action

Science, Technology, and Society

Molecular Photocopying

To learn about our human ancestors, scientists can use DNA from mummies. Well-preserved DNA can be copied using a technique called polymerase chain reaction (PCR). PCR uses enzymes called *polymerases,* which make new strands of DNA using old strands as templates. Thus, PCR is called molecular photocopying. However, scientists have to be very careful when using this process. If just one of their own skin cells falls into the PCR mixture, it will contaminate the ancient DNA with their own DNA.

Social Studies ACTIVITY

WRITING SKILL DNA analysis of mummies is helping archeologists study human history. Write a research paper about what scientists have learned about human history through DNA analysis.

Weird Science

Silly Putty™

During World War II, the supply of natural rubber was very low. So, James Wright, at General Electric, tried to make a synthetic rubber. The putty he made could be molded, stretched, and bounced. But it did not work as a rubber substitute and was ignored. Then, Peter Hodgson, a consultant for a toy company, had a brilliant idea. He marketed the putty as a toy in 1949. It was an immediate success. Hodgson created the name Silly Putty™. Although Silly Putty™ was invented more than 50 years ago, it has not changed much. More than 300 million eggs of Silly Putty have been sold since 1950.

Math ACTIVITY

In 1949, Mr. Hodgson bought 9.5 kg of putty for $147. The putty was divided into balls, each having a mass of 14 g. What was his cost for one 14 g ball of putty?

Careers

Jeannie Eberhardt

Forensic Scientist Jeannie Eberhardt says that her job as a forensic scientist is not really as glamorous as it may seem on popular TV shows. "If they bring me a garbage bag from the crime scene, then my job is to dig through the trash and look for evidence," she laughs. Jeannie Eberhardt explains that her job is to "search for, collect, and analyze evidence from crime scenes." Eberhardt says that one of the most important qualities a forensic scientist can have is the ability to be unbiased. She says that she focuses on the evidence and not on any information she may have about the alleged crime or the suspect. Eberhardt advises students who think they might be interested in a career as a forensic scientist to talk to someone who works in the field. She also recommends that students develop a broad science background. And she advises students that most of these jobs require extensive background checks. "Your actions now could affect your ability to get a job later on," she points out.

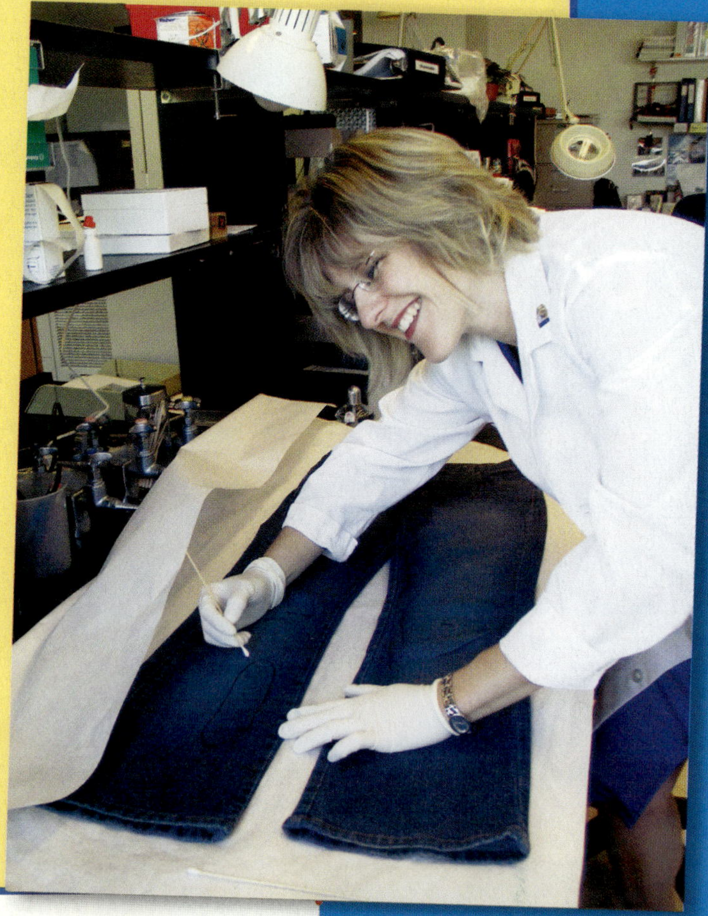

Language Arts ACTIVITY

WRITING SKILL Jeannie Eberhardt says that it is very important to be unbiased when analyzing a crime scene. Write a one-page essay explaining why it is necessary to focus on the evidence in a crime and not on personal feelings or news reports.

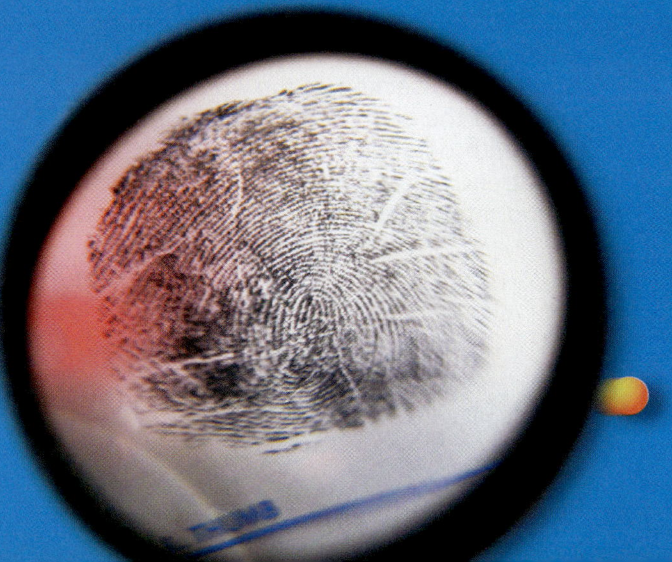

To learn more about these Science in Action topics, visit go.hrw.com and type in the keyword HP5CMPF

Current Science

Check out Current Science® articles related to this chapter by visiting go.hrw.com. Just type in the keyword HP5CS15

The Nature of Sound

About the PHOTO

Look at these dolphins swimming swiftly and silently through their watery world. Wait a minute—swiftly? Yes. Silently? No way! Dolphins use sound—clicks, squeaks, and other noises—to communicate. Dolphins also use sound to locate their food by echolocation and to find their way through murky water.

PRE-READING ACTIVITY

Graphic Organizer

Concept Map Before you read the chapter, create the graphic organizer entitled "Concept Map" described in the **Study Skills** section of the Appendix. As you read the chapter, fill in the concept map with details about each type of sound interaction.

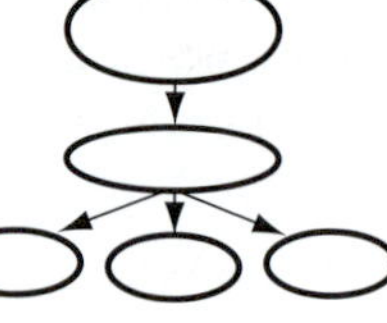

START-UP ACTIVITY

A Homemade Guitar

In this chapter, you will learn about sound. You can start by making your own guitar. It won't sound as good as a real guitar, but it will help you explore the nature of sound.

Procedure

1. Stretch a **rubber band** lengthwise around an empty **shoe box.** Place the box hollow side up. Pluck the rubber band gently. Describe what you hear.
2. Stretch **another rubber band of a different thickness** around the box. Pluck both rubber bands. Describe the differences in the sounds.
3. Put a **pencil** across the center of the box and under the rubber bands, and pluck again. Compare this sound with the sound you heard before the pencil was used.
4. Move the pencil closer to one end of the shoe box. Pluck on both sides of the pencil. Describe the differences in the sounds you hear.

Analysis

1. How did the thicknesses of the rubber bands affect the sound?
2. In steps 3 and 4, you changed the length of the vibrating part of the rubber bands. What is the relationship between the vibrating length of the rubber band and the sound that you hear?

SECTION 1

What Is Sound?

You are in a restaurant, and without warning, you hear a loud crash. A waiter dropped a tray of dishes. What a mess! But why did dropping the dishes make such a loud sound?

In this section, you'll find out what causes sound and what characteristics all sounds have in common. You'll also learn how your ears detect sound and how you can protect your hearing.

READING WARM-UP

Objectives

- Describe how vibrations cause sound.
- Explain how sound is transmitted through a medium.
- Explain how the human ear works, and identify its parts.
- Identify ways to protect your hearing.

Terms to Learn

sound wave
medium

READING STRATEGY

Prediction Guide Before reading this section, predict whether each of the following statements is true or false:

- Sound waves are made by vibrations.
- Sound waves push air particles along until they reach your ear.

Sound and Vibrations

As different as they are, all sounds have some things in common. One characteristic of sound is that it is created by vibrations. A *vibration* is the complete back-and-forth motion of an object. **Figure 1** shows one way sound is made by vibrations.

Figure 1 Sounds from a Stereo Speaker

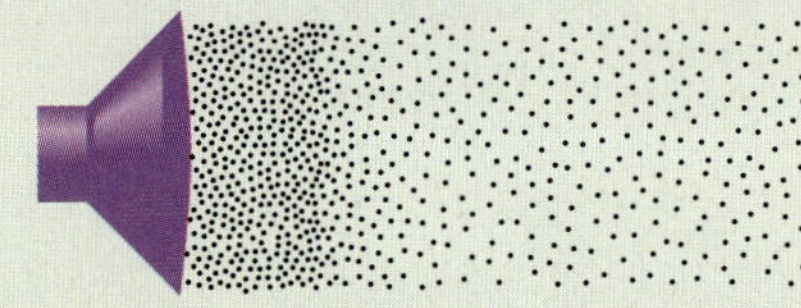

a Electrical signals make the speaker vibrate. As the speaker cone moves forward, it pushes the air particles in front of it closer together, creating a region of higher density and pressure called a *compression.*

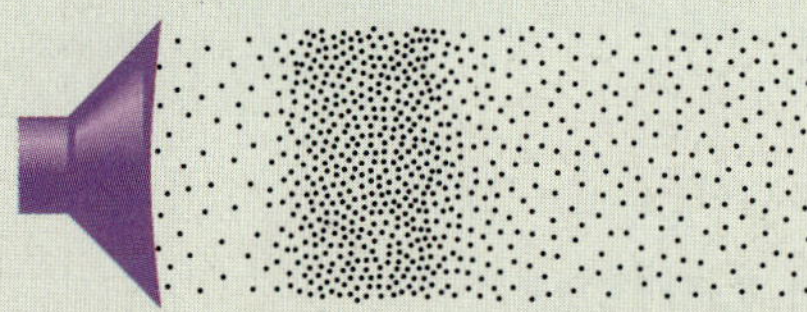

b As the speaker cone moves backward, air particles close to the cone become less crowded, creating a region of lower density and pressure called a *rarefaction.*

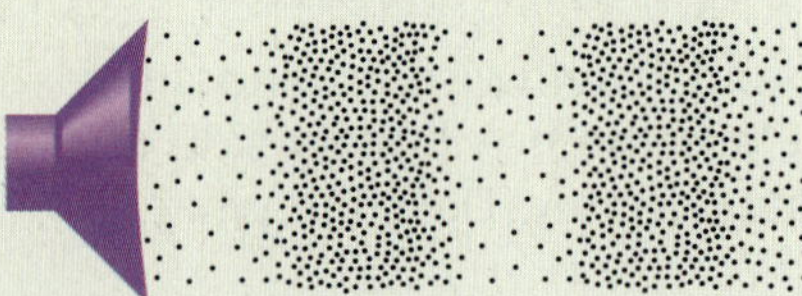

c For each vibration, a compression and a rarefaction are formed. As the compressions and rarefactions travel away from the speaker, sound is transmitted through the air.

Figure 2 *You can't actually see sound waves, but they can be represented by spheres that spread out in all directions.*

Sound Waves

Longitudinal (LAHN juh TOOD'n uhl) waves are made of compressions and rarefactions. A **sound wave** is a longitudinal wave caused by vibrations and carried through a substance. The particles of the substance, such as air particles, vibrate back and forth along the path that the sound wave travels. Sound is transmitted through the vibrations and collisions of the particles. Because the particles vibrate back and forth along the paths that sound travels, sound travels as longitudinal waves.

sound wave a longitudinal wave that is caused by vibrations and that travels through a material medium

Sound waves travel in all directions away from their source, as shown in **Figure 2.** However, air or other matter does not travel with the sound waves. The particles of air only vibrate back and forth. If air did travel with sound, wind gusts from music speakers would blow you over at a school dance!

Reading Check **What do sound waves consist of?** *(See the Appendix for answers to Reading Checks.)*

Good Vibrations

1. Gently strike a **tuning fork** on a **rubber eraser.** Watch the prongs, and listen for a sound. Describe what you see and what you hear.
2. Lightly touch the fork with your fingers. What do you feel?
3. Grasp the prongs of the fork firmly with your hand. What happens to the sound?
4. Strike the tuning fork on the eraser again, and dip the prongs in a **cup of water.** Describe what happens to the water.
5. Record your observations.

Figure 3 Tubing is connected to a pump that is removing air from the jar. As the air is removed, the ringing alarm clock sounds quieter and quieter.

Sound and Media

medium a physical environment in which phenomena occur

Another characteristic of sound is that all sound waves require a medium (plural, *media*). A **medium** is a substance through which a wave can travel. Most of the sounds that you hear travel through air at least part of the time. But sound waves can also travel through other materials, such as water, glass, and metal.

In a vacuum, however, there are no particles to vibrate. So, no sound can be made in a vacuum. This fact helps to explain the effect described in **Figure 3.** Sound must travel through air or some other medium to reach your ears and be detected.

Reading Check What does sound need in order to travel?

Vocal Sounds The vibrations that produce your voice are made inside your throat. When you speak, laugh, or sing, your lungs force air up your windpipe, causing your vocal cords to vibrate.

Do some research, and find out what role different parts of your throat and mouth play in making vocal sounds. Make a poster in which you show the different parts, and explain the role they play in shaping sound waves.

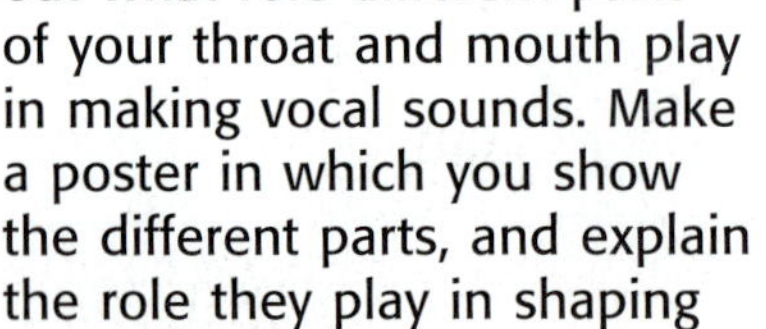

How You Detect Sound

Imagine that you are watching a suspenseful movie. Just before a door is opened, the background music becomes louder. You know that there is something scary behind that door! Now, imagine watching the same scene without the sound. You would have more difficulty figuring out what's going on if there were no sound.

Figure 4 shows how your ears change sound waves into electrical signals that allow you to hear. First, the outer ear collects sound waves. The vibrations then go to your middle ear. Very small organs increase the size of the vibrations here. These vibrations are then picked up by organs in your inner ear. Your inner ear changes vibrations into electrical signals that your brain interprets as sound.

Figure 4 How the Human Ear Works

a The **outer ear** acts as a funnel for sound waves. The *pinna* collects sound waves and directs them into the *ear canal.*

b In the **middle ear,** three bones—the *hammer, anvil,* and *stirrup*—act as levers to increase the size of the vibrations.

c In the **inner ear,** vibrations created by sound are changed into electrical signals for the brain to interpret.

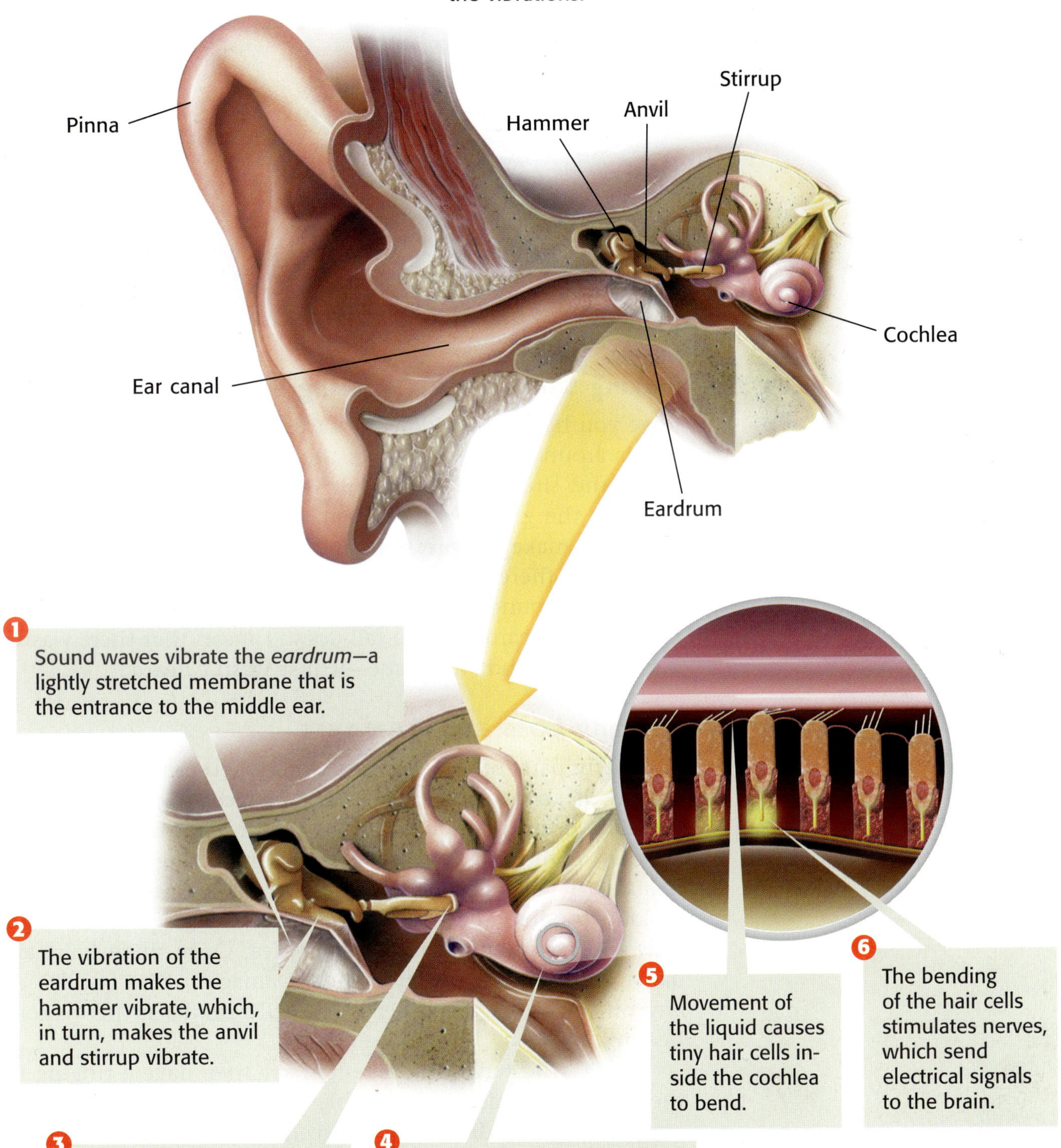

1 Sound waves vibrate the *eardrum*—a lightly stretched membrane that is the entrance to the middle ear.

2 The vibration of the eardrum makes the hammer vibrate, which, in turn, makes the anvil and stirrup vibrate.

3 The stirrup vibrates the *oval window*—the entrance to the inner ear.

4 The vibrations of the oval window create waves in the liquid inside the *cochlea.*

5 Movement of the liquid causes tiny hair cells inside the cochlea to bend.

6 The bending of the hair cells stimulates nerves, which send electrical signals to the brain.

Figure 5 *Sound is made whether or not anyone is around to hear it.*

Making Sound Versus Hearing Sound

Have you heard this riddle? If a tree falls in the forest and no one is around to hear it, does the tree make a sound? Think about the situation pictured in **Figure 5.** When a tree falls and hits the ground, the tree and the ground vibrate. These vibrations make compressions and rarefactions in the surrounding air. So, there would be a sound!

Making sound is separate from detecting sound. The fact that no one heard the tree fall doesn't mean that there wasn't a sound. A sound was made—it just wasn't heard.

INTERNET ACTIVITY

For another activity related to this chapter, go to **go.hrw.com** and type in the keyword **HP5SNDW.**

Hearing Loss and Deafness

The many parts of the ear must work together for you to hear sounds. If any part of the ear is damaged or does not work properly, hearing loss or deafness may result.

One of the most common types of hearing loss is called *tinnitus* (ti NIET us), which results from long-term exposure to loud sounds. Loud sounds can cause damage to the hair cells and nerve endings in the cochlea. Once these hairs are damaged, they do not grow back. Damage to the cochlea or any other part of the inner ear usually results in permanent hearing loss.

People who have tinnitus often say they have a ringing in their ears. They also have trouble understanding other people and hearing the difference between words that sound alike. Tinnitus can affect people of any age. Fortunately, tinnitus can be prevented.

Reading Check What causes tinnitus?

Protecting Your Hearing

Short exposures to sounds that are loud enough to be painful can cause hearing loss. Your hearing can also be damaged by loud sounds that are not quite painful, if you are exposed to them for long periods of time. There are some simple things you can do to protect your hearing. Loud sounds can be blocked out by earplugs. You can listen at a lower volume when you are using headphones, as in **Figure 6.** You can also move away from loud sounds. If you are near a speaker playing loud music, just move away from it. When you double the distance between yourself and a loud sound, the sound's intensity to your ears will be one-fourth of what it was before.

Figure 6 *Turning your radio down can help prevent hearing loss, especially when you use headphones.*

SECTION Review

Summary

- All sounds are generated by vibrations.
- Sounds travel as longitudinal waves consisting of compressions and rarefactions.
- Sound waves travel in all directions away from their source.
- Sound waves require a medium through which to travel. Sound cannot travel in a vacuum.
- Your ears convert sound into electrical impulses that are sent to your brain.
- Exposure to loud sounds can cause hearing damage.
- Using earplugs and lowering the volume of sounds can prevent hearing damage.

Using Key Terms

1. Use the following terms in the same sentence: *sound wave* and *medium.*

Understanding Key Ideas

2. Sound travels as
   a. transverse waves.
   b. longitudinal waves.
   c. shock waves.
   d. airwaves.
3. Which part of the ear increases the size of the vibrations of sound waves entering the ear?
   a. outer ear
   b. ear canal
   c. middle ear
   d. inner ear
4. Name two ways of protecting your hearing.

Critical Thinking

5. **Analyzing Processes** Explain why a person at a rock concert will not feel gusts of wind coming out of the speakers.
6. **Analyzing Ideas** If a meteorite crashed on the moon, would you be able to hear it on Earth? Why, or why not?
7. **Identifying Relationships** Recall the breaking dishes mentioned at the beginning of this section. Why was the sound that they made so loud?

Interpreting Graphics

Use the diagram of a wave below to answer the questions that follow.

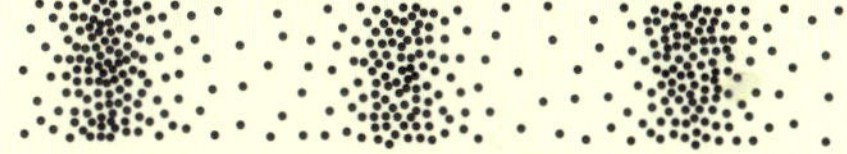

8. What kind of wave is this?
9. Draw a sketch of the diagram on a separate sheet of paper, and label the compressions and rarefactions.
10. How do vibrations make these kinds of waves?

SECTION 2

Properties of Sound

Imagine that you are swimming in a neighborhood pool. You can hear the high, loud laughter of small children and the soft splashing of the waves at the edge of the pool.

Why are some sounds loud, soft, high, or low? The differences between sounds depend on the properties of the sound waves. In this section, you will learn about properties of sound.

READING WARM-UP

Objectives

- Compare the speed of sound in different media.
- Explain how frequency and pitch are related.
- Describe the Doppler effect, and give examples of it.
- Explain how amplitude and loudness are related.
- Describe how amplitude and frequency can be "seen" on an oscilloscope.

Terms to Learn

pitch
Doppler effect
loudness
decibel

READING STRATEGY

Reading Organizer As you read this section, create an outline of the section. Use the headings from the section in your outline.

The Speed of Sound

Suppose you are standing at one end of a pool and two people from the opposite end of the pool yell at the same time. You would hear their voices at the same time. The reason is that the speed of sound depends only on the medium in which the sound is traveling. So, you would hear them at the same time—even if one person yelled louder!

How the Speed of Sound Can Change

Table 1 shows how the speed of sound varies in different media. Sound travels quickly through air, but it travels even faster in liquids and even faster in solids.

Temperature also affects the speed of sound. In general, the cooler the medium is, the slower the speed of sound. Particles of cool materials move more slowly and transmit energy more slowly than particles do in warmer materials. In 1947, pilot Chuck Yeager became the first person to travel faster than the speed of sound. Yeager flew the airplane shown in **Figure 1** at 293 m/s (about 480 mi/h) at 12,000 m above sea level. At that altitude, the temperature of the air is so low that the speed of sound is only 290 m/s.

Table 1 Speed of Sound in Different Media

Medium	Speed (m/s)
Air (0°C)	331
Air (20°C)	343
Air (100°C)	366
Water (20°C)	1,482
Steel (20°C)	5,200

Figure 1 *The X-1 airplane was the first vehicle to move faster than the speed of sound.*

Figure 2 Frequency and Pitch

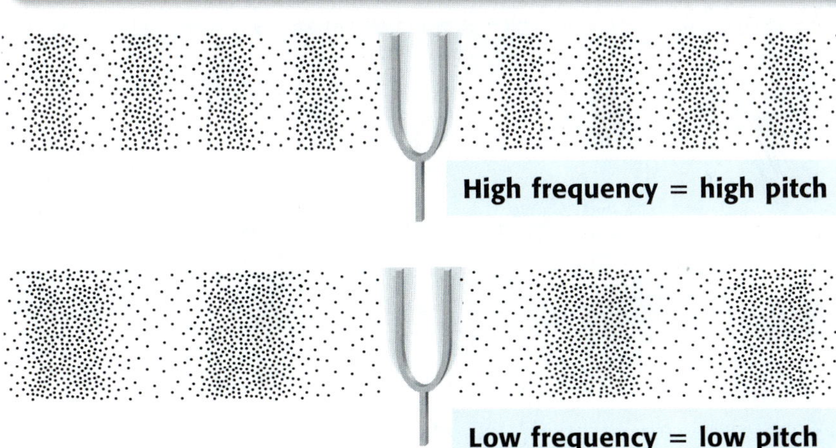

Pitch and Frequency

How low or high a sound seems to be is the **pitch** of that sound. The *frequency* of a wave is the number of crests or troughs that are made in a given time. The pitch of a sound is related to the frequency of the sound wave, as shown in **Figure 2.** Frequency is expressed in hertz (Hz), where 1 Hz = 1 wave per second. For example, the lowest note on a piano is about 40 Hz. The screech of a bat is 10,000 Hz or higher.

pitch a measure of how high or low a sound is perceived to be, depending on the frequency of the sound wave

Reading Check What is frequency? (*See the Appendix for answers to Reading Checks.*)

Frequency and Hearing

If you see someone blow a dog whistle, the whistle seems silent to you. The reason is that the frequency of the sound wave is out of the range of human hearing. But the dog hears the whistle and comes running! **Table 2** compares the range of frequencies that humans and animals can hear. Sounds that have a frequency too high for people to hear are called *ultrasonic.*

Table 2 **Frequencies Heard by Different Animals**

Animal	Frequency range (Hz)
Bat	2,000 to 110,000
Porpoise	75 to 150,000
Cat	45 to 64,000
Beluga whale	1,000 to 123,000
Elephant	16 to 12,000
Human	20 to 20,000
Dog	67 to 45,000

The Speed of Sound

The speed of sound depends on the medium through which sound is traveling and the medium's temperature. Sound travels at 343 m/s through air at a temperature of 20°C. How far will sound travel in 20°C air in 5 s?

The speed of sound in steel at 20°C is 5,200 m/s. How far can sound travel in 5 s through steel at 20°C?

Figure 3 **The Doppler Effect**

a A car with its horn honking moves toward the sound waves going in the same direction. A person in front of the car hears sound waves that are closer together.

b The car moves away from the sound waves going in the opposite direction. A person behind the car hears sound waves that are farther apart and have a lower frequency.

The Doppler Effect

Doppler effect an observed change in the frequency of a wave when the source or observer is moving

Have you ever been passed by a car with its horn honking? If so, you probably noticed the sudden change in pitch—sort of an *EEEEEOOoooowwn* sound—as the car went past you. The pitch you heard was higher as the car moved toward you than it was after the car passed. This higher pitch was a result of the Doppler effect. For sound waves, the **Doppler effect** is the apparent change in the frequency of a sound caused by the motion of either the listener or the source of the sound. **Figure 3** shows how the Doppler effect works.

In a moving sound source, such as a car with its horn honking, sound waves that are moving forward are going the same direction the car is moving. As a result, the compressions and rarefactions of the sound wave will be closer together than they would be if the sound source was not moving. To a person in front of the car, the frequency and pitch of the sound seem high. After the car passes, it is moving in the opposite direction that the sound waves are moving. To a person behind the car, the frequency and pitch of the sound seem low. The driver always hears the same pitch because the driver is moving with the car.

Loudness and Amplitude

If you gently tap a drum, you will hear a soft rumbling. But if you strike the drum with a large force, you will hear a much louder sound! By changing the force you use to strike the drum, you change the loudness of the sound that is created. **Loudness** is a measure of how well a sound can be heard.

Energy and Vibration

Look at **Figure 4.** The harder you strike a drum, the louder the boom. As you strike the drum harder, you transfer more energy to the drum. The drum moves with a larger vibration and transfers more energy to the air around it. This increase in energy causes air particles to vibrate farther from their rest positions.

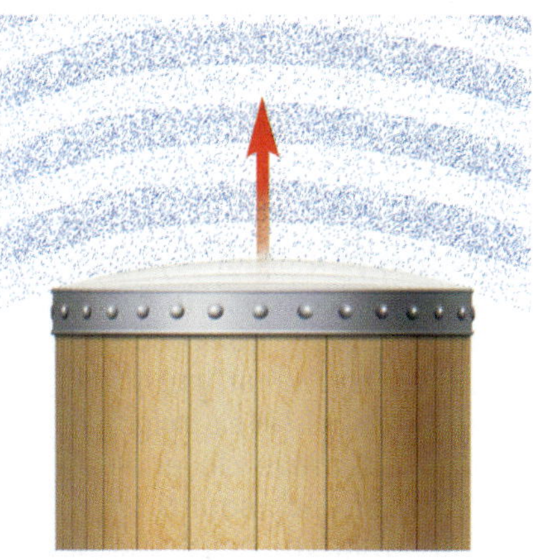

Figure 4 *When a drum is struck hard, it vibrates with a lot of energy, making a loud sound.*

Increasing Amplitude

When you strike a drum harder, you are increasing the amplitude of the sound waves being made. The *amplitude* of a wave is the largest distance the particles in a wave vibrate from their rest positions. The larger the amplitude, the louder the sound. And the smaller the amplitude, the softer the sound. One way to increase the loudness of a sound is to use an amplifier, shown in **Figure 5.** An amplifier receives sound signals in the form of electric current. The amplifier then increases the energy and makes the sound louder.

Reading Check **What is the relationship between the amplitude of a sound and its energy of vibration?**

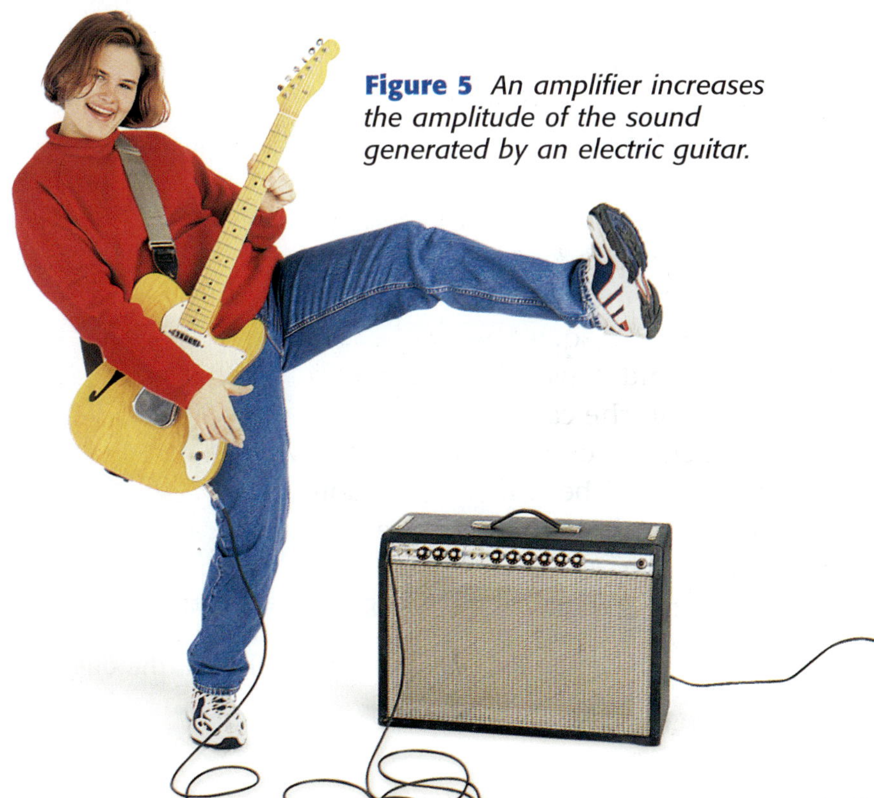

Figure 5 *An amplifier increases the amplitude of the sound generated by an electric guitar.*

loudness the extent to which a sound can be heard

Sounding Board

1. With one hand, hold a **ruler** on your **desk** so that one end of it hangs over the edge.
2. With your other hand, pull the free end of the ruler up a few centimeters, and let go.
3. Try pulling the ruler up different distances. How does the distance affect the sounds you hear? What property of the sound wave are you changing?
4. Change the length of the part that hangs over the edge. What property of the sound wave is affected? Record your answers and observations.

WRITING SKILL **Decibel Levels**

With a parent, listen for the normal sounds that happen around your house. In your **science journal,** write down some sounds and what you think their decibel levels might be. Then, move closer to the source of each sound, and write what you think the new decibel level of each is.

Table 3 Decibel Levels of Common Sounds

Decibel level	Sound
0	the softest sounds you can hear
20	whisper
25	purring cat
60	normal conversation
80	lawn mower, vacuum cleaner, truck traffic
100	chain saw, snowmobile
115	sandblaster, loud rock concert, automobile horn
120	threshold of pain
140	jet engine 30 m away
200	rocket engine 50 m away

decibel the most common unit used to measure loudness (symbol, dB)

Measuring Loudness

The most common unit used to express loudness is the **decibel** (dB). The softest sounds an average human can hear are at a level of 0 dB. Sounds that are at 120 dB or higher can be painful. **Table 3** shows some common sounds and their decibel levels.

"Seeing" Amplitude and Frequency

Sound waves are invisible. However, technology can provide a way to "see" sound waves. A device called an *oscilloscope* (uh SIL uh SKOHP) can graph representations of sound waves, as shown in **Figure 6.** Notice that the graphs look like transverse waves instead of longitudinal waves.

Reading Check What does an oscilloscope do?

Figure 6 "Seeing" Sounds

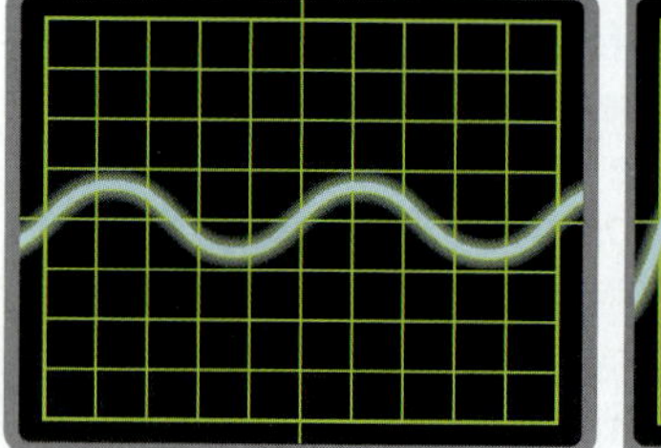

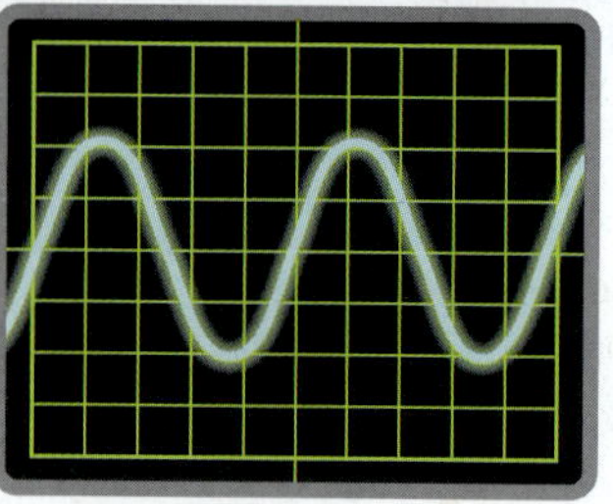

The graph on the right has a **larger amplitude** than the graph on the left. So, the sound represented on the right is **louder** than the one represented on the left.

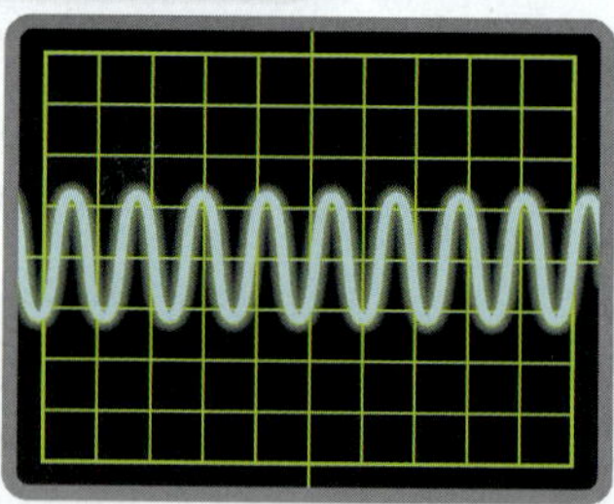

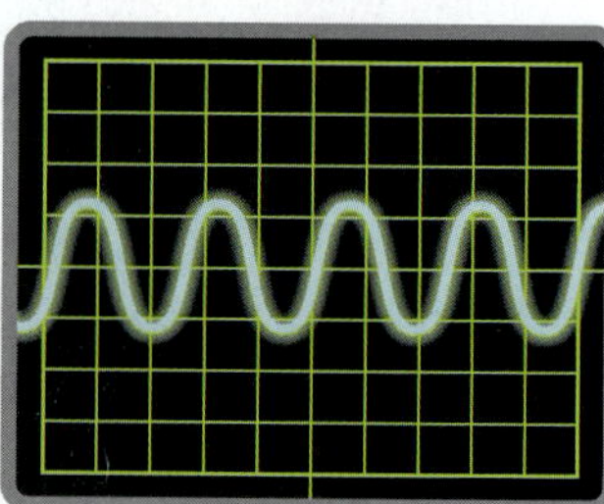

The graph on the right has a **lower frequency** than the one on the left. So, the sound represented on the right has a **lower pitch** than the one represented on the left.

From Sound to Electrical Signal

An oscilloscope is shown in **Figure 7.** A microphone is attached to the oscilloscope and changes a sound wave into an electrical signal. The electrical signal is graphed on the screen in the form of a wave. The graph shows the sound as if it were a transverse wave. So, the sound's amplitude and frequency are easier to see. The highest points (crests) of these waves represent compressions, and the lowest points (troughs) represent rarefactions. By looking at the displays on the oscilloscope, you can quickly see the differences in amplitude and frequency of different sound waves.

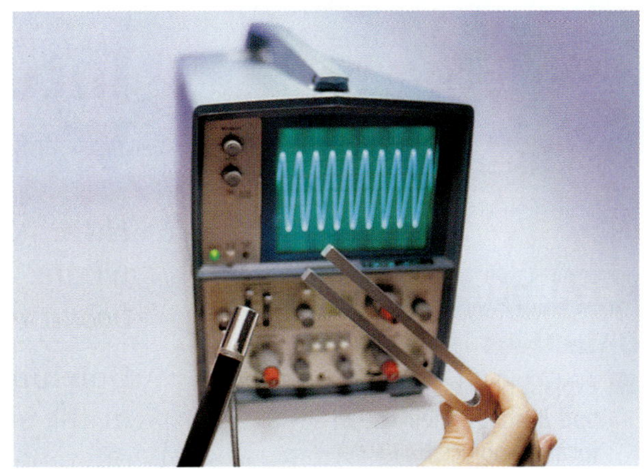

Figure 7 *An oscilloscope can be used to represent sounds.*

SECTION Review

Summary

- The speed of sound depends on the medium and the temperature.
- The pitch of a sound becomes higher as the frequency of the sound wave becomes higher. Frequency is expressed in units of Hertz (Hz), which is equivalent to waves per second.
- The Doppler effect is the apparent change in frequency of a sound caused by the motion of either the listener or the source of the sound.
- Loudness increases with the amplitude of the sound. Loudness is expressed in decibels.
- The amplitude and frequency of a sound can be measured electronically by an oscilloscope.

Using Key Terms

1. In your own words, write a definition for the term *pitch.*
2. Use the following terms in the same sentence: *loudness* and *decibel.*

Understanding Key Ideas

3. At the same temperature, in which medium does sound travel fastest?
   a. air
   b. liquid
   c. solid
   d. It travels at the same speed through all media.
4. In general, how does the temperature of a medium affect the speed of sound through that medium?
5. What property of waves affects the pitch of a sound?
6. How does an oscilloscope allow sound waves to be "seen"?

Math Skills

7. You see a distant flash of lightning, and then you hear a thunderclap 2 s later. The sound of the thunder moves at 343 m/s. How far away was the lightning?
8. In water that is near 0°C, a submarine sends out a sonar signal (a sound wave). The signal travels 1500 m/s and reaches an underwater mountain in 4 s. How far away is the mountain?

Critical Thinking

9. **Analyzing Processes** Will a listener notice the Doppler effect if both the listener and the source of the sound are traveling toward each other? Explain your answer.
10. **Predicting Consequences** A drum is struck gently, then is struck harder. What will be the difference in the amplitude of the sounds made? What will be the difference in the frequency of the sounds made?

SECTION 3

Interactions of Sound Waves

READING WARM-UP

Objectives

- Explain how echoes are made, and describe their use in locating objects.
- List examples of constructive and destructive interference of sound waves.
- Explain what resonance is.

Terms to Learn

echo	sonic boom
echolocation	standing wave
interference	resonance

READING STRATEGY

Paired Summarizing Read this section silently. In pairs, take turns summarizing the material. Stop to discuss ideas that seem confusing.

Have you ever heard of a sea canary? It's not a bird! It's a whale! Beluga whales are sometimes called sea canaries because of the many different sounds they make.

Dolphins, beluga whales, and many other animals that live in the sea use sound to communicate. Beluga whales also rely on reflected sound waves to find fish, crabs, and shrimp to eat. In this section, you will learn about reflection and other interactions of sound waves. You will also learn how bats, dolphins, and whales use sound to find food.

Reflection of Sound Waves

Reflection is the bouncing back of a wave after it strikes a barrier. You're probably already familiar with a reflected sound wave, otherwise known as an **echo.** The strength of a reflected sound wave depends on the reflecting surface. Sound waves reflect best off smooth, hard surfaces. Look at **Figure 1.** A shout in an empty gymnasium can produce an echo, but a shout in an auditorium usually does not.

The difference is that the walls of an auditorium are usually designed so that they absorb sound. If sound waves hit a flat, hard surface, they will reflect back. Reflection of sound waves doesn't matter much in a gymnasium. But you don't want to hear echoes while listening to a musical performance!

echo a reflected sound wave

Figure 1 Sound Reflection and Absorption

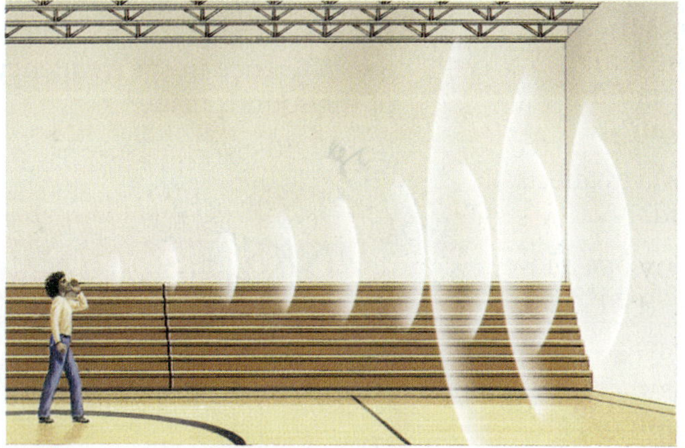

Sound waves easily reflect off the smooth, hard walls of a gymnasium. For this reason, you hear an echo.

In well-designed auditoriums, echoes are reduced by soft materials that absorb sound waves and by irregular shapes that scatter sound waves.

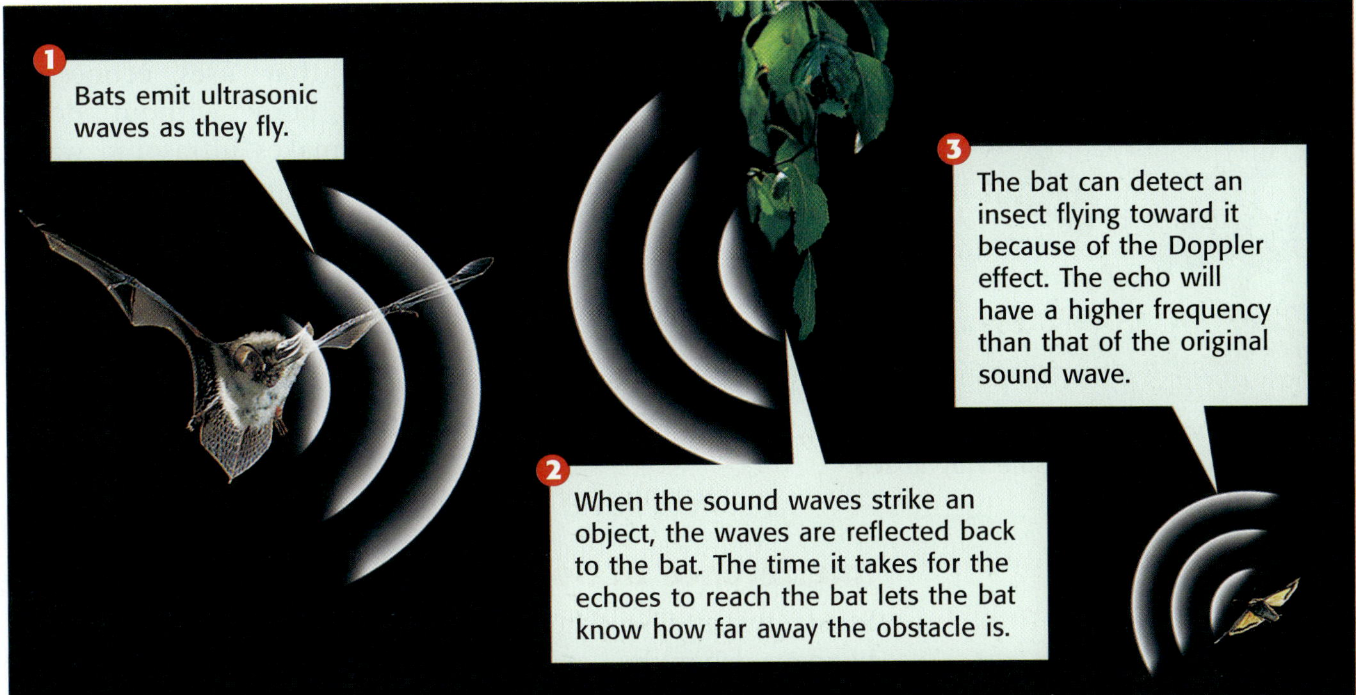

Figure 2 *Bats use echolocation to navigate around barriers and to find insects to eat.*

Echolocation

Beluga whales use echoes to find food. The use of reflected sound waves to find objects is called **echolocation.** Other animals—such as dolphins, bats, and some kinds of birds—also use echolocation to hunt food and to find objects in their paths. **Figure 2** shows how echolocation works. Animals that use echolocation can tell how far away something is based on how long it takes sound waves to echo back to their ears. Some animals, such as bats, also make use of the Doppler effect to tell if another moving object, such as an insect, is moving toward it or away from it.

echolocation the process of using reflected sound waves to find objects; used by animals such as bats

Reading Check **How is echolocation useful to some animals?** (*See the Appendix for answers to Reading Checks.*)

Echolocation Technology

People use echoes to locate objects underwater by using sonar (which stands for **so**und **n**avigation **a**nd **r**anging). *Sonar* is a type of electronic echolocation. **Figure 3** shows how sonar works. Ultrasonic waves are used because their short wavelengths give more details about the objects they reflect off. Sonar can also help navigators on ships avoid icebergs and can help oceanographers map the ocean floor.

Figure 3 *A fish finder sends ultrasonic waves down into the water. The time it takes for the echo to return helps determine the location of the fish.*

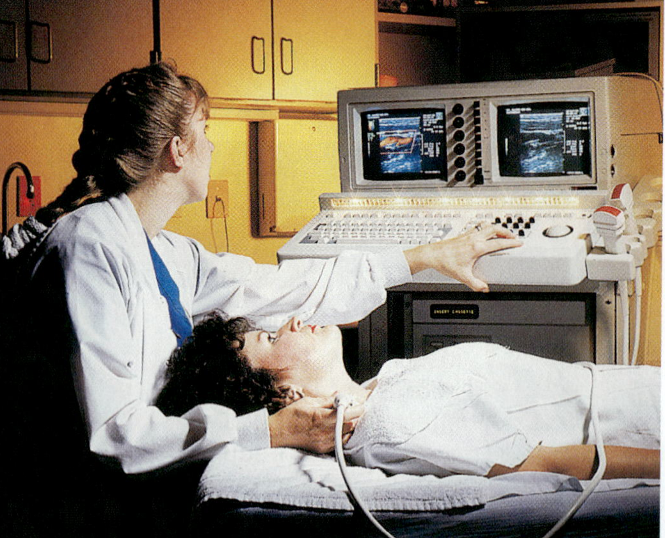

Figure 4 *Images created by ultrasonography are fuzzy, but they are a safe way to see inside a patient's body.*

Ultrasonography

Ultrasonography (UHL truh soh NAHG ruh fee) is a medical procedure that uses echoes to "see" inside a patient's body without doing surgery. A special device makes ultrasonic waves with a frequency that can be from 1 million to 10 million hertz, which reflect off the patient's internal organs. These echoes are then changed into images that can be seen on a television screen, as shown in **Figure 4.** Ultrasonography is used to examine kidneys, gallbladders, and other organs. It is also used to check the development of an unborn baby in a mother's body. Ultrasonic waves are less harmful to human tissue than X rays are.

Interference of Sound Waves

interference the combination of two or more waves that results in a single wave

sonic boom the explosive sound heard when a shock wave from an object traveling faster than the speed of sound reaches a person's ears

Sound waves also interact through interference. **Interference** happens when two or more waves overlap. **Figure 5** shows how two sound waves can combine by both constructive and destructive interference.

Orchestras and bands make use of constructive interference when several instruments of the same kind play the same notes. Interference of the sound waves causes the combined amplitude to increase, resulting in a louder sound. But destructive interference may keep some members of the audience from hearing the concert well. In certain places in an auditorium, sound waves reflecting off the walls interfere destructively with the sound waves from the stage.

Reading Check What are the two kinds of sound wave interference?

Figure 5 Constructive and Destructive Interference

Sound waves from two speakers producing sound of the same frequency combine by both constructive and destructive interference.

Constructive Interference
As the compressions of one wave overlap the compressions of another wave, the sound will be louder because the amplitude is increased.

Destructive Interference
As the compressions of one wave overlap the rarefactions of another wave, the sound will be softer because the amplitude is decreased.

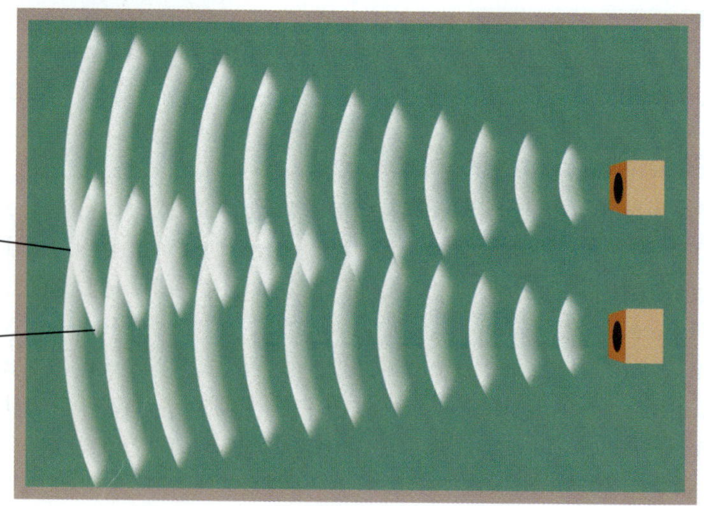

Interference and the Sound Barrier

As the source of a sound—such as a jet plane—gets close to the speed of sound, the sound waves in front of the jet plane get closer and closer together. The result is constructive interference. **Figure 6** shows what happens as a jet plane reaches the speed of sound.

For the jet in **Figure 6** to go faster than the speed of sound, the jet must overcome the pressure of the compressed sound waves. **Figure 7** shows what happens as soon as the jet reaches supersonic speeds—speeds faster than the speed of sound. At these speeds, the sound waves trail off behind the jet. At their outer edges, the sound waves combine by constructive interference to form a *shock wave.*

A **sonic boom** is the explosive sound heard when a shock wave reaches your ears. Sonic booms can be so loud that they can hurt your ears and break windows. They can even make the ground shake as it does during an earthquake.

Figure 6 *When a jet plane reaches the speed of sound, the sound waves in front of the jet combine by constructive interference. The result is a high-density compression that is called the sound barrier.*

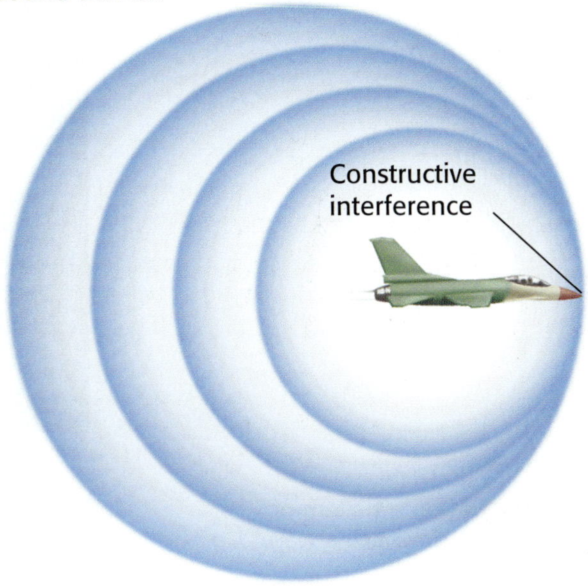

Figure 7 *When a jet travels at supersonic speeds, the sound waves it creates spread out behind it in a three-dimensional cone shape.*

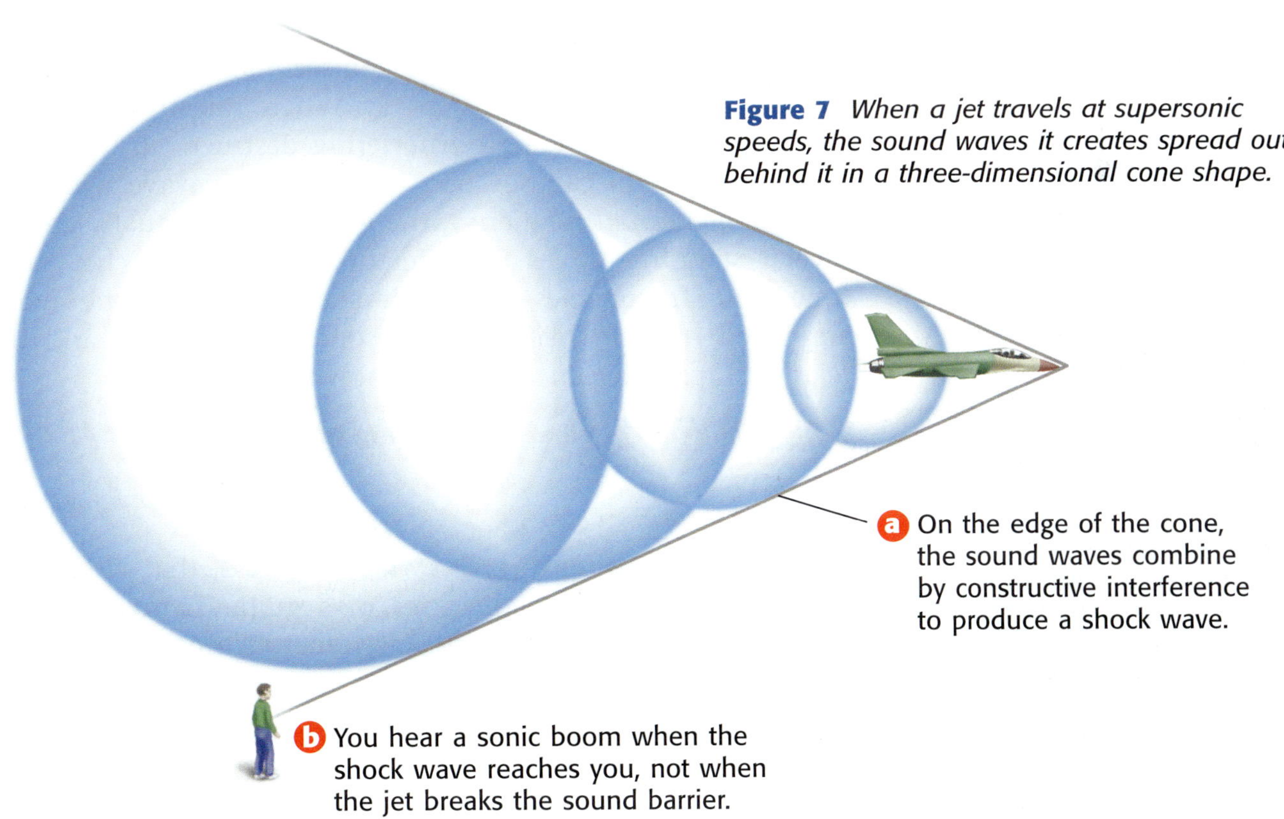

Figure 8 Resonant Frequencies of a Plucked String

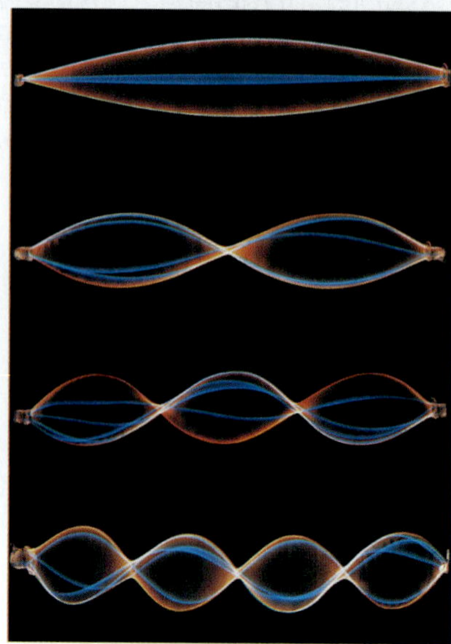

The lowest resonant frequency is called the *fundamental*.

Higher resonant frequencies are called *overtones*. The first overtone is twice the frequency of the fundamental.

The second overtone is 3 times the fundamental.

The third overtone is 4 times the fundamental.

standing wave a pattern of vibration that simulates a wave that is standing still

resonance a phenomenon that occurs when two objects naturally vibrate at the same frequency; the sound produced by one object causes the other object to vibrate

Interference and Standing Waves

When you play a guitar, you can make some pleasing sounds, and you might even play a tune. But have you ever watched a guitar string after you've plucked it? You may have noticed that the string vibrates as a standing wave. A **standing wave** is a pattern of vibration that looks like a wave that is standing still. Waves and reflected waves of the same frequency are going through the string. Where you see maximum amplitude, waves are interfering constructively. Where the string seems to be standing still, waves are interfering destructively.

Although you can see only one standing wave, which is at the *fundamental* frequency, the guitar string actually creates several standing waves of different frequencies at the same time. The frequencies at which standing waves are made are called *resonant frequencies*. Resonant frequencies and the relationships between them are shown in **Figure 8.**

Reading Check What is a standing wave?

Resonance

If you have a tuning fork, shown in **Figure 9,** that vibrates at one of the resonant frequencies of a guitar string, you can make the string make a sound without touching it. Strike the tuning fork, and hold it close to the string. The string will start to vibrate and produce a sound.

Using the vibrations of the tuning fork to make the string vibrate is an example of resonance. **Resonance** happens when an object vibrating at or near a resonant frequency of a second object causes the second object to vibrate.

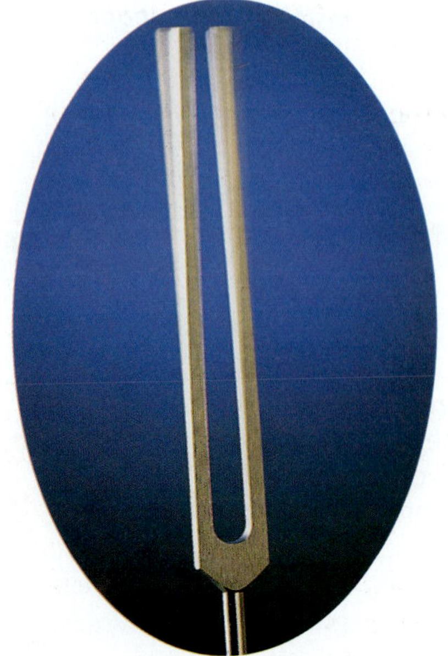

Figure 9 *When struck, a tuning fork can make another object vibrate if they both have the same resonant frequency.*

Resonance in Musical Instruments

Musical instruments use resonance to make sound. In wind instruments, vibrations are caused by blowing air into the mouthpiece. The vibrations make a sound, which is amplified when it forms a standing wave inside the instrument.

String instruments also resonate when they are played. An acoustic guitar, such as the one shown in **Figure 10,** has a hollow body. When the strings vibrate, sound waves enter the body of the guitar. Standing waves form inside the body of the guitar, and the sound is amplified.

Figure 10 *The body of a guitar resonates when the guitar is strummed.*

SECTION Review

Summary

- Echoes are reflected sound waves.
- Some animals can use echolocation to find food or to navigate around objects.
- People use echolocation technology in many underwater applications.
- Ultrasonography uses sound reflection for medical applications.
- Sound barriers and shock waves are created by interference.
- Standing waves form at an object's resonant frequencies.
- Resonance happens when a vibrating object causes a second object to vibrate at one of its resonant frequencies.

Using Key Terms

1. Use the following terms in the same sentence: *echo* and *echolocation*.

Complete each of the following sentences by choosing the correct term from the word bank.

interference   standing wave
sonic boom   resonance

2. When you pluck a string on a musical instrument, a(n) _____ forms.
3. When a vibrating object causes a nearby object to vibrate, _____ results.

Understanding Key Ideas

4. What causes an echo?
   - **a.** reflection
   - **b.** resonance
   - **c.** constructive interference
   - **d.** destructive interference
5. Describe a place in which you would expect to hear echoes.
6. How do bats use echoes to find insects to eat?
7. Give one example each of constructive and destructive interference of sound waves.

Math Skills

8. Sound travels through air at 343 m/s at 20°C. A bat emits an ultrasonic squeak and hears the echo 0.05 s later. How far away was the object that reflected it? (Hint: Remember that the sound must travel *to* the object and *back to* the bat.)

Critical Thinking

9. **Applying Concepts** Your friend is playing a song on a piano. Whenever your friend hits a certain key, the lamp on top of the piano rattles. Explain why the lamp rattles.
10. **Making Comparisons** Compare sonar and ultrasonography in locating objects.

SECTION 4

Sound Quality

Have you ever been told that the music you really like is just a lot of noise? If you have, you know that people can disagree about the difference between noise and music.

READING WARM-UP

Objectives

- Explain why different instruments have different sound qualities.
- Describe how each family of musical instruments produces sound.
- Explain how noise is different from music.

Terms to Learn

sound quality
noise

READING STRATEGY

Reading Organizer As you read this section, make a table comparing the way different instruments produce sound.

You might think of noise as sounds you don't like and music as sounds that are pleasant to hear. But the difference between music and noise does not depend on whether you like the sound. The difference has to do with sound quality.

What Is Sound Quality?

Imagine that the same note is played on a piano and on a violin. Could you tell the instruments apart without looking? The notes played have the same frequency. But you could probably tell them apart because the instruments make different sounds. The notes sound different because a single note on an instrument actually comes from several different pitches: the fundamental and several overtones. The result of the combination of these pitches is shown in **Figure 1.** The result of several pitches mixing together through interference is **sound quality.** Each instrument has a unique sound quality. **Figure 1** also shows how the sound quality differs when two instruments play the same note.

Figure 1 *Each instrument has a unique sound quality that results from the particular blend of overtones that it has.*

Sound Quality of Instruments

The difference in sound quality among different instruments comes from their structural differences. All instruments produce sound by vibrating. But instruments vary in the part that vibrates and in the way that the vibrations are made. There are three main families of instruments: string instruments, wind instruments, and percussion instruments.

sound quality the result of the blending of several pitches through interference

Reading Check **How do musical instruments differ in how they produce sound?** (*See the Appendix for answers to Reading Checks.*)

String Instruments

Violins, guitars, and banjos are examples of string instruments. They make sound when their strings vibrate after being plucked or bowed. **Figure 2** shows how two different string instruments produce sounds.

Figure 2 **String Instruments**

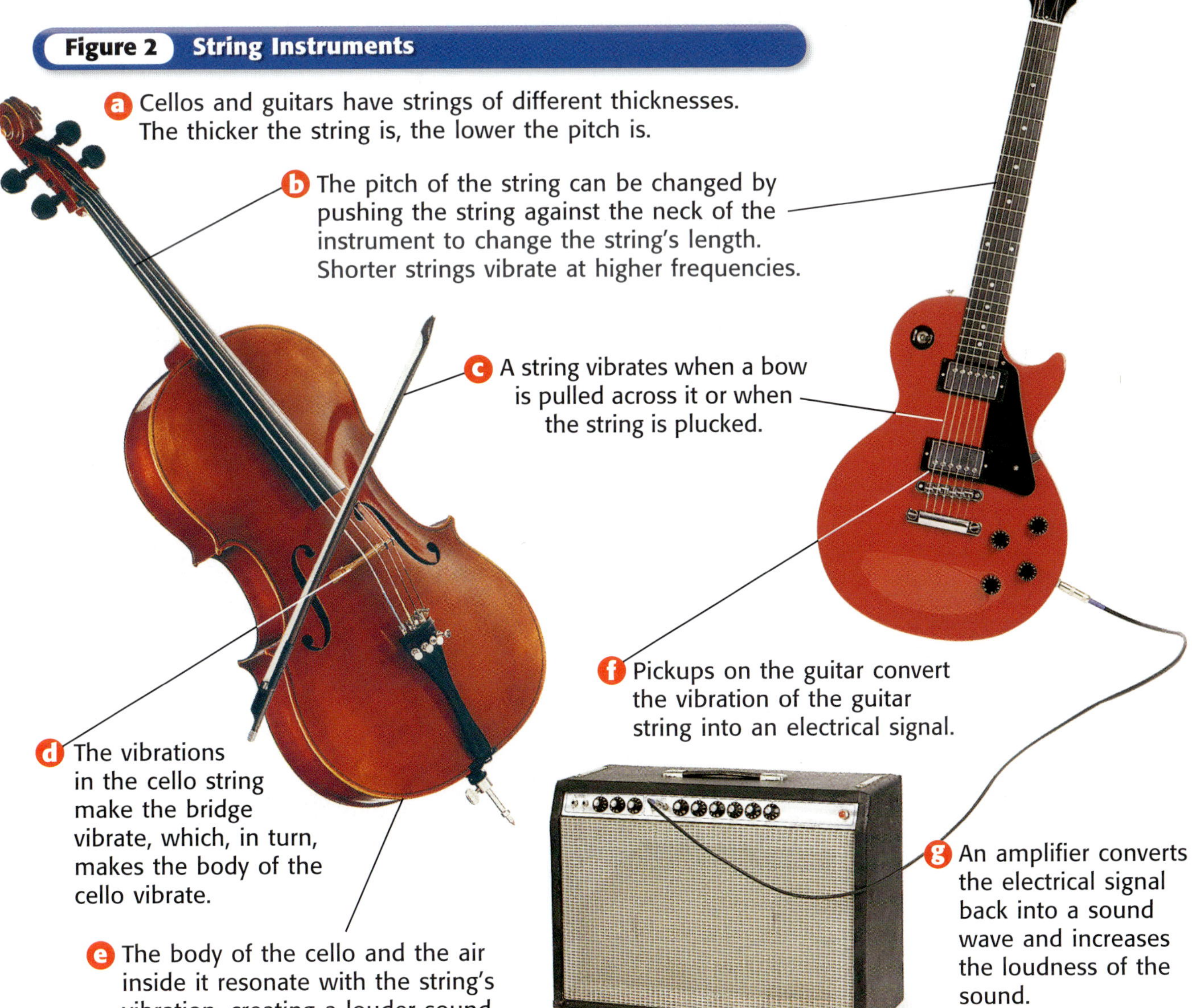

Figure 3 Wind Instruments

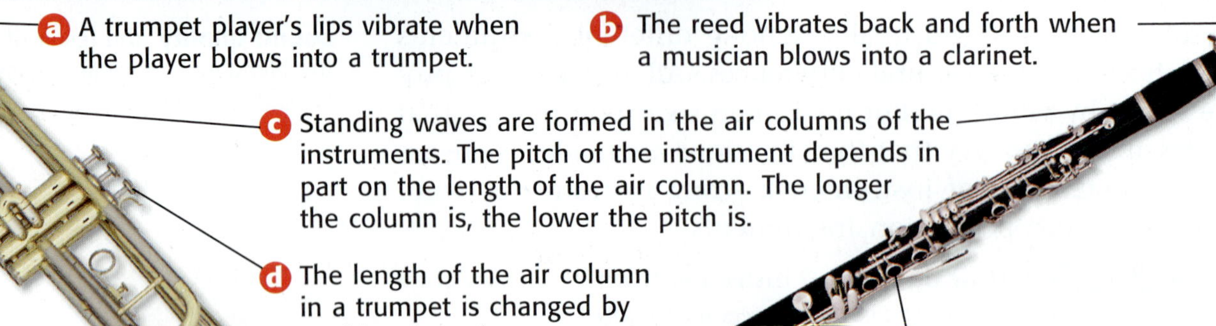

Wind Instruments

A wind instrument produces sound when a vibration is created at one end of its air column. The vibration causes standing waves inside the air column. Pitch is changed by changing the length of the air column. Wind instruments are sometimes divided into two groups—woodwinds and brass. Examples of woodwinds are saxophones, oboes, and recorders. French horns, trombones, and tubas are brass instruments. A brass instrument and a woodwind instrument are shown in **Figure 3.**

Percussion Instruments

Drums, bells, and cymbals are percussion instruments. They make sound when struck. Instruments of different sizes are used to get different pitches. Usually, the larger the instrument is, the lower the pitch is. The drums and cymbals in a trap set, shown in **Figure 4,** are percussion instruments.

Figure 4 Percussion Instruments

Music or Noise?

Most of the sounds we hear are noises. The sound of a truck roaring down the highway, the slam of a door, and the jingle of keys falling to the floor are all noises. **Noise** can be described as any sound, especially a nonmusical sound, that is a random mix of frequencies (or pitches). **Figure 5** shows on an oscilloscope the difference between a musical sound and noise.

noise a sound that consists of a random mix of frequencies

Reading Check What is the difference between music and noise?

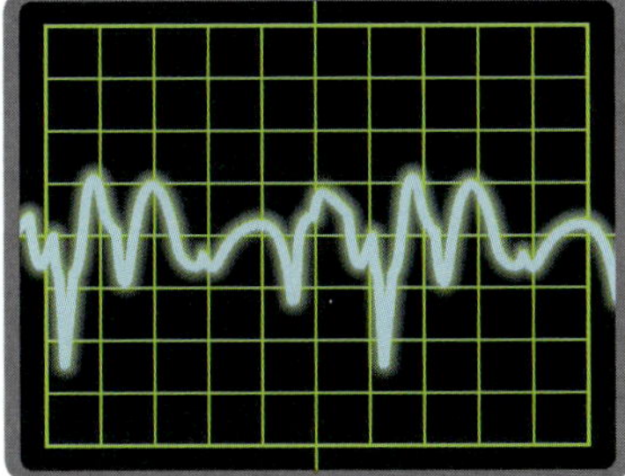

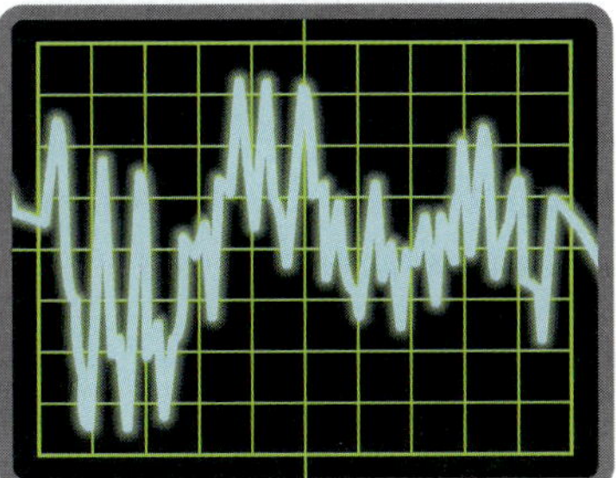

Figure 5 *A note from a French horn produces a sound wave with a repeating pattern, but noise from a clap produces complex sound waves with no regular pattern.*

SECTION Review

Summary

- Different instruments have different sound qualities.
- Sound quality results from the blending through interference of the fundamental and several overtones.
- The three families of instruments are string, wind, and percussion instruments.
- Noise is a sound consisting of a random mix of frequencies.

Using Key Terms

1. Use each of the following terms in a separate sentence: *sound quality* and *noise.*

Understanding Key Ideas

2. What interaction of sound waves determines sound quality?
   a. reflection
   b. diffraction
   c. pitch
   d. interference
3. Why do different instruments have different sound qualities?

Critical Thinking

4. **Making Comparisons** What do string instruments and wind instruments have in common in how they produce sound?
5. **Identifying Bias** Someone says that the music you are listening to is "just noise." Does the person mean that the music is a random mix of frequencies? Explain your answer.

Interpreting Graphics

6. Look at the oscilloscope screen below. Do you think the sound represented by the wave on the screen is noise or music? Explain your answer.

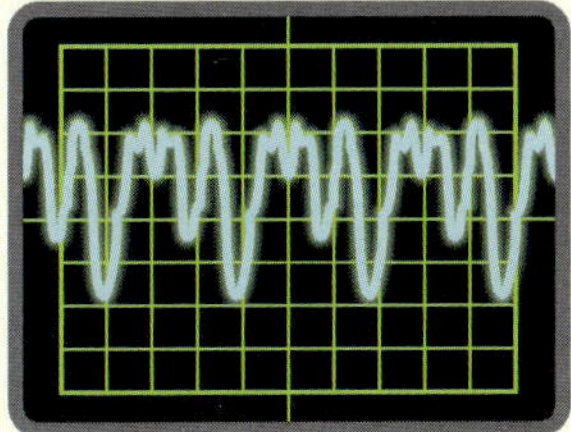

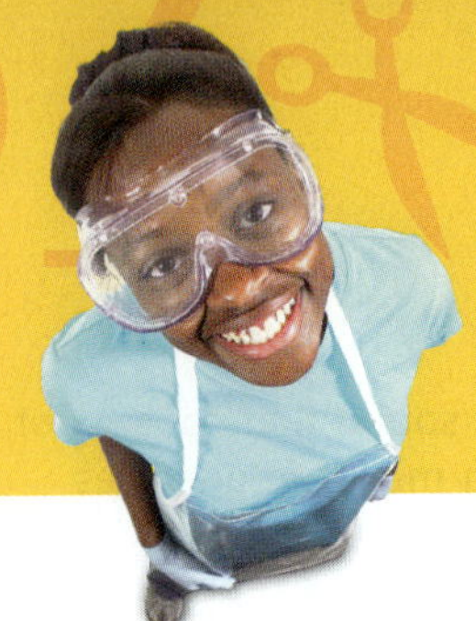

Using Scientific Methods

Skills Practice Lab

Easy Listening

Pitch describes how low or high a sound is. A sound's pitch is related to its frequency—the number of waves per second. Frequency is measured in hertz (Hz), where 1 Hz equals 1 wave per second. Most humans can hear frequencies in the range from 20 Hz to 20,000 Hz. But not everyone detects all pitches equally well at all distances. In this activity, you will collect data to see how well you and your classmates hear different frequencies at different distances.

OBJECTIVES

Measure your classmates' ability to detect different pitches at different distances.

Graph the average class data.

Form a conclusion about how easily pitches of different frequencies are heard at different distances.

MATERIALS

- eraser, hard rubber
- meterstick
- paper, graph
- tuning forks, different frequencies (4)

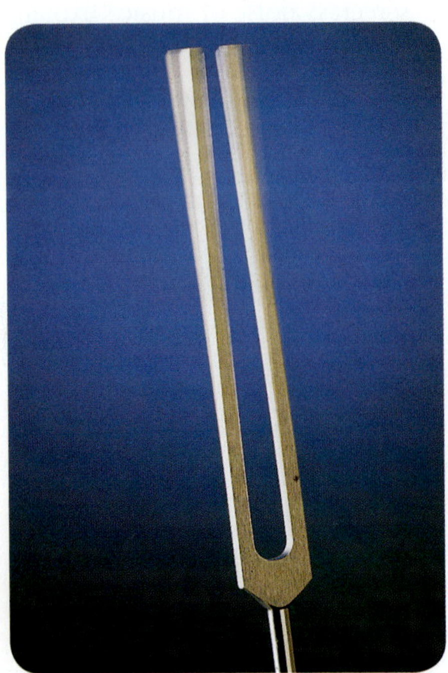

Ask a Question

1. Do most of the students in your classroom hear low-, mid-, or high-frequency sounds best?

Form a Hypothesis

2. Write a hypothesis that answers the question above. Explain your reasoning.

Test the Hypothesis

3. Choose one member of your group to be the sound maker. The others will be the listeners.
4. Copy the data table below onto another sheet of paper. Be sure to include a column for every listener in your group.

Data Collection Table

Frequency	Distance (m)			
	Listener 1	Listener 2	Listener 3	Average
1 (____Hz)				
2 (____Hz)		DO NOT WRITE IN BOOK		
3 (____Hz)				
4 (____Hz)				

5. The sound maker will choose one of the tuning forks, and record the frequency of the tuning fork in the data table.
6. The listeners should stand 1 m from the sound maker with their backs turned.

7. The sound maker will create a sound by striking the tip of the tuning fork gently with the eraser.
8. Listeners who hear the sound should take one step away from the sound maker. The listeners who do not hear the sound should stay where they are.
9. Repeat steps 7 and 8 until none of the listeners can hear the sound or the listeners reach the edge of the room.
10. Using the meterstick, the sound maker should measure the distance from his or her position to each of the listeners. All group members should record this data.
11. Repeat steps 5 through 10 with a tuning fork of a different frequency.
12. Continue until all four tuning forks have been tested.

Analyze the Results

1. **Organizing Data** Calculate the average distance for each frequency. Share your group's data with the rest of the class to make a data table for the whole class.
2. **Analyzing Data** Calculate the average distance for each frequency for the class.
3. **Constructing Graphs** Make a graph of the class results, plotting average distance (*y*-axis) versus frequency (*x*-axis).

Draw Conclusions

4. **Drawing Conclusions** Was everyone in the class able to hear all of frequencies equally? (Hint: Was the average distance for each frequency the same?)
5. **Evaluating Data** If the answer to question 4 is no, which frequency had the longest average distance? Which frequency had the shortest final distance?
6. **Analyzing Graphs** Based on your graph, do your results support your hypothesis? Explain your answer.
7. **Evaluating Methods** Do you think your class sample is large enough to confirm your hypothesis for all people of all ages? Explain your answer.

Chapter Review

USING KEY TERMS

Complete each of the following sentences by choosing the correct term from the word bank.

loudness	echoes
pitch	noise
sound quality	

1. The _____ of a sound wave depends on its amplitude.

2. Reflected sound waves are called _____.

3. Two different instruments playing the same note sound different because of _____.

UNDERSTANDING KEY IDEAS

Multiple Choice

4. If a fire engine is traveling toward you, the Doppler effect will cause the siren to sound
   - **a.** higher.
   - **b.** lower.
   - **c.** louder.
   - **d.** softer.

5. Sound travels fastest through
   - **a.** a vacuum.
   - **b.** sea water.
   - **c.** air.
   - **d.** glass.

6. If two sound waves interfere constructively, you will hear
   - **a.** a high-pitched sound.
   - **b.** a softer sound.
   - **c.** a louder sound.
   - **d.** no change in sound.

7. You will hear a sonic boom when
   - **a.** an object breaks the sound barrier.
   - **b.** an object travels at supersonic speeds.
   - **c.** a shock wave reaches your ears.
   - **d.** the speed of sound is 290 m/s.

8. Resonance can happen when an object vibrates at another object's
   - **a.** resonant frequency.
   - **b.** fundamental frequency.
   - **c.** second overtone frequency.
   - **d.** All of the above

9. A technological device that can be used to see sound waves is a(n)
   - **a.** sonar.
   - **b.** oscilloscope.
   - **c.** ultrasound.
   - **d.** amplifier.

Short Answer

10. Describe how the Doppler effect helps a beluga whale determine whether a fish is moving away from it or toward it.

11. How do vibrations cause sound waves?

12. Briefly describe what happens in the different parts of the ear.

Math Skills

13. A submarine that is not moving sends out a sonar sound wave traveling 1,500 m/s, which reflects off a boat back to the submarine. The sonar crew detects the reflected wave 6 s after it was sent out. How far away is the boat from the submarine?

CRITICAL THINKING

14 **Concept Mapping** Use the following terms to create a concept map: *sound waves, pitch, loudness, decibels, frequency, amplitude, oscilloscope, hertz,* and *interference.*

15 **Analyzing Processes** An *anechoic chamber* is a room where there is almost no reflection of sound waves. Anechoic chambers are often used to test sound equipment, such as stereos. The walls of such chambers are usually covered with foam triangles. Explain why this design eliminates echoes in the room.

16 **Applying Concepts** Would the pilot of an airplane breaking the sound barrier hear a sonic boom? Explain why or why not.

17 **Forming Hypotheses** After working in a factory for a month, a man you know complains about a ringing in his ears. What might be wrong with him? What do you think may have caused his problem? What can you suggest to him to prevent further hearing loss?

INTERPRETING GRAPHICS

Use the oscilloscope screens below to answer the questions that follow:

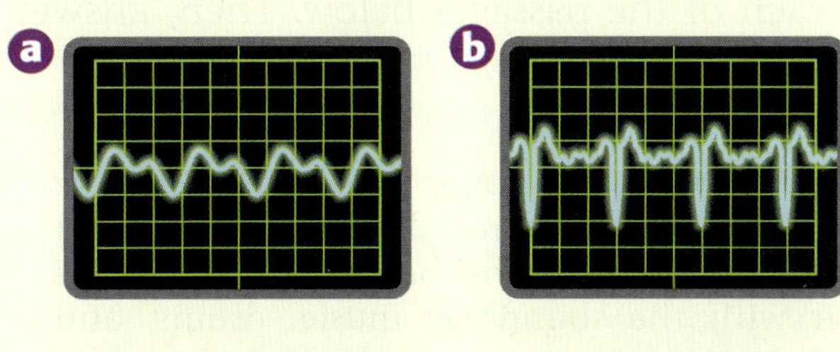

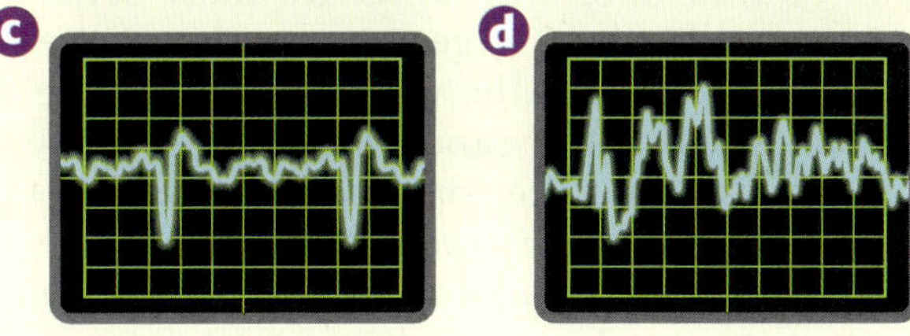

18 Which sound is noise?

19 Which represents the softest sound?

20 Which represents the sound with the lowest pitch?

21 Which two sounds were produced by the same instrument?

Standardized Test Preparation

READING

Read each of the passages below. Then, answer the questions that follow each passage.

Passage 1 Centuries ago, Marco Polo wrote about the booming sand dunes of the Asian desert. He wrote that the booming sands filled the air with the sounds of music, drums, and weapons of war. Booming sands are most often found in the middle of large deserts. They have been discovered all over the world, including the United States. Booming sands make loud, low-pitched sounds when the top layers of sand slip over the layers below, producing vibrations. The sounds have been compared to foghorns, cannon fire, and moaning. The sounds can last from a few seconds to 15 min and can be heard more than 10 km away!

1. Which is a fact in this passage?

- **A** Marco Polo loved traveling.
- **B** Booming sands always sound like moaning people.
- **C** Booming sands are the most interesting thing in Asia.
- **D** Some booming sands are found in the United States.

2. Which of the following phrases **best** describes booming sands?

- **F** found in Asia
- **G** noisy
- **H** slippery
- **I** discovered by Marco Polo

3. What causes booming sands?

- **A** vibrations caused by top layers of sand slipping over layers below
- **B** battles in the desert
- **C** animals that live beneath sand dunes
- **D** There is not enough information to determine the answer.

Passage 2 People who work in the field of architectural acoustics are concerned with controlling sound that travels in a closed space. Their goal is to make rooms and buildings quiet yet suitable for people to enjoy talking and listening to music. One major factor that affects the acoustical quality of a room is the way the room reflects sound waves. Sound waves bounce off surfaces such as doors, ceilings, and walls. Using materials that absorb sound reduces the reflection of sound waves. Materials that have small pockets of air that can trap the sound vibrations and keep them from reflecting are the most sound absorbent. Sound-absorbing floor and ceiling tiles, curtains, and upholstered furniture all help to control the reflection of sound waves.

1. The field of architectural acoustics is concerned with which of the following?

- **A** making buildings earthquake safe
- **B** controlling sound in closed spaces
- **C** designing sound-absorbing materials
- **D** making buildings as quiet as possible

2. Which of the following is a major factor in the acoustical quality of a room?

- **F** the size of the room
- **G** the furnishings in the room
- **H** the walls of the room
- **I** the noise level in the room

3. Which of the following materials is **most** likely to absorb sounds the best?

- **A** materials that have small pockets of air
- **B** surfaces such as doors, ceilings, and walls
- **C** materials that keep the room as quiet as possible
- **D** furniture that is made of wood

INTERPRETING GRAPHICS

Use the pictures of standing waves below to answer the questions that follow.

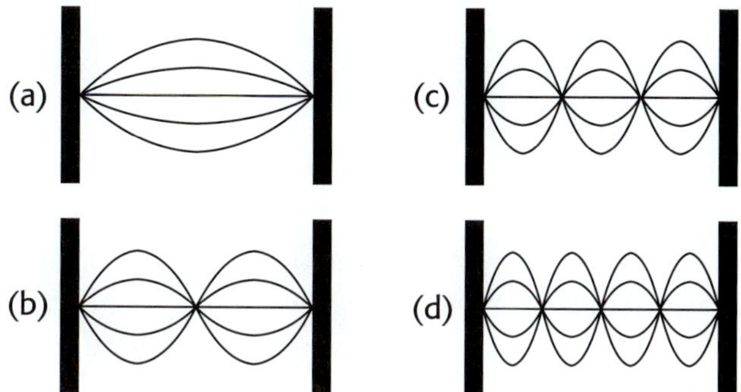

1. Which of the standing waves has the lowest frequency?
   - **A** a
   - **B** b
   - **C** c
   - **D** d

2. Which of the standing waves has the highest frequency?
   - **F** a
   - **G** b
   - **H** c
   - **I** d

3. Which of the standing waves represents the first overtone?
   - **A** a
   - **B** b
   - **C** c
   - **D** d

4. In which of the following pairs of standing waves is the frequency of the second wave twice the frequency of the first?
   - **F** a, b
   - **G** a, c
   - **H** b, c
   - **I** c, d

MATH

Read each question below, and choose the best answer.

1. The speed of sound in copper is 3,560 m/s. Which is another way to express this measure?
   - **A** 356×10^2 m/s
   - **B** 0.356×10^3 m/s
   - **C** 3.56×10^3 m/s
   - **D** 3.56×10^4 m/s

2. The speed of sound in sea water is 1,522 m/s. How far can a sound wave travel underwater in 10 s?
   - **F** 152.2 m
   - **G** 1,522 m
   - **H** 15,220 m
   - **I** 152,220 m

3. Claire likes to go swimming after work. She warms up for 120 s before she begins swimming, and it takes her an average of 55 s to swim one lap. Which equation could be used to find *w*, the number of seconds it takes for Claire to warm up and swim 15 laps?
   - **A** $w = (15 \times 120) + 55$
   - **B** $w = (15 \times 55) + 120$
   - **C** $w = 120 + 55 + 15$
   - **D** $w = (15 \times 55) \times 120$

4. The Vasquez family went bowling. They rented 6 pairs of shoes for \$3 a pair and bowled for 2 h at a rate of \$8.80/h. Which is the best estimate of the total cost of the shoes and bowling?
   - **F** \$24
   - **G** \$30
   - **H** \$36
   - **I** \$45

Science in Action

Scientific Discoveries

Jurassic Bark

Imagine you suddenly hear a loud honking sound, such as a trombone or a tuba. "Must be band tryouts," you think. You turn to find the noise and find yourself face to face with a 10 m long, 2,800 kg dinosaur with a huge tubular crest on its snout. Do you run? No—your musical friend, *Parasaurolophus,* is a vegetarian. In 1995, an almost-complete fossil skull of an adult *Parasaurolophus* was found in New Mexico. Scientists studied the noise-making qualities of *Parasaurolophus*'s crest and found that it contained many internal tubes and chambers.

Math ACTIVITY

Imagine that a standing wave with a frequency of 80 Hz is made inside the crest of a *Parasaurolophus*. What would be the frequency of the first overtone of this standing wave? the second? the third?

Science Fiction

"Ear" by Jane Yolen

Jily and her friends, Sanya and Feeny, live in a time not too far in the future. It is a time when everyone's hearing is damaged. People communicate using sign language—unless they put on their Ear. Then, the whole world is filled with sounds.

Jily and her friends visit a club called The Low Down. It is too quiet for Jily's tastes, and she wants to leave. But Sanya is dancing by herself, even though there is no music. When Jily finds Feeny, they notice some Earless kids their own age. Earless people never go to clubs, and Jily finds their presence offensive. But Feeny is intrigued.

Everyone is given an Ear at the age of 12 but has to give it up at the age of 30. Why would these kids want to go out without their Ears before the age of 30? Jily thinks the idea is ridiculous and doesn't stick around to find out the answer to such a question. But it is an answer that will change her life by the end of the next day.

Language Arts ACTIVITY

WRITING SKILL Read "Ear," by Jane Yolen, in the *Holt Anthology of Science Fiction.* Write a one-page report that discusses how the story made you think about the importance of hearing in your everyday life.

Careers

Adam Dudley

Sound Engineer Adam Dudley uses the science of sound waves every day at his job. He is the audio supervisor for the Performing Arts Center of the University of Texas at Austin. Dudley oversees sound design and technical support for campus performance spaces, including an auditorium that seats over 3,000 people.

To stage a successful concert, Dudley takes many factors into account. The size and shape of the room help determine how many speakers to use and where to place them. It is a challenge to make sure people seated in the back row can hear well enough and also to make sure that the people up front aren't going deaf from the high volume.

Adam Dudley loves his job—he enjoys working with people and technology and prefers not to wear a coat and tie. Although he is invisible to the audience, his work backstage is as crucial as the musicians and actors on stage to the success of the events.

Social Studies ACTIVITY

Research the ways in which concert halls were designed before the use of electricity for amplification. Make a model or diorama, and present it to the class, explaining the acoustical factors involved in the design.

To learn more about these Science in Action topics, visit go.hrw.com and type in the keyword HP5SNDF.

Current Science

Check out Current Science® articles related to this chapter by visiting go.hrw.com. Just type in the keyword HP5CS21.

The Nature of Light

About the PHOTO

What kind of alien life lives on this planet? Actually, this isn't a planet at all. It's an ordinary soap bubble! The brightly colored swirls on this bubble are reflections of light. Light waves combine through interference so that you see different colors on this soap bubble.

PRE-READING ACTIVITY

FOLDNOTES **Booklet** Before you read the chapter, create the FoldNote entitled "Booklet" described in the **Study Skills** section of the Appendix. Label each page of the booklet with a main idea from the chapter. As you read the chapter, write what you learn about each main idea on the appropriate page of the booklet.

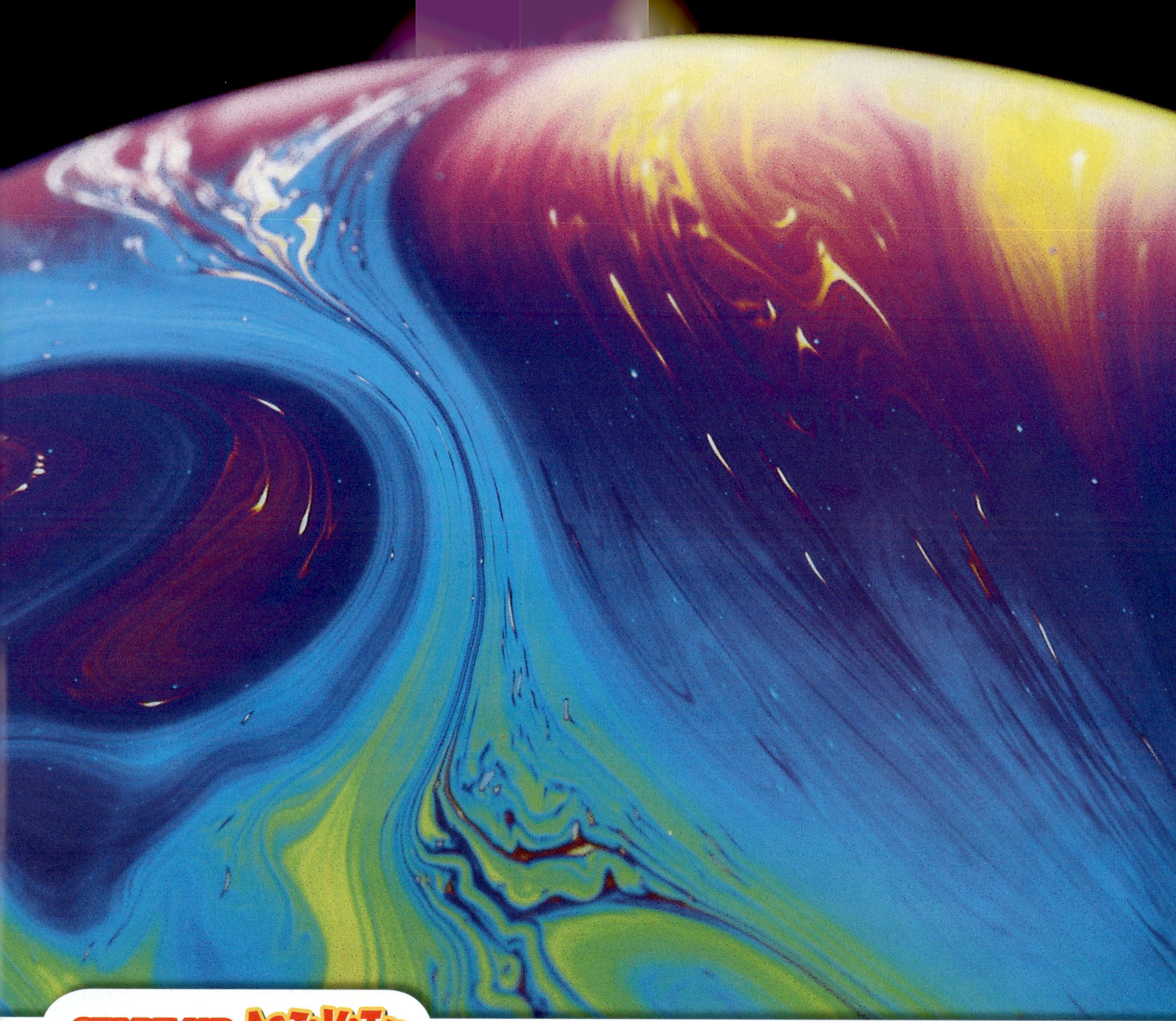

START-UP ACTIVITY

Colors of Light

Is white light really white? In this activity, you will use a spectroscope to answer that question.

Procedure

1. Your teacher will give you a **spectroscope** or instructions for making one.
2. Turn on an **incandescent light bulb.** Look at the light bulb through your spectroscope. Write a description of what you see.
3. Repeat step 2, looking at a **fluorescent light.** Again, describe what you see.

Analysis

1. Compare what you saw with the incandescent light bulb with what you saw with the fluorescent light bulb.
2. Both kinds of bulbs produce white light. What did you learn about white light by using the spectroscope?
3. Light from a flame is yellowish but is similar to white light. What do you think you would see if you used a spectroscope to look at light from a flame?

SECTION 1

What Is Light?

You can see light. It's everywhere! Light comes from the sun and from other sources, such as light bulbs. But what exactly is light?

READING WARM-UP

Objectives

- Describe light as an electromagnetic wave.
- Calculate distances traveled by light by using the speed of light.
- Explain why light from the sun is important.

Terms to Learn

electromagnetic wave
radiation

READING STRATEGY

Brainstorming The key idea of this section is light. Brainstorm words and phrases related to light.

Scientists are still studying light to learn more about it. A lot has already been discovered about light, as you will soon find out. Read on, and be enlightened!

Light: An Electromagnetic Wave

Light is a type of energy that travels as a wave. But light is different from other kinds of waves. Other kinds of waves, like sound waves and water waves, must travel through matter. Light does not require matter through which to travel. Light is an electromagnetic wave (EM wave). An **electromagnetic wave** is a wave that can travel through empty space or matter and consists of changing electric and magnetic fields.

electromagnetic wave a wave that consists of electric and magnetic fields that vibrate at right angles to each other

Fields exist around certain objects and can exert a force on another object without touching that object. For example, Earth is a source of a gravitational field. This field pulls you and all things toward Earth. But keep in mind that this field, like all fields, is not made of matter.

Figure 1 shows a diagram of an electromagnetic wave. Notice that the electric and magnetic fields are at right angles—or are *perpendicular*—to each other. These fields are also perpendicular to the direction of the wave motion.

Figure 1 *Electromagnetic waves are made of vibrating electric and magnetic fields.*

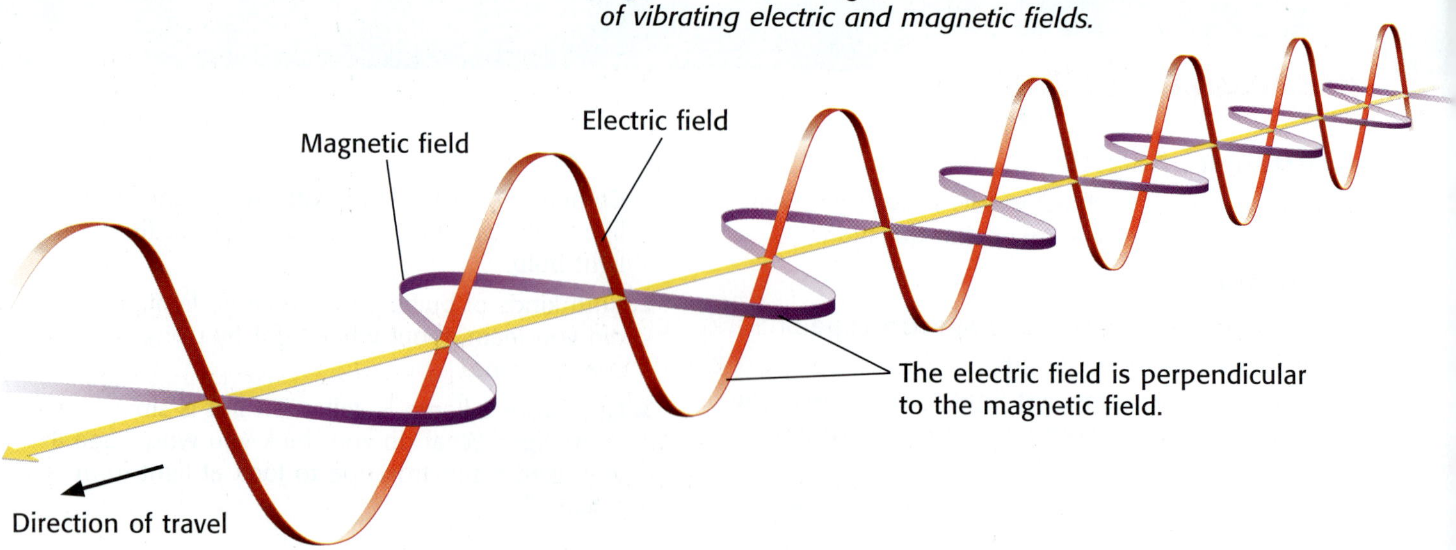

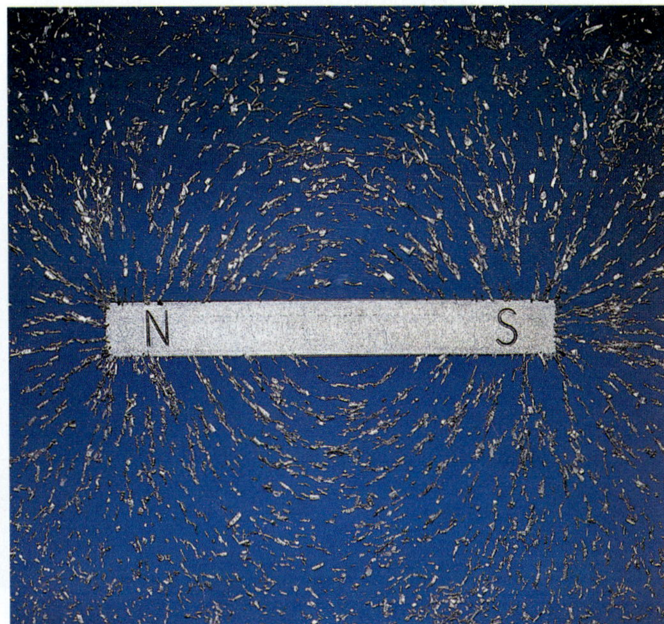

Figure 2 *The hair on the girl's head stands up because of an electric field and the iron filings form arcs around the magnet because of a magnetic field.*

Electric and Magnetic Fields

Electromagnetic waves are changing electric and magnetic fields. But what are electric and magnetic fields? An *electric field* surrounds every charged object. The electric field around a charged object pulls oppositely charged objects toward it and repels like-charged objects. You can see the effect of electric fields whenever you see objects stuck together by static electricity. **Figure 2** shows another effect of an electric field.

A *magnetic field* surrounds every magnet. Because of magnetic fields, paper clips and iron filings are pulled toward magnets. You can feel the effect of magnetic fields when you hold two magnets close together. The iron filings around the magnet in **Figure 2** form arcs in the presence of the magnet's magnetic field.

Reading Check **Where can electric fields be found?** *(See the Appendix for answers to Reading Checks.)*

How EM Waves Are Produced

An EM wave can be produced by the vibration of an electrically charged particle. When the particle vibrates, or moves back and forth, the electric field around it also vibrates. When the electric field starts vibrating, a vibrating magnetic field is created. The vibration of an electric field and a magnetic field together produces an EM wave that carries energy released by the original vibration of the particle. The transfer of energy as electromagnetic waves is called **radiation.**

CONNECTION TO Social Studies

WRITING SKILL **The Particle Model of Light**

Thinking of light as being an electromagnetic wave can explain many properties of light. But some properties of light can be explained only by using a particle model of light. In the particle model of light, light is thought of as a stream of particles called *photons*. Research the history of the particle model of light. Write a one-page paper on what you learn.

radiation transfer of energy as electromagnetic waves

Figure 3 *Thunder and lightning are produced at the same time. But you usually see lightning before you hear thunder, because light travels much faster than sound.*

The Speed of Light

Scientists have yet to discover anything that travels faster than light. In the near vacuum of space, the speed of light is about 300,000,000 m/s, or 300,000 km/s. Light waves travel slightly slower in air, glass, and other types of matter. In fact, most waves move at different speeds in different materials. (Keep in mind that even though electromagnetic waves do not need to travel through matter, they can travel through many substances.)

Believe it or not, light can travel about 880,000 times faster than sound! This fact explains the phenomenon described in **Figure 3.**

Reading Check **How does the speed of light compare with the speed of sound?**

MATH FOCUS

How Fast Is Light? The distance from Earth to the moon is 384,000 km. Calculate the time it takes for light to travel that distance.

Step 1: Write the equation for speed.

$$speed = \frac{distance}{time}$$

Step 2: Rearrange the equation by multiplying by time and dividing by speed.

$$time = \frac{distance}{speed}$$

Step 3: Replace *distance* and *speed* with the values given in the problem, and solve.

$$time = \frac{384{,}000 \text{ km}}{300{,}000 \text{ km/s}}$$

$$time = 1.28 \text{ s}$$

Now It's Your Turn

1. The distance from the sun to Venus is 108,000,000 km. Calculate the time it takes for light to travel that distance.

Light from the Sun

Even though light travels quickly, it takes about 8.3 min for light to travel from the sun to Earth. It takes this much time because Earth is 150,000,000 km away from the sun.

The sun loses energy by emitting light. The EM waves from the sun are the major source of energy on Earth. For example, plants use photosynthesis to store energy from the sun. And animals use and store energy by eating plants or by eating other animals that eat plants. Even fossil fuels, such as coal and oil, store energy from the sun. Fossil fuels are formed from the remains of plants and animals that lived millions of years ago.

Although Earth receives a large amount of energy from the sun, only a tiny fraction of the total energy given off by the sun reaches Earth. Look at **Figure 4.** The sun gives off energy as EM waves, but most of this energy travels away in space.

Figure 4 *Only a small amount of the sun's energy reaches the planets in the solar system.*

SECTION Review

Summary

- Light is an electromagnetic (EM) wave. An EM wave is a wave that consists of changing electric and magnetic fields. EM waves require no matter through which to travel.
- EM waves can be produced by the vibration of charged particles.
- The speed of light in a vacuum is about 300,000,000 m/s.
- EM waves from the sun are the major source of energy for Earth.

Using Key Terms

1. Use the following terms in the same sentence: *electromagnetic wave* and *radiation.*

Understanding Key Ideas

2. Electromagnetic waves are different from other types of waves because they can travel through
   a. air. **c.** space.
   b. glass. **d.** steel.
3. Describe light in terms of electromagnetic waves.
4. Why is light from the sun important?
5. How can electromagnetic waves be produced?

Math Skills

6. The distance from the sun to Jupiter is 778,000,000 km. How long does it take for light from the sun to reach Jupiter?

Critical Thinking

7. **Making Inferences** Why do you think the speed of light differs in water and air?
8. **Making Comparisons** How does the amount of energy produced by the sun compare with the amount of energy that reaches Earth from the sun?
9. **Applying Concepts** Explain why the energy produced by burning wood in a campfire is energy from the sun.

SECTION 2

The Electromagnetic Spectrum

READING WARM-UP

Objectives

- Identify how electromagnetic waves differ from each other.
- Describe some uses for radio waves and microwaves.
- List examples of how infrared waves and visible light are important in your life.
- Explain how ultraviolet light, X rays, and gamma rays can be both helpful and harmful.

Terms to Learn

electromagnetic spectrum

READING STRATEGY

Mnemonics As you read this section, create a mnemonic device to help you remember the kinds of EM waves.

When you look around, you can see things that reflect light to your eyes. But a bee might see the same things differently. Bees can see a kind of light—called ultraviolet light*—that you can't see!*

It might seem odd to call something you can't see *light*. The light you are most familiar with is called *visible light*. Ultraviolet light is similar to visible light. Both are kinds of electromagnetic (EM) waves. In this section, you will learn about many kinds of EM waves, including X rays, radio waves, and microwaves.

Characteristics of EM Waves

All EM waves travel at the same speed in a vacuum—300,000 km/s. How is this possible? The speed of a wave is found by multiplying its wavelength by its frequency. So, EM waves having different wavelengths can travel at the same speed as long as their frequencies are also different. The entire range of EM waves is called the **electromagnetic spectrum.** The electromagnetic spectrum is shown in **Figure 1.** The electromagnetic spectrum is divided into regions according to the length of the waves. There is no sharp division between one kind of wave and the next. Some kinds even have overlapping ranges.

Reading Check How is the speed of a wave determined? (*See the Appendix for answers to Reading Checks.*)

Figure 1 The Electromagnetic Spectrum

The electromagnetic spectrum is arranged from long to short wavelength or from low to high frequency.

Radio waves

All radio and television stations broadcast radio waves.

Microwaves

Despite their name, microwaves are not the shortest EM waves.

Infrared

Infrared means "below red."

Radio Waves

Radio waves cover a wide range of waves in the EM spectrum. Radio waves have some of the longest wavelengths and the lowest frequencies of all EM waves. In fact, radio waves are any EM waves that have wavelengths longer than 30 cm. Radio waves are used for broadcasting radio signals.

Broadcasting Radio Signals

Figure 2 shows how radio signals are broadcast. Radio stations encode sound information into radio waves by varying either the waves' amplitude or their frequency. Changing amplitude or frequency is called *modulation* (MAHJ uh LAY shuhn). You probably know that there are AM radio stations and FM radio stations. The abbreviation *AM* stands for "amplitude modulation," and the abbreviation *FM* stands for "frequency modulation."

1 A radio station converts sound into an electric current. The current produces radio waves that are sent out in all directions by the antenna.

2 A radio receives radio waves and then converts them into an electric current, which is then converted to sound.

Figure 2 *Radio waves cannot be heard, but they can carry energy that can be converted into sound.*

Comparing AM and FM Radio Waves

AM radio waves are different from FM radio waves. For example, AM radio waves have longer wavelengths than FM radio waves do. And AM radio waves can bounce off the atmosphere and thus can travel farther than FM radio waves. But FM radio waves are less affected by electrical noise than AM radio waves are. So, music broadcast from FM stations sounds better than music broadcast from AM stations.

electromagnetic spectrum all of the frequencies or wavelengths of electromagnetic radiation

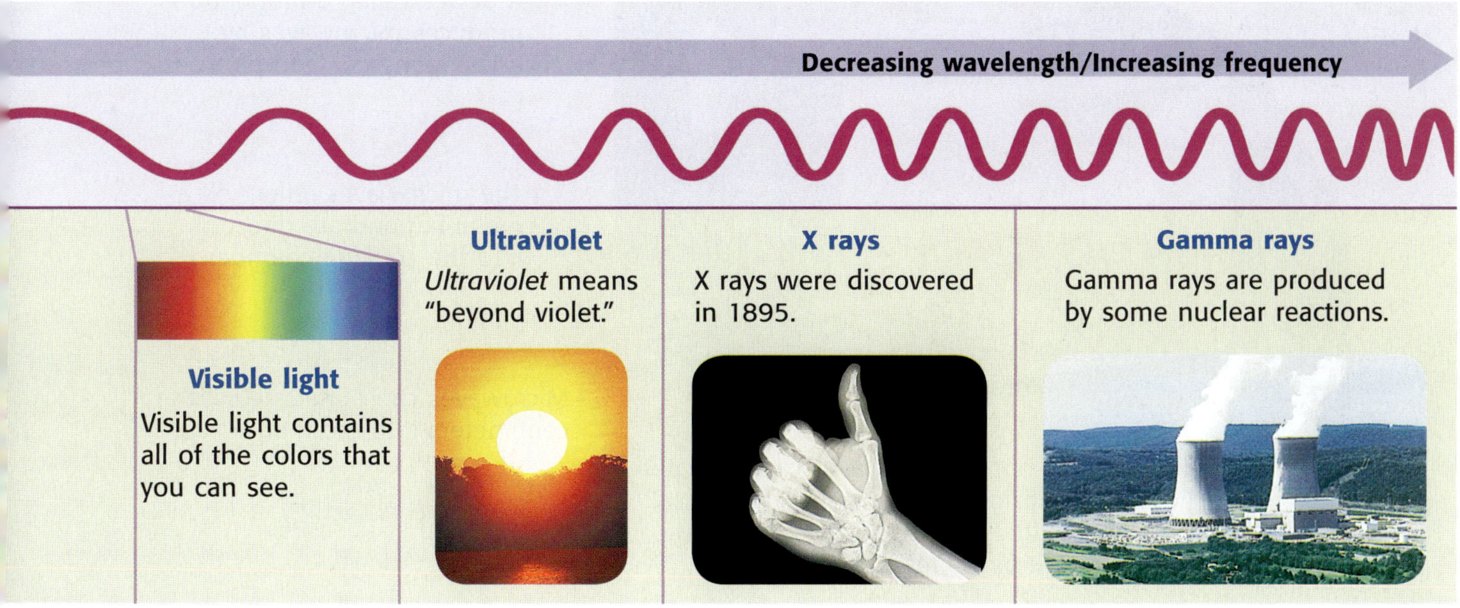

Decreasing wavelength/Increasing frequency

Visible light
Visible light contains all of the colors that you can see.

Ultraviolet
Ultraviolet means "beyond violet."

X rays
X rays were discovered in 1895.

Gamma rays
Gamma rays are produced by some nuclear reactions.

Radio Waves and Television

Television signals are also carried by radio waves. Most television stations broadcast radio waves that have shorter wavelengths and higher frequencies than those broadcast by radio stations. Like radio signals, television signals are broadcast using amplitude modulation and frequency modulation. Television stations use frequency-modulated waves to carry sound and amplitude-modulated waves to carry pictures.

Some waves carrying television signals are transmitted to artificial satellites orbiting Earth. The waves are amplified and sent to ground antennas. They then travel through cables to televisions in homes. Cable television works by this process.

Reading Check Which EM waves can carry television signals?

Microwaves

Microwaves have shorter wavelengths and higher frequencies than radio waves do. Microwaves have wavelengths between 1 mm and 30 cm. You are probably familiar with microwaves—they are created in a microwave oven, such as the one shown in **Figure 3.**

Microwaves and Communication

Like radio waves, microwaves are used to send information over long distances. For example, cellular phones send and receive signals using microwaves. And signals sent between Earth and artificial satellites in space are also carried by microwaves.

Figure 3 How a Microwave Oven Works

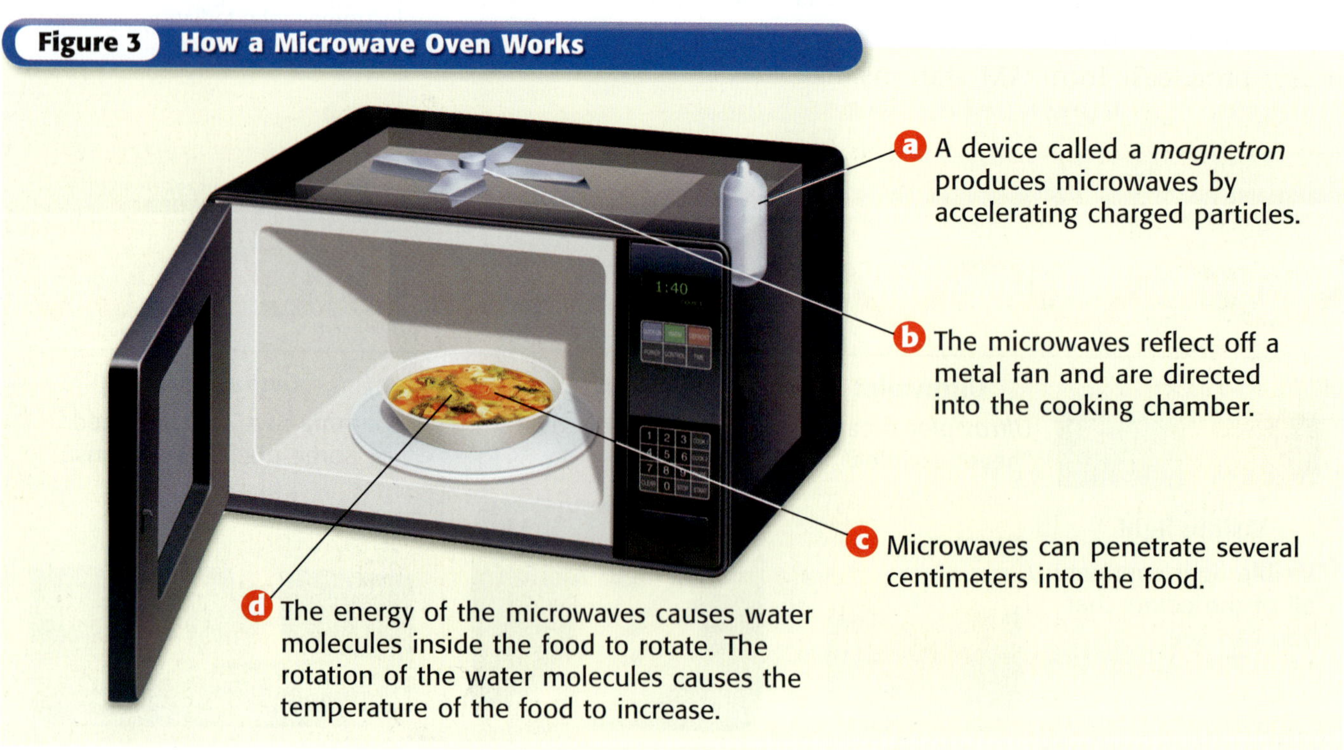

Figure 4 *Police officers use radar to detect cars going faster than the speed limit.*

Radar

Microwaves are also used in radar. *Radar* (**ra**dio **d**etection **a**nd **r**anging) is used to detect the speed and location of objects. The police officer in **Figure 4** is using radar to check the speed of a car. The radar gun sends out microwaves that reflect off the car and return to the gun. The reflected waves are used to calculate the speed of the car. Radar is also used to watch the movement of airplanes and to help ships navigate at night.

Infrared Waves

Infrared waves have shorter wavelengths and higher frequencies than microwaves do. The wavelengths of infrared waves vary between 700 nanometers and 1 mm. A nanometer (nm) is equal to 0.000000001 m.

On a sunny day, you may be warmed by infrared waves from the sun. Your skin absorbs infrared waves striking your body. The energy of the waves causes the particles in your skin to vibrate more, and you feel an increase in temperature. The sun is not the only source of infrared waves. Almost all things give off infrared waves, including buildings, trees, and you! The amount of infrared waves an object gives off depends on the object's temperature. Warmer objects give off more infrared waves than cooler objects do.

You can't see infrared waves, but some devices can detect infrared waves. For example, infrared binoculars change infrared waves into light you can see. Such binoculars can be used to watch animals at night. **Figure 5** shows a photo taken with film that is sensitive to infrared waves.

Figure 5 *In this photograph, brighter colors indicate higher temperatures.*

Figure 6 *Water droplets can separate white light into visible light of different wavelengths. As a result, you see all the colors of visible light in a rainbow.*

Visible Light

Visible light is the very narrow range of wavelengths and frequencies in the electromagnetic spectrum that human eyes respond to. Visible light waves have shorter wavelengths and higher frequencies than infrared waves do. Visible light waves have wavelengths between 400 nm and 700 nm.

Visible Light from the Sun

In addition to infrared waves, some of the energy that reaches Earth from the sun is visible light. The visible light from the sun is white light. *White light* is visible light of all wavelengths combined. Light from lamps in your home as well as from the fluorescent bulbs in your school is also white light.

Reading Check **What is white light?**

Colors of Light

Humans see the different wavelengths of visible light as different colors, as shown in **Figure 6.** The longest wavelengths are seen as red light. The shortest wavelengths are seen as violet light.

The range of colors is called the *visible spectrum.* You can see the visible spectrum in **Figure 7.** When you list the colors, you might use the imaginary name *ROY G. BiV* to help you remember their order. The capital letters in Roy's name represent the first letter of each color of visible light: **r**ed, **o**range, **y**ellow, **g**reen, **b**lue, and **v**iolet. What about the *i* in Roy's last name? You can think of *i* as standing for the color indigo. Indigo is a dark blue color.

Making a Rainbow

On a sunny day, ask a parent to use a hose or a spray bottle to make a mist of water outside. Move around until you see a rainbow in the water mist. Draw a diagram showing the positions of the water mist, the sun, the rainbow, and yourself.

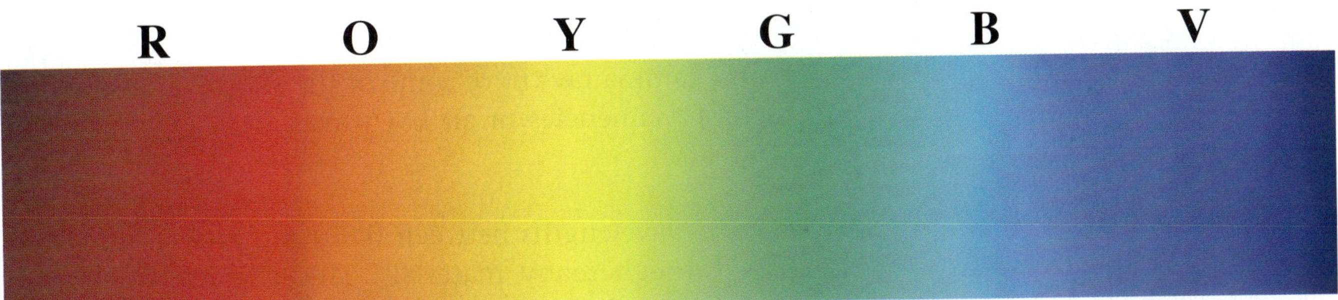

Figure 7 *The visible spectrum contains all colors of light.*

Ultraviolet Light

Ultraviolet light (UV light), like infrared and visible light, is another type of electromagnetic wave produced by the sun. Ultraviolet waves have shorter wavelengths and higher frequencies than visible light does. The wavelengths of ultraviolet light waves vary between 60 nm and 400 nm. Ultraviolet light affects your body in both bad and good ways.

Reading Check **How do ultraviolet light waves compare with visible light waves?**

Bad Effects

On the bad side, too much ultraviolet light can cause sunburn, as you can see in **Figure 8.** Too much ultraviolet light can also cause skin cancer, wrinkles, and damage to the eyes. Luckily, much ultraviolet light, like other kinds of light from the sun, does not reach Earth's surface. But you should still protect yourself against the ultraviolet light that does reach you. To do so, you should use sunscreen with a high SPF (**s**un **p**rotection **f**actor). You should also wear sunglasses that block out UV light to protect your eyes. Hats, long-sleeved shirts, and long pants can protect you, too. You need this protection even on overcast days because UV light can travel through clouds.

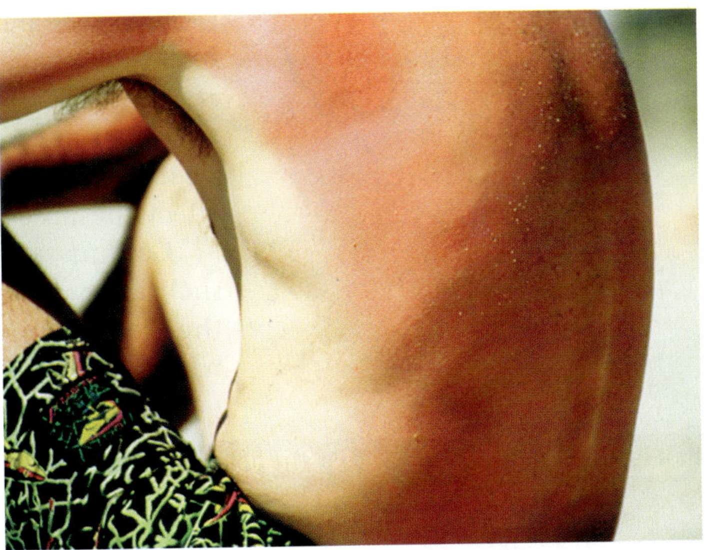

Figure 8 *Too much exposure to ultraviolet light can lead to a painful sunburn. Using sunscreen will help protect your skin.*

Good Effects

On the good side, ultraviolet waves produced by ultraviolet lamps are used to kill bacteria on food and surgical tools. In addition, small amounts of ultraviolet light are beneficial to your body. When exposed to ultraviolet light, skin cells produce vitamin D. This vitamin allows the intestines to absorb calcium. Without calcium, your teeth and bones would be very weak.

X Rays and Gamma Rays

X rays and gamma rays have some of the shortest wavelengths and highest frequencies of all EM waves.

X Rays

X rays have wavelengths between 0.001 nm and 60 nm. They can pass through many materials. This characteristic makes X rays useful in the medical field, as shown in **Figure 9.** But too much exposure to X rays can also damage or kill living cells. A patient getting an X ray may wear special aprons to protect parts of the body that do not need X-ray exposure. These aprons are lined with lead because X rays cannot pass through lead.

X-ray machines are also used as security devices in airports and other public buildings. The machines allow security officers to see inside bags and other containers without opening the containers.

Reading Check How are patients protected from X rays?

Gamma Rays

Gamma rays are EM waves that have wavelengths shorter than 0.1 nm. They can penetrate most materials very easily. Gamma rays are used to treat some forms of cancer. Doctors focus the rays on tumors inside the body to kill the cancer cells. This treatment often has good effects, but it can have bad side effects because some healthy cells may also be killed.

Gamma rays are also used to kill harmful bacteria in foods, such as meat and fresh fruits. The gamma rays do not harm the treated food and do not stay in the food. So, food that has been treated with gamma rays is safe for you to eat.

CONNECTION TO Astronomy

Gamma Ray Spectrometer In 2001, NASA put an artificial satellite called the *2001 Mars Odyssey* in orbit around Mars. The *Odyssey* is carrying a gamma ray spectrometer. A *spectrometer* is a device used to detect certain kinds of EM waves. The gamma ray spectrometer on the *Odyssey* was used to look for water and several chemical elements on Mars. Scientists hope to use this information to learn about the geology of Mars. Research the characteristics of Mars and Earth. In your **science journal,** make a chart comparing Mars and Earth.

ACTIVITY

Figure 9 How a Bone Is X Rayed

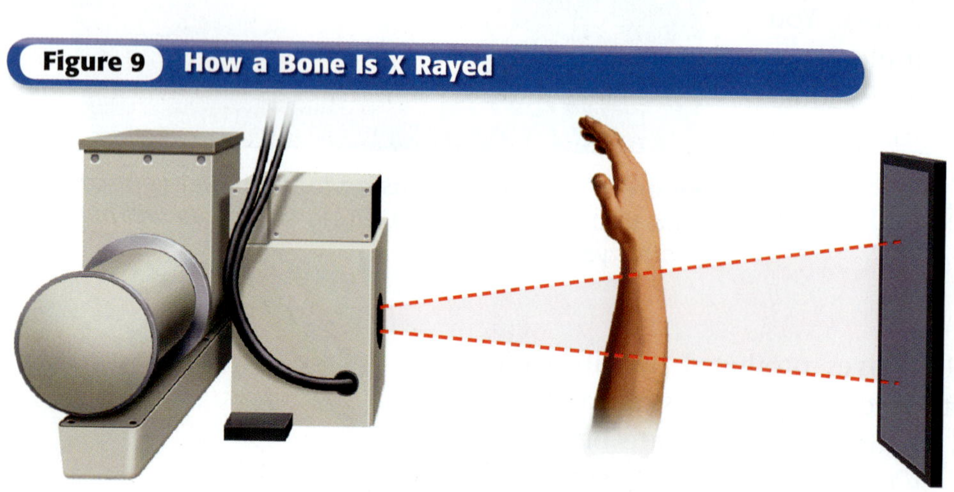

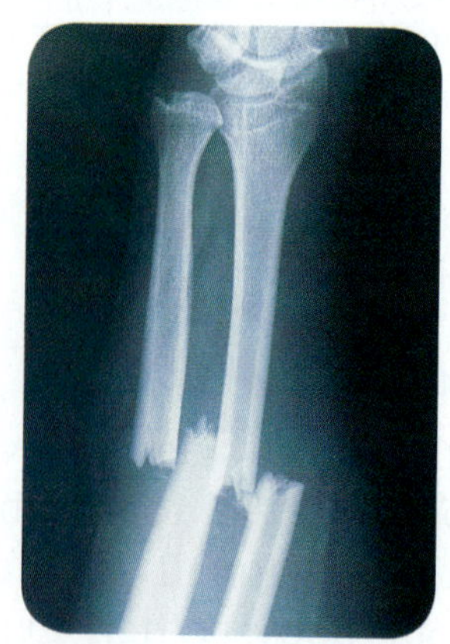

1. X rays travel easily through skin and muscle but are absorbed by bones.
2. The X rays that are not absorbed strike the film.
3. Bright areas appear on the film where X rays are absorbed by the bones.

SECTION Review

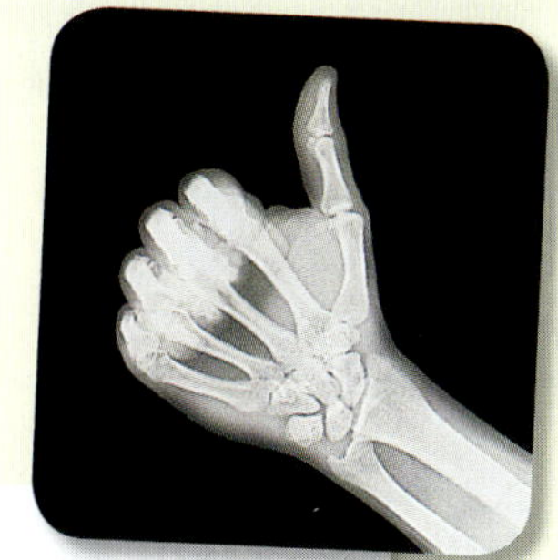

Summary

- All electromagnetic (EM) waves travel at the speed of light. EM waves differ only by wavelength and frequency.
- The entire range of EM waves is called the *electromagnetic spectrum*.
- Radio waves are used for communication.
- Microwaves are used in cooking and in radar.
- The absorption of infrared waves is felt as an increase in temperature.
- Visible light is the narrow range of wavelengths that humans can see. Different wavelengths are seen as different colors.
- Ultraviolet light is useful for killing bacteria and for producing vitamin D in the body. Overexposure to ultraviolet light can cause health problems.
- X rays and gamma rays are EM waves that are often used in medicine. Overexposure to these kinds of rays can damage or kill living cells.

Using Key Terms

1. In your own words, write a definition for the term *electromagnetic spectrum*.

Understanding Key Ideas

2. Which of the following electromagnetic waves are produced by the sun?
   a. infrared waves
   b. visible light
   c. ultraviolet light
   d. All of the above
3. How do the different kinds of EM waves differ from each other?
4. Describe two ways of transmitting information using radio waves.
5. Explain why ultraviolet light, X rays, and gamma rays can be both helpful and harmful.
6. What are two common uses for microwaves?
7. What is white light? What are two sources of white light?
8. What is the visible spectrum?

Critical Thinking

9. **Applying Concepts** Describe how three different kinds of electromagnetic waves have been useful to you today.
10. **Making Comparisons** Compare the wavelengths of infrared waves, ultraviolet light, and visible light.

Interpreting Graphics

The waves in the diagram below represent two different kinds of EM waves. Use the diagram below to answer the questions that follow.

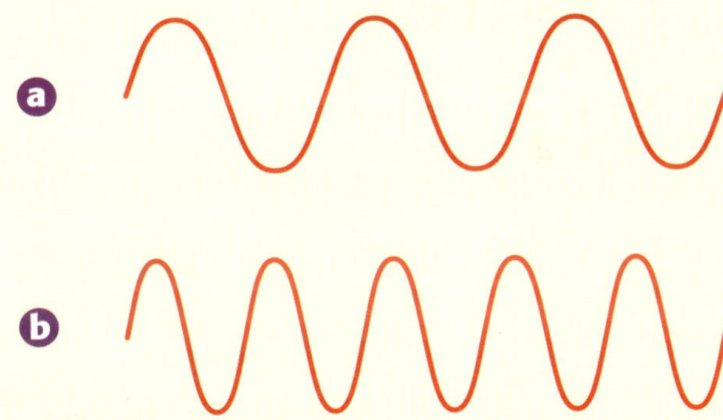

11. Which wave has the longest wavelength?
12. Suppose that one of the waves represents a microwave and one of the waves represents a radio wave. Which wave represents the microwave?

For a variety of links related to this chapter, go to www.scilinks.org

Topic: Electromagnetic Spectrum
SciLinks code: HSM0482

SECTION 3

Interactions of Light Waves

READING WARM-UP

Objectives

- Describe how reflection allows you to see things.
- Describe absorption and scattering.
- Explain how refraction can create optical illusions and separate white light into colors.
- Explain the relationship between diffraction and wavelength.
- Compare constructive and destructive interference of light.

Terms to Learn

reflection
absorption
scattering
refraction
diffraction
interference

READING STRATEGY

Reading Organizer As you read this section, make a concept map by using the terms above.

Have you ever seen a cat's eyes glow in the dark when light shines on them? Cats have a special layer of cells in the back of their eyes that reflects light.

This layer helps the cat see better by giving the eyes another chance to detect the light. Reflection is one interaction of electromagnetic waves. Because we can see visible light, it is easier to explain all wave interactions by using visible light.

Reflection

Reflection happens when light waves bounce off an object. Light reflects off objects all around you. When you look in a mirror, you are seeing light that has been reflected twice—first from you and then from the mirror. If light is reflecting off everything around you, why can't you see your image on a wall? To answer this question, you must learn the law of reflection.

The Law of Reflection

Light reflects off surfaces the same way that a ball bounces off the ground. If you throw the ball straight down against a smooth surface, it will bounce straight up. If you bounce it at an angle, it will bounce away at an angle. The *law of reflection* states that the angle of incidence is equal to the angle of reflection. *Incidence* is the arrival of a beam of light at a surface. **Figure 1** shows this law.

Reading Check What is the law of reflection? (*See the Appendix for answers to Reading Checks.*)

Figure 1 The Law of Reflection

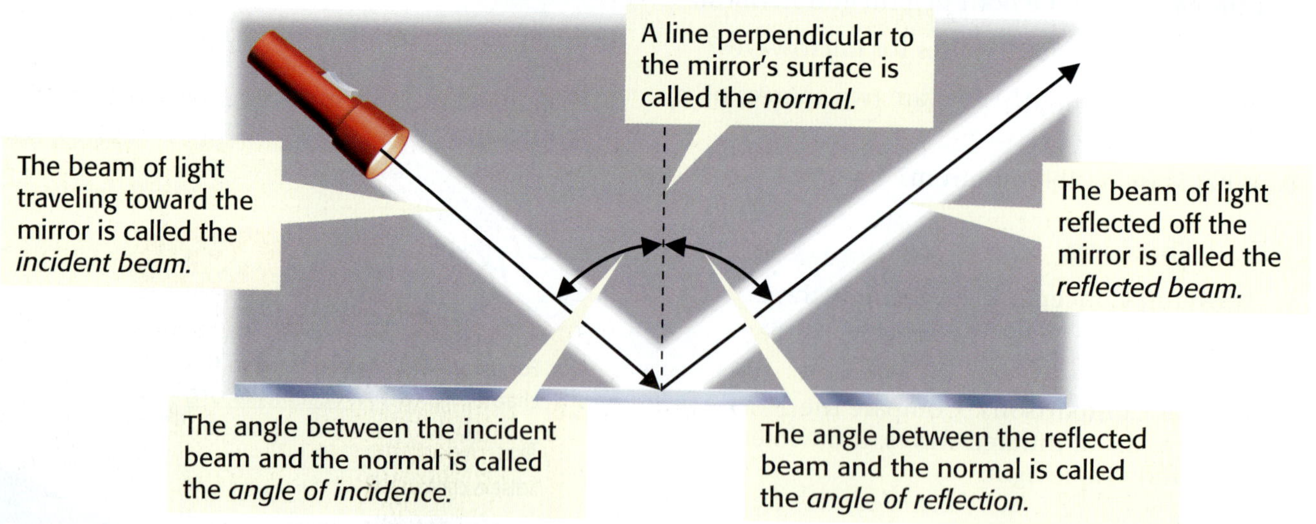

Figure 2 Regular Reflection Vs. Diffuse Reflection

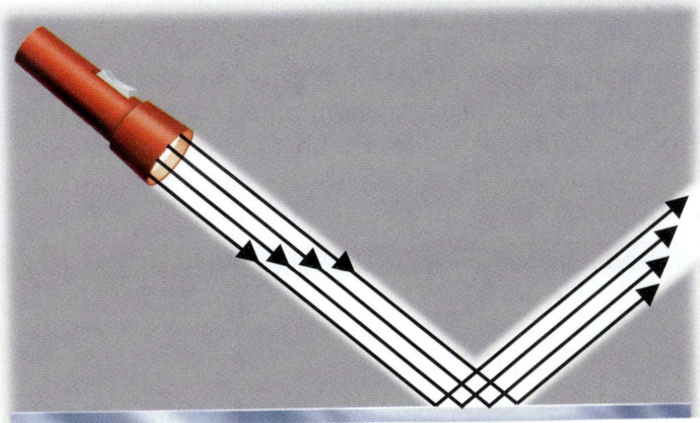

Regular reflection occurs when light beams are reflected at the same angle. When your eye detects the reflected beams, you can see a reflection on the surface.

Diffuse reflection occurs when light beams reflect at many different angles. You can't see a reflection because not all of the reflected light is directed toward your eyes.

Types of Reflection

So, why can you see your image in a mirror but not in a wall? The answer has to do with the differences between the two surfaces. A mirror's surface is very smooth. Thus, light beams reflect off all points of the mirror at the same angle. This kind of reflection is called *regular reflection.* A wall's surface is slightly rough. Light beams will hit the wall's surface and reflect at many different angles. This kind of reflection is called *diffuse reflection.* **Figure 2** shows the difference between the two kinds of reflection.

reflection the bouncing back of a ray of light, sound, or heat when the ray hits a surface that it does not go through

Light Source or Reflection?

If you look at a TV set in a bright room, you see the cabinet around the TV and the image on the screen. But if you look at the same TV in the dark, you see only the image on the screen. The difference is that the screen is a light source, but the cabinet around the TV is not.

You can see a light source even in the dark because its light passes directly into your eyes. The tail of the firefly in **Figure 3** is a light source. Flames, light bulbs, and the sun are also light sources. Objects that produce visible light are called *luminous* (LOO muh nuhs).

Most things around you are not light sources. But you can still see them because light from light sources reflects off the objects and then travels to your eyes. A visible object that is not a light source is *illuminated.*

✓ Reading Check List four different light sources.

Figure 3 *You can see the tail of this firefly because it is luminous. But you see its body because it is illuminated.*

CONNECTION TO Astronomy

Moonlight? Sometimes, the moon shines so brightly that you might think there is a lot of "moonlight." But did you know that moonlight is actually sunlight? The moon does not give off light. You can see the moon because it is illuminated by light from the sun. You see different phases of the moon because light from the sun shines only on the part of the moon that faces the sun. Make a poster that shows the different phases of the moon.

ACTIVITY

absorption in optics, the transfer of light energy to particles of matter

scattering an interaction of light with matter that causes light to change its energy, direction of motion, or both

Absorption and Scattering

Have you noticed that when you use a flashlight, the light shining on things closer to you appears brighter than the light shining on things farther away? The light is less bright the farther it travels from the flashlight. The light is weaker partly because the beam spreads out and partly because of absorption and scattering.

Absorption of Light

The transfer of energy carried by light waves to particles of matter is called **absorption.** When a beam of light shines through the air, particles in the air absorb some of the energy from the light. As a result, the beam of light becomes dim. The farther the light travels from its source, the more it is absorbed by particles, and the dimmer it becomes.

Figure 4 *A beam of light becomes dimmer partly because of scattering.*

Scattering of Light

Scattering is an interaction of light with matter that causes light to change direction. Light scatters in all directions after colliding with particles of matter. Light from the ship shown in **Figure 4** is scattered out of the beam by air particles. This scattered light allows you to see things that are outside the beam. But, because light is scattered out of the beam, the beam becomes dimmer.

Scattering makes the sky blue. Light with shorter wavelengths is scattered more than light with longer wavelengths. Sunlight is made up of many different colors of light, but blue light (which has a very short wavelength) is scattered more than any other color. So, when you look at the sky, you see a background of blue light.

Reading Check **Why can you see things outside a beam of light?**

Refraction

Imagine that you and a friend are at a lake. Your friend wades into the water. You look at her, and her feet appear to have separated from her legs! What has happened? You know her feet did not fall off, so how can you explain what you see? The answer has to do with refraction.

Refraction and Material

Refraction is the bending of a wave as it passes at an angle from one substance, or material, to another. **Figure 5** shows a beam of light refracting twice. Refraction of light waves occurs because the speed of light varies depending on the material through which the waves are traveling. In a vacuum, light travels at 300,000 km/s, but it travels more slowly through matter. When a wave enters a new material at an angle, the part of the wave that enters first begins traveling at a different speed from that of the rest of the wave.

refraction the bending of a wave as the wave passes between two substances in which the speed of the wave differs

Refraction and Lenses

A *lens* is a transparent object that refracts light to form an image. Lenses are used in cameras, telescopes, microscopes, and eyeglasses. Two kinds of lenses are convex lenses and concave lenses. *Convex lenses* are thicker in the middle than at the edges. When light beams pass through a convex lens, the beams are refracted toward each other. *Concave lenses* are thinner in the middle than at the edges. When light beams pass through a concave lens, the beams are refracted away from each other.

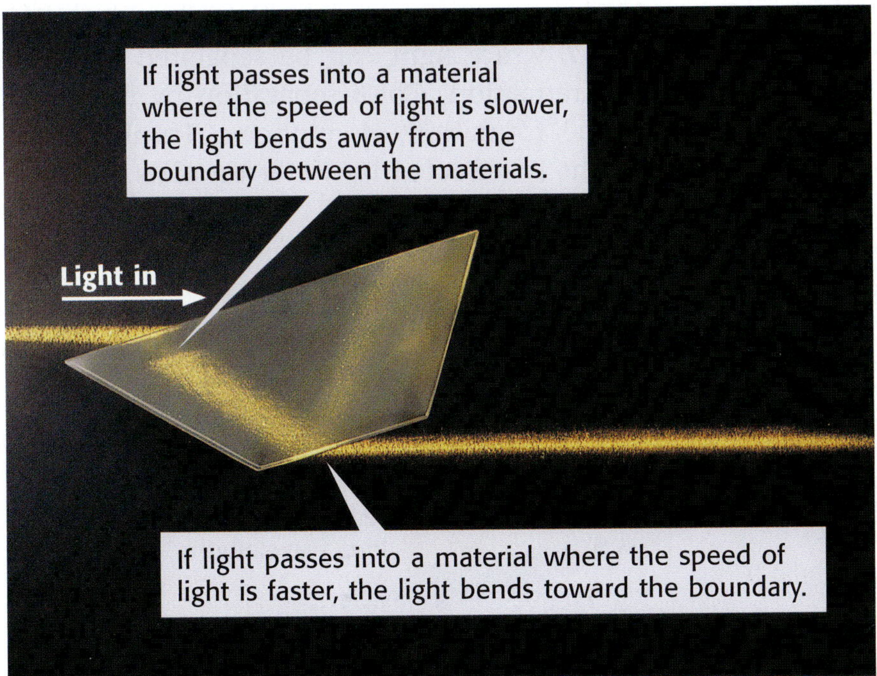

Figure 5 *Light travels more slowly through glass than it does through air. So, light refracts as it passes at an angle from air to glass or from glass to air. Notice that the light is also reflected inside the prism.*

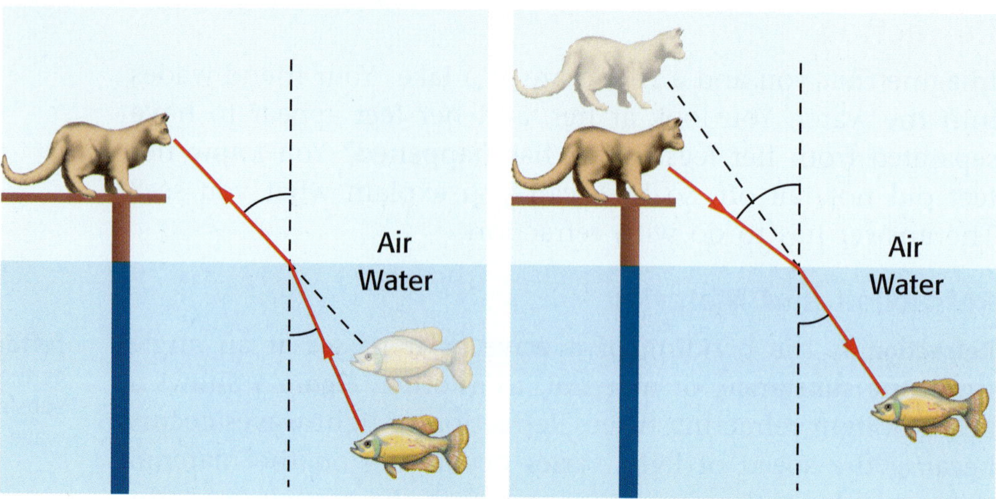

Figure 6 *Because of refraction, the cat and the fish see optical illusions. To the cat, the fish appears closer than it really is. To the fish, the cat appears farther away than it actually is.*

Refraction and Optical Illusions

Usually, when you look at an object, the light reflecting off the object travels in a straight line from the object to your eye. Your brain always interprets light as traveling in straight lines. But when you look at an object that is underwater, the light reflecting off the object does not travel in a straight line. Instead, it refracts. **Figure 6** shows how refraction creates an optical illusion. This kind of illusion causes a person's feet to appear separated from the legs when the person is wading.

Refraction and Color Separation

White light is composed of all the wavelengths of visible light. The different wavelengths of visible light are seen by humans as different colors. When white light is refracted, the amount that the light bends depends on its wavelength. Waves with short wavelengths bend more than waves with long wavelengths. As shown in **Figure 7,** white light can be separated into different colors during refraction. Color separation by refraction is responsible for the formation of rainbows. Rainbows are created when sunlight is refracted by water droplets.

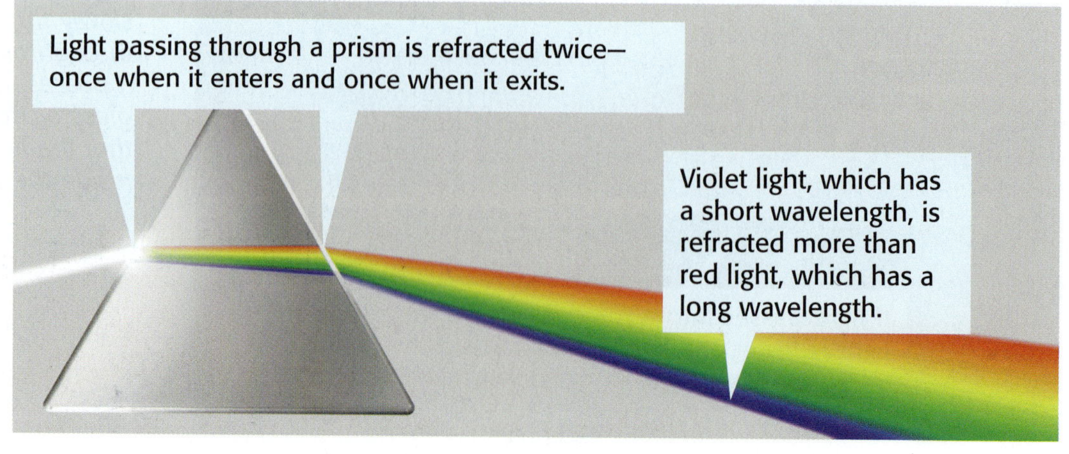

Figure 7 *A prism is a piece of glass that separates white light into the colors of visible light by refraction.*

Refraction Rainbow

1. **Tape** a **piece of construction paper** over the end of a **flashlight.** Use **scissors** to cut a slit in the paper.
2. Turn on the flashlight, and lay it on a table. Place a **prism** on end in the beam of light.
3. Slowly rotate the prism until you can see a rainbow on the surface of the table. Draw a diagram of the light beam, the prism, and the rainbow.

Diffraction

Refraction isn't the only way light waves are bent. **Diffraction** is the bending of waves around barriers or through openings. The amount a wave diffracts depends on its wavelength and the size of the barrier or the opening. The greatest amount of diffraction occurs when the barrier or opening is the same size or smaller than the wavelength.

diffraction a change in the direction of a wave when the wave finds an obstacle or an edge, such as an opening

Reading Check **The amount a wave diffracts depends on what two things?**

Diffraction and Wavelength

The wavelength of visible light is very small—about 100 times thinner than a human hair! So, a light wave cannot bend very much by diffraction unless it passes through a narrow opening, around sharp edges, or around a small barrier, as shown in **Figure 8.**

Light waves cannot diffract very much around large obstacles, such as buildings. Thus, you can't see around corners. But light waves always diffract a small amount. You can observe light waves diffracting if you examine the edges of a shadow. Diffraction causes the edges of shadows to be blurry.

Figure 8 *This diffraction pattern is made by light of a single wavelength shining around the edges of a very tiny disk.*

Interference

interference the combination of two or more waves that results in a single wave

Interference is a wave interaction that happens when two or more waves overlap. Overlapping waves can combine by constructive or destructive interference.

Constructive Interference

When waves combine by *constructive interference,* the resulting wave has a greater amplitude, or height, than the individual waves had. Constructive interference of light waves can be seen when light of one wavelength shines through two small slits onto a screen. The light on the screen will appear as a series of alternating bright and dark bands, as shown in **Figure 9.** The bright bands result from light waves combining through constructive interference.

INTERNET ACTIVITY

For another activity related to this chapter, go to **go.hrw.com** and type in the keyword **HP5LGTW.**

Reading Check What is constructive interference?

Destructive Interference

When waves combine by *destructive interference,* the resulting wave has a smaller amplitude than the individual waves had. So, when light waves interfere destructively, the result will be dimmer light. Destructive interference forms the dark bands seen in **Figure 9.**

You do not see constructive or destructive interference of white light. To understand why, remember that white light is composed of waves with many different wavelengths. The waves rarely line up to combine in total destructive interference.

Figure 9 Constructive and Destructive Interference

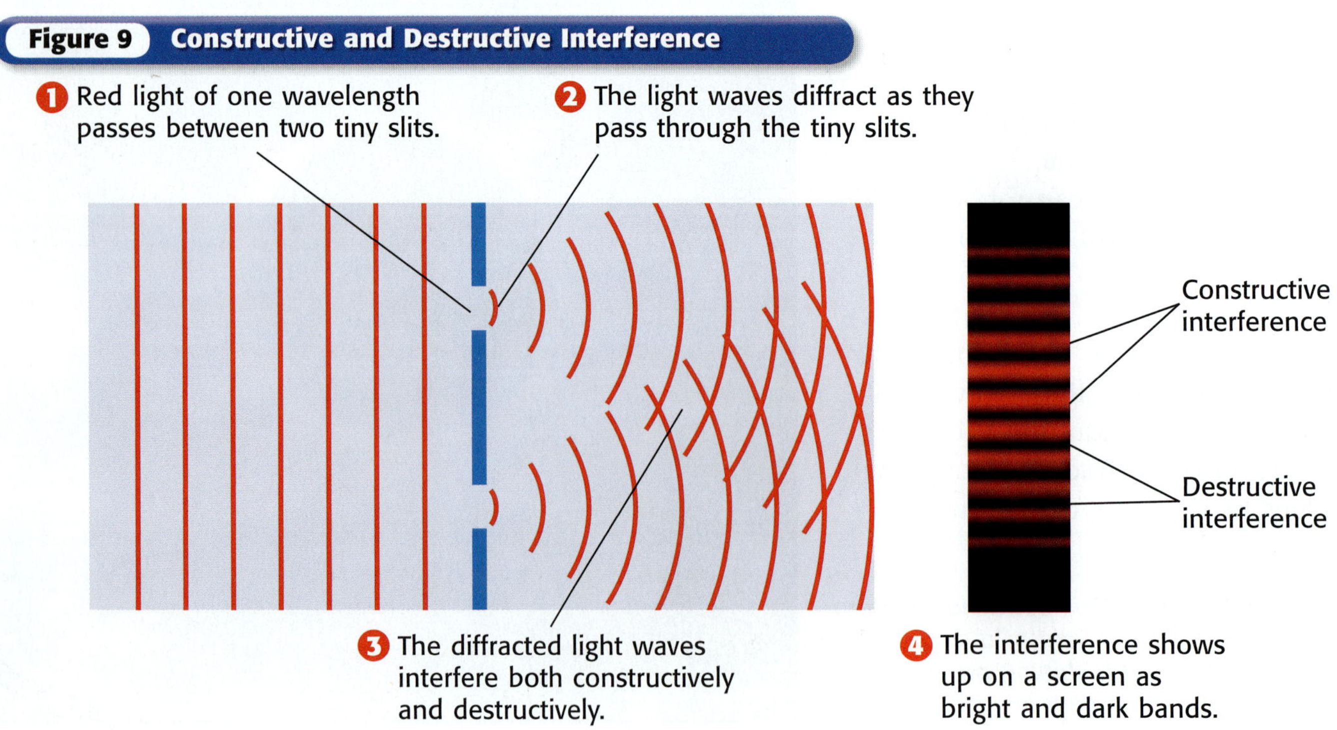

SECTION Review

- The law of reflection states that the angle of incidence is equal to the angle of reflection.
- Things that are luminous can be seen because they produce their own light. Things that are illuminated can be seen because light reflects off them.
- Absorption is the transfer of light energy to particles of matter. Scattering is an interaction of light with matter that causes light to change direction.
- Refraction of light waves can create optical illusions and can separate white light into separate colors.
- How much light waves diffract depends on the light's wavelength. Light waves diffract more when traveling through a narrow opening.
- Interference can be constructive or destructive. Interference of light waves can cause bright and dark bands.

Using Key Terms

For each pair of terms, explain how the meanings of the terms differ.

1. *refraction* and *diffraction*
2. *absorption* and *scattering*

Understanding Key Ideas

3. Which light interaction explains why you can see things that do not produce their own light?
   a. absorption
   b. reflection
   c. refraction
   d. scattering
4. Describe how absorption and scattering can affect a beam of light.
5. Why do objects that are underwater look closer than they actually are?
6. How does a prism separate white light into different colors?
7. What is the relationship between diffraction and the wavelength of light?

Critical Thinking

8. **Applying Concepts** Explain why you can see your reflection on a spoon but not on a piece of cloth.
9. **Making Inferences** The planet Mars does not produce light. Explain why you can see Mars shining like a star at night.
10. **Making Comparisons** Compare constructive interference and destructive interference.

Interpreting Graphics

Use the image below to answer the questions that follow.

11. Why doesn't the large beam of light bend like the two beams in the middle of the tank?
12. Which light interaction explains what is happening to the bottom light beam?

SECTION 4

Light and Color

Why are strawberries red and bananas yellow? How can a soda bottle be green, yet you can still see through it?

If white light is made of all the colors of light, how do things get their color from white light? Why aren't all things white in white light? Good questions! To answer these questions, you need to know how light interacts with matter.

READING WARM-UP

Objectives

- Name and describe the three ways light interacts with matter.
- Explain how the color of an object is determined.
- Explain why mixing colors of light is called *color addition*.
- Describe why mixing colors of pigments is called *color subtraction*.

Terms to Learn

transmission, transparent, translucent, opaque, pigment

READING STRATEGY

Discussion Read this section silently. Write down questions that you have about this section. Discuss your questions in a small group.

Light and Matter

When light strikes any form of matter, it can interact with the matter in three different ways—the light can be reflected, absorbed, or transmitted.

Reflection happens when light bounces off an object. Reflected light allows you to see things. Absorption is the transfer of light energy to matter. Absorbed light can make things feel warmer. **Transmission** is the passing of light through matter. You see the transmission of light all the time. All of the light that reaches your eyes is transmitted through air. Light can interact with matter in several ways at the same time. Look at **Figure 1.** Light is transmitted, reflected, and absorbed when it strikes the glass in a window.

transmission the passing of light or other form of energy through matter

Figure 1 Transmission, Reflection, and Absorption

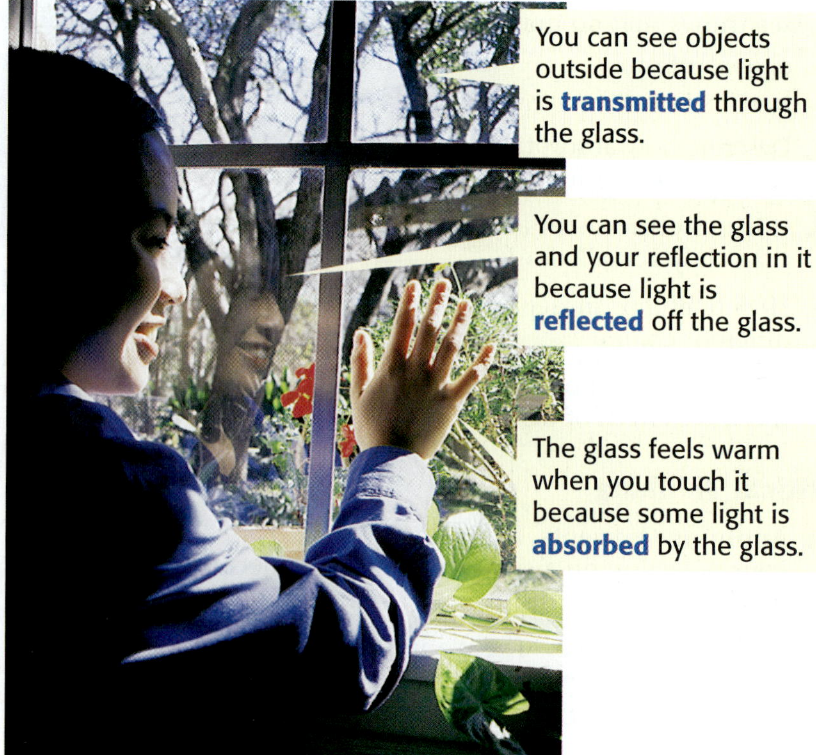

Figure 2 Transparent, Translucent, and Opaque

Transparent plastic makes it easy to see what you are having for lunch.

Translucent wax paper makes it a little harder to see exactly what's for lunch.

Opaque aluminum foil makes it impossible to see your lunch without unwrapping it.

Types of Matter

Matter through which visible light is easily transmitted is said to be **transparent.** Air, glass, and water are examples of transparent matter. You can see objects clearly when you view them through transparent matter.

Sometimes, windows in bathrooms are made of frosted glass. If you look through one of these windows, you will see only blurry shapes. You can't see clearly through a frosted window because it is translucent (trans LOO suhnt). **Translucent** matter transmits light but also scatters the light as it passes through the matter. Wax paper is an example of translucent matter.

Matter that does not transmit any light is said to be **opaque** (oh PAYK). You cannot see through opaque objects. Metal, wood, and this book are examples of opaque objects. You can compare transparent, translucent, and opaque matter in **Figure 2.**

transparent describes matter that allows light to pass through with little interference

translucent describes matter that transmits light but that does not transmit an image

opaque describes an object that is not transparent or translucent

Reading Check **List two examples of translucent objects.** *(See the Appendix for answers to Reading Checks.)*

Colors of Objects

How is an object's color determined? Humans see different wavelengths of light as different colors. For example, humans see long wavelengths as red and short wavelengths as violet. And, some colors, like pink and brown, are seen when certain combinations of wavelengths are present.

The color that an object appears to be is determined by the wavelengths of light that reach your eyes. Light reaches your eyes after being reflected off an object or after being transmitted through an object. When your eyes receive the light, they send signals to your brain. Your brain interprets the signals as colors.

Figure 3 Opaque Objects and Color

When white light shines on a strawberry, only red light is reflected. Other colors of light are absorbed. Therefore, the strawberry looks red to you.

The white hair in this cow's hide reflects all the colors of light, but the black hair absorbs all the colors.

Colors of Opaque Objects

When white light strikes a colored opaque object, some colors of light are absorbed, and some are reflected. Only the light that is reflected reaches your eyes and is detected. So, the colors of light that are reflected by an opaque object determine the color you see. For example, if a sweater reflects blue light and absorbs all other colors, you will see that the sweater is blue. Another example is shown on the left in **Figure 3.**

What colors of light are reflected by the cow shown on the right in **Figure 3**? Remember that white light includes all colors of light. So, white objects—such as the white hair in the cow's hide—appear white because all the colors of light are reflected. On the other hand, black is the absence of color. When light strikes a black object, all the colors are absorbed.

Reading Check **What happens when white light strikes a colored opaque object?**

Colors of Transparent and Translucent Objects

The color of transparent and translucent objects is determined differently than the color of opaque objects. Ordinary window glass is colorless in white light because it transmits all the colors that strike it. But some transparent objects are colored. When you look through colored transparent or translucent objects, you see the color of light that was transmitted through the material. The other colors were absorbed, as shown in **Figure 4.**

Figure 4 *This bottle is green because the plastic transmits green light.*

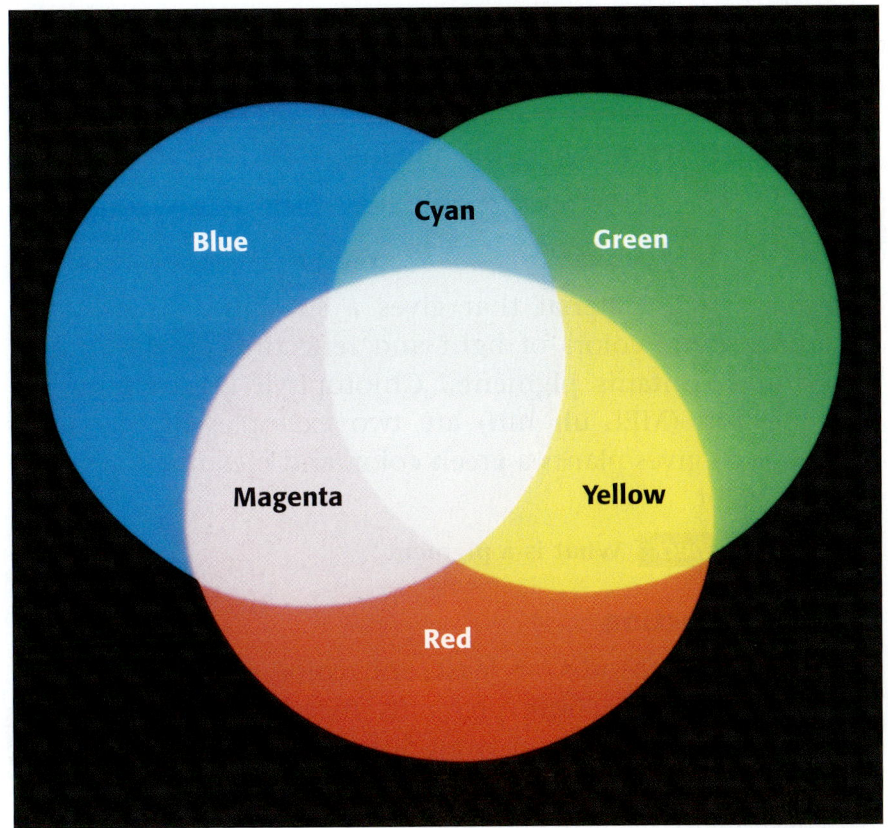

Figure 5 *Primary colors of light—written in white—combine to produce white light. Secondary colors of light—written in black—are the result of two primary colors added together.*

Mixing Colors of Light

In order to get white light, you must combine all colors of light, right? This method is one way of doing it. But you can also get light that appears white by adding just three colors of light together—red, blue, and green. The combination of these three colors is shown in **Figure 5.** In fact, these three colors can be combined in different ratios to produce many colors. Red, blue, and green are called the *primary colors of light.*

Color Addition

When colors of light combine, you see different colors. Combining colors of light is called *color addition.* When two primary colors of light are added together, you see a *secondary color of light*. The secondary colors of light are cyan (blue plus green), magenta (blue plus red), and yellow (red plus green). **Figure 5** shows how secondary colors of light are formed.

Light and Color Television

The colors on a color television are produced by color addition of the primary colors of light. A television screen is made up of groups of tiny red, green, and blue dots. Each dot will glow when the dot is hit by an electron beam. The colors given off by the glowing dots add together to produce all the different colors you see on the screen.

Television Colors

Turn on a color television. Ask an adult to carefully sprinkle a few tiny drops of water onto the television screen. Look closely at the drops of water, and discuss what you see. In your **science journal,** write a description of what you saw.

ACTIVITY

Mixing Colors of Pigment

If you have ever tried mixing paints in art class, you know that you can't make white paint by mixing red, blue, and green paint. The difference between mixing paint and mixing light is due to the fact that paint contains pigments.

Pigments and Color

pigment a substance that gives another substance or a mixture its color

A **pigment** is a material that gives a substance its color by absorbing some colors of light and reflecting others. Almost everything contains pigments. Chlorophyll (KLAWR uh FIL) and melanin (MEL uh nin) are two examples of pigments. Chlorophyll gives plants a green color, and melanin gives your skin its color.

Reading Check What is a pigment?

Color Subtraction

Each pigment absorbs at least one color of light. Look at **Figure 6.** When you mix pigments together, more colors of light are absorbed or taken away. So, mixing pigments is called *color subtraction.*

The *primary pigments* are yellow, cyan, and magenta. They can be combined to produce any other color. In fact, every color in this book was produced by using just the primary pigments and black ink. The black ink was used to provide contrast to the images. **Figure 7** shows how the four pigments combine to produce many different colors.

Rose-Colored Glasses?

1. Obtain **four plastic filters**—red, blue, yellow, and green.
2. Look through one filter at an object across the room. Describe the object's color.
3. Repeat step 2 with each of the filters.
4. Repeat step 2 with two or three filters together.
5. Why do you think the colors change when you use more than one filter?
6. Write your observations and answers.

Figure 6 *Primary pigments—written in black—combine to produce black. Secondary pigments—written in white—are the result of the subtraction of two primary pigments.*

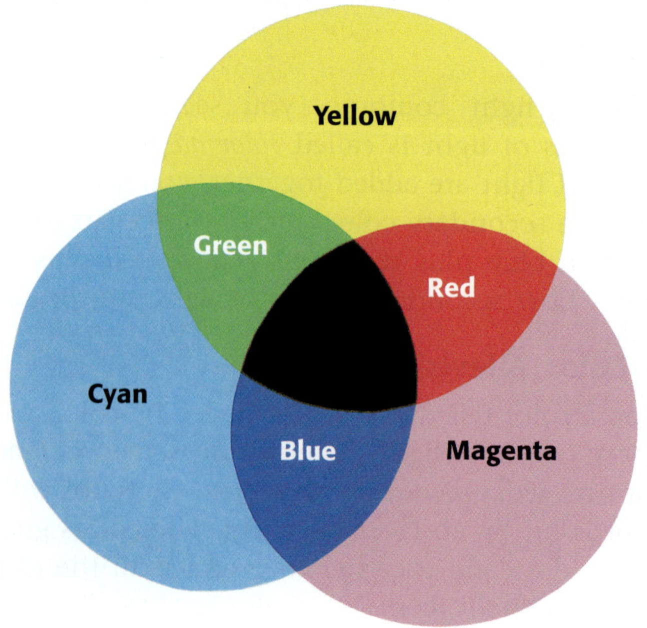

Figure 7 **Color Subtraction and Color Printing**

The picture of the balloon on the left was made by overlapping yellow ink, cyan ink, magenta ink, and black ink.

Yellow Cyan Magenta Black

SECTION Review

Summary

- Objects are transparent, translucent, or opaque, depending on their ability to transmit light.
- Colors of opaque objects are determined by the color of light that they reflect.
- Colors of translucent and transparent objects are determined by the color of light they transmit.
- White light is a mixture of all colors of light.
- Light combines by color addition. The primary colors of light are red, blue, and green.
- Pigments give objects color. Pigments combine by color subtraction. The primary pigments are magenta, cyan, and yellow.

Using Key Terms

1. Use the following terms in the same sentence: *transmission* and *transparent*.
2. In your own words, write a definition for each of the following terms: *translucent* and *opaque*.

Understanding Key Ideas

3. You can see through a car window because the window is
   a. opaque.
   b. translucent.
   c. transparent.
   d. transmitted.
4. Name and describe three different ways light interacts with matter.
5. How is the color of an opaque object determined?
6. Describe how the color of a transparent object is determined.
7. What are the primary colors of light, and why are they called *primary colors*?
8. What four colors of ink were used to print this book?

Critical Thinking

9. **Applying Concepts** What happens to the different colors of light when white light shines on an opaque violet object?
10. **Analyzing Ideas** Explain why mixing colors of light is called *color addition* but mixing pigments is called *color subtraction*.

Interpreting Graphics

11. Look at the image below. The red rose was photographed in red light. Explain why the leaves appear black and the petals appear red.

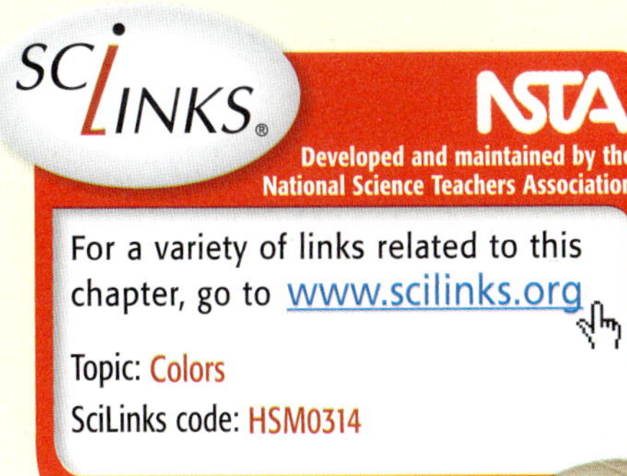

For a variety of links related to this chapter, go to www.scilinks.org

Topic: Colors
SciLinks code: HSM0314

SECTION 5

Light and Sight

When you look around, you can see objects both near and far. You can also see the different colors of the objects.

READING WARM-UP

Objectives

- Identify the parts of the human eye, and describe their functions.
- Describe three common vision problems.
- Describe surgical eye correction.

Terms to Learn

nearsightedness
farsightedness

READING STRATEGY

Reading Organizer As you read this section, make a flowchart of how the eye works.

You see objects that emit their own light because the light enters your eyes. You see all other objects because light reflected from the objects enters your eyes. But how do your eyes work, and what causes people to have vision problems?

How You Detect Light

Visible light is the part of the electromagnetic spectrum that can be detected by your eyes. Your eye gathers light to form the images that you see. The steps of this process are shown in **Figure 1.** Muscles around the lens change the thickness of the lens so that objects at different distances can be seen in focus. The light that forms the real image is detected by receptors in the retina called *rods* and *cones*. Rods can detect very dim light. Cones detect colors in bright light.

Figure 1 **How Your Eyes Work**

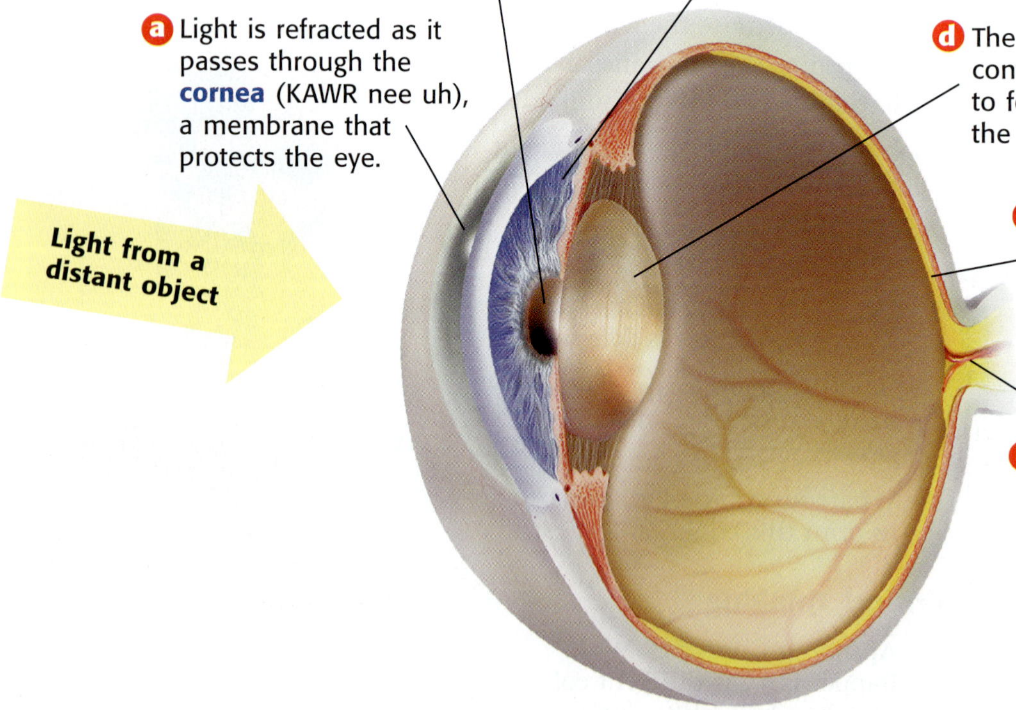

Figure 2 **Correcting Nearsightedness and Farsightedness**

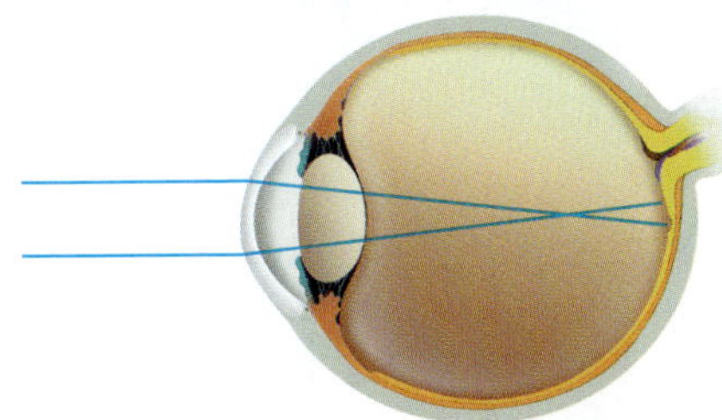

Nearsightedness happens when the eye is too long, which causes the lens to focus light in front of the retina.

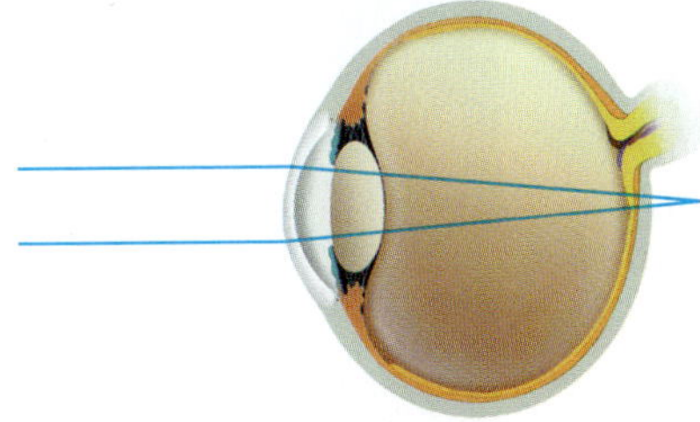

Farsightedness happens when the eye is too short, which causes the lens to focus light behind the retina.

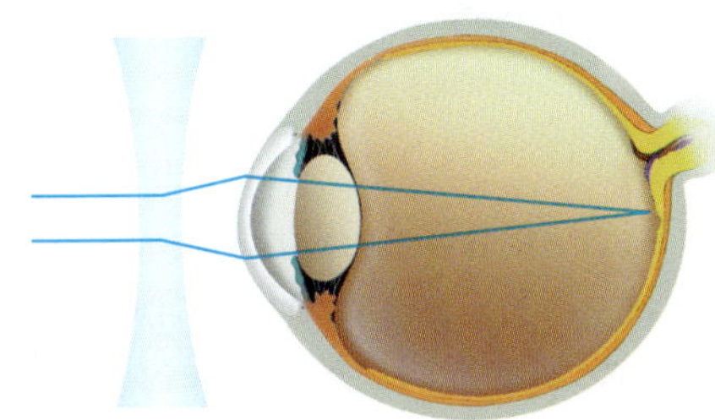

A **concave lens** placed in front of a nearsighted eye refracts the light outward. The lens in the eye can then focus the light on the retina.

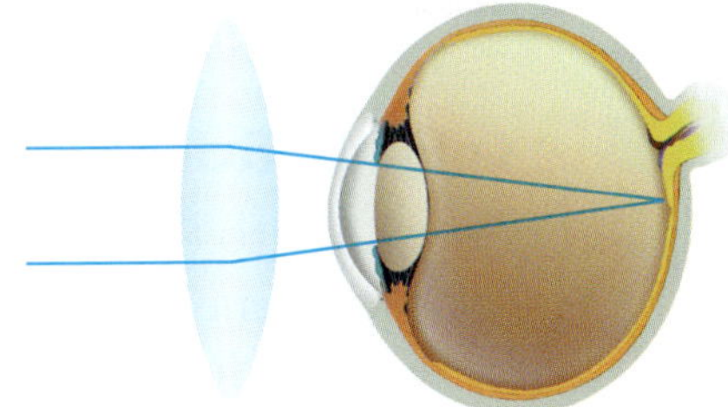

A **convex lens** placed in front of a farsighted eye focuses the light. The lens in the eye can then focus the light on the retina.

Common Vision Problems

People who have normal vision can clearly see objects that are close and objects that are far away. They can also tell the difference between all colors of visible light. But because the eye is complex, it's no surprise that many people have defects in their eyes that affect their vision.

Nearsightedness and Farsightedness

The lens of a properly working eye focuses light on the retina. So, the images formed are always clear. Two common vision problems happen when light is not focused on the retina, as shown in **Figure 2.** **Nearsightedness** happens when a person's eye is too long. A nearsighted person can see something clearly only if it is nearby. Objects that are far away look blurry. **Farsightedness** happens when a person's eye is too short. A farsighted person can see faraway objects clearly. But things that are nearby look blurry. **Figure 2** also shows how these vision problems can be corrected with glasses.

Reading Check What causes nearsightedness and farsightedness? (*See the Appendix for answers to Reading Checks.*)

nearsightedness a condition in which the lens of the eye focuses distant objects in front of rather than on the retina

farsightedness a condition in which the lens of the eye focuses distant objects behind rather than on the retina

Figure 3 *The photo on the left is what a person who has normal vision sees. The photo on the right is a simulation of what a person who has red-green color deficiency might see.*

Color Deficiency

About 5% to 8% of men and 0.5% of women in the world have *color deficiency,* or colorblindness. The majority of people who have color deficiency can't tell the difference between shades of red and green or can't tell red from green. **Figure 3** compares what a person with normal vision sees with what a person who has red-green color deficiency sees. Color deficiency cannot be corrected.

Color deficiency happens when the cones in the retina do not work properly. The three kinds of cones are named for the colors they detect most—red, green, or blue. But each kind can detect many colors of light. A person who has normal vision can see all colors of visible light. But in some people, the cones respond to the wrong colors. Those people see certain colors, such as red and green, as a different color, such as yellow.

Reading Check **What are the three kinds of cones?**

CONNECTION TO Biology

Color Deficiency and Genes The ability to see color is a sex-linked genetic trait. Certain genes control which colors of light the cones detect. If these genes are defective in a person, that person will have color deficiency. A person needs one set of normal genes to have normal color vision. Genes that control the red cones and the green cones are on the X chromosome. Women have two X chromosomes, but men have only one. So, men are more likely than women to lack a set of these genes and to have red-green color deficiency. Research two other sex-linked traits, and make a graph comparing the percentage of men and women who have the traits.

ACTIVITY

Surgical Eye Correction

Using surgery to correct nearsightedness or farsightedness is possible. Surgical eye correction works by reshaping the patient's cornea. Remember that the cornea refracts light. So, reshaping the cornea changes how light is focused on the retina.

To prepare for eye surgery, an eye doctor uses a machine to measure the patient's corneas. A laser is then used to reshape each cornea so that the patient gains perfect or nearly perfect vision. **Figure 4** shows a patient undergoing eye surgery.

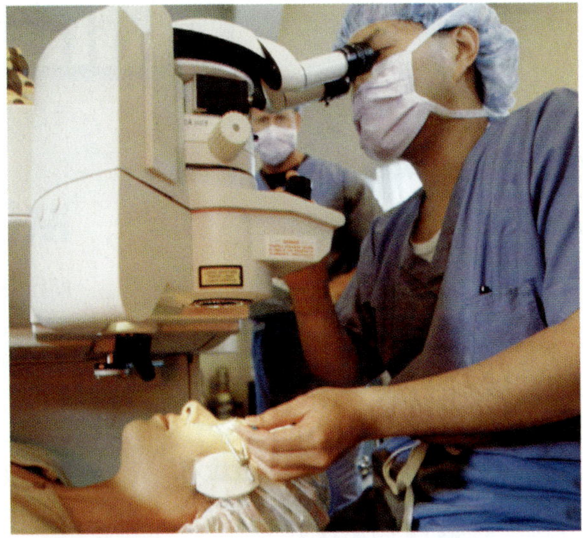

Figure 4 *An eye surgeon uses a very precise laser to reshape this patient's cornea.*

Risks of Surgical Eye Correction

Although vision-correction surgery can be helpful, it has some risks. Some patients report glares or double vision. Others have trouble seeing at night. Other patients lose vision permanently. People under 20 years old shouldn't have vision-correction surgery because their vision is still changing.

SECTION Review

Summary

- The human eye has several parts, including the cornea, the pupil, the iris, the lens, and the retina.
- Nearsightedness and farsightedness happen when light is not focused on the retina. Both problems can be corrected with glasses or eye surgery.
- Color deficiency is a condition in which cones in the retina respond to the wrong colors.
- Eye surgery can correct some vision problems.

Using Key Terms

1. Use each of the following terms in a separate sentence: *nearsightedness* and *farsightedness*.

Understanding Key Ideas

2. A person who is nearsighted will have the most trouble reading
   a. a computer screen in front of him or her.
   b. a book in his or her hands.
   c. a street sign across the street.
   d. the title of a pamphlet on a nearby table.
3. List the parts of the eye, and describe what each part does.
4. What are three common vision problems?
5. How are nearsightedness and farsightedness corrected?
6. Describe surgical eye correction.
7. What do the rods and cones in the eye do?

Math Skills

8. About 0.5% of women have a color deficiency. How many women out of 200 have a color deficiency?

Critical Thinking

9. **Forming Hypotheses** Why do you think color deficiency cannot be corrected?
10. **Expressing Opinions** Would you have surgical eye correction? Explain your reasons.

Skills Practice Lab

Mixing Colors

Mix two colors, such as red and green, and you create a new color. Is the new color brighter or darker? Color and brightness depend on the light that reaches your eye. And what reaches your eye depends on whether you are adding colors (mixing colors of light) or subtracting colors (mixing colors of pigments). In this activity, you will do both types of color formation and see the results firsthand!

OBJECTIVES

Use flashlights to mix colors of light by color addition.

Use paints to mix colors of pigments by color subtraction.

MATERIALS

Part A

- colored filters, red, green, and blue (1 of each)
- flashlights (3)
- paper, white
- tape, masking

Part B

- cups, small plastic or paper (2)
- paintbrush
- paper, white
- ruler, metric
- tape, masking
- water
- watercolor paints

SAFETY

Part A: Color Addition

Procedure

1. Tape a colored filter over each flashlight lens.
2. In a darkened room, shine the red light on a sheet of white paper. Then, shine the green light next to the red light. You should have two circles of light, one red and one green, next to each other.
3. Move the flashlights so that the circles overlap by half their diameter. What color is formed where the circles overlap? Is the mixed area brighter or darker than the single-color areas? Record your observations.

4. Repeat steps 2 and 3 with the red and blue lights.
5. Now, shine all three lights at the same point on the paper. Record your observations.

Analyze the Results

1. **Describing Events** In general, when you mixed two colors, was the result brighter or darker than the original colors?
2. **Explaining Events** In step 5, you mixed all three colors. Was the resulting color brighter or darker than when you mixed two colors? Explain your observations in terms of color addition.

Draw Conclusions

3. **Making Predictions** What do you think would happen if you mixed together all the colors of light? Explain your answer.

Part B: Color Subtraction

Procedure

1. Place a piece of masking tape on each cup. Label one cup "Clean" and the other cup "Dirty." Fill each cup about half full with water.
2. Wet the paintbrush thoroughly in the "Clean" cup. Using the watercolor paints, paint a red circle on the white paper. The circle should be approximately 4 cm in diameter.
3. Clean the brush by rinsing it first in the "Dirty" cup and then in the "Clean" cup.
4. Paint a blue circle next to the red circle. Then, paint half the red circle with the blue paint.
5. Examine the three areas: red, blue, and mixed. What color is the mixed area? Does it appear brighter or darker than the red and blue areas? Record your observations.
6. Clean the brush by repeating Step 3. Paint a green circle 4 cm in diameter, and then paint half the blue circle with green paint.

7. Examine the green, blue, and mixed areas. Record your observations.
8. Now add green paint to the mixed red-blue area so that you have an area that is a mixture of red, green, and blue paint. Clean the brush again.
9. Finally, record your observations of this new mixed area.

Analyze the Results

1. **Identifying Patterns** In general, when you mixed two colors, was the result brighter or darker than the original colors?
2. **Analyzing Results** In step 8, you mixed all three colors. Was the result brighter or darker than the result from mixing two colors? Explain what you saw in terms of color subtraction.

Draw Conclusions

3. **Drawing Conclusions** Based on your results, what do you think would happen if you mixed all the colors of paint? Explain your answer.

Chapter Review

USING KEY TERMS

Complete each of the following sentences by choosing the correct term from the word bank.

interference	radiation
scattering	opaque
translucent	transmission
electromagnetic wave	electromagnetic spectrum

1. _____ is the transfer of energy by electromagnetic waves.

2. This book is a(n) _____ object.

3. _____ is a wave interaction that occurs when two or more waves overlap and combine.

4. Light is a kind of _____ and can therefore travel through matter and space.

5. During _____, light travels through an object.

UNDERSTANDING KEY IDEAS

Multiple Choice

6. Electromagnetic waves transmit
   a. charges.
   b. fields.
   c. matter.
   d. energy.

7. Objects that transmit light easily are
   a. opaque.
   b. translucent.
   c. transparent.
   d. colored.

8. You can see yourself in a mirror because of
   a. absorption.
   b. scattering.
   c. regular reflection.
   d. diffuse reflection.

9. Shadows have blurry edges because of
   a. diffraction.
   b. scattering.
   c. diffuse reflection.
   d. refraction.

10. What color of light is produced when red light is added to green light?
    a. cyan
    b. blue
    c. yellow
    d. white

11. Prisms produce the colors of the rainbow through
    a. reflection.
    b. refraction.
    c. diffraction.
    d. interference.

12. Which kind of electromagnetic wave travels fastest in a vacuum?
    a. radio wave
    b. visible light
    c. gamma ray
    d. They all travel at the same speed.

13. A vision problem that happens when light is focused in front of the retina is
    a. farsightedness.
    b. nearsightedness.
    c. color deficiency.
    d. None of the above

Short Answer

14 How are gamma rays used?

15 What are two uses for radio waves?

16 Describe surgical eye correction. What are the risks of surgical eye correction?

Math Skills

17 Calculate the time it takes for light from the sun to reach Mercury. Mercury is 54,900,000 km away from the sun.

CRITICAL THINKING

18 **Concept Mapping** Use the following terms to create a concept map: *light, matter, reflection, absorption,* and *transmission.*

19 **Applying Concepts** A tern is a type of bird that dives underwater to catch fish. When a young tern begins learning to catch fish, the bird is rarely successful. The tern has to learn that when a fish appears to be in a certain place underwater, the fish is actually in a slightly different place. Why does the tern see the fish in the wrong place?

Air
Water

20 **Evaluating Conclusions** Imagine that you are teaching your younger brother about light. You tell him that white light is light of all the colors of the rainbow combined. But your brother says that you are wrong because mixing different colors of paint produces black and not white. Explain why your brother's conclusion is wrong.

21 **Making Inferences** If you look around a parking lot during the summer, you might see sunshades set up in the windshields of cars. How do sunshades help keep the insides of cars cool?

INTERPRETING GRAPHICS

22 Each of the pictures below shows the effects of a wave interaction of light. Identify the interaction involved.

a.

b.

c.

Standardized Test Preparation

READING

Read each of the passages below. Then, answer the questions that follow each passage.

Passage 1 Jaundice occurs in some infants when bilirubin—a pigment in healthy red blood cells—builds up in the bloodstream as blood cells break down. This excess bilirubin is deposited in the skin, giving the skin a yellowish hue. Jaundice is not dangerous if treated quickly. If left untreated, it can lead to brain damage.

The excess bilirubin in the skin is best broken down by bright blue light. For this reason, hospitals hang special blue fluorescent lights above the cribs of newborns needing treatment. The blue light is sometimes balanced with light of other colors so that doctors and nurses can be sure the baby is not blue from a lack of oxygen.

1. Which of the following is a fact in the passage?

A Jaundice is always very dangerous.

B Bilirubin in the skin of infants can be broken down with bright blue light.

C Excess bilirubin in the skin gives the skin a bright blue hue.

D Blue lights can make a baby blue from a lack of oxygen.

2. What is the purpose of this passage?

F to explain what jaundice is and how it is treated

G to warn parents about shining blue light on their babies

H to persuade light bulb manufacturers to make blue light bulbs

I to explain the purpose of bilirubin in red blood cells

Passage 2 If you have ever looked inside a toaster while toasting a piece of bread, you may have seen thin wires or bars glowing red. The wires give off energy as light when heated to a high temperature. Light produced by hot objects is called *incandescent light*. Most of the lamps in your home probably use incandescent light bulbs.

Sources of incandescent light also release a large amount of thermal energy. Thermal energy is sometimes called *heat energy*. Sometimes, thermal energy from incandescent light is used to cook food or to warm a room. But often this thermal energy is not used for anything. For example, the thermal energy given off by light bulbs is not very useful.

1. What does the word *thermal* mean, based on its use in the passage?

A light

B energy

C heat

D food

2. What is incandescent light?

F light used for cooking food

G light that is red in color

H light that is not very useful

I light produced by hot objects

3. Which of the following can be inferred from the passage?

A Sources of incandescent light are rarely found in an average home.

B A toaster uses thermal energy to toast bread.

C Incandescent light from light bulbs is often used to cook food.

D The thermal energy produced by incandescent light sources is always useful.

INTERPRETING GRAPHICS

The angles of refraction in the table were measured when a beam of light entered the material from air at a 45° angle. Use the table below to answer the questions that follow.

Material and Refraction		
Material	Index of refraction	Angle of refraction
Diamond	2.42	17°
Glass	1.52	28°
Quartz	1.46	29°
Water	1.33	32°

1. Which material has the highest index of refraction?
   - **A** diamond
   - **B** glass
   - **C** quartz
   - **D** water

2. Which material has the greatest angle of refraction?
   - **F** diamond
   - **G** glass
   - **H** quartz
   - **I** water

3. Which of the following statements **best** describes the data in the table?
   - **A** The higher the index of refraction, the greater the angle of refraction.
   - **B** The higher the index of refraction, the smaller the angle of refraction.
   - **C** The greater the angle of refraction, the higher the index of refraction.
   - **D** There is no relationship between the index of refraction and the angle of refraction.

4. Which two materials would be the most difficult to separate by observing only their angles of refraction?
   - **F** diamond and glass
   - **G** glass and quartz
   - **H** quartz and water
   - **I** water and diamond

MATH

Read each question below, and choose the best answer.

1. A square metal plate has an area of 46.3 cm^2. The length of one side of the plate is between which two values?
   - **A** 4 cm and 5 cm
   - **B** 5 cm and 6 cm
   - **C** 6 cm and 7 cm
   - **D** 7 cm and 8 cm

2. A jet was flying over the Gulf of Mexico at an altitude of 2,150 m. Directly below the jet, a submarine was at a depth of −383 m. What was the distance between the jet and the submarine?
   - **F** −2,533 m
   - **G** −1,767 m
   - **H** 1,767 m
   - **I** 2,533 m

3. The speed of light in a vacuum is exactly 299,792,458 m/s. Which of the following is a good estimate of the speed of light?
   - **A** 3.0×10^{-8} m/s
   - **B** 2.0×10^{8} m/s
   - **C** 3.0×10^{8} m/s
   - **D** 3.0×10^{9} m/s

4. The wavelength of the yellow light produced by a sodium vapor lamp is 0.000000589 m. Which of the following is equal to the wavelength of the sodium lamp's yellow light?
   - **F** -5.89×10^{7} m
   - **G** 5.89×10^{-9} m
   - **H** 5.89×10^{-7} m
   - **I** 5.89×10^{7} m

5. Amira purchased a box of light bulbs for $3.81. There are three light bulbs in the box. What is the cost per light bulb?
   - **A** $0.79
   - **B** $1.06
   - **C** $1.27
   - **D** $11.43

Science in Action

Weird Science

Fireflies Light the Way

Just as beams of light from lighthouses warn boats of approaching danger, the light of an unlikely source—fireflies—is being used by scientists to warn food inspectors of bacterial contamination.

Fireflies use an enzyme called *luciferase* to make light. Scientists have taken the gene from fireflies that tells cells how to make luciferase. They put this gene into a virus that preys on bacteria. The virus is not harmful to humans and can be mixed into meat. When the virus infects bacteria in the meat, the virus transfers the gene into the genes of the bacteria. The bacteria then produce luciferase and glow! So, if a food inspector sees glowing meat, the inspector knows that the meat is contaminated with bacteria.

Social Studies ACTIVITY

WRITING SKILL Many cultures have myths to explain certain natural phenomena. Read some of these myths. Then, write your own myth titled "How Fireflies Got Their Fire."

Science, Technology, and Society

It's a Heat Wave

In 1946, Percy Spencer visited a laboratory belonging to Raytheon—the company he worked for. When he stood near a device called a *magnetron,* he noticed that a candy bar in his pocket melted. Spencer hypothesized that the microwaves produced by the magnetron caused the candy bar to warm up and melt. To test his hypothesis, Spencer put a bag of popcorn kernels next to the magnetron. The microwaves heated the kernels, causing them to pop! Spencer's simple experiment showed that microwaves could heat foods quickly. Spencer's discovery eventually led to the development of the microwave oven—an appliance found in many kitchens today.

Math ACTIVITY

Popcorn pops when the inside of the kernel reaches a temperature of about 175°C. Convert this temperature to degrees Fahrenheit.

People in Science

Albert Einstein

A Light Pioneer When Albert Einstein was 15 years old, he asked himself, "What would the world look like if I were speeding along on a motorcycle at the speed of light?" For many years afterward, he would think about this question and about the very nature of light, time, space, and matter. He even questioned the ideas of Isaac Newton, which had been widely accepted for 200 years. Einstein was bold. And he was able to see the universe in a totally new way.

In 1905, Einstein published a paper on the nature of light. He knew from the earlier experiments of others that light was a wavelike phenomenon. But he theorized that light could also travel as particles. Scientists did not readily accept Einstein's particle theory of light. Even 10 years later, the American physicist Robert Millikan, who proved that the particle theory of light was true, was reluctant to believe his own experimental results. Einstein's theory helped pave the way for television, computers, and other important technologies. The theory also earned Einstein a Nobel Prize in physics in 1921.

Language Arts ACTIVITY

WRITING SKILL Imagine that it is 1921. You are a newspaper reporter writing an article about Albert Einstein and his Nobel Prize. Write a one-page article about Albert Einstein, his theory, and the award he won.

To learn more about these Science in Action topics, visit go.hrw.com and type in the keyword HP5LGTF.

Check out Current Science® articles related to this chapter by visiting go.hrw.com. Just type in the keyword HP5CS22.

TIMELINE

Life Science

What did you have for breakfast this morning? Your breakfast was a result of living things working together. For example, milk comes from a cow. The cow eats plants to gain energy. Bacteria help the plants obtain nutrients from the soil. And the soil has nutrients because fungi break down dead trees.

In this unit, you will study how living things interact. You will also discover ways in which all living things are similar. In addition, you will learn about organisms that cause disease and what your body does to defend itself against these organisms.

Around 2700 BCE

Si Ling-Chi, empress of China, observes silkworms in her garden and develops a process to cultivate them and make silk.

1931

The first electron microscope is developed.

1934

Dorothy Crowfoot Hodgkin uses X-ray techniques to determine the protein structure of insulin.

1970

Floppy disks for computer data storage are introduced.

1983

Dian Fossey writes *Gorillas in the Mist*, a book about her research on mountain gorillas in Africa and her efforts to save them from poachers.

Around 1000

Arab mathematician and physicist Ibn al Haytham discovers that vision is caused by the reflection of light from objects into the eye.

1684

Improvements to microscopes allow the first observation of red blood cells.

1914

His studies on agriculture and soil conservation lead George Washington Carver to perform research on peanuts.

1944

Oswald T. Avery demonstrates that DNA is the material that carries genetic properties in living organisms.

1946

ENIAC, the first entirely electronic computer, is built. It weighs 30 tons.

1967

Dr. Christiaan Barnard performs the first successful human heart transplant.

1984

A process known as DNA fingerprinting is developed by Alec Jeffreys.

1998

In China, scientists discover a fossil of a dinosaur that had feathers.

2001

A team of scientists led by Philippa Uwins announces that tiny nanobes that are 20 to 150 nanometers wide have been found in Australia. Scientists debate whether these particles are living.

It's Alive!! Or Is It?

About the PHOTO

What does it mean to say something is *alive*? Machines have some of the characteristics of living things, but machines do not have all of these characteristics. This amazing robot insect can respond to changes in its environment. It can walk over obstacles. It can perform some tasks. But it is still not alive. How is it like and unlike a living insect?

PRE-READING ACTIVITY

Graphic Organizer

Concept Map Before you read the chapter, create the graphic organizer entitled "Concept Map" described in the **Study Skills** section of the Appendix. As you read the chapter, fill in the concept map with details about the characteristics of living things.

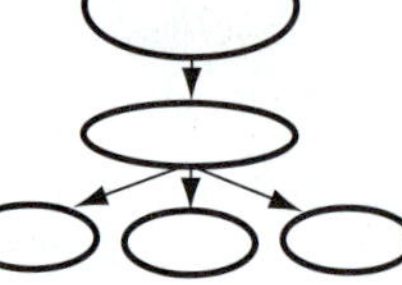

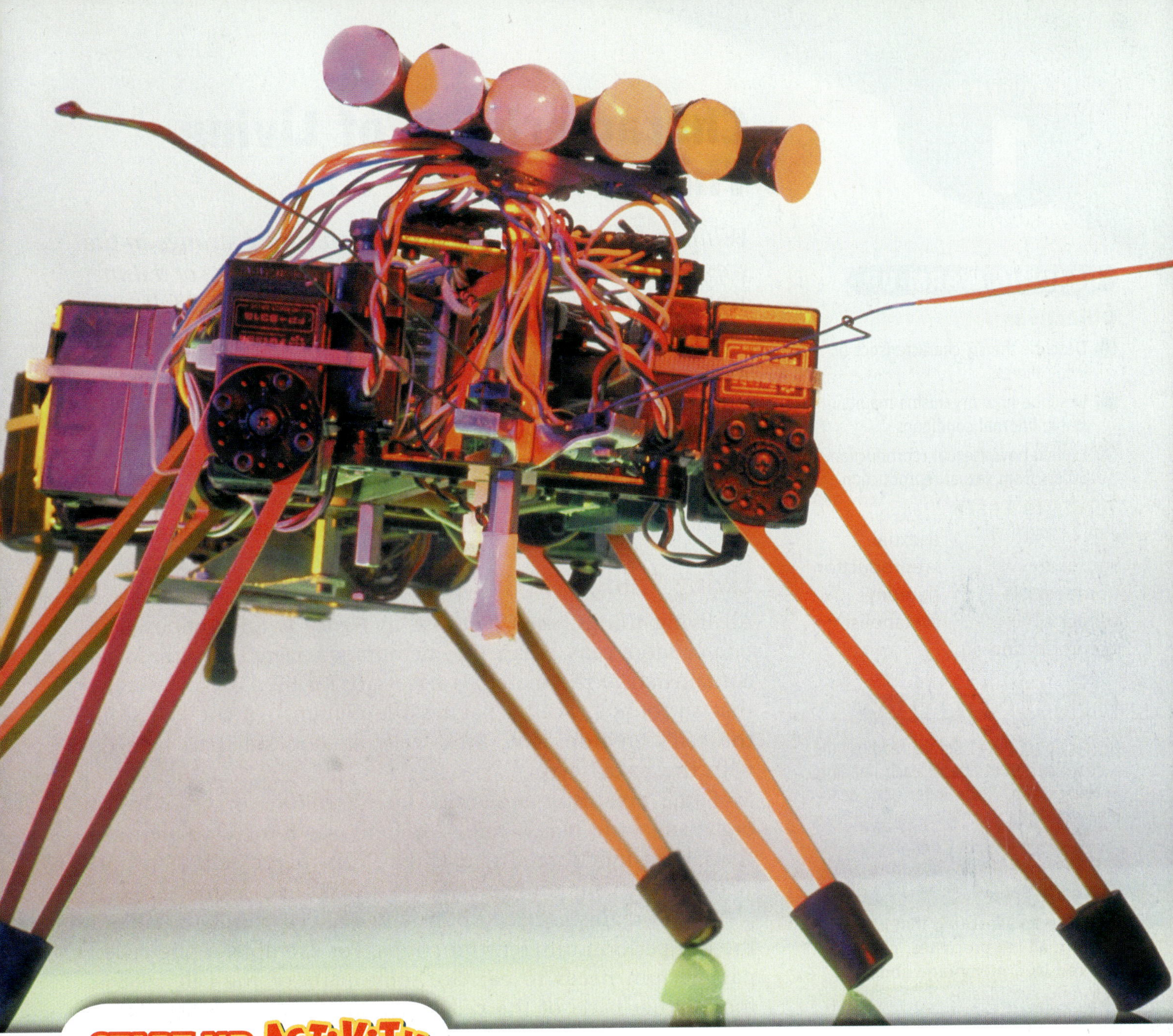

START-UP ACTIVITY

Lights On!

In this activity, you will work with a partner to see how eyes react to changes in light.

Procedure

1. Observe a classmate's eyes in a lighted room. Note the size of your partner's pupils.
2. Have your partner keep both eyes open. Ask him or her to cover each one with a cupped hand. Wait about one minute.
3. Instruct your partner to pull away both hands quickly. Immediately, look at your partner's pupils. Record what happens.
4. Now, briefly shine a **flashlight** into your partner's eyes. Record how this affects your partner's pupils. **Caution:** Do not use the sun as the source of the light.
5. Change places with your partner, and repeat steps 1–4 so that your partner can observe your eyes.

Analysis

1. How did your partner's eyes respond to changes in the level of light?
2. How did changes in the size of your pupils affect your vision? What does this tell you about why pupils change size?

SECTION 1

Characteristics of Living Things

READING WARM-UP

Objectives

- Describe the six characteristics of living things.
- Describe how organisms maintain stable internal conditions.
- Explain how asexual reproduction differs from sexual reproduction.

Terms to Learn

cell
stimulus
homeostasis
sexual reproduction
asexual reproduction
heredity
metabolism

READING STRATEGY

Prediction Guide Before reading this section, write the title of each heading in this section. Next, under each heading, write what you think you will learn.

While outside one day, you notice something strange in the grass. It's slimy, bright yellow, and about the size of a dime. You have no idea what it is. Is it a plant part that fell from a tree? Is it alive? How can you tell?

An amazing variety of living things exists on Earth. But living things are all alike in several ways. What does a dog have in common with a bacterium? What does a fish have in common with a mushroom? And what do *you* have in common with a slimy, yellow blob, known as a *slime mold*? Read on to find out about the six characteristics that all organisms share.

Living Things Have Cells

All living things, such as those in **Figure 1**, are composed of one or more cells. A **cell** is a membrane-covered structure that contains all of the materials necessary for life. The membrane that surrounds a cell separates the contents of the cell from the cell's environment. Most cells are too small to be seen with the naked eye.

Some organisms are made up of trillions of cells. In an organism with many cells, different kinds of cells perform specialized functions. For example, your nerve cells transport signals, and your muscle cells are specialized for movement.

In an organism made up of only one cell, different parts of the cell perform different functions. For example, a one-celled paramecium needs to eat. So, some parts of the cell take in food. Other parts of the cell break down the food. Still other parts of the cell excrete wastes.

cell the smallest unit that can perform all life processes; cells are covered by a membrane and have DNA and cytoplasm

Figure 1 *Some organisms, such as the protists on the right, are made of one cell or a few cells. The monkeys on the left are made up of trillions of cells.*

Figure 2 *The touch of an insect triggers the Venus' flytrap to close its leaves quickly.*

Living Things Sense and Respond to Change

All organisms have the ability to sense change in their environment and to respond to that change. When your pupils are exposed to light, they respond by becoming smaller. A change that affects the activity of the organism is called a **stimulus** (plural, *stimuli*).

Stimuli can be chemicals, gravity, light, sounds, hunger, or anything that causes organisms to respond in some way. A gentle touch causes a response in the plant shown in **Figure 2.**

stimulus anything that causes a reaction or change in an organism or any part of an organism

homeostasis the maintenance of a constant internal state in a changing environment

Homeostasis

Even though an organism's outside environment may change, conditions inside an organism's body must stay the same. Many chemical reactions keep an organism alive. These reactions can take place only when conditions are exactly right, so an organism must maintain stable internal conditions to survive. The maintenance of a stable internal environment is called **homeostasis** (HOH mee OH STAY sis).

Responding to External Changes

Your body maintains a temperature of about 37°C. When you get hot, your body responds by sweating. When you get cold, your muscles twitch in an attempt to warm you up. This twitching is called *shivering*. Whether you are sweating or shivering, your body is trying to return itself to normal.

Other animals also need to have stable internal conditions. But many cannot respond the way you do. They have to control their body temperature by moving from one environment to another. If they get too warm, they move to the shade. If they get too cool, they move out into the sunlight.

Reading Check How do some animals maintain homeostasis? (*See the Appendix for answers to Reading Checks.*)

CONNECTION TO Physics

Temperature Regulation Your body temperature does not change very much throughout the day. When you exercise, you sweat. Sweating helps keep your body temperature stable. As your sweat evaporates, your skin cools. Given this information, why do you think you feel cooler faster when you stand in front of a fan?

Figure 3 *Like most animals, bears produce offspring by sexual reproduction.*

Figure 4 *The hydra can reproduce asexually by forming buds that break off and grow into new individuals.*

Living Things Reproduce

Organisms make other organisms similar to themselves. They do so in one of two ways: by sexual reproduction or by asexual reproduction. In **sexual reproduction,** two parents produce offspring that will share characteristics of both parents. Most animals and plants reproduce in this way. The bear cubs in **Figure 3** were produced sexually by their parents.

In **asexual reproduction,** a single parent produces offspring that are identical to the parent. **Figure 4** shows an organism that reproduces asexually. Most single-celled organisms reproduce in this way.

Living Things Have DNA

The cells of all living things contain the molecule **d**eoxyribo**n**ucleic (dee AHKS uh RIE boh noo KLEE ik) **a**cid, or DNA. *DNA* controls the structure and function of cells. When organisms reproduce, they pass copies of their DNA to their offspring. Passing DNA ensures that offspring resemble parents. The passing of traits from one generation to the next is called **heredity.**

Living Things Use Energy

Organisms use energy to carry out the activities of life. These activities include such things as making food, breaking down food, moving materials into and out of cells, and building cells. An organism's **metabolism** (muh TAB uh LIZ uhm) is the total of all of the chemical activities that the organism performs.

✓ Reading Check **Name four chemical activities in living things that require energy.**

sexual reproduction reproduction in which the sex cells from two parents unite, producing offspring that share traits from both parents

asexual reproduction reproduction that does not involve the union of sex cells and in which one parent produces offspring identical to itself

heredity the passing of genetic traits from parent to offspring

metabolism the sum of all chemical processes that occur in an organism

Living Things Grow and Develop

All living things, whether they are made of one cell or many cells, grow during periods of their lives. In a single-celled organism, the cell gets larger and divides, making other organisms. In organisms made of many cells, the number of cells gets larger, and the organism gets bigger.

In addition to getting larger, living things may develop and change as they grow. Just like the organisms in **Figure 5,** you will pass through different stages in your life as you develop into an adult.

Figure 5 *Over time, acorns develop into oak seedlings, which become oak trees.*

SECTION Review

Summary

- Organisms are made of one or more cells.
- Organisms detect and respond to stimuli.
- Organisms make more organisms like themselves by reproducing either asexually or sexually.
- Organisms have DNA.
- Organisms use energy to carry out the chemical activities of life.
- Organisms grow and develop.

Using Key Terms

Complete each of the following sentences by choosing the correct term from the word bank.

cells	stimulus
homeostasis	metabolism

1. Sunlight can be a ___.
2. Living things are made of ___.

Understanding Key Ideas

3. Homeostasis means maintaining
   a. stable internal conditions.
   b. varied internal conditions.
   c. similar offspring.
   d. varied offspring.
4. Explain the difference between asexual and sexual reproduction.
5. Describe the six characteristics of living things.

Math Skills

6. Bacteria double every generation. One bacterium is in the first generation. How many are in the sixth generation?

Critical Thinking

7. **Applying Concepts** How do you respond to some stimuli in your environment?
8. **Identifying Relationships** What does the fur coat of a bear have to do with homeostasis?

For a variety of links related to this chapter, go to www.scilinks.org

Topic: Characteristics of Living Things
SciLinks code: HSM0258

SECTION 2

The Necessities of Life

Would it surprise you to learn that you have the same basic needs as a tree, a frog, and a fly?

In fact, almost every organism has the same basic needs: water, air, a place to live, and food.

READING WARM-UP

Objectives

- Explain why organisms need food, water, air, and living space.
- Describe the chemical building blocks of cells.

Terms to Learn

producer
consumer
decomposer
protein
carbohydrate
lipid
phospholipid
ATP
nucleic acid

READING STRATEGY

Discussion Read this section silently. Write down questions that you have about this section. Discuss your questions in a small group.

Water

You may know that your body is made mostly of water. In fact, your cells and the cells of almost all living organisms are approximately 70% water. Most of the chemical reactions involved in metabolism require water.

Organisms differ greatly in terms of how much water they need and how they get it. You could survive for only about three days without water. You get water from the fluids you drink and the food you eat. The desert-dwelling kangaroo rat never drinks. It gets all of its water from its food.

Air

Air is a mixture of several different gases, including oxygen and carbon dioxide. Most living things use oxygen in the chemical process that releases energy from food. Organisms living on land get oxygen from the air. Organisms living in water either take in dissolved oxygen from the water or come to the water's surface to get oxygen from the air. The European diving spider in **Figure 1** goes to great lengths to get oxygen.

Green plants, algae, and some bacteria need carbon dioxide gas in addition to oxygen. These organisms produce food and oxygen by using photosynthesis (FOHT oh SIN thuh sis). In *photosynthesis*, green organisms convert the energy in sunlight to energy stored in food.

Reading Check What process do plants use to make food? (*See the Appendix for answers to Reading Checks.*)

Figure 1 *This spider surrounds itself with an air bubble that provides the spider with a source of oxygen underwater.*

A Place to Live

All organisms need a place to live that contains all of the things they need to survive. Some organisms, such as elephants, require a large amount of space. Other organisms may live their entire life in one place.

Space on Earth is limited. So, organisms often compete with each other for food, water, and other necessities. Many animals, including the warbler in **Figure 2,** will claim a particular space. After claiming a space, they try to keep other animals away.

Figure 2 *A warbler's song is more than just a pretty tune. The warbler is protecting its home by telling other warblers to stay out of its territory.*

Food

All living things need food. Food gives organisms energy and the raw materials needed to carry on life processes. Organisms use nutrients from food to replace cells and build body parts. But not all organisms get food in the same way. In fact, organisms can be grouped into three different groups based on how they get their food.

Making Food

Some organisms, such as plants, are called producers. **Producers** can make their own food. Like most producers, plants use energy from sunlight to make food, in the form of sugars, from water and carbon dioxide. Some producers get energy and food from the chemicals in their environment.

producer an organism that can make its own food by using energy from its surroundings

consumer an organism that eats other organisms or organic matter

decomposer an organism that gets energy by breaking down the remains of dead organisms or animal wastes and consuming or absorbing the nutrients

Taking Food

Other organisms are called **consumers** because they must eat (consume) other organisms to get food. The frog in **Figure 3** is an example of a consumer. It gets the energy it needs by eating insects and other organisms.

Some consumers are decomposers. **Decomposers** are organisms that get their food by breaking down the nutrients in dead organisms or animal wastes. The mushroom in **Figure 3** is a decomposer.

Figure 3 *The frog is a consumer. The mushroom is a decomposer. The green plants are producers.*

Pen a Menu

WRITING SKILL With a parent, write a menu for a favorite meal. Using Nutrition Facts labels, find out which items on your menu include proteins, carbohydrates, and fats. Try making the meal.

Putting It All Together

Some organisms make their own food. Some organisms get food from eating other organisms. But all organisms need to break down food in order to use the nutrients in it.

Nutrients are made up of molecules. A *molecule* is a substance made when two or more atoms combine. Molecules made of different kinds of atoms are *compounds*. Molecules found in living things are usually made of different combinations of six elements: carbon, hydrogen, nitrogen, oxygen, phosphorus, and sulfur. These elements combine to form proteins, carbohydrates, lipids, ATP, and nucleic acids.

Proteins

Almost all of the life processes of a cell involve proteins. **Proteins** are large molecules that are made up of smaller molecules called *amino acids*.

protein a molecule that is made up of amino acids and that is needed to build and repair body structures and to regulate processes in the body

Making Proteins

The nutrients in food can be used as building materials. For example, organisms break down the proteins in food to supply their cells with amino acids. These amino acids are then linked together to form new proteins. Some proteins are made up of only a few amino acids, but others contain more than 10,000 amino acids.

Proteins in Action

Proteins have many different functions. Some proteins form structures that are easy to see, such as those in **Figure 4.** Other proteins are very small and help cells do their jobs. Inside red blood cells, the protein hemoglobin (HEE moh GLOH bin) binds to oxygen to deliver and release oxygen throughout the body. Other proteins, called *enzymes* (EN ZIEMZ), start or speed up chemical reactions in cells.

Figure 4 *Spider webs, hair, horns, and feathers are all made from proteins.*

Figure 5 *The extra sugar in a potato plant is stored in the potato as starch, a complex carbohydrate.*

Carbohydrates

Molecules made of sugars are called **carbohydrates.** Cells use carbohydrates as a source of energy and for energy storage. An organism's cells break down carbohydrates to release the energy stored in them. There are two kinds of carbohydrates—simple carbohydrates and complex carbohydrates.

carbohydrate a class of energy-giving nutrients that includes sugars, starches, and fiber; contains carbon, hydrogen, and oxygen

Simple Carbohydrates

Simple carbohydrates are made up of one sugar molecule or a few sugar molecules linked together. Table sugar and the sugar in fruits are examples of simple carbohydrates.

Complex Carbohydrates

When an organism has more sugar than it needs, its extra sugar may be stored as complex carbohydrates. *Complex carbohydrates* are made of hundreds of sugar molecules linked together. Plants, such as the potato plant in **Figure 5,** store extra sugar as starch. When you eat mashed potatoes, you are eating a potato plant's stored starch. Your body then breaks down this complex carbohydrate to release the energy stored in the potato.

Reading Check **What is the difference between simple carbohydrates and complex carbohydrates?**

How Much Oxygen?

Each red blood cell carries about 250 million molecules of hemoglobin. How many molecules of oxygen could a single red blood cell deliver throughout the body if every hemoglobin molecule attached to four oxygen molecules?

Starch Search

1. Obtain several **food samples** from your teacher.
2. Put **a few drops of iodine** on each sample. Record your observations. **Caution:** Iodine can stain clothing.
3. When iodine comes into contact with starch, a black substance appears. Which samples contain starch?

Figure 6 **Phospholipid Membranes**

The head of a phospholipid molecule is attracted to water, but the tail is not.

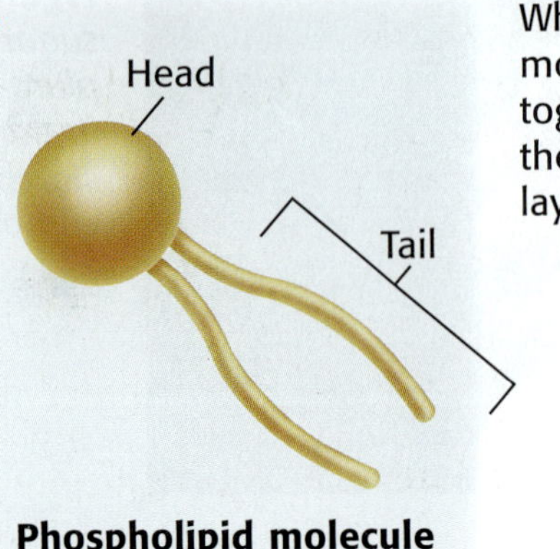

Phospholipid molecule

When phospholipid molecules come together in water, they form two layers.

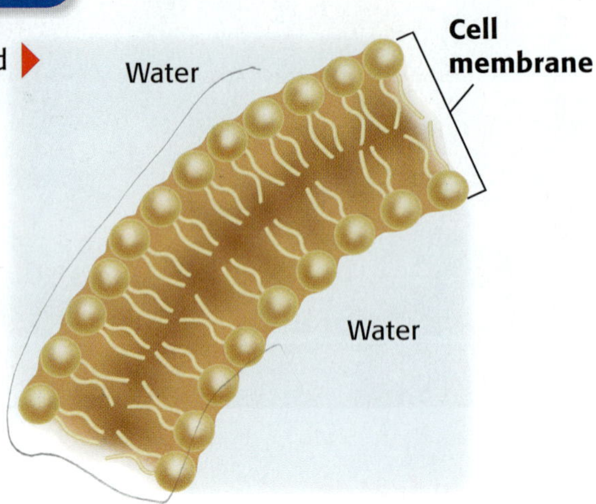

lipid a type of biochemical that does not dissolve in water; fats and steroids are lipids

phospholipid a lipid that contains phosphorus and that is a structural component in cell membranes

ATP **a**denosine **t**ri**p**hosphate, a molecule that acts as the main energy source for cell processes

Lipids

Lipids are compounds that cannot mix with water. Lipids have many important jobs in the cell. Like carbohydrates, some lipids store energy. Other lipids form the membranes of cells.

Phospholipids

All cells are surrounded by a cell membrane. The cell membrane helps protect the cell and keep the internal conditions of the cell stable. **Phospholipids** (FAHS foh LIP idz) are the molecules that form much of the cell membrane. The head of a phospholipid molecule is attracted to water. The tail is not. Cells are mostly water. When phospholipids are in water, the tails come together, and the heads face out into the water. **Figure 6** shows how phospholipid molecules form two layers in water.

Fats and Oils

Fats and oils are lipids that store energy. When an organism has used up most of its carbohydrates, it can get energy from these lipids. The structures of fats and oils are almost the same, but at room temperature, most fats are solid, and most oils are liquid. Most of the lipids stored in plants are oils. Most of the lipids stored in animals are fats.

Reading Check What is one difference between oils and fats?

ATP

Adenosine **trip**hosphate (uh DEN uh SEEN trie FAHS FAYT), also called ATP, is another important molecule. **ATP** is the major energy-carrying molecule in the cell. The energy in carbohydrates and lipids must be transferred to ATP, which then provides fuel for cellular activities.

CONNECTION TO Social Studies

Whaling In the 1900s, whales were hunted and killed for their oil. Whale oil was often used as fuel for oil lamps. Most of the oil taken from whales was taken from their fat, or *blubber*. Some whales had blubber over 18 in. thick, producing over 40 barrels of oil per whale. Research whether anyone still hunts whales or uses whale oil. Make a presentation to the class on your findings.

Nucleic Acids

Nucleic acids are sometimes called the blueprints of life because they have all the information needed for a cell to make proteins. **Nucleic acids** are large molecules made up of molecules called *nucleotides* (NOO klee oh TIEDZ). A nucleic acid may have thousands of nucleotides. The order of those nucleotides stores information. DNA is a nucleic acid. A DNA molecule is like a recipe book entitled *How to Make Proteins.* When a cell needs to make a certain protein, the cell gets information from the order of the nucleotides in DNA. This order of nucleotides tells the cell the order of the amino acids that are linked together to make that protein.

nucleic acid a molecule made up of subunits called *nucleotides*

For another activity related to this chapter, go to **go.hrw.com** and type in the keyword **HL5ALVW.**

SECTION Review

Summary

- Organisms need water for cellular processes.
- Organisms need oxygen to release the energy contained in their food.
- Organisms must have a place to live.
- Cells store energy in carbohydrates, which are made of sugars.
- Proteins are made up of amino acids. Some proteins are enzymes.
- Fats and oils store energy and make up cell membranes.
- Cells use molecules of ATP to fuel their activities.
- Nucleic acids, such as DNA, are made up of nucleotides.

Using Key Terms

For each pair of terms, explain how the meanings of the terms differ.

1. *producer* and *consumer*
2. *lipid* and *phospholipid*

Understanding Key Ideas

3. Plants store extra sugar as
   a. proteins.
   b. starch.
   c. nucleic acids.
   d. phospholipids.
4. Explain why organisms need food, water, air, and living space.
5. Describe the chemical building blocks of cells.
6. Why are decomposers categorized as consumers? How do they differ from producers?
7. What are the subunits of proteins?

Math Skills

8. Protein A is a chain of 660 amino acids. Protein B is a chain of 11 amino acids. How many times more amino acids does protein A have than protein B?

Critical Thinking

9. **Making Inferences** Could life as we know it exist on Earth if air contained only oxygen? Explain.
10. **Identifying Relationships** How might a cave, an ant, and a lake each meet the needs of an organism?
11. **Predicting Consequences** What would happen to the supply of ATP in your cells if you did not eat enough carbohydrates? How would this affect your cells?
12. **Applying Concepts** Which resource do you think is most important to your survival: water, air, a place to live, or food? Explain your answer.

For a variety of links related to this chapter, go to www.scilinks.org

Topic: The Necessities of Life
SciLinks code: HSM1018

SECTION 3

The Organization of Living Things

READING WARM-UP

Objectives

- List three advantages of being multicellular.
- Describe the four levels of organization in living things.
- Explain the relationship between the structure and function of a part of an organism.

Terms to Learn

tissue	organism
organ	structure
organ system	function

READING STRATEGY

Paired Summarizing Read this section silently. In pairs, take turns summarizing the material. Stop to discuss ideas that seem confusing.

In some ways, organisms are like machines. Some machines have just one part. But most machines have many parts. Some organisms exist as a single cell. Other organisms have many—even trillions—of cells.

Most cells are smaller than the period that ends this sentence. Yet, every cell in every organism performs all the processes of life. So, are there any advantages to having many cells?

The Benefits of Being Multicellular

You are a *multicellular organism.* This means that you are made of many cells. The cells in your body are continually dividing to make more cells. New cells replace those cells that die as a normal part of life. Multicellular organisms grow by making more small cells, not by making their cells larger. For example, an elephant is bigger than you are, but its cells are about the same size as yours. An elephant just has more cells than you do. Some benefits of being multicellular are the following:

- **Larger Size** Many multicellular organisms are small. But they are usually larger than single-celled organisms. Larger organisms are prey for fewer predators. Larger predators can eat a wider variety of prey.
- **Longer Life** The life span of a multicellular organism is not limited to the life span of any single cell.
- **Specialization** Each type of cell has a particular job. Specialization makes the organism more efficient. For example, the cardiac muscle cell in **Figure 1** is a specialized muscle cell. Heart muscle cells contract and make the heart pump blood.

Reading Check **List three advantages of being multicellular.** (*See the Appendix for answers to Reading Checks.*)

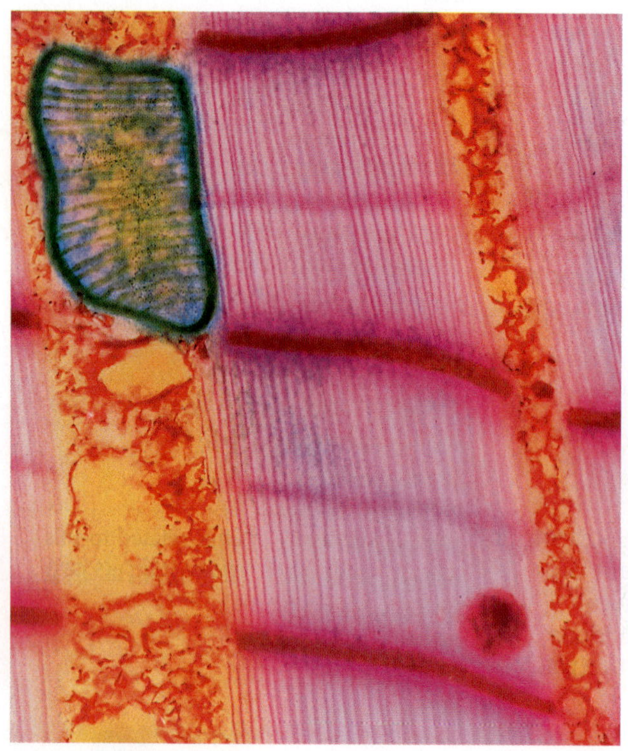

Figure 1 *This photomicrograph shows a small part of one heart muscle cell. The green line surrounds one of many mitochondria, the powerhouses of the cell. The pink areas are muscle filaments.*

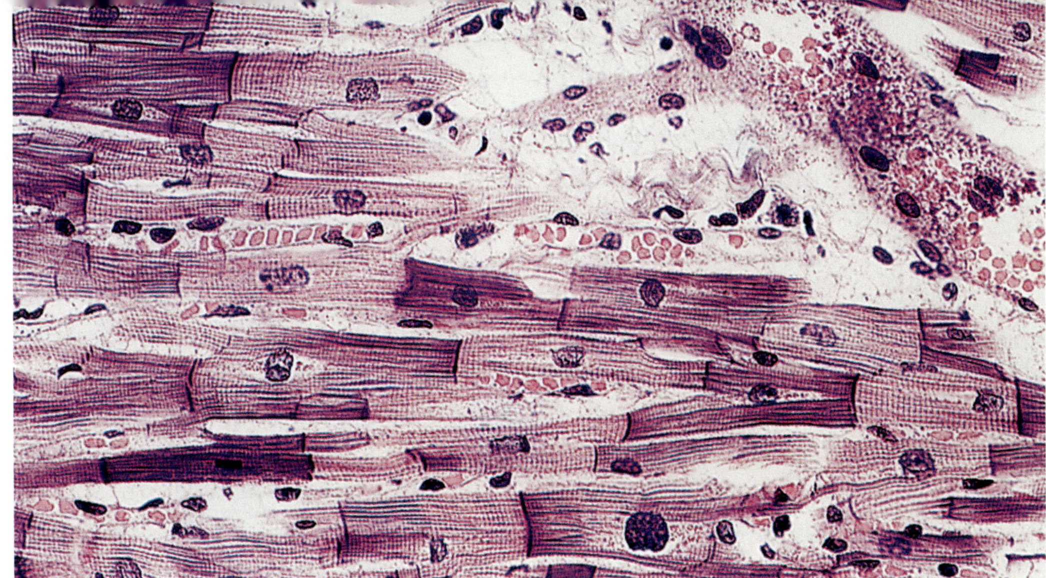

Figure 2 *This photomicrograph shows cardiac muscle tissue. Cardiac muscle tissue is made up of many cardiac cells.*

Cells Working Together

A **tissue** is a group of cells that work together to perform a specific job. The material around and between the cells is also part of the tissue. The cardiac muscle tissue, shown in **Figure 2,** is made of many cardiac muscle cells. Cardiac muscle tissue is just one type of tissue in a heart.

Animals have four basic types of tissues: nerve tissue, muscle tissue, connective tissue, and protective tissue. In contrast, plants have three types of tissues: transport tissue, protective tissue, and ground tissue. Transport tissue moves water and nutrients through a plant. Protective tissue covers the plant. It helps the plant retain water and protects the plant against damage. Photosynthesis takes place in ground tissue.

tissue a group of similar cells that perform a common function

organ a collection of tissues that carry out a specialized function of the body

Tissues Working Together

A structure that is made up of two or more tissues working together to perform a specific function is called an **organ.** For example, your heart is an organ. It is made mostly of cardiac muscle tissue. But your heart also has nerve tissue and tissues of the blood vessels that all work together to make your heart the powerful pump that it is.

Another organ is your stomach. It also has several kinds of tissue. In the stomach, muscle tissue makes food move in and through the stomach. Special tissues make chemicals that help digest your food. Connective tissue holds the stomach together, and nervous tissue carries messages back and forth between the stomach and the brain. Other organs include the intestines, brain, and lungs.

Plants also have different kinds of tissues that work together as organs. A leaf is a plant organ that contains tissue that traps light energy to make food. Other examples of plant organs are stems and roots.

Reading Check What is an organ?

A Pet Protist

Imagine that you have a tiny box-shaped protist for a pet. To care for your pet protist properly, you have to figure out how much to feed it. The dimensions of your protist are roughly 25 µm × 20 µm × 2 µm. If seven food particles per second can enter through each square micrometer of surface area, how many particles can your protist eat in 1 min?

WRITING SKILL **Feeding Your Cells**

With a parent, research how tissues, such as blood, and organ systems, such as the digestive system, provide your body cells with the nutrients and fuel they need to survive. Write a summary of what you learn in your **science journal.**

Organs Working Together

A group of organs working together to perform a particular function is called an **organ system.** Each organ system has a specific job to do in the body.

For example, the digestive system is made up of several organs, including the stomach and intestines. The digestive system's job is to break down food into small particles. Other parts of the body then use these small particles as fuel. In turn, the digestive system depends on the respiratory and cardiovascular systems for oxygen. The cardiovascular system, shown in **Figure 3,** includes organs and tissues such as the heart and blood vessels. Plants also have organ systems. They include leaf systems, root systems, and stem systems.

Reading Check List the levels of organization in living things.

Organisms

Anything that can perform life processes by itself is an **organism.** An organism made of a single cell is called a *unicellular organism.* Bacteria, most protists, and some kinds of fungi are unicellular. Although some of these organisms live in colonies, they are still unicellular. When unicellular organisms live together, all of the cells in the colony are the same. Each cell must carry out all life processes in order for that cell to survive. In contrast, even the simplest multicellular organism has specialized cells that depend on each other for the organism to survive.

organ system a group of organs that work together to perform body functions

organism a living thing; anything that can carry out life processes independently

structure the arrangement of parts in an organism

function the special, normal, or proper activity of an organ or part

Figure 3 Levels of Organization in the Cardiovascular System

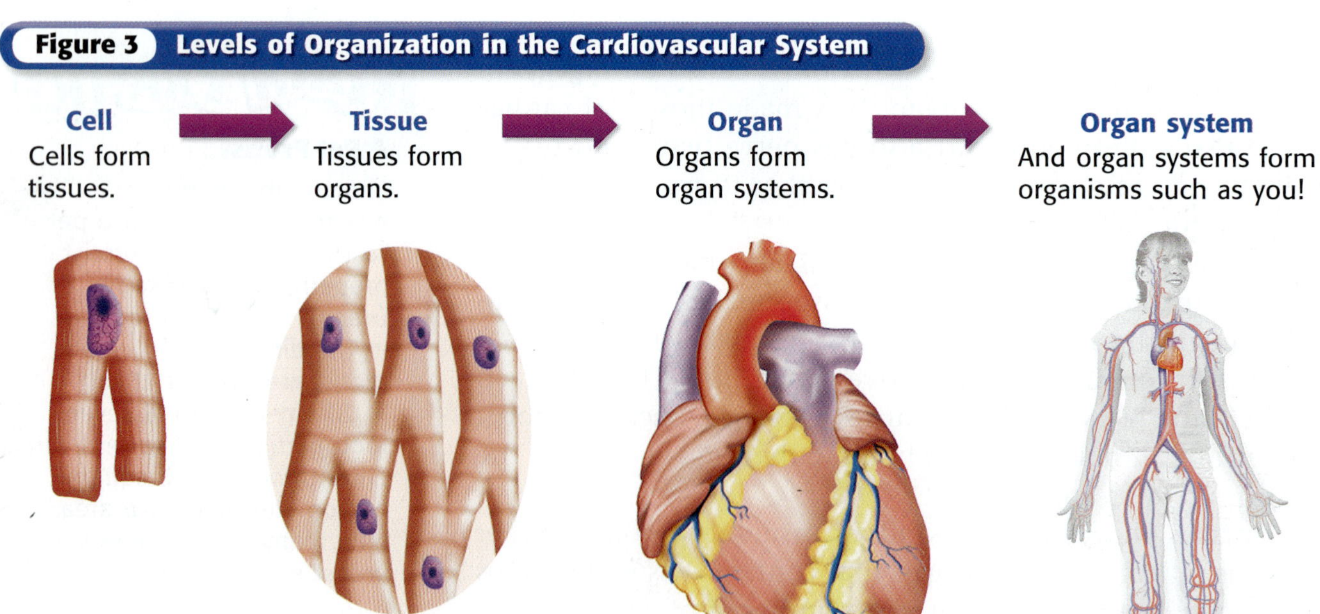

Structure and Function

In organisms, structure and function are related. **Structure** is the arrangement of parts in an organism. It includes the shape of a part and the material of which the part is made. **Function** is the job the part does. For example, the structure of the lungs is a large, spongy sac. In the lungs, there are millions of tiny air sacs called *alveoli*. Blood vessels wrap around the alveoli, as shown in **Figure 4.** Oxygen from air in the alveoli enters the blood. Blood then brings oxygen to body tissues. Also, in the alveoli, carbon dioxide leaves the blood and is exhaled.

The structures of alveoli and blood vessels enable them to perform a function. Together, they bring oxygen to the body and get rid of its carbon dioxide.

Figure 4 The Structure and Function of Alveoli

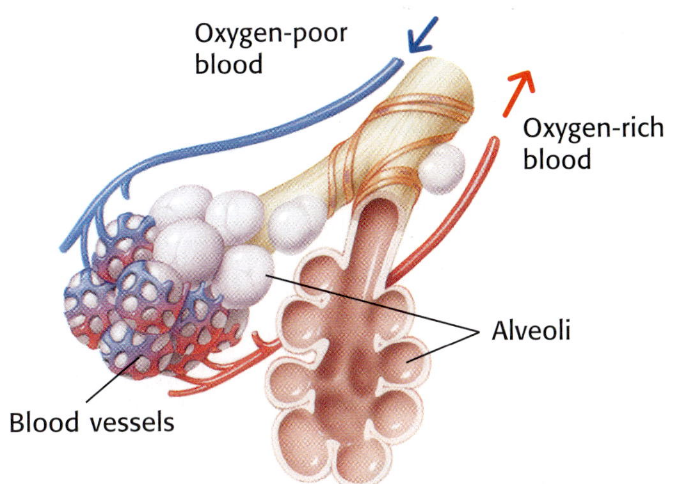

SECTION Review

Summary

- Advantages of being multicellular are larger size, longer life, and cell specialization.
- Four levels of organization are cell, tissue, organ, and organ system.
- A *tissue* is a group of cells working together. An *organ* is two or more tissues working together. An *organ system* is two or more organs working together.
- In organisms, a part's structure and function are related.

Using Key Terms

1. Use each of the following terms in a separate sentence: *tissue, organ,* and *function.*

Understanding Key Ideas

2. What are the four levels of organization in living things?
   - **a.** cell, multicellular, organ, organ system
   - **b.** single cell, multicellular, tissue, organ
   - **c.** larger size, longer life, specialized cells, organs
   - **d.** cell, tissue, organ, organ system

Math Skills

3. One multicellular organism is a cube. Each of its sides is 3 cm long. Each of its cells is 1 cm^3. How many cells does it have? If each side doubles in length, how many cells will it then have?

Critical Thinking

4. **Applying Concepts** Explain the relationship between structure and function. Use alveoli as an example. Be sure to include more than one level of organization.
5. **Making Inferences** Why can multicellular organisms be more complex than unicellular organisms? Use the three advantages of being multicellular to help explain your answer.

SECTION 4

Natural Selection

The relationship between structure and function can explain why certain organisms look the way they do. Natural selection is also useful for understanding the world around you.

READING WARM-UP

Objectives

- Give three examples of natural selection in action.
- Outline the process of speciation.

Terms to Learn

generation time
speciation

READING STRATEGY

Prediction Guide Before reading this section, write the title of each heading in this section. Next, under each heading, write what you think you will learn.

Have you ever had to take an antibiotic? Antibiotics are supposed to kill bacteria. But sometimes, bacteria are not killed. Why? A population of bacteria might develop an adaptation through natural selection. Most bacteria are killed by the chemicals in antibiotics. But in some cases, a few bacteria are naturally *resistant* to the chemicals, so they are not killed. These survivors are then able to pass this adaptation to their offspring. This is an example of how natural selection works.

Changes in Populations

The theory of natural selection explains how a population changes in response to its environment. If natural selection is always taking place, a population will tend to be well adapted to its environment. But not all individuals are the same. The individuals that are likely to survive and reproduce are those that are best adapted at the time.

Adaptation to Hunting

Changes in populations are sometimes observed when a new force affects the survival of individuals. In Uganda, scientists think that hunting is affecting the elephant population. In 1930, about 99% of the male elephants in one area had tusks. Only 1% of the elephants were born without tusks. Today, 85% of the male elephants in Uganda have tusks. What happened?

A male African elephant that has tusks is shown in **Figure 1.** People hunt the elephants for their tusks. As a result, fewer of the elephants that have tusks survive to reproduce, and more of the tuskless elephants survive. When the tuskless elephants reproduce, they pass the tuskless trait to their offspring.

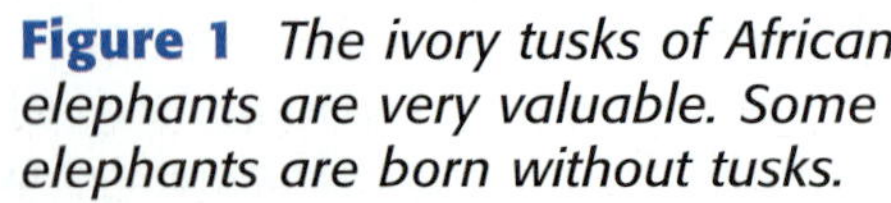

Figure 1 *The ivory tusks of African elephants are very valuable. Some elephants are born without tusks.*

Figure 2 Natural Selection of Insecticide Resistance

1 An insecticide will kill most insects, but a few may survive. These survivors have genes that make them resistant to the insecticide.

2 The survivors then reproduce, passing the insecticide-resistance genes to their offspring.

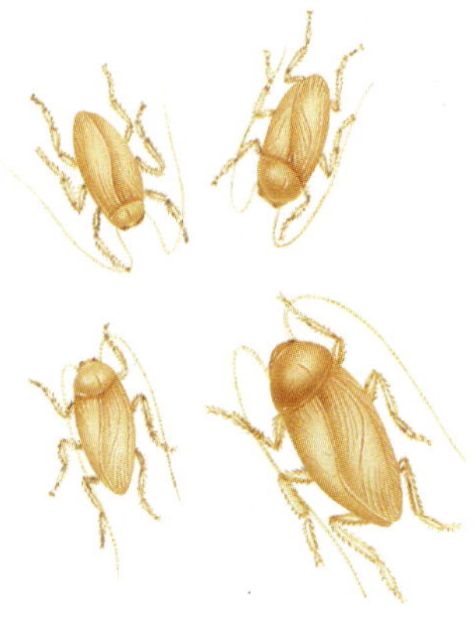

3 In time, the replacement population of insects is made up mostly of individuals that have the insecticide-resistance genes.

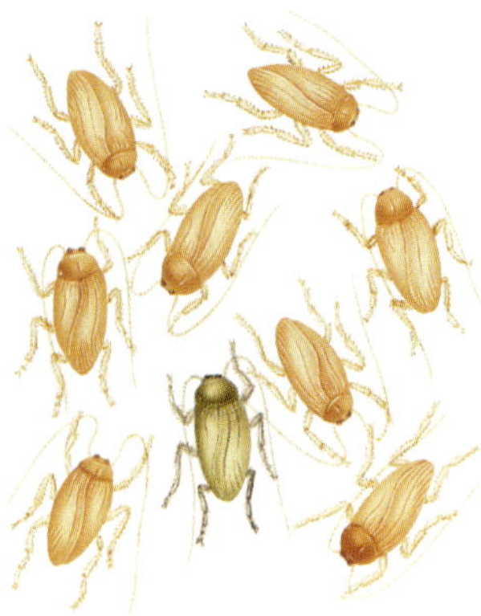

4 When the same kind of insecticide is used on the insects, only a few are killed because most of them are resistant to that insecticide.

Insecticide Resistance

People have always wanted to control the insect populations around their homes and farms. Many insecticides are used to kill insects. But some chemicals that used to work well do not work as well anymore. Some individual insects within the population are resistant to certain insecticides. **Figure 2** shows how a population of insects might become resistant to common insecticides.

More than 500 kinds of insects are now resistant to certain insecticides. Insects can quickly develop resistance because they often produce many offspring and have short generation times. **Generation time** is the average time between one generation of offspring and the next.

generation time the period between the birth of one generation and the birth of the next generation

Reading Check **Why do insects quickly develop resistance to insecticides?** (*See the Appendix for answers to Reading Checks.*)

Competition for Mates

In the process of evolution, survival is simply not enough. Natural selection is at work when individuals reproduce. In organisms that reproduce sexually, finding a mate is part of the struggle to reproduce. Many species have so much competition for mates that interesting adaptations result. For example, the females of many bird species prefer to mate with males that have certain types of colorful feathers.

Forming a New Species

Sometimes, drastic changes that can form a new species take place. In the animal kingdom, a *species* is a group of organisms that can mate with each other to produce fertile offspring. A new species may form after a group becomes separated from the original population. This group forms a new population. Over time, the two populations adapt to their different environments. Eventually, the populations can become so different that they can't mate anymore. Each population may then be considered a new species. The formation of a new species as a result of evolution is called **speciation** (SPEE shee AY shuhn). **Figure 3** shows how new species of Galápagos finches may have formed. Speciation may happen in other ways as well.

speciation the formation of new species as a result of evolution

Separation

Speciation often begins when a part of a population becomes separated from the rest. The process of separation can happen in several ways. For example, a newly formed canyon, mountain range, or lake can divide the members of a population.

Reading Check How can parts of a population become separated?

Figure 3 The Evolution of Galápagos Finch Species

1. Some finches left the mainland and reached one of the islands (separation).

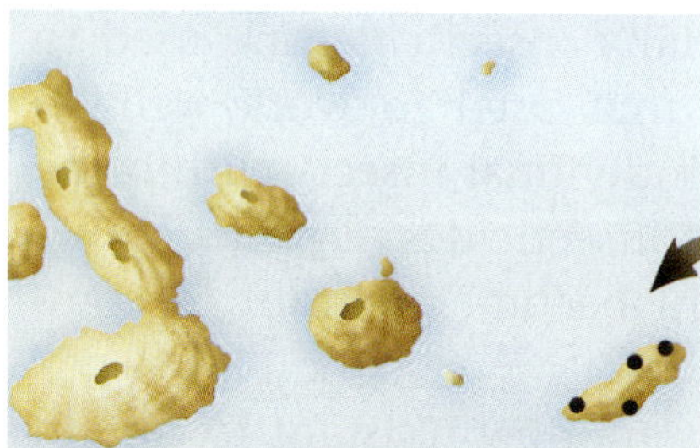

2. The finches reproduced and adapted to the environment (adaptation).

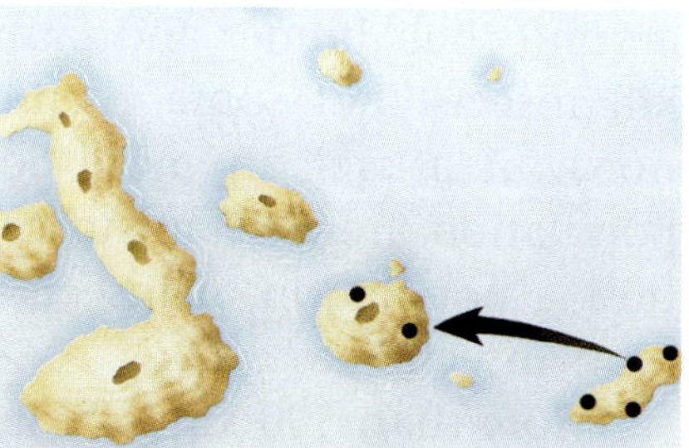

3. Some finches flew to a second island (separation).

4. The finches reproduced and adapted to the different environment (adaptation).

5. Some finches flew back to the first island but could no longer interbreed with the finches there (division).

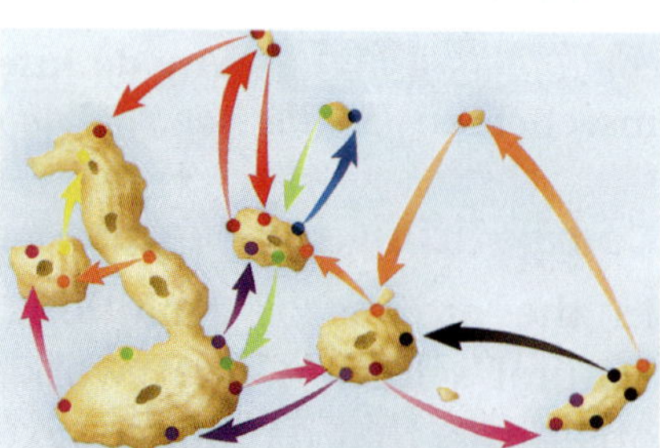

6. This process may have occurred over and over again as the finches flew to the other islands.

Adaptation

Populations constantly undergo natural selection. After two groups have separated, natural selection may act on each group in different ways. Over many generations, the separated groups may evolve different sets of traits. If the environmental conditions for each group differ, the adaptations in the groups will also differ.

Division

Over many generations, two separated groups of a population may become very different. Even if a geographical barrier is removed, the groups may not be able to interbreed anymore. At this point, the two groups are no longer the same species.

Figure 4 shows another way that populations may stop interbreeding. Leopard frogs and pickerel frogs probably had the same ancestor species. Then, at some point, some of these frogs began to mate at different times during the year.

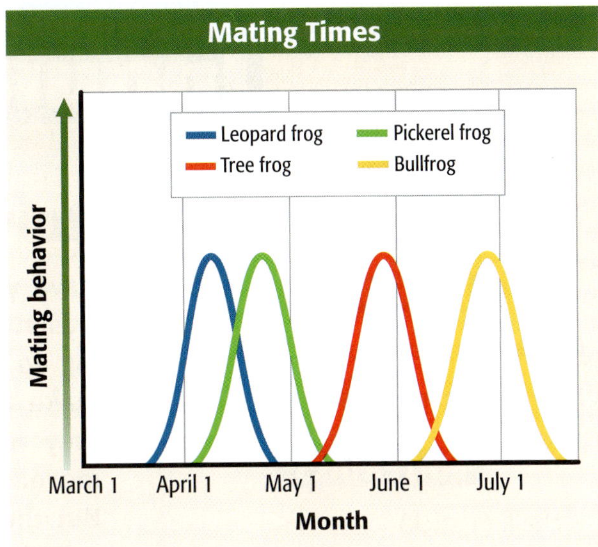

Figure 4 *The leopard frog and the pickerel frog are similar species. However, leopard frogs do not search for mates at the same time of year that pickerel frogs do.*

SECTION Review

Summary

- Natural selection explains how populations adapt to changes in their environment. A variety of examples of such adaptations can be found.
- Natural selection also explains how one species may evolve into another. Speciation occurs as populations undergo separation, adaptation, and division.

Using Key Terms

1. In your own words, write a definition for the term *speciation*.

Understanding Key Ideas

2. Two populations have evolved into two species when
   a. the populations are separated.
   b. the populations look different.
   c. the populations can no longer interbreed.
   d. the populations adapt.
3. Explain why the number of tuskless elephants in Uganda may be increasing.

Math Skills

4. A female cockroach can produce 80 offspring at a time. If half of the offspring produced by a certain female are female and each female produces 80 offspring, how many cockroaches are there in the third generation?

Critical Thinking

5. **Forming Hypotheses** Most kinds of cactus have leaves that grow in the form of spines. The stems or trunks become thick, juicy pads or barrels. Explain how these cactus parts might have evolved.
6. **Making Comparisons** Suggest an organism other than an insect that might evolve an adaptation to human activities.

Using Scientific Methods

Inquiry Lab

Roly-Poly Races

OBJECTIVES

Observe responses to stimuli.

Analyze responses to stimuli.

MATERIALS

- chalk (1 stick)
- container, plastic, small, with lid
- gloves, protective
- isopod (4)
- potato, raw (1 small slice)
- ruler, metric
- soil (8 oz)
- stopwatch

SAFETY

Have you ever watched a bug run? Did you wonder why it was running? The bug you saw running was probably reacting to a stimulus. In other words, something happened to make the bug run! One characteristic of living things is that they respond to stimuli. In this activity, you will study the movement of roly-polies. Roly-polies are also called *pill bugs*. But they are not really bugs; they are land-dwelling animals called *isopods*. Isopods live in dark, moist areas under rocks or wood. You will provide stimuli to determine how fast your isopod can move and what affects its speed and direction. Remember that isopods are living things and must be treated gently and respectfully.

Ask a Question

1. Ask a question such as, "Which stimuli cause pill bugs to run?"

Form a Hypothesis

2. Using your question as a guide, form a hypothesis. For example, you could form the following hypothesis: "Light, sound, and touch stimulate pill bugs to run."

Test the Hypothesis

3. Choose a partner, and decide together how you will run your roly-poly race. Discuss some gentle ways to stimulate your isopods to move. Choose five or six things that might cause movement, such as a gentle nudge or a change in temperature, sound, or light. Check your choices with your teacher.
4. Make a data table similar to the table below. Label the columns with the stimuli that you've chosen. Label the rows "Isopod 1," "Isopod 2," "Isopod 3," and "Isopod 4."

Isopod Responses			
	Stimulus 1	Stimulus 2	Stimulus 3
Isopod 1			
Isopod 2			
Isopod 3	DO NOT WRITE IN BOOK		
Isopod 4			

5. Place a layer of soil that is 1 cm or 2 cm deep in a small plastic container. Add a small slice of potato and a piece of chalk. Your isopods will eat these items.
6. Place four isopods in your container. Observe them for a minute or two before you perform your tests. Record your observations.
7. Decide which stimulus you want to test first. Carefully arrange the isopods at the "starting line." The starting line can be an imaginary line at one end of the container.
8. Gently stimulate each isopod at the same time and in the same way. In your data table, record the isopods' responses to the stimulus. Be sure to record the distance that each isopod travels. Don't forget to time the race.
9. Repeat steps 7–8 for each stimulus. Be sure to wait at least 2 min between trials.

Analyze the Results

1. **Describing Events** Describe the way that isopods move. Do their legs move together?
2. **Analyzing Results** Did your isopods move before or between the trials? Did the movement seem to have a purpose, or were the isopods responding to a stimulus? Explain.

Draw Conclusions

3. **Interpreting Information** Did any of the stimuli make the isopods move faster or go farther? Explain.

Applying Your Data

Like isopods and all other living things, humans react to stimuli. Describe three stimuli that might cause humans to run.

Chapter Review

USING KEY TERMS

Complete each of the following sentences by choosing the correct term from the word bank.

tissue	generation time
consumer	carbohydrate
homeostasis	producer

1. The process of maintaining a stable internal environment is known as ___.
2. A group of cells working together to perform a specific function is a(n) ___.
3. A ___ obtains food by eating other organisms.
4. Starch is a ___ and is made up of sugars.
5. Populations of bacteria can evolve quickly because they have a short ___.

UNDERSTANDING KEY IDEAS

Multiple Choice

6. Which of the following statements about cells is true?
   - **a.** Cells are the structures that contain all of the materials necessary for life.
   - **b.** Cells are found in all organisms.
   - **c.** Cells are sometimes specialized for particular functions.
   - **d.** All of the above
7. A change in an organism's environment that affects the organism's activities is a
   - **a.** response.
   - **b.** stimulus.
   - **c.** metabolism.
   - **d.** producer.
8. Organisms must have food because
   - **a.** food is a source of energy.
   - **b.** food supplies cells with oxygen.
   - **c.** organisms never make their own food.
   - **d.** All of the above
9. The benefits of being multicellular include
   - **a.** smaller size, longer life, and cell specialization.
   - **b.** generalized cells and longer life.
   - **c.** larger size and more enemies.
   - **d.** longer life, larger size, and specialized cells.
10. Organisms store energy in
    - **a.** nucleic acids.
    - **b.** phospholipids.
    - **c.** lipids.
    - **d.** water.
11. The molecule that contains the information about how to make proteins is
    - **a.** ATP.
    - **b.** a carbohydrate.
    - **c.** DNA.
    - **d.** a phospholipid.
12. The subunits of nucleic acids are
    - **a.** nucleotides.
    - **b.** oils.
    - **c.** sugars.
    - **d.** amino acids.

Short Answer

13 What is the difference between asexual reproduction and sexual reproduction?

14 In one or two sentences, explain why living things must have air.

15 Outline an example of the process of speciation.

CRITICAL THINKING

16 **Concept Mapping** Use the following terms to create a concept map: *cell, carbohydrates, protein, enzymes, DNA, sugars, lipids, nucleotides, amino acids,* and *nucleic acid.*

17 **Identifying Relationships.** Explain how the structure and function of an organism's parts are related. Give an example.

18 **Applying Concepts** Explain how populations of organisms are held in check and changed by environmental factors.

19 **Evaluating Hypotheses** Your friend tells you that the stimulus of music makes his goldfish swim faster. How would you design a controlled experiment to test your friend's claim?

INTERPRETING GRAPHICS

The pictures below show the same plant over a period of 3 days. Use the pictures below to answer the questions that follow.

20 What is the plant doing?

21 What characteristic(s) of living things is the plant exhibiting?

Standardized Test Preparation

READING

Read each of the passages below. Then, answer the questions that follow each passage.

Passage 1 Organisms make other organisms similar to themselves. They do so in one of two ways: by sexual reproduction or by asexual reproduction. In sexual reproduction, two parents produce offspring that will share characteristics of both parents. Most animals and plants reproduce in this way. In asexual reproduction, a single parent produces offspring that are identical to the parent. Most single-celled organisms reproduce in this way.

1. In the passage, what does the term *asexual reproduction* mean?

A A single parent produces offspring.
B Two parents make identical offspring.
C Plants make offspring.
D Animals make offspring.

2. What is characteristic of offspring produced by sexual reproduction?

F They are identical to both parents.
G They share the traits of both parents.
H They are identical to one parent.
I They are identical to each other.

3. What is characteristic of offspring produced by asexual reproduction?

A They are identical to both parents.
B They share the traits of both parents.
C They are identical to one parent.
D They are usually plants.

4. What is the difference between sexual and asexual reproduction?

F the number of offspring produced
G the number of parents needed to produce offspring
H the number of traits produced
I the number of offspring that survive

Passage 2 In 1996, a group of researchers led by NASA scientists studied a 3.8-billion-year-old meteorite named ALH84001. These scientists agree that ALH84001 is a potato-sized piece of the planet Mars. They also agree that it fell to Earth about 13,000 years ago. It was discovered in Antarctica in 1984. According to the NASA team, ALH84001 brought with it evidence that life once existed on Mars.

Scientists found certain kinds of organic molecules (molecules containing carbon) on the surface of ALH84001. These molecules are similar to those left behind when living things break down substances for food. When these scientists examined the interior of the meteorite, they found the same organic molecules throughout. Because these molecules were spread throughout the meteorite, scientists concluded that the molecules were not contamination from Earth. The NASA team believes that these organic compounds are strong evidence that tiny organisms similar to bacteria lived, ate, and died on Mars millions of years ago.

1. How old is the meteorite named ALH84001?

A 13,000 years old
B millions of years old
C 3.8 billion years old
D 3.8 trillion years old

2. Which of the following would best support a claim that life might have existed on Mars?

F remains of organisms
G water
H meteorite temperatures similar to Earth temperatures
I oxygen

INTERPRETING GRAPHICS

The graph below shows an ill person's body temperature. Use the graph below to answer the questions that follow.

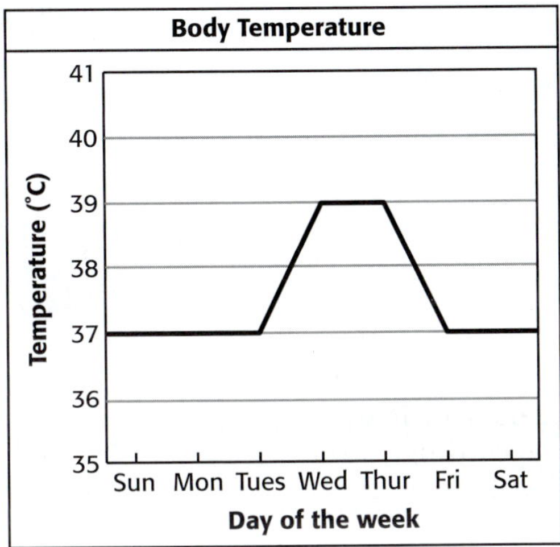

1. A fever is a spike in temperature. On which day does this person have a fever?
   A Sunday
   B Monday
   C Wednesday
   D Saturday

2. A body with a fever is often fighting an infection. Fevers help eliminate the pathogens that cause the infection. According to the chart, when does this person probably have the highest fever?
   F Sunday
   G Monday
   H Wednesday
   I Saturday

3. What is the highest temperature that this fever reaches?
   A 37°C
   B 38°C
   C 39°C
   D 40°C

4. What is probably this person's normal body temperature?
   F 37°C
   G 38°C
   H 39°C
   I 40°C

MATH

Read each question below, and choose the best answer.

1. An aquarium is a place where fish can live. What is the volume of the aquarium shown below?

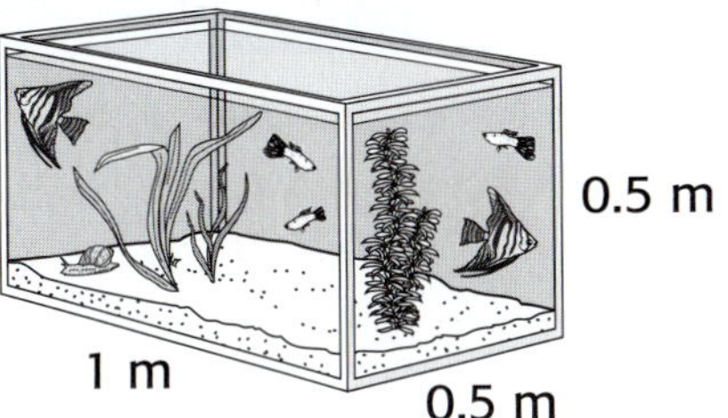

   A 0.25 m
   B 0.25 m^2
   C 0.25 m^3
   D 0.52 m^3

2. The cost of admission to a natural history museum is $7 per adult. What is the total cost of admission for a group of five adults?
   F $25
   G $35
   H $45
   I $55

3. Lee biked 25.3 km on Monday, 20.7 km on Tuesday, and 15.6 km on Wednesday. How many kilometers did Lee bike during those three days?
   A 66.1 km
   B 61.6 km
   C 51.6 km
   D 16.6 km

4. Laura collected 24 leaves. One-third of the leaves were oak leaves. How many oak leaves did Laura collect?
   F 6
   G 8
   H 12
   I 24

Science in Action

Science, Technology, and Society

Chess-Playing Computers

Computers can help us explore how humans think. One way to explore how humans think is to study how people and computers play chess against each other.

A computer's approach to chess is straightforward. By calculating each piece's possible board position for the next few moves, a computer creates what is called a *position tree*. A position tree shows how each move can lead to other moves. This way of playing requires millions of calculations.

Human chess champions play differently. Humans calculate only three or four moves every minute. Even so, human champions are still a match for computer opponents. By studying the ways that people and computers play chess, scientists are learning how people think and make choices.

Math Activity

A chess-playing computer needs to evaluate 3 million positions before a move. If you could evaluate two positions in 1 min, how long would it take you to evaluate 3 million possible positions?

Science Fiction

"They're Made Out of Meat" by Terry Bisson

Two space explorers millions of light-years from home are visiting an uncharted sector of the universe to find signs of life. Their mission is to contact, welcome, and log any and all beings in this part of the universe.

During their mission, they encounter a life-form quite unlike anything they have ever seen before. It looked too strange and, well, disgusting. The explorers have very strong doubts about adding this new organism to the list. But the explorers' official duty is to contact and welcome all life-forms no matter how ugly they are. Can the explorers bring themselves to perform their duty?

You'll find out by reading "They're Made Out of Meat," a short story by Terry Bisson. This story is in the *Holt Anthology of Science Fiction*.

Language Arts Activity

WRITING SKILL Write a story about what happens when the explorers next meet the creatures on the star in G445 zone.

People in Science

Janis Davis-Street

NASA Nutritionist Do astronauts eat shrimp cocktail in space? Yes, they do! Shrimp cocktail is nutritious and tastes so good that it is one of the most popular foods in the space program. And eating a proper diet helps astronauts stay healthy while they are in space.

But who figures out what astronauts need to eat? Janis Davis-Street is a nutritionist and laboratory supervisor for the Nutritional Biochemistry Laboratory at the Johnson Space Center in Houston, Texas. She was born in Georgetown, Guyana, on the northeastern coast of South America. She was educated in Canada.

Davis-Street is part of a team that uses their knowledge of nutrition, biology, and chemistry to figure out the nutritional requirements for spaceflight. For example, they determine how many calories and other nutrients each astronaut needs per day during spaceflight.

The Nutritional Biochemistry Laboratory's work on the space shuttle missions and *Mir* space station developed into tests that allow NASA to help ensure astronaut health before, during, and after flight. These tests are important for understanding how the human body adapts to long space missions, and for determining whether treatments for preventing bone and muscle loss during spaceflight are working.

Social Studies ACTIVITY

Scientists from more than 30 countries have been on space missions. Research which countries have provided astronauts or cosmonauts for space missions. Using a map, place self-stick notes on countries that have provided scientists for space missions. Write the names of the appropriate scientists on the self-stick notes.

To learn more about these Science in Action topics, visit go.hrw.com and type in the keyword HL5ALVF.

Check out Current Science® articles related to this chapter by visiting go.hrw.com. Just type in the keyword HL5CS02.

The Cell in Action

About the PHOTO

This adult katydid is emerging from its last immature, or nymph, stage. As the katydid changed from a nymph to an adult, every structure of its body changed. To grow and change, an organism must produce new cells. When a cell divides, it makes a copy of its genetic material.

PRE-READING ACTIVITY

Tri-Fold Before you read the chapter, create the FoldNote entitled "Tri-Fold" described in the **Study Skills** section of the Appendix. Write what you know about the actions of cells in the column labeled "Know." Then, write what you want to know in the column labeled "Want." As you read the chapter, write what you learn about the actions of cells in the column labeled "Learn."

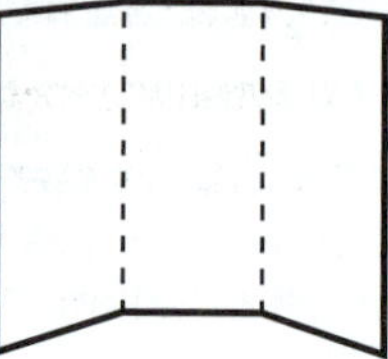

START-UP ACTIVITY

Cells in Action

Yeast are single-celled fungi that are an important ingredient in bread. Yeast cells break down sugar molecules to release energy. In the process, carbon dioxide gas is produced, which causes bread dough to rise.

Procedure

1. Add **4 mL of a sugar solution** to **10 mL of a yeast-and-water mixture**. Use a **stirring rod** to thoroughly mix the two liquids.
2. Pour the stirred mixture into a small test tube.
3. Place a slightly **larger test tube** over the **small test tube.** The top of the small test tube should touch the bottom of the larger test tube.
4. Hold the test tubes together, and quickly turn both test tubes over. Place the test tubes in a test-tube rack.
5. Use a **ruler** to measure the height of the fluid in the large test tube. Wait 20 min, and then measure the height of the liquid again.

Analysis

1. What is the difference between the first height measurement and the second height measurement?
2. What do you think caused the change in the fluid's height?

SECTION 1

Exchange with the Environment

READING WARM-UP

Objectives

- Explain the process of diffusion.
- Describe how osmosis occurs.
- Compare passive transport with active transport.
- Explain how large particles get into and out of cells.

Terms to Learn

diffusion
osmosis
passive transport
active transport
endocytosis
exocytosis

READING STRATEGY

Reading Organizer As you read this section, make a table comparing active transport and passive transport.

What would happen to a factory if its power were shut off or its supply of raw materials never arrived? What would happen if the factory couldn't get rid of its garbage?

Like a factory, an organism must be able to obtain energy and raw materials and get rid of wastes. An organism's cells perform all of these functions. These functions keep cells healthy so that they can divide. Cell division allows organisms to grow and repair injuries.

The exchange of materials between a cell and its environment takes place at the cell's membrane. To understand how materials move into and out of the cell, you need to know about diffusion.

What Is Diffusion?

What happens if you pour dye on top of a layer of gelatin? At first, it is easy to see where the dye ends and the gelatin begins. But over time, the line between the two layers will blur, as shown in **Figure 1.** Why? Everything, including the gelatin and the dye, is made up of tiny moving particles. Particles travel from where they are crowded to where they are less crowded. This movement from areas of high concentration (crowded) to areas of low concentration (less crowded) is called **diffusion** (di FYOO zhuhn). Dye particles diffuse from where they are crowded (near the top of the glass) to where they are less crowded (in the gelatin). Diffusion also happens within and between living cells. Cells do not need to use energy for diffusion.

diffusion the movement of particles from regions of higher density to regions of lower density

Figure 1 *The particles of the dye and the gelatin slowly mix by diffusion.*

Figure 2 Osmosis

1 The side that holds only pure water has the higher concentration of water particles.

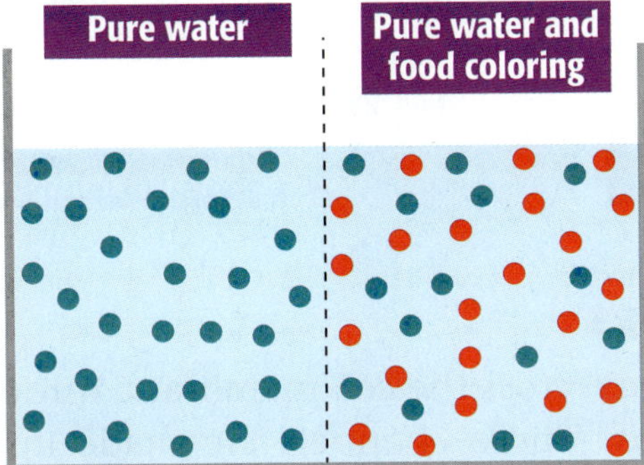

2 During osmosis, water particles move to where they are less concentrated.

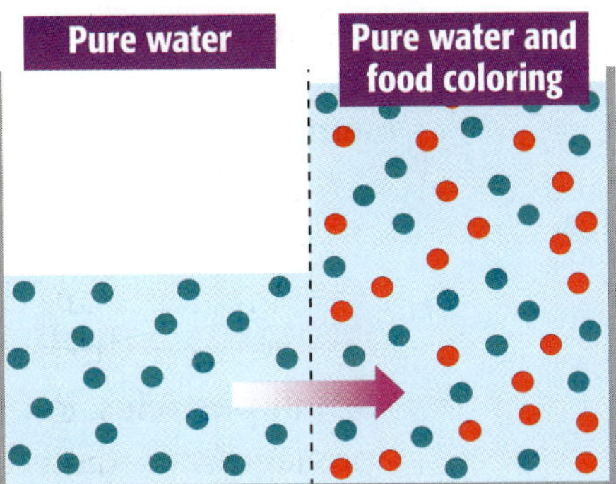

Diffusion of Water

The cells of organisms are surrounded by and filled with fluids that are made mostly of water. The diffusion of water through cell membranes is so important to life processes that it has been given a special name—**osmosis** (ahs MOH sis).

osmosis the diffusion of water through a semipermeable membrane

Water is made up of particles, called *molecules*. Pure water has the highest concentration of water molecules. When you mix something, such as food coloring, sugar, or salt, with water, you lower the concentration of water molecules. **Figure 2** shows how water molecules move through a membrane that is semipermeable (SEM i PUHR mee uh buhl). *Semipermeable* means that only certain substances can pass through. The picture on the left in **Figure 2** shows liquids that have different concentrations of water. Over time, the water molecules move from the liquid with the high concentration of water molecules to the liquid with the lower concentration of water molecules.

The Cell and Osmosis

Osmosis is important to cell functions. For example, red blood cells are surrounded by plasma. Plasma is made up of water, salts, sugars, and other particles. The concentration of these particles is kept in balance by osmosis. If red blood cells were in pure water, water molecules would flood into the cells and cause them to burst. When red blood cells are put into a salty solution, the concentration of water molecules inside the cell is higher than the concentration of water outside. This difference makes water move out of the cells, and the cells shrivel up. Osmosis also occurs in plant cells. When a wilted plant is watered, osmosis makes the plant firm again.

Reading Check Why would red blood cells burst if you placed them in pure water? (*See the Appendix for answers to Reading Checks.*)

Bead Diffusion

1. Put three groups of **colored beads** on the bottom of a **plastic bowl.** Each group should be made up of five beads of the same color.
2. Stretch some **clear plastic wrap** tightly over the top of the bowl. Gently shake the bowl for 10 seconds while watching the beads.
3. How is the scattering of the beads like the diffusion of particles? How is it different from the diffusion of particles?

Figure 3 *In passive transport, particles travel through proteins to areas of lower concentration. In active transport, cells use energy to move particles, usually to areas of higher concentration.*

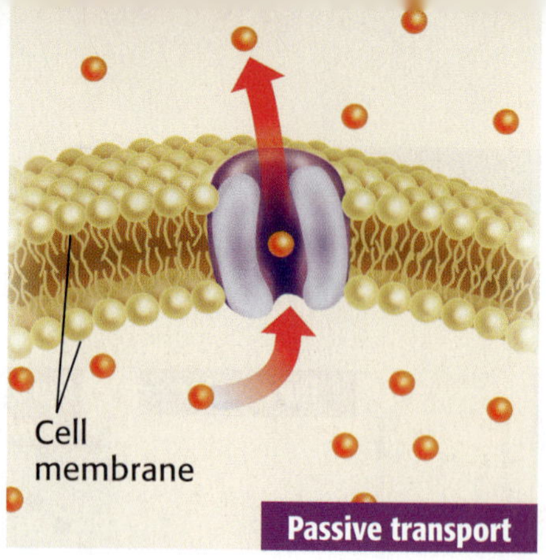

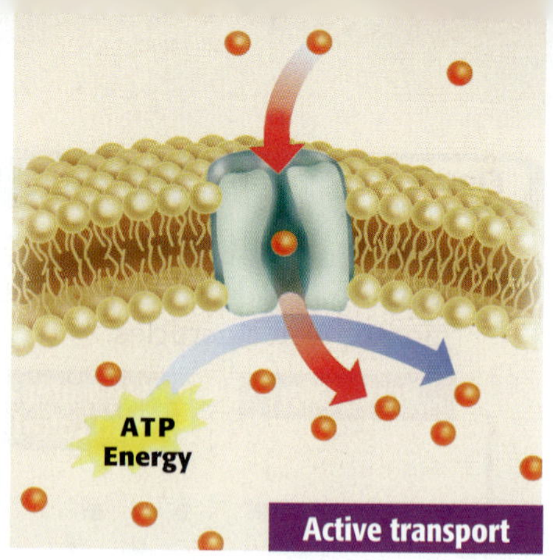

Moving Small Particles

Small particles, such as sugars, cross the cell membrane through passageways called *channels*. These channels are made up of proteins in the cell membrane. Particles travel through these channels by either passive or active transport. The movement of particles across a cell membrane without the use of energy by the cell is called **passive transport**, and is shown in **Figure 3.** During passive transport, particles move from an area of high concentration to an area of low concentration. Diffusion and osmosis are examples of passive transport.

A process of transporting particles that requires the cell to use energy is called **active transport.** Active transport usually involves the movement of particles from an area of low concentration to an area of high concentration.

passive transport the movement of substances across a cell membrane without the use of energy by the cell

active transport the movement of substances across the cell membrane that requires the cell to use energy

endocytosis the process by which a cell membrane surrounds a particle and encloses the particle in a vesicle to bring the particle into the cell

Moving Large Particles

Small particles cross the cell membrane by diffusion, passive transport, and active transport. Large particles move into and out of the cell by processes called *endocytosis* and *exocytosis*.

Endocytosis

The active-transport process by which a cell surrounds a large particle, such as a large protein, and encloses the particle in a vesicle to bring the particle into the cell is called **endocytosis** (EN doh sie TOH sis). *Vesicles* are sacs formed from pieces of cell membrane. **Figure 4** shows endocytosis.

Figure 4 **Endocytosis**

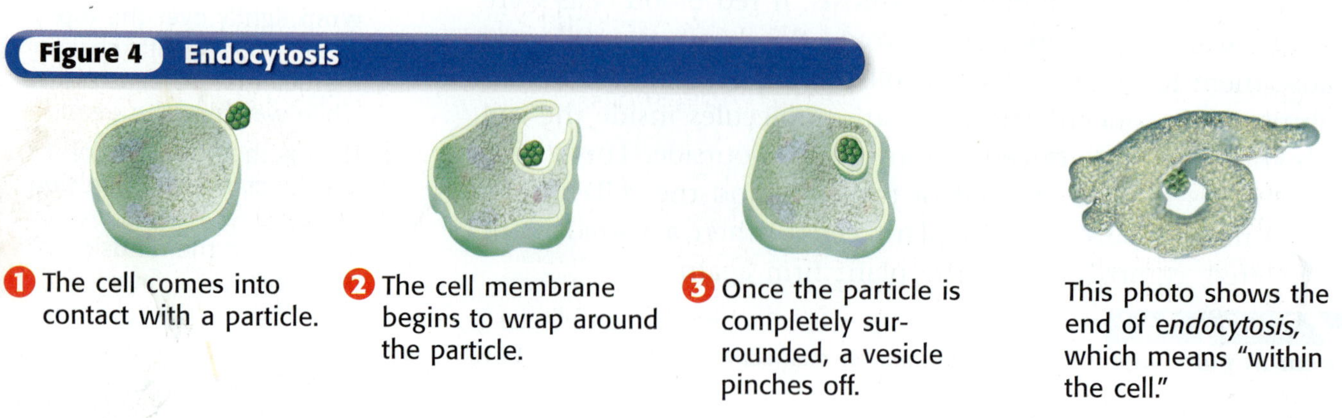

1. The cell comes into contact with a particle.
2. The cell membrane begins to wrap around the particle.
3. Once the particle is completely surrounded, a vesicle pinches off.

This photo shows the end of *endocytosis,* which means "within the cell."

Figure 5 **Exocytosis**

1 Large particles that must leave the cell are packaged in vesicles.

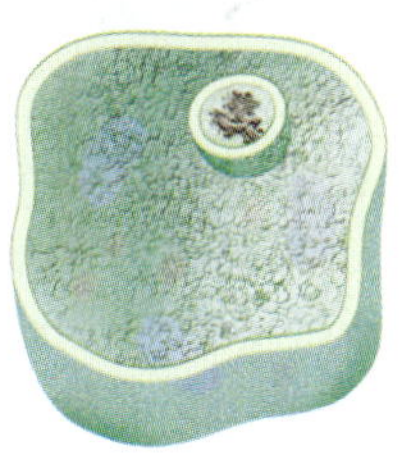

2 The vesicle travels to the cell membrane and fuses with it.

3 The cell releases the particle to the outside of the cell.

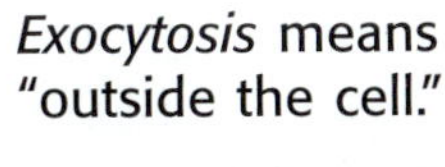

Exocytosis means "outside the cell."

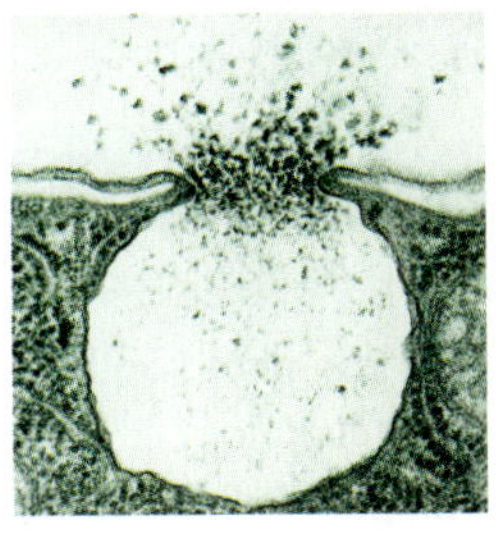

Exocytosis

When large particles, such as wastes, leave the cell, the cell uses an active-transport process called **exocytosis** (EK soh sie TOH sis). During exocytosis, a vesicle forms around a large particle within the cell. The vesicle carries the particle to the cell membrane. The vesicle fuses with the cell membrane and releases the particle to the outside of the cell. **Figure 5** shows exocytosis.

exocytosis the process in which a cell releases a particle by enclosing the particle in a vesicle that then moves to the cell surface and fuses with the cell membrane

Reading Check What is exocytosis?

SECTION Review

Summary

- Diffusion is the movement of particles from an area of high concentration to an area of low concentration.
- Osmosis is the diffusion of water through a semipermeable membrane.
- Cells move small particles by diffusion, which is an example of passive transport, and by active transport.
- Large particles enter the cell by endocytosis, and exit the cell by exocytosis.

Using Key Terms

For each pair of terms, explain how the meanings of the terms differ.

1. *diffusion* and *osmosis*
2. *active transport* and *passive transport*
3. *endocytosis* and *exocytosis*

Understanding Key Ideas

4. The movement of particles from a less crowded area to a more crowded area requires
   - **a.** sunlight.
   - **b.** energy.
   - **c.** a membrane.
   - **d.** osmosis.
5. What structures allow small particles to cross cell membranes?

Math Skills

6. The area of particle 1 is 2.5 mm^2. The area of particle 2 is 0.5 mm^2. The area of particle 1 is how many times as big as the area of particle 2?

Critical Thinking

7. **Predicting Consequences** What would happen to a cell if its channel proteins were damaged and unable to transport particles? What would happen to the organism if many of its cells were damaged in this way? Explain your answer.
8. **Analyzing Ideas** Why does active transport require energy?

SECTION 2

Cell Energy

Why do you get hungry? Feeling hungry is your body's way of telling you that your cells need energy.

READING WARM-UP

Objectives

- Describe photosynthesis and cellular respiration.
- Compare cellular respiration with fermentation.

Terms to Learn

photosynthesis
cellular respiration
fermentation

READING STRATEGY

Discussion Read this section silently. Write down questions that you have about this section. Discuss your questions in a small group.

Cells need energy because they carry out the basic functions of living things. For example, cells get energy from food and get rid of wastes. Cells do similar work in all living things. But cells get energy in different ways. Plant cells get their energy from the sun. Most animal cells get energy from food.

From Sun to Cell

Nearly all of the energy that fuels life comes from the sun. Plants capture energy from the sun and change it into food through a process called **photosynthesis.** The food that plants make supplies them with energy. This food also becomes a source of energy for the organisms that eat the plants.

photosynthesis the process by which plants, algae, and some bacteria use sunlight, carbon dioxide, and water to make food

Photosynthesis

Plant cells have molecules that absorb light energy. These molecules are called *pigments*. Chlorophyll (KLAWR uh FIL), the main pigment used in photosynthesis, gives plants their green color. Chlorophyll is found in chloroplasts.

Plants use the energy captured by chlorophyll to change carbon dioxide and water into food. The food they make can be used immediately or stored. Food is stored as simple sugars, such as starch, sucrose, and glucose. These sugars are carbohydrates. Plants also produce oxygen as they make food through photosynthesis. Photosynthesis is summarized in **Figure 1.**

Photosynthesis

$$6CO_2 + 6H_2O + \text{Light energy} \rightarrow C_6H_{12}O_6 + 6O_2$$

Carbon dioxide, Water, Glucose, Oxygen

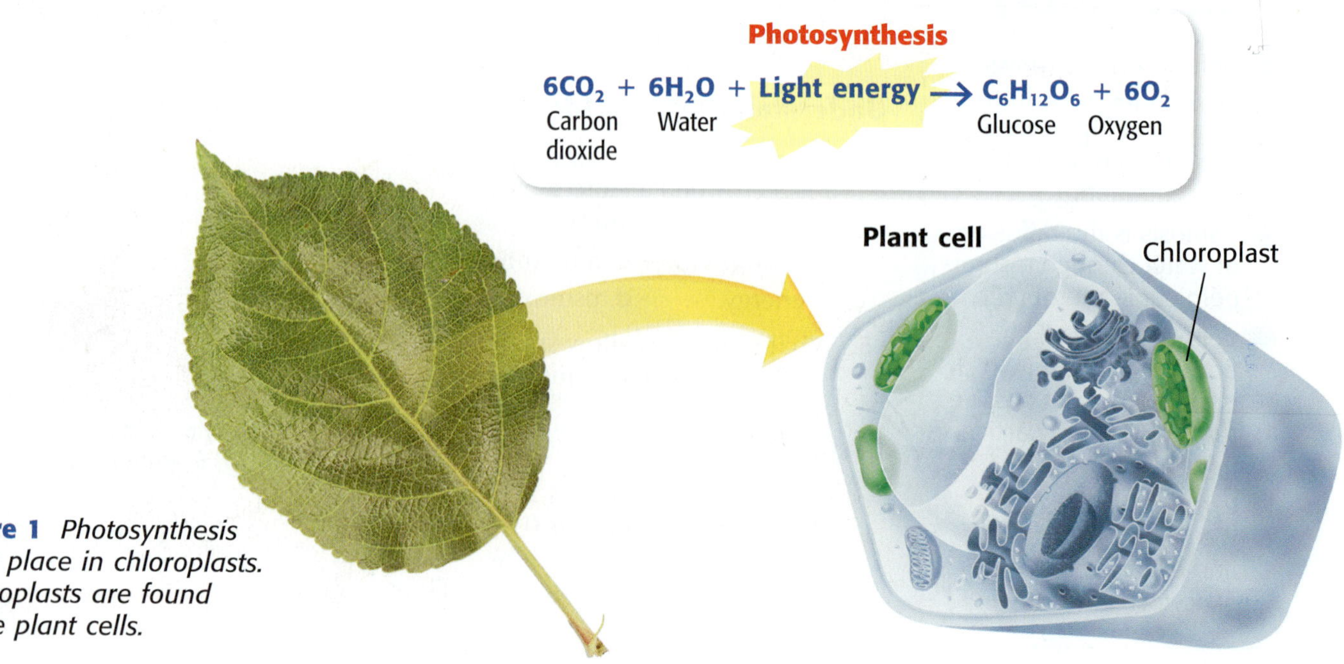

Figure 1 *Photosynthesis takes place in chloroplasts. Chloroplasts are found inside plant cells.*

Getting Energy from Food

Animal cells have different ways of getting energy from food. But each way of getting energy from food provides the cells with fuel for its activities. And each way produces the building materials that are needed for an animal to grow.

Cellular Respiration

Using oxygen to break down food is called **cellular respiration.** The word *respiration* means "breathing," but cellular respiration is different from breathing. Breathing supplies the oxygen needed for cellular respiration. Breathing also removes carbon dioxide, which is a waste product of cellular respiration. But cellular respiration is a chemical process that occurs in cells.

Most complex organisms, such as the cow in **Figure 2,** obtain energy through cellular respiration. During cellular respiration, food (such as glucose) is broken down into CO_2 and H_2O, and energy is released. Most of the energy released maintains body temperature. Some of the energy is used to form adenosine triphosphate (ATP). ATP supplies energy that fuels cell activities.

Most of the process of cellular respiration takes place in the cell membrane of prokaryotic cells. But in the cells of eukaryotes, cellular respiration takes place mostly in the mitochondria. The process of cellular respiration is summarized in **Figure 2.** Does the equation in the figure remind you of the equation for photosynthesis? **Figure 3** on the next page shows how photosynthesis and respiration are related.

Reading Check **What is the difference between cellular respiration and breathing?** (*See the Appendix for answers to Reading Checks.*)

CONNECTION TO Chemistry

Earth's Early Atmosphere

Scientists think that Earth's early atmosphere lacked oxygen. Because of this lack of oxygen, early organisms used fermentation to get energy from food. When organisms began to photosynthesize, the oxygen they produced entered the atmosphere. How do you think this oxygen changed how other organisms got energy?

cellular respiration the process by which cells use oxygen to produce energy from food

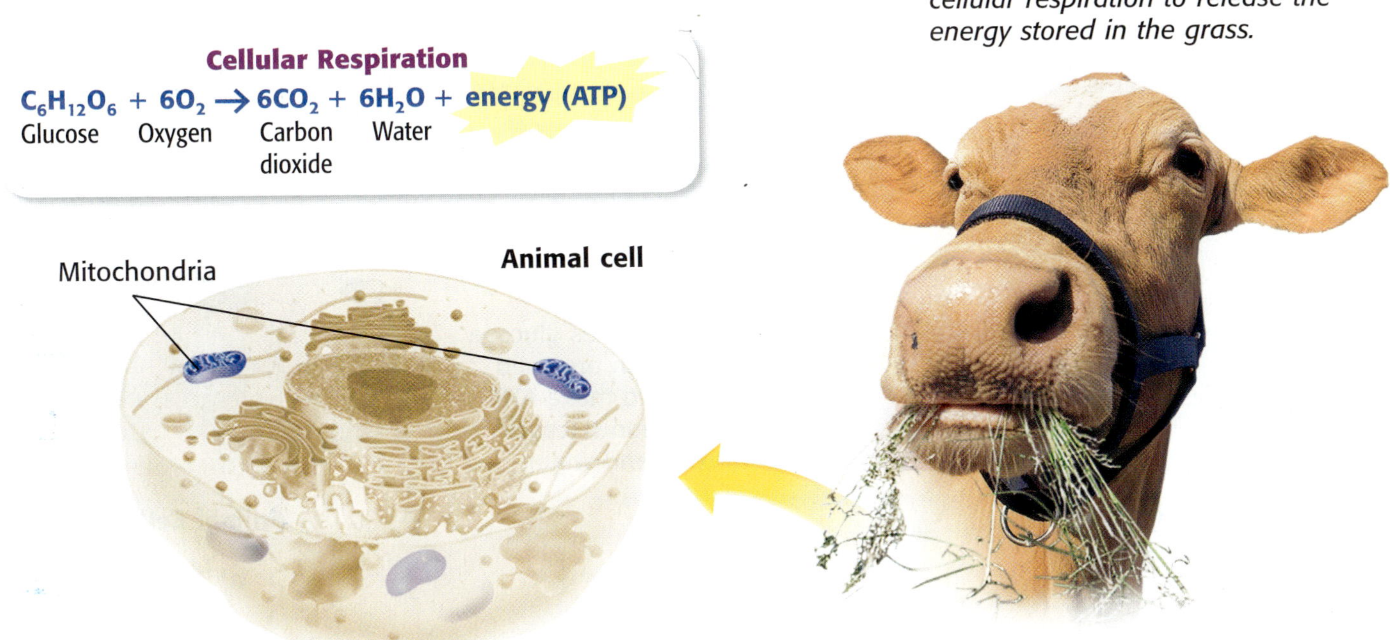

Figure 2 *The mitochondria in the cells of this cow will use cellular respiration to release the energy stored in the grass.*

Figure 3 The Connection Between Photosynthesis and Respiration

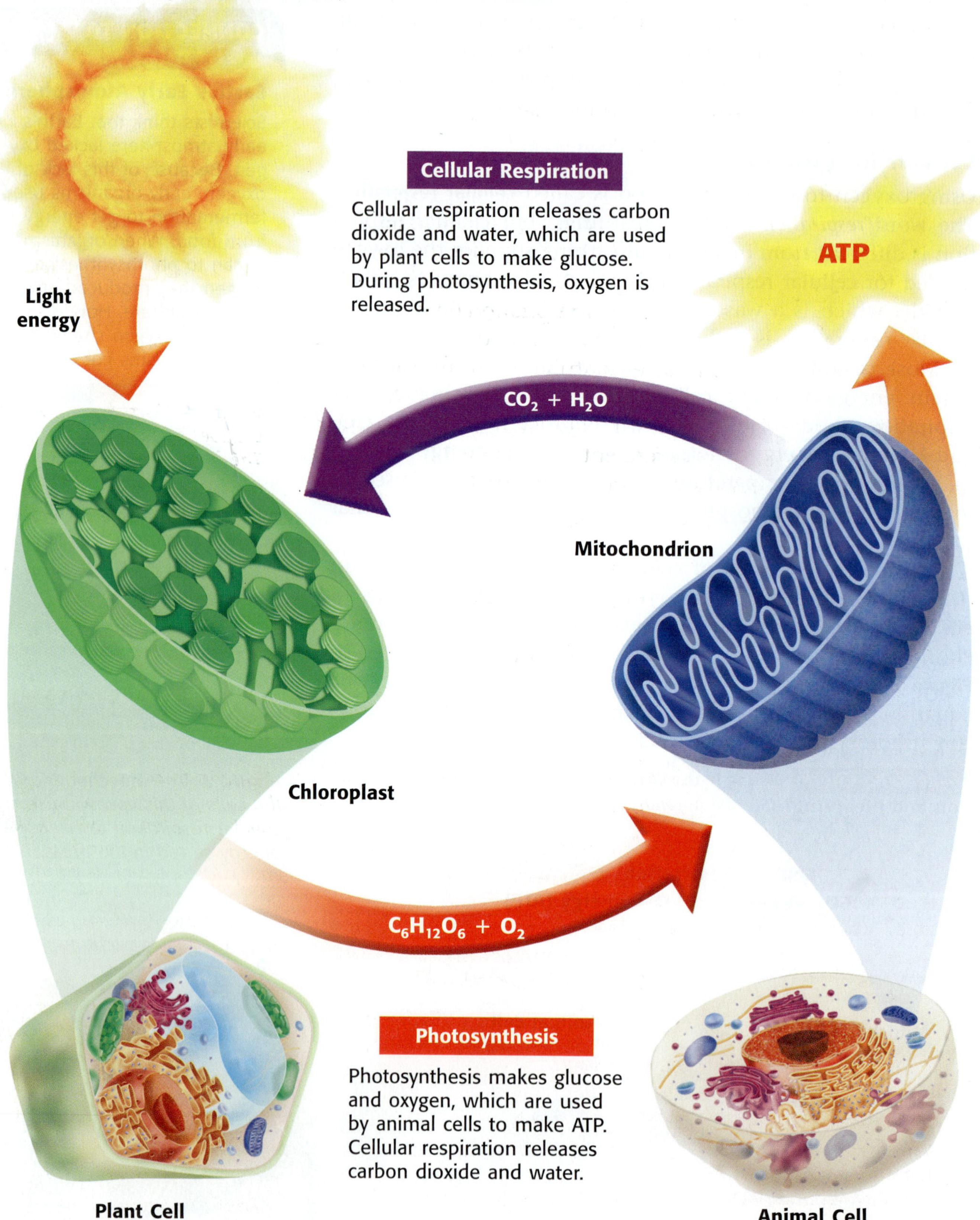

Connection Between Photosynthesis and Respiration

As shown in **Figure 3,** photosynthesis transforms energy from the sun into glucose. During photosynthesis, cells use CO_2 to make glucose, and the cells release O_2. During cellular respiration, cells use O_2 to break down glucose and release energy and CO_2. Each process makes the materials that are needed for the other process to occur elsewhere.

fermentation the breakdown of food without the use of oxygen

Fermentation

When cells can't get the oxygen needed for cellular respiration, they use fermentation to get energy. **Fermentation** is a way of getting energy without using oxygen. Fermentation releases less energy from a given food than cellular respiration does. One kind of fermentation happens in your muscles and produces lactic acid. The buildup of lactic acid contributes to muscle fatigue and causes a burning sensation. This kind of fermentation also happens in the muscle cells of other animals and in some fungi and bacteria. Another type of fermentation occurs in some types of bacteria and in yeast as described in **Figure 4.**

Figure 4 *Yeast forms carbon dioxide during fermentation. The bubbles of CO_2 gas cause the dough to rise.*

Reading Check What are two kinds of fermentation?

SECTION Review

Summary

- Most of the energy that fuels life processes comes from the sun.
- The sun's energy is converted into food by the process of photosynthesis.
- Cellular respiration breaks down glucose into water, carbon dioxide, and energy.
- Fermentation is a way that cells get energy from their food without using oxygen.

Using Key Terms

1. In your own words, write a definition for the term *fermentation*.

Understanding Key Ideas

2. O_2 is released during
   a. cellular respiration.
   b. photosynthesis.
   c. breathing.
   d. fermentation.
3. How are photosynthesis and cellular respiration related?
4. How are respiration and fermentation similar? How are they different?

Math Skills

5. Cells of plant A make 120 molecules of glucose an hour. Cells of plant B make half as much glucose as plant A does. How much glucose does plant B make every minute?

Critical Thinking

6. **Analyzing Relationships** Why are plants important to the survival of all other organisms?
7. **Applying Concepts** You have been given the job of restoring life to a barren island. What types of organisms would you put on the island? If you want to have animals on the island, what other organisms must you bring? Explain your answer.

SECTION 3

The Cell Cycle

In the time that it takes you to read this sentence, your body will have made millions of new cells! Making new cells allows you to grow and replace cells that have died.

READING WARM-UP

Objectives

- Explain how cells produce more cells.
- Describe the process of mitosis.
- Explain how cell division differs in animals and plants.

Terms to Learn

cell cycle
chromosome
homologous chromosomes
mitosis
cytokinesis

READING STRATEGY

Paired Summarizing Read this section silently. In pairs, take turns summarizing the material. Stop to discuss ideas that seem confusing.

The environment in your stomach is so acidic that the cells lining your stomach must be replaced every few days. Other cells are replaced less often, but your body is constantly making new cells.

The Life of a Cell

As you grow, you pass through different stages in life. Your cells also pass through different stages in their life cycle. The life cycle of a cell is called the **cell cycle.**

The cell cycle begins when the cell is formed and ends when the cell divides and forms new cells. Before a cell divides, it must make a copy of its deoxyribonucleic acid (DNA). DNA is the hereditary material that controls all cell activities, including the making of new cells. The DNA of a cell is organized into structures called **chromosomes.** Copying chromosomes ensures that each new cell will be an exact copy of its parent cell. How does a cell make more cells? It depends on whether the cell is prokaryotic (with no nucleus) or eukaryotic (with a nucleus).

cell cycle the life cycle of a cell

chromosome in a eukaryotic cell, one of the structures in the nucleus that are made up of DNA and protein; in a prokaryotic cell, the main ring of DNA

Making More Prokaryotic Cells

Prokaryotic cells are less complex than eukaryotic cells are. Bacteria, which are prokaryotes, have ribosomes and a single, circular DNA molecule but don't have membrane-enclosed organelles. Cell division in bacteria is called *binary fission,* which means "splitting into two parts." Binary fission results in two cells that each contain one copy of the circle of DNA. A few of the bacteria in **Figure 1** are undergoing binary fission.

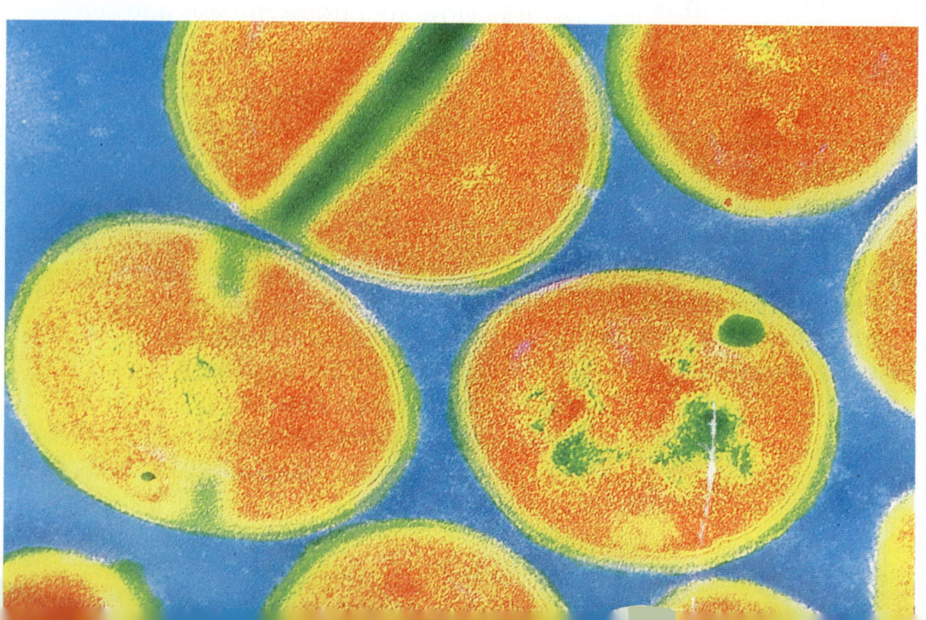

Figure 1 *Bacteria reproduce by binary fission.*

Eukaryotic Cells and Their DNA

Eukaryotic cells are more complex than prokaryotic cells are. The chromosomes of eukaryotic cells contain more DNA than those of prokaryotic cells do. Different kinds of eukaryotes have different numbers of chromosomes. Comparing the number of chromosomes can show how eukaryotes are related. So, scientists can use number of chromosomes to classify organisms. More-complex eukaryotes do not necessarily have more chromosomes than simpler eukaryotes do. For example, fruit flies have 8 chromosomes, potatoes have 48, and humans have 46. **Figure 2** shows the 46 chromosomes of a human body cell lined up in pairs. These pairs are made up of similar chromosomes known as **homologous chromosomes** (hoh MAHL uh guhs KROH muh SOHMZ).

Reading Check **Do more-complex organisms always have more chromosomes than simpler organisms do?** (*See the Appendix for answers to Reading Checks.*)

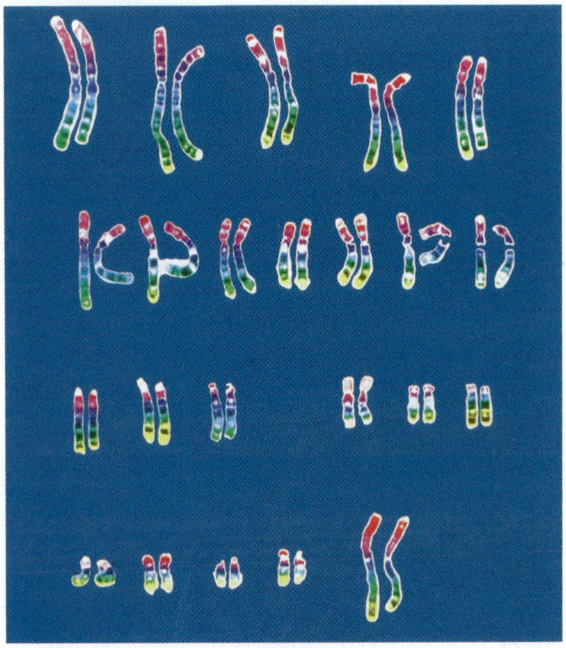

Figure 2 *Human body cells have 46 chromosomes, or 23 pairs of chromosomes.*

Making More Eukaryotic Cells

The eukaryotic cell cycle includes three stages. In the first stage, called *interphase,* the cell grows and copies its organelles and chromosomes. After each chromosome is duplicated, the two copies are called *chromatids*. Chromatids are held together at a region called the *centromere*. The joined chromatids twist and coil and condense into an X shape, as shown in **Figure 3.**

In the second stage, the chromatids separate. The process of chromosome separation is called **mitosis.** Mitosis ensures that each new cell receives a copy of each chromosome. In the third stage, the cell splits into two cells. These cells are identical to each other and to the original cell.

homologous chromosomes chromosomes that have the same sequence of genes and the same structure

mitosis in eukaryotic cells, a process of cell division that forms two new nuclei, each of which has the same number of chromosomes

Figure 3 *This duplicated chromosome consists of two chromatids. The chromatids are joined at the centromere.*

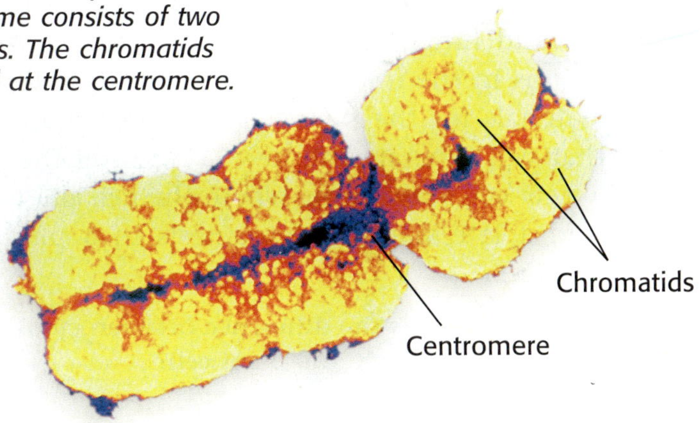

Chromatids

Centromere

CONNECTION TO Language Arts

Picking Apart Vocabulary

Brainstorm what words are similar to the parts of the term *binary fission*. What can you guess about the meaning of the term's root words? Look up the roots of the words, and explain how they help describe the concept.

ACTIVITY

Figure 4 **The Cell Cycle**

Copying DNA (Interphase)

Before mitosis begins, chromosomes are copied. Each chromosome is then two chromatids.

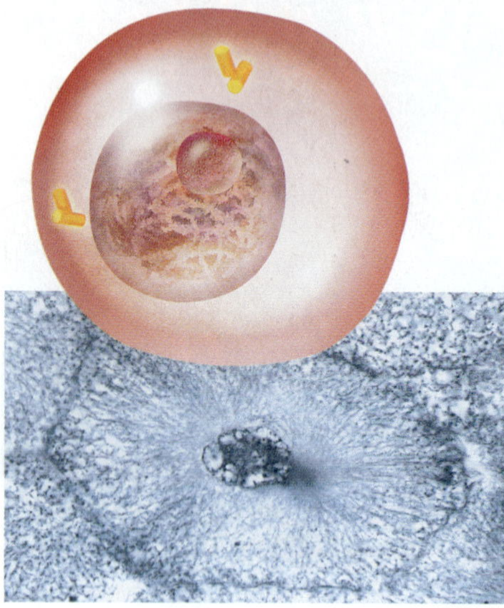

Mitosis Phase 1 (Prophase)

Mitosis begins. The nuclear membrane dissolves. Chromosomes condense into rodlike structures.

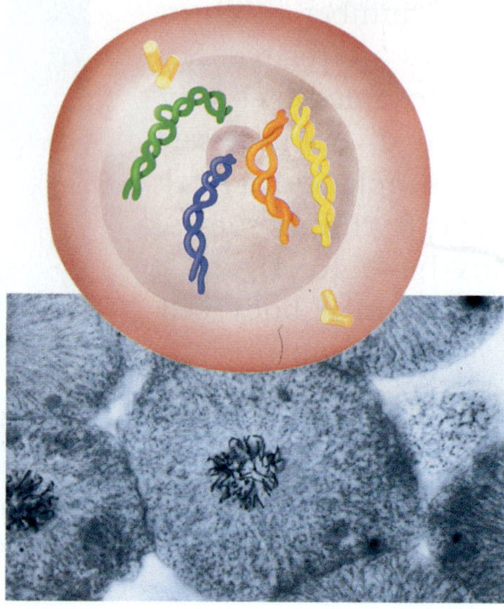

Mitosis Phase 2 (Metaphase)

The chromosomes line up along the equator of the cell. Homologous chromosomes pair up.

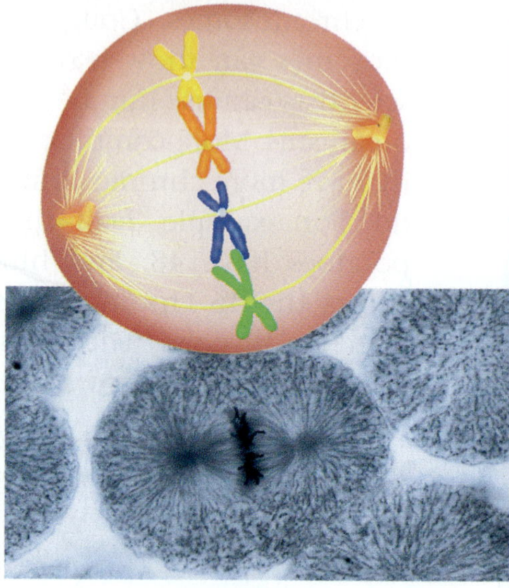

cytokinesis the division of the cytoplasm of a cell

Mitosis and the Cell Cycle

Figure 4 shows the cell cycle and the phases of mitosis in an animal cell. Mitosis has four phases that are shown and described above. This diagram shows only four chromosomes to make it easy to see what's happening inside the cell.

Cytokinesis

In animal cells and other eukaryotes that do not have cell walls, division of the cytoplasm begins at the cell membrane. The cell membrane begins to pinch inward to form a groove, which eventually pinches all the way through the cell, and two daughter cells form. The division of cytoplasm is called **cytokinesis** and is shown at the last step of **Figure 4.**

Eukaryotic cells that have a cell wall, such as the cells of plants, algae, and fungi, reproduce differently. In these cells, a *cell plate* forms in the middle of the cell. The cell plate contains the materials for the new cell membranes and the new cell walls that will separate the new cells. After the cell splits into two, a new cell wall forms where the cell plate was. The cell plate and a late stage of cytokinesis in a plant cell are shown in **Figure 5.**

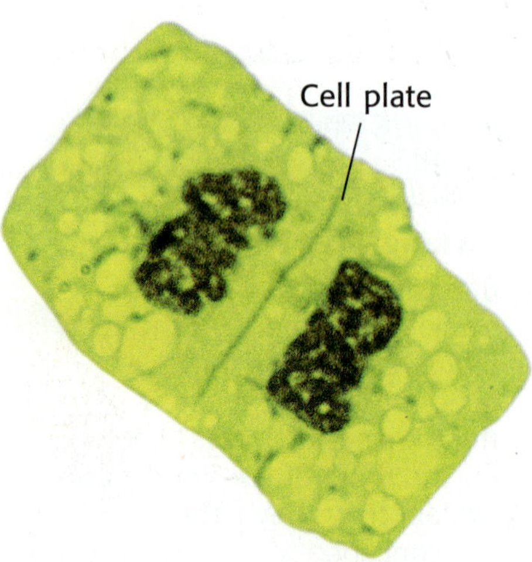

Figure 5 *When a plant cell divides, a cell plate forms and the cell splits into two cells.*

Reading Check **What is the difference between cytokinesis in an animal cell and cytokinesis in a plant cell?**

Mitosis Phase 3 (Anaphase)

The chromatids separate and move to opposite sides of the cell.

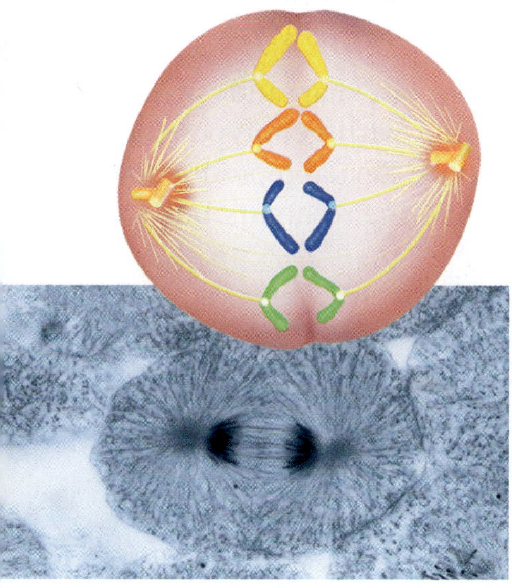

Mitosis Phase 4 (Telophase)

A nuclear membrane forms around each set of chromosomes, and the chromosomes unwind. Mitosis is complete.

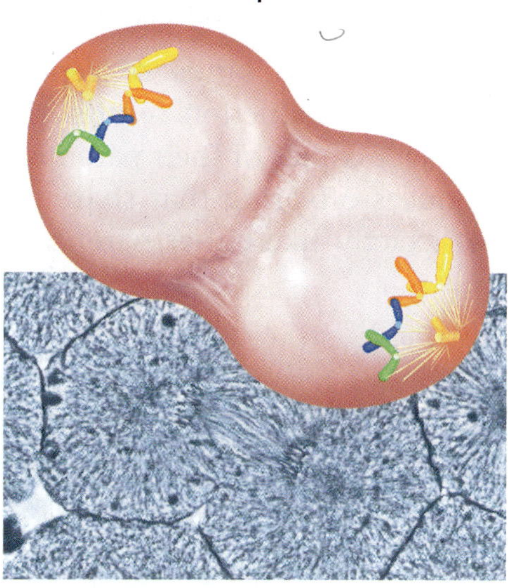

Cytokinesis

In cells that lack a cell wall, the cell pinches in two. In cells that have a cell wall, a cell plate forms between the two new cells.

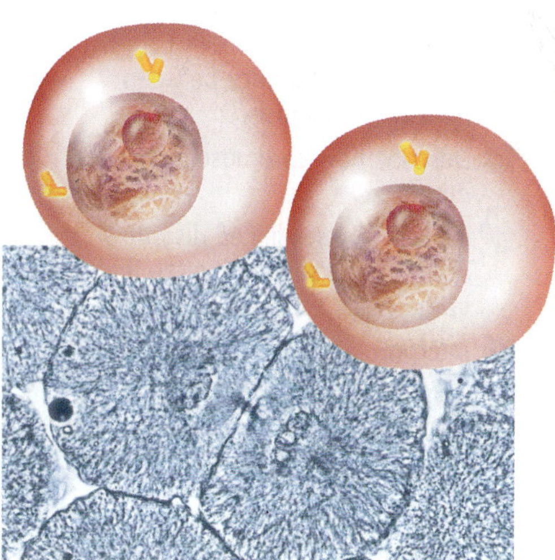

SECTION Review

Summary

- A cell produces more cells by first copying its DNA.
- Eukaryotic cells produce more cells through the four phases of mitosis.
- Mitosis produces two cells that have the same number of chromosomes as the parent cell.
- At the end of mitosis, a cell divides the cytoplasm by cytokinesis.
- In plant cells, a cell plate forms between the two new cells during cytokinesis.

Using Key Terms

1. In your own words, write a definition for each of the following terms: *cell cycle* and *cytokinesis*.

Understanding Key Ideas

2. Eukaryotic cells
   a. do not divide.
   b. undergo binary fission.
   c. undergo mitosis.
   d. have cell walls.
3. Why is it important for chromosomes to be copied before cell division?
4. Describe mitosis.

Math Skills

5. Cell A takes 6 h to complete division. Cell B takes 8 h to complete division. After 24 h, how many more copies of cell A would there be than cell B?

Critical Thinking

6. **Predicting Consequences** What would happen if cytokinesis occurred without mitosis?
7. **Applying Concepts** How does mitosis ensure that a new cell is just like its parent cell?
8. **Making Comparisons** Compare the processes that animal cells and plant cells use to make new cells. How are the processes different?

SECTION 4

Meiosis

How do traits pass from parents to offspring? Genes *are sets of instructions for inherited traits. But where are genes located, and how do they pass information? Understanding reproduction is the first step to finding the answers.*

READING WARM-UP

Objectives

- Explain the difference between mitosis and meiosis.
- Describe how chromosomes determine sex.
- Explain why sex-linked disorders occur in one sex more often than in the other.
- Interpret a pedigree.

Terms to Learn

meiosis
sex chromosome
pedigree

READING STRATEGY

Reading Organizer As you read this section, make a flowchart of the steps of meiosis.

There are two kinds of reproduction: asexual and sexual. Asexual reproduction results in offspring with genotypes that are exact copies of their parent's genotype. Sexual reproduction produces offspring that share traits with their parents but are not exactly like either parent.

Asexual Reproduction

In *asexual reproduction,* only one parent cell is needed. The structures inside the cell are copied, and then the parent cell divides, making two exact copies. This type of cell reproduction is known as *mitosis.* Most of the cells in your body and most single-celled organisms reproduce in this way.

Sexual Reproduction

In sexual reproduction, two parent cells join together to form offspring that are different from both parents. The parent cells are called *sex cells.* A female sex cell joins with a male sex cell to form a fertilized egg which has genetic information from each parent. This egg multiplies to form a new organism.

Sex cells are different from ordinary body cells because they have only 23 chromosomes—half the usual number. Ordinary body cells have 46 chromosomes that can be arranged in pairs of homologous chromosomes, as shown in **Figure 1.** Each sex cell has only one of the chromosomes from each homologous pair of chromosomes.

meiosis a process in cell division during which the number of chromosomes decreases to half the original number by two divisions of the nucleus, which results in the production of sex cells

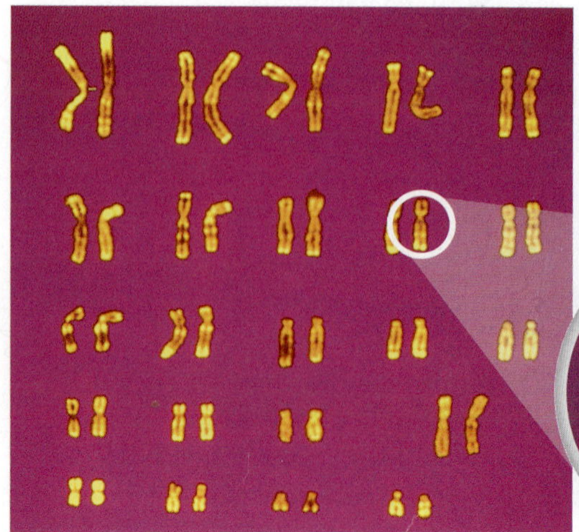
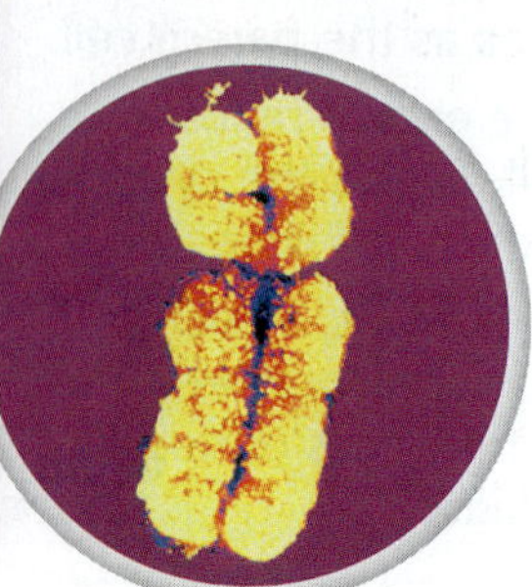

Figure 1 *One member of a pair of homologous chromosomes from a human body cell is shown below.*

Meiosis

Sex cells are made during meiosis (mie OH sis). **Meiosis** is a copying process that produces cells with half the usual number of chromosomes. Each sex cell receives one-half of each homologous pair. For example, a human egg cell has 23 chromosomes, and a sperm cell has 23 chromosomes. The new cell that forms when an egg cell and a sperm cell join has 46 chromosomes.

Reading Check **How many chromosomes does a human egg cell have?** *(See the Appendix for answers to Reading Checks.)*

Genes and Chromosomes

What does all of this have to do with the location of genes? A graduate student named Walter Sutton made an important observation about chromosomes. Sutton was studying sperm cells in grasshoppers. Sutton knew of earlier studies which showed that the egg and sperm must each contribute the same amount of information to the offspring. That was the only way the 3:1 ratio found in the second generation could be explained. Sutton also knew from his own studies that although eggs and sperm were different, they did have something in common: Their chromosomes were located inside a nucleus. Using his observations of meiosis, his understanding of earlier studies on inheritance, and some creative thinking, Sutton proposed something very important:

Genes are located on chromosomes!

Understanding meiosis was critical to finding the location of genes. Before you learn about meiosis, review mitosis, shown in **Figure 2.** Meiosis is outlined in **Figure 3** on the next two pages.

CONNECTION TO Language Arts

Greek Roots The word *mitosis* is related to a Greek word that means "threads." Threadlike spindles are visible during mitosis. The word *meiosis* comes from a Greek word that means "to make smaller." How do you think meiosis got its name?

Figure 2 Mitosis Revisited

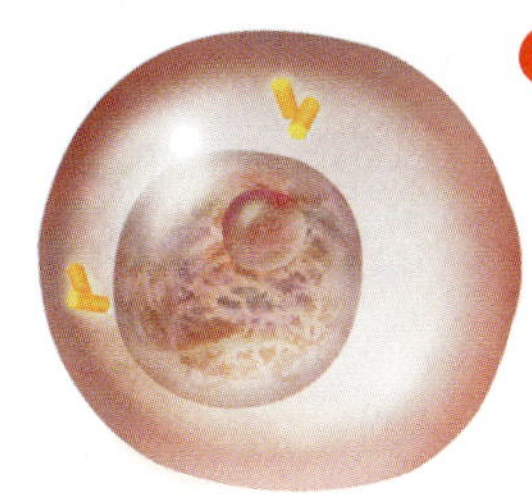

1. Each chromosome is copied.

2. The chromosomes thicken and shorten. Each chromosome consists of two identical copies, called *chromatids.*

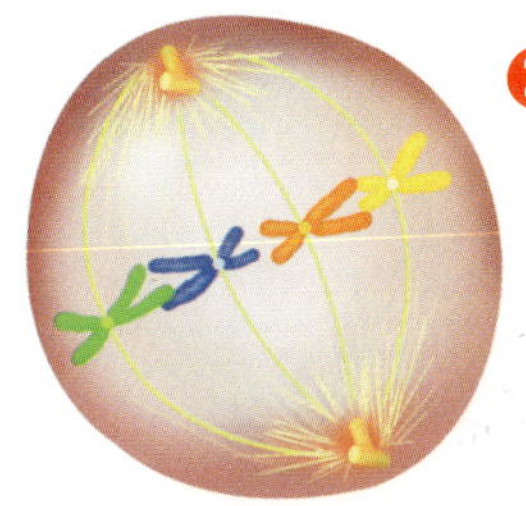

3. The nuclear membrane dissolves. The chromatids line up along the equator (center) of the cell.

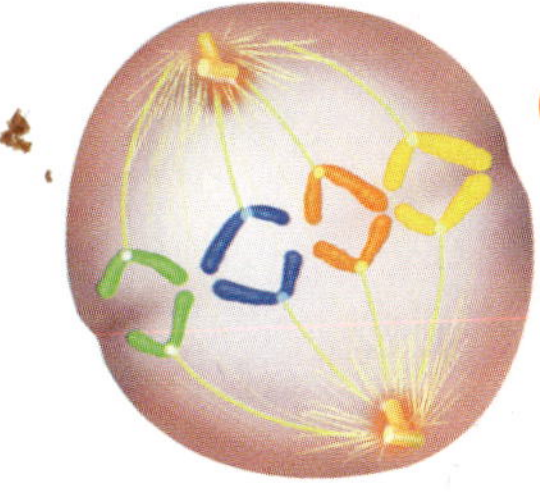

4. The chromatids pull apart.

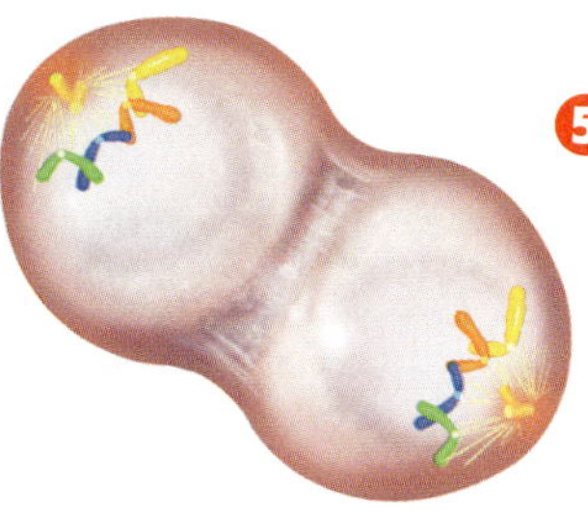

5. The nuclear membrane forms around the separated chromatids. The chromosomes unwind, and the cell divides.

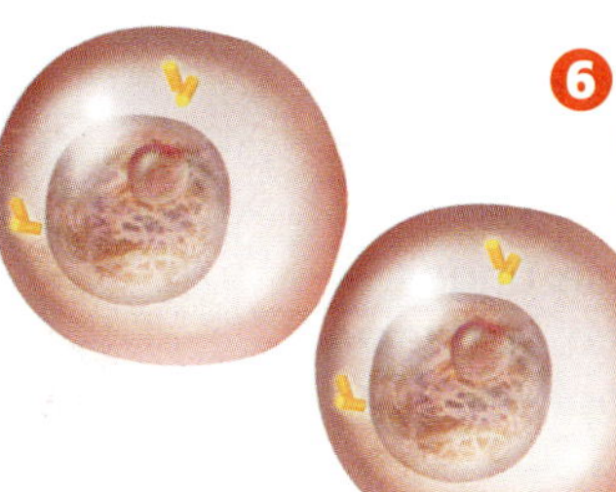

6. The result is two identical copies of the original cell.

The Steps of Meiosis

During mitosis, chromosomes are copied once, and then the nucleus divides once. During meiosis, chromosomes are copied once, and then the nucleus divides twice. The resulting sperm and eggs have half the number of chromosomes of a normal body cell. **Figure 3** shows all eight steps of meiosis. Read about each step as you look at the figure. Different types of living things have different numbers of chromosomes. In this illustration, only four chromosomes are shown.

Reading Check **How many cells are made from one parent cell during meiosis?**

Figure 3 **Steps of Meiosis**

Read about each step as you look at the diagram. Different types of living things have different numbers of chromosomes. In this diagram, only four chromosomes are shown.

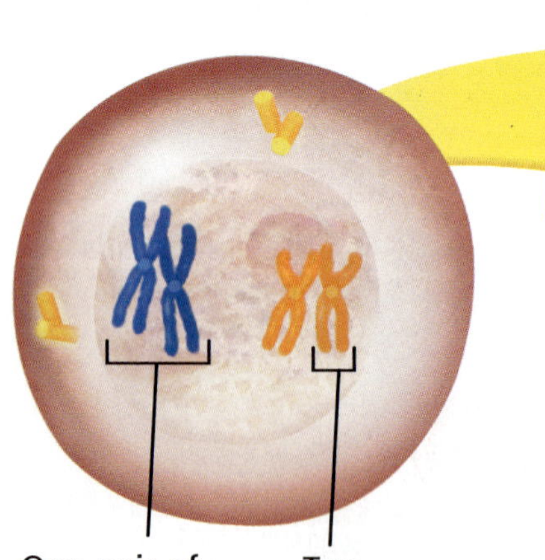

1 Before meiosis begins, the chromosomes are in a threadlike form. Each chromosome makes an exact copy of itself, forming two halves called *chromatids*. The chromosomes then thicken and shorten into a form that is visible under a microscope. The nuclear membrane disappears.

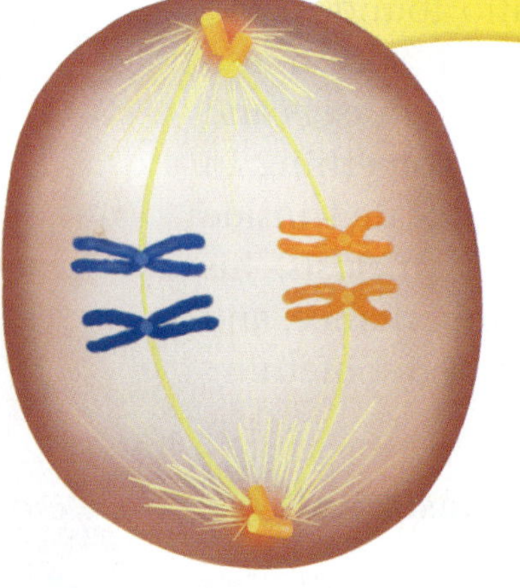

2 Each chromosome is now made up of two identical chromatids. Similar chromosomes pair with one another, and the paired homologous chromosomes line up at the equator of the cell.

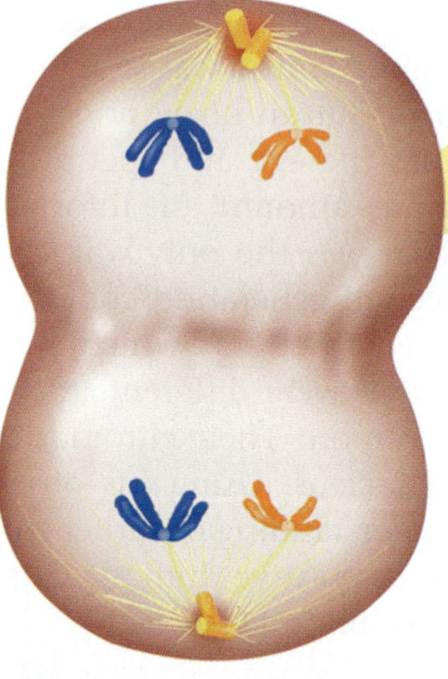

3 The chromosomes separate from their homologous partners and then move to opposite ends of the cell.

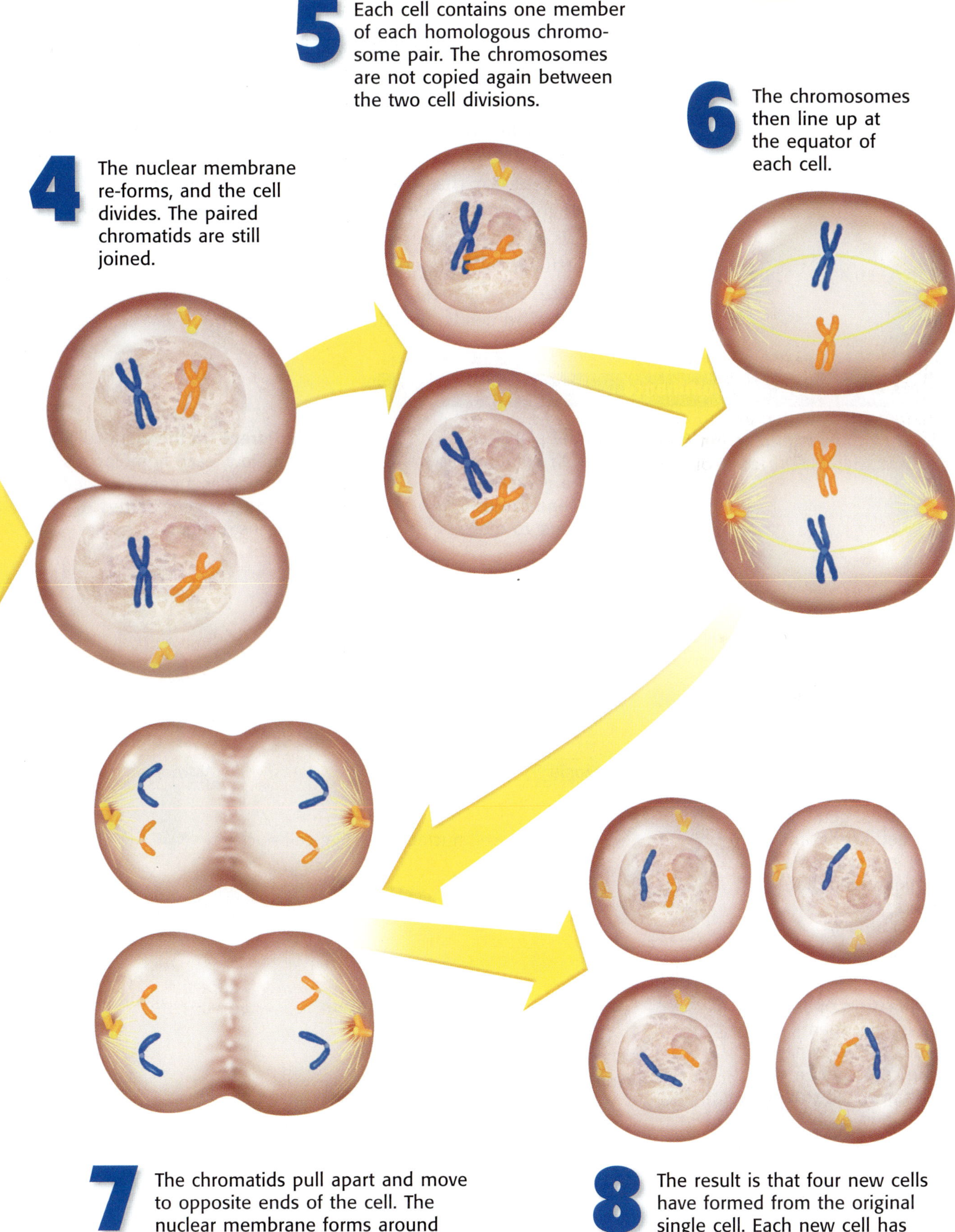
5
Each cell contains one member of each homologous chromosome pair. The chromosomes are not copied again between the two cell divisions.
6
The chromosomes then line up at the equator of each cell.
4
The nuclear membrane re-forms, and the cell divides. The paired chromatids are still joined.
7
The chromatids pull apart and move to opposite ends of the cell. The nuclear membrane forms around the separated chromosomes, and the cells divide.
8
The result is that four new cells have formed from the original single cell. Each new cell has half the number of chromosomes present in the original cell.

Inheriting Traits

The steps of meiosis help explain how traits are inherited. Individuals inherit a homologous chromosome from each parent. The genes on these chromosomes determine what traits appear. Some traits are dominant, and others are recessive. *Dominant* traits appear even if only one homologous chromosome carries the gene for that trait. *Recessive* traits appear only if both homologous chromosomes carry that gene. **Figure 4** shows what happens to a pair of homologous chromosomes during meiosis and fertilization. The cross shown is between a plant that has two round seed genes and a plant that has two wrinkled seed genes. Round seeds are the dominant trait.

Figure 4 **Meiosis and Dominance**

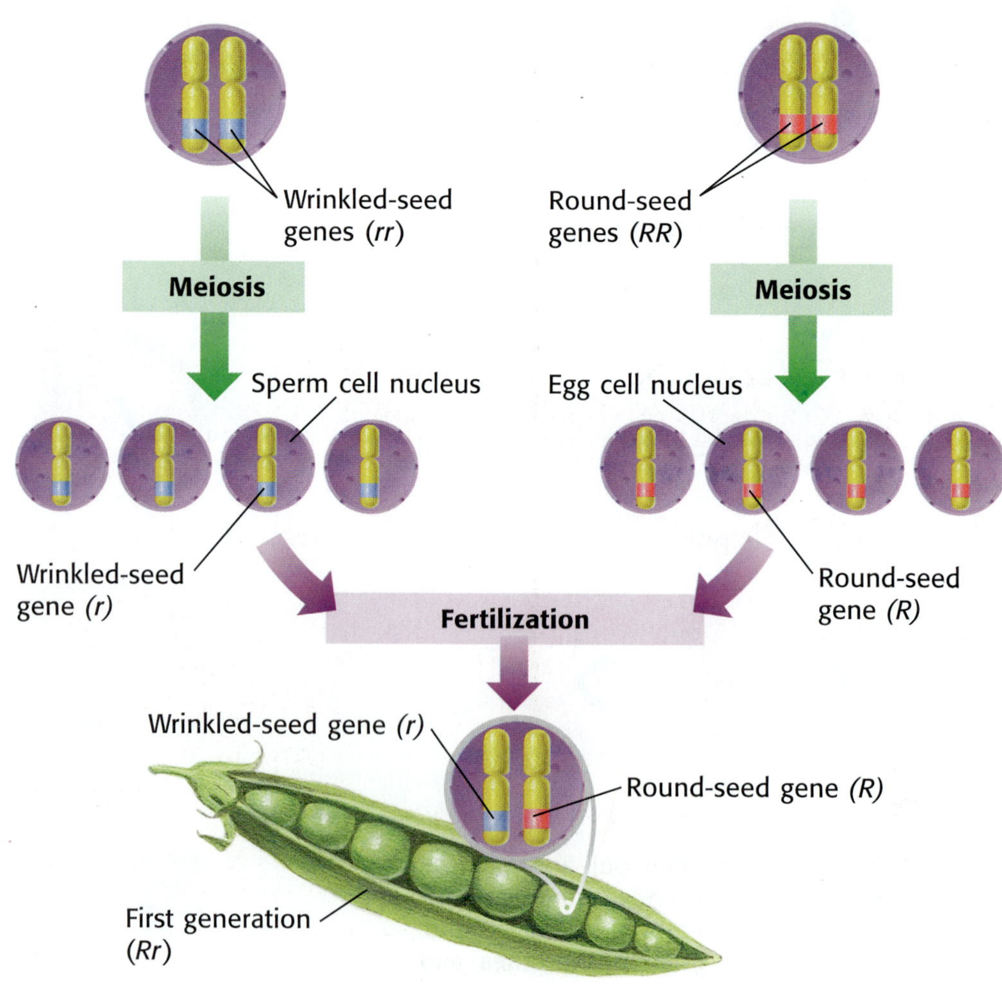

a Following **meiosis,** each sperm cell has a recessive gene for wrinkled seeds, and each egg cell has a dominant gene for round seeds.

b **Fertilization** of any egg by any sperm results in the same gene combination (*Rr*). The same trait (round) appears in each offspring.

Sex Chromosomes

Information contained on chromosomes determines many kinds of traits. **Sex chromosomes** carry genes that determine sex. In humans, females have two X chromosomes. But human males have one X chromosome and one Y chromosome.

During meiosis, one of each of the chromosome pairs ends up in a sex cell. Females have two X chromosomes in each body cell. When meiosis produces the egg cells, each egg gets one X chromosome. Males have both an X chromosome and a Y chromosome in each body cell. Meiosis produces sperm with either an X or a Y chromosome. An egg fertilized by a sperm with an X chromosome will produce a female. If the sperm contains a Y chromosome, the offspring will be male, as shown in **Figure 5.**

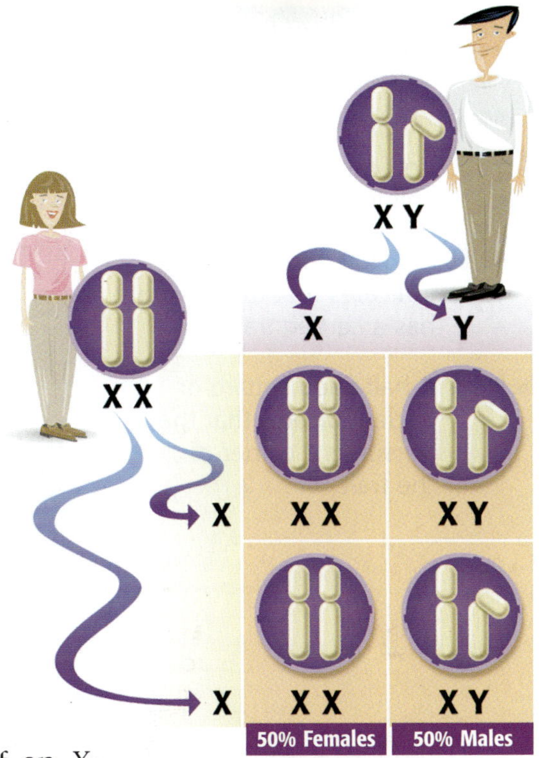

Figure 5 *Egg and sperm combine to form either the XX or XY combination.*

Sex-Linked Disorders

The Y chromosome does not carry all of the genes of an X chromosome. Females have two X chromosomes, so they carry two copies of each gene found on the X chromosome. This makes a backup gene available if one becomes damaged. Males have only one copy of each gene on their one X chromosome. The genes for certain disorders, such as colorblindness, are carried on the X chromosome. These disorders are called *sex-linked disorders*. Because the gene for such disorders is recessive, men are more likely to have sex-linked disorders.

sex chromosome one of the pair of chromosomes that determine the sex of an individual

People who are colorblind can have trouble distinguishing between shades of red and green. To help the colorblind, some cities have added shapes to their street lights, as shown in **Figure 6.** Hemophilia (HEE moh FIL ee uh) is another sex-linked disorder. Hemophilia prevents blood from clotting, and people with hemophilia bleed for a long time after small cuts. Hemophilia can be fatal.

Figure 6 *This stoplight in Canada is designed to help the colorblind see signals easily. This photograph was taken over a few minutes to show all three shapes.*

Figure 7 **Pedigree for a Recessive Disease**

Males — Females

Vertical lines connect children to their parents.

A solid square or circle indicates that the person has a certain trait.

A half-filled square or circle indicates that the person is a carrier of the trait.

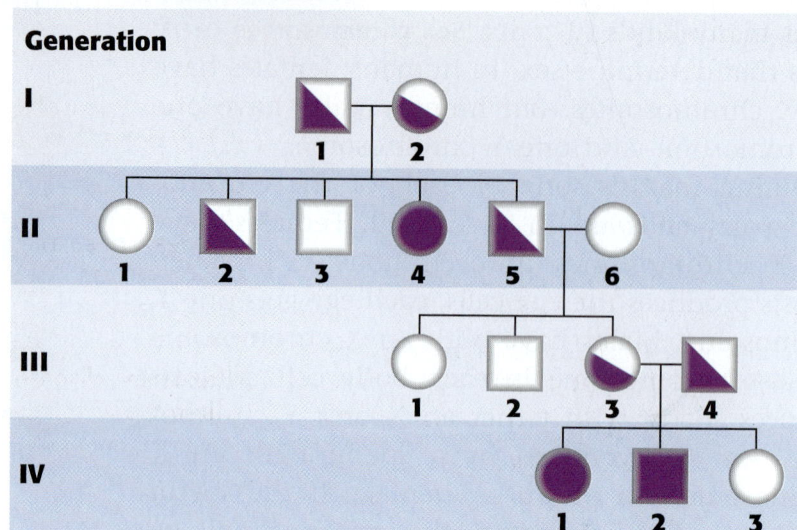

Genetic Counseling

Hemophilia and other genetic disorders can be traced through a family tree. If people are worried that they might pass a disease to their children, they may consult a genetic counselor. These counselors often make use of a diagram known as a **pedigree,** which is a tool for tracing a trait through generations of a family. By making a pedigree, a counselor can often predict whether a person is a carrier of a hereditary disease. The pedigree shown in **Figure 7** traces a disease called *cystic fibrosis* (SIS tik FIE broh sis). Cystic fibrosis causes serious lung problems. People with this disease have inherited two copies of the recessive gene. Both parents need to be carriers of the gene for the disease to show up in their children.

pedigree a diagram that shows the occurrence of a genetic trait in several generations of a family

Pedigrees can be drawn up to trace any trait through a family tree. You could even draw a pedigree that would show how you inherited your hair color. Many different pedigrees could be drawn for a typical family.

Selective Breeding

For thousands of years, humans have seen the benefits of the careful breeding of plants and animals. In *selective breeding,* organisms with desirable characteristics are mated. You have probably enjoyed the benefits of selective breeding, although you may not have realized it. For example, you have probably eaten an egg from a chicken that was bred to produce more eggs. Your pet dog may be a result of selective breeding. Roses, like the one shown in **Figure 8,** have been selectively bred to produce large flowers. Wild roses are much smaller and have fewer petals than roses that you could buy at a nursery.

Figure 8 *Roses have been selectively bred to create large, bright flowers.*

SECTION Review

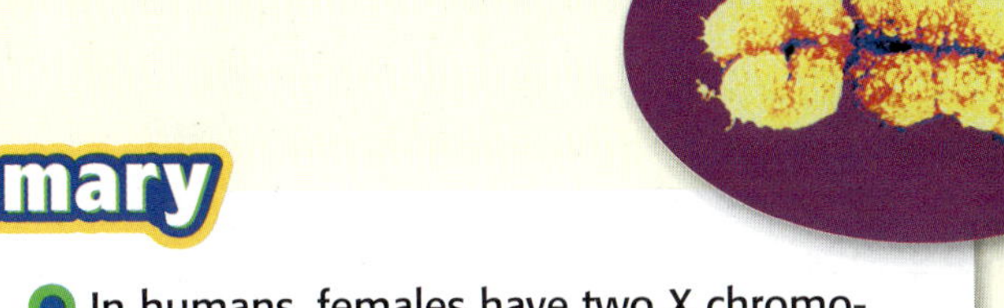

- In mitosis, chromosomes are copied once, and then the nucleus divides once. In meiosis, chromosomes are copied once, and then the nucleus divides twice.
- The process of meiosis produces sex cells, which have half the number of chromosomes. These two halves combine during reproduction.
- In humans, females have two X chromosomes. So, each egg contains one X chromosome. Males have both an X and a Y chromosome. So, each sperm cell contains either an X or a Y chromosome.
- Sex-linked disorders occur in males more often than in females. Colorblindness and hemophilia are examples of sex-linked disorders.
- A pedigree is a diagram used to trace a trait through many generations of a family.

Using Key Terms

1. Use each of the following terms in the same sentence: *meiosis* and *sex chromosomes.*

In each of the following sentences, replace the incorrect term with the correct term from the word bank.

pedigree	homologous chromosomes
meiosis	mitosis

2. During fertilization, chromosomes are copied, and then the nucleus divides twice.
3. A Punnett square is used to show how inherited traits move through a family.
4. During meiosis, sex cells line up in the middle of the cell.

Understanding Key Ideas

5. Genes are found on
   a. chromosomes.
   b. proteins.
   c. alleles.
   d. sex cells.
6. If there are 14 chromosomes in pea plant cells, how many chromosomes are present in a sex cell of a pea plant?
7. Draw the eight steps of meiosis. Label one chromosome, and show its position in each step.

Interpreting Graphics

Use this pedigree to answer the question below.

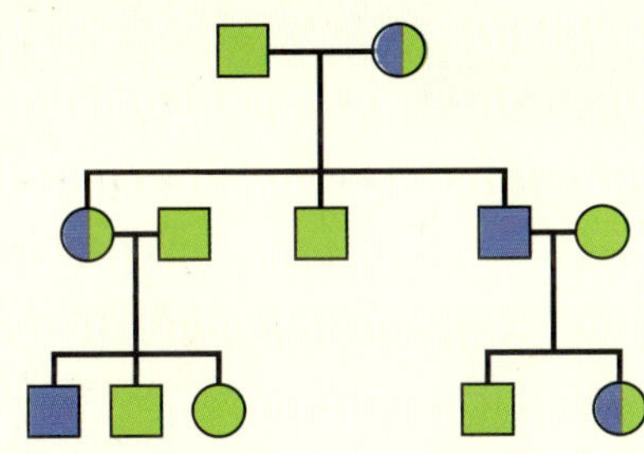

8. Is this disorder sex linked? Explain your reasoning.

Critical Thinking

9. **Identifying Relationships** Put the following in order of smallest to largest: chromosome, gene, and cell.
10. **Applying Concepts** A pea plant has purple flowers. What genes for flower color could the sex cells carry?

Using Scientific Methods

Inquiry Lab

The Perfect Taters Mystery

OBJECTIVES

Examine osmosis in potato cells.

Design a procedure that will give the best results.

MATERIALS

- cups, clear plastic, small
- potato pieces, freshly cut
- potato samples (A, B, and C)
- salt
- water, distilled

SAFETY

You are the chief food detective at Perfect Taters Food Company. The boss, Mr. Fries, wants you to find a way to keep his potatoes fresh and crisp before they are cooked. His workers have tried several methods, but these methods have not worked. Workers in Group A put the potatoes in very salty water, and something unexpected happened to the potatoes. Workers in Group B put the potatoes in water that did not contain any salt, and something else happened! Workers in Group C didn't put the potatoes in any water, and that didn't work either. Now, you must design an experiment to find out what can be done to make the potatoes stay crisp and fresh.

- Before you plan your experiment, review what you know. You know that potatoes are made of cells. Plant cells contain a large amount of water. Cells have membranes that hold water and other materials inside and keep some things out. Water and other materials must travel across cell membranes to get into and out of the cell.
- Mr. Fries has told you that you can obtain as many samples as you need from the workers in Groups A, B, and C. Your teacher will have these samples ready for you to observe.
- Make a data table like the one below. List your observations in the data table. Make as many observations as you can about the potatoes tested by workers in Groups A, B, and C.

Observations	
Group A	
Group B	DO NOT WRITE IN BOOK
Group C	

Ask a Question

1. Now that you have made your observations, state Mr. Fries's problem in the form of a question that can be answered by your experiment.

Form a Hypothesis

2. Form a hypothesis based on your observations and your questions. The hypothesis should be a statement about what causes the potatoes not to be crisp and fresh. Based on your hypothesis, make a prediction about the outcome of your experiment. State your prediction in an if-then format.

Test the Hypothesis

3. Once you have made a prediction, design your investigation. Check your experimental design with your teacher before you begin. Mr. Fries will give you potato pieces, water, salt, and no more than six containers.

4. Keep very accurate records. Write your plan and procedure. Make data tables. To be sure your data is accurate, measure all materials carefully and make drawings of the potato pieces before and after the experiment.

Analyze the Results

1. **Explaining Events** Explain what happened to the potato cells in Groups A, B, and C in your experiment. Include a discussion of the cell membrane and the process of osmosis.

Draw Conclusions

2. **Analyzing Results** Write a letter to Mr. Fries that explains your experimental method, results, and conclusion. Then, make a recommendation about how he should handle the potatoes so that they will stay fresh and crisp.

USING KEY TERMS

1. Use the following terms in the same sentence: *diffusion* and *osmosis*.

2. In your own words, write a definition for each of the following terms: *exocytosis* and *endocytosis*.

Complete each of the following sentences by choosing the correct term from the word bank.

cellular respiration
photosynthesis
fermentation

3. Plants use ___ to make glucose.

4. During ___, oxygen is used to break down food molecules releasing large amounts of energy.

For each pair of terms, explain how the meanings of the terms differ.

5. *meiosis* and *mitosis*

6. *homologous chromosomes* and *sex chromosomes*

7. *cellular respiration* and *fermentation*

UNDERSTANDING KEY IDEAS

Multiple Choice

8. The process in which particles move through a membrane from a region of low concentration to a region of high concentration is
   - **a.** diffusion.
   - **b.** passive transport.
   - **c.** active transport.
   - **d.** fermentation.

9. What is the result of mitosis and cytokinesis?
   - **a.** two identical cells
   - **b.** two nuclei
   - **c.** chloroplasts
   - **d.** two different cells

10. Before the energy in food can be used by a cell, the energy must first be transferred to molecules of
    - **a.** proteins.
    - **b.** carbohydrates.
    - **c.** DNA.
    - **d.** ATP.

11. Which of the following cells would form a cell plate during the cell cycle?
    - **a.** a human cell
    - **b.** a prokaryotic cell
    - **c.** a plant cell
    - **d.** All of the above

Short Answer

12. Are exocytosis and endocytosis examples of active or passive transport? Explain your answer.

13. Name the cell structures that are needed for photosynthesis and the cell structures that are needed for cellular respiration.

14. Describe the three stages of the cell cycle of a eukaryotic cell.

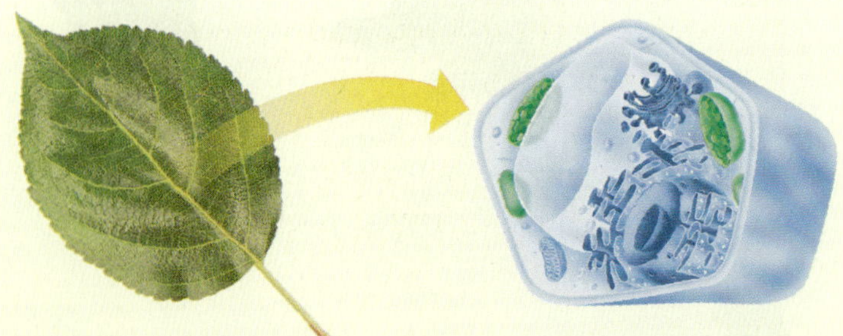

CRITICAL THINKING

15 **Concept Mapping** Use the following terms to create a concept map: *chromosome duplication, cytokinesis, prokaryote, mitosis, cell cycle, binary fission,* and *eukaryote.*

16 **Making Inferences** Which one of the plants pictured below was given water mixed with salt, and which one was given pure water? Explain how you know, and be sure to use the word *osmosis* in your answer.

17 **Identifying Relationships** Why would your muscle cells need to be supplied with more food when there is a lack of oxygen than when there is plenty of oxygen present?

18 **Applying Concepts** Explain how two people who are not colorblind could have a child who is colorblind. Would this child be a boy or a girl?

19 **Predicting Consequences** Suppose that a gardener bred two round-seed pea plants that each had one round-seed gene and one winkle-seed gene. Do you think all the offspring would have round seeds? Why or why not?

INTERPRETING GRAPHICS

The picture below shows a cell. Use the picture below to answer the questions that follow.

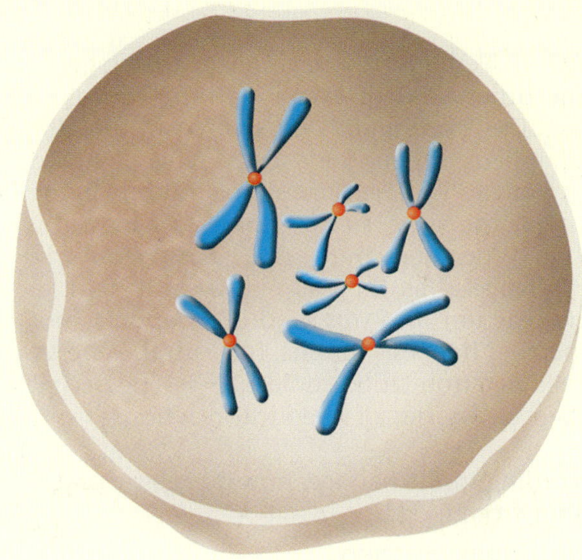

20 Is the cell prokaryotic or eukaryotic?

21 Which stage of the cell cycle is this cell in?

22 How many chromatids are present? How many pairs of homologous chromosomes are present?

23 How many chromosomes will be present in each of the new cells after the cell divides?

Standardized Test Preparation

READING

Read each of the passages below. Then, answer the questions that follow each passage.

Passage 1 Perhaps you have heard that jogging or some other kind of exercise "burns" a lot of Calories. The word *burn* is often used to describe what happens when your cells release stored energy from food. The burning of food in living cells is not the same as the burning of logs in a campfire. When logs burn, the energy stored in wood is released as thermal energy and light in a single reaction. But this kind of reaction is not the kind that happens in cells. Instead, the energy that cells get from food molecules is released at each step of a series of chemical reactions.

1. According to the passage, how do cells release energy from food?

- **A** in a single reaction
- **B** as thermal energy and light
- **C** in a series of reactions
- **D** by burning

2. Which of the following statements is a fact in the passage?

- **F** Wood burns better than food does.
- **G** Both food and wood have stored energy.
- **H** Food has more stored energy than wood does.
- **I** When it is burned, wood releases only thermal energy.

3. According to the passage, why might people be confused between what happens in a living cell and what happens in a campfire?

- **A** The word *burn* may describe both processes.
- **B** Thermal energy is released during both processes.
- **C** Wood can be burned and broken down by living cells.
- **D** Jogging and other exercises use energy.

Passage 2 The word *respiration* means "breathing," but cellular respiration is different from breathing. Breathing supplies your cells with the oxygen that they need for cellular respiration. Breathing also rids your body of carbon dioxide, which is a waste product of cellular respiration. Cellular respiration is the chemical process that releases energy from food. Most organisms obtain energy from food through cellular respiration. During cellular respiration, oxygen is used to break down food (glucose) into CO_2 and H_2O, and energy is released. In humans, most of the energy released is used to maintain body temperature.

1. According to the passage, what is glucose?

- **A** a type of chemical process
- **B** a type of waste product
- **C** a type of organism
- **D** a type of food

2. According to the passage, how does cellular respiration differ from breathing?

- **F** Breathing releases carbon dioxide, but cellular respiration releases oxygen.
- **G** Cellular respiration is a chemical process that uses oxygen to release energy from food, but breathing supplies cells with oxygen.
- **H** Cellular respiration requires oxygen, but breathing does not.
- **I** Breathing rids your body of waste products, but cellular respiration stores wastes.

3. According to the passage, how do humans use most of the energy released?

- **A** to break down food
- **B** to obtain oxygen
- **C** to maintain body temperature
- **D** to get rid of carbon dioxide

INTERPRETING GRAPHICS

The graph below shows the cell cycle. Use this graph to answer the questions that follow.

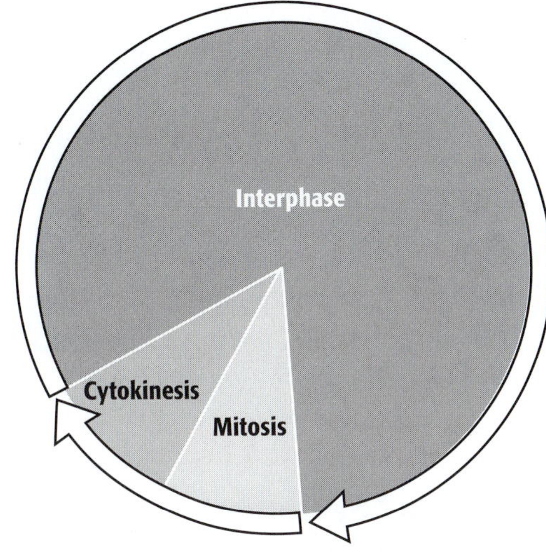

1. Which part of the cell cycle lasts longest?
   - **A** interphase
   - **B** mitosis
   - **C** cytokinesis
   - **D** There is not enough information to determine the answer.

2. Which of the following lists the parts of the cell cycle in the proper order?
   - **F** mitosis, cytokinesis, mitosis
   - **G** interphase, cytokinesis, mitosis
   - **H** interphase, mitosis, interphase
   - **I** mitosis, cytokinesis, interphase

3. Which part of the cell cycle is the briefest?
   - **A** interphase
   - **B** cell division
   - **C** cytokinesis
   - **D** There is not enough information to determine the answer.

4. Why is the cell cycle represented by a circle?
   - **F** The cell cycle is a continuous process that begins again after it finishes.
   - **G** The cell cycle happens only in cells that are round.
   - **H** The cell cycle is a linear process.
   - **I** The cell is in interphase for more than half of the cell cycle.

MATH

Read each question below, and choose the best answer.

1. A normal cell spends 90% of its time in interphase. How is 90% expressed as a fraction?
   - **A** 3/4
   - **B** 4/5
   - **C** 85/100
   - **D** 9/10

2. If a cell lived for 3 weeks and 4 days, how many days did it live?
   - **F** 7
   - **G** 11
   - **H** 21
   - **I** 25

3. How is $2 \times 3 \times 3 \times 3 \times 3$ expressed in exponential notation?
   - **A** 3×2^4
   - **B** 2×3^3
   - **C** 3^4
   - **D** 2×3^4

4. Cell A has 3 times as many chromosomes as cell B has. After cell B's chromosomes double during mitosis, cell B has 6 chromosomes. How many chromosomes does cell A have?
   - **F** 3
   - **G** 6
   - **H** 9
   - **I** 18

5. If $x + 2 = 3$, what does $x + 1$ equal?
   - **A** 4
   - **B** 3
   - **C** 2
   - **D** 1

6. If $3x + 2 = 26$, what does $x + 1$ equal?
   - **F** 7
   - **G** 8
   - **H** 9
   - **I** 10

Science in Action

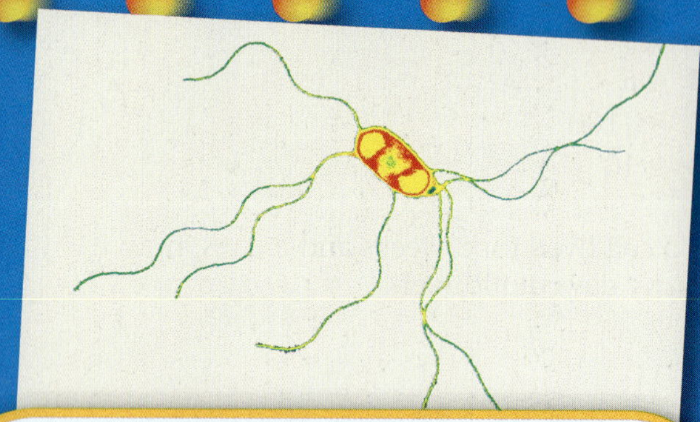

Scientific Discoveries

Electrifying News About Microbes

Your car is out of fuel, and there isn't a service station in sight. This is not a problem! Your car's motor runs on electricity supplied by trillions of microorganisms. Some chemists think that "living" batteries will someday operate everything from watches to entire cities. A group of scientists at King's College in London have demonstrated that microorganisms can convert food into usable electrical energy. The microorganisms convert foods such as table sugar and molasses most efficiently. An efficient microorganism can convert more than 90% of its food into compounds that will fuel an electric reaction. A less efficient microbe will only convert 50% of its food into these types of compounds.

Math ACTIVITY

An efficient microorganism converts 90% of its food into fuel compounds, and an inefficient microorganism converts only 50%. If the inefficient microorganism makes 60 g of fuel out of a possible 120 g of food, how much fuel would an efficient microorganism make out of the same amount of food?

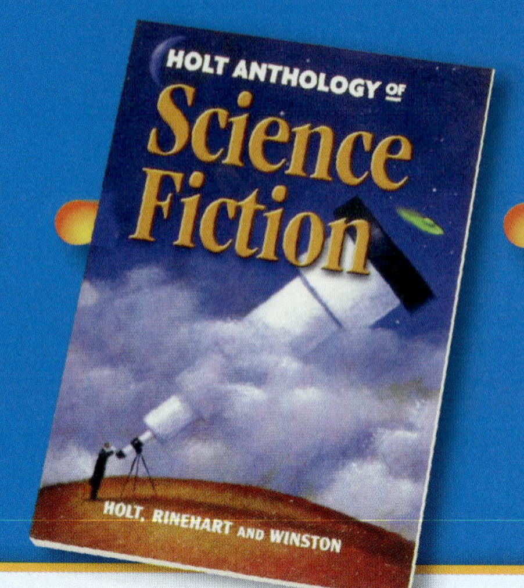

Science Fiction

"Contagion" by Katherine MacLean

A quarter mile from their spaceship, the *Explorer,* a team of doctors walk carefully along a narrow forest trail. Around them, the forest looks like a forest on Earth in the fall—the leaves are green, copper, purple, and fiery red. But it isn't fall. And the team is not on Earth.

Minos is enough like Earth to be the home of another colony of humans. But Minos might also be home to unknown organisms that could cause severe illness or death among the crew of *Explorer*. These diseases might be enough like diseases on Earth to be contagious, but they might be different enough to be very difficult to treat.

Something large moves among the shadows—it looks like a man. What happens next? Read Katherine's MacLean's "Contagion" in the *Holt Anthology of Science Fiction* to find out.

Language Arts ACTIVITY

WRITING SKILL Write two to three paragraphs that describe what you think might happen next in the story.

Careers

Jerry Yakel

Neuroscientist Jerry Yakel credits a sea slug for making him a neuroscientist. In a college class studying neurons, or nerve cells, Yakel got to see firsthand how ions move across the cell membrane of *Aplysia californica,* also known as a sea hare. He says, "I was totally hooked. I knew that I wanted to be a neurophysiologist then and there. I haven't wavered since."

Today, Yakel is a senior investigator for the National Institutes of Environmental Health Sciences, which is part of the U.S. government's National Institutes of Health. "We try to understand how the normal brain works," says Yakel of his team. "Then, when we look at a diseased brain, we train to understand where the deficits are. Eventually, someone will have an idea about a drug that will tweak the system in this or that way."

Yakel studies the ways in which nicotine affects the human brain. "It is one of the most prevalent and potent neurotoxins in the environment," says Yakel. "I'm amazed that it isn't higher on the list of worries for the general public."

Social Studies ACTIVITY

WRITING SKILL Research a famous or historical figure in science. Write a short report that outlines how he or she became interested in science.

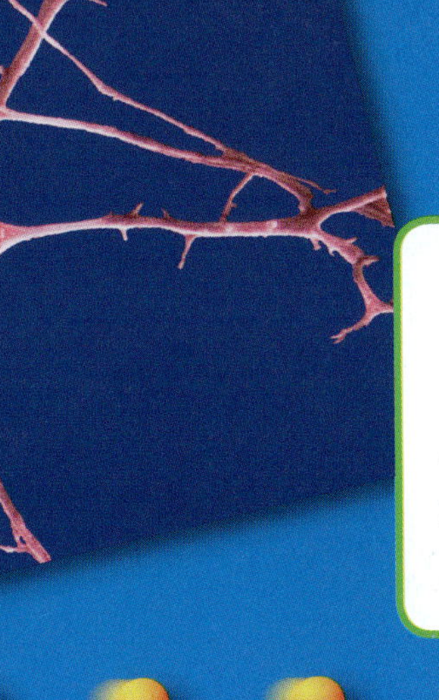

To learn more about these Science in Action topics, visit go.hrw.com and type in the keyword **HL5ACTF.**

Check out Current Science® articles related to this chapter by visiting go.hrw.com. Just type in the keyword HL5CS04.

Bacteria and Viruses

About the PHOTO

Bacteria are everywhere. Some provide us with medicines, and some make foods we eat. Others, such as the one pictured here, can cause illness. This bacterium is a kind of *Salmonella,* and it can cause food poisoning. *Salmonella* can live inside chickens and other birds. Cooking eggs and chicken properly helps make sure that you don't get sick from *Salmonella*.

PRE-READING ACTIVITY

FOLDNOTES

Double Door Before you read the chapter, create the FoldNote entitled "Double Door" described in the **Study Skills** section of the Appendix. Write "Bacteria" on one flap of the double door and "Viruses" on the other flap. As you read the chapter, compare the two topics, and write characteristics of each on the inside of the appropriate flap.

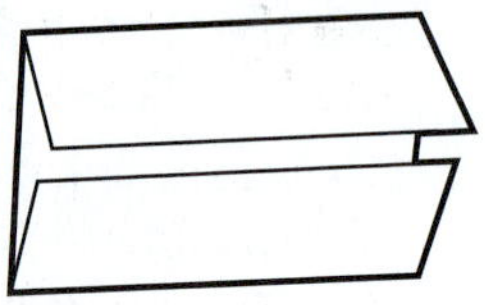

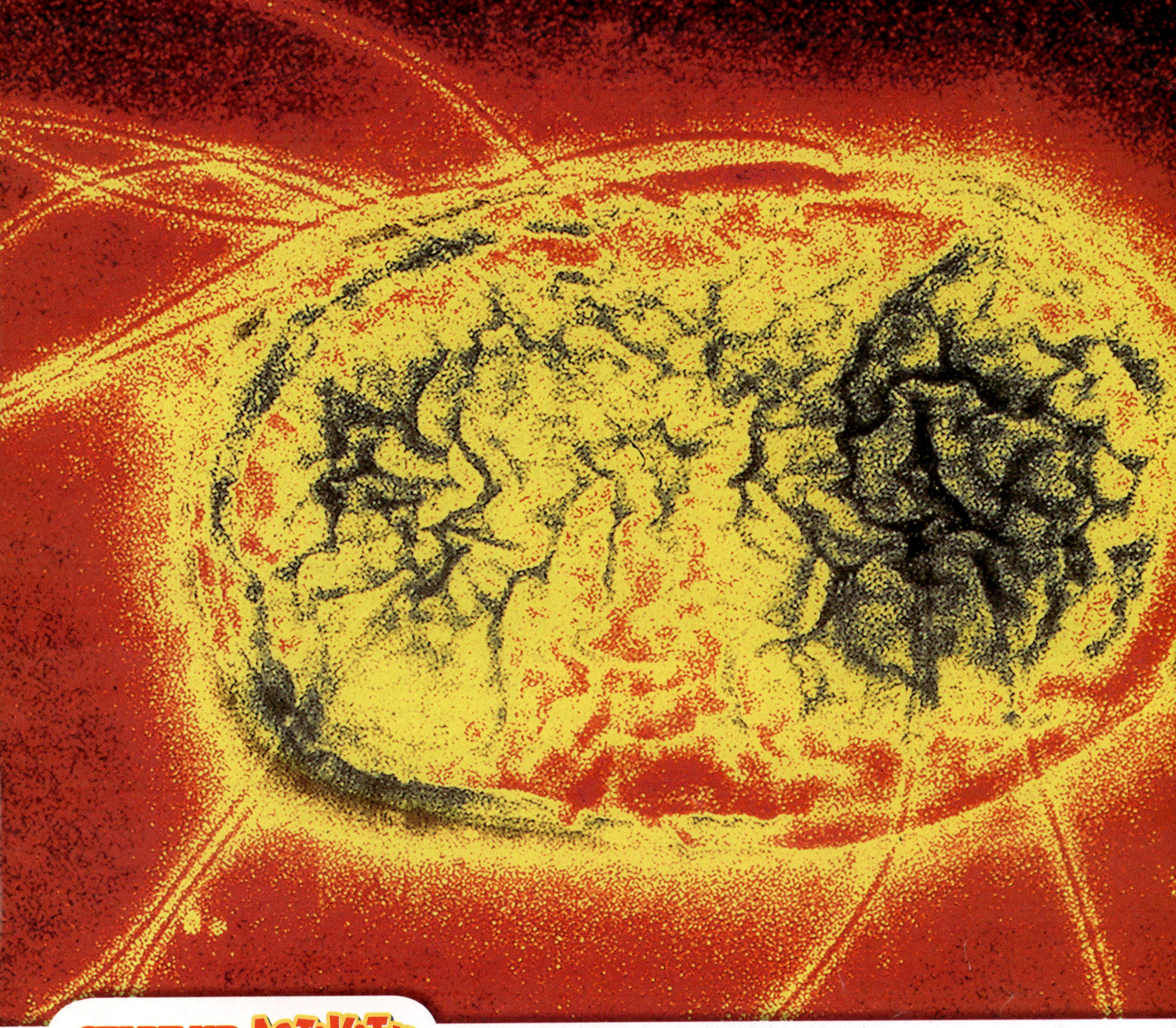

START-UP ACTIVITY

Our Constant Companions

Bacteria are in the soil, in the air, and even inside your body. When grown in a laboratory, microscopic bacteria form colonies that you can see. In this activity, you will observe some of the bacteria that share your world.

Procedure

1. Get **three plastic Petri dishes containing nutrient agar** from your teacher. Label one dish "Hand," another "Breath," and another "Soil."
2. Wipe your finger across the agar in the dish labeled "Hand." Breathe into the dish labeled "Breath." Place a **small amount of soil** in the dish labeled "Soil."
3. Secure the Petri dish lids with **transparent tape.** Wash your hands. Keep the dishes upside down in a warm, dark place for about one week. **Caution:** Do not open the Petri dishes after they are sealed.
4. Observe the Petri dishes each day. What do you see? Record your observations.

Analysis

1. How does the appearance of the colonies growing on the agar in each dish differ? What do bacterial colonies look like?
2. Which source caused the most bacterial growth—your hand, your breath, or the soil? Why do you think this source caused the most growth?

SECTION 1

Bacteria

READING WARM-UP

Objectives

- Describe the characteristics of bacteria.
- Explain how bacteria reproduce.
- Compare and contrast eubacteria and archaebacteria.

Terms to Learn

prokaryote
binary fission
endospore

READING STRATEGY

Prediction Guide Before reading this section, predict whether each of the following statements is true or false:

- There are only a few kinds of bacteria.
- Most bacteria are too small to see.

How many bacteria are in a handful of soil? Would you believe that a single gram of soil—which is about the mass of a pencil eraser—may have more than 2.5 billion bacteria? A handful of soil may contain trillions of bacteria!

There are more types of bacteria on Earth than all other living things combined. Most bacteria are too small to be seen without a microscope. But not all bacteria are the same size. In fact, the largest known bacteria are 1,000 times larger than the average bacterium. One of these giant bacteria was first found inside a surgeonfish and is shown in **Figure 1.**

Characteristics of Bacteria

All living things fit into one of six kingdoms: Protista, Plantae, Fungi, Animalia, Eubacteria, or Archaebacteria. Bacteria make up the kingdoms Eubacteria (YOO bak TIR ee uh) and Archaebacteria (AHR kee bak TIR ee uh). These two kingdoms contain the oldest forms of life on Earth. All bacteria are single-celled organisms. Bacteria are usually one of three main shapes: bacilli, cocci, or spirilla.

Reading Check **What two kingdoms are made up of bacteria?** *(See the Appendix for answers to Reading Checks.)*

Figure 1 *The giant bacteria inside this fish are 0.6 mm long, which is big enough to see without a microscope.*

Figure 2 The Most Common Shapes of Bacteria

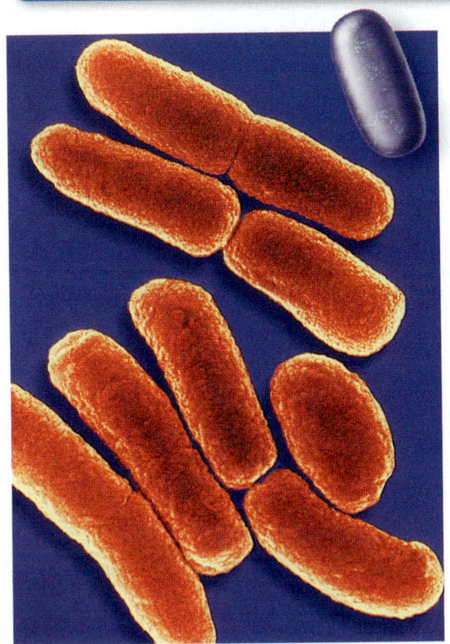

Bacilli (buh SIL IE) are rod shaped. They have a large surface area, which helps them take in nutrients. But a large surface area can cause them to dry out easily.

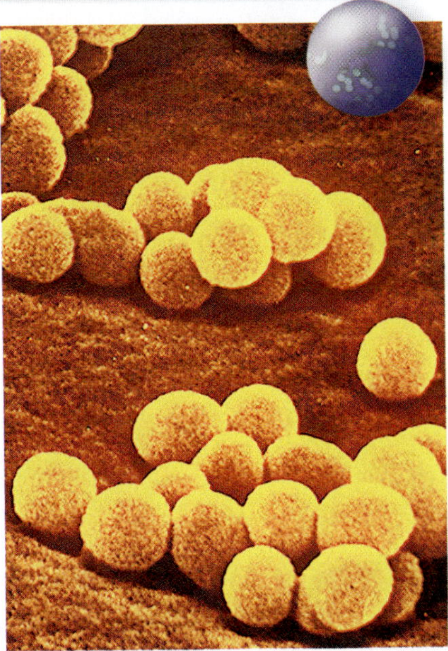

Cocci (KAHK SIE) are spherical. They do not dry out as quickly as rod-shaped bacteria.

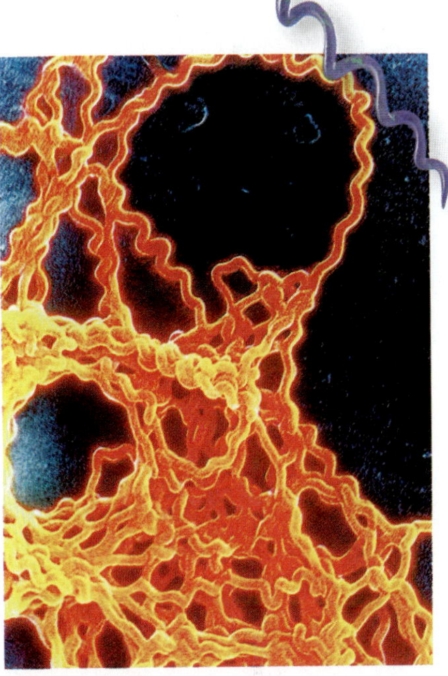

Spirilla (spie RIL uh) are long and spiral shaped. They use flagella at both ends to move like a corkscrew.

The Shape of Bacteria

Most bacteria have a rigid cell wall that gives them their shape. **Figure 2** shows the three most common shapes of bacteria. Bacilli (buh SIL IE) are rod shaped. Cocci (KAHK SIE) are spherical. Spirilla (spie RIL uh) are long and spiral shaped. Each shape helps bacteria in a different way.

Some bacteria have hairlike parts called *flagella* (fluh JEL uh) that help them move around. Flagella spin to push a bacterium through water or other liquids.

prokaryote an organism that consists of a single cell that does not have a nucleus

No Nucleus!

All bacteria are single-celled organisms that do not have a nucleus. An organism that does not have a nucleus is called a **prokaryote** (proh KAR ee OHT). A prokaryote is able to move, get energy, and reproduce like cells that have a nucleus, which are called *eukaryotes* (yoo KAR ee OHTZ).

Prokaryotes function as independent organisms. Some bacteria stick together to form strands or films, but each bacterium is still functioning as a single organism. Most prokaryotes are much simpler and smaller than eukaryotes. Prokaryotes also reproduce differently than eukaryotes do.

Spying on Spirilla

1. Using a **microscope,** observe prepared **slides of bacteria.** Draw each type of bacteria you see.
2. What different shapes do you see? What are these shapes called?

Figure 3 Binary Fission

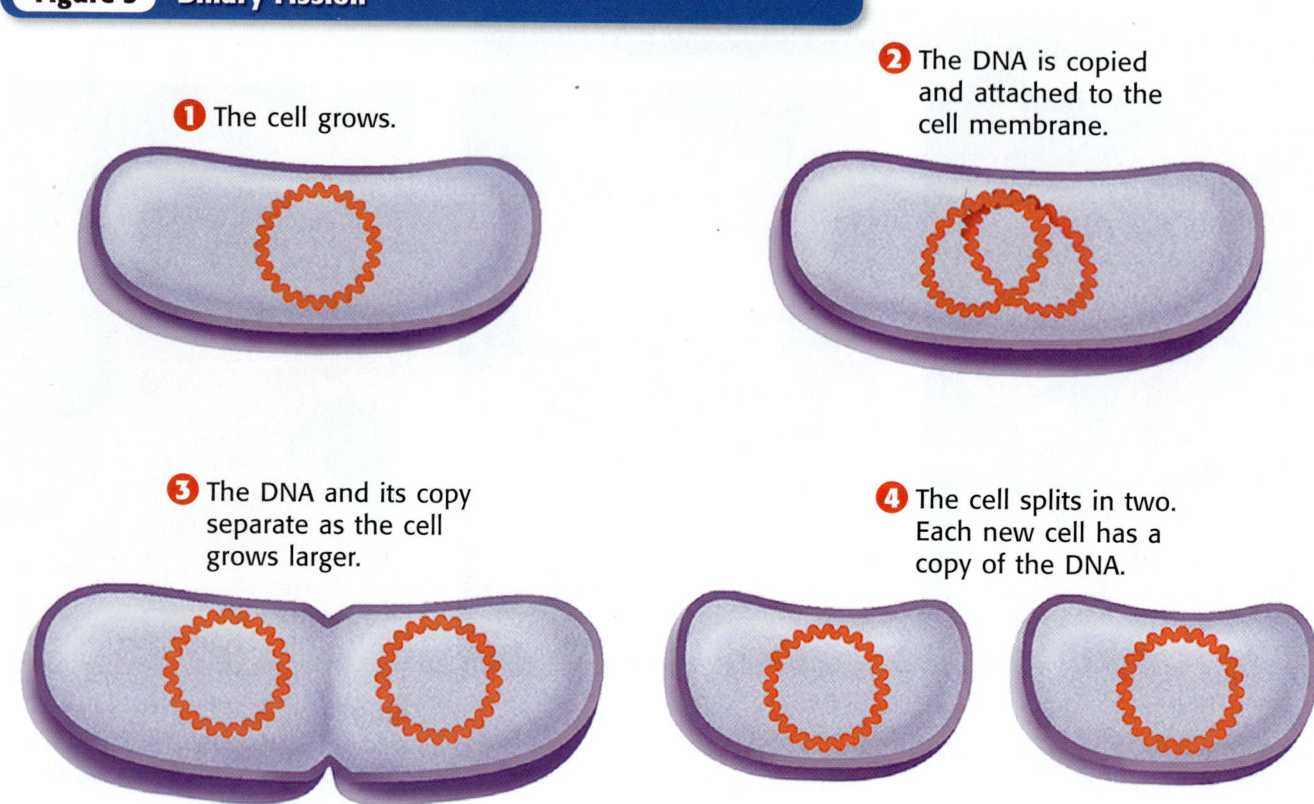

Bacterial Reproduction

binary fission a form of asexual reproduction in single-celled organisms by which one cell divides into two cells of the same size

Bacteria reproduce by the process shown in **Figure 3.** This process is called binary fission (BIE nuh ree FISH uhn). **Binary fission** is reproduction in which one single-celled organism splits into two single-celled organisms.

Prokaryotes have no nucleus, so their DNA is not surrounded by a membrane. The DNA of bacteria is in circular loops. In the first step of binary fission, the cell's DNA is copied. The DNA and its copy then bind to different places on the inside of the cell membrane. As the cell and its membrane grow bigger, the loops of DNA separate. Finally, when the cell is about double its original size, the membrane pinches inward as shown in **Figure 4.** A new cell wall forms and separates the two new cells. Each new cell has one exact copy of the parent cell's DNA.

Reading Check What is binary fission?

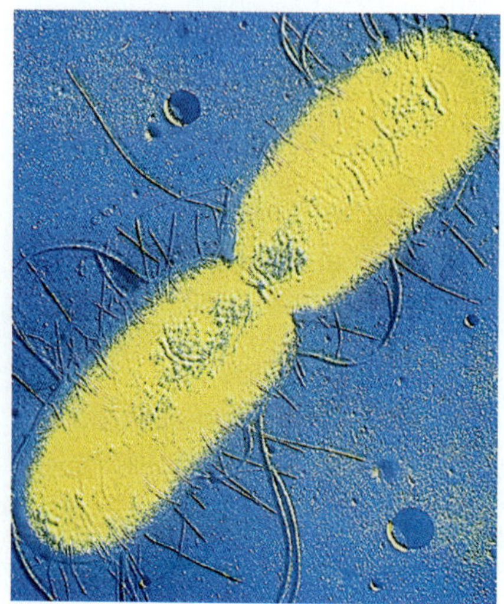

Figure 4 *This bacterium is about to complete binary fission.*

Endospores

Most species of bacteria do well in warm, moist places. In dry or cold surroundings, some species of bacteria will die. In these conditions, other bacteria become inactive and form endospores (EN doh SPAWRZ). An **endospore** contains genetic material and proteins and is covered by a thick, protective coat. Many endospores can survive in hot, cold, and very dry places. When conditions improve, the endospores break open, and the bacteria become active again. Scientists found endospores inside an insect that was preserved in amber for 30 million years. When the endospores were moistened in a laboratory, bacteria began to grow! A similar piece of amber can be seen in **Figure 5.**

Figure 5 *Endospores found in a preserved insect like this one showed scientists that bacteria can survive for millions of years.*

Kingdom Eubacteria

Most bacteria are eubacteria. The kingdom Eubacteria has more individuals than all of the other five kingdoms combined. Scientists think that eubacteria have lived on Earth for more than 3.5 billion years.

Eubacteria Classification

Eubacteria are classified by the way they get food. Most eubacteria, such as those breaking down the leaf in **Figure 6,** are consumers. Consumers get their food by eating other organisms. Many bacteria are decomposers, which feed on dead organisms. Other consumer bacteria live in or on the body of another organism. Eubacteria that make their own food are called *producers*. Like plants, producer bacteria use the energy from sunlight to make food. These bacteria are often green.

endospore a thick-walled protective spore that forms inside a bacterial cell and resists harsh conditions

Figure 6 *Decomposers, such as the ones helping to decay this leaf, return nutrients to the soil for other living things to use.*

Cyanobacteria

Cyanobacteria (SIE uh noh bak TIR ee uh) are producers. Cyanobacteria usually live in water. These bacteria contain the green pigment chlorophyll. Chlorophyll is important to photosynthesis (the process of making food from the energy in sunlight). Many cyanobacteria have other pigments as well. Some have a blue pigment that helps in photosynthesis. This pigment gives those cyanobacteria a blue tint. Other cyanobacteria have red pigment. Flamingos get their pink color from eating red cyanobacteria.

Some scientists think that billions of years ago, bacteria similar to cyanobacteria began to live inside larger cells. According to this theory, the bacteria made food, and the cells provided protection. This combination may have given rise to the first plants on Earth.

CONNECTION TO Language Arts

Colorful Names *Cyanobacteria* means "blue bacteria." Many other names also refer to colors. You might not recognize these colors because the words for the colors are in another language. Look at the list of Greek color words below. Write down two English words that have one of the color roots in them. (Hint: Many words have the color as the first part of the word.)

melano = black
chloro = green
erythro = red
leuko = white

Kingdom Archaebacteria

The three main types of archaebacteria are *heat lovers, salt lovers,* and *methane makers*. Heat lovers live in ocean vents and hot springs. They live in very hot water, usually from 60°C to 80°C, but they can survive temperatures of more than 250°C. Salt lovers live in environments that have high levels of salt, such as the Dead Sea and Great Salt Lake. Methane makers give off methane gas and live in swamps and animal intestines. **Figure 7** shows one type of methane maker found in the mud of swamps.

Figure 7 *These bacteria are methane makers. This micrograph shows two bacteria sliced across their narrow side and a dividing bacterium sliced lengthwise.*

Harsh Environments

Archaebacteria often live where nothing else can. Most archaebacteria prefer environments where there is little or no oxygen. Scientists have found them in the hot springs at Yellowstone National Park and beneath 430 m of ice in Antarctica. Archaebacteria have even been found living 8 km below the Earth's surface! Even though they are often found in these harsh environments, many archaebacteria can also be found in moderate environments in Earth's oceans.

Archaebacteria are very different from eubacteria. Not all archaebacteria have cell walls. When they do have them, the cell walls are chemically different from those of eubacteria.

For another activity related to this chapter, go to **go.hrw.com** and type in the keyword **HL5VIRW.**

SECTION Review

Summary

- Bacteria are single-celled organisms that are the smallest and simplest living things on Earth.
- Most bacteria have a rigid cell wall that gives them their shape. The main shapes of bacteria are rod shaped (bacilli), spherical (cocci), and spiral shaped (spirilla).
- Bacteria reproduce by binary fission. In binary fission, one cell divides into two cells.
- Eubacteria have cell walls and are either producers (bacteria that make their own food) or consumers (bacteria that get food from other organisms).
- Archaebacteria often live in harsh environments.

Using Key Terms

The statements below are false. For each statement, replace the underlined term to make a true statement.

1. Bacteria are <u>eukaryotes.</u>
2. Bacteria reproduce by <u>primary fission.</u>

Understanding Key Ideas

3. The structure that helps some bacteria survive harsh conditions is called a(n)
   a. endospore. **c.** exospore.
   b. shell. **d.** exoskeleton.
4. How are eubacteria and archaebacteria different?
5. Draw and label the four stages of binary fission.
6. Describe one advantage of each shape of bacteria.
7. What two things do producer bacteria and plants have in common?

Math Skills

8. An ounce (oz) is equal to about 28 g. If 1 g of soil contains 2.5 billion bacteria, how many bacteria are in 1 oz of soil?

Critical Thinking

9. **Applying Concepts** Many bacteria cannot reproduce in cooler temperatures and are destroyed at high temperatures. How do humans take advantage of this fact when preparing and storing food?
10. **Making Comparisons** Scientists are studying cold and dry environments on Earth that are like the environment on Mars. What kind of bacteria do you think they might find in these environments on Earth? Explain.
11. **Forming Hypotheses** You are studying a lake and the bacteria that live in it. What conditions of the lake would you measure to form a hypothesis about what kind of bacteria may live there?

SECTION 2

Bacteria's Role in the World

Have you ever had strep throat or a cavity in your tooth? Did you know that both are caused by bacteria?

Bacteria live in our water, our food, and our bodies. Much of what we know about bacteria was learned by scientists fighting bacterial diseases. But of the thousands of types of bacteria, only a few hundred cause disease. Many bacteria do things that are important and even helpful to us.

READING WARM-UP

Objectives

- Explain how life on Earth depends on bacteria.
- List three ways bacteria are useful to people.
- Describe two ways in which bacteria can be harmful to people.

Terms to Learn

bioremediation
antibiotic
pathogenic bacteria

READING STRATEGY

Reading Organizer As you read this section, create an outline of the section. Use the headings from the section in your outline.

Good for the Environment

Life as we know it could not exist without bacteria. Bacteria are very important to the health of Earth. They help recycle dead animals and plants. Bacteria also play an important role in the nitrogen cycle.

Nitrogen Fixation

Most living things depend on plants. Plants need nitrogen to grow. Nitrogen gas makes up about 78% of the air, but most plants cannot use nitrogen directly from the air. They need to take in a different form of nitrogen. Nitrogen-fixing bacteria take in nitrogen from the air and change it to a form that plants can use. This process, called *nitrogen fixation,* is described in **Figure 1.**

✓ Reading Check **What is nitrogen fixation?** (*See the Appendix for answers to Reading Checks.*)

Figure 1 Bacteria's Role in the Nitrogen Cycle

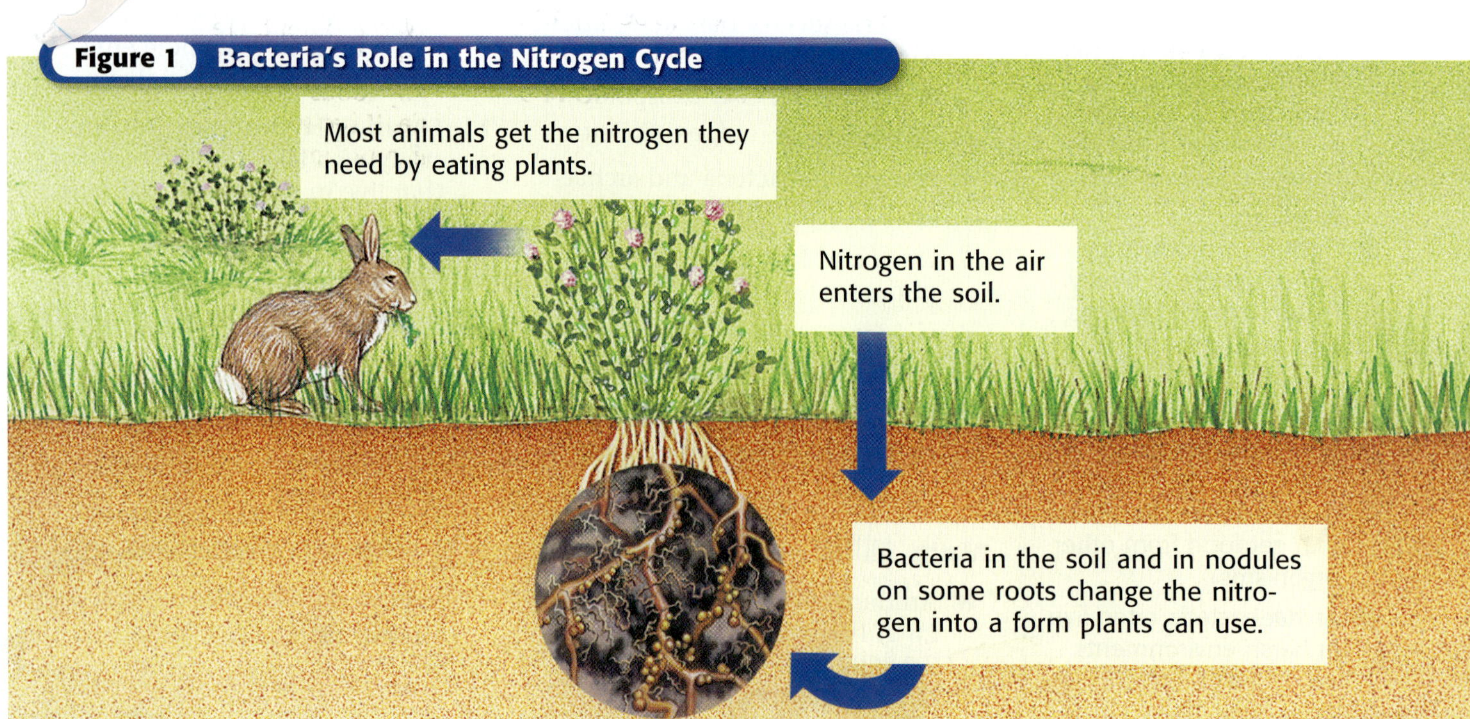

Recycling

Have you ever seen dead leaves and twigs on a forest floor? These leaves and twigs are recycled over time with the help of bacteria. Decomposer bacteria break down dead plant and animal matter. Breaking down dead matter makes nutrients available to other living things.

Cleaning Up

Bacteria and other microorganisms are also used to fight pollution. **Bioremediation** (BIE oh ri MEE dee AY shuhn) means using microorganisms to change harmful chemicals into harmless ones. Bioremediation is used to clean up hazardous waste from industries, farms, and cities. It is also used to clean up oil spills. The workers in **Figure 2** are using bacteria to remove pollutants from the soil.

Figure 2 *Bioremediating bacteria are added to soil to eat pollutants. The bacteria then release the pollutants as harmless waste.*

Good for People

Bacteria do much more than help keep our environment clean. Bacteria also help produce many of the foods we eat every day. They even help make important medicines.

bioremediation the biological treatment of hazardous waste by living organisms

Bacteria in Your Food

Believe it or not, people raise bacteria for food! Every time you eat cheese, yogurt, buttermilk, or sour cream, you are also eating bacteria. Lactic acid-producing bacteria break down the sugar in milk, which is called *lactose*. In the process, the bacteria change lactose into lactic acid. Lactic acid preserves and adds flavor to the food. All of the foods shown in **Figure 3** were made with the help of bacteria.

Make a Meal Plan

With a parent, create a week's meal plan without any foods made with bacteria. What would your diet be like without bacteria?

Figure 3 *Bacteria are used to make many kinds of foods.*

Figure 4 *Genes from the* Xenopus *frog were used to produce the first genetically engineered bacteria.*

antibiotic medicine used to kill bacteria and other microorganisms

pathogenic bacteria bacteria that cause disease

Making Medicines

What's the best way to fight disease-causing bacteria? Would you believe that the answer is to use other bacteria? **Antibiotics** are medicines used to kill bacteria and other microorganisms. Many antibiotics are made by bacteria.

Insulin

The human body needs insulin to break down and use sugar and carbohydrates. People who have diabetes do not make enough insulin. In the 1970s, scientists discovered how to put genes into bacteria so that the bacteria would make human insulin. The insulin can then be separated from the bacteria and given to people who have diabetes.

Genetic Engineering

When scientists change the genes of bacteria, or any other living thing, the process is called *genetic engineering.* Scientists have been genetically engineering bacteria since 1973. In that year, researchers put genes from a frog like the one in **Figure 4** into the bacterium *Escherichia coli* (ESH uh RIK ee uh KOH LIE). The bacterium then started making copies of the frog genes. Scientists can now engineer bacteria to make many products, such as insecticides, cleansers, and adhesives.

Reading Check What is genetic engineering?

Harmful Bacteria

Figure 5 *Vaccines can protect you from bacterial diseases such as tetanus and diptheria.*

Humans couldn't live without bacteria, but bacteria can also cause harm. Scientists learned in the 1800s that some bacteria are pathogenic (PATH uh JEN ik). **Pathogenic bacteria** are bacteria that cause disease. Pathogenic bacteria get inside a host organism and take nutrients from the host's cells. In the process, they harm the host. Today, we are protected from many bacterial diseases by vaccination, as shown in **Figure 5.** Many bacterial diseases can also be treated with antibiotics.

Diseases in Other Organisms

Bacteria cause diseases in other organisms as well as in people. Have you ever seen a plant with odd-colored spots or soft rot? If so, you've seen bacterial damage to plants. Pathogenic bacteria attack plants, animals, protists, fungi, and even other bacteria. They can cause damage to grain, fruit, and vegetable crops. The branch of the pear tree in **Figure 6** shows the effects of pathogenic bacteria. Plants are sometimes treated with antibiotics. Scientists have also genetically engineered certain plants to be resistant to disease-causing bacteria.

Figure 6 *This branch of a pear tree has a bacterial disease called* fire blight.

SECTION Review

Summary

- Bacteria are important to life on Earth because they fix nitrogen and decompose dead matter.
- Bacteria are useful to people because they help make foods and medicines.
- Scientists have genetically engineered bacteria to make medicines.
- Pathogenic bacteria are harmful to people. Bacteria can also harm the crops we grow for food.

Using Key Terms

1. In your own words, write a definition for the term *bioremediation.*
2. Use the following terms in the same sentence: *pathogenic bacteria* and *antibiotic.*

Understanding Key Ideas

3. What are two ways that bacteria affect plants?
4. How can bacteria both cause and cure diseases?
5. Explain two ways in which bacteria are crucial to life on Earth.
6. Describe two ways your life was affected by bacteria today.

Math Skills

7. Nitrogen makes up about 78% of air. If you have 2 L of air, how many liters of nitrogen are in the air?

Critical Thinking

8. **Identifying Relationships** Legumes, which include peas and beans, are efficient nitrogen fixers. Legumes are also a good source of amino acids. What chemical element would you expect to find in amino acids?
9. **Applying Concepts** Design a bacterium that will be genetically engineered. What do you want it to do? How would it help people or the environment?

SECTION 3

Viruses

One day, you discover red spots on your skin. More and more spots appear, and they begin turning into itchy blisters. What do you have?

The spots could be chickenpox. Chickenpox is a disease caused by a virus. A **virus** is a microscopic particle that gets inside a cell and often destroys the cell. Many viruses cause diseases, such as the common cold, flu, and acquired immune deficiency syndrome (AIDS).

READING WARM-UP

Objectives

- Explain how viruses are similar to and different from living things.
- List the four major virus shapes.
- Describe the two kinds of viral reproduction.

Terms to Learn

virus
host

READING STRATEGY

Discussion Read this section silently. Write down questions that you have about this section. Discuss your questions in a small group.

It's a Small World

Viruses are tiny. They are smaller than the smallest bacteria. About 5 billion virus particles could fit in a single drop of blood. Viruses can change rapidly. So, a virus's effect on living things can also change. Because viruses are so small and change so often, scientists don't know exactly how many types exist. These properties also make them difficult to fight.

Are Viruses Living?

Like living things, viruses contain protein and genetic material. But viruses, such as the ones shown in **Figure 1,** don't act like living things. They can't eat, grow, break down food, or use oxygen. In fact, a virus cannot function on its own. A virus can reproduce only inside a living cell that serves as a host. A **host** is a living thing that a virus or parasite lives on or in. Using a host's cell as a tiny factory, the virus forces the host to make viruses rather than healthy new cells.

virus a microscopic particle that gets inside a cell and often destroys the cell

host an organism from which a parasite takes food or shelter

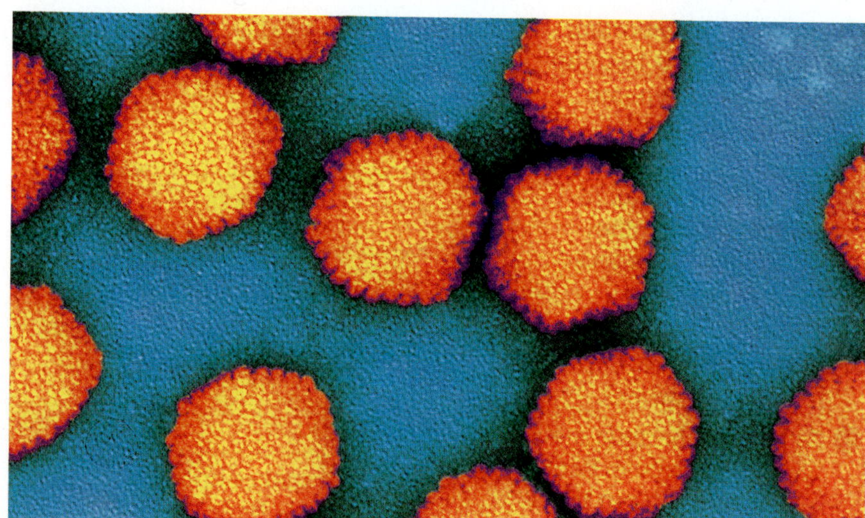

Figure 1 *Viruses are not cells. They do not have cytoplasm or organelles.*

Figure 2 **The Basic Shapes of Viruses**

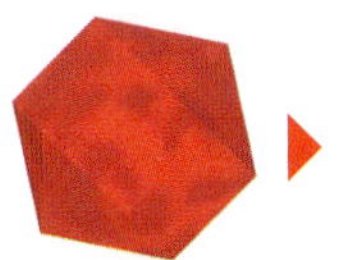

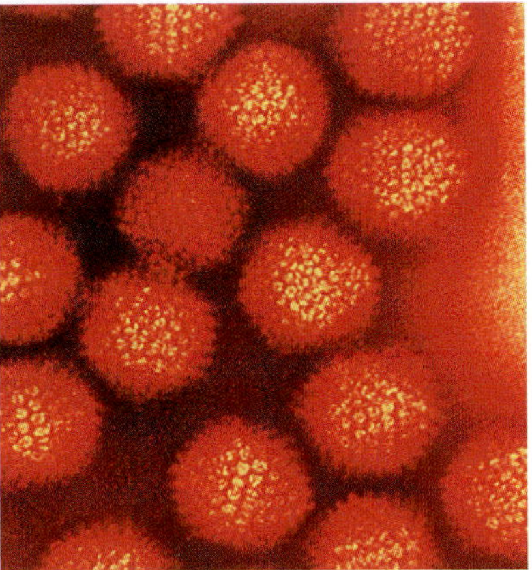

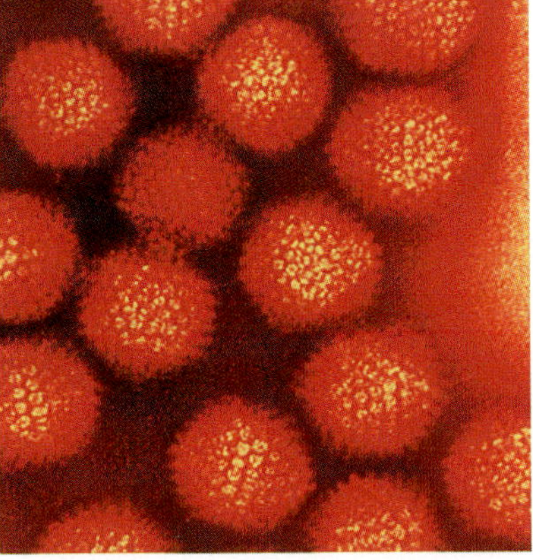

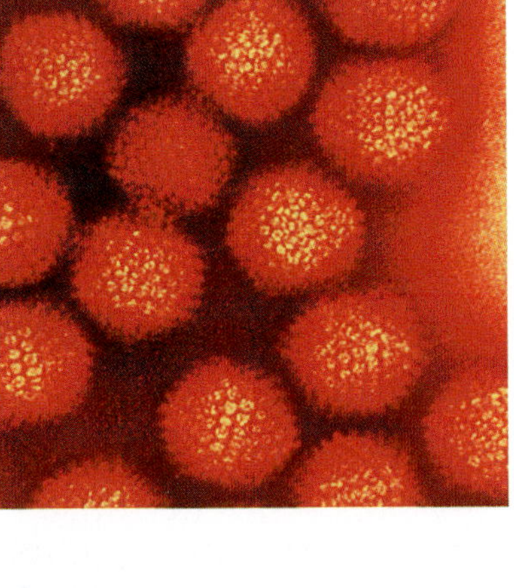

Crystals
The polio virus is shaped like the crystals shown here.

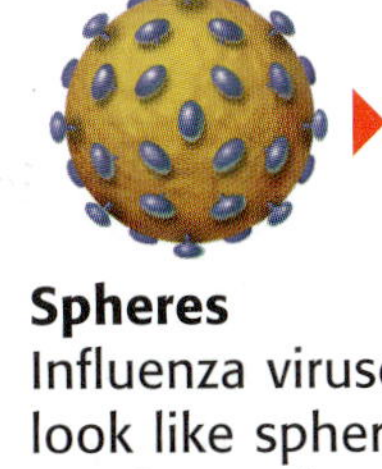

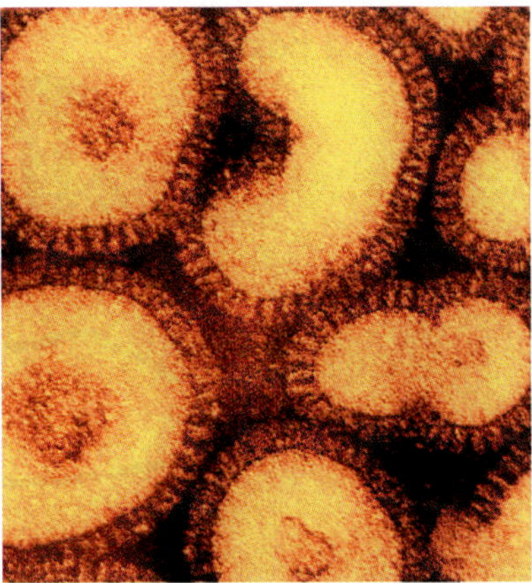

Spheres
Influenza viruses look like spheres. HIV is another virus that has this structure.

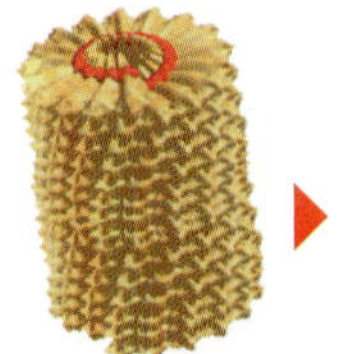

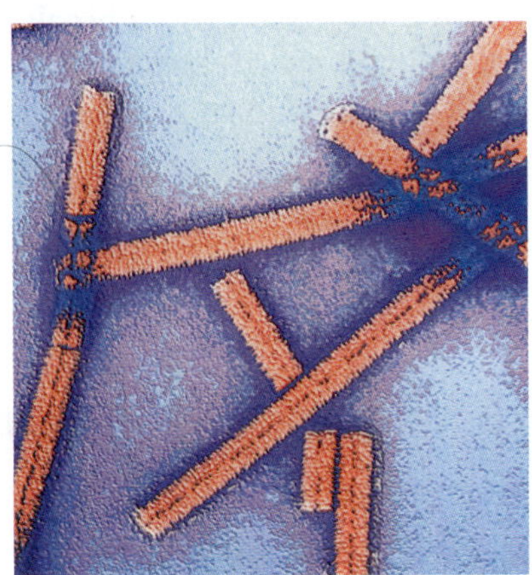

Cylinders
The tobacco mosaic virus is shaped like a cylinder and attacks tobacco plants.

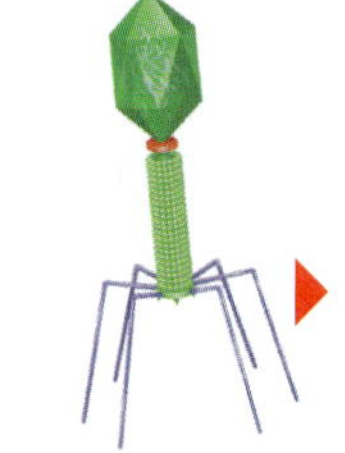

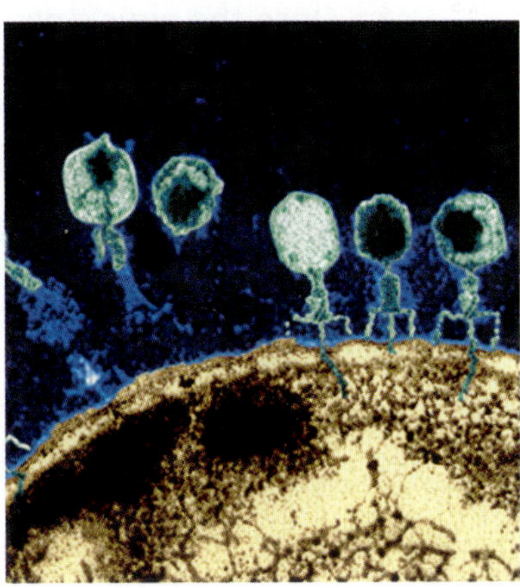

Spacecraft
One group of viruses attacks only bacteria. Many of these look almost like spacecraft.

Classifying Viruses

Viruses can be grouped by their shape, the type of disease they cause, their life cycle, or the kind of genetic material they contain. The four main shapes of viruses are shown in **Figure 2.** Every virus is made up of genetic material inside a protein coat. The protein coat protects the genetic material and helps a virus enter a host cell. Many viruses have a protein coat that matches characteristics of their specific host.

The genetic material in viruses is either DNA or RNA. Most RNA is made up of one strand of nucleotides. Most DNA is made up of two strands of nucleotides. Both DNA and RNA contain information for making proteins. The viruses that cause warts and chickenpox contain DNA. The viruses that cause colds and the flu contain RNA. The virus that causes AIDS, which is called the *human immunodeficiency virus* (HIV), also contains RNA.

Reading Check **What are two ways in which viruses can be classified?** (*See the Appendix for answers to Reading Checks.*)

Sizing Up a Virus

If you enlarged an average virus 600,000 times, it would be about the size of a small pea. How tall would you be if you were enlarged 600,000 times?

Figure 3 The Lytic Cycle

1 The virus finds and joins itself to a host cell.

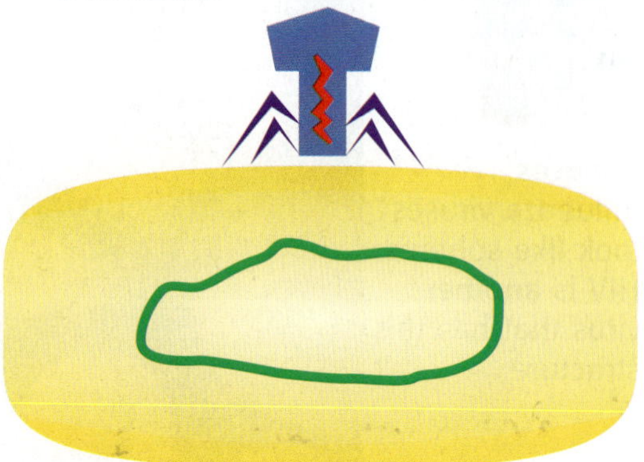

2 The virus enters the cell, or the virus's genetic material is injected into the cell.

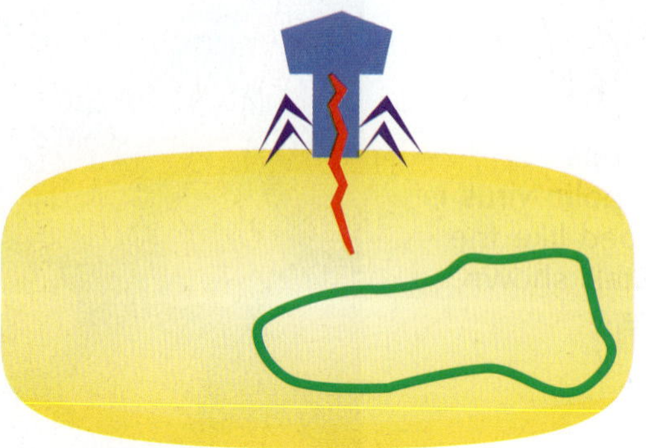

3 Once the virus's genes are inside, they take over the direction of the host cell and turn it into a virus factory.

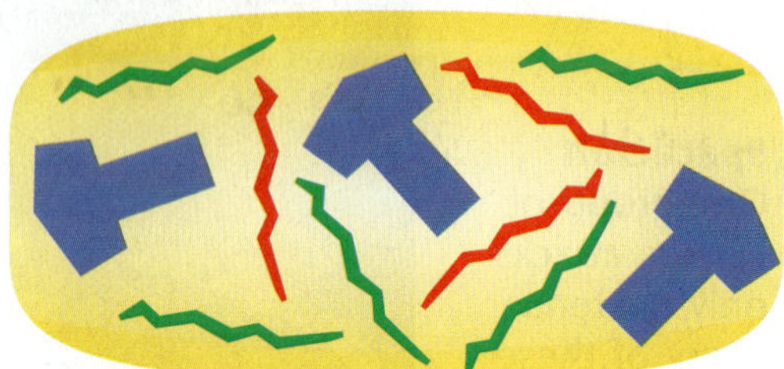

4 The new viruses break out of the host cell, which kills the host cell. The cycle begins again.

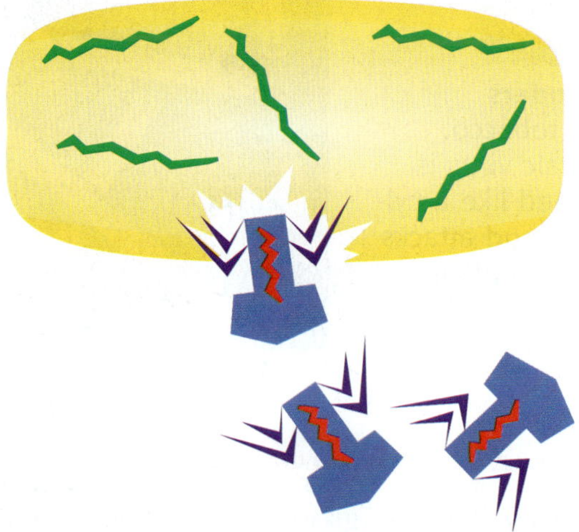

A Destructive House Guest

The one thing that viruses do that living things also do is make more of themselves. Viruses attack living cells and turn them into virus factories. This cycle is called the *lytic cycle* (LIT ik SIE kuhl), and it is shown in **Figure 3.**

Reading Check What is the lytic cycle?

A Time Bomb

Some viruses don't go straight into the lytic cycle. These viruses also put their genetic material into the host cell. But new viruses are not made right away. In the lysogenic (LIE soh JEN ik) cycle, each new cell gets a copy of the virus's genes when the host cell divides. The genes can stay inactive for a long time. When the genes do become active, they begin the lytic cycle and make copies of the virus.

CONNECTION TO Chemistry

Viral Crystals Many viruses can form into crystals. Scientists can study X rays of these crystals to learn about the structure of viruses. Why do you think scientists want to learn more about viruses?

Treating a Virus

Antibiotics do not kill viruses. But scientists have recently developed antiviral (AN tie VIE ruhl) medications. Many of these medicines stop viruses from reproducing. Because many viral diseases do not have cures, it is best to prevent a viral infection from happening in the first place. Childhood vaccinations give your immune system a head start in fighting off viruses. Having current vaccinations can prevent you from getting a viral infection. It is also a good practice to wash your hands often and never to touch wild animals. If you do get sick from a virus, like the boy in **Figure 4,** it is often best to rest and drink extra fluids. As with any sickness, you should tell your parents or a doctor.

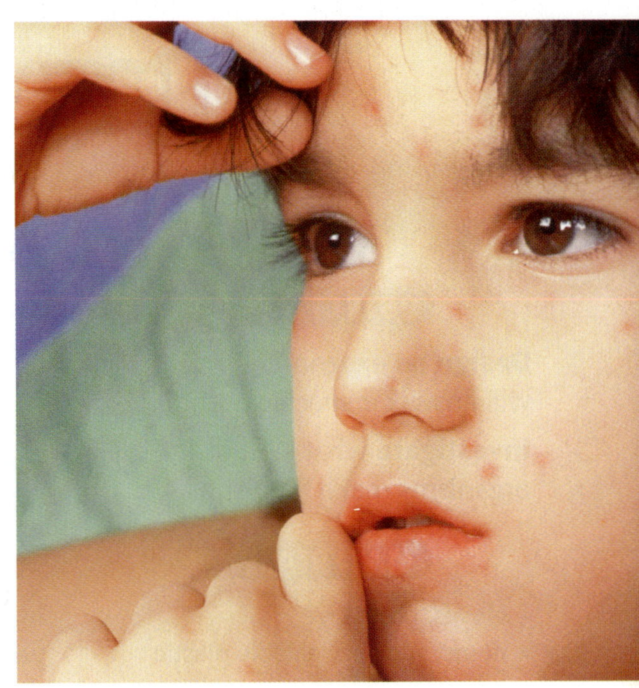

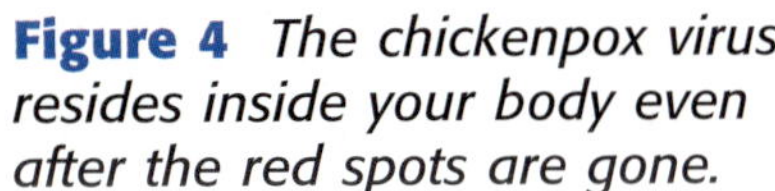

Figure 4 *The chickenpox virus resides inside your body even after the red spots are gone.*

SECTION Review

Summary

- Viruses have characteristics of living and nonliving things. They reproduce in living cells.
- Viruses may be classified by their shape, the kind of disease they cause, or their life cycle.
- To reproduce, a virus must enter a cell, reproduce itself, and then break open the cell. This is called the lytic cycle.
- In the lysogenic cycle, the genes of a virus are incorporated into the genes of the host cell.

Using Key Terms

1. Use the following terms in the same sentence: *virus* and *host.*

Understanding Key Ideas

2. One characteristic viruses have in common with living things is that they
   - **a.** eat.
   - **b.** reproduce.
   - **c.** sleep.
   - **d.** grow.
3. Describe the four steps in the lytic cycle.
4. Explain how the lytic cycle and the lysogenic cycle are different.

Math Skills

5. A bacterial cell infected by a virus divides every 20 min. After 10,000 divisions, the new viruses are released from their host cell. About how many weeks will this process take?

Critical Thinking

6. **Making Inferences** Do you think modern transportation has had an effect on the way viruses spread? Explain.
7. **Identifying Relationships** What characteristics of viruses do you think have made finding drugs to attack them difficult?
8. **Expressing Opinions** Do you think that vaccinations are important even in areas where a virus is not found?

Using Scientific Methods

Inquiry Lab

Aunt Flossie and the Intruder

Aunt Flossie is a really bad housekeeper! She never cleans the refrigerator, and things get really gross in there. Last week she pulled out a plastic bag that looked like it was going to explode! The bag was full of gas that she did not put there! Aunt Flossie remembered from her school days that gases are released from living things as waste products. Something had to be alive in the bag!

Aunt Flossie became very upset that there was an intruder in her refrigerator. She refuses to bake another cookie until you determine the nature of the intruder.

OBJECTIVES

Design an experiment that will answer a specific question.

Investigate what kind of organisms make food spoil.

MATERIALS

- gloves, protective
- items, such as sealable plastic bags, food samples, a scale, or a thermometer, to be determined by the students and approved by the teacher as needed for each experiment

SAFETY

Ask a Question

1. How did gas get into Aunt Flossie's bag?

Form a Hypothesis

2. Write a hypothesis which answers the question above. Explain your reasoning.

Test the Hypothesis

3. Design an experiment that will determine how gas got into Aunt Flossie's bag. Make a list of the materials you will need, and prepare all the data tables you will need for recording your observations.
4. Get your teacher's approval of your experimental design and your list of materials before you begin.
5. Dispose of your materials according to your teacher's instructions at the end of your experiment. **Caution:** Do not open any bags of spoiled food or allow any of the contents to escape.

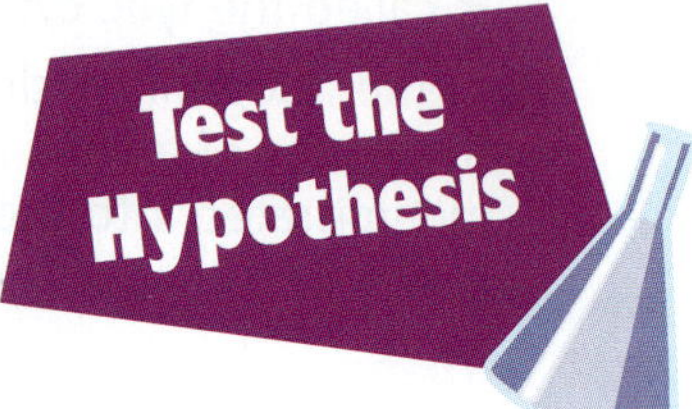

Analyze the Results

1. **Organizing Data** What data did you collect from your experiment?

Draw Conclusions

2. **Drawing Conclusions** What conclusions can you draw from your investigation? Where did the gas come from?

3. **Evaluating Methods** If you were going to perform another investigation, what would you change in the experiment to give better results? Explain your answer.

Communicating Your Data

WRITING SKILL Write a letter to Aunt Flossie describing your experiment. Explain what produced the gas in the bag and your recommendations for preventing these intruders in her refrigerator in the future.

Chapter Review

USING KEY TERMS

1. In your own words, write a definition for the term *pathogenic bacteria*.

Complete each of the following sentences by choosing the correct term from the word bank.

binary fission	endospore
antibiotic	bioremediation
virus	bacteria

2. Most bacteria reproduce by ___.

3. Bacterial infections can be treated with ___.

4. A(n) ___ needs a host to reproduce.

UNDERSTANDING KEY IDEAS

Multiple Choice

5. Bacteria are used for all of the following EXCEPT
   a. making certain foods.
   b. making antibiotics.
   c. cleaning up oil spills.
   d. preserving fruit.

6. In the lytic cycle, the host cell
   a. is destroyed.
   b. destroys the virus.
   c. becomes a virus.
   d. undergoes cell division.

7. A bacterial cell
   a. is an endospore.
   b. has a loop of DNA.
   c. has a distinct nucleus.
   d. is a eukaryote.

8. Eubacteria
   a. include methane makers.
   b. include decomposers.
   c. all have chlorophyll.
   d. are rod-shaped.

9. Cyanobacteria
   a. are consumers.
   b. are parasites.
   c. contain chlorophyll.
   d. are decomposers.

10. Archaebacteria
   a. are a special type of eubacteria.
   b. live only in places without oxygen.
   c. are lactic acid-producing bacteria.
   d. can live in hostile environments.

11. Viruses
   a. are about the same size as bacteria.
   b. have nuclei.
   c. can reproduce only within a host cell.
   d. do not infect plants.

12. Bacteria are important to the planet as
   a. decomposers of dead organic matter.
   b. processors of nitrogen.
   c. makers of medicine.
   d. All of the above

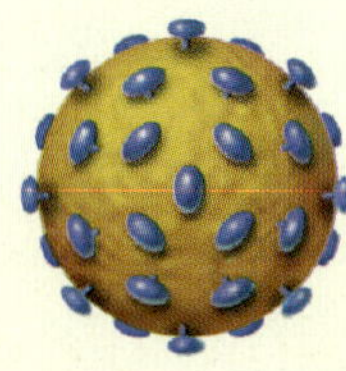

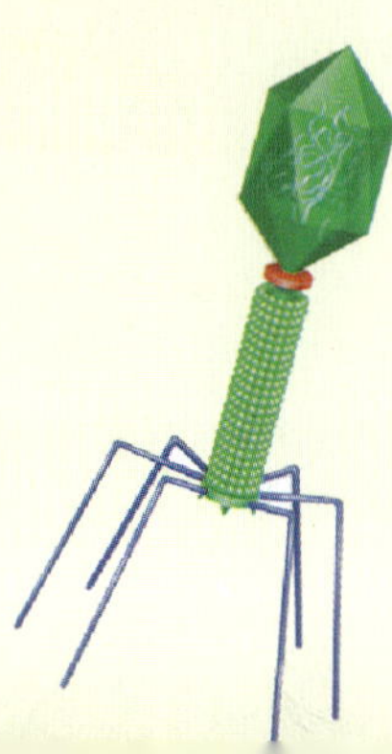

Short Answer

13. How are the functions of nitrogen-fixing bacteria and decomposers similar?

14. Which cycle takes more time, the lytic cycle or the lysogenic cycle?

15. Describe two ways in which viruses do not act like living things.

16. What is bioremediation?

17. Describe how doctors can treat a viral infection.

CRITICAL THINKING

18. **Concept Mapping** Use the following terms to create a concept map: *eubacteria, bacilli, cocci, spirilla, consumers, producers,* and *cyanobacteria.*

19. **Predicting Consequences** Describe some of the problems you think bacteria might face if there were no humans.

20. **Applying Concepts** Many modern soaps contain chemicals that kill bacteria. Describe one good outcome and one bad outcome of the use of antibacterial soaps.

21. **Identifying Relationships** Some people have digestive problems after they take a course of antibiotics. Why do you think these problems happen?

INTERPRETING GRAPHICS

The diagram below illustrates the stages of binary fission. Match each statement with the correct stage.

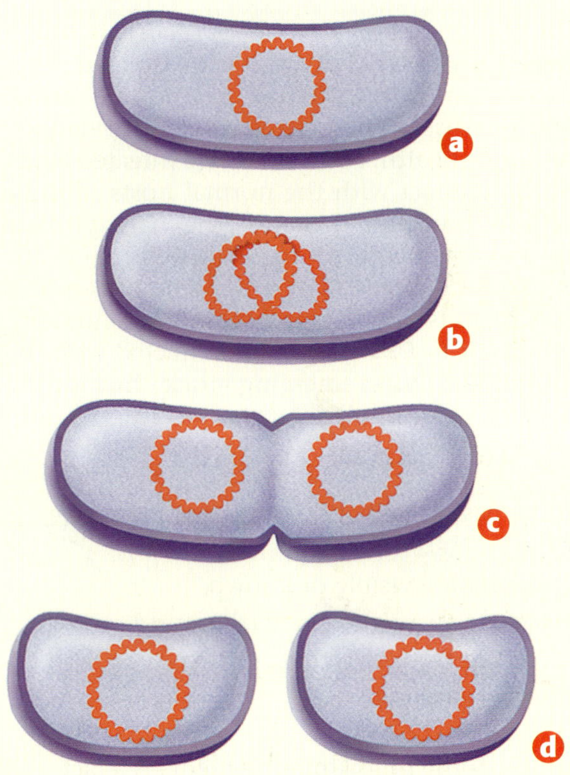

22. The DNA loops separate.

23. The DNA loop replicates.

24. The parent cell starts to expand.

25. The DNA attaches to the cell membrane.

Standardized Test Preparation

READING

Read each of the passages below. Then, answer the questions that follow each passage.

Passage 1 Viruses that evolve in isolated areas and that can infect human beings are called emerging viruses. These new viruses are dangerous to public health. People become infected when they have contact with the normal hosts of these viruses. In the United States, the hantavirus is considered an emerging virus. First detected in the southwestern United States, the hantavirus occurs in wild rodents and can infect and kill humans. Roughly 40% to 50% of humans infected with the hantavirus die. Other emerging viruses include the Ebola (Africa), Lassa (Africa), and Machupo (South America) viruses.

1. In the passage, what does the word *emerging* mean?

A to become visible or known
B to fade away into the background
C to melt from two things into one
D to become urgent

2. Which of the following statements is a fact from the passage?

F Hantavirus causes death in more than 40% of its victims.
G Hantavirus causes death in more than 50% of its victims.
H Hantavirus causes death in fewer than 30% of its victims.
I Hantavirus causes death in fewer than 40% of its victims.

3. Which of the following is an emergent virus in South America?

A Ebola virus
B Lassa virus
C SARS virus
D Machupo virus

Passage 2 Less than 100 years ago, people had no way to treat bacterial infections. But in 1928, a Scottish scientist named Alexander Fleming discovered the first antibiotic, or bacteria-killing drug. This first antibiotic was called *penicillin*. The discovery of antibiotics improved healthcare dramatically. However, scientists are now realizing that many bacteria are becoming resistant to existing antibiotics. Scientists are hoping that a particular type of virus called a bacteriophage (bak TIR ee uh FAHJ) might hold the key to fighting bacteria in the future. Bacteriophages destroy bacteria cells. Each kind of bacteriophage can infect only a particular species of bacteria.

1. In what year was penicillin discovered?

A 1905
B 1928
C 1969
D 1974

2. According to the passage, what might be the key to fighting bacteria in the future?

F antibiotics
G bacteriophages
H penicillin
I antibiotic-resistant bacteria

3. According to the passage, what can each kind of bacteriophage infect?

A viruses that cause disease
B only antibiotic-resistant bacteria
C all kinds of bacteria
D only a particular species of bacteria

INTERPRETING GRAPHICS

The images below show the four main shapes of viruses. Use these pictures to answer the questions that follow.

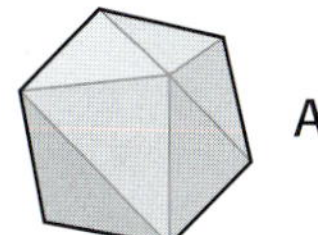

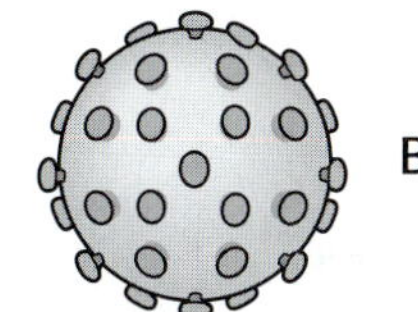

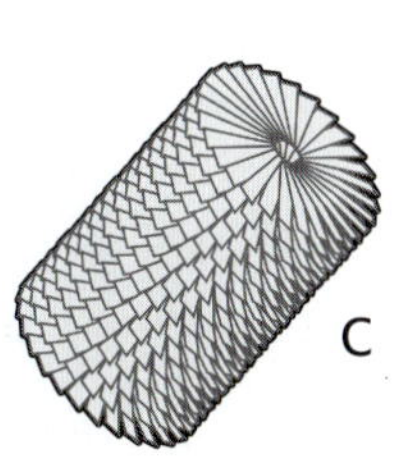

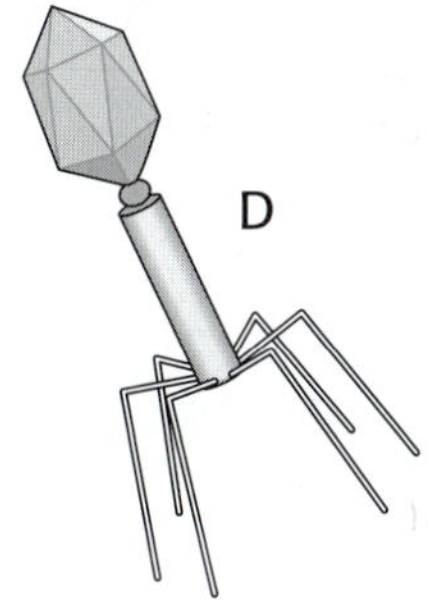

1. Which viral shape attacks only bacteria?

A virus A
B virus B
C virus C
D virus D

2. Which viral shape is the cylinder?

F virus A
G virus B
H virus C
I virus D

3. Which viral shape would you expect to have the largest surface area–to-volume ratio?

A virus A
B virus B
C virus C
D virus D

MATH

Read each question below, and choose the best answer.

1. Reagan spent \$26 for four equally priced CDs. Which of the following equations could be used to find how much each CD costs?

A $4 \times \$26 = n$
B $n = \$26 - 4$
C $4 \times n = \$26$
D $n \times \$26 = 4$

2. What is $5 + (-8)$ equal to?

F -13
G -3
H 3
I 13

3. What is $-9 - 2$ equal to?

A -11
B -7
C -4
D 7

4. What is the solution to $45 \div 0.009$?

F 5,000
G 500
H 50
I 5

5. What is $-9 + 2$ equal to?

A -11
B -7
C -4
D 7

6. Jennifer, Beth, and Sienna live 8 km, 2.2 km, and 7.4 km from the school. Which of the following is a reasonable estimate of the average distance these friends live from the school?

F 6 km
G 7.4 km
H 9 km
I 18 km

Science in Action

Science, Technology, and Society

Edible Vaccines

Vaccines protect you from life-threatening diseases. But vaccinations are expensive, and the people who give them must go through extensive training. These and other factors often prevent people in developing countries from getting vaccinations. But help may be on the way. Scientists are developing edible vaccines. Imagine eating a banana and getting the same protection you would from several painful injections. These vaccines are made from DNA that encodes a protein in the disease-causing particles. This DNA can then be inserted into the banana's genes. Researchers are still working on safe and effective edible vaccines.

Language Arts ACTIVITY

WRITING SKILL Write an advertisement for an edible vaccine. Be sure to describe the benefits of vaccinations.

Scientific Discoveries

Spanish Flu and the Flu Cycle

In 1918, a version of the influenza (the flu) virus killed millions of people worldwide. This disease, mistakenly called the Spanish Flu (it probably started in China), was one of the worst epidemics in history. Doctors and scientists realized that the large movement of people during the First World War probably made it easier for the Spanish Flu to spread. But the question of how this common disease could become so deadly remained unknown. One important factor is that the influenza virus is constantly changing. Many scientists now think that the influenza virus mutates into a more deadly form about every 30 years. There were flu epidemics in 1918, 1957, and 1968, which leads some scientists to believe that we are overdue for another flu epidemic.

Social Studies ACTIVITY

WRITING SKILL Conduct an interview with an older member of your family. Ask them how the flu, smallpox, tuberculosis, or polio has affected their lives. Write a report that includes information on how doctors deal with the disease today.

People in Science

Laytonville Middle School

Composting Project In 1973, Mary Appelhof tried an experiment. She knew that bacteria can help break down dead organic matter. In her basement, she set up a bin with worms and dumped her food scraps in there. Her basement didn't smell like garbage because her worms were eating the food scraps! Composting uses heat, bacteria, and, sometimes, worms to break down food wastes. Composting turns these wastes into fertilizer.

Binet Payne, a teacher at the Laytonville Middle School in California, decided to try Appelhof's composting system. Ms. Payne asked her students to separate their school cafeteria's trash into different categories: veggie wastes (worm food), protein foods (meat, milk, and cheese), bottles, cans, bags (to be recycled), and "yucky trash" (napkins and other nonrecyclables). The veggie waste was placed into the worm bins, and the protein foods were used to feed a local farm's chickens and pigs. In the first year, the Garden Project saved the school $6,000, which otherwise would have been used to dump the garbage into a landfill.

If the school saved $6,000 the first year, how much money did the school save each day of the year?

To learn more about these Science in Action topics, visit go.hrw.com and type in the keyword HL5VIRF.

Check out Current Science® articles related to this chapter by visiting go.hrw.com. Just type in the keyword HL5CS10.

Interactions of Living Things

About the PHOTO

A chameleon is about to grab an insect using its long tongue. A chameleon's body can change color to match its surroundings. Blending in helps the chameleon sneak up on its prey and also keeps the chameleon safe from animals that would like to make a snack out of a chameleon.

FOLDNOTES **Tri-Fold** Before you read the chapter, create the FoldNote entitled "Tri-Fold" described in the **Study Skills** section of the Appendix. Write what you know about the interactions of living things in the column labeled "Know." Then, write what you want to know in the column labeled "Want." As you read the chapter, write what you learn about the interactions of living things in the column labeled "Learn."

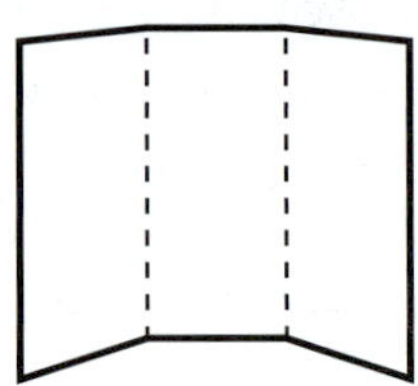

START-UP ACTIVITY

Who Eats Whom?

In this activity, you will learn how organisms interact when finding (or becoming) the next meal.

Procedure

1. On each of **five index cards,** print the name of one of the following organisms: killer whale, cod fish, krill shrimp, algae, and leopard seal.
2. On your desk, arrange the cards in a chain to show who eats whom.
3. Record the order of your cards.
4. In nature, would you expect to see more killer whales or cod? Arrange the cards in order of most individuals in an organism group to fewest.

Analysis

1. What might happen to the other organisms if algae were removed from this group? What might happen if the killer whales were removed?
2. Are there any organisms in this group that eat more than one kind of food? (Hint: What else might a seal, a fish, or a killer whale eat?) How could you change the order of your cards to show this information? How could you use pieces of string to show these relationships?

SECTION 1

Everything Is Connected

An alligator drifts in a weedy Florida river, watching a long, thin fish called a gar. *The gar swims too close to the alligator. Then, in a rush of murky water, the alligator swallows the gar whole and slowly swims away.*

READING WARM-UP

Objectives

- Distinguish between the biotic and abiotic parts of the environment.
- Explain how populations and communities are related.
- Describe how the abiotic parts of the environment affect ecosystems.

Terms to Learn

ecology
biotic
abiotic
population
community
ecosystem
biosphere

READING STRATEGY

Reading Organizer As you read this section, create an outline of the section. Use the headings from the section in your outline.

It is clear that two organisms have interacted when one eats the other. But organisms have many interactions other than simply "who eats whom." For example, alligators dig underwater holes to escape from the heat. After the alligators abandon these holes, fish and other aquatic organisms live in the holes during the winter dry period.

Studying the Web of Life

All living things are connected in a web of life. Scientists who study the web of life specialize in the science of ecology. **Ecology** is the study of the interactions of organisms with one another and with their environment.

The Two Parts of an Environment

An organism's environment consists of all the things that affect the organism. These things can be divided into two groups. All of the organisms that live together and interact with one another make up the **biotic** part of the environment. The **abiotic** part of the environment consists of the nonliving factors, such as water, soil, light, and temperature. How many biotic parts and abiotic parts do you see in **Figure 1?**

Figure 1 *The alligator affects, and is affected by, many organisms in its environment.*

Organization in the Environment

At first glance, the environment may seem disorganized. However, the environment can be arranged into different levels, as shown in **Figure 2.** The first level is made of an individual organism. The second level is larger and is made of similar organisms, which form a population. The third level is made of different populations, which form a community. The fourth level is made of a community and its abiotic environment, which form an ecosystem. The fifth and final level contains all ecosystems, which form the biosphere.

ecology the study of the interactions of living organisms with one another and with their environment

biotic describes living factors in the environment

abiotic describes the nonliving part of the environment, including water, rocks, light, and temperature

Figure 2 The Five Levels of Environmental Organization

Meeting the Neighbors

1. Explore two or three blocks of your neighborhood.
2. Draw a map of the area's biotic and abiotic features. For example, map the location of sidewalks, large rocks, trees, water features, and any animals you see. Remember to approach all plants and animals with caution. Use your map to answer the following questions.
3. How are the biotic factors affected by the abiotic factors?
4. How are the abiotic factors affected by the biotic factors?

Populations

population a group of organisms of the same species that live in a specific geographical area

community all the populations of species that live in the same habitat and interact with each other

A salt marsh, such as the one shown in **Figure 3,** is a coastal area where grasslike plants grow. Within the salt marsh are animals. Each animal is a part of a **population,** or a group of individuals of the same species that live together. For example, all of the seaside sparrows that live in the same salt marsh are members of a population. The individuals in the population often compete with one another for food, nesting space, and mates.

Communities

A **community** consists of all of the populations of species that live and interact in an area. The animals and plants you see in **Figure 3** form a salt-marsh community. The populations in a community depend on each other for food, shelter, and many other things.

Figure 3 *Examine the picture of a salt marsh. Try to find examples of each level of organization in this environment.*

Ecosystems

An **ecosystem** is made up of a community of organisms and the abiotic environment of the community. An ecologist studying the ecosystem could examine how organisms interact as well as how temperature, precipitation, and soil characteristics affect the organisms. For example, the rivers that empty into the salt marsh carry nutrients, such as nitrogen, from the land. These nutrients affect the growth of the cordgrass and algae.

ecosystem a community of organisms and their abiotic environment

biosphere the part of Earth where life exists

The Biosphere

The **biosphere** is the part of Earth where life exists. It extends from the deepest parts of the ocean to high in the air where plant spores drift. Ecologists study the biosphere to learn how organisms interact with the abiotic environment—Earth's atmosphere, water, soil, and rock. The water in the abiotic environment includes fresh water and salt water as well as water that is frozen in polar icecaps and glaciers.

For another activity related to this chapter, go to **go.hrw.com** and type in the keyword **HL5INTW.**

Reading Check What is the biosphere? (*See the Appendix for answers to Reading Checks.*)

SECTION Review

Summary

- All living things are connected in a web of life.
- The biotic part of an environment is made up of all of the living things found within it.
- The abiotic part of an environment is made up of all of the nonliving things found within it, such as water and light.
- An ecosystem is made up of a community of organisms and its abiotic environment.

Using Key Terms

1. In your own words, write a definition for the term *ecology.*
2. Use the following terms in the same sentence: *biotic* and *abiotic.*

Understanding Key Ideas

3. Which one of the following is the highest level of environmental organization?
   a. ecosystem
   b. community
   c. population
   d. organism
4. What makes up a community?
5. Give two examples of how abiotic factors can affect an ecosystem.

Math Skills

6. From sea level, the biosphere goes up about 9 km and down about 19 km. What is the thickness of the biosphere in meters?

Critical Thinking

7. **Analyzing Relationships** What would happen to the other organisms in the salt-marsh ecosystem if the cordgrass suddenly died?
8. **Identifying Relationships** Explain in your own words what people mean when they say that everything is connected.
9. **Analyzing Ideas** Do ecosystems have borders? Explain your answer.

For a variety of links related to this chapter, go to www.scilinks.org

Topic: Biotic and Abiotic Factors; Organization in the Environment

SciLinks code: HSM0164; HSM1079

SECTION 2

Living Things Need Energy

Do you think you could survive on only water and vitamins? Eating food satisfies your hunger because it provides something you cannot live without—energy.

READING WARM-UP

Objectives

- Describe the functions of producers, consumers, and decomposers in an ecosystem.
- Distinguish between a food chain and a food web.
- Explain how energy flows through a food web.
- Describe how the removal of one species affects the entire food web.

Terms to Learn

herbivore
carnivore
omnivore
food chain
food web
energy pyramid

READING STRATEGY

Reading Organizer As you read this section, make a table comparing producers, consumers, and decomposers.

Living things need energy to survive. For example, black-tailed prairie dogs, which live in the grasslands of North America, eat grass and seeds to get the energy they need. Everything a prairie dog does requires energy. The same is true for the plants that grow in the grasslands where the prairie dogs live.

The Energy Connection

Organisms, in a prairie or any community, can be divided into three groups based on how they get energy. These groups are producers, consumers, and decomposers. Examine **Figure 1** to see how energy passes through an ecosystem.

Producers

Organisms that use sunlight directly to make food are called *producers*. They do this by using a process called *photosynthesis*. Most producers are plants, but algae and some bacteria are also producers. Grasses are the main producers in a prairie ecosystem. Examples of producers in other ecosystems include cordgrass and algae in a salt marsh and trees in a forest. Algae are the main producers in the ocean.

Figure 1 *Living things get their energy either from the sun or from eating other organisms.*

Energy Sunlight is the source of energy for almost all living things.

Producer Plants use the energy in sunlight to make food.

Consumer The black-tailed prairie dog (herbivore) eats seeds and grass in the grasslands of western North America.

Consumer All of the prairie dogs in a colony watch for enemies, such as coyotes (carnivore), hawks, and badgers. Occasionally, a prairie dog is killed and eaten by a coyote.

Consumers

Organisms that eat other organisms are called *consumers*. They cannot use the sun's energy to make food like producers can. Instead, consumers eat producers or other animals to obtain energy. There are several kinds of consumers. A consumer that eats only plants is called a **herbivore.** Herbivores found in the prairie include grasshoppers, prairie dogs, and bison. A **carnivore** is a consumer that eats animals. Carnivores in the prairie include coyotes, hawks, badgers, and owls. Consumers known as **omnivores** eat both plants and animals. The grasshopper mouse is an example of an omnivore. It eats insects, lizards, and grass seeds.

Scavengers are omnivores that eat dead plants and animals. The turkey vulture is a scavenger in the prairie. A vulture will eat what is left after a coyote has killed and eaten an animal. Scavengers also eat animals and plants that have died from natural causes.

Reading Check **What are organisms that eat other organisms called?** (*See the Appendix for answers to Reading Checks.*)

herbivore an organism that eats only plants

carnivore an organism that eats animals

omnivore an organism that eats both plants and animals

A Chain Game

With the help of your parent, make a list of the foods you ate at your most recent meal. Trace the energy of each food back to the sun. Which foods on your list were consumers? How many were producers?

Decomposers

Organisms that get energy by breaking down dead organisms are called *decomposers*. Bacteria and fungi are decomposers. These organisms remove stored energy from dead organisms. They produce simple materials, such as water and carbon dioxide, which can be used by other living things. Decomposers are important because they are nature's recyclers.

Consumer A turkey vulture (scavenger) may eat some of the coyote's leftovers. A scavenger can pick bones completely clean.

Decomposer Any prairie dog remains not eaten by the coyote or the turkey vulture are broken down by bacteria (decomposer) and fungi that live in the soil.

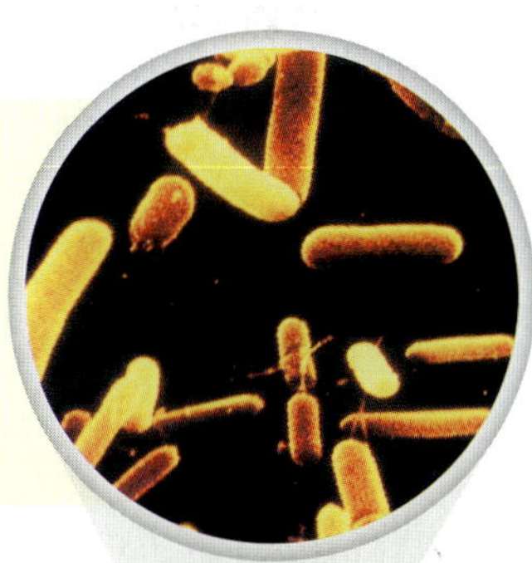

food chain the pathway of energy transfer through various stages as a result of the feeding patterns of a series of organisms

food web a diagram that shows the feeding relationships between organisms in an ecosystem

Food Chains and Food Webs

Figure 1 on the previous page, shows a food chain. A **food chain** is a diagram that shows how energy in food flows from one organism to another. Because few organisms eat just one kind of food, simple food chains are rare.

The energy connections in nature are more accurately shown by a food web than by a food chain. A **food web** is a diagram that shows the feeding relationships between organisms in an ecosystem. **Figure 2** shows a simple food web. Notice that an arrow goes from the prairie dog to the coyote, showing that the prairie dog is food for the coyote. The prairie dog is also food for the mountain lion. Energy moves from one organism to the next in a one-way direction, even in a food web. Any energy not immediately used by an organism is stored in its tissues. Only the energy stored in an organism's tissues can be used by the next consumer. There are two interconnected global food webs on Earth: a land food web and an aquatic food web.

Figure 2 *The green arrows show how energy moves when one organism eats another. Most consumers eat a variety of foods and can be eaten by a variety of other consumers.*

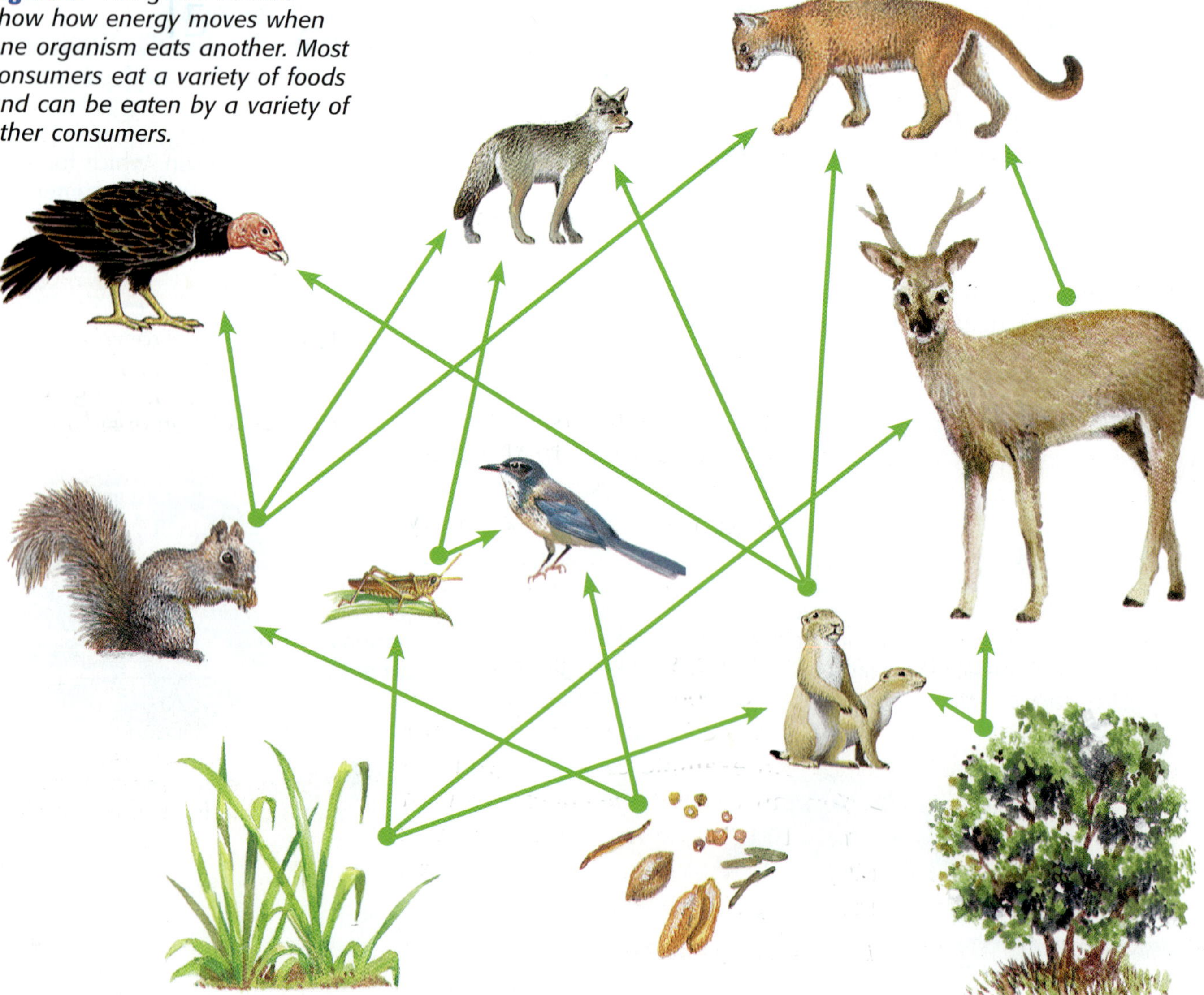

Figure 3 *The pyramid represents energy. As you can see, more energy is available at the base of the pyramid than at its top.*

Energy Pyramids

Grass uses most of the energy it gets from sunlight for its own life processes. But some of the energy is stored in the grass' tissues. This energy is used by the prairie dogs and other animals that eat the grass. Prairie dogs use most of the energy they get from eating grass and store only a little in their tissues. Therefore, a population of prairie dogs can support only a few coyotes. In the community, there must be more grass than prairie dogs and more prairie dogs than coyotes.

The energy at each level of the food chain can be seen in an energy pyramid. An **energy pyramid** is a diagram that shows an ecosystem's loss of energy. An example of an energy pyramid is shown in **Figure 3.** You can see that the energy pyramid has a large base and a small top. Less energy is available at higher levels because only energy stored in the tissues of an organism can be transferred to the next level.

energy pyramid a triangular diagram that shows an ecosystem's loss of energy, which results as energy passes through the ecosystem's food chain

Reading Check What is an energy pyramid?

Figure 4 *As the wilderness was settled, the gray wolf population in the United States declined.*

Wolves and the Energy Pyramid

One species can be very important to the flow of energy in an environment. Gray wolves, which are shown in **Figure 4,** are consumers that control the populations of many other animals. The diet of gray wolves can include anything from a lizard to an elk. Because gray wolves are predators that prey on large animals, their place is at the top of the food pyramid.

Once common throughout much of the United States, gray wolves were almost wiped out as the wilderness was settled. Without wolves, some species, such as elk, were no longer controlled. The overpopulation of elk in some areas led to overgrazing. The overgrazing left too little grass to support the elk and other populations who depended on the grass for food. Soon, almost all of the populations in the area were affected by the loss of the gray wolves.

Reading Check **How were other animals affected by the disappearance of the gray wolf?**

Gray Wolves and the Food Web

Gray wolves were brought back to Yellowstone National Park in 1995. The reintroduced wolves soon began to breed. **Figure 5** shows a wolf caring for pups. The U.S. Fish and Wildlife Service thinks the return of the wolves will restore the natural energy flow in the area, bring populations back into balance, and help restore the park's natural integrity.

Not everyone approves, however. Ranchers near Yellowstone are concerned about the safety of their livestock. Cows and sheep are not the natural prey of wolves. However, the wolves will eat cows and sheep if they are given the chance.

Figure 5 *In small wolf packs, only one female has pups. They are cared for by all of the males and females in the pack.*

Balance in Ecosystems

As wolves become reestablished in Yellowstone National Park, they kill the old, injured, and diseased elk. This process is reducing the number of elk. The smaller elk population is letting more plants grow. So, the numbers of animals that eat the plants, such as snowshoe hares, and the animals that eat the hares, such as foxes, are increasing.

All organisms in a food web are important for the health and balance of all other organisms in the food web. But the debate over the introduction of wolves to Yellowstone National Park will most likely continue for years to come.

Energy Pyramids

Draw an energy pyramid for a river ecosystem that contains four levels—aquatic plants, insect larvae, bluegill fish, and a largemouth bass. The plants obtain 10,000 units of energy from sunlight. If each level uses 90% of the energy it receives from the previous level, how many units of energy are available to the bass?

SECTION Review

Summary

- Producers use the energy in sunlight to make their own food.
- Consumers eat producers and other organisms to gain energy.
- Food chains represent how energy flows from one organism to another.
- All organisms are important to maintain the balance of energy in the food web.
- Energy pyramids show how energy is lost at each food chain level.

Using Key Terms

1. Use each of the following terms in a separate sentence: *herbivores, carnivores,* and *omnivores.*
2. In your own words, write a definition for each of the following terms: *food chain, food web,* and *energy pyramid.*

Understanding Key Ideas

3. Herbivores, carnivores, and scavengers are all examples of
   a. producers.
   b. decomposers.
   c. consumers.
   d. omnivores.
4. Explain the importance of decomposers in an ecosystem.
5. Describe how producers, consumers, and decomposers are linked in a food chain.
6. Describe how energy flows through a food web.

Math Skills

7. The plants in each square meter of an ecosystem obtained 20,810 Calories of energy from sunlight per year. The herbivores in that ecosystem ate all the plants but obtained only 3,370 Calories of energy. How much energy did the plants use?

Critical Thinking

8. **Identifying Relationships** Draw two food chains, and depict how they link together to form a food web.
9. **Applying Concepts** Are consumers found at the top or bottom of an energy pyramid? Explain your answer.
10. **Predicting Consequences** What would happen if a species disappeared from an ecosystem?

SECTION 3

Types of Interactions

READING WARM-UP

Objectives

- Explain the relationship between carrying capacity and limiting factors.
- Describe the two types of competition.
- Distinguish between mutualism, commensalism, and parasitism. Give an example of coevolution.

Terms to Learn

carrying capacity	mutualism
prey	commensalism
predator	parasitism
symbiosis	coevolution

READING STRATEGY

Reading Organizer As you read this section, make a concept map by using the terms above.

Look at the seaweed forest shown in **Figure 1** *below. How many fish do you see? How many seaweed plants do you count? Why do you think there are more members of the seaweed population than members of the fish population?*

In natural communities, the sizes of populations of different organisms can vary greatly. This variation happens because everything in the environment affects every other thing. Populations also affect every other population.

Interactions with the Environment

Most living things produce more offspring than will survive. A female frog, for example, might lay hundreds of eggs in a small pond. In a few months, the population of frogs in that pond will be about the same as it was the year before. Why won't the pond become overrun with frogs? An organism, such as a frog, interacts with biotic and abiotic factors in its environment that can control the size of its population.

Limiting Factors

Populations cannot grow without stopping, because the environment contains a limited amount of food, water, living space, and other resources. A resource that is so scarce that it limits the size of a population is called a *limiting factor*. For example, food becomes a limiting factor when a population becomes too large for the amount of food available. Any single resource can be a limiting factor to a population's size.

Figure 1 *This seaweed forest is home to a large number of interacting species.*

Carrying Capacity

The largest population that an environment can support is known as the **carrying capacity.** When a population grows larger than its carrying capacity, limiting factors in the environment cause individuals to die off or leave. As individuals die or leave, the population decreases.

carrying capacity the largest population that an environment can support at any given time

For example, after a rainy season, plants may produce a large crop of leaves and seeds. This large amount of food may cause an herbivore population to grow. If the next year has less rainfall, there won't be enough food to support the large herbivore population. In this way, a population may become larger than the carrying capacity, but only for a little while. A limiting factor will cause the population to die back. The population will return to a size that the environment can support.

Interactions Between Organisms

Populations contain individuals of a single species that interact with one another, such as a group of rabbits feeding in the same area. Communities contain interacting populations, such as a coral reef with many species of corals trying to find living space. Ecologists have described four main ways that species and individuals affect each other: competition, predators and prey, symbiotic relationships, and coevolution.

Reading Check **What are four main ways organisms affect one another?** (*See the Appendix for answers to Reading Checks.*)

Competition

When two or more individuals or populations try to use the same resource, such as food, water, shelter, space, or sunlight, it is called *competition*. Because resources are in limited supply in the environment, their use by one individual or population decreases the amount available to other organisms.

Competition happens between individuals *within* a population. The elks in Yellowstone National Park are herbivores that compete with each other for the same food plants in the park. This competition is a big problem in winter when many plants die.

Competition also happens *between* populations. The different species of trees in **Figure 2** are competing with each other for sunlight and space.

Figure 2 *Some of the trees in this forest grow tall to reach sunlight, which reduces the amount of sunlight available to shorter trees nearby.*

Predators and Prey

Many interactions between species consist of one organism eating another. The organism that is eaten is called the **prey.** The organism that eats the prey is called the **predator.** When a bird eats a worm, the worm is prey and the bird is the predator.

prey an organism that is killed and eaten by another organism

predator an organism that eats all or part of another organism

Predator Adaptations

To survive, predators must be able to catch their prey. Predators have a wide variety of methods and abilities for doing so. The cheetah, for example, is able to run very quickly to catch its prey. The cheetah's speed gives it an advantage over other predators competing for the same prey.

Other predators, such as the goldenrod spider, shown in **Figure 3,** ambush their prey. The goldenrod spider blends in so well with the goldenrod flower that all it has to do is wait for its next insect meal to arrive.

Figure 3 *The goldenrod spider is difficult for its insect prey to see. Can you see it?*

Prey Adaptations

Prey have their own methods and abilities to keep from being eaten. Prey are able to run away, stay in groups, or camouflage themselves. Some prey are poisonous. They may advertise their poison with bright colors to warn predators to stay away. The fire salamander, shown in **Figure 4,** sprays a poison that burns. Predators quickly learn to recognize its *warning coloration.*

Many animals run away from predators. Prairie dogs run to their underground burrows when a predator approaches. Many small fishes, such as anchovies, swim in groups called *schools.* Antelopes and buffaloes stay in herds. All the eyes, ears, and noses of the individuals in the group are watching, listening, and smelling for predators. This behavior increases the likelihood of spotting a potential predator.

Figure 4 *Many predators know better than to eat the fire salamander! This colorful animal will make a predator very sick.*

Camouflage

One way animals avoid being eaten is by being hard to see. A rabbit often freezes so that its natural color blends into a background of shrubs or grass. Blending in with the background is called *camouflage*. Many animals mimic twigs, leaves, stones, bark, or other materials in their environment. One insect, called a walking stick, looks just like a twig. Some walking sticks even sway a bit, as though a breeze were blowing.

Reading Check **What is camouflage, and how does it prevent an animal from being eaten?**

Defensive Chemicals

The spines of a porcupine clearly signal trouble to a potential predator, but other defenses may not be as obvious. Some animals defend themselves with chemicals. The skunk and the bombardier beetle both spray predators with irritating chemicals. Bees, ants, and wasps inject a powerful acid into their attackers. The skin of both the poison arrow frog and a bird called the *hooded pitohui* contains a deadly toxin. Any predator that eats, or tries to eat, one of these animals will likely die.

CONNECTION TO Environmental Science

Pretenders Some animals are pretenders. They don't have defensive chemicals. But they use warning coloration to their advantage. The Scarlet king snake has colored stripes that make it look like the poisonous coral snake. Even though the Scarlet king snake is harmless, predators see its bright colors and leave it alone. What might happen if there were more pretenders than there were animals with real defensive chemicals?

Warning Coloration

Animals that have a chemical defense need a way to warn predators that they should look elsewhere for a meal. Their chemical weapons are often advertised by warning colors, as shown in **Figure 5.** Predators will avoid any animal that has the colors and patterns they associate with pain, illness, or unpleasant experiences. The most common warning colors are bright shades of red, yellow, orange, black, and white.

Figure 5 *The warning coloration of the yellow jacket (left) and the pitohui (above) warns predators that they are dangerous.*

Symbiosis

symbiosis a relationship in which two different organisms live in close association with each other

mutualism a relationship between two species in which both species benefit

commensalism a relationship between two organisms in which one organism benefits and the other is unaffected

Some species have very close interactions with other species. **Symbiosis** is a close, long-term association between two or more species. The individuals in a symbiotic relationship can benefit from, be unaffected by, or be harmed by the relationship. Often, one species lives in or on the other species. The thousands of symbiotic relationships in nature are often classified into three groups: mutualism, commensalism, and parasitism.

Mutualism

A symbiotic relationship in which both organisms benefit is called **mutualism** (MYOO choo uhl IZ uhm). For example, you and a species of bacteria that lives in your intestines benefit each other! The bacteria get food from you, and you get vitamins that the bacteria produce.

Another example of mutualism happens between corals and algae. Coral near the surface of the water provide a home for algae. The algae produce food for the coral by photosynthesis. When a coral dies, its skeleton is used by other corals. Over a long time, these skeletons build up large formations that lie under the surface of warm seas, as shown in **Figure 6.**

Figure 6 *In the smaller photo above, you can see the gold-colored algae inside the coral.*

Reading Check Which organism benefits in mutualism?

Commensalism

A symbiotic relationship in which one organism benefits and the other is unaffected is called **commensalism.** One example of commensalism is the relationship between sharks and smaller fish called *remoras*. **Figure 7** shows a shark with a remora attached to its body. Remoras "hitch a ride" and feed on scraps of food left by sharks. The remoras benefit from this relationship, while sharks are unaffected.

Figure 7 *The remora attached to the shark benefits from the relationship. The shark neither benefits from nor is harmed by the relationship.*

Figure 8 *The tomato hornworm is being parasitized by young wasps. Do you see their cocoons?*

Parasitism

A symbiotic association in which one organism benefits while the other is harmed is called **parasitism** (PAR uh SIT IZ uhm). The organism that benefits is called the *parasite*. The organism that is harmed is called the *host*. The parasite gets nourishment from its host while the host is weakened. Sometimes, a host dies. Parasites, such as ticks, live outside the host's body. Other parasites, such as tapeworms, live inside the host's body.

Figure 8 shows a bright green caterpillar called a *tomato hornworm*. A female wasp laid tiny eggs on the caterpillar. When the eggs hatch, each young wasp will burrow into the caterpillar's body. The young wasps will actually eat the caterpillar alive! In a short time, the caterpillar will be almost completely eaten and will die. When that happens, the adult wasps will fly away.

In this example of parasitism, the host dies. Most parasites, however, do not kill their hosts. Most parasites don't kill their hosts because parasites depend on their hosts. If a parasite were to kill its host, the parasite would have to find a new host.

parasitism a relationship between two species in which one species, the parasite, benefits from the other species, the host, which is harmed

coevolution the evolution of two species that is due to mutual influence, often in a way that makes the relationship more beneficial to both species

Coevolution

Relationships between organisms change over time. Interactions can also change the organisms themselves. When a long-term change takes place in two species because of their close interactions with one another, the change is called **coevolution.**

The ant and the acacia tree shown in **Figure 9** have a mutualistic relationship. The ants protect the tree by attacking other organisms that come near the tree. The tree has special structures that make food for the ants. The ants and the acacia tree may have coevolved through interactions between the two species. Coevolution can take place between any organisms that live close together. But changes happen over a very long period of time.

Figure 9 *Ants collect food made by the acacia tree and store the food in their shelter, which is also made by the tree.*

Coevolution and Flowers

A *pollinator* is an organism that carries pollen from one flower to another. Pollination is necessary for reproduction in most plants.

Flowers have changed over millions of years to attract pollinators. Pollinators such as bees, bats, and hummingbirds can be attracted to a flower because of its color, odor, or nectar. Flowers pollinated by hummingbirds make nectar with the right amount of sugar for the bird. Hummingbirds have long beaks, which help them drink the nectar.

Some bats, such as the one shown in **Figure 10,** changed over time to have long, thin tongues and noses to help them reach the nectar in flowers. As the bat feeds on the nectar, its nose becomes covered with pollen. The next flower it eats from will be pollinated with the pollen it is gathering from this flower. The long nose helps it to feed and also makes it a better pollinator.

Because flowers and their pollinators have interacted so closely over millions of years, there are many examples of coevolution between them.

Reading Check Why do flowers need to attract pollinators?

CONNECTION TO Social Studies

Rabbits in Australia In 1859, settlers released 12 rabbits in Australia. There was plenty of food and no natural predators for the rabbits. The rabbit population increased so fast that the country was soon overrun by rabbits. Then, the Australian government introduced a rabbit virus to control the population. The first time the virus was used, more than 99% of the rabbits died. The survivors reproduced, and the rabbit population grew large again. The second time the virus was used, about 90% of the rabbits died. Once again, the rabbit population increased. The third time the virus was used, only about 50% of the rabbits died. Suggest what changes might have occurred in the rabbits and the virus.

Figure 10 *This bat is drinking nectar with its long, skinny tongue. The bat has coevolved with the flower over millions of years.*

SECTION Review

Summary

- Limiting factors in the environment keep a population from growing without limit.
- Two or more individuals or populations trying to use the same resource is called *competition.*
- A predator is an organism that eats all or part of another organism. The organism that is eaten is called *prey.*
- Prey have developed features such as camouflage, chemical defenses, and warning coloration, to protect them from predators.
- Symbiosis occurs when two organisms form a very close relationship with one another over time.
- Close relationships over a very long time can result in coevolution. For example, flowers and their pollinators have evolved traits that benefit both.

Using Key Terms

1. In your own words, write a definition for the term *carrying capacity*.
2. Use each of the following terms in a separate sentence: *mutualism, commensalism,* and *parasitism*.

Understanding Key Ideas

3. Which of the following is NOT a prey adaptation?
   a. camouflage
   b. chemical defenses
   c. warning coloration
   d. parasitism
4. Identify two things organisms compete with one another for.
5. Briefly describe one example of a predator-prey relationship. Identify the predator and the prey.

Critical Thinking

6. **Making Comparisons** Compare coevolution with symbiosis.
7. **Identifying Relationships** Explain the probable relationship between the giant *Rafflesia* flower, which smells like rotting meat, and the carrion flies that buzz around it. (Hint: *Carrion* means "rotting flesh.")
8. **Predicting Consequences** Predict what might happen if all of the ants were removed from an acacia tree.

Interpreting Graphics

The population graph below shows the growth of a species of *Paramecium* (single-celled microorganism) over 18 days. Food was added to the test tube occasionally. Use this graph to answer the questions that follow.

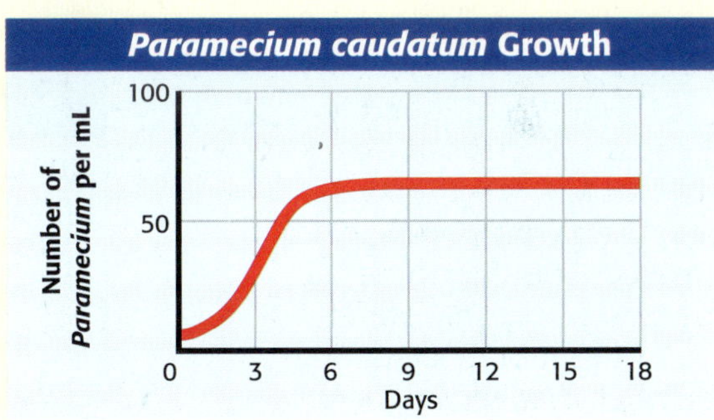

9. What is the carrying capacity of the test tube as long as food is added?
10. Predict what will happen if no more food is added?
11. What keeps the number of *Paramecium* at a steady level?

For a variety of links related to this chapter, go to www.scilinks.org

Topic: Predator/Prey; Coevolution
SciLinks code: HSM1205; HSM0309

Skills Practice Lab

Capturing the Wild Bean

OBJECTIVES

Estimate the size of a "population" of beans.

Calculate the difference between your estimation and the actual number of beans.

MATERIALS

- bag, paper lunch, small
- beans, pinto
- calculator (optional)
- marker, permanent

When wildlife biologists study a group of organisms in an area, they need to know how many organisms live there. Sometimes, biologists worry that a certain organism is outgrowing the environment's carrying capacity. Other times, scientists need to know if an organism is becoming rare so that steps can be taken to protect it. However, animals can be difficult to count because they can move around and hide. Because of this challenge, biologists have developed methods to estimate the number of animals in a specific area. One of these counting methods is called the *mark-recapture method*.

In this activity, you will enter the territory of the wild pinto bean to estimate the number of beans that live in the paper-bag habitat.

Procedure

1. Prepare a data table like the one below.

Mark-Recapture Data Table

Number of animals in first capture	Total number of animals in recapture	Number of marked animals in recapture	Calculated estimate of population	Actual total population
	DO NOT WRITE IN BOOK			

2. Your teacher will provide you with a paper bag containing an unknown number of beans. Carefully reach into the bag, and remove a handful of beans.

3. Count the number of beans you have "captured." Record this number in your data table under "Number of animals in first capture."
4. Use the permanent marker to carefully mark each bean that you have just counted. Allow the marks to dry completely. When all the marks are dry, place the marked beans back into the bag.
5. Gently mix the beans in the bag so that the marks won't rub off. Once again, reach into the bag. "Capture" and remove a handful of beans.
6. Count the number of beans in your "recapture." Record this number in your data table under "Total number of animals in recapture."
7. Count the beans in your recapture that have marks from the first capture. Record this number in your data table under "Number of marked animals in recapture."
8. Calculate your estimation of the total number of beans in the bag by using the following equation:

$$\frac{\text{number of beans in recapture} \times \text{number of beans marked}}{\text{number of marked beans in recapture}} = \text{calculated estimate of population}$$

   Enter this number in your data table under "Calculated estimate of population."
9. Place all the beans in the bag. Then empty the bag on your work table. Be careful that no beans escape! Count each bean as you place them one at a time back into the bag. Record the number in your data table under "Actual total population."

Analyze the Results

1. **Evaluating Results** How close was your estimate to the actual number of beans?

Draw Conclusions

2. **Evaluating Methods** If your estimate was not close to the actual number of beans, how might you change your mark-recapture procedure? If you did not recapture any marked beans, what might be the cause?

Applying Your Data

How could you use the mark-recapture method to estimate the population of turtles in a small pond? Explain your procedure.

Chapter Review

USING KEY TERMS

1. Use each of the following terms in a separate sentence: *symbiosis, mutualism, commensalism,* and *parasitism.*

Complete each of the following sentences by choosing the correct term from the word bank.

biotic	abiotic
ecosystem	community

2. The environment includes _____ factors including water, rocks, and light.

3. The environment also includes _____, or living, factors.

4. A community of organisms and their environment is called a(n) _____.

For each pair of terms, explain how the meanings of the terms differ.

5. *community* and *population*

6. *ecosystem* and *biosphere*

7. *producers* and *consumers*

UNDERSTANDING KEY IDEAS

Multiple Choice

8. A tick sucks blood from a dog. In this relationship, the tick is the _____ and the dog is the _____.
   - **a.** parasite, prey
   - **b.** predator, host
   - **c.** parasite, host
   - **d.** host, parasite

9. Resources such as water, food, or sunlight are likely to be limiting factors
   - **a.** when population size is decreasing.
   - **b.** when predators eat their prey.
   - **c.** when the population is small.
   - **d.** when a population is approaching the carrying capacity.

10. Nature's recyclers are
    - **a.** predators.
    - **b.** decomposers.
    - **c.** producers.
    - **d.** omnivores.

11. A beneficial association between coral and algae is an example of
    - **a.** commensalism.
    - **b.** parasitism.
    - **c.** mutualism.
    - **d.** predation.

12. The process by which energy moves through an ecosystem can be represented by
    - **a.** food chains.
    - **b.** energy pyramids.
    - **c.** food webs.
    - **d.** All of the above

13. Which organisms does the base of an energy pyramid represent?
    - **a.** producers
    - **b.** carnivores
    - **c.** herbivores
    - **d.** scavengers

14. Which of the following is the correct order in a food chain?
    - **a.** sun→producers→herbivores→scavengers→carnivores
    - **b.** sun→consumers→predators→parasites→hosts
    - **c.** sun→producers→decomposers→consumers→omnivores
    - **d.** sun→producers→herbivores→carnivores→scavengers

15. Remoras and sharks have a relationship that is best described as
   - **a.** mutualism.
   - **b.** commensalism.
   - **c.** predator and prey.
   - **d.** parasitism.

Short Answer

16. Describe how energy flows through a food web.

17. Explain how the food web changed when the gray wolf disappeared from Yellowstone National Park.

18. How are the competition between two trees of the same species and the competition between two different species of trees similiar?

19. How do limiting factors affect the carrying capacity of an environment?

20. What is coevolution?

CRITICAL THINKING

21. **Concept Mapping** Use the following terms to create a concept map: *herbivores, organisms, producers, populations, ecosystems, consumers, communities, carnivores,* and *biosphere.*

22. **Identifying Relationships** Could a balanced ecosystem contain producers and consumers but not decomposers? Why or why not?

23. **Predicting Consequences** Some biologists think that certain species, such as alligators and wolves, help maintain biological diversity in their ecosystems. Predict what might happen to other organisms, such as gar fish or herons, if alligators were to become extinct in the Florida Everglades.

24. **Expressing Opinions** Do you think there is a carrying capacity for humans? Why or why not?

INTERPRETING GRAPHICS

Use the energy pyramid below to answer the questions that follow.

25. According to the energy pyramid, are there more prairie dogs or plants?

26. What level has the most energy?

27. Would an energy pyramid such as this one exist in nature?

28. How could you change this pyramid to look like one representing a real ecosystem?

Standardized Test Preparation

READING

Read each of the passages below. Then, answer the questions that follow each passage.

Passage 1 Two or more individuals trying to use the same resource, such as food, water, shelter, space, or sunlight is called *competition.* Because resources are in limited supply in the environment, the use of them by one individual or population decreases the amount available to other organisms. Competition also occurs between individuals within a population. The elk in Yellowstone National Park are herbivores that compete with each other for the same food plants in the park.

1. According to the passage, competition occurs between which of the following?

- **A** individuals trying to use the same resource
- **B** elk and carnivores
- **C** food and shelter
- **D** individuals trying to use different resources

2. According to the passage, food, water, shelter, space, and sunlight are examples of

- **F** populations.
- **G** things found in Yellowstone National Park.
- **H** competition.
- **I** resources.

3. Based on the passage, which of the following statements is a fact?

- **A** Competition occurs only between individuals of different populations.
- **B** Competition occurs between individuals within a population and between individuals of different populations.
- **C** Competition increases the amount of resources available to individuals.
- **D** Because resources are abundant in the environment, competition rarely happens between individuals of different populations.

Passage 2 In the deserts of northern Africa and the Middle East, water is a scarce and valuable resource. In this area, no permanent streams flow except for the Nile. More than 1.6 million square kilometers of this region typically have no rainfall for years at a time. However, much of this area has large aquifers. The water that these aquifers contain dates back to much wetter times thousands of years ago. Occasionally, water reaches the surface to form an oasis. Wells supply the rest of the water used throughout the region. In some regions of Saudi Arabia and Kuwait, wells drilled for water more often strike oil.

1. According to the passage, an aquifer contains what resource?

- **A** oil
- **B** water
- **C** wells
- **D** oasis

2. Based on the passage, which of the following statements is a fact?

- **F** The Nile no longer flows through northern Africa and the Middle East.
- **G** The water found in aquifers is from recent rainfall.
- **H** Wells drilled in Saudi Arabia and Kuwait are more likely to strike oil than water.
- **I** The desert regions of northern Africa and the Middle East receive rainfall almost every day.

3. According to the passage, an oasis forms under what conditions?

- **A** when water stays beneath the surface
- **B** when water is drilled from a well
- **C** when it rains
- **D** when water reaches the surface

INTERPRETING GRAPHICS

The graphs below show the population growth for two populations. Use these graphs to answer the questions that follow.

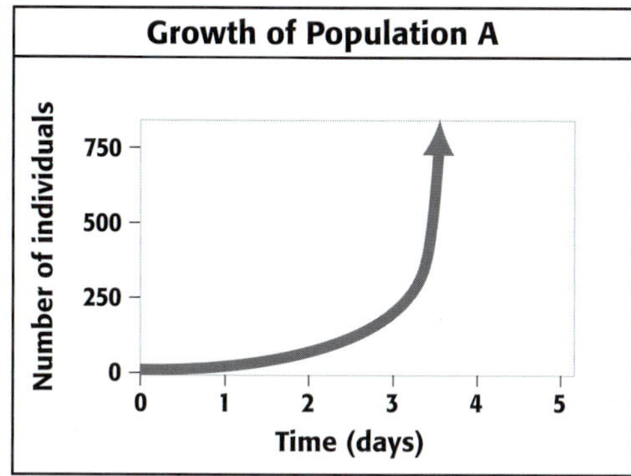

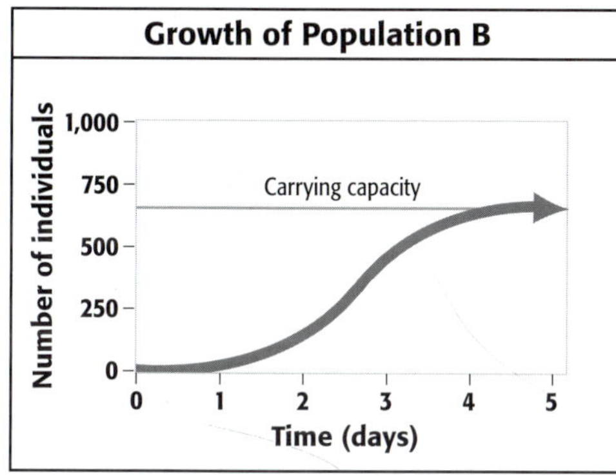

1. After 2 days, which population has more individuals?

A Population A has more individuals.
B Population B has more individuals.
C The populations are the same.
D There is not enough information to determine the answer.

2. After 5 days, which population has more individuals?

F Population A has more individuals.
G Population B has more individuals.
H The populations are the same.
I There is not enough information to determine the answer.

3. On day 10, which statement is probably true?

A Population B is larger than population A.
B Population A is the same as it was on day 5.
C Population A and B are the same.
D Population B is the same as it was on day 5.

MATH

Read each question below, and choose the best answer.

1. The figure below is a map of a forest ecosystem. What is the area of this ecosystem?

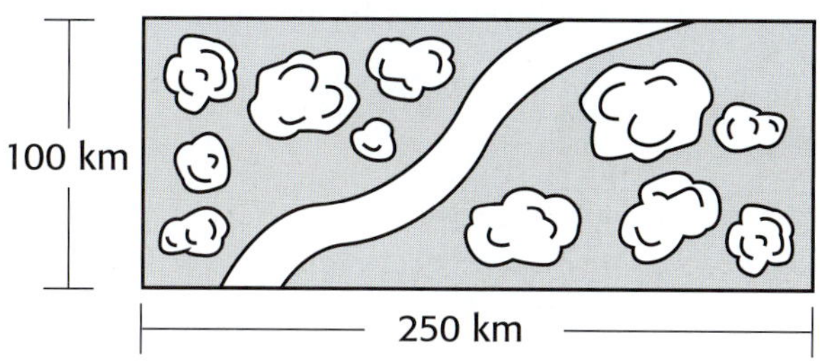

A 25,000 km^2
B 32,000 km
C 1,200 km^2
D 2,500 km

2. If an antelope eats 7 kg of vegetation in 2 days, how many kilograms of vegetation does it eat per day?

F 2/7 kg
G 3/5 kg
H 3 1/2 kg
I 7 1/2 kg

3. If $x = 3$ and $y = x + 2$, what is y?

A 2
B 4
C 5
D 8

4. If $x = 4$ and $y = x + 2$, what is y?

F 2
G 5
H 6
I 8

Science in Action

Scientific Debate

How Did Dogs Become Pets?

Did humans change dogs to be the social and helpful creatures they are today? Or were dogs naturally social? Did dogs start moving closer to our campfires long ago? Or did humans find dogs and bring them into our homes? The way in which dogs became our friends, companions, and helpers is still a question. Some scientists think humans and training are the reasons for many of our dogs' best features. Other scientists think dogs and humans have both changed over time to form their strong and unique bond.

Math ACTIVITY

Scientists have found fossils of dogs that are 15,000 years old. Generation time is the time between the birth of one generation and the next. If the generation time for dogs is 1.5 years, how many generations have there been in the last 15,000 years?

Weird Science

Follicle Mites

What has a tiny tubelike body and short stumpy legs and lives in your eyebrows and eyelashes? Would you believe a small animal lives there? It's called a follicle mite, and humans are its host. Studies show that more than 97% of adults have these mites. Except in rare cases, follicle mites are harmless.

Like all large animals, human beings are hosts to a variety of smaller creatures that live in or on our bodies and share our bodies' resources. Bacteria that live in our lower digestive tracks help to produce vitamins such as folic acid and vitamin K. Other bacteria may help maintain proper pH levels in our bodies.

Language Arts ACTIVITY

WRITING SKILL Imagine that you were shrunk to the size of a follicle mite. How would you get food? Where would you sleep? Write a short story describing one day in your new, tiny life.

Careers

Dalton Dockery

Horticulture Specialist Did you know that instead of using pesticides to get rid of insects that are eating the plants in your garden, you can use other insects? "It is a healthy way of growing vegetables without the use of chemicals and pesticides, and it reduces the harmful effects pesticides have on the environment," says Dalton Dockery, a horticulture specialist in North Carolina. Some insects, such as ladybugs and praying mantises, are natural predators of many insects that are harmful to plants. They will eat other bugs but leave your precious plants in peace. Using bugs to drive off pests is just one aspect of natural gardening. Natural gardening takes advantage of relationships that already exist in nature and uses these interactions to our benefit. For Dockery, the best parts about being a horticultural specialist are teaching people how to preserve the environment, getting to work outside regularly, and having the opportunity to help people on a daily basis.

Social Studies ACTIVITY

WRITING SKILL Research gardening or farming techniques in other cultures. Do other cultures use any of the same aspects of natural gardening as horticultural specialists? Write a short report describing your findings.

To learn more about these Science in Action topics, visit go.hrw.com and type in the keyword HL5INTF.

Current Science

Check out Current Science® articles related to this chapter by visiting go.hrw.com. Just type in the keyword HL5CS18.

Environmental Problems and Solutions

About the PHOTO

After an oil spill, volunteers try to capture oil-covered penguins. The oil affects the penguins' ability to float. So, oil-covered penguins often won't go into the water to get food. The penguins may also swallow oil, harming their stomach, kidneys, and lungs. Once captured, the penguins are fed activated charcoal. The charcoal helps the penguins get rid of any oil they have swallowed. Then, the birds are washed to remove oil from their feathers.

PRE-READING ACTIVITY

FOLDNOTES

Two-Panel Flip Chart

Before you read the chapter, create the FoldNote entitled "Two-Panel Flip Chart" described in the **Study Skills** section of the Appendix. Label the flaps of the two-panel flip chart with "Environmental problems" and "Environmental solutions." As you read the chapter, write information you learn about each category under the appropriate flap.

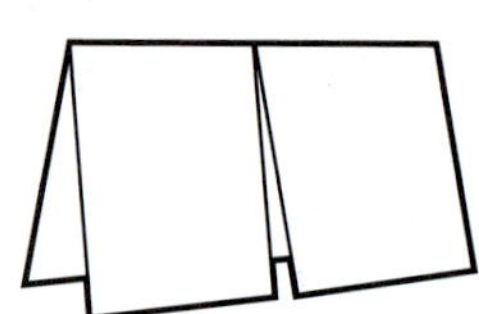

START-UP ACTIVITY

Recycling Paper

In this activity, you will be making paper without cutting down trees. You will be reusing paper that has already been made.

Procedure

1. Tear **two sheets of old newspaper** into small pieces, and put them in a **blender.** Add **1 L of water.** Cover and blend until the mixture is soupy.
2. Fill a **square pan** with **water** to a depth of 2 cm to 3 cm. Place a **wire screen** in the pan. Pour 250 mL of the paper mixture onto the screen, and spread the mixture evenly.
3. Lift the screen out of the water with the paper on it. Drain excess water into the pan.
4. Place the screen inside a **section of newspaper.** Close the newspaper, and turn it over so that the screen is on top of the paper mixture.
5. Cover the newspaper with a **flat board.** Press on the board to squeeze out excess water.
6. Open the newspaper, and let your paper mixture dry overnight. Use your recycled paper to write a note to a friend!

Analysis

1. How is your paper like regular paper? How is it different?
2. What could you do to improve your papermaking methods?

SECTION 1

Environmental Problems

Maybe you've heard warnings about dirty air, water, and soil. Or you've heard about the destruction of rain forests. Do these warnings mean our environment is in trouble?

In the late 1700s, the Industrial Revolution began. People started to rely more and more on machines. As a result, more harmful substances entered the air, water, and soil.

READING WARM-UP

Objectives

- List five kinds of pollutants.
- Distinguish between renewable and nonrenewable resources.
- Describe the impact of exotic species.
- Explain why human population growth has increased.
- Describe how habitat destruction affects biodiversity.
- Give two examples of how pollution affects humans.

Terms to Learn

pollution
renewable resource
nonrenewable resource
overpopulation
biodiversity

READING STRATEGY

Reading Organizer As you read this section, make a concept map by using the terms above.

Pollution

Today, machines don't produce as much pollution as they once did. But there are more sources of pollution today than there once were. **Pollution** is an unwanted change in the environment caused by substances, such as wastes, or forms of energy, such as radiation. Anything that causes pollution is called a *pollutant*. Some pollutants are produced by natural events, such as volcanic eruptions. Many pollutants are human-made. Pollutants may harm plants, animals, and humans.

pollution an unwanted change in the environment caused by substances or forms of energy

Garbage

The average American throws away more trash than the average person in any other nation—about 12 kg of trash a week. This trash often goes to a landfill like the one in **Figure 1.** Other landfills contain medical waste, lead paint, and other hazardous wastes. *Hazardous waste* includes wastes that can catch fire; corrode, or eat through metal; explode; or make people sick. Many industries, such as paper mills and oil refineries, produce hazardous wastes.

Reading Check What is hazardous waste? *(See the Appendix for answers to Reading Checks.)*

Figure 1 *Every year, Americans throw away about 200 million metric tons of garbage.*

Figure 2 *Fertilizer promotes the growth of algae. As dead algae decompose, oxygen in the water is used up. So, fish die because they cannot get oxygen.*

Chemicals

People need and use many chemicals. Some chemicals are used to treat diseases. Other chemicals are used in plastics and preserved foods. Sometimes, the same chemicals that help people may harm the environment. As shown in **Figure 2,** fertilizers and pesticides may pollute soil and water.

CFCs and PCBs are two groups of harmful chemicals. Ozone protects Earth from harmful ultraviolet light. CFCs destroy ozone. CFCs were used in aerosols, refrigerators, and plastics. The second group, PCBs, was once used in appliances and paints. PCBs are poisonous and may cause cancer. Today, the use of CFCs and PCBs is banned. But CFCs are still found in the atmosphere. And PCBs are still found in even the most remote areas on Earth.

CONNECTION TO Social Studies

WRITING SKILL **Technology** Write a report describing how technologies having to do with food production, sanitation, and disease prevention have changed how people live and work. How have changes in these technologies affected the growth of the human population?

High-Powered Wastes

Nuclear power plants provide electricity to many homes and businesses. The plants also produce radioactive wastes. *Radioactive wastes* are hazardous wastes that give off radiation. Some of these wastes take thousands of years to become harmless.

Gases

Earth's atmosphere is made up of a mixture of gases, including carbon dioxide. The atmosphere acts as a protective blanket. It keeps Earth warm enough for life to exist. Since the Industrial Revolution, however, the amount of carbon dioxide in the atmosphere has increased. Carbon dioxide and other air pollutants act like a greenhouse, trapping heat around the Earth. Many scientists think the increase in carbon dioxide has increased global temperatures. If temperatures continue to rise, the polar icecaps could melt. Then, the level of the world's oceans would rise. Coastal areas could flood as a result.

MATH PRACTICE

Water Depletion

In one day, millions of liters of water are removed from a water supply. Of this volume, 30 million liters cannot be replaced naturally. Today, the water supply has 60 billion liters of water. In years, how long would the water supply last if water continued to be removed at this rate? If water were removed at the same rate as it was replaced, how long would the water supply last?

Noise

Some pollutants affect the senses. These pollutants include loud noises. Too much noise is not just annoying. Noise pollution affects your ability to hear and think clearly. And it may damage your hearing. People who work in noisy environments, such as in construction zones, must protect their ears.

Resource Depletion

Some of Earth's resources are renewable. But other resources are nonrenewable. A **renewable resource** is one that can be used over and over or has an unlimited supply. Solar and wind energy are renewable resources, as are some kinds of trees. A **nonrenewable resource** is one that cannot be replaced or that can be replaced only over thousands or millions of years. Most minerals and fossil fuels, such as oil and coal, are nonrenewable resources.

Nonrenewable resources cannot last forever. These resources will become more expensive as they become harder to find. The removal of some materials from the Earth also carries a high price tag. This removal may lead to oil spills, loss of habitat, and damage from mining, as shown in **Figure 3.**

renewable resource a natural resource that can be replaced at the same rate at which the resource is consumed

nonrenewable resource a resource that forms at a rate that is much slower than the rate at which it is consumed

Renewable or Nonrenewable?

Some resources once thought to be renewable are becoming nonrenewable. For example, scientists used to think that fresh water was a renewable resource. However, in some areas, water supplies are being used faster than they are being replaced. Eventually, these areas may run out of fresh water. So, scientists are working on ways to keep these water supplies from being used up.

Figure 3 *This area has been mined for iron using a method called* strip mining.

Exotic Species

People are always on the move. Without knowing it, people carry other species with them. Plant seeds, animal eggs, and adult organisms are carried from one part of the world to another. An organism that makes a home for itself in a new place outside its native home is an *exotic species*. Exotic species often thrive in new places. One reason is that they are free from the predators found in their native homes.

Exotic species can become pests and compete with native species. In 2002, the northern snakehead fish was found in a Maryland pond. This fish, shown in **Figure 4,** is from Asia. Scientists are concerned because the northern snakehead eats other fish, amphibians, small birds, and some mammals. It can also move across land. The northern snakehead could invade more lakes and ponds.

Reading Check What are exotic species?

Figure 4 *Northern snakehead fish can move across land in search of water. These fish can survive out of water for up to four days!*

Human Population Growth

Look at **Figure 5.** In 1800, there were 1 billion people on Earth. By 2000, there were more than 6 billion people. Advances in medicine, such as immunizations, and advances in farming have made human population growth possible. Overall, these advances are beneficial. But some people argue that there may eventually be too many people on Earth. **Overpopulation** happens when the number of individuals becomes so large that the individuals can't get the resources they need to survive. However, many scientists think that human population growth will slow down or level off before it reaches that point.

overpopulation the presence of too many individuals in an area for the available resources

Figure 5 *Recently, the human population has been doubling every few decades.*

Figure 6 *Deforestation can leave soil exposed to erosion.*

Habitat Destruction

People need homes. People also need food and building materials. But when land is cleared for construction, crops, mines, or lumber, the topsoil may erode. Chemicals may pollute nearby streams and rivers. The organisms that were living in these areas may be left without food and shelter. These organisms may die.

An organism's *habitat* is where it lives. Every habitat has its own number and variety of organisms, or **biodiversity.** If a habitat is damaged or destroyed, biodiversity is lost.

biodiversity the number and variety of organisms in a given area during a specific period of time

Forest Habitats

Trees provide humans with oxygen, lumber, food, rubber, and paper. For some of these products, such as lumber and paper, trees must be cut down. *Deforestation* is the clearing of forest lands, as shown in **Figure 6.** At one time, many of these cleared forests were not replanted. Today, lumber companies often plant new trees to replace the trees that were cut down. However, some biodiversity is still lost.

Tropical rain forests, the most diverse habitats on Earth, are sometimes cleared for farmland, roads, and lumber. But after a tropical rain forest is cleared, the area cannot grow to be as diverse as it once was. Also, thin tropical soils are often badly damaged.

CONNECTION TO Social Studies

Wood Identify a country that is a major exporter of wood. List some of the ways this wood is used. Research the impact this exportation is having on that country's forests. Make a poster describing your findings.

ACTIVITY

Marine Habitats

Many people think of oil spills when they think of pollution in marine habitats. This is an example of *point-source pollution*, or pollution that comes from one source. Spilled oil pollutes both open waters and coastal habitats.

A second kind of water pollution is *nonpoint-source pollution*. This kind of pollution comes from many different sources. Nonpoint-source pollution often happens when chemicals on land are washed into rivers, lakes, and oceans. These chemicals can harm or kill many of the organisms that live in marine habitats.

In addition to oil and chemicals, plastics are also sometimes dumped into marine habitats. Animals may mistake plastics for food. Or animals may become tangled in plastics. Dumping plastics into the ocean is against the law. However, this law is difficult to enforce.

Reading Check What are point-source and nonpoint-source pollution?

Effects on Humans

Trees and marine life are not the only organisms affected by pollution and habitat destruction. Pollution and habitat destruction affect humans, too. Sometimes, the effect is immediate. Polluted air affects people with respiratory problems. If you drink polluted water, you may get sick. Sometimes, the damage is not apparent right away. Some chemicals cause cancers many years after a person is exposed to them. Over time, natural resources may be hard to find or used up. Your children or grandchildren may have to deal with these problems.

Anything that harms other organisms may eventually harm people, too. Caring for the environment means being aware of what is happening now and looking ahead to the future.

SECTION Review

Summary

- Pollutants include garbage, chemicals, high-energy wastes, gases, and noise.
- Renewable resources can be used over and over. Nonrenewable resources cannot be replaced or are replaced over thousands or millions of years.
- Exotic species can become pests and compete with native species.
- Overpopulation happens when a population is so large that it can't get what it needs to survive.
- Habitat destruction can lead to soil erosion, water pollution, and decreased biodiversity.
- In addition to harming the environment, pollution can harm humans.

Using Key Terms

The statements below are false. For each statement, replace the underlined term to make a true statement.

1. Coal is a renewable resource.
2. Overpopulation is the number and variety of organisms in an area.

Understanding Key Ideas

3. Which of the following can cause pollution?
   a. noise
   b. garbage
   c. chemicals
   d. All of the above
4. Pollution
   a. does not affect humans.
   b. can make humans sick.
   c. makes humans sick only after many years.
   d. None of the above
5. Compare renewable and nonrenewable resources.
6. Why has human population growth increased?
7. What is an exotic species?
8. How does habitat destruction affect biodiversity?

Math Skills

9. Jodi's family produces 48 kg of garbage each week. What is the percentage decrease if they reduce the amount of garbage to 40 kg per week?

Critical Thinking

10. **Applying Concepts** Explain how each of the following can help people but harm the environment: hospitals, old refrigerators, and road construction.
11. **Making Inferences** Explain how human population growth is related to pollution problems.
12. **Predicting Consequences** How can the pollution of marine habitats affect humans?

SECTION 2

Environmental Solutions

As the human population grows, it will need more resources. People will need food, healthcare, transportation, and waste disposal. What does this mean for the Earth?

READING WARM-UP

Objectives

- Explain the importance of conservation.
- Describe the three Rs.
- Explain how biodiversity can be maintained.
- List five environmental strategies.

Terms to Learn

conservation
recycling

READING STRATEGY

Discussion Read this section silently. Write down questions that you have about this section. Discuss your questions in a small group.

All of these needs will have an impact on the Earth. If people don't use resources wisely, people will continue to pollute the air, soil, and water. More natural habitats could be lost. Many species could die out as a result. But there are many things people can do to protect the environment.

Conservation

One way to care for the Earth is conservation (KAHN suhr VAY shuhn). **Conservation** is the preservation and wise use of natural resources. You can ride your bike to conserve fuel. At the same time, you prevent air pollution. You can use organic compost instead of chemical fertilizer in your garden. Doing so conserves the resources needed to make the fertilizer. Also, you may reduce soil and water pollution.

Practicing conservation means using fewer natural resources. Conservation helps reduce waste and pollution. Also, conservation can help prevent habitat destruction. The three Rs are shown in **Figure 1.** They describe three ways to conserve resources: Reduce, Reuse, and Recycle.

conservation the preservation and wise use of natural resources

Reading Check What are the three Rs? (*See the Appendix for answers to Reading Checks.*)

Figure 1 *By reducing, reusing, and recycling, these teens are conserving resources.*

Reduce

What is the best way to conserve the Earth's natural resources? Use less of them! Doing so also helps reduce pollution.

Reducing Waste and Pollution

As much as one-third of the waste produced by some countries is packaging material. Products can be wrapped in less paper and plastic to reduce waste. For example, fast-food restaurants used to serve sandwiches in large plastic containers. Today, sandwiches are usually wrapped in thin paper instead. This paper is more biodegradable than plastic. Something that is *biodegradable* can be broken down by living organisms, such as bacteria. Scientists, such as the ones in **Figure 2,** are working to make biodegradable plastics.

Many people and companies are using less-hazardous materials in making their products. For example, some farmers don't use synthetic chemicals on their crops. Instead, they practice organic farming. They use mulch, compost, manure, and natural pest control. Agricultural specialists are also working on farming techniques that are better for the environment.

Figure 2 *These scientists are studying ways to make biodegradable plastics.*

Reducing the Use of Nonrenewable Resources

Some scientists are looking for sources of energy that can replace fossil fuels. For example, solar energy can be used to power homes, such as the home shown in **Figure 3.** Scientists are studying power sources such as wind, tides, and falling water. Car companies have developed electric and hydrogen-fueled automobiles. Driving these cars uses fewer fossil fuels and produces less pollution than driving gas-fueled cars does.

Figure 3 *The people who live in this home use solar panels to get energy from the sun.*

Figure 4 *This home was built with reused tires and aluminum cans.*

Reuse

Do you get hand-me-down clothes from an older sibling? Do you try to fix broken sports equipment instead of throwing it away? If so, you are helping conserve resources by *reusing* products.

Reusing Products

Every time you reuse a plastic bag, one bag fewer needs to be made. Reusing the plastic bag at the grocery store is just one way to reuse the bag. Reusing products is an important way to conserve resources.

You might be surprised at how many materials can be reused. For example, building materials can be reused. Wood, bricks, and tiles can be used in new structures. Old tires can be reused, too. They can be reused for playground surfaces. As shown in **Figure 4,** some tires are even reused to build new homes!

Figure 5 *This golf course is being watered with reclaimed water.*

Reusing Water

About 100 billion liters of water are used each day in American homes. Most of this water goes down the drain. Many communities are experiencing water shortages. Some of these communities are experimenting with reusing, or reclaiming, wastewater.

One way to reclaim water is to use organisms to clean the water. These organisms include plants and filter-feeding animals, such as clams. Often, reclaimed water isn't pure enough to drink. But it can be used to water crops, lawns, and golf courses, such as the one shown in **Figure 5.** Sometimes, reclaimed water is returned to underground water supplies.

✓ *Reading Check* **Describe how water is reused.**

Recycle

Another example of reuse is recycling. **Recycling** is the recovery of materials from waste. Sometimes, recyclable items, such as paper, are used to make the same kinds of products. Other recyclable items are made into different products. For example, yard clippings can be recycled into a natural fertilizer.

recycling the process of recovering valuable or useful materials from waste or scrap

Recycling Trash

Plastics, paper, aluminum, wood, glass, and cardboard are examples of materials that can be recycled. Every week, about half a million trees are used to make Sunday newspapers. Recycling newspapers could save millions of trees. Recycling aluminum saves 95% of the energy needed to change raw ore into aluminum. Glass can be recycled over and over again to make new bottles and jars.

Many communities make recycling easy. Some cities provide containers for glass, plastic, aluminum, and paper. People can leave these containers on the curb. Each week, the materials are picked up for recycling, as shown in **Figure 6.** Other cities have centers where people can take materials for recycling.

Figure 6 *In some communities, recyclable materials are picked up each week.*

Recycling Resources

Waste that can be burned can also be used to generate electricity. Electricity is generated in waste-to-energy plants, such as the one shown in **Figure 7.** Using garbage to make electricity is an example of *resource recovery.* Some companies are beginning to make electricity with their own waste. Doing so saves the companies money and conserves resources.

About 16% of the solid waste in the United States is burned in waste-to-energy plants. But some people are concerned that these plants pollute the air. Other people worry that the plants reduce recycling.

Figure 7 *A waste-to-energy plant can provide electricity to many homes and businesses.*

Figure 8 *What could happen if a fungus attacks a banana field? Biodiversity is low in fields of crops such as bananas.*

Maintaining Biodiversity

You know the three Rs. What else can you do to help the environment? You can help maintain biodiversity! So, how does biodiversity help the environment?

Imagine a forest with only one kind of tree. If a disease hit that species, the entire forest might die. Now, imagine a forest with 10 species of trees. If a disease hits one species, 9 other species will remain. Bananas, shown in **Figure 8,** are an important crop. But banana fields are not very diverse. Fungi threaten the survival of bananas. Farmers often use chemicals to control fungi. Growing other plants among the bananas, or increasing biodiversity, can also prevent the spread of fungi.

Biodiversity is also important because each species has a unique role in an ecosystem. Losing one species could disrupt an entire ecosystem. For example, if an important predator is lost, its prey will multiply. The prey might eat the plants in an area, keeping other animals from getting food. Eventually, even the prey won't have food. So, the prey will starve.

Figure 9 *Thanks to captive-breeding programs, the California condor population is increasing.*

Protecting Species

One way to maintain biodiversity is to protect individual species. In the United States, a law called the *Endangered Species Act* was designed to do just that. Endangered species are put on a special list. The law forbids activities that would harm a species on this list. The law also requires the development of recovery programs for each endangered species. Some endangered species, such as the California condor in **Figure 9,** are now increasing in number.

Anyone can ask the government to add a species to or remove a species from the endangered species list. This process can take years to complete. The government must study the species and its habitat before making a decision.

Protecting Habitats

Waiting until a species is almost extinct to begin protecting it is like waiting until your teeth are rotting to begin brushing them. Scientists want to prevent species from becoming endangered and from becoming extinct.

Plants, animals, and microorganisms depend on each other. Each organism is part of a huge, interconnected web of organisms. The entire web should be protected to protect these organisms. To protect the web, complete habitats, not just individual species, must be preserved. Nature preserves, such as the one shown in **Figure 10,** are one way to protect entire habitats.

Figure 10 *Setting aside public lands for wildlife is one way to protect habitats.*

Environmental Strategies

Laws have been passed to help protect the Earth's environment. By following those laws, people can help the environment. People can also use the following environmental strategies:

- **Reduce pollution.** Recycle as much as possible, and buy recycled products. Don't dump wastes on farmland, in forests, or into rivers, lakes, and oceans. Participate in a local cleanup project.
- **Reduce pesticide use.** Use only pesticides that are targeted specifically for harmful insects. Avoid pesticides that might harm beneficial insects, such as ladybugs or spiders. Use natural pesticides that interfere with how certain insects grow, develop, and reproduce.
- **Protect habitats.** Preserve entire habitats. Conserve wetlands. Reduce deforestation. Use resources at a rate that allows them to be replenished naturally.
- **Learn about local issues.** Attend local meetings about laws and projects that may affect your local environment. Research the impact of the project, and let people know about your concerns.
- **Develop alternative energy sources.** Increase the use of renewable energy, such as solar power and wind power.

For another activity related to this chapter, go to **go.hrw.com** and type in the keyword **HL5ENVW.**

The *Environmental Protection Agency* (EPA) is a government organization that helps protect the environment. The EPA works to help people have a clean environment in which to live, work, and play. The EPA keeps people informed about environmental issues and helps enforce environmental laws.

What is the EPA?

What You Can Do

Reduce, reuse, and recycle. Protect the Earth. These are jobs for everyone. Children as well as adults can help clean up the Earth. By doing so, people can improve their environment. And they can improve their quality of life.

The list in **Figure 11** offers some suggestions for how *you* can help. How many of these things do you already do? What can you add to the list?

Figure 11 **How You Can Help the Environment**

1. Volunteer at a local preserve or nature center, and help other people learn about conservation.
2. Give away your old toys.
3. Use recycled paper.
4. Fill up both sides of a sheet of paper.
5. Start an environmental awareness club at your school or in your neighborhood.
6. Recycle glass, plastics, paper, aluminum, and batteries.
7. Don't buy any products made from an endangered plant or animal.
8. Turn off electrical devices when you are not using them.
9. Wear hand-me-downs.
10. Share books with friends, or use the library.
11. Walk, ride a bicycle, or use public transportation.
12. Carry a reusable cloth shopping bag to the store.
13. Use a lunch box, or reuse your paper lunch bags.
14. Turn off the water while you brush your teeth.
15. Buy products made from biodegradable and recycled materials.
16. Use cloth napkins and kitchen towels.
17. Buy things in packages that can be recycled.
18. Use rechargeable batteries.
19. Make a compost heap.

SECTION Review

Summary

- Conservation is the preservation and wise use of natural resources. Conservation helps reduce pollution, ensures that resources will be available in the future, and protects habitats.
- The three Rs are Reduce, Reuse, and Recycle. Reducing means using fewer resources. Reusing means using materials and products over and over. Recycling is the recovery of materials from waste.
- Biodiversity is vital for maintaining healthy ecosystems. A loss of one species can affect an entire ecosystem.
- Biodiversity can be preserved by protecting endangered species and entire habitats.
- Environmental strategies include reducing pollution, reducing pesticide use, protecting habitats, enforcing the Endangered Species Act, and developing alternative energy resources.

Using Key Terms

1. Use each of the following terms in a separate sentence: *conservation* and *recycling*.

Understanding Key Ideas

2. Which of the following is NOT a strategy to protect the environment?
   - **a.** preserving entire habitats
   - **b.** using pesticides that target all insects
   - **c.** reducing deforestation
   - **d.** increasing the use of solar power
3. Conservation
   - **a.** has little effect on the environment.
   - **b.** is the use of more natural resources.
   - **c.** involves using more fossil fuels.
   - **d.** can prevent pollution.
4. Describe the three Rs.
5. Describe why biodiversity is important. How can biodiversity be protected?

Critical Thinking

6. **Applying Concepts** Liza rode her bike to the store. She bought items that had little packaging and put her purchases into her backpack. Describe how Liza practiced conservation.
7. **Identifying Relationships** How does conservation of resources also reduce pollution and protect habitats?

Interpreting Graphics

Use the pie graph below to answer the questions that follow.

Land Use in the United States

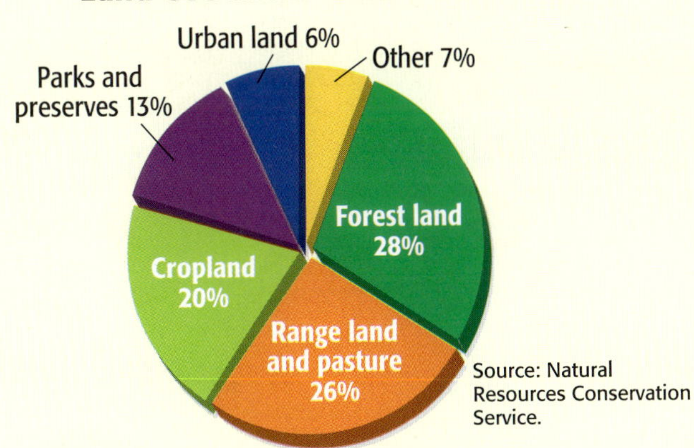

8. If half of the forest land were made into preserves, what percentage of total land would be parks and preserves?
9. If 10% of the cropland were not planted, what percentage of land would be used for crops?

Using Scientific Methods

Inquiry Lab

Biodiversity—What a Disturbing Thought!

OBJECTIVES

Examine biodiversity in your community.

Identify which areas in your community have the greatest biodiversity.

MATERIALS

- items to be determined by the students and approved by the teacher (Possible field equipment includes a meterstick, binoculars, a magnifying lens, and forceps.)
- stakes (4)
- twine

SAFETY

Biodiversity is important for the stability of an ecosystem. Microorganisms, plants, and animals all have a role in an ecosystem. In this activity, you will investigate areas outside your school to determine which areas contain the greatest biodiversity.

Ask a Question

1. Based on your understanding of biodiversity, do you expect a forest or an area planted with crops to be more diverse?

Form a Hypothesis

2. Select an area that is highly disturbed (such as a yard) and an area that is relatively undisturbed (such as a vacant lot). Make a hypothesis about which area contains the greater biodiversity. Get your teacher's approval of your selected locations.

Test the Hypothesis

3. Design a procedure to determine which area contains the greater biodiversity. Have your plan approved by your teacher before you begin.

Prairie

Wheat Field

4. To discover smaller organisms, measure off a square meter, set stakes at the corners, and mark the area with twine. Use a magnifying lens to observe organisms. When you record your observations, refer to organisms in the following way: Ant A, Ant B, and so on. Make note of any visits by larger organisms.

5. Create any data tables that you might need for recording your data. If you observe your areas on more than one occasion, make data tables for each observation period. Organize your data into clear and understandable categories.

Analyze the Results

1. **Explaining Events** What factors did you consider before deciding which habitats were disturbed or undisturbed?

2. **Constructing Maps** Draw a map of the land around your school. Label areas of high biodiversity and those of lower biodiversity.

3. **Analyzing Data** What problems did you have while making observations and recording data for each habitat? How did you solve these problems?

Draw Conclusions

4. **Drawing Conclusions** Review your hypothesis. Did your data support your hypothesis? Explain your answer.

5. **Evaluating Methods** Describe possible errors in your investigation. What are ways you could improve your procedure to eliminate errors?

6. **Applying Conclusions** Do you think that the biodiversity around your school increased or decreased since the school was built? Explain your answer.

Applying Your Data

The photographs of the prairie and of the wheat field on this page are beautiful. One of these areas, however, is very low in biodiversity. Describe each photograph, and explain the difference in biodiversity.

Chapter Review

USING KEY TERMS

Complete each of the following sentences by choosing the correct term from the word bank.

conservation
pollution
recycling
biodiversity
overpopulation
renewable resource
nonrenewable resource

1. A(n) ___ is a resource that is replaced at a much slower rate than it is used.

2. The presence of too many individuals in a population for available resources is called ___.

3. ___ is an unwanted change in the environment caused by wastes.

4. The preservation and wise use of natural resources is called ___.

5. ___ is the number and variety of organisms in an area.

UNDERSTANDING KEY IDEAS

Multiple Choice

6. Preventing habitat destruction is important because
   a. organisms do not live independently of each other.
   b. protection of habitats is a way to promote biodiversity.
   c. the balance of nature could be disrupted if habitats were destroyed.
   d. All of the above

7. Exotic species
   a. do not affect native species.
   b. are species that make a home for themselves in a new place.
   c. are not introduced by human activity.
   d. do not take over an area.

8. A renewable resource
   a. is a natural resource that can be replaced as quickly as it is used.
   b. is a natural resource that takes thousands or millions of years to be replaced.
   c. includes fossil fuels, such as coal or oil.
   d. will eventually run out.

Short Answer

9. Describe how you can use the three Rs to conserve resources.

10. What are five kinds of pollutants?

11. Explain why human population growth has increased.

12. What are two things that can be done to maintain biodiversity?

13. List five environmental strategies.

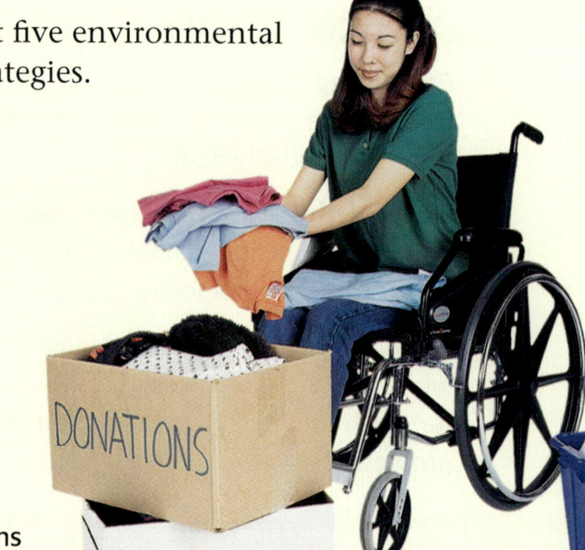

CRITICAL THINKING

14. **Concept Mapping** Use the following terms to create a concept map: *pollution, radioactive wastes, gases, pollutants, CFCs, PCBs, hazardous wastes, chemicals, noise,* and *garbage.*

15. **Analyzing Ideas** How may deforestation have contributed to the extinction of some species?

16. **Predicting Consequences** Imagine that the supply of fossil fuels is going to run out in 50 years. What will happen if people are not prepared when the supply runs out? What might be done to prepare for such an event?

17. **Evaluating Conclusions** A scientist thinks that farms should be planted with many different kinds of crops instead of a single crop. Based on what you learned about biodiversity, evaluate the scientist's conclusion. What problems might this cause?

18. **Applying Concepts** Imagine that a new species has moved into a local habitat. The species feeds on some of the same plants that the native species do, but it has no natural predators. Describe what might happen to local habitats as a result.

19. **Making Inferences** Many scientists think that forests are nonrenewable resources. Explain why they might have this opinion.

INTERPRETING GRAPHICS

The line graph below shows the concentration of carbon dioxide in the atmosphere between 1958 and 1994. Use this graph to answer the questions that follow.

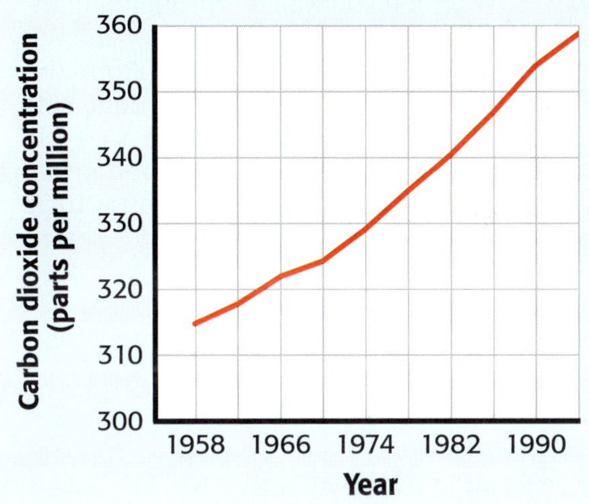

20. What was the concentration of carbon dioxide in parts per million in 1960? in 1994?

21. What is the average change in carbon dioxide concentration every 4 years?

22. If the concentration of carbon dioxide continues to change at the rate shown in the graph, what will the concentration be in 2010?

Standardized Test Preparation

READING

Read the passages below. Then, answer the questions that follow each passage.

Passage 1 The scientist woke up and jogged over to the rain forest. There she observed the water-recycling experiment. She took a swim in the ocean, after which she walked through a mangrove forest on her way home. At home, she ate lunch and went to the computer lab. From the lab, she could monitor the sensors that would alert her if any part of the ecosystem failed to cycle properly. This monitoring was very important to the scientist and her research team because their lives depended on the health of their sealed environment.

1. Based on the passage, the reader can conclude which of the following?

- **A** The scientist lives in an artificial environment.
- **B** The scientist lives by herself.
- **C** The scientist and her research team are studying a newly discovered island.
- **D** The scientist does not rely on the health of her environment.

2. Which of the following statements is a fact in the passage?

- **F** The scientist is scared that her environment is being destroyed.
- **G** The scientist depends on sensors to alert her to trouble.
- **H** The scientist lives in an open environment.
- **I** The scientist eats lunch at home every day.

3. Based on the passage, which of the following events happened first?

- **A** The scientist walked through the mangrove forest.
- **B** The scientist checked the water-recycling experiment.
- **C** The scientist swam in the ocean.
- **D** The scientist ate lunch.

Passage 2 All along the Gulf Coast, marine scientists and Earth scientists are trying to find methods to reduce or eliminate the dead zone. They have made models of the Mississippi River ecosystem that have accurately predicted the data that have since been collected. The scientists have changed the models to see what happens. For example, wetlands are one of nature's best filters. Wetlands take up a lot of the chemicals present in water. Scientists predict that adding wetlands to the Mississippi River watershed could reduce the chemicals reaching the Gulf of Mexico, possibly reducing the dead zone.

1. Based on the passage, what can you conclude about the dead zone?

- **A** It is found in the Mississippi River.
- **B** It may be prevented by adding wetlands to the Mississippi River watershed.
- **C** It reduces the chemicals reaching the Gulf of Mexico.
- **D** It is not caused by chemicals.

2. Based on the passage, which of the following statements about models is true?

- **F** Models do not accurately predict data.
- **G** Scientists do not change models.
- **H** Scientists use models to make predictions.
- **I** Models are always used for research.

3. Based on the passage, why did the scientists change their models?

- **A** to predict the effects of adding wetlands to the Mississippi River watershed
- **B** to find out why the dead zone happened
- **C** to eliminate the dead zone
- **D** to predict why there are a lot of chemicals in the Gulf of Mexico

INTERPRETING GRAPHICS

The table below shows the change in ozone levels between 1960 and 1990 above Halley Bay, Antarctica. Use the table to answer the questions that follow.

October Ozone Levels Above Halley Bay, Antarctica, in Dobson Units (DU)

Year	Ozone level (DU)
1960	300
1970	280
1980	235
1990	190

1. According to the table, which of the following is the most likely ozone level for October 2000?

A 120 DU
B 150 DU
C 235 DU
D 280 DU

2. According to the table, the ozone level above Halley Bay is doing which of the following?

F It steadily increased between 1960 and 1990.
G It fell by 37% between 1960 and 1990.
H It decreased by an average of 37 DU per year.
I It decreased by about 25% every 10 years.

3. What is the percent decrease in ozone level between 1980 and 1990?

A 16%
B 19%
C 24%
D 81%

4. What is the average loss of ozone level per year in DU?

F 4 DU
G 6 DU
H 37 DU
I 63 DU

MATH

Read each question below, and choose the best answer.

1. About 15 m of topsoil covers the western plains of the United States. If topsoil forms at the rate of 2.5 cm per 500 years, how long did it take for 15 m of topsoil to form?

A 3,000 years
B 7,500 years
C 18,750 years
D 300,000 years

2. The dimensions of a habitat are 16 km by 6 km. If these dimensions are decreased by 50%, what will the area of the habitat be?

F 22 km^2
G 24 km^2
H 48 km^2
I 96 km^2

3. If each person in a city of 500,000 people throws away 12 kg of trash each week, how many metric tons of trash does the city produce per year? (There are 1,000 kg in a metric ton.)

A 6,000 metric tons
B 26,000 metric tons
C 312,000 metric tons
D 312,000,000 metric tons

4. Producing one ton of new glass creates about 175 kg of mining waste. Using 50% recycled glass cuts this rate by 75%. Which of the following equations calculates *y*, the mass of mining waste produced using 50% recycled glass?

F $y = 175 \times 0.25$
G $y = 175 \times 0.75$
H $y = 175 \times 0.5$
I $y = 175 \div 0.75$

Science in Action

Scientific Debate

Where Should the Wolves Roam?

The U.S. Fish and Wildlife Service once listed the gray wolf as an endangered species and devised a plan to reintroduce the wolf to parts of the U.S. The goal was to establish a population of at least 100 wolves at each location. In April 2003, gray wolves were reclassified as a threatened species in much of the United States. Eventually, gray wolves may be removed from the endangered species list entirely. But some ranchers and hunters are uneasy about the reintroduction of gray wolves, and some environmentalists and wolf enthusiasts think the plan doesn't go far enough to protect wolves.

Math Activity

Scientists tried to establish a population of 100 wolves in Idaho. But the population grew to 285 wolves. By what percentage did the population exceed expectations?

Science, Technology, and Society

Hydrogen-Fueled Automobiles

Can you imagine a car that purrs quieter than a kitten and gives off water vapor instead of harmful pollutants? These cars may sound like science fiction. But such cars already exist! They run on one of the most common elements in the world—hydrogen. Some car companies are already speculating that one day all cars will run on hydrogen. The U.S. government has also taken notice. In 2003, President George W. Bush promised $1.2 billion to help research and develop hydrogen-fueled cars.

Language Arts Activity

WRITING SKILL Research hydrogen-fueled cars. Then, write a letter to a car company, your senator, or the President expressing your opinion about the development of hydrogen-fueled cars.

People in Science

Phil McCrory

Hairy Oil Spills Phil McCrory, a hairdresser in Huntsville, Alabama, asked a brilliant question when he saw an otter whose fur was drenched with oil from the *Exxon Valdez* oil spill. If the otter's fur soaked up all the oil, why wouldn't human hair do the same? McCrory gathered hair from the floor of his salon and took it home to perform his own experiments. He stuffed hair into a pair of his wife's pantyhose and tied the ankles together to form a bagel-shaped bundle. McCrory floated the bundle in his son's wading pool and poured used motor oil into the center of the ring. When he pulled the ring closed, not a drop of oil remained in the water!

McCrory approached the National Aeronautics and Space Administration (NASA) with his discovery. Based on tests performed by NASA, scientists estimated that 64 million kilograms of hair in reusable mesh pillows could have cleaned up all of the oil spilled by the *Exxon Valdez* within a week! Unfortunately, the $2 billion spent on the cleanup removed only about 12% of the oil.

Social Studies ACTIVITY

Make a map of an oil spill. Show the areas that were affected. Indicate some of the animal populations affected by the spill, such as penguins.

To learn more about these Science in Action topics, visit go.hrw.com and type in the keyword HL5ENVF.

Check out Current Science® articles related to this chapter by visiting go.hrw.com. Just type in the keyword HL5CS21.

The Digestive and Urinary Systems

About the PHOTO

Is this a giant worm? No, it's an X ray of a healthy large intestine! Your large intestine helps your body preserve water. As mostly digested food passes through your large intestine, water is drawn out of the food. This water is returned to the bloodstream. The gray shadow behind the intestine is the spinal column. The areas that look empty are actually filled with organs. A special liquid helps this large intestine show up on the X ray.

PRE-READING ACTIVITY

Graphic Organizer

Chain-of-Events Chart Before you read the chapter, create the graphic organizer entitled "Chain-of-Events Chart" described in the **Study Skills** section of the Appendix. As you read the chapter, fill in the chart with details about each step of the processes that your body uses to digest food.

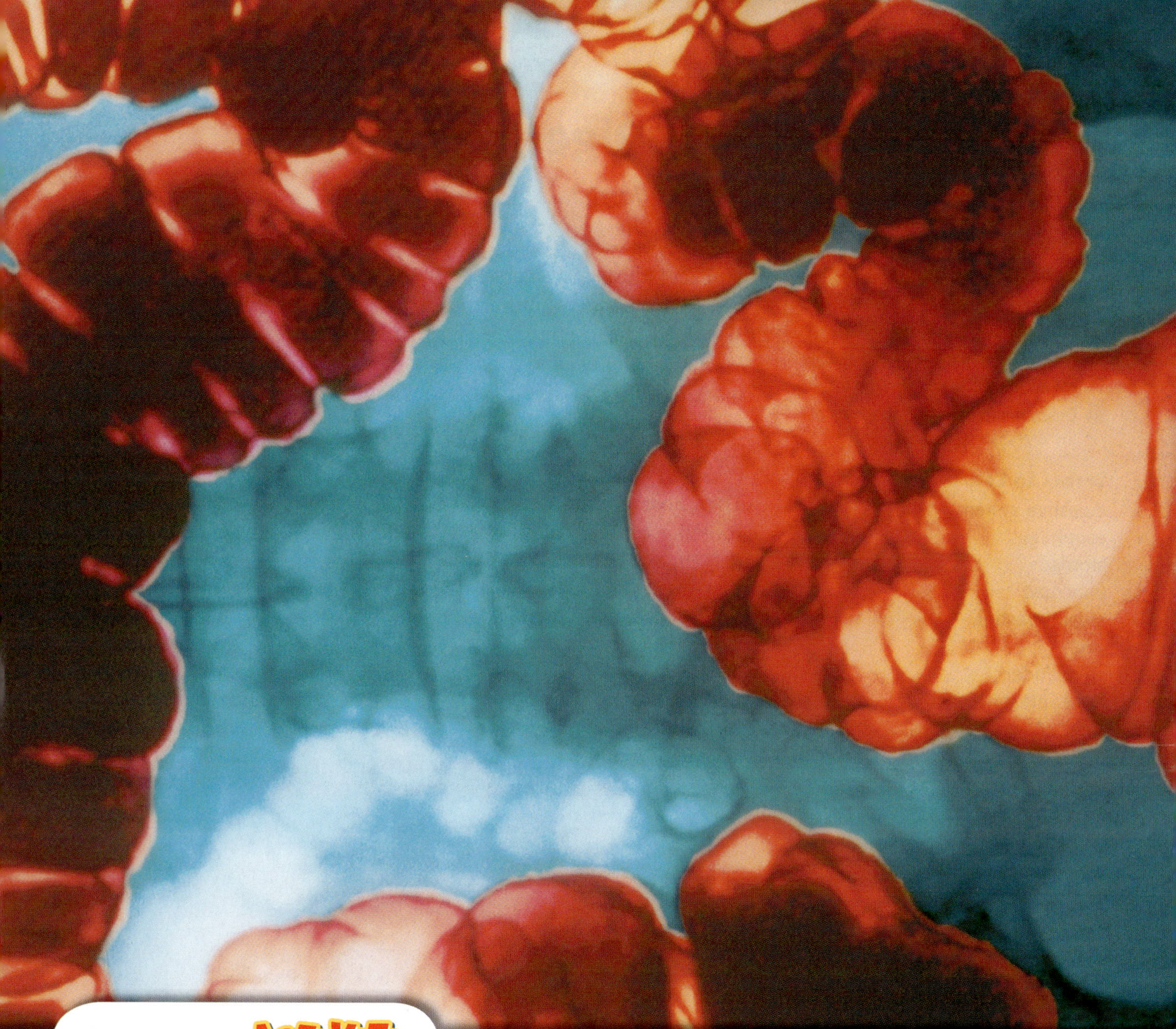

START-UP ACTIVITY

Changing Foods

The stomach breaks down food by, in part, squeezing the food. You can model the action of the stomach in the following activity.

Procedure

1. Add **200 mL of flour** and **100 mL of water** to a **resealable plastic bag.**
2. Mix **100 mL of vegetable oil** with the flour and water.
3. Seal the plastic bag.
4. Shake the bag until the flour, water, and oil are well mixed.
5. Remove as much air from the bag as you can, and reseal the bag carefully.
6. Knead the bag carefully with your hands for 5 min. Be careful to keep the bag sealed.

Analysis

1. Describe the mixture before and after you kneaded the bag.
2. How might the changes you saw in the mixture relate to how your stomach digests food?
3. Do you think this activity is a good model of how your stomach works? Explain your answer.

SECTION 1

The Digestive System

It's your last class before lunch, and you're starving! Finally, the bell rings, and you get to eat!

READING WARM-UP

Objectives

- Compare mechanical digestion with chemical digestion.
- Describe the parts and functions of the digestive system.

Terms to Learn

digestive system
esophagus
stomach
pancreas
small intestine
liver
gallbladder
large intestine

READING STRATEGY

Prediction Guide Before reading this section, write the title of each heading in this section. Next, under each heading, write what you think you will learn.

You feel hungry because your brain receives signals that your cells need energy. But eating is only the beginning of the story. Your body must change a meal into substances that you can use. Your **digestive system,** shown in **Figure 1,** is a group of organs that work together to digest food so that it can be used by the body.

digestive system the organs that break down food so that it can be used by the body

Digestive System at a Glance

The most obvious part of your digestive system is a series of tubelike organs called the *digestive tract.* Food passes through the digestive tract. The digestive tract includes your mouth, pharynx, esophagus, stomach, small intestine, large intestine, rectum, and anus. The human digestive tract can be more than 9 m long! The liver, gallbladder, pancreas, and salivary glands are also part of the digestive system. But food does not pass through these organs.

Figure 1 The Digestive System

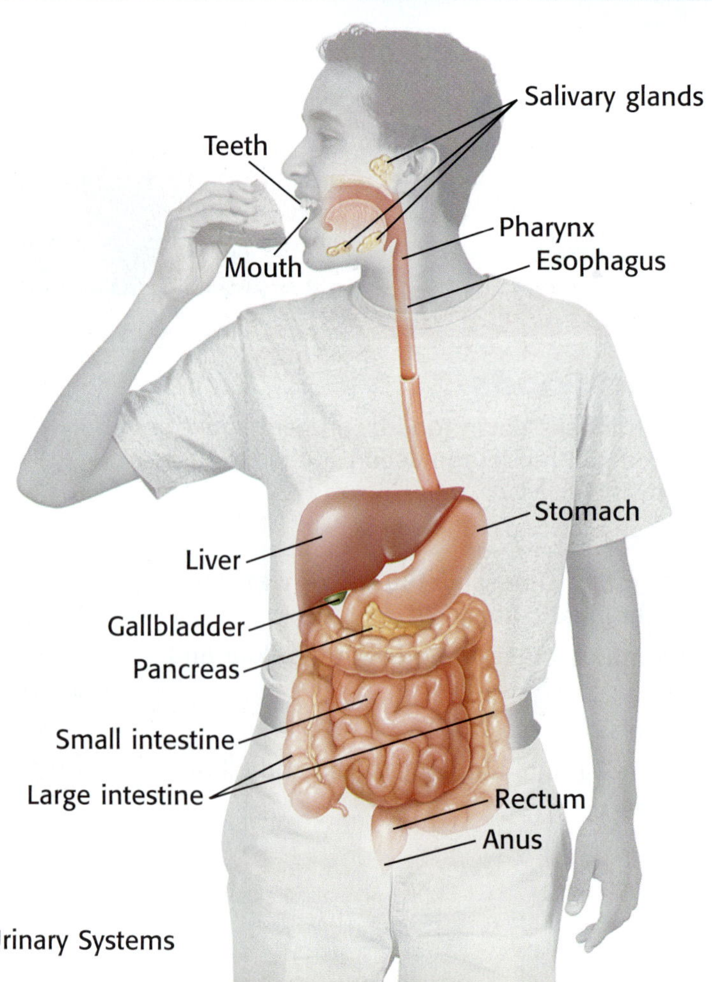

Breaking Down Food

Digestion is the process of breaking down food, such as a peanut butter and jelly sandwich, into a form that can pass from the digestive tract into the bloodstream. There are two types of digestion—mechanical and chemical. The breaking, crushing, and mashing of food is called *mechanical digestion.* In *chemical digestion,* large molecules are broken down into nutrients. Nutrients are substances in food that the body needs for normal growth, maintenance, and repair.

Three major types of nutrients—carbohydrates, proteins, and fats—make up most of the food you eat. In fact, a peanut butter and jelly sandwich contains all three of these nutrients. Substances called *enzymes* break some nutrients into smaller particles that the body can use. For example, proteins are chains of smaller molecules called *amino acids.* Proteins are too large to be absorbed into the bloodstream. So, enzymes cut up the chain of amino acids. The amino acids are small enough to pass into the bloodstream. This process is shown in **Figure 2.**

Reading Check **How do enzymes help digestion?** (*See the Appendix for answers to Reading Checks.*)

Figure 2 **The Role of Enzymes in Protein Digestion**

1. Enzymes act as chemical scissors to cut the long chains of amino acids into small chains.

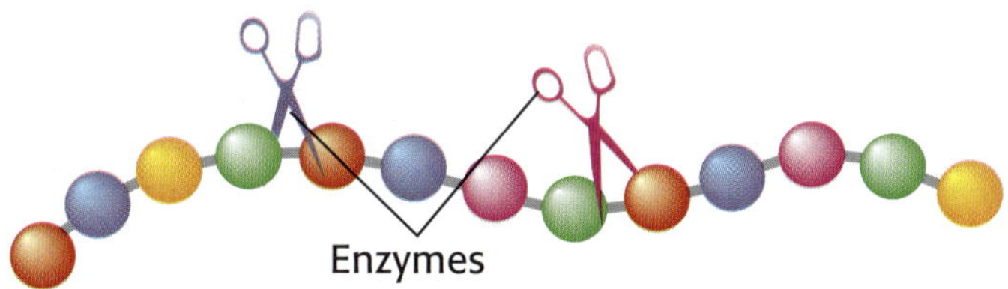

2. The small chains are split by other enzymes.

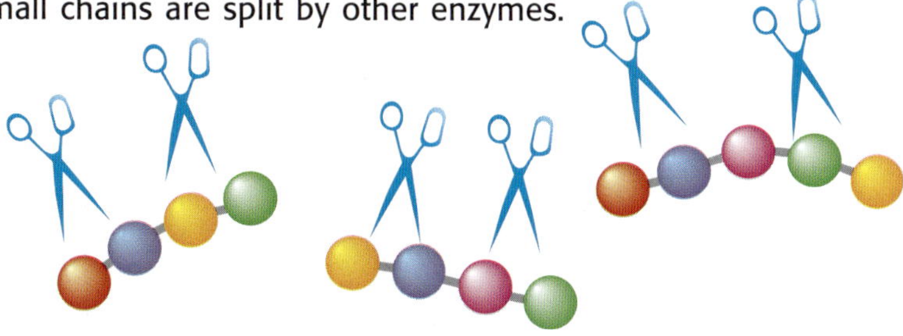

3. Individual amino acids are small enough to enter the bloodstream, where they can be used to make new proteins.

Quick Lab

Break It Up!

1. Drop **one piece of hard candy** into a **clear plastic cup of water.**
2. Wrap an **identical candy** in a **towel,** and crush the candy with a **hammer.** Drop the candy into a **second clear cup of water.**
3. The next day, examine both cups. What is different about the two candies?
4. What type of digestion is represented by breaking the hard candy?
5. How does chewing your food help the process of digestion?

Digestion Begins in the Mouth

Chewing is important for two reasons. First, chewing creates small, slippery pieces of food that are easier to swallow than big, dry pieces are. Second, small pieces of food are easier to digest.

Teeth

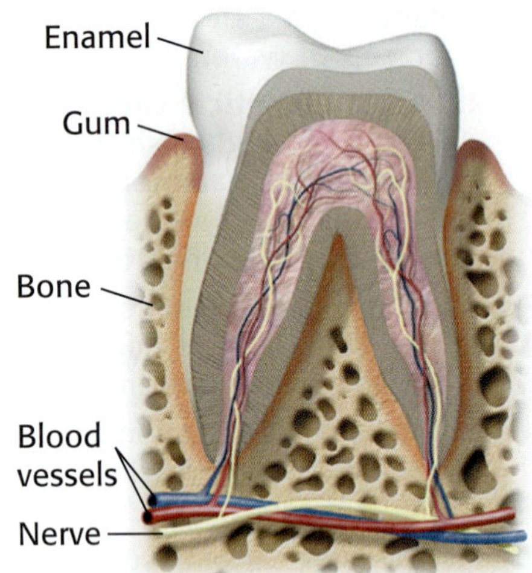

Figure 3 *A tooth, such as this molar, is made of many kinds of tissue.*

Teeth are very important organs for mechanical digestion. With the help of strong jaw muscles, teeth break and grind food. The outermost layer of a tooth, the *enamel,* is the hardest material in the body. Enamel protects nerves and softer material inside the tooth. **Figure 3** shows a cross section of a tooth.

Have you ever noticed that your teeth have different shapes? Look at **Figure 4** to locate the different kinds of teeth. The molars are well suited for grinding food. The *premolars* are perfect for mashing food. The sharp teeth at the front of your mouth, the *incisors* and *canines,* are for shredding food.

Saliva

As you chew, the food mixes with a liquid called *saliva*. Saliva is made in salivary glands located in the mouth. Saliva contains an enzyme that begins the chemical digestion of carbohydrates. Saliva changes complex carbohydrates into simple sugars.

Leaving the Mouth

esophagus a long, straight tube that connects the pharynx to the stomach

Once the food has been reduced to a soft mush, the tongue pushes it into the throat, which leads to a long, straight tube called the **esophagus** (i SAHF uh guhs). The esophagus squeezes the mass of food with rhythmic muscle contractions called *peristalsis* (PER uh STAL sis). Peristalsis forces the food into the stomach.

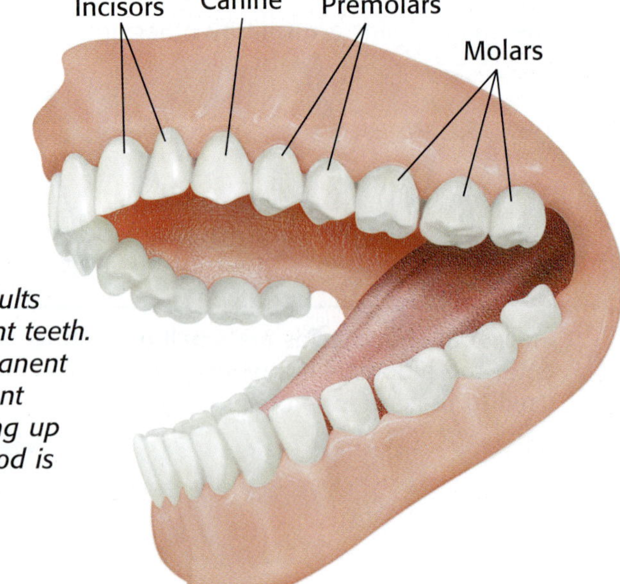

Figure 4 *Most adults have 32 permanent teeth. Each type of permanent tooth has a different function in breaking up food before the food is swallowed.*

Figure 5 **The Stomach**

The stomach squeezes and mixes food for hours before it releases the mixture into the small intestine.

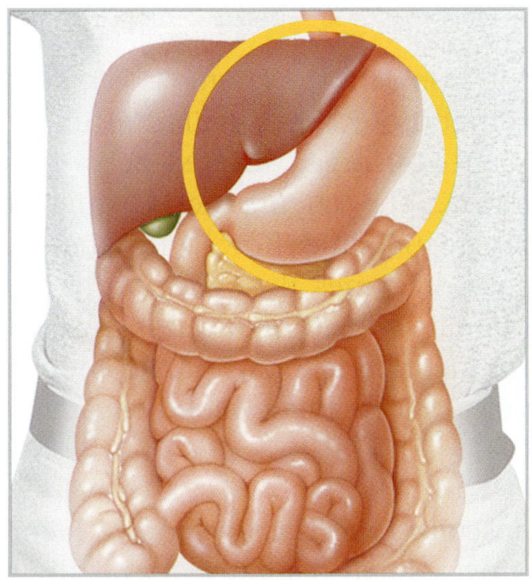

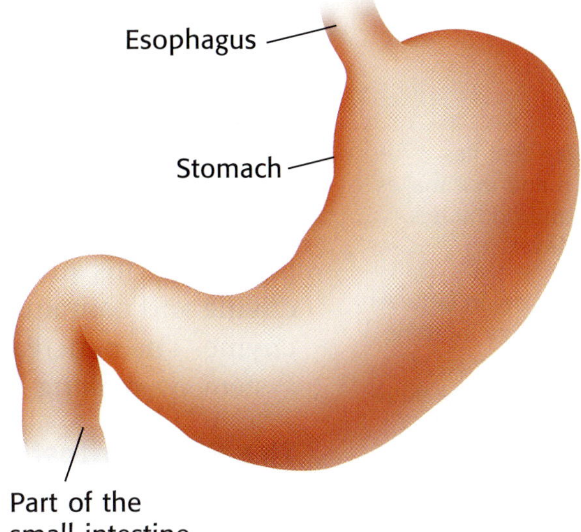

The Harsh Environment of the Stomach

The **stomach** is a muscular, saclike, digestive organ attached to the lower end of the esophagus. The stomach is shown in **Figure 5.** The stomach continues the mechanical digestion of your meal by squeezing the food with muscular contractions. While this squeezing is taking place, tiny glands in the stomach produce enzymes and acid. The enzymes and acid work together to break food into nutrients. Stomach acid also kills most bacteria that you might swallow with your food. After a few hours of combined mechanical and chemical digestion, your peanut butter and jelly sandwich has been reduced to a soupy mixture called *chyme* (KIEM).

stomach the saclike, digestive organ between the esophagus and the small intestine that breaks down food into a liquid by the action of muscles, enzymes, and acids

Reading Check What is chyme?

Leaving the Stomach

The stomach slowly releases the chyme into the small intestine through a small ring of muscle that works like a valve. This valve keeps food in the stomach until the food has been thoroughly mixed with digestive fluids. Each time the valve opens and closes, it lets a small amount of chyme into the small intestine. Because the stomach releases chyme slowly, the intestine has more time to mix the chyme with fluids from the liver and pancreas. These fluids help digest food and stop the harsh acids in chyme from hurting the small intestine.

Tooth Truth

Young children get a first set of 20 teeth called *baby teeth*. These teeth usually fall out and are replaced by 32 permanent teeth. How many more permanent teeth than baby teeth does a person have? What is the ratio of baby teeth to permanent teeth? Be sure to express the ratio in its most reduced form.

The Pancreas and Small Intestine

Most chemical digestion takes place after food leaves the stomach. Proteins, carbohydrates, and fats in the chyme are digested by the small intestine and fluids from the pancreas.

The Pancreas

When the chyme leaves the stomach, the chyme is very acidic. The pancreas makes fluids that protect the small intestine from the acid. The **pancreas** is an oval organ located between the stomach and small intestine. The chyme never enters the pancreas. Instead, the pancreatic fluid flows into the small intestine. This fluid contains enzymes that chemically digest chyme and contains bicarbonate, which neutralizes the acid in chyme. The pancreas also functions as a part of the endocrine system by making hormones that regulate blood sugar.

pancreas the organ that lies behind the stomach and that makes digestive enzymes and hormones that regulate sugar levels

small intestine the organ between the stomach and the large intestine where most of the breakdown of food happens and most of the nutrients from food are absorbed

The Small Intestine

The **small intestine** is a muscular tube that is about 2.5 cm in diameter. Other than having a small diameter, it is really not that small. In fact, if you stretched the small intestine out, it would be longer than you are tall—about 6 m! If you flattened out the surface of the small intestine, it would be larger than a tennis court! How is this possible? The inside wall of the small intestine is covered with fingerlike projections called *villi,* shown in **Figure 6.** The surface area of the small intestine is very large because of the villi. The villi are covered with tiny, nutrient-absorbing cells. Once the nutrients are absorbed, they enter the bloodstream.

CONNECTION TO Social Studies

WRITING SKILL **Parasites** Intestinal parasites are organisms, such as roundworms and hookworms, that infect people and live in their digestive tract. Worldwide, intestinal parasites infect more than 1 billion people. Some parasites can be deadly. Research intestinal parasites in a library or on the Internet. Then, write a report on a parasite, including how it spreads, what problems it causes, how many people have it, and what can be done to stop it.

Figure 6 **The Small Intestine and Villi**

The highly folded lining of the small intestine has many fingerlike projections called *villi.*

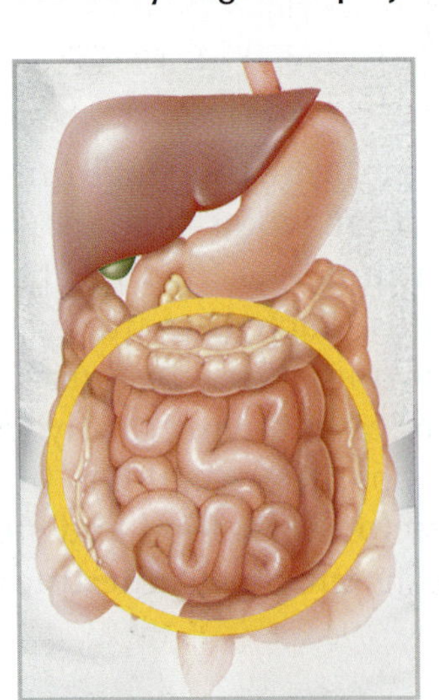

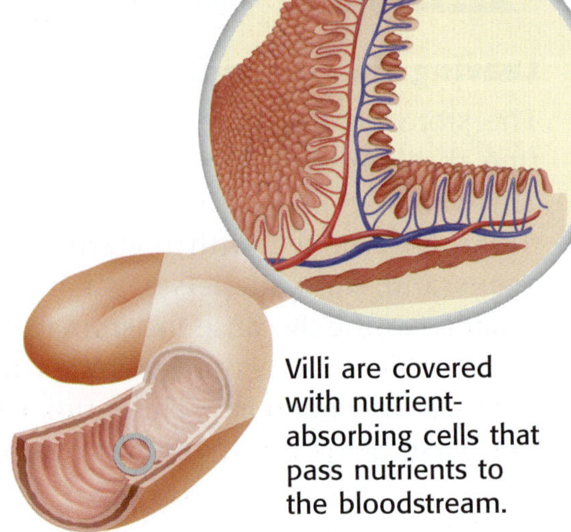

Villi are covered with nutrient-absorbing cells that pass nutrients to the bloodstream.

Figure 7 The Liver and the Gallbladder

Food does not move through the liver, gallbladder, and pancreas even though these organs are linked to the small intestine.

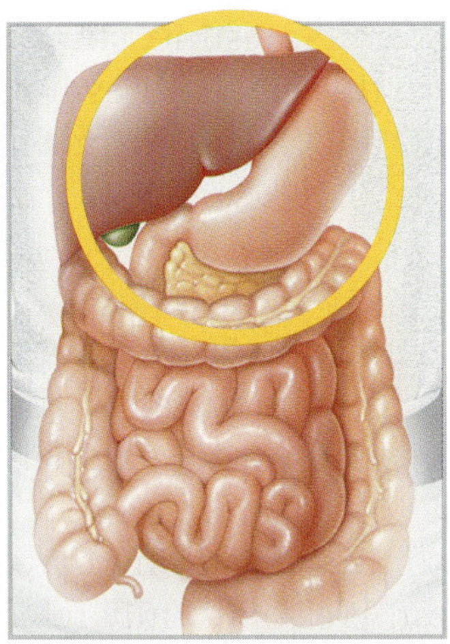

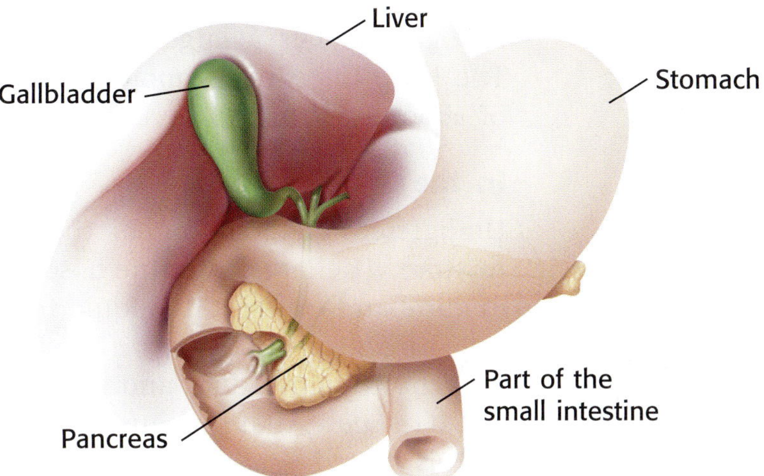

The Liver and Gallbladder

The **liver** is a large, reddish brown organ that helps with digestion. A human liver can be as large as a football. Your liver is located toward your right side, slightly higher than your stomach, as shown in **Figure 7.** The liver helps with digestion in the following ways:

- It makes bile to break up fat.
- It stores nutrients.
- It breaks down toxins.

liver the largest organ in the body; it makes bile, stores and filters blood, and stores excess sugars as glycogen

gallbladder a sac-shaped organ that stores bile produced by the liver

Breaking Up Fat

Although bile is made by the liver, bile is temporarily stored in a small, saclike organ called the **gallbladder,** shown in **Figure 7.** Bile is squeezed from the gallbladder into the small intestine, where the bile breaks large fat droplets into very small droplets. This mechanical process allows more fat molecules to be exposed to digestive enzymes.

Reading Check How does bile help digest fat?

Storing Nutrients and Protecting the Body

After nutrients are broken down, they are absorbed into the bloodstream and carried through the body. Nutrients that are not needed right away are stored in the liver. The liver then releases the stored nutrients into the bloodstream as needed. The liver also captures and detoxifies many chemicals in the body. For instance, the liver produces enzymes that break down alcohol and many other drugs.

Bile Model

You can model the way bile breaks down fat and oil by using dish soap. At home with a parent, put a small amount of water in a small jar. Then, add a few drops of vegetable oil to the water. Notice that the two liquids separate. Draw a picture of the jar and its contents. Next, add a few drops of dishwashing soap to the water, tighten the lid securely onto the jar, and shake the jar. What happened to the three liquids in the jar? Draw another picture of the jar and its contents.

The End of the Line

large intestine the wider and shorter portion of the intestine that removes water from mostly digested food and that turns the waste into semisolid feces, or stool

Material that can't be absorbed into the blood is pushed into the large intestine. The **large intestine** is the organ of the digestive system that stores, compacts, and then eliminates indigestible material from the body. The large intestine, shown in **Figure 8,** has a larger diameter than the small intestine. The large intestine is about 1.5 m long, and has a diameter of about 7.5 cm.

In the Large Intestine

Undigested material enters the large intestine as a soupy mixture. The large intestine absorbs most of the water in the mixture and changes the liquid into semisolid waste materials called *feces,* or *stool.*

Whole grains, fruits, and vegetables contain a carbohydrate, called *cellulose,* that humans cannot digest. We commonly refer to this material as *fiber*. Fiber keeps the stool soft and keeps material moving through the large intestine.

Reading Check How does eating fiber help digestion?

Leaving the Body

The *rectum* is the last part of the large intestine. The rectum stores feces until they can be expelled. Feces pass to the outside of the body through an opening called the *anus*. It has taken your sandwich about 24 hours to make this journey through your digestive system.

CONNECTION TO Environmental Science

Waste Away Feces and other human wastes contain microorganisms and other substances that can contaminate drinking water. Every time you flush a toilet, the water and wastes go through the sewer to a wastewater treatment plant. At the wastewater treatment plant, the disease-causing microorganisms are removed, and the clean water is released back to rivers, lakes, and streams. Find out where the wastewater treatment plants are in your area. Report to your class where their wastewater goes.

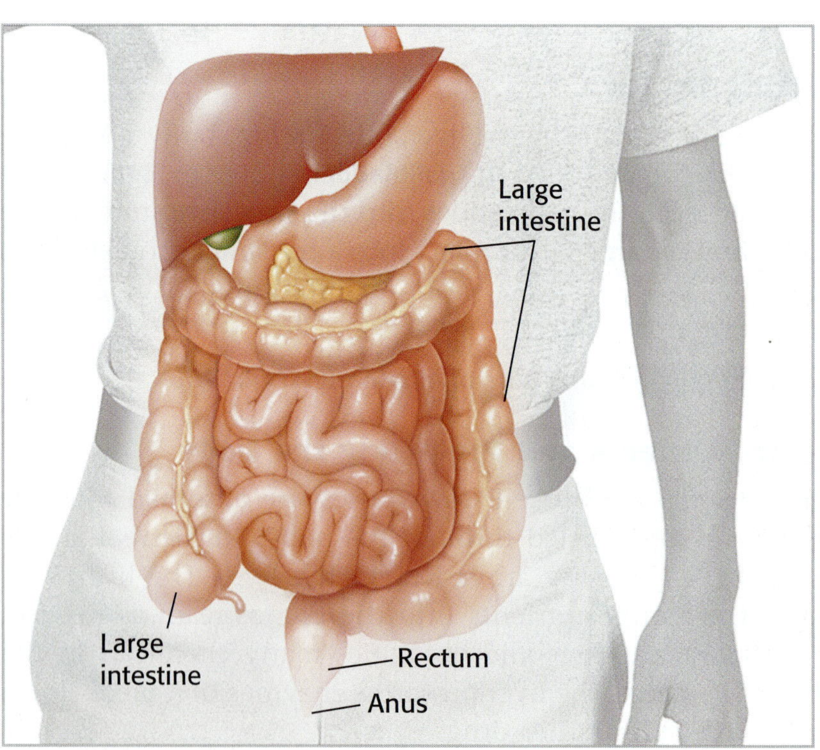

Figure 8 *The large intestine is the final organ of digestion.*

SECTION Review

Summary

- Your digestive system is a group of organs that work together to digest food so that the nutrients from food can be used by the body.
- The breaking and mashing of food is called *mechanical digestion*. Chemical digestion is the process that breaks large food molecules into simpler molecules.
- The stomach mixes food with acid and enzymes that break down nutrients. The mixture is called *chyme.*
- In the small intestine, pancreatic fluid and bile are mixed with chyme.
- From the small intestine, nutrients enter the bloodstream and are circulated to the body's cells.
- The liver makes bile, stores nutrients, and breaks down toxins.
- The large intestine absorbs water, changing liquid waste into semisolid stool, or feces.

Using Key Terms

1. Use each of the following terms in a separate sentence: *digestive system, large intestine,* and *small intestine.*

Understanding Key Ideas

2. Which of the following is NOT a function of the liver?
   - **a.** to secrete bile
   - **b.** to store nutrients
   - **c.** to detoxify chemicals
   - **d.** to compact wastes
3. What is the difference between mechanical digestion and chemical digestion?
4. What happens to the food that you eat when it gets to your stomach?
5. Describe the role of the liver, gallbladder, and pancreas in digestion.
6. Put the following steps of digestion in order.
   - **a.** Food is chewed by the teeth in the mouth.
   - **b.** Water is absorbed by the large intestine.
   - **c.** Food is reduced to chyme in the stomach.
   - **d.** Food moves down the esophagus.
   - **e.** Nutrients are absorbed by the small intestine.
   - **f.** The pancreas releases enzymes.

Critical Thinking

7. **Evaluating Conclusions** Explain the following statement: "Digestion begins in the mouth."
8. **Identifying Relationships** How would the inability to make saliva affect digestion?

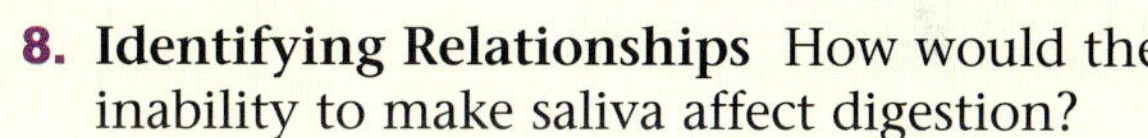

Interpreting Graphics

9. Label and describe the function of each of the organs in the diagram below.

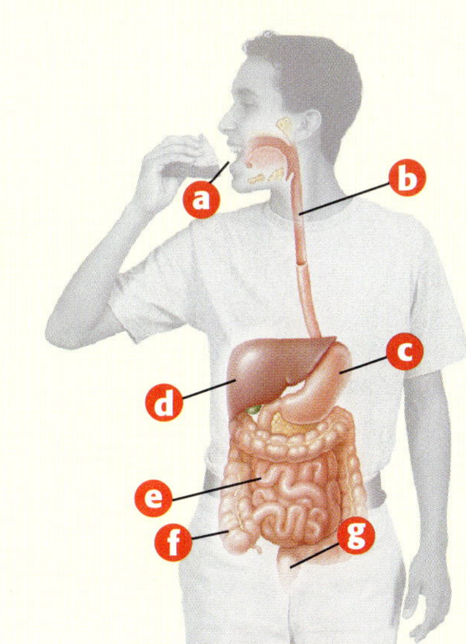

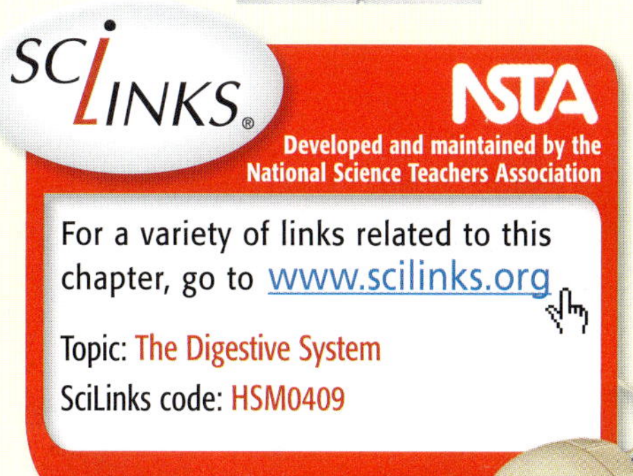

SECTION 2

The Urinary System

As blood travels through the tissues, it picks up waste produced by the body's cells. Your blood is like a train that comes to town to drop off supplies and take away garbage. If the waste is not removed, your body can actually be poisoned.

READING WARM-UP

Objectives

- Describe the parts and functions of the urinary system.
- Explain how the kidneys filter blood.
- Describe three disorders of the urinary system.

Terms to Learn

urinary system
kidney
nephron

READING STRATEGY

Reading Organizer As you read this section, create an outline of the section. Use the headings from the section in your outline.

Excretion is the process of removing waste products from the body. Three of your body systems have a role in excretion. Your integumentary system releases waste products and water when you sweat. Your respiratory system releases carbon dioxide and water when you exhale. Finally, the **urinary system** contains the organs that remove waste products from your blood.

urinary system the organs that produce, store, and eliminate urine

Cleaning the Blood

As your body performs the chemical activities that keep you alive, waste products, such as carbon dioxide and ammonia, are made. Your body has to get rid of these waste products to stay healthy. The urinary system, shown in **Figure 1,** removes these waste products from the blood.

Figure 1 Urinary System

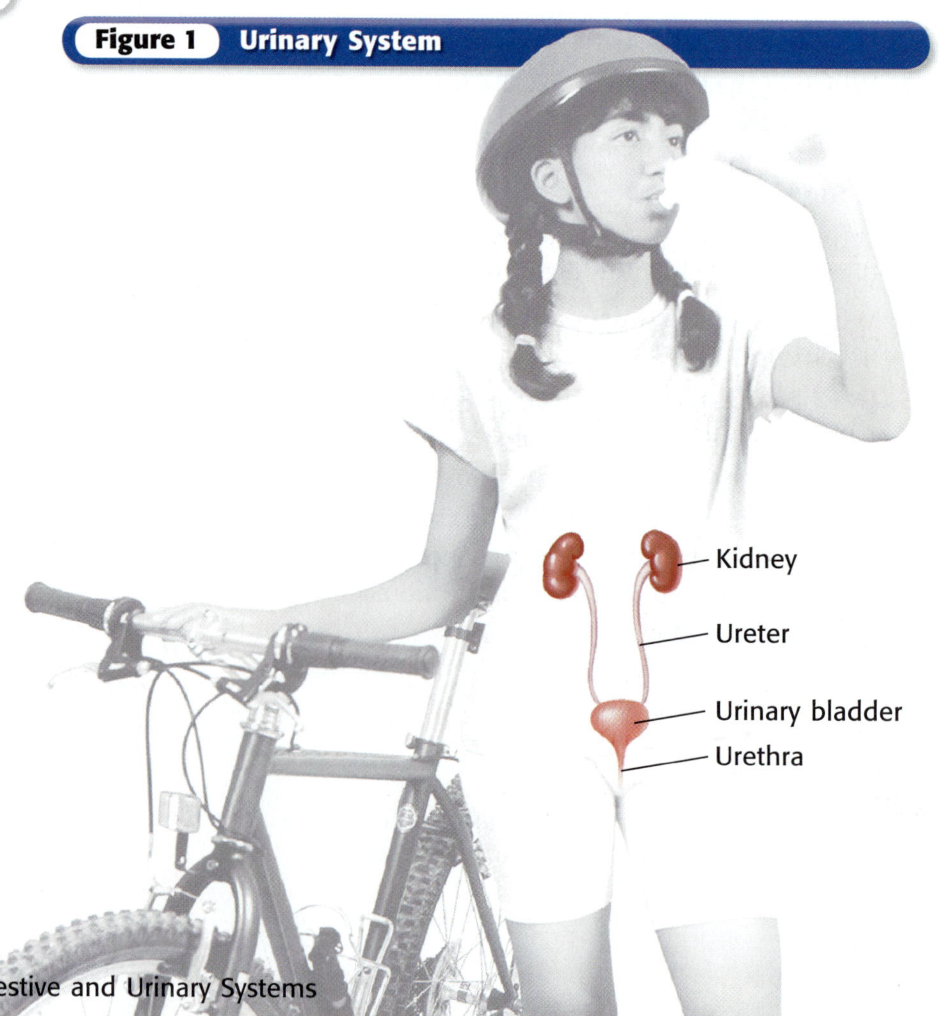

The Kidneys as Filters

The **kidneys** are a pair of organs that constantly clean the blood. Your kidneys filter about 2,000 L of blood each day. Your body holds only 5.6 L of blood, so your blood cycles through your kidneys about 350 times per day!

Inside each kidney, shown in **Figure 2,** are more than 1 million nephrons. **Nephrons** are microscopic filters in the kidney that remove wastes from the blood. Nephrons remove many harmful substances. One of the most important substances removed by nephrons is urea (yoo REE uh), which contains nitrogen and is formed when cells use protein for energy.

kidney one of the pair of organs that filter water and wastes from the blood and that excrete products as urine

nephron the unit in the kidney that filters blood

Reading Check **How are nephrons related to kidneys?** (*See the Appendix for answers to Reading Checks.*)

Figure 2 **How the Kidneys Filter Blood**

1. A large artery brings blood into each kidney.
2. Tiny blood vessels branch off the main artery and pass through part of each nephron.
3. Water and other small substances, such as glucose, salts, amino acids, and urea, are forced out of the blood vessels and into the nephrons.
4. As these substances flow through the nephrons, most of the water and some nutrients are moved back into blood vessels that wrap around the nephrons. A concentrated mixture of waste materials is left behind in the nephrons.
5. The cleaned blood, which has slightly less water and much less waste material, leaves each kidney in a large vein to recirculate in the body.
6. The yellow fluid that remains in the nephrons is called *urine.* Urine leaves each kidney through a slender tube called the *ureter* and flows into the *urinary bladder,* where urine is stored.
7. Urine leaves the body through another tube called the *urethra. Urination* is the process of expelling urine from the body.

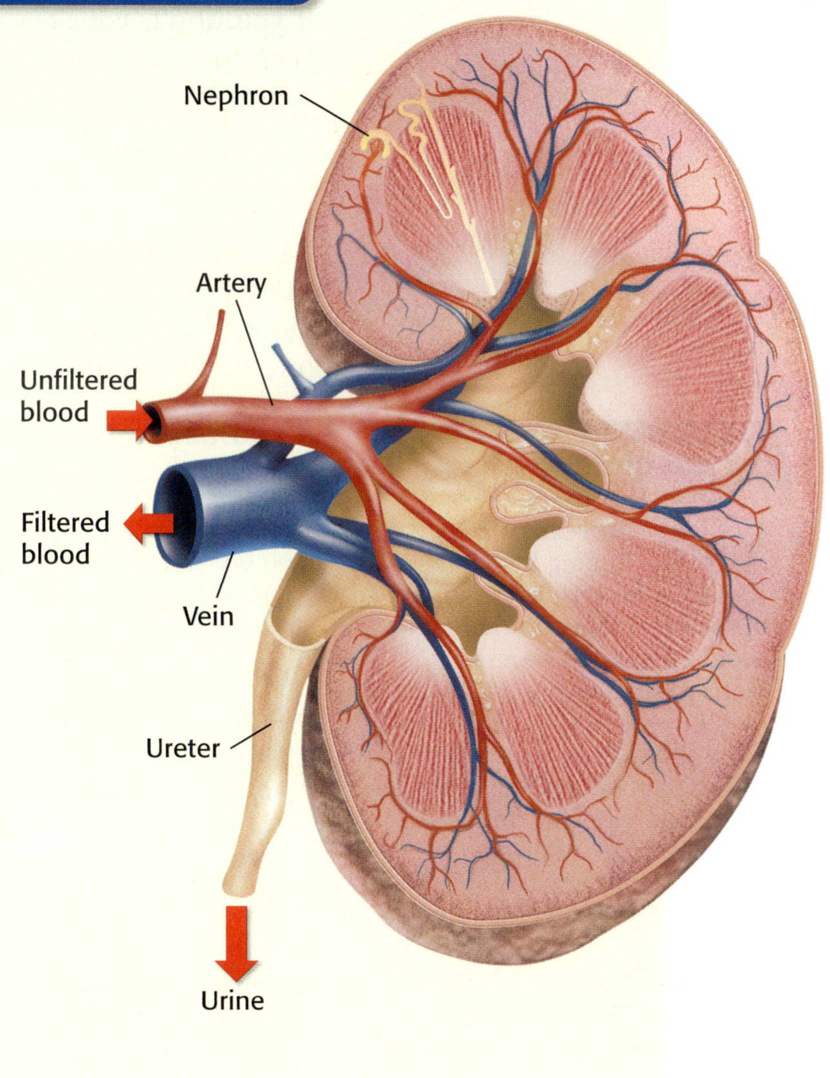

Figure 3 *Drinking water when you exercise helps replace the water you lose when you sweat.*

Water In, Water Out

You drink water every day. You lose water every day in sweat and urine. You need to get rid of as much water as you drink. If you don't, your body will swell up. So, how does your body keep the water levels in balance? The balance of fluids is controlled by chemical messengers in the body called *hormones*.

Sweat and Thirst

When you are too warm, as the boy in **Figure 3** is, you lose a lot of water in the form of sweat. The evaporation of water from your skin cools you down. As the water content of the blood drops, the salivary glands produce less saliva. This is one of the reasons you feel thirsty.

Antidiuretic Hormone

When you get thirsty, other parts of your body react to the water shortage, too. A hormone called *antidiuretic hormone* (AN tee DIE yoo RET ik HAWR MOHN), or ADH, is released. ADH signals the kidneys to take water from the nephrons. The nephrons return the water to the bloodstream. Thus, the kidneys make less urine. When your blood has too much water, small amounts of ADH are released. The kidneys react by allowing more water to stay in the nephrons and leave the body as urine.

Diuretics

Some beverages contain caffeine, which is a *diuretic* (DIE yoo RET ik). Diuretics cause the kidneys to make more urine, which decreases the amount of water in the blood. When you drink a beverage that contains water and caffeine, the caffeine increases fluid loss. So, your body gets to use less of the water from the caffeinated beverage than from a glass of water.

Reading Check What are diuretics?

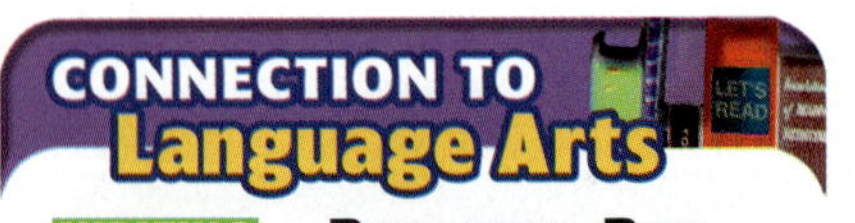

CONNECTION TO Language Arts

WRITING SKILL **Beverage Ban** During football season, a football coach insists that all members of the team avoid caffeinated beverages. Many of the players are upset by the news. Pretend that you are the coach. Write a letter to the members of the team explaining why it is better for them to drink water than to drink beverages that contain caffeine. Read the letter aloud to members of your family. Ask them how you could make your letter more convincing.

Urinary System Problems

The urinary system regulates body fluids and removes wastes from the blood. Any problems with water regulation can become dangerous for your body. Some common urinary system problems are described below.

- **Bacterial Infections** Bacteria can get into the bladder and ureters through the urethra and cause painful infections. Infections should be treated early, before they spread to the kidneys. Infections in the kidneys can permanently damage the nephrons.
- **Kidney Stones** Sometimes, salts and other wastes collect inside the kidneys and form kidney stones like the one in **Figure 4.** Some kidney stones interfere with urine flow and cause pain. Most kidney stones pass naturally from the body, but sometimes they must be removed by a doctor.
- **Kidney Disease** Damage to nephrons can prevent normal kidney functioning and can lead to kidney disease. If a person's kidneys do not function properly, a kidney machine can be used to filter waste from the blood.

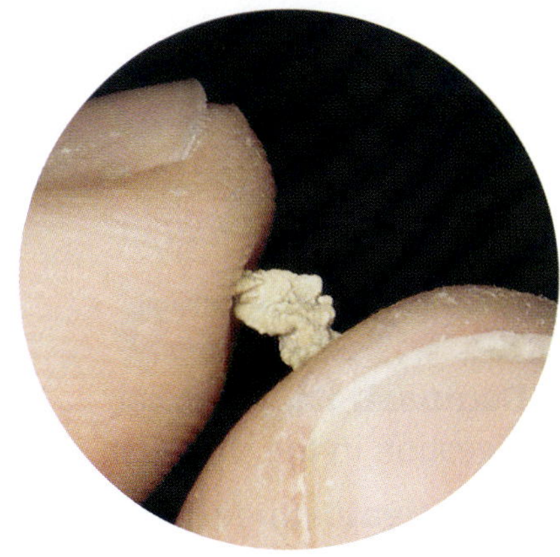

Figure 4 *This kidney stone had to be removed from a patient's urinary system.*

SECTION Review

Summary

- The urinary system removes liquid waste as urine. The filtering structures in the kidney are called *nephrons.*
- Most of the water in the blood is returned to the bloodstream. Urine passes through the ureter, into the bladder, and out of the body through the urethra.
- Disorders of the urinary system include infections, kidney stones, and kidney disease.

Using Key Terms

1. In your own words, write a definition for the term *urinary system.*

Understanding Key Ideas

2. Which event happens first?
   a. Water is absorbed into blood.
   b. A large artery brings blood into the kidney.
   c. Water enters the nephrons.
   d. The nephron separates water from wastes.
3. How do kidneys filter blood?
4. Describe three disorders of the urinary system.

Math Skills

5. A study has shown that 75% of teenage boys drink 34 oz of soda per day. How many 12 oz cans of soda would a boy drink in a week if he drank 34 oz per day?

Critical Thinking

6. **Applying Concepts** Which of the following contains more water: the blood going into the kidney or the blood leaving it?
7. **Predicting Consequences** When people have one kidney removed, their other kidney can often keep their blood clean. But the remaining kidney often changes. Predict how the remaining kidney may change to do the work of two kidneys.

Using Scientific Methods

Skills Practice Lab

As the Stomach Churns

The stomach, as you know, performs not only mechanical digestion but also chemical digestion. As the stomach churns, which moves the food particles around, the digestive fluids—acid and enzymes—are added to begin protein digestion.

Commercially prepared meat tenderizers contain enzymes from plants that break down, or digest, proteins. Two types of meat tenderizer are commonly available at grocery stores. One type of tenderizer contains an enzyme called *papain,* from papaya. Another type of tenderizer contains an enzyme called *bromelain,* from pineapple. In this lab, you will test the effects of these two types of meat tenderizers on beef stew meat.

OBJECTIVES

Demonstrate chemical digestion in the stomach.

Investigate three forms of chemical digestion.

MATERIALS

- beef stew meat, 1 cm cubes (3)
- eyedropper
- gloves, protective
- graduated cylinder, 25 mL
- hydrochloric acid, very dilute, 0.1 M
- measuring spoon, 1/4 tsp
- meat tenderizer, commercially prepared, containing bromelain
- meat tenderizer, commercially prepared, containing papain
- tape, masking
- test tubes (4)
- test-tube marker
- test-tube rack
- water

SAFETY

Ask a Question

1. Determine which question you will answer through your experiment. That question may be one of the following: Which meat tenderizer will work faster? Which one will make the meat more tender? Will the meat tenderizers change the color of the meat or water? What might these color changes, if any, indicate?

Form a Hypothesis

2. Form a hypothesis from the question you formed in step 1. **Caution:** Do not taste any of the materials in this activity.

Test the Hypothesis

3. Identify all variables and controls present in your experiment. In your notebook, make a data table that includes these variables and controls. Use this data table to record your observations and results.

4. Label one test tube with the name of one tenderizer, and label the other test tube with the name of the other tenderizer. Label the third test tube "Control." What will the test tube labeled "Control" contain?

5. Pour 20 mL of water into each test tube.
6. Use the eyedropper to add four drops of very dilute hydrochloric acid to each test tube. **Caution:** Hydrochloric acid can burn your skin. If any acid touches your skin, rinse the area with running water and tell your teacher immediately.
7. Use the measuring spoon to add 1/4 tsp of each meat tenderizer to its corresponding test tube.
8. Add one cube of beef to each test tube.
9. Record your observations for each test tube immediately, after 5 min, after 15 min, after 30 min, and after 24 h.

Analyze the Results

1. **Describing Events** Did you immediately notice any differences in the beef in the three test tubes? At what time interval did you notice a significant difference in the appearance of the beef in the test tubes? Explain the differences.
2. **Examining Data** Did one meat tenderizer perform better than the other? Explain how you determined which performed better.

Draw Conclusions

3. **Evaluating Results** Was your hypothesis supported? Explain your answer.
4. **Applying Conclusions** Many animals that sting have venom composed of proteins. Explain how applying meat tenderizer to the wound helps relieve the pain of such a sting.

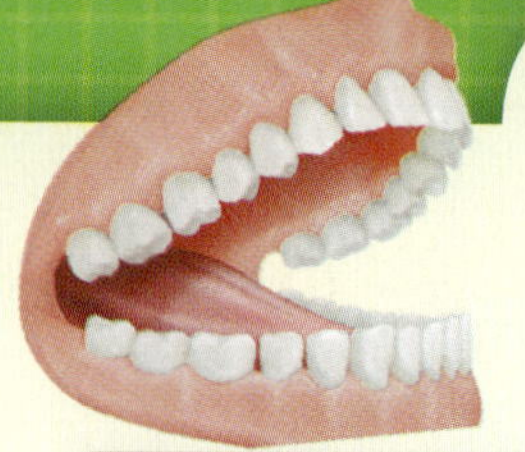

Chapter Review

USING KEY TERMS

Complete each of the following sentences by choosing the correct term from the word bank.

pancreas	digestive system
large intestine	stomach
kidney	small intestine
nephron	urinary system

1. The ____ secretes juices into the small intestine.
2. The saclike organ at the end of the esophagus is called the ____.
3. The ___ is an organ that contains millions of nephrons.
4. A group of organs that removes waste from the blood and excretes it from the body is called the ____.
5. The ____ is a group of organs that work together to break down food.
6. Indigestible material is formed into feces in the ____.

UNDERSTANDING KEY IDEAS

Multiple Choice

7. The hormone that signals the kidneys to make less urine is
   a. urea.
   b. caffeine.
   c. ADH.
   d. ATP.
8. Which of the following organs aids digestion by producing bile?
   a. stomach
   b. pancreas
   c. small intestine
   d. liver
9. The part of the kidney that filters the blood is the
   a. artery.
   b. ureter.
   c. nephron.
   d. urethra.
10. The fingerlike projections that line the small intestine are called
    a. emulsifiers.
    b. fats.
    c. amino acids.
    d. villi.
11. Which of the following is NOT part of the digestive tract?
    a. mouth
    b. kidney
    c. stomach
    d. rectum
12. The soupy mixture of food, enzymes, and acids in the stomach is called
    a. chyme.
    b. villi.
    c. urea.
    d. vitamins.
13. The stomach helps with
    a. storing food.
    b. chemical digestion.
    c. physical digestion.
    d. All of the above
14. The gall bladder stores
    a. food.
    b. urine.
    c. bile.
    d. villi.
15. The esophagus connects the
    a. pharynx to the stomach.
    b. stomach to the small intestine.
    c. kidneys to the nephrons.
    d. stomach to the large intestine.

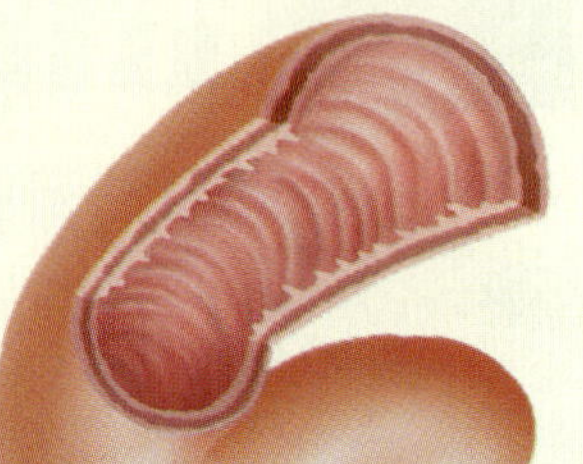

Short Answer

16 Why is it important for the pancreas to release bicarbonate into the small intestine?

17 How does the structure of the small intestine help the small intestine absorb nutrients?

18 What is a kidney stone?

CRITICAL THINKING

19 **Concept Mapping** Use the following terms to create a concept map: *teeth, stomach, digestion, bile, saliva, mechanical digestion, gallbladder,* and *chemical digestion.*

20 **Predicting Consequences** How would digestion be affected if the liver were damaged?

21 **Analyzing Processes** When you put a piece of carbohydrate-rich food, such as bread, a potato, or a cracker, into your mouth, the food tastes bland. But if this food sits on your tongue for a while, the food will begin to taste sweet. What digestive process causes this change in taste?

22 **Making Comparisons** The recycling process for one kind of plastic begins with breaking the plastic into small pieces. Next, chemicals are used to break the small pieces of plastic down to its building blocks. Then, those building blocks are used to make new plastic. How is this process both like and unlike human digestion?

INTERPRETING GRAPHICS

The bar graph below shows how long the average meal spends in each portion of your digestive tract. Use the graph below to answer the questions that follow.

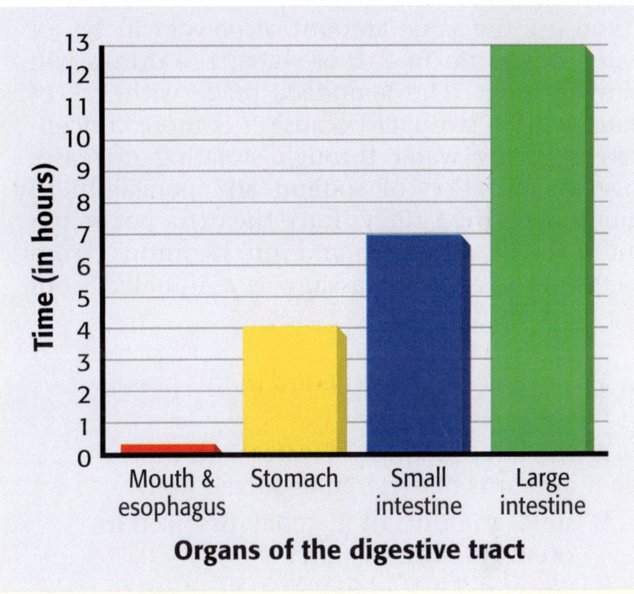

23 In which part of your digestive tract does the food spend the longest amount of time?

24 On average, how much longer does food stay in the small intestine than in the stomach?

25 Which organ mixes food with special substances to make chyme? Approximately how long does food remain in this organ?

26 Bile breaks large fat droplets into very small droplets. How long is the food in your body before it comes into contact with bile?

Standardized Test Preparation

READING

Read the passage below. Then, read each question that follows the passage. Decide which is the best answer to each question.

Passage 1 When you lose water, your blood becomes more concentrated. Think about how you make a powdered drink, such as lemonade. If you use the same amount of powder in 1 L of water as you do in 2 L of water, the drinks will taste different. The lemonade made with 1 L of water will be stronger because it is more concentrated. Losing water through sweating increases the concentration of sodium and potassium in your blood. The kidneys force the extra potassium out of the blood stream and into nephrons. From the nephrons, the potassium is eliminated from the body in urine.

1. The words *more concentrated* in this passage refer to

A the same amount of water with different amounts of material dissolved in it.

B small amounts of material dissolved in small amounts of water.

C large amounts of material dissolved in large amounts of water.

D a given amount of material dissolved in a smaller amount of water.

2. Which of the following statements is a fact from the passage?

F Blood contains both potassium and sodium.

G Losing too much sodium is dangerous.

H Potassium and sodium can be replaced by drinking an exercise drink.

I Tears contain sodium.

Passage 2 Three major types of nutrients—carbohydrates, proteins, and fats—make up most of the food you eat. Chemical substances called *enzymes* break these nutrients into smaller particles for the body to use. For example, proteins, which are chains of smaller molecules called *amino acids,* are too large to be absorbed into the bloodstream. So, enzymes cut the chain of amino acids. These amino acids are small enough to pass into the bloodstream to be used by the body.

1. According to the passage, what is a carbohydrate?

A an enzyme

B a substance made of amino acids

C a nutrient

D the only substance in a healthy diet

2. Which of the following statements is a fact from the passage?

F Carbohydrates, fats, and proteins are three major types of nutrients.

G Proteins are made of fats and carbohydrates.

H Some enzymes create chains of proteins.

I Fats are difficult to digest.

3. Which of the following can be inferred from the passage?

A To be useful to the body, nutrients must be small enough to enter the bloodstream.

B Carbohydrates are made of amino acids.

C Amino acids are made of proteins.

D Without enough protein, the body cannot grow.

INTERPRETING GRAPHICS

Use the figure below to answer the questions that follow.

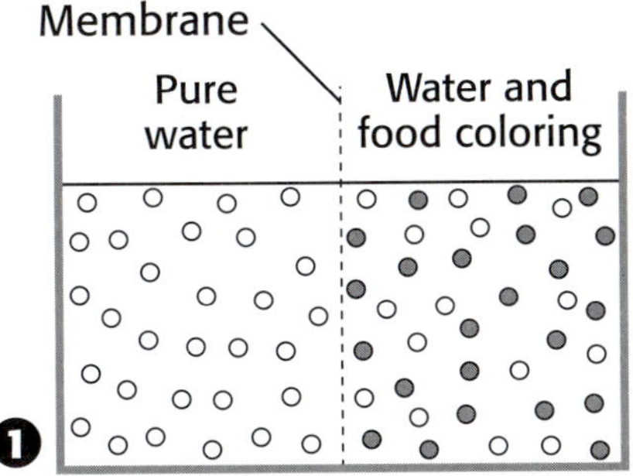

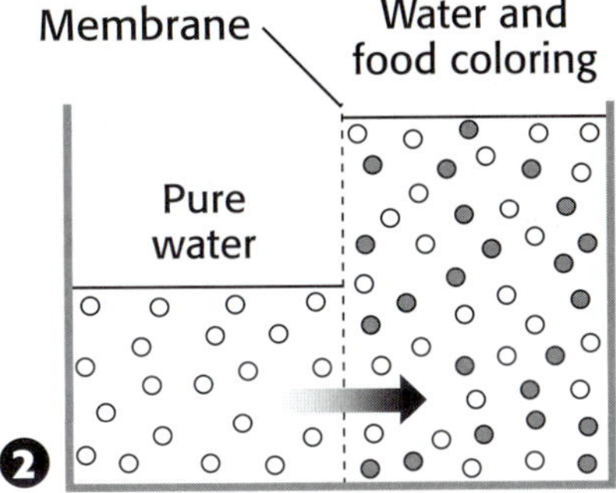

1. The container is divided by a membrane. What can you conclude from the diagram?

A Water molecules can pass through the membrane.

B Food-coloring molecules can pass through the membrane.

C Both water molecules and food-coloring molecules can pass through the membrane.

D Neither water molecules nor food-coloring molecules can pass through the membrane.

2. If the membrane has holes that separate molecules by size,

F food-coloring molecules are larger than water molecules.

G water molecules are larger than food-coloring molecules.

H water molecules and food-coloring molecules are the same size.

I the holes are smaller than both water molecules and food-coloring molecules.

3. The concentration of food-coloring molecules in the columns labeled "Water and food coloring"

A is greater in 2 than in 1.

B is greater in 1 than in 2.

C is the same in 1 and 2.

D cannot change.

MATH

Read each question below, and choose the best answer.

1. Cora is 1.5 m tall. Cora's small intestine is 6 m long. How many times longer is Cora's small intestine than her height?

A 3 times longer

B 4 times longer

C 5 times longer

D 6 times longer

2. During a water-balance study that was performed for one day, a woman drank 1,500 mL of water. The food she ate contained 750 mL of water, and her body produced 250 mL of water internally during normal body processes. She lost 900 mL of water in sweat, 1,500 mL in urine, and 100 mL in feces. Overall, how much water did she gain or lose during the day?

F She gained 1,500 mL of water.

G She lost 900 mL of water.

H She gained as much water as she lost.

I She lost twice as much water as she gained.

3. There are 6 blue marbles, 2 red marbles, and 4 green marbles in a bag. If someone selects 1 marble at random from the bag, what is the probability that the marble will be blue?

A 1/5

B 1/4

C 1/3

D 1/2

Science in Action

Weird Science

Tapeworms

What if you found out that you had a constant mealtime companion who didn't want just a bite but wanted it all? And what if that companion never asked for your permission? This mealtime companion might be a tapeworm. Tapeworms are invertebrate flatworms. These flatworms are parasites. A parasite is an organism that obtains its food by living in or on another organism. A tapeworm doesn't have a digestive tract of its own. Instead, a tapeworm absorbs the nutrients digested by the host. Some tape worms can grow to be over 10 m long. Cooking beef, pork, and fish properly can help prevent people from getting tapeworms. People or animals who get tapeworms can be treated with medicines.

Social Studies

WRITING SKILL The World Health Organization and the Pan American Health Organization have made fighting intestinal parasites in children a high priority. Conduct library or Internet research on Worm Busters, which is a program for fighting parasites. Write a brief report of your findings.

Science, Technology, and Society

Pill Cameras

Open wide and say "Ahhhh." When you have a problem with your mouth or teeth, doctors can examine you pretty easily. But when people have problems that are further down their digestive tract, examination becomes more difficult. So, some doctors have recently created a tiny, disposable camera that patients can swallow. As the camera travels down the digestive tract, the camera takes pictures and sends them to a tiny recorder that patients wear on their belt. The camera takes about 57,000 images during its trip. Later, doctors can review the pictures and see the pictures of the patient's entire digestive tract.

Math Activity

If a pill camera takes 57,000 images while it travels through the digestive system and takes about two pictures per second, how many hours is the camera in the body?

Careers

Christy Krames

Medical Illustrator Christy Krames is a medical illustrator. For 19 years, she has created detailed illustrations of the inner workings of the human body. Medical illustrations allow doctors and surgeons to share concepts, theories, and techniques with colleagues and allow students to learn about the human body.

Medical illustrators often draw tiny structures or body processes that would be difficult or impossible to photograph. For example, a photograph of a small intestine can show the entire organ. But a medical illustrator can add to the photograph an enlarged drawing of the tiny villi inside the intestine. Adding details helps to better explain how small parts of organs work together so that the organs can function.

Medical illustration requires knowledge of both art and science. So, Christy Krames studied both art and medicine in college. Often, Krames must do research before she draws a subject. Her research may include reading books, observing surgical procedures, or even dissecting a pig's heart. This research results in accurate and educational drawings of the inner body.

Language Arts ACTIVITY

WRITING SKILL Pretend you are going to publish an atlas of the human body. Write a classified advertisement to hire medical illustrators. Describe the job, and describe the qualities that the best candidates will have. As you write the ad, remember you are trying to persuade the best illustrators to contact you.

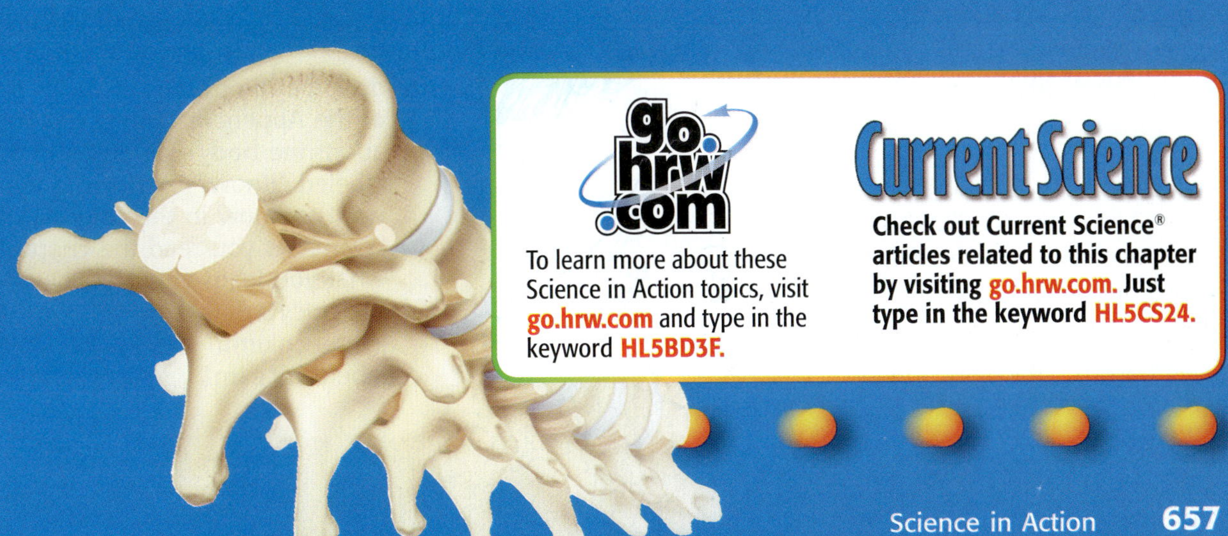

go.hrw.com

To learn more about these Science in Action topics, visit go.hrw.com and type in the keyword HL5BD3F.

Current Science

Check out Current Science® articles related to this chapter by visiting go.hrw.com. Just type in the keyword HL5CS24.

Understanding Diseases

About the PHOTO

Does this photograph look like a bunch of pink roof tiles? It is a photograph of human skin cells. Just as roof tiles protect a house from environmental factors such as bad weather, human skin cells protect our bodies from dangerous environmental factors, including millions of tiny living things that cause disease.

PRE-READING ACTIVITY

FOLDNOTES **Tri-Fold** Before you read the chapter, create the FoldNote entitled "Tri-Fold" described in the **Study Skills** section of the Appendix. Write what you know about the body's defenses in the column labeled "Know." Then, write what you want to know in the column labeled "Want." As you read the chapter, write what you learn about the body's defenses in the column labeled "Learn."

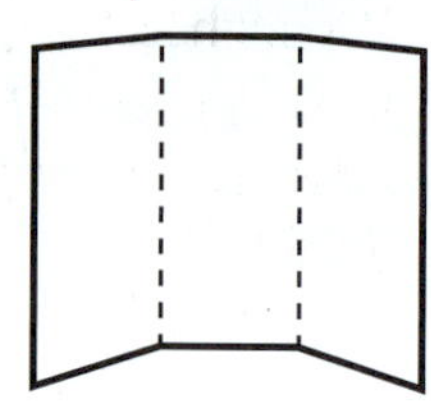

START-UP ACTIVITY

Invisible Invaders

In this activity, you will see tiny organisms grow.

Procedure

1. Obtain **two Petri dishes containing nutrient agar.** Label them "Washed" and "Unwashed."
2. Rub **two marbles** between the palms of your hands. Observe the appearance of the marbles.
3. Roll one marble in the Petri dish labeled "Unwashed."
4. Put on a pair of **disposable gloves.** Wash the other marble with **soap** and **warm water** for 4 min.
5. Roll the washed marble in the Petri dish labeled "Washed."
6. Secure the lids of the Petri dishes with **transparent tape.** Place the dishes in a warm, dark place. Do not open the Petri dishes after they are sealed.
7. Record changes in the Petri dishes for 1 week.

Analysis

1. How did the washed and unwashed marbles compare? How did the Petri dishes differ after several days?
2. Why is it important to wash your hands before eating?

SECTION 1

Disease

You've probably heard it before: "Cover your mouth when you sneeze!" "Wash your hands!" "Don't put that in your mouth!"

What is all the fuss about? When people say these things to you, they are concerned about the spread of disease.

READING WARM-UP

Objectives

- Explain the difference between infectious diseases and noninfectious diseases.
- Identify five ways that you might come into contact with a pathogen.
- Discuss how medical discoveries have changed the way that diseases are treated.

Terms to Learn

noninfectious disease
infectious disease
pathogen

READING STRATEGY

Paired Summarizing Read this section silently. In pairs, take turns summarizing the material. Stop to discuss ideas that seem confusing.

Causes of Disease

When you have a *disease,* your normal body functions are disrupted. Some diseases, such as most cancers and heart disease, are not spread from one person to another. They are called **noninfectious diseases.**

Noninfectious diseases can be caused by a variety of factors. For example, a genetic disorder causes the disease hemophilia (HEE moh FIL ee uh), in which a person's blood does not clot properly. Smoking, lack of physical activity, and a high-fat diet can greatly increase a person's chances of getting certain noninfectious diseases. Avoiding harmful habits may help you avoid noninfectious diseases.

A disease that can be passed from one living thing to another is an **infectious disease.** Infectious diseases are caused by agents called **pathogens.** Viruses and some bacteria, fungi, protists, and worms may all cause diseases. **Figure 1** shows some enlarged images of common pathogens.

noninfectious disease a disease that cannot spread from one individual to another

infectious disease a disease that is caused by a pathogen and that can be spread from one individual to another

pathogen a virus, microorganism, or other organism that causes disease

Figure 1 Pathogens

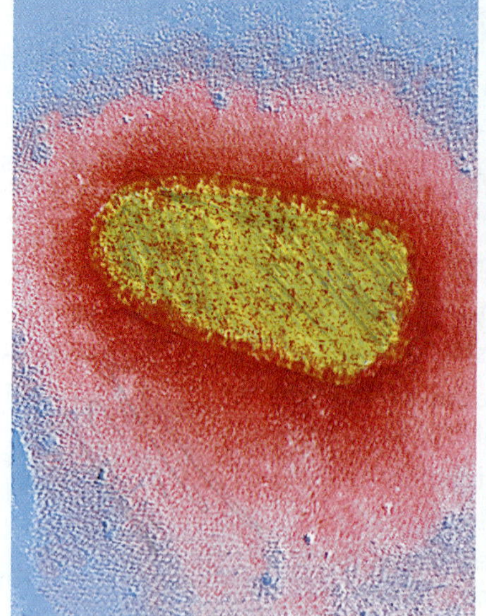

This virus causes rabies.

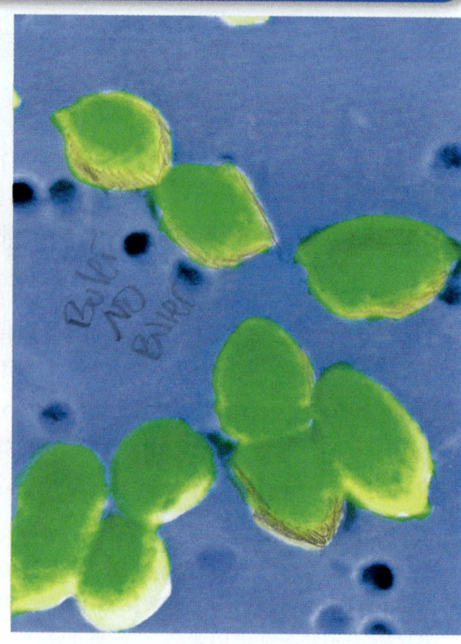

Streptococcus bacteria can cause strep throat.

Pathways to Pathogens

There are many ways pathogens can be passed from one person to another. Being aware of them can help you stay healthy.

Air

Some pathogens travel through the air. For example, a single sneeze, such as the one shown in **Figure 2,** releases thousands of tiny droplets of moisture that can carry pathogens.

Figure 2 *A sneeze can force thousands of pathogen-carrying droplets out of your body at up to 160 km/h.*

Contaminated Objects

You may already know that if you drink from a glass that an infected person has just used, you could become infected with a pathogen. A person who is sick may leave bacteria or viruses on many other objects, too. For example, contaminated doorknobs, keyboards, combs, and towels can pass pathogens.

Person to Person

Some pathogens are spread by direct person-to-person contact. You can become infected with some illnesses by kissing, shaking hands, or touching the sores of an infected person.

Animals

Some pathogens are carried by animals. For example, humans can get a fungus called *ringworm* from handling an infected dog or cat. Also, ticks may carry bacteria that cause Lyme disease or Rocky Mountain spotted fever.

Food and Water

Drinking water in the United States is generally safe. But water lines can break, or treatment plants can become flooded. These problems may allow microorganisms to enter the public water supply. Bacteria growing in foods and beverages can cause illness, too. For example, meat, fish, and eggs that are not cooked enough can still contain dangerous bacteria or parasites. Even leaving food out at room temperature can give bacteria such as salmonella the chance to grow and produce toxins in the food. Refrigerating foods can slow the growth of many of these pathogens. Because bacteria grow in food, washing all used cooking surfaces and tools is also important.

Reading Check **Why must you cook meat and eggs thoroughly?** *(See the Appendix for answers to Reading Checks.)*

CONNECTION TO Social Studies

Disease and History Many diseases have shaped history. For example, yellow fever, which is caused by a virus that is spread by mosquitoes, was one of the obstacles in building the Panama Canal. Only after people learned how to prevent the spread of the yellow fever virus could the canal be completed.

Use information from Internet and library research to create a poster describing how one infectious disease affected history.

ACTIVITY

Putting Pathogens in Their Place

Until the 20th century, patients often died of bacterial infections. People did not know that bacteria and other germs caused diseases. For this reason, attempts at treating diseases were not very successful. But doctors eventually learned that simple cleanliness could help prevent the spread of many diseases. Today, hospitals and clinics use a variety of technologies to prevent the spread of pathogens.

Medicine

With your family, look at the safety information on some over-the-counter medicines that your family uses. Together, come up with a list of safety rules for using medicines wisely. Make a poster about using medicines safely, and share it with the class.

Early Attempts

Throughout history, people have tried to explain what caused disease. Many people thought that diseases were caused by spiritual forces and needed spiritual cures. For centuries, many people believed that an imbalance of body fluids caused disease. In these cases, doctors treated disease by making patients vomit or bleed. After people understood that germs can cause disease, doctors could develop more-effective treatments.

Medical Discoveries

Scientists are constantly discovering new ways to fight diseases. Antibiotics are one discovery that has had a major impact on fighting pathogens. An *antibiotic* is a substance that can kill or slow the growth of microorganisms, such as bacteria. **Figure 3** shows a fungus that makes one common antibiotic. Antibiotics may also be used to treat infections caused by small, parasitic fungi and protists. However, viruses, such as those that cause colds, are not affected by antibiotics. Antibiotics can kill only living things, and viruses are not alive. The only way to destroy viruses in your body is to locate and kill the cells that they have invaded.

Recent medical discoveries allow people to treat diseases that were untreatable even just a few years ago. Success at constantly finding new ways to treat diseases gives people hope that future discoveries will lead to cures for diseases that are currently not well understood.

Figure 3 *Some* penicillium *fungi make penicillin, which is used as an antibiotic.*

Using Medicine Wisely

Because medicines have such powerful effects on the body, people who use medicines must be very careful. Following a doctor's instructions for taking medicines is important. Following these instructions will keep you from taking too much or too little of a medicine. Taking too much medicine could make you sick. Taking too little medicine could keep the medicine from working properly. Always follow instructions when taking medicines.

Using medicines responsibly includes understanding how they work. Taking medicines carelessly can encourage pathogens to fight back. When an antibiotic kills the bacteria that cause an infection, a few bacteria may survive. These bacteria may have a natural resistance to the drug. The resistant bacteria may multiply quickly, and the antibiotic wouldn't be able to stop them. To make sure that antibiotics do not become ineffective, people should use antibiotics only when prescribed by a doctor.

Reading Check When should people use antibiotics?

Epidemic!

You catch a cold and return to your school while sick. Your friends don't have immunity to your cold. On the first day, you expose five friends to your cold. The next day, each of those friends passes the virus to five more people. If this pattern continues for 5 more days, how many people will be exposed to the virus?

SECTION Review

Summary

- Noninfectious diseases cannot be spread between people. Infectious diseases are caused by pathogens that are passed between living things.
- Pathogens can travel through the air or can be spread by contact with other people, contaminated objects, animals, food, or water.
- Researchers have developed medicines to fight pathogens. People must use these medicines carefully.

Using Key Terms

1. In your own words, write a definition for each of the following terms: *infectious disease* and *noninfectious disease.*

Understanding Key Ideas

2. Which of the following is an infectious disease?
   a. cancer
   b. bacterial infection
   c. heart disease
   d. hemophilia
3. How can you be sure that you are using medicine safely?
4. Name five ways that pathogens spread.

Math Skills

5. If 10 people who have a virus each expose 25 more people to the virus, how many people will be exposed to the virus?

Critical Thinking

6. **Applying Concepts** Many foods have expiration dates that state when you should no longer eat the foods. Why do you think foods are not safe to eat after the expiration date?
7. **Analyzing Methods** Explain what may happen if a doctor does not wear gloves when treating patients. Explain what may happen if a doctor does not put on fresh gloves for each patient.

SECTION 2

The Germ Theory

How often do you brush you teeth, wash your hands, or put something in the refrigerator? You may not realize that all of these actions protect you from germs.

READING WARM-UP

Objectives

- Describe the germ theory.
- Explain how pasteurization prevents spoilage.
- Explain how vaccination works.
- Describe how the germ theory has affected modern health practices.

Terms to Learn

pasteurization
vaccine
immunity

READING STRATEGY

Discussion Read this section silently. Write down questions that you have about this section. Discuss your questions in a small group.

Brushing your teeth and washing your hands remove germs from your body. How did people realize that getting rid of these tiny organisms could protect their health? In the 1800s, a French scientist named Louis Pasteur did experiments that led people to accept the germ theory. The *germ theory* states that microorganisms can cause disease.

Louis Pasteur

Louis Pasteur's experiments have led to improvements in the lives of people around the world. Pasteur, shown in the painting in **Figure 1,** conducted many experiments that showed how microorganisms affected food, drinks, animals, and people. *Microorganisms* are tiny living things, such as bacteria, small fungi, and protists. Pasteur showed that these living things grow in our food, our drinks, and our bodies. Most importantly, his work led people to understand that some of these living things can cause disease.

After scientists understood the cause of many diseases, they began researching ways to prevent and fight disease. Since then, medical science has grown very powerful. Billions of lives have been saved or improved by medical discoveries.

Figure 1 *Louis Pasteur studied microorganisms, including how they affect people.*

Pasteurization

Pasteur did not just show how microorganisms affect people. He developed ways to prevent the damage that they cause. One method of controlling these organisms is named after Pasteur. This method is called *pasteurization* (PAS tuhr i ZAY shuhn). **Pasteurization** is the process of using heat to kill microorganisms in foods.

Figure 2 *These cows have foot and mouth disease, which is caused by a virus.*

Spoiled Milk

Pasteur's work on pasteurization began with a question about what causes milk and wine to spoil. Pasteur showed that microorganisms enter products from the air, multiply quickly, and make waste products. These wastes are sometimes desirable, such as when yeast waste makes bread rise. But often the wastes cause products to spoil. For example, acetic acid spoils wine. Pasteur found that heating a liquid to a temperature that would kill the microorganisms could keep the liquid from spoiling.

After realizing that microorganisms from the air affected milk and wine, Pasteur looked for their effects in other places. He found that microorganisms played a role in many animal diseases. This discovery led other scientists to experiment further. They found that specific kinds of organisms cause specific diseases, such as the foot and mouth disease affecting the cows in **Figure 2.**

pasteurization the process of using heat to kill microorganisms in foods by heating the food to a certain temperature for a specified period of time

Healthy Groceries

Pasteurization is still used today. This practice protects dairy products and many other foods from spoiling. Because of pasteurization, you can buy food or drinks at a grocery store and trust that you will not get sick when you eat or drink them. The milk that the girl in **Figure 3** is drinking has been pasteurized.

Reading Check **How does pasteurization keep milk and wine from spoiling?** (*See the Appendix for answers to Reading Checks.*)

Figure 3 *Today, pasteurization is used to kill pathogens in many types of food, including dairy products, shellfish, and juices.*

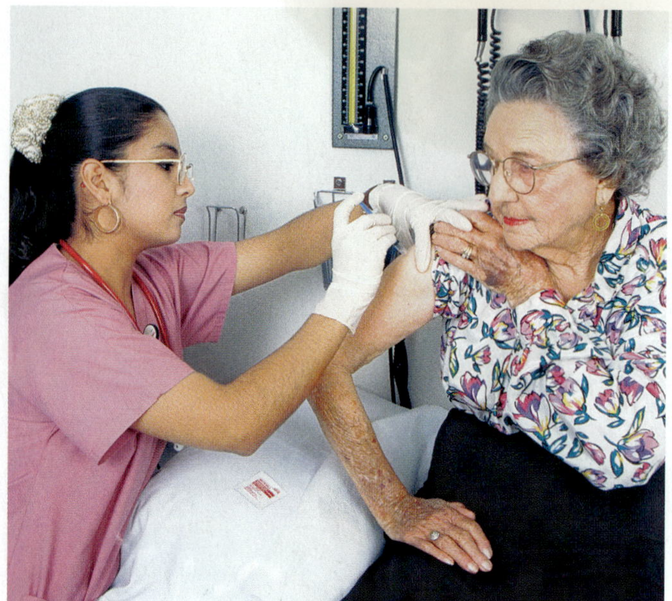

Figure 4 *Today, vaccinations are commonly used to protect people from viral diseases.*

Vaccines

Pasteur's work also supported Edward Jenner's work on vaccinations. In the late 1700s, Jenner studied a disease called *smallpox*. He observed that people who had been infected with cowpox seemed to have protection against smallpox. Jenner's work led to the first modern vaccine. A **vaccine** is a substance that helps the body develop resistance to a disease. Jenner vaccinated people with cowpox to protect them from smallpox.

Jenner's work was not widely accepted in the 1700s. At that time, people did not know what a pathogen was. But Pasteur's work on microorganisms helped explain why vaccinations worked. Now, the vaccination process, shown in **Figure 4,** is used regularly.

vaccine a substance prepared from killed or weakened pathogens or genetic material and that is introduced into a body to produce immunity

immunity the ability to resist an infectious disease

Building Immunity

Pasteur realized that vaccines work because the body learns to recognize a harmless version of a pathogen. If the pathogen enters the body again, the body recognizes and fights it quickly. The ability to resist an infectious disease is called **immunity.** People can build immunity to a disease without getting sick.

Defeating Diseases

Today, vaccines are used to fight many serious diseases. Vaccines have even controlled smallpox, as shown in **Figure 5.** Modern vaccines contain pathogens that are killed or treated so that they can't make you very sick. The vaccine is enough like the pathogen to allow your body to build immunity.

Figure 5 *A large vaccination campaign in schools in the late 1940s helped control the smallpox virus.*

Controlling Germs

The acceptance of the germ theory changed public health practices dramatically. Many healthy behaviors that seem obvious now were not understood before the germ theory. The importance of sanitation practices, such as cleaning our bodies and our environments regularly, was not well understood before Pasteur's time. The government has created inspection regulations to ensure food and water safety, as shown in **Figure 6.** Quarantine policies stop serious diseases from spreading to large populations by keeping people thought to be infected separated from uninfected people. Aseptic surgical techniques prevent microorganisms from entering the body during surgery.

Modern health practices have led to a great ability to control germs and stop disease. But germs are always changing, and many germs are still major threats. Scientists today are still researching better ways to use chemicals and processes to destroy the microorganisms that cause disease.

✓ Reading Check **What are four modern health practices that developed as a result of the germ theory?**

Figure 6 *Federal regulations require that food inspectors check that food will not spread disease to consumers.*

SECTION Review

Summary

- The germ theory states that microorganisms can cause disease.
- Pasteurization is the process of using heat to kill microorganisms that cause liquids to spoil.
- Vaccinations protect people from disease by helping the body build immunity.
- Sanitation, food regulations, quarantine, and aseptic technologies are a few ways that the germ theory has affected public health practices.

Using Key Terms

1. In your own words, write a definition for each of the following terms: *pasteurization, vaccine,* and *immunity.*

Understanding Key Ideas

2. What do vaccines contain?

a. treated pathogens
b. heat
c. antibiotics
d. pasteurization

3. What is the germ theory?

4. How does pasteurization work?

5. How does vaccination work?

Math Skills

6. If vaccination campaign workers give 250 vaccinations each day for a week in each of 20 locations, how many people will be vaccinated during the week?

Critical Thinking

7. Identifying Relationships Why might the risk of infectious disease be high in a community that has no water treatment facility?

8. Applying Concepts Use your knowledge of the germ theory to write a paragraph explaining to a younger person why taking baths or showers and keeping the kitchen and other rooms clean are important.

SECTION 3

Your Body's Defenses

Bacteria and viruses can be in the air, in the water, and on all the surfaces around you.

READING WARM-UP

Objectives

- Describe how your body keeps out pathogens.
- Explain how the immune system fights infections.
- Describe four challenges to the immune system.

Terms to Learn

immune system
macrophage
T cell
B cell
antibody
memory B cell
allergy
autoimmune disease
cancer

READING STRATEGY

Reading Organizer As you read this section, make a flowchart of the steps of how your body responds to a virus.

Your body must constantly protect itself against pathogens that are trying to invade it. But how does your body do that? Luckily, your body has its own built-in defense system.

First Lines of Defense

For a pathogen to harm you, it must attack a part of your body. Usually, though, very few of the pathogens around you make it past your first lines of defense.

Many organisms that try to enter your eyes or mouth are destroyed by special enzymes. Pathogens that enter your nose are washed down the back of your throat by mucus. The mucus carries the pathogens to your stomach, where most are quickly digested.

Your skin is made of many layers of flat cells. The outermost layers are dead. As a result, many pathogens that land on your skin have difficulty finding a live cell to infect. As **Figure 1** shows, the dead skin cells are constantly dropping off your body as new skin cells grow from beneath. As the dead skin cells flake off, they carry away viruses, bacteria, and other microorganisms. In addition, glands secrete oil onto your skin's surface. The oil contains chemicals that kill many pathogens.

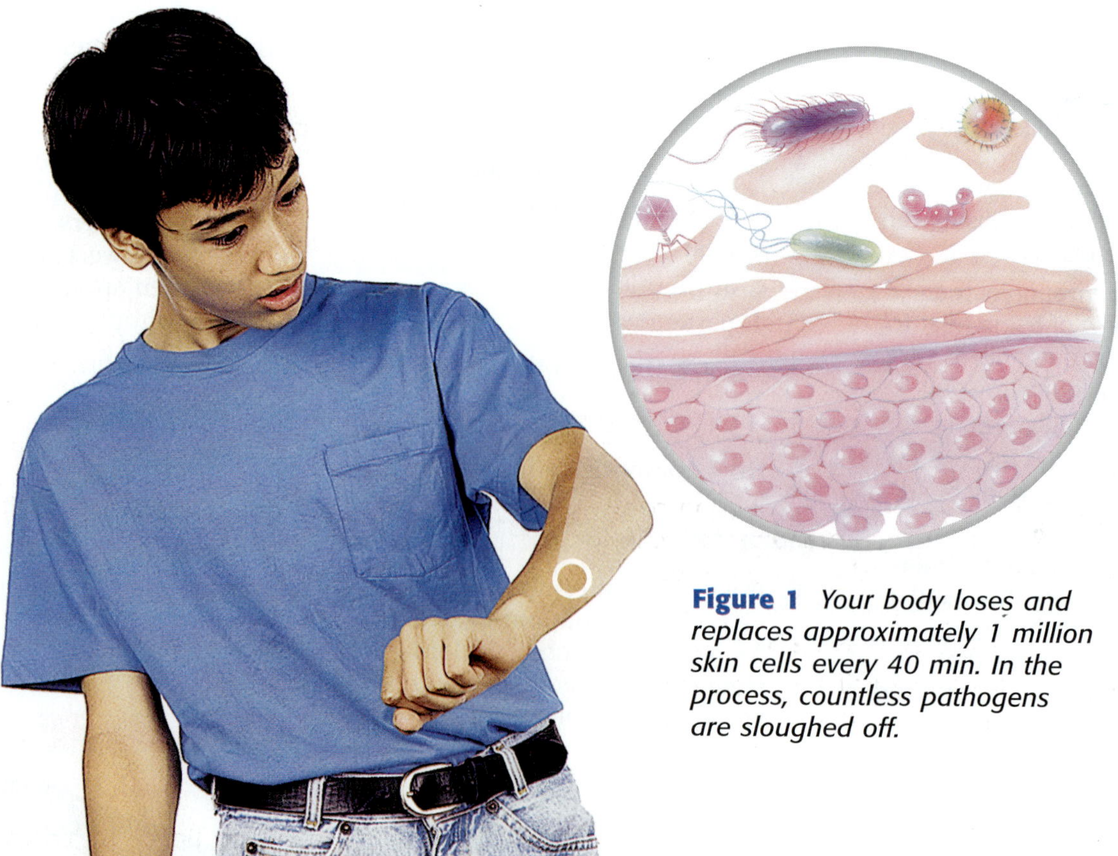

Figure 1 *Your body loses and replaces approximately 1 million skin cells every 40 min. In the process, countless pathogens are sloughed off.*

Failure of First Lines

Sometimes, skin is cut or punctured and pathogens can enter the body. The body acts quickly to keep out as many pathogens as possible. Blood flow to the injured area increases. Cell parts in the blood called *platelets* help seal the open wound so that no more pathogens can enter.

The increased blood flow also brings cells that belong to the **immune system,** the body system that fights pathogens. The immune system is not localized in any one place in your body. It is not controlled by any one organ, such as the brain. Instead, it is a team of individual cells, tissues, and organs that work together to keep you safe from invading pathogens.

immune system the cells and tissues that recognize and attack foreign substances in the body

macrophage an immune system cell that engulfs pathogens and other materials

T cell an immune system cell that coordinates the immune system and attacks many infected cells

B cell a white blood cell that makes antibodies

antibody a protein made by B cells that binds to a specific antigen

Cells of the Immune System

The immune system consists mainly of three kinds of cells. One kind is the macrophage (MAK roh FAYJ). **Macrophages** engulf and digest many microorganisms or viruses that enter your body. If only a few microorganisms or viruses have entered a wound, the macrophages can easily stop them.

The other two main kinds of immune system cells are T cells and B cells. **T cells** coordinate the immune system and attack many infected cells. **B cells** are immune system cells that make antibodies. **Antibodies** are proteins that attach to specific antigens. *Antigens* are substances that stimulate an immune response. Your body is capable of making billions of different antibodies. Each antibody usually attaches to only one kind of antigen, as illustrated in **Figure 2.**

Reading Check How do macrophages help fight disease? (*See the Appendix for answers to Reading Checks.*)

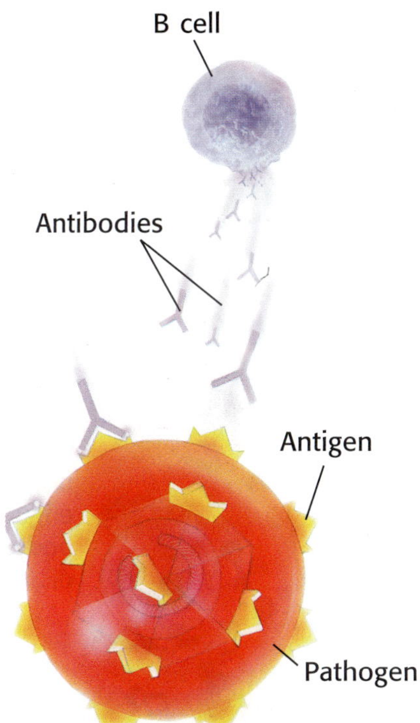

Figure 2 *An antibody's shape is very specialized. It matches an antigen like a key fits a lock.*

Only Skin Deep

1. Cut an **apple** in half.
2. Place **plastic wrap** over each half. The plastic wrap will act as skin.
3. Use **scissors** to cut the plastic wrap on one of the apple halves, and then use an **eyedropper** to drip **food coloring** on each apple half. The food coloring represents pathogens coming into contact with your body.
4. What happened to each apple half?
5. How is the plastic wrap similar to skin?
6. How is the plastic wrap different from skin?

Responding to a Virus

If virus particles enter your body, some of the particles may pass into body cells and begin to replicate. Other virus particles will be engulfed and broken up by macrophages. This is just the beginning of the immune response. The process your immune system uses to fight an invading virus is summarized in the figure below.

Reading Check **What are two things that can happen to virus particles when they enter the body?**

INTERNET ACTIVITY

For another activity related to this chapter, go to **go.hrw.com** and type in the keyword **HL5BD6W.**

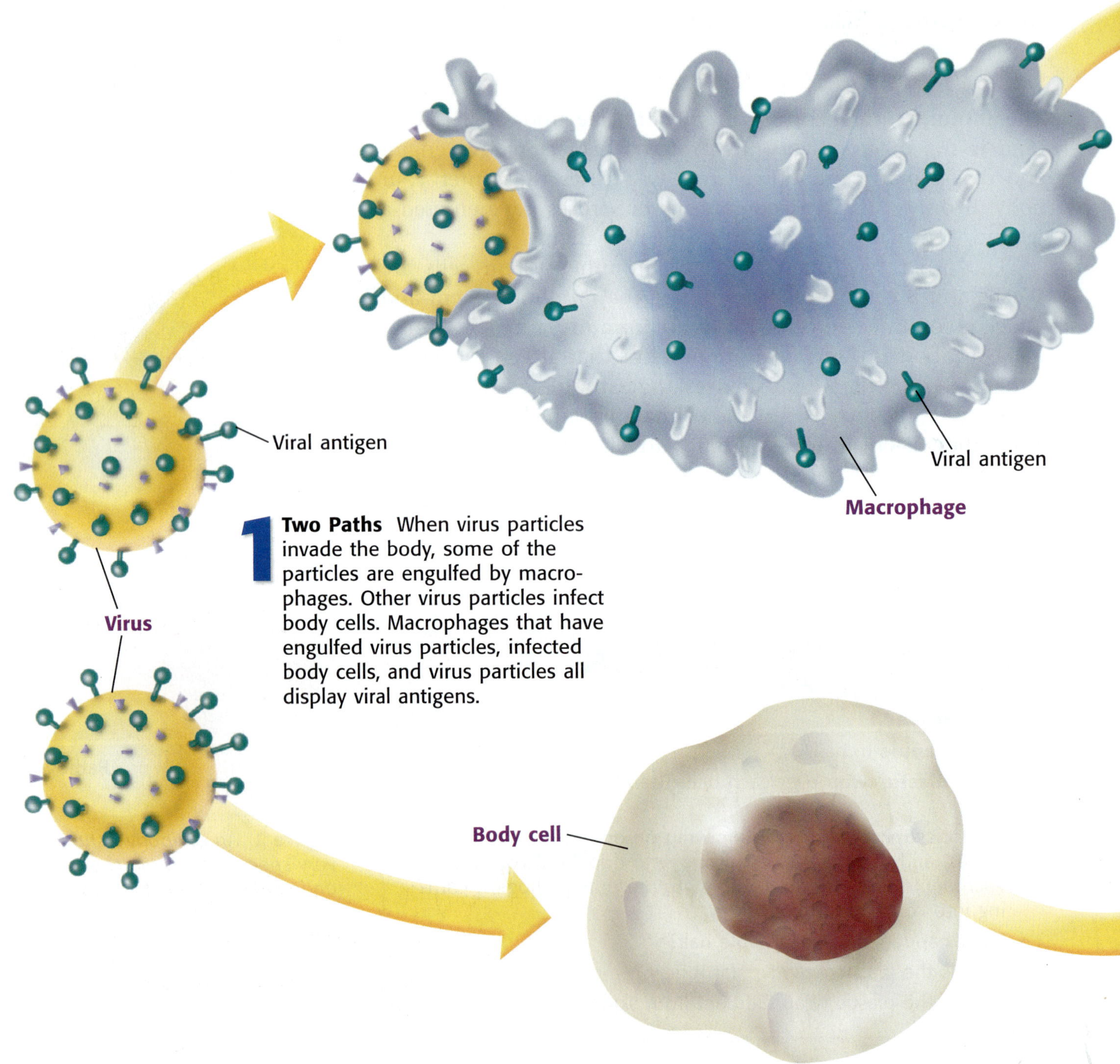

1 Two Paths When virus particles invade the body, some of the particles are engulfed by macrophages. Other virus particles infect body cells. Macrophages that have engulfed virus particles, infected body cells, and virus particles all display viral antigens.

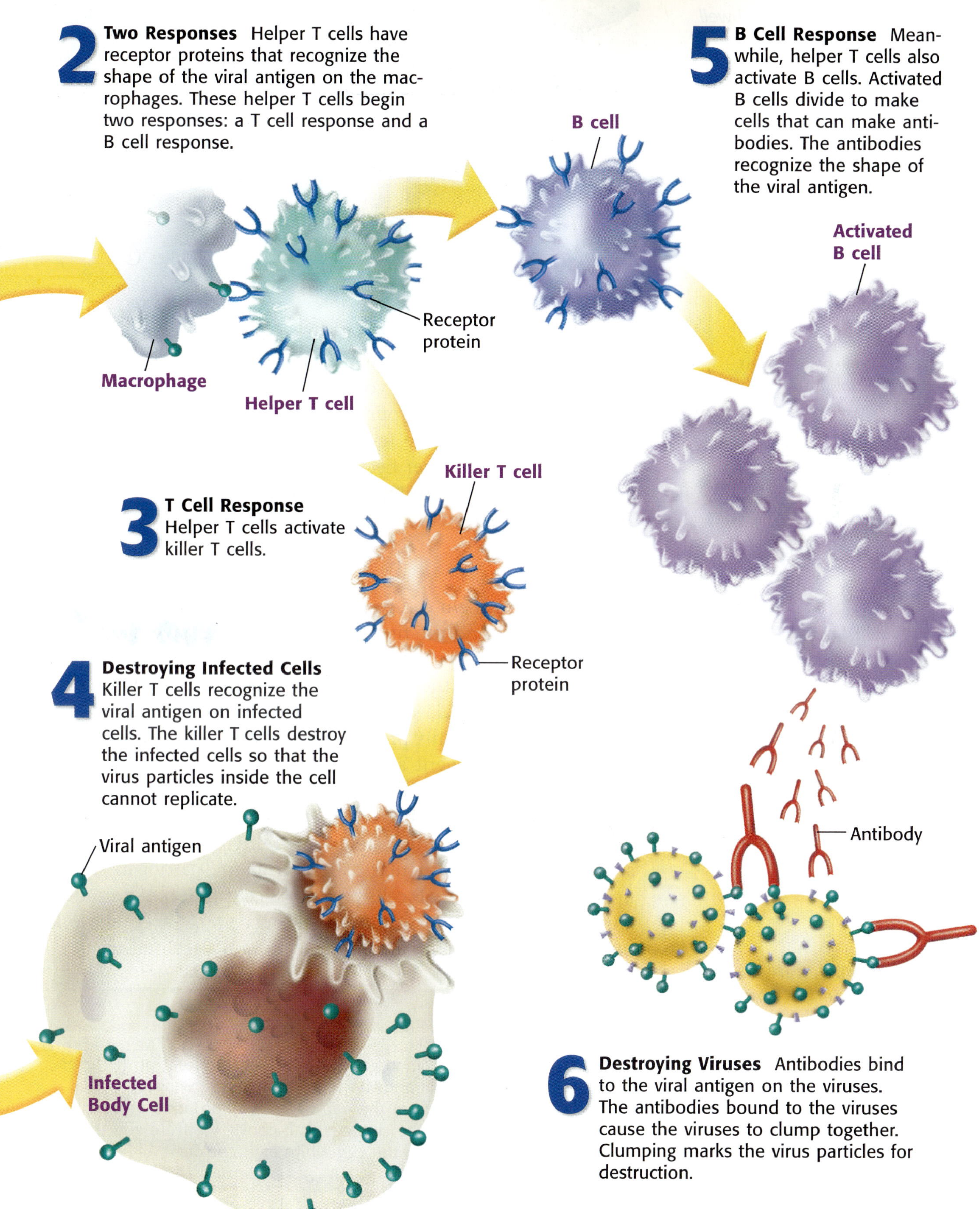
2 Two Responses Helper T cells have receptor proteins that recognize the shape of the viral antigen on the macrophages. These helper T cells begin two responses: a T cell response and a B cell response.
Macrophage
Helper T cell
Receptor protein
B cell
5 B Cell Response Meanwhile, helper T cells also activate B cells. Activated B cells divide to make cells that can make antibodies. The antibodies recognize the shape of the viral antigen.
Activated B cell
Killer T cell
3 T Cell Response Helper T cells activate killer T cells.
Receptor protein
4 Destroying Infected Cells Killer T cells recognize the viral antigen on infected cells. The killer T cells destroy the infected cells so that the virus particles inside the cell cannot replicate.
Viral antigen
Infected Body Cell
Antibody
6 Destroying Viruses Antibodies bind to the viral antigen on the viruses. The antibodies bound to the viruses cause the viruses to clump together. Clumping marks the virus particles for destruction.

Figure 3 *You may not feel well when you have a fever. But a fever is one way that your body fights infections.*

Fevers

The man in **Figure 3** is sick and has a fever. What is a fever? When macrophages activate the helper T cells, they send a chemical signal that tells your brain to turn up the thermostat. In a few minutes, your body's temperature can rise several degrees. A moderate fever of one or two degrees actually helps you get well faster because it slows the growth of some pathogens. As shown in **Figure 4,** a fever also helps B cells and T cells multiply faster.

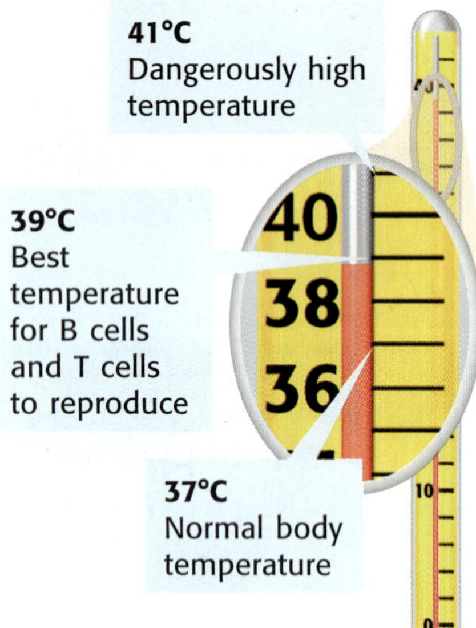

Figure 4 *A slight fever helps immune cells reproduce. But a fever of more than a few degrees can become dangerous.*

Memory Cells

Your immune system can respond to a second encounter faster than it can respond the first time. B cells must have had previous contact with a pathogen before they can make the correct antibodies. During the first encounter with a new pathogen, specialized B cells make antibodies that are effective against that particular invader. This process takes about 2 weeks, which is far too long to prevent an infection. Therefore, the first time you are infected, you usually get sick.

A few of the B cells become memory B cells. **Memory B cells** are cells in your immune system that "remember" how to make an antibody for a particular pathogen. If the pathogen shows up again, the memory B cells produce B cells that make enough antibodies in just 3 or 4 days to protect you. Vaccines work because B cells recognize pathogens from a vaccination.

memory B cell a B cell that responds to an antigen more strongly when the body is reinfected with an antigen than it does during its first encounter with the antigen

CONNECTION TO Chemistry

Bent out of Shape When you have a fever, the heat of the fever changes the shape of viral or bacterial proteins, slowing or preventing the reproduction of the pathogen. With an adult present, observe how an egg white changes as it cooks. What do you think happens to the protein in the egg white as it cooks?

ACTIVITY

Challenges to the Immune System

The immune system is a very effective body-defense system, but it is not perfect. The immune system is unable to deal with some diseases. There are also conditions in which the immune system does not work properly.

allergy a reaction to a harmless or common substance by the body's immune system

autoimmune disease a disease in which the immune system attacks the organism's own cells

Allergies

Sometimes, the immune system overreacts to antigens that are not dangerous to the body. This inappropriate reaction is called an **allergy.** Allergies may be caused by many things, including certain foods and medicines. Many people have allergic reactions to pollen, shown in **Figure 5.** Symptoms of allergic reactions range from a runny nose and itchy eyes to more serious conditions, such as asthma.

Doctors are not sure why the immune system overreacts in some people. Scientists think allergies might be useful because the mucus draining from your nose carries away pollen, dust, and microorganisms.

Figure 5 *Pollen is one substance that can cause allergic reactions.*

Autoimmune Diseases

A disease in which the immune system attacks the body's own cells is called an **autoimmune disease.** In an autoimmune disease, immune system cells mistake body cells for pathogens. One autoimmune disease is rheumatoid arthritis (ROO muh TOYD ahr THRIET is), in which the immune system attacks the joints. A common location for rheumatoid arthritis is the joints of the hands, as shown in **Figure 6.** Other autoimmune diseases include type 1 diabetes, multiple sclerosis, and lupus.

Reading Check **Name four autoimmune diseases.**

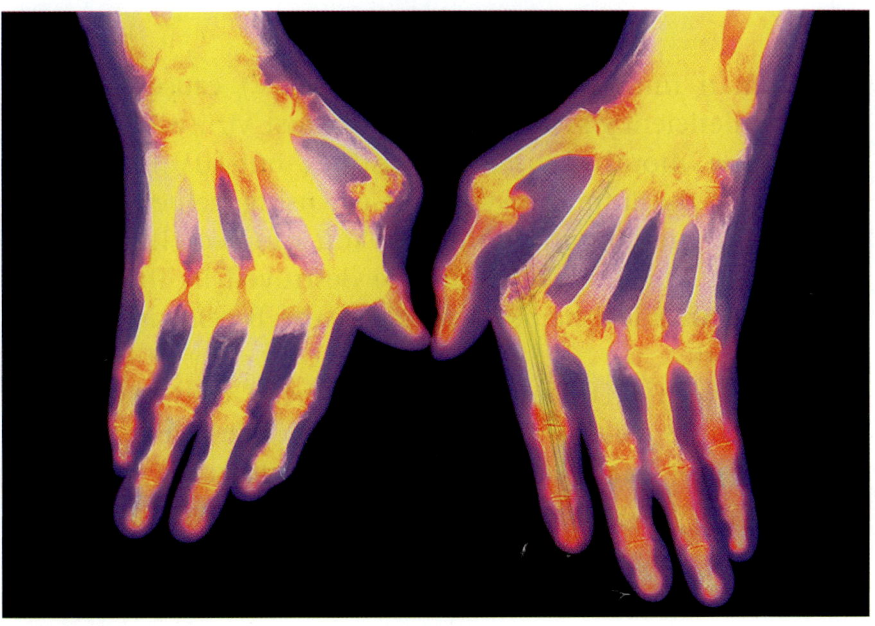

Figure 6 *In rheumatoid arthritis, immune system cells cause joint-tissue swelling, which can lead to joint deformities.*

Figure 7 **Immune Cells Fighting Cancer**

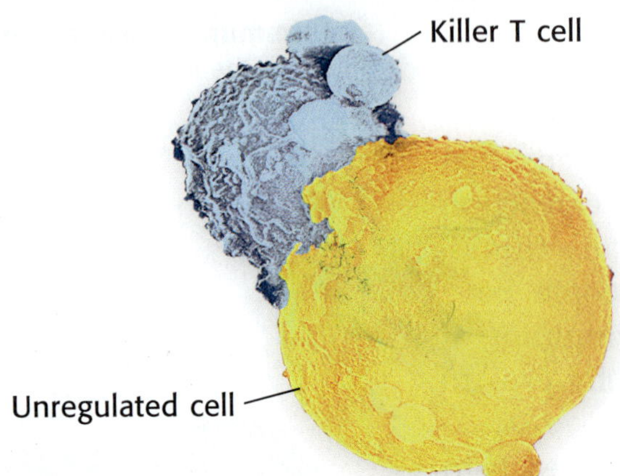

Cancer

Healthy cells divide at a carefully regulated rate. Occasionally, a cell doesn't respond to the body's regulation and begins dividing at an uncontrolled rate. As can be seen in **Figure 7,** killer T cells destroy this type of cell. Sometimes, the immune system cannot control the division of these cells. **Cancer** is the condition in which cells divide at an uncontrolled rate.

cancer a disease in which the cells begin dividing at an uncontrolled rate and become invasive

Many cancers will invade nearby tissues. They can also enter the cardiovascular system or lymphatic system. Cancers can then be transported to other places in the body. Cancers disrupt the normal activities of the organs they have invaded, sometimes leading to death. Today, though, there are many treatments for cancer. Surgery, radiation, and certain drugs can be used to remove or kill cancer cells or slow their division.

AIDS

The human immunodeficiency virus (HIV) causes acquired immune deficiency syndrome (AIDS). Most viruses infect cells in the nose, mouth, lungs, or intestines, but HIV is different. HIV infects the immune system itself, using helper T cells as factories to produce more viruses. You can see HIV particles in **Figure 8.** The helper T cells are destroyed in the process. Remember that the helper T cells put the B cells and killer T cells to work.

People with AIDS have very few helper T cells, so nothing activates the B cells and killer T cells. Therefore, the immune system cannot attack HIV or any other pathogen. People with AIDS don't usually die of AIDS itself. They die of other diseases that they are unable to fight off.

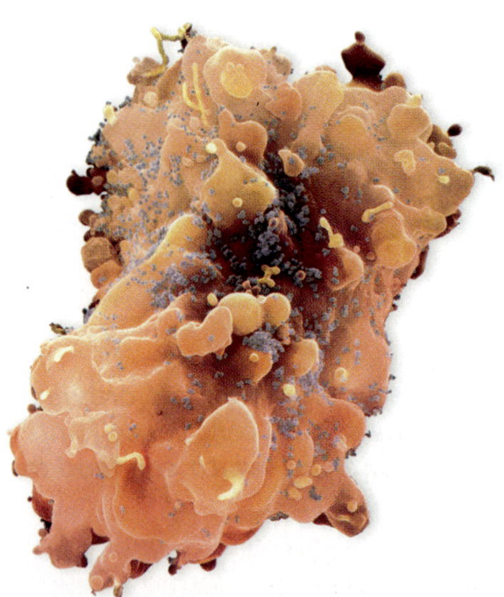

Figure 8 *The blue particles on this helper T cell are human immunodeficiency viruses. They replicated inside the T cell.*

Reading Check What virus causes AIDS?

SECTION Review

- Macrophages engulf pathogens, display antigens on their surface, and activate helper T cells. The helper T cells put the killer T cells and B cells to work.
- Killer T cells kill infected cells. B cells make antibodies.
- Fever helps speed immune-cell growth and slow pathogen growth.
- Memory B cells remember how to make an antibody for a pathogen that the body has previously fought.
- An allergy is the overreaction of the immune system to a harmless antigen.
- Autoimmune diseases are responses in which the immune system attacks healthy tissue.
- Cancer cells are cells that undergo uncontrolled division.
- AIDS is a disease that results when the human immunodeficiency virus kills helper T cells.

Using Key Terms

For each pair of terms, explain how the meanings of the terms differ.

1. *B cell* and *T cell*
2. *autoimmune disease* and *allergy*

Understanding Key Ideas

3. Your body's first line of defense against pathogens includes
   a. skin. c. T cells.
   b. macrophages. d. B cells.
4. List three ways your body defends itself against pathogens.
5. Name three different cells in the immune system, and describe how they respond to pathogens.
6. Describe four challenges to the immune system.
7. What characterizes a cancer cell?

Critical Thinking

8. **Identifying Relationships** Can your body make antibodies for pathogens that you have never been in contact with? Why or why not?
9. **Applying Concepts** If you had chickenpox at age 7, what might prevent you from getting chickenpox again at age 8?

Interpreting Graphics

10. Look at the graph below. Over time, people with AIDS become very sick and are unable to fight off infection. Use the information in the graph below to explain why this occurs.

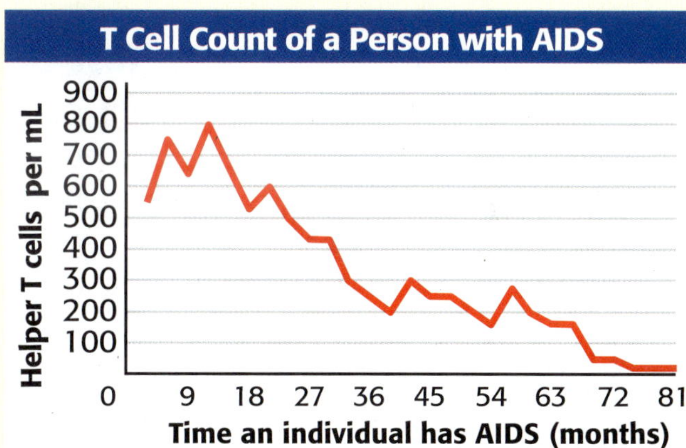

Using Scientific Methods

Skills Practice Lab

Passing the Cold

OBJECTIVES

Investigate how diseases spread.

Analyze data about how diseases spread.

MATERIALS

- beaker or a cup, 200 mL
- eyedropper
- gloves, protective
- solution, unknown, 50 mL

SAFETY

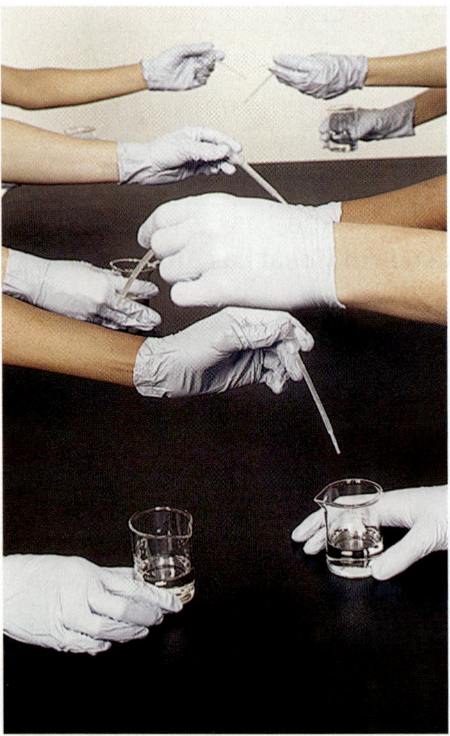

There are more than 100 viruses that cause the symptoms of the common cold. Any of the viruses can be passed from person to person—through the air or through direct contact. Although you can't catch the same cold virus twice, there are many viruses that cause the cold, so you can get a cold many times. In this activity, you will track the progress of an outbreak in your class.

Ask a Question

1. With other members of your group, form a question about the spread of disease. For example "How are cold viruses passed from person to person?" or "How can the progress of an outbreak be modeled?"

Form a Hypothesis

2. Form a hypothesis based on the question you asked.

Test the Hypothesis

3. Obtain an empty cup or beaker, an eyedropper, and 50 mL of one of the solutions from your teacher. Only one student will have the "cold virus" solution. You will see a change in your solution when you have become "infected."
4. Your teacher will divide the class into two equal groups. If there is an extra student, that person will record data on the board. Otherwise, the teacher will act as the recorder.
5. The two groups should form straight lines, facing each other.
6. Each time your teacher says the word *mix,* fill your eyedropper with your solution, and place 10 drops of your solution in the beaker of the person in the line opposite you without touching your eyedropper to the other liquid.
7. Gently stir the liquid in your cup with your eyedropper. Do not put your eyedropper in anyone else's solution.
8. If your solution changes color, raise your hand so that the recorder can record the number of students who have been "infected."

Results of Experiment			
Trial	Number of infected people	Total number of people	Percentage of infected people
1			
2			
3			
4			
5			
6			
7			
8			
9			
10			

9. Your teacher will instruct one line to move one person to the right. Then, the person at the end of the line without a partner should go to the other end of the line.
10. Repeat steps 5–9 nine more times for a total of 10 trials.
11. Return to your desk, and create a data table in your notebook similar to the table above. The column with the title "Total number of people" will remain the same in every row. Enter the data from the board into your data table.
12. Find the percentage of infected people for the last column by dividing the number of infected people by the total number of people and multiplying by 100 in each line.

Analyze the Results

1. **Describing Events** Did you become infected? If so, during which trial did you become infected?
2. **Examining Data** Did everyone eventually become infected? If so, how many trials were necessary to infect everyone?

Draw Conclusions

3. **Interpreting Information** Explain at least one reason why this simulation may underestimate the number of people who might have been infected in real life.
4. **Applying Conclusions** Use your results to make a line graph showing the change in the infection percentage per trial.

Applying Your Data

Do research in the library or on the Internet to find out some of the factors that contribute to the spread of a cold virus. What is the best and easiest way to reduce your chances of catching a cold? Explain your answer.

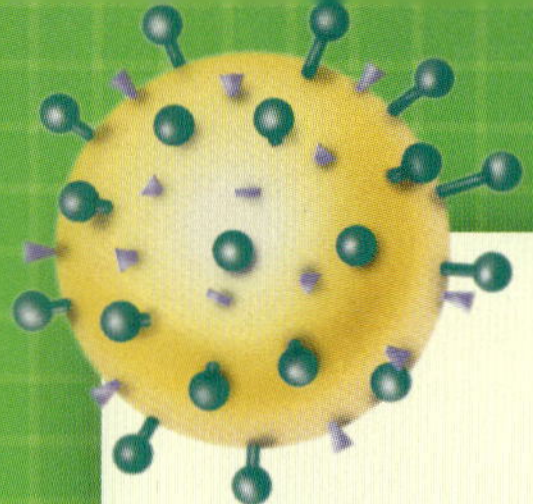

Chapter Review

USING KEY TERMS

Complete each of the following sentences by choosing the correct term from the word bank.

antibody	cancer
infectious disease	B cells
vaccines	T cells
pathogen	allergy

1. ____ can prevent disease.

2. Antibiotics can be used to kill a(n) ____.

3. Macrophages attract helper ____.

4. A(n) ____ binds to an antigen.

5. An immune system overreaction to a harmless substance is a(n) ____.

6. ____ is the unregulated growth of cells.

UNDERSTANDING KEY IDEAS

Multiple Choice

7. Pathogens are
   a. all viruses and microorganisms.
   b. viruses and microorganisms that cause disease.
   c. noninfectious organisms.
   d. all bacteria that live in water.

8. Pasteurization can kill germs that are living in
   a. animals.
   b. liquids.
   c. viruses.
   d. people.

9. The skin keeps pathogens out by
   a. staying warm enough to kill them.
   b. releasing killer T cells onto the surface.
   c. shedding dead cells and secreting oils.
   d. All of the above

10. Memory B cells
    a. kill pathogens.
    b. activate killer T cells.
    c. activate killer B cells.
    d. produce B cells that make antibodies.

11. The germ theory states that
    a. microorganisms can cause disease.
    b. heating milk will kill germs.
    c. people can build immunity.
    d. All of the above

12. Macrophages
    a. make antibodies.
    b. release helper T cells.
    c. live in the gut.
    d. engulf pathogens.

Short Answer

13. Explain how macrophages start an immune response.

14. Describe the role of helper T cells in responding to an infection.

15. Name two ways that you come into contact with pathogens.

CRITICAL THINKING

16 **Concept Mapping** Use the following terms to create a concept map: *macrophages, helper T cells, B cells, antibodies, antigens, killer T cells,* and *memory B cells.*

17 **Identifying Relationships** Why does the disappearance of helper T cells in AIDS patients damage the immune system?

18 **Predicting Consequences** Many people take fever-reducing drugs as soon as their temperature exceeds 37°C. Why might it not be a good idea to reduce a fever immediately with drugs?

19 **Evaluating Data** The risk of dying from a whooping cough vaccine is about one in 1 million. In contrast, the risk of dying from whooping cough is about one in 500. Discuss the pros and cons of this vaccination.

20 **Applying Concepts** Some people who live on farms drink milk fresh from the cow without pasteurizing the milk. Do you think drinking unpasteurized fresh milk could be safe? Why could drinking that milk be safer than drinking unpasteurized milk from a grocery store?

INTERPRETING GRAPHICS

The graph below compares the concentration of antibodies in the blood the first time you are exposed to a pathogen with the concentration of antibodies the next time you are exposed to the pathogen. Use the graph below to answer the questions that follow.

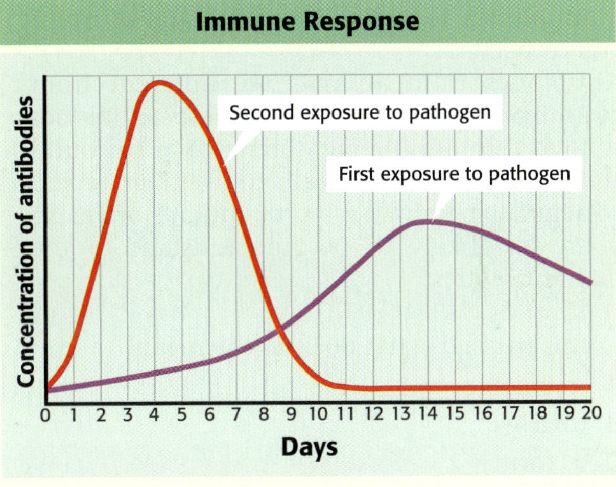

21 Are there more antibodies present during the first week of the first exposure or the first week of the second exposure? Why do you think this is so?

22 What is the difference in recovery time between the first exposure and second exposure? Why?

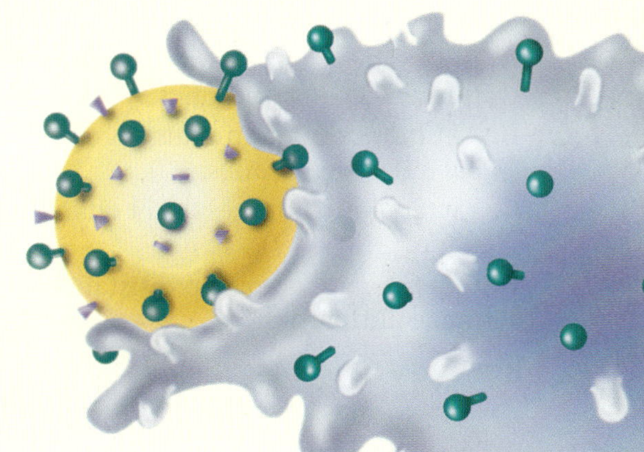

Standardized Test Preparation

READING

Read each of the passages below. Then, answer the questions that follow each passage.

Passage 1 Bacteria are becoming resistant to many human-made antibiotics, which means that the drugs no longer affect the bacteria. Scientists now face the challenge of developing new antibiotics that can overcome the resistant strains of bacteria.

Antibiotics from animals are different from some human-made antibiotics. These antibiotics bore holes through the membranes that surround bacterial cells, causing the cells to disintegrate and die. Bacterial membranes don't mutate often, so they are less likely to become resistant to the animal antibiotics.

1. In this passage, what does *mutate* mean?

A to change
B to grow
C to form
D to degrade

2. Based on the passage, which of the following statements is a fact?

F Bacterial membranes are on the inside of the bacterial cell.
G Bacterial membranes are on the outside of the bacterial cell.
H All strains of bacteria mutate.
I Bacterial membranes never change.

3. Based on the passage, which of the following sentences is false?

A Antibiotics from animals are different from human-made antibiotics.
B Antibiotics from animals bore holes in bacterial membranes.
C Bacterial membranes don't change very often.
D Bacteria rarely develop resistance to human-made antibiotics.

Passage 2 Drinking water in the United States is generally safe, but water lines can break, or treatment plants can become flooded, allowing microorganisms to enter the public water supply. Bacteria growing in foods and beverages can cause illness, too. Refrigerating foods can slow the growth of many of these pathogens, but meat, fish, and eggs that are not cooked enough can still contain dangerous bacteria or parasites. Leaving food out at room temperature can give bacteria such as *salmonella* time to grow and produce toxins in the food. For these reasons, it is important to wash all used cooking tools.

1. Which of the following statements can you infer from this passage?

A Treatment plants help keep drinking water safe.
B Treatment plants never become flooded.
C Eliminating treatment plants would help keep water safe.
D New treatment plants are better than old ones.

2. Which of the following statements can you infer from the passage?

F Bacteria that live in food produce more toxins than molds produce.
G Cooking food thoroughly kills bacteria living in the food.
H Some bacteria are helpful to humans.
I Illnesses caused by bacteria living in food are seldom serious.

3. According to this passage, what do pathogens cause?

A disease
B flooding
C water-line breaks
D water supplies

INTERPRETING GRAPHICS

The graph below shows the reported number of people living with HIV/AIDS. Use the graph to answer the questions that follow.

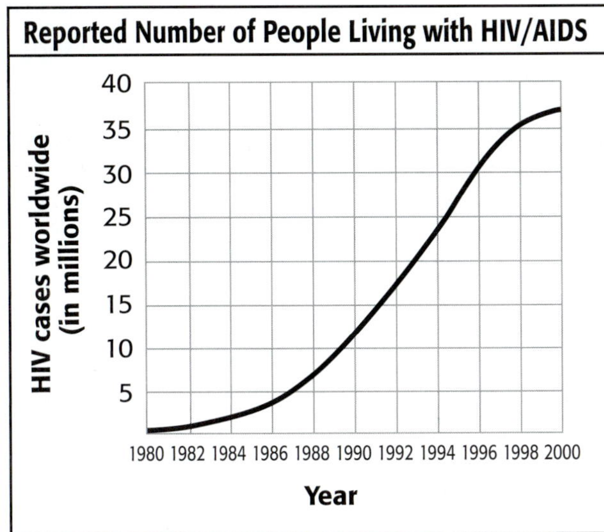

Source: Joint United Nations Program on HIV/AIDS

1. When did the number of people living with HIV/AIDS reach 5 million?
   - **A** 1985
   - **B** 1986
   - **C** 1987
   - **D** 1988

2. When did the number of people living with HIV/AIDS reach 30 million?
   - **F** 1996
   - **G** 1997
   - **H** 1998
   - **I** 1999

3. When was the rate of increase of people with HIV/AIDS the **greatest**?
   - **A** from 1980 to 1982
   - **B** from 1984 to 1986
   - **C** from 1988 to 1990
   - **D** from 1998 to 2000

4. What percentage of the people who are infected with HIV do not yet have AIDS?
   - **F** 10%
   - **G** 24%
   - **H** 75%
   - **I** There is not enough information to determine the answer.

5. If the virus continued to spread as the graph indicates, in the year 2002, about how many people would be infected with HIV?
   - **A** 30 million
   - **B** 35 million
   - **C** 39 million
   - **D** 60 million

6. Which part of the graph indicates the rate of infection?
   - **F** x-axis
   - **G** y-axis
   - **H** slope of the line being graphed
   - **I** number of years in the sample

MATH

Read each question below, and choose the best answer.

1. Suppose you have 50,000 flu viruses on your fingers and you rub your eyes. Only 20,000 viruses enter your eyes, 10,000 dissolve in chemicals, and 10,000 are washed down into your nose. Of those, you sneeze out 2,000. How many viruses are left to wash down the back of your throat and possibly start an infection?
   - **A** 50,000
   - **B** 10,000
   - **C** 8,000
   - **D** 5,000

2. In which of the following lists are the numbers in order from smallest to greatest?
   - **F** 0.027, 0.072, 0.270, 0.720
   - **G** 0.270, 0.072, 0.720, 0.270
   - **H** 0.072, 0.027, 0.270, 0.720
   - **I** 0.720, 0.270, 0.072, 0.027

Science in Action

Weird Science

Frogs in the Medicine Cabinet?

Frog skin, mouse intestines, cow lungs, and shark stomachs are all being tested to make more effective medicines to combat harmful bacteria. In 1896, a biologist named Michael Zasloff was studying African clawed frogs. He noticed that cuts in the frogs' skin healed quickly and never became infected. Zasloff decided to investigate further. He found that when a frog was cut, its skin released a liquid antibiotic that killed invading bacteria. Furthermore, sand sharks, moths, pigs, mice, and cows also contain chemicals that kill bacteria and other microorganisms. These useful antibiotics are even found in the small intestines of humans!

Social Studies ACTIVITY

Many medicines were discovered in plants or animals by people living near those plants or animals. Research the origin of one or two common medicines discovered this way. Make a poster showing a world map and the location of the medicines that you researched.

Scientific Discoveries

Medicine for Peanut Allergies

Scientists estimate that 1.5 million people in the United States suffer from peanut allergies. Every year 50 to 100 people in the United States die from an allergic reaction to peanuts. Peanuts and peanut oil are used to make many foods. People who have a peanut allergy sometimes mistakenly eat these foods and suffer severe reactions. A new drug has been discovered to help people control severe reactions. The drug is called TNX-901. The drug is actually an antibody that binds to the antibodies that the body makes during the allergic reaction to the peanuts. By binding these antibodies, the drug controls the allergic response.

Math ACTIVITY

During the testing of the new drug, 84 people were given four injections over the course of 4 months. One-fourth of the people participating received injections of a control that had no medicine in it. The rest of the people participating received different doses of the drug. How many people received the control? How many people received medicine? How many shots containing medicine were administered during the 4-month test?

Careers

Terrel Shepherd III

Nurse Terrel Shepherd III is a registered nurse (RN) at Texas Children's Hospital in Houston, Texas. RNs have many responsibilities. These responsibilities include giving patients their medications, assessing patients' health, and establishing intravenous access. Nurses also serve as a go-between for the patient and the doctor. Although most nurses work in hospitals or clinics, some nurses work for corporations. Pediatric nurses such as Shepherd work specifically with infants, children, and adolescents. The field of nursing offers a wide variety of job opportunities including home-care nurses, traveling nurses, and flight nurses. The hospital alone has many areas of expertise for nurses, including geriatrics (working with the elderly), intensive care, administration, and surgery. Traditionally, nursing has been considered to be a woman's career. However, since nursing began as a profession, men and women have practiced nursing. A career in nursing is possible for anyone who does well in science, enjoys people, and wants to make a difference in people's lives.

Language Arts ACTIVITY

WRITING SKILL Create a brochure that persuades people to consider a career in nursing. Describe nursing as a career, the benefits of becoming a nurse, and the education needed to be a nurse. Illustrate the brochure with pictures of nurses from the Internet or from magazines.

To learn more about these Science in Action topics, visit go.hrw.com and type in the keyword HL5BD6F.

Current Science

Check out Current Science® articles related to this chapter by visiting go.hrw.com. Just type in the keyword HL5CS27.

Staying Healthy

About the PHOTO

What do you see in this photo? Sure, you can see five students facing the camera, but what else does the picture tell you? The bright eyes, happy smiles, and shiny hair show radiant health. Having a clear mind and a long, active life depend on having a healthy body. Keeping your body healthy depends on eating well; avoiding drugs, cigarettes, and alcohol; and staying safe.

PRE-READING ACTIVITY

FOLDNOTES

Booklet Before you read the chapter, create the FoldNote entitled "Booklet" described in the **Study Skills** section of the Appendix. Label each page of the booklet with a main idea from the chapter. As you read the chapter, write what you learn about each main idea on the appropriate page of the booklet.

START-UP ACTIVITY

Conduct a Survey

How healthy are the habits of your classmates? Find out for yourself.

Procedure

1. Copy and answer yes or no to each of the five questions at right. Do not put your name on the survey.

Analysis

1. As a class, record the data from the completed surveys in a chart. For each question, calculate the percentage of your class that answered yes.
2. What good and bad habits do your classmates have?

1. Do you exercise at least three times a week?
2. Do you wear a seat belt every time you ride in a car?
3. Do you eat five or more servings of fruits and vegetables every day?
4. Do you use sunscreen to protect your skin when you are outdoors?
5. Do you eat a lot of high-fat foods?

SECTION 1

Good Nutrition

Does the saying "You are what you eat" mean that you are pizza? No, but substances in pizza help build your body.

Protein in the cheese may become part of your hair. Carbohydrates in the crust can give you energy for your next race.

READING WARM-UP

Objectives

- Identify the six groups of nutrients, and explain their importance to good health.
- Describe the Food Guide Pyramid.
- Understand how to read Nutrition Facts labels.
- Explain the dangers of various nutritional disorders.

Terms to Learn

nutrient
carbohydrate
protein
fat
mineral
vitamin
malnutrition

READING STRATEGY

Reading Organizer As you read this section, create an outline of the section. Use the headings from the section in your outline.

Nutrients

Are you more likely to have potato chips or broccoli for a snack? If you eat many foods that are high in fat, such as potato chips, your food choices probably are not as healthy as they could be. Broccoli is a healthier food than potato chips. But eating only broccoli, as the person in **Figure 1** is doing, does not give you a balanced diet.

To stay healthy, you need to take in **nutrients,** or substances that provide the materials needed for life processes. Nutrients are grouped into six classes: *carbohydrates, proteins, fats, water, vitamins,* and *minerals.* Carbohydrates, proteins, and fats provide energy for the body in units called *Calories* (Cal).

nutrient a substance in food that provides energy or helps form body tissues and that is necessary for life and growth

Carbohydrates

Carbohydrates are your body's main source of energy. A **carbohydrate** is a chemical composed of simple sugars. There are two types of carbohydrates: simple and complex. *Simple carbohydrates* are sugars. They are easily digested and give you quick energy. *Complex carbohydrates* are made up of many sugar molecules linked together. They are digested slowly and give you long-lasting energy. Some complex carbohydrates are good sources of fiber. Fiber is a part of a healthy diet and is found in whole-grain foods, such as brown rice and whole-wheat bread. Many fruits and vegetables also contain fiber.

carbohydrate a class of energy-giving nutrients that includes sugars, starches, and fiber

Figure 1 *Eating only one food, even a healthy food, will not give you all the substances your body needs.*

Protein

Proteins are found in body fluids, muscle, bone, and skin. **Proteins** are nutrients used to build and repair your body. Your body makes the proteins it needs, but it must have the necessary building blocks, called *amino acids*. Your digestive system breaks down protein into individual amino acids that are then used to make new proteins. Some foods, such as poultry, fish, milk, and eggs, provide all of the amino acids your body needs. Foods that contain all of these essential amino acids are called *complete proteins*. *Incomplete proteins* contain only some of the essential amino acids. Most plant foods contain incomplete protein, but eating a variety of plant foods will provide all of the amino acids your body needs.

Reading Check What is an incomplete protein? (*See the Appendix for answers to Reading Checks.*)

Figure 2 *This sample meal provides many of the nutrients a growing teenager needs.*

Fats

Another class of nutrients that is important to a healthy meal, such as the meal shown in **Figure 2,** is fat. **Fats** are energy-storage nutrients. Fats are needed to store and transport vitamins, produce hormones, keep skin healthy, and provide insulation. Fats also provide more energy than either proteins or carbohydrates. There are two types of fats: saturated and unsaturated. *Saturated fats* are found in meat, dairy products, coconut oil, and palm oil. Saturated fats raise blood cholesterol levels. Although *cholesterol* is a fat-like substance found naturally in the body, high levels can increase the risk of heart disease. *Unsaturated fats* and foods high in fiber may help reduce blood cholesterol levels. Your body cannot make unsaturated fats. They must come from vegetable oils and fish in your diet. The body needs both kinds of fats.

protein a molecule that is made up of amino acids and that is needed to build and repair body structures and to regulate processes in the body

fat an energy-storage nutrient that helps the body store some vitamins

Water

You cannot survive for more than a few days without water. Your body is about 70% water. Water is in every cell of your body. The main functions of water are to transport substances, regulate body temperature, and provide lubrication. Some scientists think you should drink at least eight glasses of water a day. When you exercise you need more water, as shown in **Figure 3.** You also get water from other liquids you drink and the foods you eat. Fresh fruits and vegetables, juices, soups, and milk are good sources of water.

Figure 3 *When you exercise, you need to drink more water.*

Table 1 Some Essential Vitamins		
Vitamin	**What it does**	**Where you get it**
A	keeps skin and eyes healthy; builds strong bones and teeth	yellow and orange fruits and vegetables, leafy greens, meats, and milk
B (various forms)	helps body use carbohydrates; helps blood, nerves, and heart function	meats, whole grains, beans, peas, nuts, and seafood
C	strengthens tissues; helps the body absorb iron, fight disease	citrus fruits, leafy greens, broccoli, peppers, and cabbage
D	builds strong bones and teeth; helps the body use calcium and phosphorus	sunlight, enriched milk, eggs, and fish
E	protects red blood cells from destruction; keeps skin healthy	oils, fats, eggs, whole grains, wheat germ, liver, and leafy greens
K	assists with blood clotting	leafy greens, tomatoes, and potatoes

Minerals

If you eat a balanced diet, you should get all of the vitamins and minerals you need. **Minerals** are elements that are essential for good health. You need six minerals in large amounts: calcium, chloride, magnesium, phosphorus, potassium, and sodium. There are at least 12 minerals that are required in very small amounts. These include fluorine, iodine, iron, and zinc. Calcium is necessary for strong bones and teeth. Magnesium and sodium help the body use proteins. Potassium is needed to regulate your heartbeat and produce muscle movement, and iron is necessary for red blood cell production.

mineral a class of nutrients that are chemical elements that are needed for certain body processes

vitamin a class of nutrients that contain carbon and that are needed in small amounts to maintain health and allow growth

Vitamins

Vitamins are another class of nutrients. **Vitamins** are compounds that control many body functions. Only vitamin D can be made by the body, so you have to get most vitamins from food. **Table 1** provides information about six essential vitamins.

CONNECTION TO Oceanography

Nutritious Seaweed Kelp, a type of seaweed, is a good source of iodine. This nutritious food is grown on special farms off the coasts of China and Japan. What other nutritious foods come from the sea?

Eating for Good Health

You have learned which nutrients you need for good health. But how can you know how much to eat? Keep in mind that the number of Calories you need depends on your age, gender, activity level, and heredity. For example, most teenage girls need about 2,200 Cal per day, and most boys need about 2,800 Cal. You should also remember that *where* you get your Calories is as important as *how many* Calories you get. The Food Guide Pyramid, shown in **Figure 4,** can help you make good food choices.

✓ Reading Check **Using the Food Guide Pyramid below, design a healthy lunch that includes one food from each food group.**

Brown Bag Test

1. Cut a **brown paper bag** into squares.
2. Gather a variety of **foods.** Place a piece of each food on a different square and leave overnight.
3. What do you see when you remove the food?
4. Which food had the most oil? Which food had the least?

Figure 4 The Food Guide Pyramid

The U.S. Department of Agriculture and the Department of Health and Human Services developed the Food Guide Pyramid to help Americans make healthy food choices. The Food Guide Pyramid divides foods into six groups. It shows how many servings you need daily from each group and gives examples of foods for each. This pyramid also provides sample serving sizes for each group. Within each group, the food choices are up to you.

Fats, oils, and sweets
Use sparingly.

Milk, yogurt, and cheese 2 to 3 servings
- 1 cup of milk or yogurt
- 1 1/2 oz of natural cheese
- 2 oz of processed cheese

Meat, poultry, fish, beans, eggs, and nuts 2 to 3 servings
- 2 to 3 oz of cooked poultry, fish, or lean meat
- 1/2 cup of cooked dried beans
- 1 egg

Vegetables 3 to 5 servings
- 1/2 cup of chopped vegetables
- 1 cup of raw, leafy vegetables
- 3/4 cup of cooked vegetables

Fruits 2 to 4 servings
- 1 medium apple, banana, or orange
- 1/2 cup of chopped, cooked, or canned fruit
- 3/4 cup of fruit juice

Bread, cereal, rice, and pasta 6 to 11 servings
- 1 slice of bread
- 1 oz of ready-to-eat cereal
- 1/2 cup of rice or pasta
- 1/2 cup of cooked cereal

Nutrition Facts	
Serving Size 1/2 cup (120 ml)	
Servings per Container 2.5	
Amount per Serving	**Prepared**
Calories	70
Calories from Fat	25
	% Daily Value
Total Fat 2.5 g	4%
Saturated Fat 1 g	5%
Cholesterol 15 mg	5%
Sodium 960 mg	40%
Total Carbohydrate 8 g	3%
Dietary Fiber less than 1 g	4%
Sugars 1 g	
Protein 3 g	
Vitamin A	15%
Vitamin C	0%
Calcium	0%
Iron	4%

Serving information

Number of Calories per serving

Percentage of daily values

*Percent Daily Values are based on a 2,000 Calorie diet. Your daily values may be higher or lower depending on your Calorie needs:

	Calories	2,000	2,500
Total Fat	Less than	65g	80g
Sat Fat	Less than	20g	25g
Cholesterol	Less than	300mg	300mg
Sodium	Less than	2,400mg	2,400mg
Total Carbohydrate		300g	375g
Dietary Fiber		25g	30g
Protein		50g	60g

Figure 5 *Nutrition Facts labels provide a lot of information.*

Reading Food Labels

Packaged foods must have Nutrition Facts labels. **Figure 5** shows a Nutrition Facts label for chicken noodle soup. Nutrition Facts labels show what amount of each nutrient is in one serving of the food. You can tell whether a food is high or low in a nutrient by looking at its daily value. Reading food labels can help you make healthy eating choices. The percentage of daily values shown is based on a diet that consists of 2,000 Cal per day. Most teenagers need more than 2,000 Cal per day. The number of Calories needed depends on factors such as height, weight, age, and level of activity. Playing sports and exercising use up Calories that need to be replaced for you to grow.

Reading Check **For what nutrients does chicken noodle soup provide more than 10% of the daily value?**

Nutritional Disorders

Unhealthy eating habits can cause nutritional disorders. **Malnutrition** occurs when someone does not eat enough of the nutrients needed by the body. Malnutrition can result from eating too few or too many Calories or not taking in enough of the right nutrients. Malnutrition affects how one looks and how quickly one's body can repair damage and fight illness.

malnutrition a disorder of nutrition that results when a person does not consume enough of each of the nutrients that are needed by the human body

Anorexia Nervosa and Bulimia Nervosa

Anorexia nervosa (AN uh REKS ee uh nuhr VOH suh) is an eating disorder characterized by self-starvation and an intense fear of gaining weight. Anorexia nervosa can lead to severe malnutrition.

Bulimia nervosa (boo LEE mee uh nuhr VOH suh) is a disorder characterized by binge eating followed by induced vomiting. Sometimes, people suffering from bulimia nervosa use laxatives or diuretics to rid their bodies of food and water. Bulimia nervosa can damage teeth and the digestive system and can lead to kidney and heart failure.

Both anorexia and bulimia can cause weak bones, low blood pressure, and heart problems. These eating disorders can be fatal if not treated. If you are worried that you or someone you know may have an eating disorder, talk to an adult.

What Percentage?

Use the Nutrition Facts label above to answer the following question. The recommended daily value of fat is 72 g for teenage girls and 90 g for teenage boys. What percentage of the daily recommended fat value is provided in one cup of soup?

Obesity

Eating too much food that is high in fat and low in other nutrients, such as junk food and fast food, can lead to malnutrition. *Obesity* (oh BEE suh tee) is having an extremely high percentage of body fat. People suffering from obesity may not be eating a variety of foods that provide them with the correct balance of essential nutrients. Having an inactive lifestyle can also contribute to obesity.

Obesity increases the risk of high blood pressure, heart disease, and diabetes. Eating a more balanced diet and exercising regularly can help reduce obesity. Obesity may also be caused by other factors. Scientists are studying the links between obesity and heredity.

SECTION Review

Summary

- A healthy diet has a balance of carbohydrates, proteins, fats, water, vitamins, and minerals.
- The Food Guide Pyramid is a good guide for healthy eating.
- Nutrition Facts labels provide information needed to plan a healthy diet.
- Anorexia nervosa and bulimia nervosa cause malnutrition and damage to many body systems.
- Obesity can lead to heart disease and diabetes.

Using Key Terms

1. In your own words, write a definition for each of the following terms: *nutrient, mineral,* and *vitamin.*

Understanding Key Ideas

2. Malnutrition can be caused by
   a. obesity.
   b. bulimia nervosa.
   c. anorexia nervosa.
   d. All of the above
3. What information is found on a Nutrition Facts label?
4. Give an example of a carbohydrate, a protein, and a fat.
5. If vitamins and minerals do not supply energy, why are they important to a healthy diet?
6. How do anorexia nervosa and bulimia nervosa differ?
7. How can someone who is obese suffer from malnutrition?

Math Skills

8. If you eat 2,500 Cal per day and 20% are from fat, 30% are from protein, and 50% are from carbohydrates, how many Calories of each nutrient do you eat?

Critical Thinking

9. **Applying Concepts** Name some of the nutrients that can be found in a glass of milk.
10. **Identifying Relationships** Explain how eating a variety of foods can help ensure good nutrition.
11. **Predicting Consequences** How would your growth be affected if your diet consistently lacked important nutrients?
12. **Applying Concepts** Explain how you can use the Nutrition Facts label to choose food that is high in calcium.

SECTION 2

Risks of Alcohol and Other Drugs

You see them in movies and on television and read about them in magazines. But what are drugs?

You are exposed to information, and misinformation, about drugs every day. So, how can you make the best decisions?

READING WARM-UP

Objectives

- Describe the difference between psychological and physical dependence.
- Explain the hazards of tobacco, alcohol, and illegal drugs.
- Distinguish between the positive and negative uses of drugs.

Terms to Learn

drug
addiction
nicotine
alcoholism
narcotic

READING STRATEGY

Reading Organizer As you read this section, make a table comparing the positive and negative uses of drugs.

What Is a Drug?

Any chemical substance that causes a physical or psychological change is called a **drug.** Drugs come in many forms, as shown in **Figure 1.** Some drugs enter the body through the skin. Other drugs are swallowed, inhaled, or injected. Drugs are classified by their effects. *Analgesics* (AN'l JEE ziks) relieve pain. *Antibiotics* (AN tie bie AHT iks) fight bacterial infections, and *antihistamines* (AN tie HIS tuh MEENZ) control cold and allergy symptoms. *Stimulants* speed up the central nervous system, and *depressants* slow it down. When used correctly, legal drugs can help your body heal. When used illegally or improperly, however, drugs can do great harm.

Dependence and Addiction

The body can develop *tolerance* to a drug. Tolerance means that larger and larger doses of the drug are needed to get the same effect. The body can also form a *physical dependence* or need for a drug. If the body doesn't receive a drug that it is physically dependent on, withdrawal symptoms occur. Withdrawal symptoms include nausea, vomiting, pain, and tremors.

Addiction is the loss of control of drug-taking behavior. Once addicted, a person finds it very hard to stop taking a drug. Sometimes, the need for a drug is not due only to physical dependence. Some people also form *psychological dependence* on a drug, which means that they feel powerful cravings for the drug.

Figure 1 *All of these products contain drugs.*

Types of Drugs

There are many kinds of drugs. Some drugs are made from plants, and some are made in a lab. You can buy some drugs at the grocery store, while others can be prescribed only by a doctor. Some drugs are illegal to buy, sell, or possess.

drug any substance that causes a change in a person's physical or psychological state

addiction a dependence on a substance, such as alcohol or another drug

Herbal Medicines

Information about herbal medicines has been handed down for centuries, and some herbs contain chemicals with important healing properties. The tea in **Figure 2** contains chamomile and is made from a plant. Chamomile has chemicals in it that can help you sleep. However, herbs are drugs and should be used carefully. The Federal Drug Administration does not regulate herbal medicines or teas and cannot guarantee their safety.

Figure 2 *Some herbs can be purchased in health-food stores. Medicinal herbs should always be used with care.*

Over-the-Counter and Prescription Drugs

Over-the-counter drugs can be bought without a prescription. A prescription is written by a doctor and describes the drug, directions for use, and the amount of the drug to be taken.

Many over-the-counter and prescription drugs are powerful healing agents. However, some drugs also produce unwanted side effects. *Side effects* are uncomfortable symptoms, such as nausea, headaches, drowsiness, or more serious problems.

Whether purchased with or without a prescription, all drugs must be used with care. Information on proper use can be found on the label. **Figure 3** shows some general drug safety tips.

Reading Check **What is the difference between an over-the-counter drug and a prescription drug?** (*See the Appendix for answers to Reading Checks.*)

Figure 3 **Drug Safety Tips**

- *Never take another person's prescription medicine.*
- *Read the label before each use. Always follow the instructions on the label and those provided by your doctor or pharmacist.*
- *Do not take more or less medication than prescribed.*
- *Consult a doctor if you have any side effects.*
- *Throw away leftover and out-of-date medicines.*

Figure 4 **Effects of Smoking**

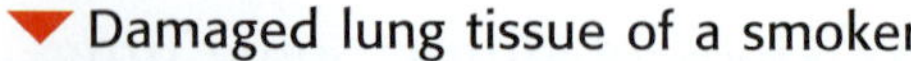

Healthy lung tissue of a nonsmoker

Damaged lung tissue of a smoker

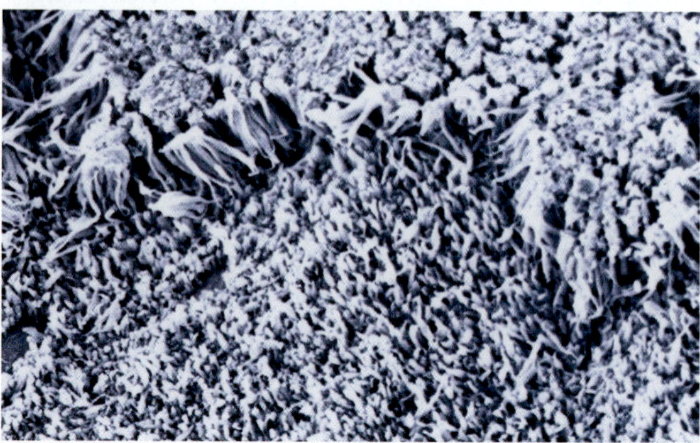

Tobacco

nicotine a toxic, addictive chemical that is found in tobacco and that is one of the major contributors to the harmful effects of smoking

alcoholism a disorder in which a person repeatedly drinks alcoholic beverages in an amount that interferes with the person's health and activities

Cigarettes are addictive, and smoking has serious health effects. **Nicotine** (NIK uh TEEN) is a chemical in tobacco that increases heart rate and blood pressure and is extremely addictive. Smokers experience a decrease in physical endurance. **Figure 4** shows the effects of smoking on the cilia of your lungs. Cilia clean the air you breathe and prevent debris from entering your lungs. Smoking increases the chances of lung cancer, and it has been linked to other cancers, emphysema, chronic bronchitis, and heart disease. Experts estimate that there are more than 430,000 deaths related to smoking each year in the United States. Secondhand smoke also poses significant health risks.

Like cigarettes, smokeless, or chewing, tobacco is addictive and can cause health problems. Nicotine is absorbed through the lining of the mouth. Smokeless tobacco increases the risk of several cancers, including mouth and throat cancer. It also causes gum disease and yellowing of the teeth.

Figure 5 *This car was in an accident involving a drunk driver.*

Alcohol

It is illegal in most of the United States for people under the age of 21 to use alcohol. Alcohol slows down the central nervous system and can cause memory loss. Excessive use of alcohol can damage the liver, pancreas, brain, nerves, and cardiovascular system. In very large quantities, alcohol can cause death. Alcohol is a factor in more than half of all suicides, murders, and accidental deaths. **Figure 5** shows the results of one alcohol-related accident. Alcohol also affects decision making and can lead you to take unhealthy risks.

People can suffer from **alcoholism,** which means that they are physically and psychologically dependent on alcohol. Alcoholism is considered a disease, and genetic factors are thought to influence the development of alcoholism in some people.

Figure 6 *Smoking marijuana can make your health and dreams go up in smoke.*

Marijuana

Marijuana is an illegal drug that comes from the Indian hemp plant. Marijuana affects different people in different ways. It may increase anxiety or cause feelings of paranoia. Marijuana slows reaction time, impairs thinking, and causes a loss of coordination. Regular use of marijuana can affect many areas of your life, as described in **Figure 6.**

narcotic a drug that is derived from opium and that relieves pain and induces sleep

Cocaine

Cocaine and its more purified form, crack, are made from the coca plant. Both drugs are illegal and highly addictive. Users can become addicted to them in a very short time. Cocaine can produce feelings of intense excitement followed by anxiety and depression. Both drugs increase heart rate and blood pressure and can cause heart attacks, even among first-time users.

Reading Check **What are two dangers to users of cocaine?**

Narcotics and Designer Drugs

Drugs made from the opium plant are called **narcotics.** Some narcotics are used to treat severe pain. Narcotics are illegal unless prescribed by a doctor. Some narcotics are never legal. For example, heroin is one of the most addictive narcotics and is always illegal. Heroin is usually injected, and users often share needles. Therefore, heroin users have a high risk of becoming infected with diseases such as hepatitis and AIDS. Heroin users can also die of an overdose of the drug.

Other illegal drugs include inhalants, barbiturates (bahr BICH uhr itz), amphetamines (am FET uh MEENZ), and *designer drugs.* Designer drugs are made by making small changes to existing drugs. Ecstasy, or "X," is a designer drug that causes feelings of well-being. Over time, the drug causes lesions (LEE zhuhnz), or holes, in a user's brain, as shown in **Figure 7.** Ecstasy users are also more likely to develop depression.

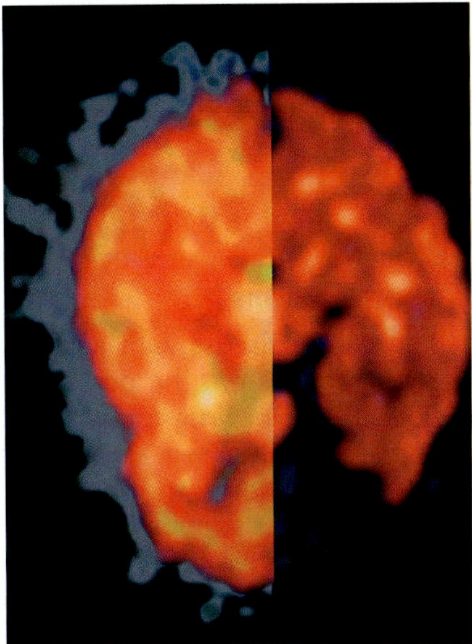

Figure 7 *The brain scan on the left shows a healthy brain. The scan on the right is from a teenager who has regularly used Ecstasy.*

Figure 8 *Drug abuse can leave you depressed and feeling alone.*

Figure 9 Drug Myths

Myth ***"It's only alcohol, not drugs."***

Reality Alcohol is a mood-altering and mind-altering drug. It affects the central nervous system and is addictive.

Myth ***"I won't get hooked on one or two cigarettes a day."***

Reality Addiction is not related to the amount of a drug used. Some people become addicted after using a drug once or twice.

Myth ***"I can quit any time I want."***

Reality Addicts may quit and return to drug usage many times. Their inability to stay drug-free shows how powerful the addiction is.

Hallucinogens

Hallucinogens (huh LOO si nuh juhnz) distort the senses and cause mood changes. Users have hallucinations, which means that they see and hear things that are not real. LSD and PCP are powerful, illegal hallucinogens. Sniffing glue or solvents can also cause hallucinations and serious brain damage.

Drug Abuse

A drug user takes a drug to prevent or improve a medical condition. The drug user obtains the drug legally and uses the drug properly. A drug abuser does not take a drug to relieve a medical condition. An abuser may take drugs for the temporary good feelings they produce, to escape from problems, or to belong to a group. The drug is often obtained illegally, and it is often taken without knowledge of the drug's dangers.

Reading Check **What is the difference between drug use and drug abuse?**

How Drug Abuse Starts

Nicotine, alcohol, and marijuana are sometimes called *gateway drugs* because they are often the first drugs a person abuses. The abuse of other, more dangerous drugs may follow the abuse of gateway drugs. Peer pressure is often the reason that young people begin to use drugs. Teenagers may drink, smoke, or try marijuana to make friends or avoid being teased. Because drug abusers often stand out, it can sometimes be hard to see that many teenagers do not abuse drugs.

Many teenagers begin using illegal drugs to feel part of a group, but drug abuse has many serious consequences. Drug abuse can lead to problems with friends, family, school, and handling money. These problems often lead to depression and social isolation, as shown in **Figure 8.**

Many people who start using drugs do not recognize the dangers. Misinformation about drugs is everywhere. Several common drug myths are discussed in **Figure 9.**

Good Reasons

WRITING SKILL Discuss with your parent the possible effects of drug abuse on your family. Then, write yourself a letter giving reasons why you should stay drug-free. Put your letter in a safe place. If you ever find yourself thinking about using drugs, take out your letter and read it.

ACTIVITY

Getting Off Drugs

People who abuse drugs undergo emotional and physical changes. Teenagers who had few problems often begin to have problems with school, family, and money when they start to use drugs.

The first step to quitting drugs is to admit to abusing drugs and to decide to stop. It is important for the addicted person to get the proper medical treatment. There are drug treatment centers, like the one shown in **Figure 10,** available to help. Getting off drugs can be extremely difficult. Withdrawal symptoms are often painful, and powerful cravings for a drug can continue long after a person quits. But people who stop abusing drugs lead happier and healthier lives.

Figure 10 *Drug treatment centers help people get off drugs and back on track to healthier, happier lives.*

SECTION Review

Summary

- Physical dependence causes withdrawal symptoms when a person stops using a drug. Psychological dependence causes powerful cravings.
- There are many types of drugs, including over-the-counter, prescription, and herbal medicines.
- Tobacco contains the highly addictive chemical nicotine.
- Abuse of alcohol can lead to alcoholism.
- Illegal drugs include marijuana, cocaine, hallucinogens, designer drugs, and many narcotics.
- Getting off drugs requires proper medical treatment.

Using Key Terms

1. In your own words, write a definition for the terms *drug, addiction,* and *narcotic.*

Understanding Key Ideas

2. Which of the following products does NOT contain a drug?
   a. cola
   b. fruit juice
   c. herbal tea
   d. cough syrup
3. Describe the difference between physical and psychological dependence.
4. What is the difference between drug use and drug abuse?
5. How does addiction occur, and what are two consequences of drug addiction?
6. Name two different kinds of illegal drugs, and give examples of each.

Math Skills

7. If 2,200 people between the ages of 16 and 20 die every year in alcohol-related car crashes, how many die every day?

Critical Thinking

8. **Analyzing Relationships** How are nicotine, alcohol, heroin, and cocaine similar? How are they different?
9. **Analyzing Ideas** What are two ways that a person who abuses drugs can get in trouble with the law?
10. **Predicting Consequences** How can drug abuse damage family relationships?
11. **Making Inferences** Driving a car while under the influence of drugs can put others in danger. Describe another situation in which one person's drug abuse could put other people in danger.

For a variety of links related to this chapter, go to www.scilinks.org

Topic: Drug and Alcohol Abuse
SciLinks code: HSM0428

SECTION 3

Healthy Habits

Do you like playing sports or acting in plays? How does your health affect your favorite activities?

Whatever you do, the better your health is, the better you can perform. Keeping yourself healthy is a daily responsibility.

READING WARM-UP

Objectives

- Describe three important aspects of good hygiene.
- Explain why exercise and sleep are important to good health.
- Describe methods of handling stress.
- List three ways to stay safe at home, on the road, and outdoors.
- Plan what you would do in the case of an accident.

Terms to Learn

hygiene
aerobic exercise
stress

READING STRATEGY

Prediction Guide Before reading this section, write the title of each heading in this section. Next, under each heading, write what you think you will learn.

Taking Care of Your Body

The science of preserving and protecting your health is known as **hygiene.** It sounds simple, but washing your hands is the best way to prevent the spread of disease and infection. You should always wash your hands after using the bathroom and before and after handling food. Taking care of your skin, hair, and teeth is important for good hygiene. Good hygiene includes regularly using sunscreen, shampooing your hair, and brushing and flossing your teeth daily.

hygiene the science of health and ways to preserve health

Good Posture

Posture is also important to health. Good posture helps you look and feel your best. Bad posture strains your muscles and ligaments and makes breathing difficult. To have good posture, imagine a vertical line passing through your ear, shoulder, hip, knee, and ankle when you stand, as shown in **Figure 1.** When working at a desk, you should maintain good posture by pulling your chair forward and planting your feet firmly on the floor.

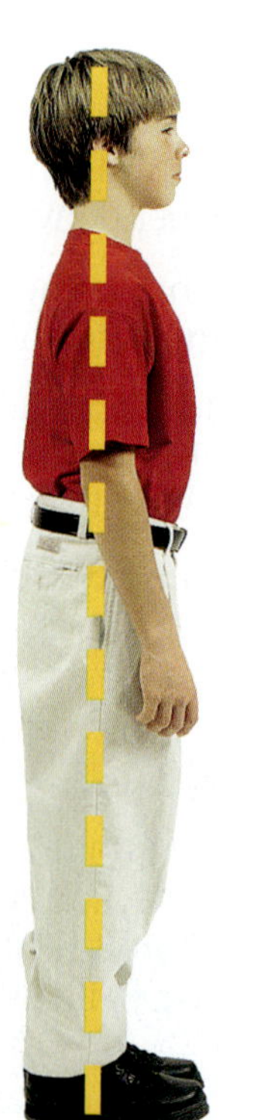

When you have good posture, your ear, shoulder, hip, knee, and ankle are in a straight line.

Bad posture strains your muscles and ligaments and can make breathing difficult.

Figure 1 *A slumped posture strains your lower back.*

Exercise

Aerobic exercise at least three times a week is essential to good health. **Aerobic exercise** is vigorous, constant exercise of the whole body for 20 minutes or more. Walking, running, swimming, and biking are all examples of aerobic exercise. **Figure 2** shows another popular aerobic exercise—basketball.

Aerobic exercise increases the heart rate. As a result, more oxygen is taken in and distributed throughout the body. Over time, aerobic exercise strengthens the heart, lungs, and bones. It burns Calories, helps your body conserve some nutrients, and aids digestion. It also gives you more energy and stamina and improves music tone. Aerobic exercise protects your physical and mental health.

Figure 2 *Aerobic exercise can be fun if you choose an activity you enjoy.*

Reading Check **What are two benefits of regular exercise?** *(See the Appendix for answers to Reading Checks.)*

Sleep

Believe it or not, teenagers actually need more sleep than younger children. Do you ever fall asleep in class, like the girl in **Figure 3,** or feel tired in the middle of the afternoon? If so, you may not be getting enough sleep. Scientists say that teenagers need about 9.5 hours of sleep each night.

At night, the body goes through several cycles of progressively deeper sleep, with periods of lighter sleep in between. If you do not sleep long enough, you will not enter the deepest, most restful period of sleep.

aerobic exercise physical exercise intended to increase the activity of the heart and lungs to promote the body's use of oxygen

Figure 3 *If you fall asleep easily during the day, you are probably not getting enough sleep.*

CONNECTION TO Language Arts

Dreamy Poetry

You are not wrong, who deem
That my days have been a dream;
Yet if hope has flown away
In a night, or in a day,
In a vision, or in none,
Is it therefore the less gone?
All that we see or seem
Is but a dream within a dream.

(Edgar Allan Poe, "A Dream Within a Dream")

What do you think Poe means by "a dream within a dream?" Why do you think there are many poems written about dreams or sleep?

Coping with Stress

You have a big soccer game tomorrow. Are you excited and ready for action? You got a low grade on your English paper. Are you upset or angry? The game and the test are causing you stress. **Stress** is the physical and mental response to pressure.

Some stress is a normal part of life. Stress stimulates your body to prepare for difficult or dangerous situations. However, sometimes you may have no outlet for the stress, and it builds up. Many things are causing stress for the girl shown in **Figure 4.** Excess stress is harmful to your health and can decrease your ability to carry out your daily activities.

You may not even realize you are stressed until your body reacts. Perhaps you get a headache, have an upset stomach, or lie awake at night. You might feel tired all the time or begin an old nervous habit, such as nail-biting. You may become irritable or resentful. All of these things can be signs of too much stress.

Figure 4 *Can you identify all of the things in this picture that could cause stress?*

stress a physical or mental response to pressure

Dealing with Stress

Different people are stressed by different things. Once you identify the source of the stress, you can find ways to deal with it. If you cannot remove the cause of stress, here are some ideas for handling stress.

- Share your problems. Talk things over with someone you trust, such as a parent, friend, teacher, or school counselor.
- Make a list of all the things you would like to get done, and rank the things in order of importance. Do the most important things first.
- Exercise regularly, and get enough sleep.
- Pet a friendly animal.
- Spend some quiet time alone, or practice deep breathing or other relaxation techniques.

For another activity related to this chapter, go to **go.hrw.com** and type in the keyword **HL5BD7W.**

Injury Prevention

Have you ever fallen off your bike or sprained your ankle? Accidents happen, and they can cause injury and even death. It is impossible to prevent all accidents, but you can decrease your risk by using your common sense and following basic safety rules.

Safety Outdoors

Always dress appropriately for the weather and for the activity. Never hike or camp alone. Tell someone where you are going and when you expect to return. If you do not bring water from home, be sure to purify any water you drink in the wilderness.

Learn how to swim. It could save your life! Never swim alone, and do not dive into shallow water or water of unknown depth. When in a boat, wear a life jacket. If a storm threatens, get out of the water and seek shelter.

Reading Check Name three safety tips for the outdoors.

Safety at Home

Many accidents can be avoided. **Figure 5** shows tips for safety around the house.

Figure 5 **Home Safety Tips**

Figure 6 *It is always important to use the appropriate safety equipment.*

Safety on the Road

In the car, always wear a seat belt, even if you are traveling only a short distance. Never ride in a car with someone who has been drinking. Safety equipment and common sense are your best defense against injury. When riding a bicycle, always wear a helmet like those shown in **Figure 6.** Ride with traffic, and obey all traffic rules. Be sure to signal when stopping or turning.

Safety in Class

Accidents can happen in school, especially in a lab class or during woodworking class. To avoid hurting yourself and others, always follow your teacher's instructions, and wear the proper safety equipment at all times.

When Accidents Happen

No matter how well you practice safety measures, accidents can still happen. What should you do if a friend chokes on food and cannot breathe? What if a friend is stung by a bee and has a violent allergic reaction?

Figure 7 *When calling 911, stay calm and listen carefully to what the dispatcher tells you.*

Call for Help

Once you've checked for other dangers, call for medical help immediately, as the person shown in **Figure 7** is doing. In most communities, you can dial 911. Speak slowly and clearly. Give the complete address and a description of the location. Describe the accident, the number of people injured, and the types of injuries. Ask what to do, and listen carefully to the instructions. Let the other person hang up first to be sure there are no more questions or instructions for you.

Learn First Aid

If you want to learn more about what to do in an emergency, you can take a first-aid or CPR course, such as the one shown in **Figure 8.** *CPR* can revive a person who is not breathing and has no heartbeat. If you are over 12 years old, you can become certified in both CPR and first aid. Some baby-sitting classes also provide information on first aid. The American Red Cross, community organizations, and local hospitals offer these classes. However, you should not attempt any lifesaving procedure unless you have been trained.

Reading Check What is CPR, and how can you learn it?

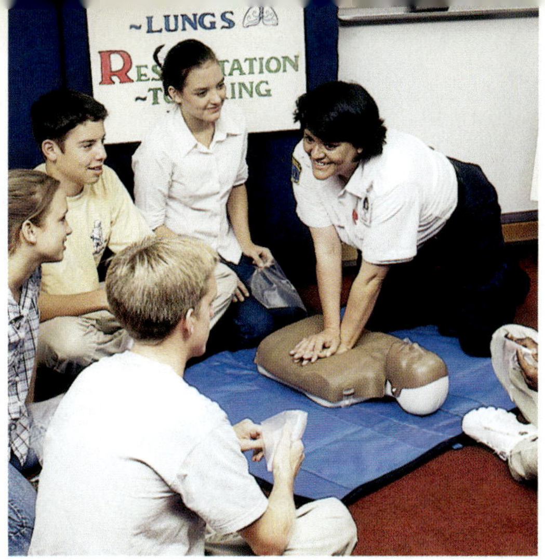

Figure 8 *These teenagers are taking a CPR course to prepare themselves for emergency situations.*

SECTION Review

Summary

- Good hygiene includes taking care of your skin, hair, and teeth.
- Good posture is important to health.
- Exercise keeps your heart, lungs, and bones healthy.
- Teenagers need more than 9 hours of sleep to stay rested and healthy.
- Coping with stress is an important part of staying physically and emotionally healthy.
- It is important to be aware of the possible hazards around your home, outdoors, and at school. Using the appropriate safety equipment can also help keep you safe.

Using Key Terms

Complete each of the following sentences by choosing the correct term from the word bank.

hygiene	aerobic exercise
sleep	stress

1. The science of protecting your health is called ___.
2. ___ strengthens your heart, lungs, and bones.
3. ___ is the physical and mental response to pressure.

Understanding Key Ideas

4. Which of the following is important for good health?
   a. irregular exercise
   b. getting your hair cut
   c. taking care of your teeth
   d. getting plenty of sun
5. List two things you should do when calling for help in a medical emergency.
6. List three ways to stay safe when you are outside, and three ways to stay safe at home.
7. How do seat belts and safety equipment protect you?

Math Skills

8. It is estimated that only 65% of adults wear their seat belts. If there are 10,000 people driving in your area right now, how many of them are wearing their seat belts?

Critical Thinking

9. **Applying Concepts** What situations cause you stress? What can you do to help relieve the stress you are feeling?
10. **Making Inferences** According to the newspaper, the temperature outside is 61°F right now. Later, it will be 90°F outside. If you and your friends want to play soccer in the park, what should you wear? What should you bring with you?

Skills Practice Lab

Keep It Clean

OBJECTIVES

Investigate how well antibacterial soap works.

Practice counting bacterial colonies.

MATERIALS

- incubator
- pencil, wax
- Petri dishes, nutrient agar–filled, sterile (3)
- scrub brush, new
- soap, liquid antibacterial
- stopwatch
- tape, transparent

SAFETY

One of the best ways to prevent the spread of bacterial and viral infections is to frequently wash your hands with soap and water. Many companies advertise that their soap ingredients can destroy bacteria normally found on the body. In this activity, you will investigate how effective antibacterial soaps are at killing bacteria.

Procedure

1. Keeping the agar plates closed at all times, use the wax pencil to label the bottoms of three agar plates. Label one plate "Control," one plate "No soap," and one plate "Soap."
2. Without washing your hands, carefully press several surfaces of your hands on the agar plate marked "Control." Have your partner immediately put the cover back on the plate. After you touch the agar, do not touch anything with either hand.
3. Hold your right hand under running water for 2 min. Ask your partner to scrub all surfaces of your right hand with the scrub brush throughout these 2 min. Be sure that he or she scrubs under your fingernails. After scrubbing, your partner should turn off the water and open the plate marked "No soap." Touching only the agar, carefully press on the "No soap" plate with the same surfaces of your right hand that you used to press on the "Control" plate.

4. Repeat step 3, but use your left hand instead of your right. This time, ask your partner to scrub your left hand with liquid antibacterial soap and the scrub brush. Use the plate marked "Soap" instead of the plate marked "No soap."

5. Secure the lid of each plate to its bottom half with transparent tape. Place the plates upside down in the incubator. Incubate all three plates overnight at 37°C.

6. Remove the plates from the incubator, and turn them right side up. Check each plate for the presence of bacterial colonies, and count the number of colonies present on each plate. Record this information. **Caution:** Do not remove the lids on any of the plates.

Analyze the Results

1. **Examining Data** Compare the bacterial growth on the plates. Which plate contained the most growth? Which contained the least?

Draw Conclusions

2. **Drawing Conclusions** Does water alone effectively kill bacteria? Explain.

Applying Your Data

Repeat this experiment, but scrub with regular, not antibacterial, liquid soap. Describe how the results of the two experiments differ.

Chapter Review

USING KEY TERMS

Complete each of the following sentences by choosing the correct term from the word bank.

nutrients	Food Guide Pyramid
addiction	drug
malnutrition	

1. Carbohydrates, proteins, fats, vitamins, minerals, and water are the six categories of ___.

2. The ___ divides foods into six groups and gives a recommended number of servings for each group.

3. Both bulimia nervosa and anorexia nervosa cause ___.

4. A physical or psychological dependence on a drug can lead to ___.

5. A(n) ___ is any substance that causes a change in a person's physical or psychological state.

UNDERSTANDING KEY IDEAS

Multiple Choice

6. Which of the following statements about drugs is true?
   - **a.** A child cannot become addicted to drugs.
   - **b.** Smoking just one or two cigarettes is safe for anyone.
   - **c.** Alcohol is not a drug.
   - **d.** Withdrawal symptoms may be painful.

7. What does alcohol do to the central nervous system (CNS)?
   - **a.** It speeds the CNS up.
   - **b.** It slows the CNS down.
   - **c.** It keeps the CNS regulated.
   - **d.** It has no effect on the CNS.

8. To keep your teeth healthy,
   - **a.** brush your teeth as hard as you can.
   - **b.** use a toothbrush until it is worn out.
   - **c.** brush at least twice a day.
   - **d.** floss at least once a week.

9. According to the Food Guide Pyramid, what foods should you eat most?
   - **a.** meats
   - **b.** milk, yogurt, and cheese
   - **c.** fruits and vegetables
   - **d.** bread, cereal, rice, and pasta

10. Which of the following can help you deal with stress?
    - **a.** ignoring your homework
    - **b.** drinking a caffeinated drink
    - **c.** talking to a friend
    - **d.** watching television

11. Tobacco use increases the risk of
    - **a.** lung cancer.
    - **b.** car accidents.
    - **c.** liver damage.
    - **d.** depression.

Short Answer

12. Are all narcotics illegal? Explain.

13. What are three dangers of tobacco and alcohol use?

14. What are the three types of nutrients that provide energy in Calories, and what is the main function of each type in the body?

15. Name two conditions that can lead to malnutrition.

16. Explain why you should always wear safety equipment when you ride your bicycle.

CRITICAL THINKING

17. **Concept Mapping** Use the following terms to create a concept map: *carbohydrates, water, proteins, nutrients, fats, vitamins, minerals, saturated fats,* and *unsaturated fats.*

18. **Applying Concepts** You have recently become a vegetarian, and you worry that you are not getting enough protein. Name two foods that you could eat to get more protein.

19. **Analyzing Ideas** Your two-year-old cousin will be staying with your family. Name three things that you can do to make sure that the house is safe for a young child.

INTERPRETING GRAPHICS

Look at the photos below. The people in the photos are not practicing safe habits. List the unsafe habits shown in these photos. For each unsafe habit, tell what the corresponding safe habit is.

20. 

21. 

Standardized Test Preparation

READING

Read each of the passages below. Then, answer the questions that follow each passage.

Passage 1 A chronic disease is a disease that, once developed, is always present and will not go away. Chronic bronchitis is a disease that causes the airways in the lungs to become swollen. This irritation causes a lot of mucus to form in the lungs. As a result, a person who has chronic bronchitis coughs a lot. Another chronic condition is emphysema. Emphysema destroys the tiny air sacs and the walls in the lungs. The holes in the air sacs cannot heal. Eventually, the lung tissue dies, and the lungs can no longer work. Cigarette smoking causes more than 80% of all cases of chronic bronchitis and emphysema.

1. In the passage, what does the word *chronic* mean?

A disappearing
B temporary
C always present
D mucus filled

2. According to the passage, what disease destroys the tiny air sacs and walls of the lungs?

F chronic bronchitis
G emphysema
H chronic cough
I cigarette smoking

3. Which of the following is a true statement according to the passage?

A Holes in the air sacs of lungs heal very quickly.
B Cigarette smoking causes more than 80% of all cases of chronic bronchitis and emphysema.
C Cigarette smoking does not cause chronic bronchitis or emphysema.
D Chronic bronchitis will go away after a person stops smoking cigarettes.

Passage 2 Each body reacts differently to alcohol. Several factors affect how a body reacts to alcohol. A person who has several drinks in a short time is likely to be affected more than a person who has a single drink in the same amount of time. Food in a drinker's stomach can also slow alcohol absorption into the blood. Finally, the way that women absorb and process alcohol differs from the way that men do. If a man and a woman drink the same amount of alcohol, the woman's blood alcohol content (BAC) will be higher than the man's. As BAC increases, mental and physical abilities decline. Muscle coordination, which is especially important for walking and driving, decreases. Vision becomes blurred. Speech and memory are impaired. A high BAC can cause a person to pass out or even die.

1. According to the passage, what does *BAC* stand for?

A blood alcohol content
B blood alcohol contaminant
C blurred alcohol capacity
D blood alcoholic coordination

2. According to the passage, which of the following factors can affect BAC?

F time of day
G food in the stomach
H age
I physical activity

3. Which of the following is a fact according to the passage?

A Alcohol does not affect mood or mental abilities.
B Men absorb alcohol in the same way that women do.
C Alcohol decreases muscle coordination.
D Everybody reacts to alcohol in the same way.

INTERPRETING GRAPHICS

The figure below shows a sample prescription drug label. Use this figure to answer the questions that follow.

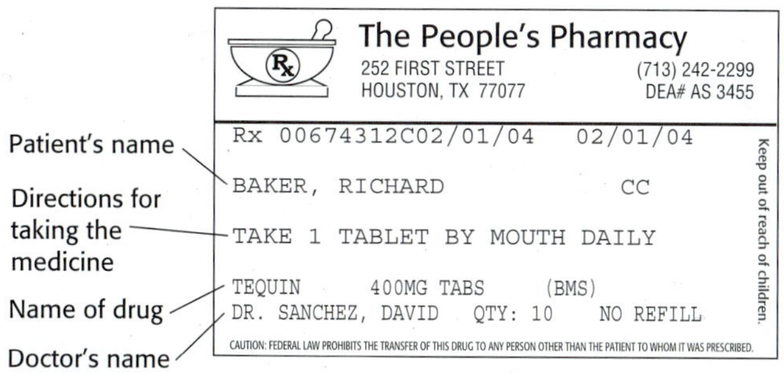

1. According to the label, what is the patient's name?
   - **A** Richard Baker
   - **B** Baker Richard
   - **C** David Sanchez
   - **D** James Beard

2. According to the label, how often should the medication be taken?
   - **F** once a day
   - **G** twice a day
   - **H** three times a day
   - **I** once a week

3. According to the label, how many refills remain on the prescription?
   - **A** 0
   - **B** 1
   - **C** 2
   - **D** 3

4. If this patient follows the directions exactly, how long will he need to take this medicine?
   - **F** 1 day
   - **G** 5 days
   - **H** 10 days
   - **I** There is not enough data to determine the answer.

MATH

Read each question below, and choose the best answer.

1. Which of the following ratios is equal to 2/4?
   - **A** 1/2
   - **B** 17/18
   - **C** 5/2
   - **D** 7/2

2. If 1 gal = 3.79 L, how many liters are in 3 gal?
   - **F** 3.79 L
   - **G** 7.58 L
   - **H** 11.37 L
   - **I** 15.16 L

3. Approximately how many liters are in 5 gal?
   - **A** 5 L
   - **B** 10 L
   - **C** 20 L
   - **D** 30 L

4. Ada has just built a car for a Pinewood Derby. She wants to find the average speed of her new car. During her first test run, she goes 5 mi/h. During her second run, she goes 4 mi/h, and in her third run, she goes 6 mi/h. What is her average speed?
   - **F** 4 mi/h
   - **G** 5 mi/h
   - **H** 6 mi/h
   - **I** 7 mi/h

5. Which of the following numbers is largest?
   - **A** 1×10^2
   - **B** 1×10^5
   - **C** 3×10^5
   - **D** 5×10^4

6. On Saturday, Mae won a goldfish at the school carnival. On the way home, Mae and her mother bought a fishbowl for $10.25, a container of fish food for $3.75, and a plastic coral for $8.15. How much money did Mae and her mother spend?
   - **F** $11.90
   - **G** $18.40
   - **H** $22.15
   - **I** $30.30

Science in Action

Bones can become severely weakened by the female athlete triad.

Science, Technology, and Society

Meatless Munching

Recent studies suggest that a vegetarian diet may reduce the risk of heart disease, adult-onset diabetes, and some forms of cancer. However, a vegetarian diet takes careful planning. Vegetarians must ensure that they get the proper balance of protein and vitamins in their diet. New foods that can help vegetarians remain healthy are being developed constantly. Meat substitutes are now made from soybeans, textured vegetable protein, and tofu. One new food, which is shown above, is made of a fungus that is a relative of mushrooms and truffles.

Social Studies ACTIVITY

WRITING SKILL Research a culture that has a mostly vegetarian diet, such as Hindu or Buddhist. What kinds of food do the people eat? Why don't they eat animals? Write a short report on your findings.

Scientific Discoveries

Female Athlete Triad

Getting enough exercise is an important part of staying healthy. But in 1992, doctors learned that too much exercise can be harmful for women. When a girl or woman exercises too much, three things can happen. She may lose too much weight. She may stop having her period. And her bones may become very weak. These three symptoms form the female athlete triad. To prevent this condition, female athletes need to take in enough Calories. Women who exercise heavily and try to lose weight may have a reduction in estrogen. Estrogen is the hormone that helps regulate the menstrual cycle. Low levels of estrogen and inadequate nutrition can cause bones to become weak and brittle. The photo above shows bone that has been weakened greatly.

Math ACTIVITY

Some scientists recommend that teenagers get 1,200 to 1,500 mg of calcium every day. A cup of milk has 300 mg of calcium, and a serving of yogurt has 400 mg of calcium. Calculate two combinations of milk and yogurt that would give you the recommended 1,500 mg of calcium.

Careers

Russell Selger

Guidance Counselor Guidance counselors help students think about their future by helping them discover their interests. After focusing their interests, a guidance counselor helps students plan a good academic schedule. A guidance counselor might talk to you about taking an art or computer science class that may help you discover a hidden talent. Many skills are vital to being a good guidance counselor. The job requires empathy, which is the ability to understand and sympathize with another person's feelings. Counselors also need patience, good listening skills, and a love of helping young people. Russell Selger, a guidance counselor at Timberlane Middle School, has a great respect for middle school students. "The kids are just alive. They want to learn. There's something about the spark that they have, and it's so much fun to guide them through all of this stuff," he explains.

WRITING SKILL Visit the guidance counselor's office at your school. What services does your guidance counselor offer? Conduct an interview with a guidance counselor. Ask why he or she became a counselor. Write an article for the school paper about your findings.

To learn more about these Science in Action topics, visit go.hrw.com and type in the keyword HL5BD7F.

Current Science

Check out Current Science® articles related to this chapter by visiting go.hrw.com. Just type in the keyword HL5CS28.

LabBook

Contents

Using Scientific Methods

Skills Practice Lab

Crystal Growth

Magma forms deep below the Earth's surface at depths of 25 km to 160 km and at extremely high temperatures. Some magma reaches the surface and cools quickly. Other magma gets trapped in cracks or magma chambers beneath the surface and cools very slowly. When magma cools slowly, large, well-developed crystals form. But when magma erupts onto the surface, it cools more quickly. There is not enough time for large crystals to grow. The size of the crystals found in igneous rocks gives geologists clues about where and how the rocks formed.

In this experiment, you will demonstrate how the rate of cooling affects the size of crystals in igneous rocks by cooling crystals of magnesium sulfate at two different rates.

MATERIALS

- aluminum foil
- basalt
- beaker, 400 mL
- gloves, heat-resistant
- granite
- hot plate
- laboratory scoop, pointed
- magnesium sulfate ($MgSO_4$) (Epsom salts)
- magnifying lens
- marker, dark
- pumice
- tape, masking
- test tube, medium-sized
- thermometer, Celsius
- tongs, test-tube
- watch (or clock)
- water, distilled
- water, tap, 200 mL

SAFETY

Ask a Question

1. How does temperature affect the formation of crystals?

Form a Hypothesis

2. Suppose you have two solutions that are identical in every way except for temperature. How will the temperature of a solution affect the size of the crystals and the rate at which they form?

Test the Hypothesis

3. Put on your gloves, apron, and goggles.
4. Fill the beaker halfway with tap water. Place the beaker on the hot plate, and let it begin to warm. The temperature of the water should be between 40°C and 50°C. **Caution:** Make sure the hot plate is away from the edge of the lab table.
5. Examine two or three crystals of the magnesium sulfate with your magnifying lens. On a separate sheet of paper, describe the color, shape, luster, and other interesting features of the crystals.
6. On a separate sheet of paper, draw a sketch of the magnesium sulfate crystals.

7. Use the pointed laboratory scoop to fill the test tube about halfway with the magnesium sulfate. Add an equal amount of distilled water.
8. Hold the test tube in one hand, and use one finger from your other hand to tap the test tube gently. Observe the solution mixing as you continue to tap the test tube.
9. Place the test tube in the beaker of hot water, and heat it for approximately 3 min. **Caution:** Be sure to direct the opening of the test tube away from you and other students.
10. While the test tube is heating, shape your aluminum foil into two small boatlike containers by doubling the foil and turning up each edge.
11. If all the magnesium sulfate is not dissolved after 3 min, tap the test tube again, and heat it for 3 min longer. **Caution:** Use the test-tube tongs to handle the hot test tube.

12. With a marker and a piece of masking tape, label one of your aluminum boats "Sample 1," and place it on the hot plate. Turn the hot plate off.
13. Label the other aluminum boat "Sample 2," and place it on the lab table.
14. Using the test-tube tongs, remove the test tube from the beaker of water, and evenly distribute the contents to each of your foil boats. Carefully pour the hot water in the beaker down the drain. Do not move or disturb either of your foil boats.
15. Copy the table below onto a separate sheet of paper. Using the magnifying lens, carefully observe the foil boats. Record the time it takes for the first crystals to appear.

Crystal-Formation Table

Crystal formation	Time	Size and appearance of crystals	Sketch of crystals
Sample 1			
Sample 2		DO NOT WRITE IN BOOK	

16. If crystals have not formed in the boats before class is over, carefully place the boats in a safe place. You may then record the time in days instead of in minutes.
17. When crystals have formed in both boats, use your magnifying lens to examine the crystals carefully.

Analyze the Results

1. Was your prediction correct? Explain.
2. Compare the size and shape of the crystals in Samples 1 and 2 with the size and shape of the crystals you examined in step 5. How long do you think the formation of the original crystals must have taken?

Draw Conclusions

3. Granite, basalt, and pumice are all igneous rocks. The most distinctive feature of each is the size of its crystals. Different igneous rocks form when magma cools at different rates. Examine a sample of each with your magnifying lens.
4. Copy the table below onto a separate sheet of paper, and sketch each rock sample.
5. Use what you have learned in this activity to explain how each rock sample formed and how long it took for the crystals to form. Record your answers in your table.

Igneous Rock Observations

	Granite	Basalt	Pumice
Sketch			
How did the rock sample form?			
Rate of cooling			

Communicating Your Data

Describe the size and shape of the crystals you would expect to find when a volcano erupts and sends material into the air and when magma oozes down the volcano's slope.

Model-Making Lab

Metamorphic Mash

Metamorphism is a complex process that takes place deep within the Earth, where the temperature and pressure would turn a human into a crispy pancake. The effects of this extreme temperature and pressure are obvious in some metamorphic rocks. One of these effects is the reorganization of mineral grains within the rock. In this activity, you will investigate the process of metamorphism without being charred, flattened, or buried.

MATERIALS

- cardboard (or plywood), very stiff, small pieces
- clay, modeling
- knife, plastic
- sequins (or other small flat objects)

SAFETY

Procedure

1. Flatten the clay into a layer about 1 cm thick. Sprinkle the surface with sequins.
2. Roll the corners of the clay toward the middle to form a neat ball.
3. Carefully use the plastic knife to cut the ball in half. On a separate sheet of paper, describe the position and location of the sequins inside the ball.
4. Put the ball back together, and use the sheets of cardboard or plywood to flatten the ball until it is about 2 cm thick.
5. Using the plastic knife, slice open the slab of clay in several places. Describe the position and location of the sequins in the slab.

Analyze the Results

1. What physical process does flattening the ball represent?
2. Describe any changes in the position and location of the sequins that occurred as the clay ball was flattened into a slab.

Draw Conclusions

3. How are the sequins oriented in relation to the force you put on the ball to flatten it?
4. Do you think the orientation of the mineral grains in a foliated metamorphic rock tells you anything about the rock? Defend your answer.

Applying Your Data

Suppose you find a foliated metamorphic rock that has grains running in two distinct directions. Use what you have learned in this activity to offer a possible explanation for this observation.

Skills Practice Lab

Power of the Sun

The sun radiates energy in every direction. Like the sun, the energy radiated by a light bulb spreads out in all directions. But how much energy an object receives depends on how close that object is to the source. As you move farther from the source, the amount of energy you receive decreases. For example, if you measure the amount of energy that reaches you from a light and then move three times farther away, you will discover that nine times less energy will reach you at your second position. Energy from the sun travels as light energy. When light energy is absorbed by an object it is converted into thermal energy. Power is the rate at which one form of energy is converted to another, and it is measured in watts. Because power is related to distance, nearby objects can be used to measure the power of far-away objects. In this lab you will calculate the power of the sun using an ordinary 100-watt light bulb.

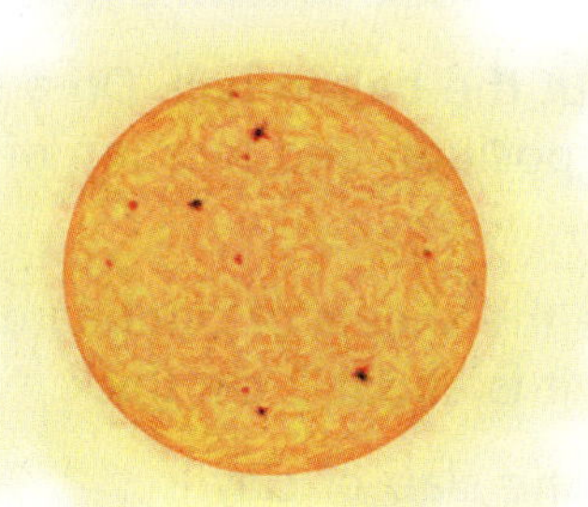

MATERIALS

- aluminum strip, 2 × 8 cm
- calculator, scientific
- clay, modeling
- desk lamp with a 100 W bulb and removable shade
- gloves, protective
- marker, black permanent
- mason jar, cap, and lid with hole in center
- pencil
- ruler, metric
- thermometer, Celsius
- watch (or clock) that indicates seconds

Procedure

1. Gently shape the piece of aluminum around a pencil so that it holds on in the middle and has two wings, one on either side of the pencil.
2. Bend the wings outward so that they can catch as much sunlight as possible.
3. Use the marker to color both wings on one side of the aluminum strip black.
4. Remove the pencil and place the aluminum snugly around the thermometer near the bulb. **Caution:** Do not press too hard—you do not want to break the thermometer! Wear protective gloves when working with the thermometer and the aluminum.
5. Carefully slide the top of the thermometer through the hole in the lid. Place the lid on the jar so that the thermometer bulb is inside the jar, and screw down the cap.
6. Secure the thermometer to the jar lid by molding clay around the thermometer on the outside of the lid. The aluminum wings should be in the center of the jar.
7. Read the temperature on the thermometer. Record this as room temperature.
8. Place the jar on a windowsill in the sunlight. Turn the jar so that the black wings are angled toward the sun.
9. Watch the thermometer until the temperature reading stops rising. Record the temperature.
10. Remove the jar from direct sunlight, and allow it to return to room temperature.
11. Remove any shade or reflector from the lamp. Place the lamp at one end of a table.
12. Place the jar about 30 cm from the lamp. Turn the jar so that the wings are angled toward the lamp.

13. Turn on the lamp, and wait about 1 minute.
14. Move the jar a few centimeters toward the lamp until the temperature reading starts to rise. When the temperature stops rising, compare it with the reading you took in step 9.
15. Repeat step 14 until the temperature matches the temperature you recorded in step 9.
16. If the temperature reading rises too high, move the jar away from the lamp and allow it to cool. Once the reading has dropped to at least 5°C below the temperature you recorded in step 9, you may begin again at step 12.
17. When the temperature in the jar matches the temperature you recorded in step 9, record the distance between the center of the light bulb and the thermometer bulb.

Analyze the Results

1. The thermometer measured the same amount of energy absorbed by the jar at the distance you measured to the lamp. In other words, your jar absorbed as much energy from the sun at a distance of 150 million kilometers as it did from the 100 W light bulb at the distance you recorded in step 17.
2. Use the following formula to calculate the power of the sun (be sure to show your work):

$$\frac{\text{power of the sun}}{(\text{distance to the sun})^2} = \frac{\text{power of the lamp}}{(\text{distance to the lamp})^2}$$

Hint: $(\text{distance})^2$ means that you multiply the distance by itself. If you found that the lamp was 5 cm away from the jar, for example, the $(\text{distance})^2$ would be 25.

Hint: Convert 150,000,000 km to 15,000,000,000,000 cm.

3. Review the discussion of scientific notation in the Math Refresher found in the Appendix at the back of this book. You will need to understand this technique for writing large numbers in order to compare your calculation with the actual figure. For practice, convert the distance to the sun given above in step 2 of Analyze the Results to scientific notation.

15,000,000,000,000 cm = $1.5 \times 10^{\underline{?}}$ cm

Draw Conclusions

4. The sun emits 3.7×10^{26} W of power. Compare your answer in step 2 with this value. Was this a good way to calculate the power of the sun? Explain.

Using Scientific Methods

Model-Making Lab

Oh, the Pressure!

When scientists want to understand natural processes, such as mountain formation, they often make models to help them. Models are useful in studying how rocks react to the forces of plate tectonics. A model can demonstrate in a short amount of time geological processes that take millions of years. Do the following activity to find out how folding and faulting occur in the Earth's crust.

MATERIALS

- can, soup (or rolling pin)
- clay, modeling, 4 colors
- knife, plastic
- newspaper
- pencils, colored
- poster board, 5 cm × 5 cm squares (2)
- poster board, 5 cm × 15 cm strip

SAFETY

Ask a Question

1. How do synclines, anticlines, and faults form?

Form a Hypothesis

2. On a separate piece of paper, write a hypothesis that is a possible answer to the question above. Explain your reasoning.

Test the Hypothesis

3. Use modeling clay of one color to form a long cylinder, and place the cylinder in the center of the glossy side of the poster-board strip.
4. Mold the clay to the strip. Try to make the clay layer the same thickness all along the strip; you can use the soup can or rolling pin to even it out. Pinch the sides of the clay so that the clay is the same width and length as the strip. Your strip should be at least 15 cm long and 5 cm wide.

5. Flip the strip over on the newspaper your teacher has placed across your desk. Carefully peel the strip from the modeling clay.
6. Repeat steps 3–5 with the other colors of modeling clay. Each person should have a turn molding the clay. Each time you flip the strip over, stack the new clay layer on top of the previous one. When you are finished, you should have a block of clay made of four layers.
7. Lift the block of clay, and hold it parallel to and just above the tabletop. Push gently on the block from opposite sides, as shown below.

8. Use the colored pencils to draw the results of step 6. Use the terms *syncline* and *anticline* to label your diagram. Draw arrows to show the direction that each edge of the clay was pushed.
9. Repeat steps 3–6 to form a second block of clay.
10. Cut the second block of clay in two at a 45° angle as seen from the side of the block.

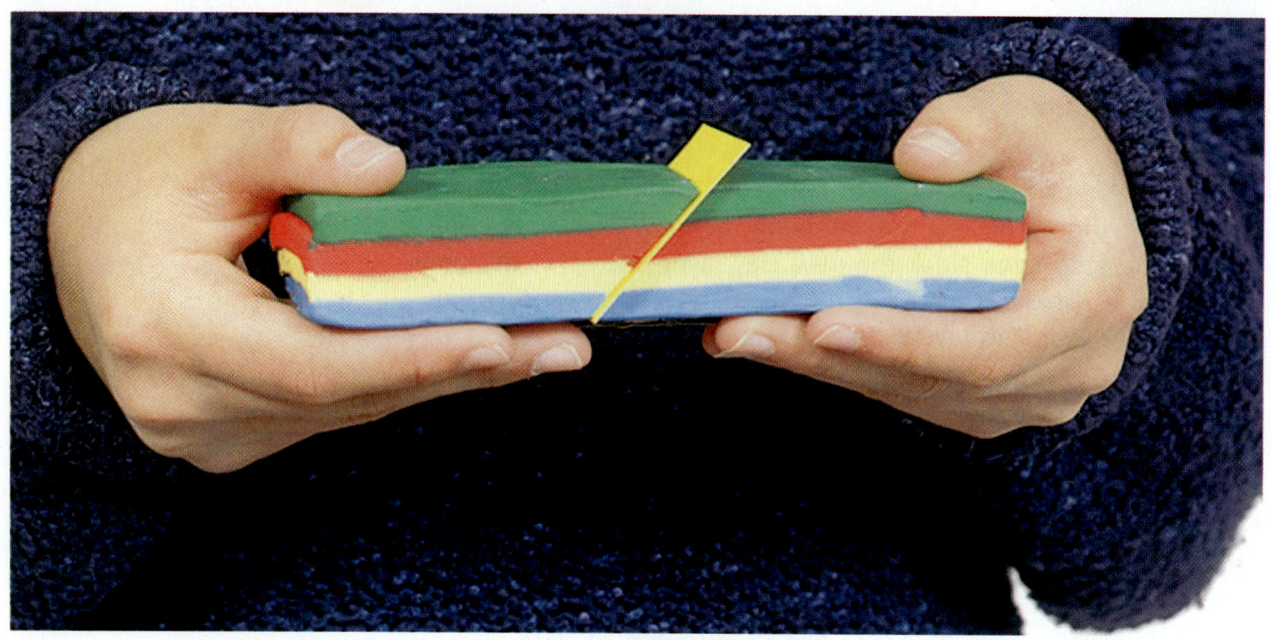

11. Press one poster-board square on the angled end of each of the block's two pieces. The poster board represents a fault. The two angled ends represent a hanging wall and a footwall. The model should resemble the one in the photograph above.
12. Keeping the angled edges together, lift the blocks, and hold them parallel to and just above the tabletop. Push gently on the two blocks until they move. Record your observations.
13. Now, hold the two pieces of the clay block in their original position, and slowly pull them apart, allowing the hanging wall to move downward. Record your observations.

Analyze the Results

1. What happened to the first block of clay in step 7? What kind of force did you apply to the block of clay?
2. What happened to the pieces of the second block of clay in step 12? What kind of force did you apply to them?
3. What happened to the pieces of the second block of clay in step 13? Describe the forces that acted on the block and the way the pieces of the block reacted.

Draw Conclusions

4. Summarize how the forces you applied to the blocks of clay relate to the way tectonic forces affect rock layers. Be sure to use the terms *fold, fault, anticline, syncline, hanging wall, footwall, tension,* and *compression* in your summary.

Skills Practice Lab

Some Go "Pop," Some Do Not

Volcanic eruptions range from mild to violent. When volcanoes erupt, the materials left behind provide information to scientists studying the Earth's crust. Mild, or nonexplosive, eruptions produce thin, runny lava that is low in silica. During nonexplosive eruptions, lava simply flows down the side of the volcano. Explosive eruptions, on the other hand, do not produce much lava. Instead, the explosions hurl ash and debris into the air. The materials left behind are light in color and high in silica. These materials help geologists determine the composition of the crust underneath the volcanoes.

MATERIALS

- paper, graph (1 sheet)
- pencils (or markers), red, yellow, and orange
- ruler, metric

Procedure

1. Copy the map below onto graph paper. Take care to line the grid up properly.
2. Locate each volcano from the list on the next page by drawing a circle with a diameter of about 2 mm in the proper location on your copy of the map. Use the latitude and longitude grids to help you.
3. Review all the eruptions for each volcano. For each explosive eruption, color the circle red. For each quiet volcano, color the circle yellow. For volcanoes that have erupted in both ways, color the circle orange.

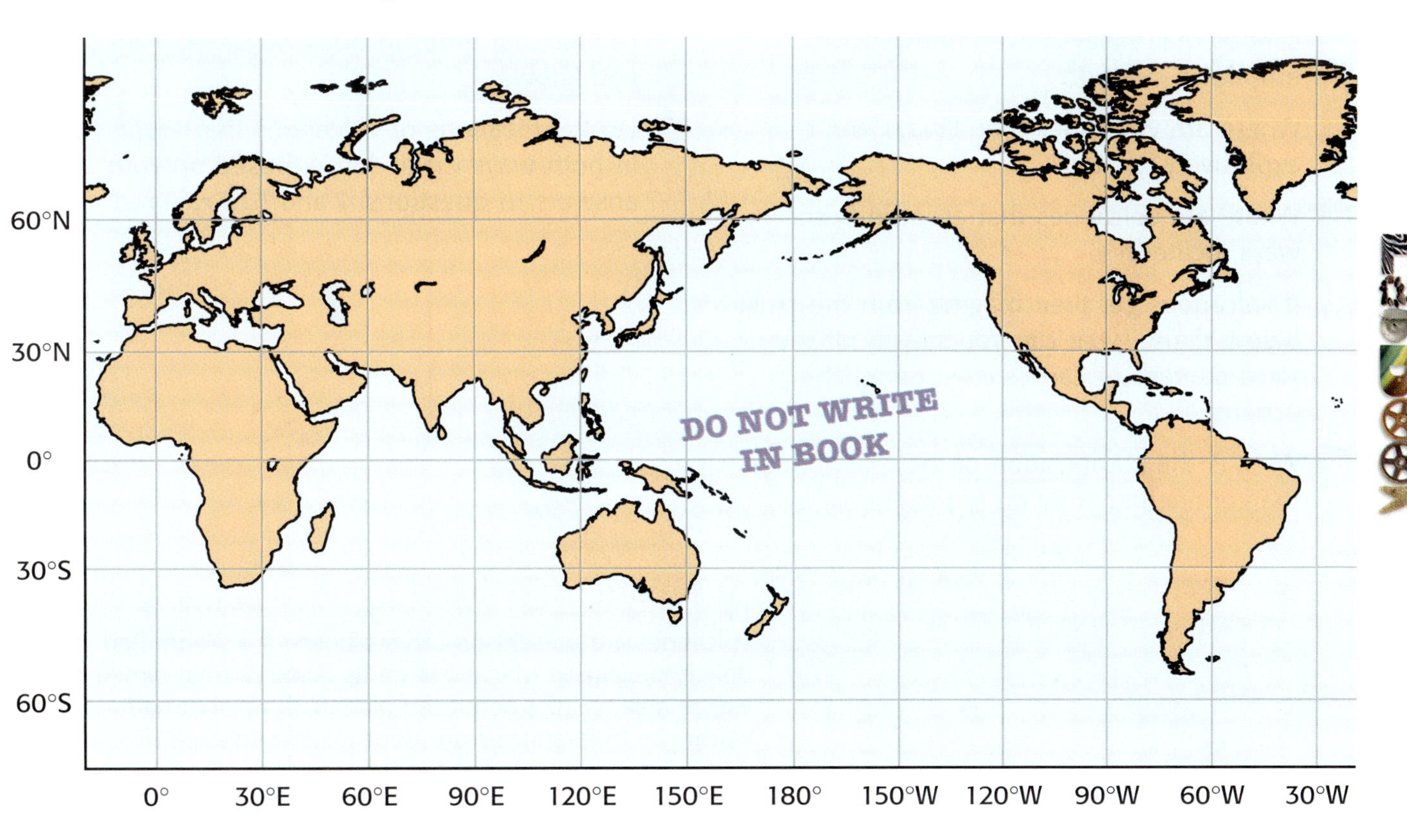

Volcanic Activity Chart		
Volcano name	**Location**	**Description**
Mount St. Helens	46°N 122°W	An explosive eruption blew the top off the mountain. Light-colored ash covered thousands of square kilometers. Another eruption sent a lava flow down the southeast side of the mountain.
Kilauea	19°N 155°W	One small eruption sent a lava flow along 12 km of highway.
Rabaul caldera	4°S 152°E	Explosive eruptions have caused tsunamis and have left 1–2 m of ash on nearby buildings.
Popocatépetl	19°N 98°W	During one explosion, Mexico City closed the airport for 14 hours because huge columns of ash made it too difficult for pilots to see. Eruptions from this volcano have also caused damaging avalanches.
Soufriere Hills	16°N 62°W	Small eruptions have sent lava flows down the hills. Other explosive eruptions have sent large columns of ash into the air.
Long Valley caldera	37°N 119°W	Explosive eruptions have sent ash into the air.
Okmok	53°N 168°W	Recently, there have been slow lava flows from this volcano. Twenty-five hundred years ago, ash and debris exploded from the top of this volcano.
Pavlof	55°N 161°W	Eruption clouds have been sent 200 m above the summit. Eruptions have sent ash columns 10 km into the air. Occasionally, small eruptions have caused lava flows.
Fernandina	42°N 12°E	Eruptions have ejected large blocks of rock from this volcano.
Mount Pinatubo	15°N 120°E	Ash and debris from an explosive eruption destroyed homes, crops, and roads within 52,000 km^2 around the volcano.

Analyze the Results

1. According to your map, where are volcanoes that always have nonexplosive eruptions located?
2. Where are volcanoes that always erupt explosively located?
3. Where are volcanoes that erupt in both ways located?
4. If volcanoes get their magma from the crust below them, what can you say about the silica content of Earth's crust under the oceans?
5. What is the composition of the crust under the continents? How do we know?

Draw Conclusions

6. What is the source of materials for volcanoes that erupt in both ways? How do you know?
7. Do the locations of volcanoes that erupt in both ways make sense, based on your answers to questions 4 and 5? Explain.

Applying Your Data

Volcanoes are present on other planets. If a planet had only nonexplosive volcanoes on its surface, what would we be able to infer about the planet? If a planet had volcanoes that ranged from nonexplosive to explosive, what might that tell us about the planet?

Model-Making Lab

Dune Movement

Wind moves the sand by a process called *saltation.* The sand skips and bounces along the ground in the same direction as the wind is blowing. As sand is blown across a beach, the dunes change. In this activity, you will investigate the effect wind has on a model sand dune.

MATERIALS

- bag, paper, large enough to hold half the box
- box, cardboard, shallow
- hair dryer
- marker
- mask, filter
- ruler, metric
- sand, fine

SAFETY

Procedure

1. Use the marker to draw and label vertical lines 5 cm apart along one side of the box.
2. Fill the box about halfway with sand. Brush the sand into a dune shape about 10 cm from the end of the box.
3. Use the lines you drew along the edge of the box to measure the location of the dune's peak to the nearest centimeter.
4. Slide the box into the paper bag until only about half the box is exposed, as shown below.
5. Put on your safety goggles and filter mask. Hold the hair dryer so that it is level with the peak of the dune and about 10–20 cm from the open end of the box.
6. Turn on the hair dryer at the lowest speed, and direct the air toward the model sand dune for 1 min.
7. Record the new location of the model dune.
8. Repeat steps 5 and 6 three times. After each trial, measure and record the location of the dune's peak.

Analyze the Results

1. How far did the dune move during each trial?
2. How far did the dune move overall?

Draw Conclusions

3. How might the dune's movement be affected if you were to turn the hair dryer to the highest speed?

Applying Your Data

Flatten the sand. Place a barrier, such as a rock, in the sand. Position the hair dryer level with the top of the sand's surface. How does the rock affect the dune's movement?

Using Scientific Methods

Skills Practice Lab

Creating a Kettle

As glaciers recede, they leave huge amounts of rock material behind. Sometimes receding glaciers form moraines by depositing some of the rock material in ridges. At other times, glaciers leave chunks of ice that form depressions called *kettles*. As the ice melts, these depressions may form ponds or lakes. In this activity, you will discover how kettles are formed by creating your own.

MATERIALS

- ice, cubes of various sizes (4–5)
- ruler, metric
- sand
- tub, small

Ask a Question

1. How are kettles formed?

Form a Hypothesis

2. Write a hypothesis that could answer the question above.

Test the Hypothesis

3. Fill the tub three-quarters full with sand.
4. Describe the size and shape of each ice cube.
5. Push the ice cubes to various depths in the sand.
6. Put the tub where it won't be disturbed overnight.
7. Closely observe the sand around the area where you left each ice cube.
8. What happened to the ice cubes?
9. Use a metric ruler to measure the depth and diameter of the indentation left by each ice cube.

Analyze the Results

1. How does this model relate to the size and shape of a natural kettle?
2. In what ways are your model kettles similar to real ones? How are they different?

Draw Conclusions

3. Based on your model, what can you conclude about the formation of kettles by receding glaciers?

Skills Practice Lab

Global Impact

For years, scientists have debated the topic of global warming. Is the temperature of the Earth actually getting warmer? In this activity, you will examine a table to determine if the data indicate any trends. Be sure to notice how much the trends seem to change as you analyze different sets of data.

MATERIALS

- pencils, colored (4)
- ruler, metric

Procedure

1. The table below shows average global temperatures recorded over the last 100 years.
2. Draw a graph. Label the horizontal axis "Time." Mark the grid in 5-year intervals. Label the vertical axis "Temperature (°C)," with values ranging from 13°C to 15°C.
3. Starting with 1900, use the numbers in red to plot the temperature in 20-year intervals. Connect the dots with straight lines.
4. Using a ruler, estimate the average slope for the temperatures. Draw a red line to represent the slope.
5. Using different colors, plot the temperatures at 10-year intervals and 5-year intervals on the same graph. Connect each set of dots, and draw the average slope for each set.

Analyze the Results

1. Examine your completed graph, and explain any trends you see in the graphed data. Was there an increase or a decrease in average temperature over the last 100 years?
2. What similarities and differences did you see between each set of graphed data?

Draw Conclusions

3. What conclusions can you draw from the data you graphed in this activity?
4. What would happen if your graph were plotted in 1-year intervals? Try it!

Average Global Temperatures

Year	°C	Year	°C	Year	°C	Year	°C	Year	°C	Year	°C
1900	**14.0**	1917	13.6	1934	14.0	1951	14.0	1968	13.9	1985	14.1
1901	13.9	1918	13.6	**1935**	**13.9**	1952	14.0	1969	14.0	1986	14.2
1902	13.8	1919	13.8	1936	14.0	1953	14.1	**1970**	**14.0**	1987	14.3
1903	13.6	**1920**	**13.8**	1937	14.1	1954	13.9	1971	13.9	1988	14.4
1904	13.5	1921	13.9	1938	14.1	**1955**	**13.9**	1972	13.9	1989	14.2
1905	**13.7**	1922	13.9	1939	14.0	1956	13.8	1973	14.2	**1990**	**14.5**
1906	13.8	1923	13.8	**1940**	**14.1**	1957	14.1	1974	13.9	1991	14.4
1907	13.6	1924	13.8	1941	14.1	1958	14.1	**1975**	**14.0**	1992	14.1
1908	13.7	**1925**	**13.8**	1942	14.1	1959	14.0	1976	13.8	1993	14.2
1909	13.7	1926	14.1	1943	14.0	**1960**	**14.0**	1977	14.2	1994	14.3
1910	**13.7**	1927	14.0	1944	14.1	1961	14.1	1978	14.1	**1995**	**14.5**
1911	13.7	1928	14.0	**1945**	**14.0**	1962	14.0	1979	14.1	1996	14.4
1912	13.7	1929	13.8	1946	14.0	1963	14.0	**1980**	**14.3**	1997	14.4
1913	13.8	**1930**	**13.9**	1947	14.1	1964	13.7	1981	14.4	1998	14.5
1914	14.0	1931	14.0	1948	14.0	**1965**	**13.8**	1982	14.1	1999	14.5
1915	14.0	1932	14.0	1949	13.9	1966	13.9	1983	14.3	**2000**	**14.5**
1916	13.8	1933	13.9	**1950**	**13.8**	1967	14.0	1984	14.1	2001	14.5

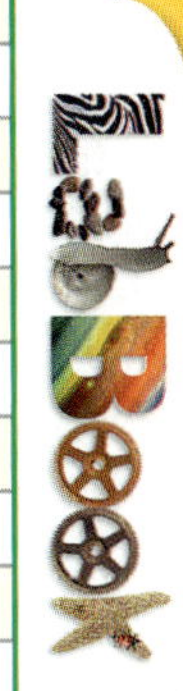

Using Scientific Methods

Skills Practice Lab

For the Birds

You and a partner have a new business building birdhouses. But your first clients have told you that birds do not want to live in the birdhouses you have made. The clients want their money back unless you can solve the problem. You need to come up with a solution right away!

You remember reading an article about microclimates in a science magazine. Cities often heat up because the pavement and buildings absorb so much solar radiation. Maybe the houses are too warm! How can the houses be kept cooler?

You decide to investigate the roofs; after all, changing the roofs would be a lot easier than building new houses. In order to help your clients and the birds, you decide to test different roof colors and materials to see how these variables affect a roof's ability to absorb the sun's rays.

One partner will test the color, and the other partner will test the materials. You will then share your results and make a recommendation together.

MATERIALS

- cardboard (4 pieces)
- paint, black, white, and light blue tempera
- rubber, beige or tan
- thermometers, Celsius (4)
- watch (or clock)
- wood, beige or tan

SAFETY

Part A: Color Test

Ask a Question

1. What color would be the best choice for the roof of a birdhouse?

Form a Hypothesis

2. Write down the color you think will keep a birdhouse coolest.

Test the Hypothesis

3. Paint one piece of cardboard black, another piece white, and a third light blue.
4. After the paint has dried, take the three pieces of cardboard outside, and place a thermometer on each piece.
5. In an area where there is no shade, place each piece at the same height so that all three receive the same amount of sunlight. Leave the pieces in the sunlight for 15 min.
6. Leave a fourth thermometer outside in the shade to measure the temperature of the air.
7. Record the reading of the thermometer on each piece of cardboard. Also, record the outside temperature.

Analyze the Results

1. Did each of the three thermometers record the same temperature after 15 min? Explain.
2. Were the temperature readings on each of the three pieces of cardboard the same as the reading for the outside temperature? Explain.

Draw Conclusions

3. How do your observations compare with your hypothesis?

Part B: Material Test

Ask a Question

1. Which material would be the best choice for the roof of a birdhouse?

Form a Hypothesis

2. Write down the material you think will keep a birdhouse coolest.

Test the Hypothesis

3. Take the rubber, wood, and the fourth piece of cardboard outside, and place a thermometer on each.
4. In an area where there is no shade, place each material at the same height so that they all receive the same amount of sunlight. Leave the materials in the sunlight for 15 min.
5. Leave a fourth thermometer outside in the shade to measure the temperature of the air.
6. Record the temperature of each material. Also, record the outside temperature. After you and your partner have finished your investigations, take a few minutes to share your results.

Analyze the Results

1. Did each of the thermometers on the three materials record the same temperature after 15 min? Explain.
2. Were the temperature readings on the rubber, wood, and cardboard the same as the reading for the outside temperature? Explain.

Draw Conclusions

3. How do your observations compare with your hypothesis?
4. Which material would you use to build the roofs for your birdhouses? Why?
5. Which color would you use to paint the new roofs? Why?

Applying Your Data

Make three different-colored samples for each of the three materials. When you measure the temperatures for each sample, how do the colors compare for each material? Is the same color best for all three materials? How do your results compare with what you concluded in steps 4 and 5 under Draw Conclusions of this activity? What's more important, color or material?

Using Scientific Methods

Skills Practice Lab

I See the Light!

How do you find the distance to an object you can't reach? You can do it by measuring something you can reach, finding a few angles, and using mathematics. In this activity, you'll practice measuring the distances of objects here on Earth. When you get used to it, you can take your skills to the stars!

MATERIALS

- calculator, scientific
- meterstick
- pencil, sharp
- poster board, 16 × 16 cm
- protractor
- ruler, metric
- scissors
- string, 30 cm
- tape measure, metric
- tape, transparent

SAFETY

Ask a Question

1. How can you measure the distance to a star?

Form a Hypothesis

2. Write a hypothesis that might answer this question. Explain your reasoning.

Test the Hypothesis

3. Draw a line 4 cm away from the edge of one side of the piece of poster board. Fold the poster board along this line.
4. Tape the protractor to the poster board with its flat edge against the fold, as shown in the photo below.

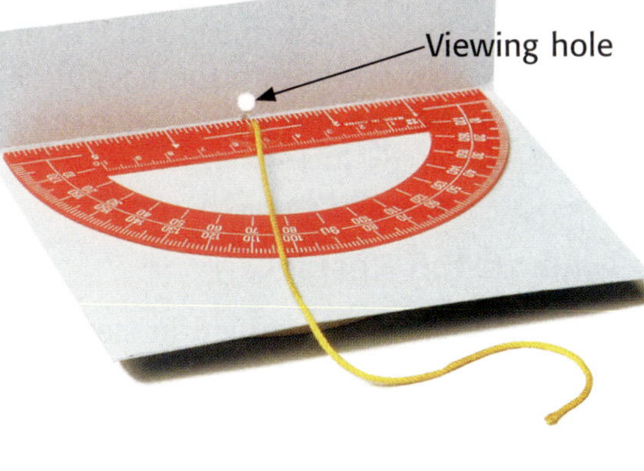

5. Use a pencil to carefully punch a hole through the poster board along its folded edge at the center of the protractor.
6. Thread the string through the hole, and tape one end to the underside of the poster board. The other end should be long enough to hang off the far end of the poster board.
7. Carefully punch a second hole in the smaller area of the poster board halfway between its short sides. The hole should be directly above the first hole and should be large enough for the pencil to fit through. This hole is the viewing hole of your new parallax device. This device will allow you to measure the distance of faraway objects.
8. Find a location that is at least 50 steps away from a tall, narrow object, such as the school's flagpole or a tall tree. (This object will represent background stars.) Set the meterstick on the ground with one of its long edges facing the flagpole.
9. Ask your partner, who represents a nearby star, to take 10 steps toward the flagpole, starting at the left end of the meterstick. You will be the observer. When you stand at the left end of the meterstick, which represents the location of the sun, your partner's nose should be lined up with the flagpole.

10. Move to the other end of the meterstick, which represents the location of Earth. Does your partner appear to the left or right of the flagpole? Record your observations.
11. Hold the string so that it runs straight from the viewing hole to the 90° mark on the protractor. Using one eye, look through the viewing hole along the string, and point the device at your partner's nose.
12. Holding the device still, slowly move your head until you can see the flagpole through the viewing hole. Move the string so that it lines up between your eye and the flagpole. Make sure the string is taut, and hold it tightly against the protractor.
13. Read and record the angle made by the string and the string's original position at 90° (count the number of degrees between 90° and the string's new position).
14. Use the measuring tape to find and record the distance from the left end of the meterstick to your partner's nose.
15. Now, find a place outside that is at least 100 steps away from the flagpole. Set the meterstick on the ground as before, and repeat steps 9–14.

Analyze the Results

1. The angle you recorded in step 13 is called the *parallax angle.* The distance from one end of the meterstick to the other is called the *baseline.* With this angle and the length of your baseline, you can calculate the distance to your partner.
2. To calculate the distance (d) to your partner, use the following equation:

$$d = b/\tan A$$

In this equation, A is the parallax angle, and b is the length of the baseline (1 m). (Tan A means the tangent of angle A, which you will learn more about in math classes.)

3. To find d, enter 1 (the length of your baseline in meters) into the calculator, press the division key, enter the value of A (the parallax angle you recorded), then press the tan key. Finally, press the equals key.

4. Record this result. It is the distance in meters between the left end of the meterstick and your partner. You may want to use a table like the one below.
5. How close is this calculated distance to the distance you measured?
6. Repeat steps 1–3 under Analyze the Results using the angle you found when the flagpole was 100 steps away.

Draw Conclusions

7. At which position, 50 steps or 100 steps from the flagpole, did your calculated distance better match the actual distance as measured?
8. What do you think would happen if you were even farther from the flagpole?
9. When astronomers use parallax, their "flagpoles" are distant stars. Might this affect the accuracy of their parallax readings?

Distance by Parallax Versus Measuring Tape

	At 50 steps	At 100 steps
Parallax		
angle		
Distance (calculated)		
Distance (measured)		

Skills Practice Lab

Built for Speed

Imagine that you are an engineer at GoCarCo, a toy-vehicle company. GoCarCo is trying to beat the competition by building a new toy vehicle. Several new designs are being tested. Your boss has given you one of the new toy vehicles and instructed you to measure its speed as accurately as possible with the tools you have. Other engineers (your classmates) are testing the other designs. Your results could decide the fate of the company!

MATERIALS

- meterstick
- stopwatch
- tape, masking
- toy vehicle

SAFETY

Procedure

1. How will you accomplish your goal? Write a paragraph to describe your goal and your procedure for this experiment. Be sure that your procedure includes several trials.
2. Show your plan to your boss (teacher). Get his or her approval to carry out your procedure.
3. Perform your stated procedure. Record all data. Be sure to express all data in the correct units.

Analyze the Results

1. What was the average speed of your vehicle? How does your result compare with the results of the other engineers?
2. Compare your technique for determining the speed of your vehicle with the techniques of the other engineers. Which technique do you think is the most effective?
3. Was your toy vehicle the fastest? Explain why or why not.

Applying Your Data

Think of several conditions that could affect your vehicle's speed. Design an experiment to test your vehicle under one of those conditions. Write a paragraph to explain your procedure. Be sure to include an explanation of how that condition changes your vehicle's speed.

Skills Practice Lab

Relating Mass and Weight

Why do objects with more mass weigh more than objects with less mass? All objects have weight on Earth because their mass is affected by Earth's gravitational force. Because the mass of an object on Earth is constant, the relationship between the mass of an object and its weight is also constant. You will measure the mass and weight of several objects to verify the relationship between mass and weight on the surface of Earth.

MATERIALS

- balance, metric
- classroom objects, small
- paper, graph
- scissors
- spring scale (force meter)
- string

SAFETY

Procedure

1. Copy the table below.

Mass and Weight Measurements

Object	Mass (g)	Weight (N)
	DO NOT WRITE IN BOOK	

2. Using the metric balance, find the mass of five or six small classroom objects designated by your teacher. Record the masses.
3. Using the spring scale, find the weight of each object. Record the weights. (You may need to use the string to create a hook with which to hang some objects from the spring scale, as shown at right.)

Analyze the Results

1. Using your data, construct a scatterplot of weight (*y*-axis) versus mass (*x*-axis). Draw a line that best fits all your data points.
2. Does the graph confirm the relationship between mass and weight on Earth? Explain your answer.

Using Scientific Methods

Skills Practice Lab

Science Friction

In this experiment, you will investigate three types of friction—static, sliding, and rolling—to determine which is the largest force and which is the smallest force.

Ask a Question

1. Which type of friction is the largest force—static, sliding, or rolling? Which is the smallest?

Form a Hypothesis

2. Write a statement or statements that answer the questions above. Explain your reasoning.

Test the Hypothesis

3. Cut a piece of string, and tie it in a loop that fits in the textbook, as shown on the next page. Hook the string to the spring scale.
4. Practice the next three steps several times before you collect data.
5. To measure the static friction between the book and the table, pull the spring scale very slowly. Record the largest force on the scale before the book starts to move.
6. After the book begins to move, you can determine the sliding friction. Record the force required to keep the book sliding at a slow, constant speed.
7. Place two or three rods under the book to act as rollers. Make sure the rollers are evenly spaced. Place another roller in front of the book so that the book will roll onto it. Pull the force meter slowly. Measure the force needed to keep the book rolling at a constant speed.

MATERIALS

- rods, wood or metal (3–4)
- scissors
- spring scale (force meter)
- string
- textbook (covered)

SAFETY

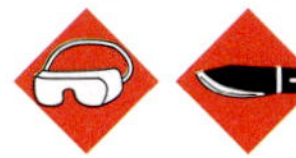

Analyze the Results

1. Which type of friction was the largest? Which was the smallest?
2. Do the results support your hypothesis? If not, how would you revise or retest your hypothesis?

Draw Conclusions

3. Compare your results with those of another group. Are there any differences? Working together, design a way to improve the experiment and resolve possible differences.

Using Scientific Methods

Skills Practice Lab

A Marshmallow Catapult

Catapults use projectile motion to launch objects. In this lab, you will build a simple catapult and determine the angle at which the catapult will launch an object the farthest.

MATERIALS

- marshmallows, miniature (2)
- meterstick
- protractor
- spoon, plastic
- tape, duct
- wood block, 3.5 cm × 3.5 cm × 1 cm

SAFETY

Ask a Question

1. At what angle, from 10° to 90°, will a catapult launch a marshmallow the farthest?

Form a Hypothesis

2. Write a hypothesis that is a possible answer to your question.

Angle	Distance 1 (cm)	Distance 2 (cm)	Average distance	Data Collection
10°	DO NOT WRITE IN BOOK			

Test the Hypothesis

3. Copy the table above. In your table, add one row each for 20°, 30°, 40°, 50°, 60°, 70°, 80°, and 90° angles.
4. Using duct tape, attach the plastic spoon to the 1 cm side of the block. Use enough tape to attach the spoon securely.
5. Place one marshmallow in the center of the spoon, and tape it to the spoon. This marshmallow serves as a ledge to hold the marshmallow that will be launched.
6. Line up the bottom corner of the block with the bottom center of the protractor, as shown in the photograph. Start with the block at 10°.
7. Place a marshmallow in the spoon, on top of the taped marshmallow. Pull the spoon back lightly, and let go. Measure and record the distance from the catapult that the marshmallow lands. Repeat the measurement, and calculate an average.
8. Repeat step 7 for each angle up to 90°.

Analyze the Results

1. At what angle did the catapult launch the marshmallow the farthest? Explain any differences from your hypothesis.

Draw Conclusions

2. At what angle should you throw a ball or shoot an arrow so that it will fly the farthest? Why? Support your answer with your data.

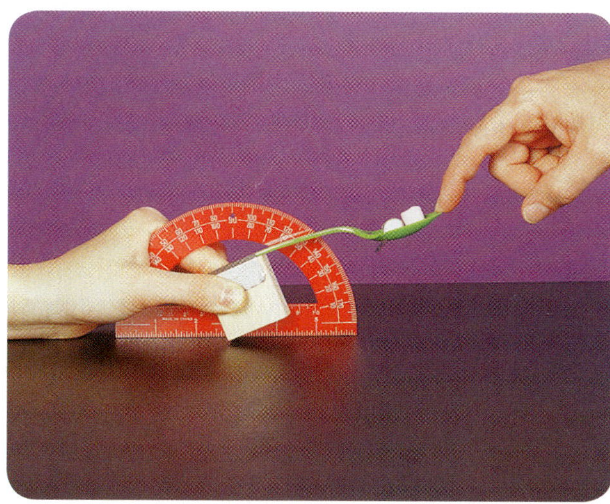

Model-Making Lab

Blast Off!

You have been hired as a rocket scientist for NASA. Your job is to design a rocket that will have a controlled flight while carrying a payload. Keep in mind that Newton's laws will have a powerful influence on your rocket.

MATERIALS

- balloon, long, thin
- cup, paper, small
- fishing line, 3 m
- meterstick
- pencil
- pennies
- straw, straight plastic
- string, 15 cm (2)
- tape, masking
- twist tie

SAFETY

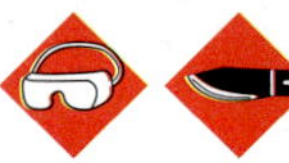

Procedure

1. When you begin your experiment, your teacher will tape one end of the fishing line to the ceiling.
2. Use a pencil to poke a small hole in each side of the cup near the top. Place a 15 cm piece of string through each hole, and tape down the ends inside.
3. Inflate the balloon, and use the twist tie to hold it closed.
4. Tape the free ends of the strings to the sides of the balloon near the bottom. The cup should hang below the balloon. Your model rocket should look like a hot-air balloon.
5. Thread the fishing line that is hanging from the ceiling through the straw. Tape the balloon securely to the straw. Tape the loose end of the fishing line to the floor.
6. Untie the twist tie while holding the end of the balloon closed. When you are ready, release the end of the balloon. Mark and record the maximum height of the rocket.
7. Repeat the procedure, adding a penny to the cup each time until your rocket cannot lift any more pennies.

Analyze the Results

1. In a paragraph, describe how all three of Newton's laws influenced the flight of your rocket.

Draw Conclusions

2. Draw a diagram of your rocket. Label the action and reaction forces.

Applying Your Data

Brainstorm ways to modify your rocket so that it will carry the most pennies to the maximum height. Select the best design. When your teacher has approved all the designs, build and launch your rocket. Which variable did you modify? How did this variable affect your rocket's flight?

Skills Practice Lab

Quite a Reaction

Catapults have been used for centuries to throw objects great distances. According to Newton's third law of motion (whenever one object exerts a force on a second object, the second object exerts an equal and opposite force on the first), when an object is launched, something must also happen to the catapult. In this activity, you will build a kind of catapult that will allow you to observe the effects of Newton's third law of motion and the law of conservation of momentum.

MATERIALS

- cardboard rectangles, 10 cm × 15 cm (3)
- glue
- marble
- meterstick
- pushpins (3)
- rubber band
- scissors
- straws, plastic (6)
- string

SAFETY

Procedure

1. Glue the cardboard rectangles together to make a stack of three.
2. Push two of the pushpins into the cardboard stack near the corners at one end, as shown below. These pushpins will be the anchors for the rubber band.
3. Make a small loop of string.
4. Put the rubber band through the loop of string, and then place the rubber band over the two pushpin anchors. The rubber band should be stretched between the two anchors with the string loop in the middle.
5. Pull the string loop toward the end of the cardboard stack opposite the end with the anchors, and fasten the loop in place with the third pushpin.
6. Place the six straws about 1 cm apart on a tabletop or on the floor. Then, carefully center the catapult on top of the straws.
7. Put the marble in the closed end of the V formed by the rubber band.
8. Use scissors to cut the string holding the rubber band, and observe what happens. (Be careful not to let the scissors touch the cardboard catapult when you cut the string.)

9. Reset the catapult with a new piece of string. Try launching the marble several times to be sure that you have observed everything that happens during a launch. Record all your observations.

Analyze the Results

1. Which has more mass, the marble or the catapult?
2. What happened to the catapult when the marble was launched?
3. How far did the marble fly before it landed?
4. Did the catapult move as far as the marble did?

Draw Conclusions

5. Explain why the catapult moved backward.
6. If the forces that made the marble and the catapult move apart are equal, why didn't the marble and the catapult move apart the same distance? (Hint: The fact that the marble can roll after it lands is not the answer.)
7. The momentum of an object depends on the mass and velocity of the object. What is the momentum of the marble before it is launched? What is the momentum of the catapult? Explain your answers.
8. Using the law of conservation of momentum, explain why the marble and the catapult move in opposite directions after the launch.

Applying Your Data

How would you modify the catapult if you wanted to keep it from moving backward as far as it did? (It still has to rest on the straws.) Using items that you can find in the classroom, design a catapult that will move backward less than the one originally designed.

Model-Making Lab

Finding a Balance

Usually, balancing a chemical equation involves just writing. But in this activity, you will use models to practice balancing chemical equations, as shown below. By following the rules, you will soon become an expert equation balancer!

MATERIALS

- envelopes, each labeled with an unbalanced equation

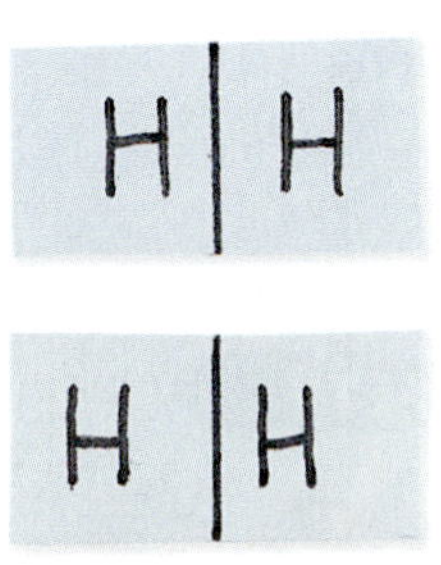

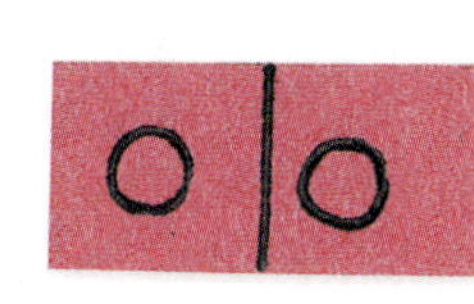

Example

$\_H_2 + \_O_2 \rightarrow \_H_2O$

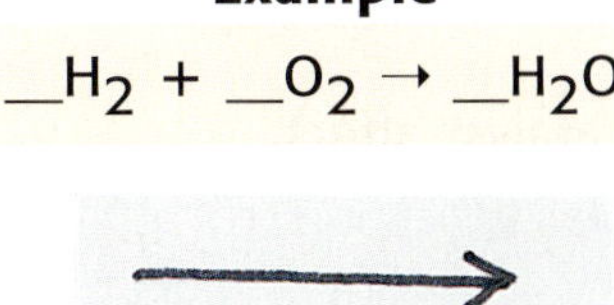

Balanced Equation

$2H_2 + O_2 \rightarrow 2H_2O$

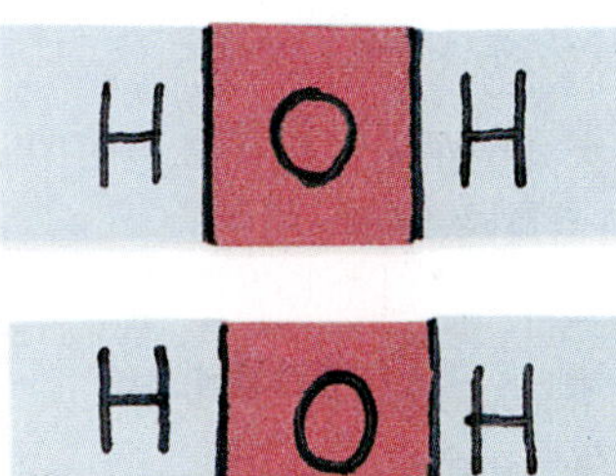

Procedure

1. The rules are as follows:
   a. Reactant-molecule models may be placed only to the left of the arrow.
   b. Product-molecule models may be placed only to the right of the arrow.
   c. You may use only complete molecule models.
   d. At least one of each of the reactant and product molecules shown in the equation must be included in the model when you are finished.
2. Select one of the labeled envelopes. Copy the unbalanced equation written on the envelope.
3. Open the envelope, and pull out the molecule models and the arrow. Place the arrow in the center of your work area.
4. Put one model of each molecule that is a reactant on the left side of the arrow and one model of each product on the right side.
5. Add one reactant-molecule or product-molecule model at a time until the number of each of the different-colored squares on each side of the arrow is the same. Remember to follow the rules.
6. When the equation is balanced, count the number of each of the molecule models you used. Write these numbers as coefficients, as shown in the balanced equation above.
7. Select another envelope, and repeat the steps until you have balanced all of the equations.

Analyze the Results

1. The rules specify that you are allowed to use only complete molecule models. How are these rules similar to what occurs in a real chemical reaction?
2. In chemical reactions, energy is either released or absorbed. Devise a way to improve the model to show energy being released or absorbed.

Using Scientific Methods

Skills Practice Lab

Cata-what? Catalyst!

Catalysts increase the rate of a chemical reaction without being changed during the reaction. In this experiment, hydrogen peroxide, H_2O_2, decomposes into oxygen, O_2, and water, H_2O. An enzyme present in liver cells acts as a catalyst for this reaction. You will investigate the relationship between the amount of the catalyst and the rate of the decomposition reaction.

MATERIALS

- beaker, 600 mL
- funnel
- graduated cylinder, 10 mL
- hydrogen peroxide, 3% solution
- liver cubes, small (2)
- mortar and pestle
- tape, masking
- test tubes, 10 mL (3)
- tweezers
- water, hot

SAFETY

Ask a Question

1. How does the amount of a catalyst affect reaction rate?

Form a Hypothesis

2. Write a statement that answers the question above. Explain your reasoning.

Test the Hypothesis

3. Put a small piece of masking tape near the top of each test tube, and label the tubes "1," "2," and "3."
4. Create a hot-water bath by filling the beaker half full with hot water.
5. Using the funnel and graduated cylinder, measure 5 mL of the hydrogen peroxide solution into each test tube. Place the test tubes in the hot-water bath for 5 min.
6. While the test tubes warm up, grind one liver cube with the mortar and pestle.
7. After 5 min, use the tweezers to place the cube of liver in test tube 1. Place the ground liver in test tube 2. Leave test tube 3 alone.
8. Observe the reaction rate (the amount of bubbling) in all three test tubes, and record your observations.

Analyze the Results

1. Does liver appear to be a catalyst? Explain your answer.
2. Which type of liver (whole or ground) produced a faster reaction? Why?
3. What is the purpose of test tube 3?

Draw Conclusions

4. How do your results support or disprove your hypothesis?
5. Why was a hot-water bath used? (Hint: Look in your book for a definition of *activation energy*.)

Skills Practice Lab

Putting Elements Together

A synthesis reaction is a reaction in which two or more substances combine to form a single compound. The resulting compound has different chemical and physical properties than the substances from which it is composed. In this activity, you will synthesize, or create, copper(II) oxide from the elements copper and oxygen.

- balance, metric
- Bunsen burner (or portable burner)
- copper powder
- evaporating dish
- gauze, wire
- gloves, protective
- igniter
- paper, weighing
- ring stand and ring
- tongs

SAFETY

Procedure

1. Copy the table below.

Data Collection Table	
Object	**Mass (g)**
Evaporating dish	
Copper powder	DO NOT WRITE IN BOOK
Copper + evaporating dish after heating	
Copper(II) oxide	

2. Use the metric balance to measure the mass (to the nearest 0.1 g) of the empty evaporating dish. Record this mass in the table.
3. Place a piece of weighing paper on the metric balance, and measure approximately 10 g of copper powder. Record the mass (to the nearest 0.1 g) in the table. **Caution:** Wear protective gloves when working with copper powder.
4. Use the weighing paper to place the copper powder in the evaporating dish. Spread the powder over the bottom and up the sides as much as possible. Discard the weighing paper.

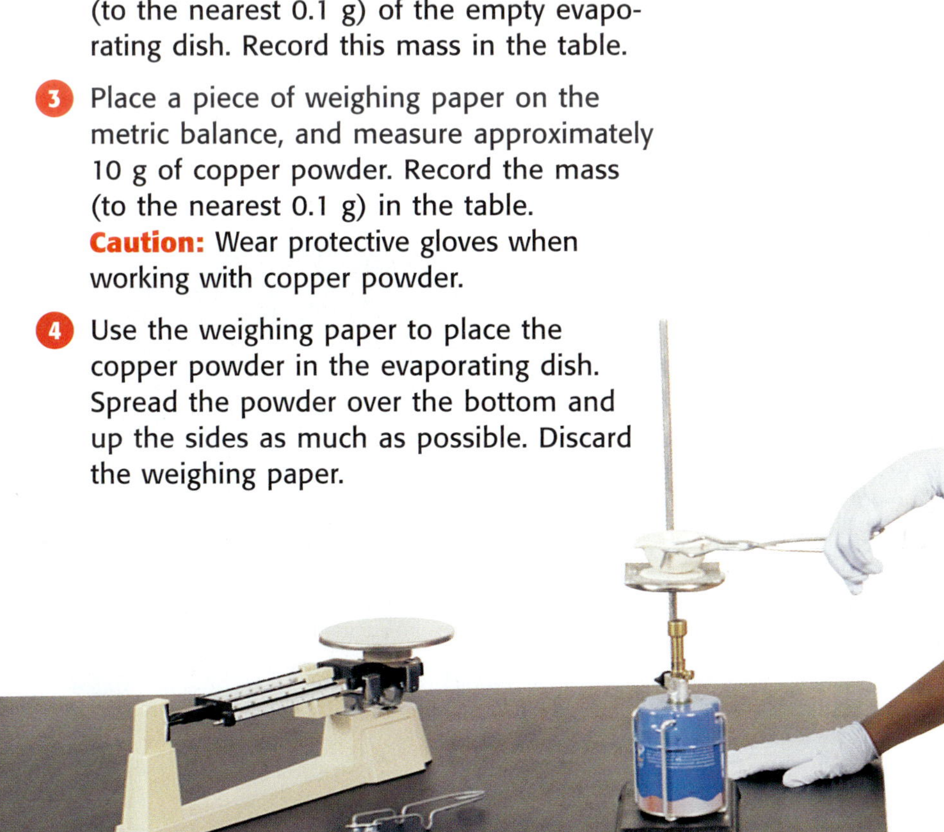

5. Set up the ring stand and ring. Place the wire gauze on top of the ring. Carefully place the evaporating dish on the wire gauze.
6. Place the Bunsen burner under the ring and wire gauze. Use the igniter to light the Bunsen burner. **Caution:** Use extreme care when working near an open flame.
7. Heat the evaporating dish for 10 min.
8. Turn off the burner, and allow the evaporating dish to cool for 10 min. Use tongs to remove the evaporating dish and to place it on the balance to determine the mass. Record the mass in the table.
9. Determine the mass of the reaction product—copper(II) oxide—by subtracting the mass of the evaporating dish from the mass of the evaporating dish and copper powder after heating. Record this mass in the table.

Analyze the Results

1. What evidence of a chemical reaction did you observe after the copper was heated?
2. Explain why there was a change in mass.
3. How does the change in mass support the idea that this reaction is a synthesis reaction?

Draw Conclusions

4. Why was powdered copper used rather than a small piece of copper? (Hint: How does surface area affect the rate of the reaction?)
5. Why was the copper heated? (Hint: Look in your book for the discussion of activation energy.)
6. The copper bottoms of cooking pots can turn black when used. How is that similar to the results you obtained in this lab?

Applying Your Data

Rust, shown below, is iron(III) oxide—the product of a synthesis reaction between iron and oxygen. How does painting a car help prevent this type of reaction?

Using Scientific Methods

Skills Practice Lab

Making Salt

A neutralization reaction between an acid and a base produces water and a salt. In this lab, you will react an acid with a base and then let the water evaporate. You will then examine what is left for properties that tell you that it is indeed a salt.

MATERIALS

- beaker, 100 mL
- eyedroppers (2)
- evaporating dish
- gloves, protective
- graduated cylinder, 100 mL
- hydrochloric acid
- magnifying lens
- phenolphthalein solution in a dropper bottle
- stirring rod, glass
- sodium hydroxide
- water, distilled

SAFETY

Ask a Question

1. Write a question about reactions between acids and bases.

Form a Hypothesis

2. Write a hypothesis that may answer the question you asked in the step above.

Test the Hypothesis

3. Put on protective gloves. Carefully measure 25 mL of hydrochloric acid in a graduated cylinder, and then pour it into the beaker. Carefully rinse the graduated cylinder with distilled water to clean out any leftover acid. **Caution:** Hydrochloric acid is corrosive. If any should spill on you, immediately flush the area with water, and notify your teacher.
4. Add 3 drops of phenolphthalein indicator to the acid in the beaker. You will not see anything happen yet because this indicator won't show its color unless too much base is present.
5. Measure 20 mL of sodium hydroxide (base) in the graduated cylinder, and add it slowly to the beaker with the acid. Use the stirring rod to mix the substances completely. **Caution:** Sodium hydroxide is also corrosive. If any should spill on you, immediately flush the area with water, and notify your teacher.
6. Use an eyedropper to add more base, a few drops at a time, to the acid-base mixture in the beaker. Be sure to stir the mixture after each few drops. Continue adding drops of base until the mixture remains colored after stirring.

7. Use another eyedropper to add acid to the beaker, 1 drop at a time, until the color just disappears after stirring.
8. Pour the mixture carefully into an evaporating dish, and place the dish where your teacher tells you to allow the water to evaporate overnight.
9. The next day, examine your evaporating dish, and with a magnifying lens, study the crystals that were left. Identify the color, shape, and other properties of the crystals.

Analyze the Results

1. The following equation is for the reaction that occurred in this experiment:

$$HCl + NaOH \longrightarrow H_2O + NaCl$$

   NaCl is ordinary table salt and forms very regular cubic crystals that are white. Did you find white cubic crystals?
2. The phenolphthalein indicator changes color in the presence of a base. Why did you add more acid in step 7 until the color disappeared?

Applying Your Data

Another neutralization reaction occurs between hydrochloric acid and potassium hydroxide, KOH. The equation for this reaction is as follows:

$$HCl + KOH \longrightarrow H_2O + KCl$$

What are the products of this neutralization reaction? How do they compare with those you discovered in this experiment?

Inquiry Lab

The Speed of Sound

In the chapter entitled "The Nature of Sound," you learned that the speed of sound in air is 343 m/s at 20°C (approximately room temperature). In this lab, you'll design an experiment to measure the speed of sound yourself—and you'll determine if you're "up to speed"!

MATERIALS

- items to be determined by the students and approved by the teacher

Procedure

1. Brainstorm with your teammates to come up with a way to measure the speed of sound. Consider the following as you design your experiment:

   a. You must have a method of making a sound. Some simple examples include speaking, clapping your hands, and hitting two boards together.

   b. Remember that speed is equal to distance divided by time. You must devise methods to measure the distance that a sound travels and to measure the amount of time it takes for that sound to travel that distance.

   c. Sound travels very rapidly. A sound from across the room will reach your ears almost before you can start recording the time! You may wish to have the sound travel a long distance.

   d. Remember that sound travels in waves. Think about the interactions of sound waves. You might be able to include these interactions in your design.

2. Discuss your experimental design with your teacher, including any equipment you need. Your teacher may have questions that will help you improve your design.

3. Once your design is approved, carry out your experiment. Be sure to perform several trials. Record your results.

Analyze the Results

1. Was your result close to the value given in the introduction to this lab? If not, what factors may have caused you to get such a different value?

2. Why was it important for you to perform several trials in your experiment?

Draw Conclusions

3. Compare your results with those of your classmates. Determine which experimental design provided the best results. Explain why you think this design was so successful.

Skills Practice Lab

Tuneful Tube

If you have seen a singer shatter a crystal glass simply by singing a note, you have seen an example of resonance. For the glass to shatter, the note has to match the resonant frequency of the glass. A column of air within a cylinder can also resonate if the air column is the proper length for the frequency of the note. In this lab, you will investigate the relationship between the length of an air column, the frequency, and the wavelength during resonance.

MATERIALS

- eraser, pink, rubber
- graduated cylinder, 100 mL
- paper, graph
- plastic tube, supplied by your teacher
- ruler, metric
- tuning forks, different frequencies (4)
- water

SAFETY

Procedure

1. Copy the data table below.

Data Collection Table				
Frequency (Hz)				
Length (cm)				

2. Fill the graduated cylinder with water.
3. Hold a plastic tube in the water so that about 3 cm is above the water.
4. Record the frequency of the first tuning fork. Gently strike the tuning fork with the eraser, and hold the tuning fork so that the prongs are just above the tube, as shown at right. Slowly move the tube and fork up and down until you hear the loudest sound.
5. Measure the distance from the top of the tube to the water. Record this length in your data table.
6. Repeat steps 3–5 using the other three tuning forks.

Analyze the Results

1. Calculate the wavelength (in centimeters) of each sound wave by dividing the speed of sound in air (343 m/s at 20°C) by the frequency and multiplying by 100.
2. Make the following graphs: air column length versus frequency and wavelength versus frequency. On both graphs, plot the frequency on the *x*-axis.
3. Describe the trend between the length of the air column and the frequency of the tuning fork.
4. How are the pitches you heard related to the wavelengths of the sounds?

Skills Practice Lab

The Energy of Sound

In the chapter entitled "The Nature of Sound," you learned about various properties and interactions of sound. In this lab, you will perform several activities that will demonstrate that the properties and interactions of sound all depend on one thing—the energy carried by sound waves.

MATERIALS

- cup, plastic, small, filled with water
- eraser, pink, rubber
- rubber band
- string, 50 cm
- tuning forks, same frequency (2), different frequency (1)

SAFETY

Part A: Sound Vibrations

Procedure

1. Lightly strike a tuning fork with the eraser. Slowly place the prongs of the tuning fork in the plastic cup of water. Record your observations.

Part B: Resonance

Procedure

1. Strike a tuning fork with the eraser. Quickly pick up a second tuning fork in your other hand, and hold it about 30 cm from the first tuning fork.
2. Place the first tuning fork against your leg to stop the tuning fork's vibration. Listen closely to the second tuning fork. Record your observations, including the frequencies of the two tuning forks.
3. Repeat steps 1 and 2, using the remaining tuning fork as the second tuning fork.

Part C: Interference

Procedure

1. Use the two tuning forks that have the same frequency, and place a rubber band tightly over the prongs near the base of one tuning fork, as shown at right. Strike both tuning forks against the eraser. Hold the stems of the tuning forks against a table, 3 cm to 5 cm apart. If you cannot hear any differences, move the rubber band up or down the prongs. Strike again. Record your observations.

Part D: The Doppler Effect

Procedure

1. Your teacher will tie the piece of string securely to the base of one tuning fork. Your teacher will then strike the tuning fork and carefully swing the tuning fork in a circle overhead. Record your observations.

Analyze the Results

1. How do your observations demonstrate that sound waves are carried through vibrations?
2. Explain why you can hear a sound from the second tuning fork when the frequencies of the tuning forks used are the same.
3. When using tuning forks of different frequencies, would you expect to hear a sound from the second tuning fork if you strike the first tuning fork harder? Explain your reasoning.
4. Did you notice the sound changing back and forth between loud and soft? A steady pattern like this one is called a *beat frequency.* Explain this changing pattern of loudness and softness in terms of interference (both constructive and destructive).
5. Did the tuning fork make a different sound when your teacher was swinging it than when he or she was holding it? If yes, explain why.
6. Is the actual pitch of the tuning fork changing when it is swinging? Explain.

Draw Conclusions

7. Explain how your observations from each part of this lab verify that sound waves carry energy from one point to another through a vibrating medium.
8. Particularly loud thunder can cause the windows of your room to rattle. How is this evidence that sound waves carry energy?

Using Scientific Methods

Skills Practice Lab

What Color of Light Is Best for Green Plants?

Plants grow well outdoors under natural sunlight. However, some plants are grown indoors under artificial light. A variety of colored lights are available for helping plants grow indoors. In this experiment, you'll test several colors of light to discover which color best meets the energy needs of green plants.

MATERIALS

- bean seedlings
- colored lights, supplied by your teacher
- marker, felt-tip
- paper towels
- Petri dishes with covers
- tape, masking
- water

SAFETY

Ask a Question

1. Which color of light is best for growing green plants?

Form a Hypothesis

2. Write a hypothesis that answers the question above. Explain your reasoning.

Test the Hypothesis

3. Use the masking tape and marker to label the side of each Petri dish with your name and the type of light under which you will place the dish.
4. Place a moist paper towel in each Petri dish. Place 5 seedlings on top of the paper towel. Cover each dish.
5. Record your observations of the seedlings, such as length, color, and number of leaves.
6. Place each dish under the appropriate light.
7. Observe the Petri dishes every day for at least 5 days. Record your observations.

Analyze the Results

1. Based on your results, which color of light is best for growing green plants? Which color of light is worst?

Draw Conclusions

2. Remember that the color of an opaque object (such as a plant) is determined by the colors the object reflects. Use this information to explain your answer to question 1 above.
3. Would a purple light be good for growing purple plants? Explain.

Skills Practice Lab

Which Color Is Hottest?

Will a navy blue hat or a white hat keep your head warmer in cool weather? Colored objects absorb energy, which can make the objects warmer. How much energy is absorbed depends on the object's color. In this experiment, you will test several colors under a bright light to determine which colors absorb the most energy.

MATERIALS

- light source
- paper, colored, squares
- paper, graph
- paper towels
- pencils or pens, colored
- tape, transparent
- thermometer
- water, room-temperature

SAFETY

Procedure

1. Copy the table below. Be sure to have one column for each color of paper you use and enough rows to end at 3 min.

Data Collection Table

Time (s)	White	Red	Blue	Black
0				
15				
30				
45				
etc.				

DO NOT WRITE IN BOOK

2. Tape a piece of colored paper around the bottom of a thermometer, and hold it under the light source. Record the temperature every 15 s for 3 min.
3. Cool the thermometer by removing the piece of paper and placing the thermometer in the cup of room-temperature water. After 1 min, remove the thermometer, and dry it with a paper towel.
4. Repeat steps 2 and 3 with each color, making sure to hold the thermometer at the same distance from the light source.

Analyze the Results

1. Prepare a graph of temperature (*y*-axis) versus time (*x*-axis). Using a different colored pencil or pen for each set of data, plot all data on one graph.
2. Rank the colors you used in order from hottest to coolest.

Draw Conclusions

3. Compare the colors, based on the amount of energy each absorbs.
4. In this experiment, a white light was used. How would your results be different if you used a red light? Explain.
5. Use the relationship between color and energy absorbed to explain why different colors of clothing are used for different seasons.

Skills Practice Lab

The Best-Bread Bakery Dilemma

The chief baker at the Best-Bread Bakery thinks that the yeast the bakery received may be dead. Yeast is a central ingredient in bread. Yeast is a living organism, a member of the kingdom Fungi, and it undergoes the same life processes as other living organisms. When yeast grows in the presence of oxygen and other nutrients, yeast produces carbon dioxide. The gas forms bubbles that cause bread dough to rise. Thousands of dollars may be lost if the yeast is dead.

The Best-Bread Bakery has requested that you test the yeast. The bakery has furnished samples of live yeast and some samples of the yeast in question.

MATERIALS

- beaker, 250 mL
- flour
- gloves, heat-resistant
- graduated cylinder
- hot plate
- magnifying lens
- scoopula (or small spoon)
- stirring sticks, wooden (3)
- sugar
- test-tube rack
- test tubes (3) (or clear plastic cups)
- thermometer, Celsius, with clip
- water, 125 mL
- yeast samples (live, A, and B)

SAFETY

Procedure

1. Make a data table similar to the one below. Leave plenty of room to write your observations.
2. Examine each yeast sample with a magnifying lens. You may want to sniff the samples to determine the presence of an odor. (Your teacher will demonstrate the appropriate way to detect odors in this lab.) Record your observations in the data table.
3. Label three test tubes or plastic cups "Live Yeast," "Sample A Yeast," and "Sample B Yeast."
4. Fill a beaker with 125 mL of water, and place the beaker on a hot plate. Use a thermometer to be sure the water does not get warmer than 32°C. Attach the thermometer to the side of the beaker with a clip so the thermometer doesn't touch the bottom of the beaker. Turn off the hot plate when the water temperature reaches 32°C.

Yeast sample	Observations	0 min	5 min	10 min	15 min	20 min	25 min	Dead or alive?
Live								
Sample A								
Sample B								

5. Add a small scoop (about 1/2 tsp) of each yeast sample to the correctly labeled container. Add a small scoop of sugar to each container.
6. Add 10 mL of the warm water to each container, and stir.
7. Add a small scoop of flour to each container, and stir again. The flour will help make the process more visible but is not necessary as food for the yeast.
8. Observe the samples carefully. Look for bubbles. Make observations at 5 min intervals. Write your observations in the data table.
9. In the last column of the data table, write "alive" or "dead" based on your observations during the experiment.

Analyze the Results

1. Describe any differences in the yeast samples before the experiment.
2. Describe the appearance of the yeast samples at the conclusion of the experiment.
3. Why was a sample of live yeast included in the experiment?
4. Why was sugar added to the samples?
5. Based on your observations, is either Sample A or Sample B alive?

Draw Conclusions

6. Write a letter to the Best-Bread Bakery stating your recommendation to use or not use the yeast samples. Give reasons for your recommendation.

Applying Your Data

Based on your observations of the nutrient requirements of yeast, design an experiment to determine the ideal combination of nutrients. Vary the amount of nutrients, or examine different energy sources.

Skills Practice Lab

Stayin' Alive!

Every second of your life, your body's trillions of cells take in, use, and store energy. They repair themselves, reproduce, and get rid of waste. Together, these processes are called *metabolism.* Your cells use the food that you eat to provide the energy you need to stay alive.

Your Basal Metabolic Rate (BMR) is a measurement of the energy that your body needs to carry out all the basic life processes while you are at rest. These processes include breathing, keeping your heart beating, and keeping your body's temperature stable. Your BMR is influenced by your gender, your age, and many other things. Your BMR may be different from everyone else's, but it is normal for you. In this activity, you will find the amount of energy, measured in Calories, you need every day in order to stay alive.

MATERIALS

- bathroom scale
- tape measure

Procedure

1. Find your weight on a bathroom scale. If the scale measures in pounds, you must convert your weight in pounds to your mass in kilograms. To convert your weight in pounds (lb) to mass in kilograms (kg), multiply the number of pounds by 0.454.

Example: If Carlos weighs 125 lb, his mass in kilograms is:

$$\begin{array}{r} 125\ \text{lb} \\ \times\ \underline{0.454} \\ 56.75\ \text{kg} \end{array}$$

2. Use a tape measure to find your height. If the tape measures in inches, convert your height in inches to height in centimeters. To convert your height in inches (in.) to your height in centimeters (cm), multiply the number of inches by 2.54.

If Carlos is 62 in. tall, his height in centimeters is:

$$\begin{array}{r} 62\ \text{in.} \\ \times\ \underline{2.54} \\ 157.48\ \text{cm} \end{array}$$

3. Now that you know your height and mass, use the appropriate formula below to get a close estimate of your BMR. Your answer will give you an estimate of the number of Calories your body needs each day just to stay alive.

Calculating Your BMR	
Females	**Males**
65 + (10 × your mass in kilograms) + (1.8 × your height in centimeters) − (4.7 × your age in years)	66 + (13.5 × your mass in kilograms) + (5 × your height in centimeters) − (6.8 × your age in years)

4. Your metabolism is also influenced by how active you are. Talking, walking, and playing games all take more energy than being at rest. To get an idea of how many Calories your body needs each day to stay healthy, select the lifestyle that best describes yours from the table at right. Then multiply your BMR by the activity factor.

Activity Factors

Activity lifestyle	Activity factor
Moderately inactive (normal, everyday activities)	1.3
Moderately active (exercise 3 to 4 times a week)	1.4
Very active (exercise 4 to 6 times a week)	1.6
Extremely active (exercise 6 to 7 times a week)	1.8

Analyze the Results

1. In what way could you compare your whole body to a single cell? Explain.
2. Does an increase in activity increase your BMR? Does an increase in activity increase your need for Calories? Explain your answers.

Draw Conclusions

3. If you are moderately inactive, how many more Calories would you need if you began to exercise every day?

Applying Your Data

The best energy sources are those that supply the correct amount of Calories for your lifestyle and also provide the nutrients you need. Research in the library or on the Internet to find out which kinds of foods are the best energy sources for you. How does your list of best energy sources compare with your diet?

List everything you eat and drink in 1 day. Find out how many Calories are in each item, and find the total number of Calories you have consumed. How does this number of Calories compare with the number of Calories you need each day for all your activities?

Model-Making Lab

Viral Decorations

Although viruses are made of only protein and nucleic acids, their structures have many different shapes that help them attach to and invade living cells. One viral shape can be constructed from the template provided by your teacher. In this activity, you will construct and modify a model of a virus.

MATERIALS

- glue (or tape)
- markers, colored
- paper, construction
- pipe cleaners, twist ties, buttons, string, plastic wrap, and other scrap materials for making variations of the virus
- scissors
- virus model template

SAFETY

Procedure

1. Obtain a virus model template from your teacher. Carefully copy the template on a piece of construction paper. You may make the virus model as large as your teacher allows.
2. Plan how you will modify your virus. For example, you might want to add the tail and tail fibers of a bacteriophage or wrap the model in plastic to represent the envelope that surrounds the protein coat in HIV.
3. Color your virus model, and cut it out by cutting on the solid black lines. Then, fold the virus model along the dotted lines.
4. Glue or tape each lettered tab under the corresponding lettered triangle. For example, glue or tape the large *Z* tab under the *Z*-shaded triangle. When you are finished, you should have a closed box with 20 sides.
5. Apply the modifications that you planned. Give your virus a name, and write it on the model. Decorate your classroom with your virus and those of your classmates.

Analyze the Results

1. Describe the modifications you made to your virus model, and explain how the virus might use them.
2. If your virus causes disease, explain what disease it causes, how it reproduces, and how the virus is spread.

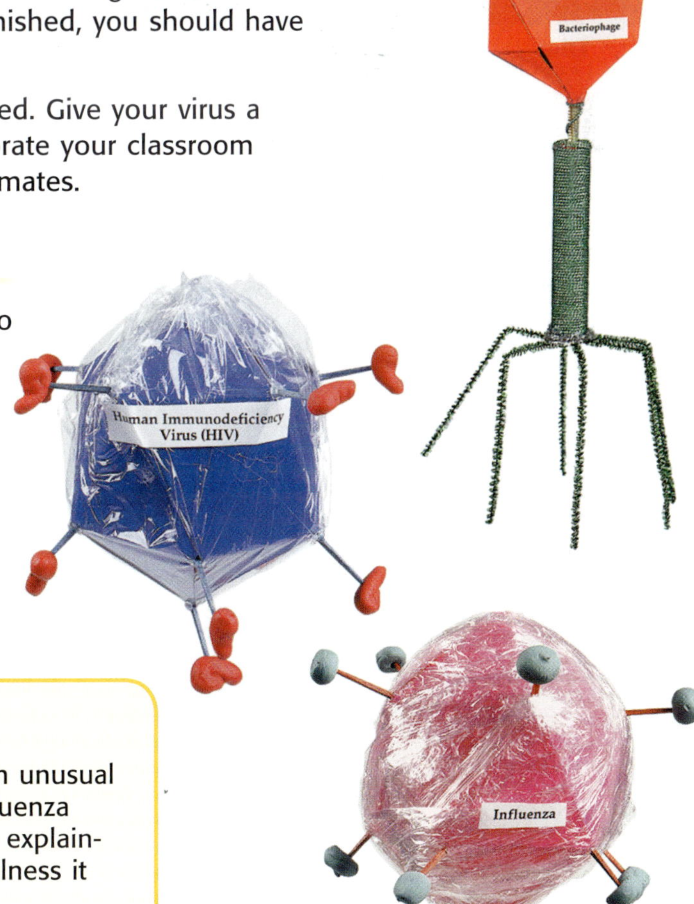

Communicating Your Data

Research in the library or on the Internet an unusual virus that causes an illness, such as the influenza virus, HIV, or Ebola virus. Write a paragraph explaining what is unusual about the virus, what illness it causes, and how it might be avoided.

Model-Making Lab

Adaptation: It's a Way of Life

Since the beginning of life on Earth, species have had special characteristics called *adaptations* that have helped them survive changes in environmental conditions. Changes in a species' environment include climate changes, habitat destruction, or the extinction of prey. These things can cause a species to die out unless the species has a characteristic that helps it survive. For example, a species of bird may have an adaptation for eating sunflower seeds and ants. If the ant population dies out, the bird can still eat seeds and can therefore survive.

In this activity, you will explore several adaptations and design an organism with adaptations you choose. Then, you will describe how these adaptations help the organism survive.

MATERIALS

- arts-and-crafts materials, various
- markers, colored
- magazines for cutouts
- poster board
- scissors

SAFETY

Procedure

1. Study the chart below. Choose one adaptation from each column. For example, an organism might be a scavenger that burrows underground and has spikes on its tail!

Adaptations		
Diet	**Type of transportation**	**Special adaptation**
carnivore	flies	uses sensors to detect heat
herbivore	glides through the air	is active only at night and has excellent night vision
omnivore	burrows underground	changes colors to match its surroundings
scavenger	runs fast	has armor
decomposer	swims	has horns
	hops	can withstand extreme temperature changes
	walks	secretes a terrible and sickening scent
	climbs	has poison glands
	floats	has specialized front teeth
	slithers	has tail spikes
		stores oxygen in its cells so it does not have to breathe continuously
		one of your own invention

2. Design an organism that has the three adaptations you have chosen. Use poster board, colored markers, picture cutouts, or craft materials of your choosing to create your organism.
3. Write a caption on your poster describing your organism. Describe its appearance, its habitat, its niche, and the way its adaptations help it survive. Give your organism a two-part "scientific" name that is based on its characteristics.
4. Display your creation in your classroom. Share with classmates how you chose the adaptations for your organism.

Analyze the Results

1. What does your imaginary organism eat?
2. In what environment or habitat would your organism be most likely to survive—in the desert, tropical rain forest, plains, icecaps, mountains, or ocean? Explain your answer.
3. Is your creature a mammal, a reptile, an amphibian, a bird, or a fish? What modern organism (on Earth today) or ancient organism (extinct) is your imaginary organism most like? Explain the similarities between the two organisms. Do some research outside the lab, if necessary, to find out about a real organism that may be similar to your imaginary organism.

Draw Conclusions

4. If there were a sudden climate change, such as daily downpours of rain in a desert, would your imaginary organism survive? What adaptations for surviving such a change does it have?

Applying Your Data

Call or write to an agency such as the U.S. Fish and Wildlife Service to get a list of endangered species in your area. Choose an organism on that list. Describe the organism's niche and any special adaptations it has that help it survive. Find out why it is endangered and what is being done to protect it.

Examine the illustration of the animal at right. Based on its physical characteristics, describe its habitat and niche. Is this a real animal?

Skills Practice Lab

Deciding About Environmental Issues

You make hundreds of decisions every day. Some of them are complicated, but many of them are very simple, such as what to wear or what to eat for lunch. Deciding what to do about an environmental issue can be very difficult. There are many different factors that must be considered. How will a certain solution affect people's lives? How much will it cost? Is it ethically right?

In this activity, you will analyze an issue in four steps to help you make a decision about it. Find out about environmental issues that are being discussed in your area. Examine newspapers, magazines, and other publications to find out what the issues are. Choose one local issue to evaluate. For example, you could evaluate whether the city should spend the money to provide recycling bins and special trucks for picking up recyclable trash.

MATERIALS

- newspapers, magazines, and other publications containing information about environmental issues

A Four-Step Decision-Making Model

Gather Information

Consider Values

Explore Consequences

Make a Decision

Procedure

1. Write a statement about an environmental issue.
2. Read about your issue in several publications. On a separate sheet of paper, summarize important facts.
3. The values of an issue are the things that you consider important. Examine the diagram below. Several values are given. Which values do you think apply most to the environmental issue you are considering? Are there other values that you believe will help you make a decision about the issue? Consider at least four values in making your decision.

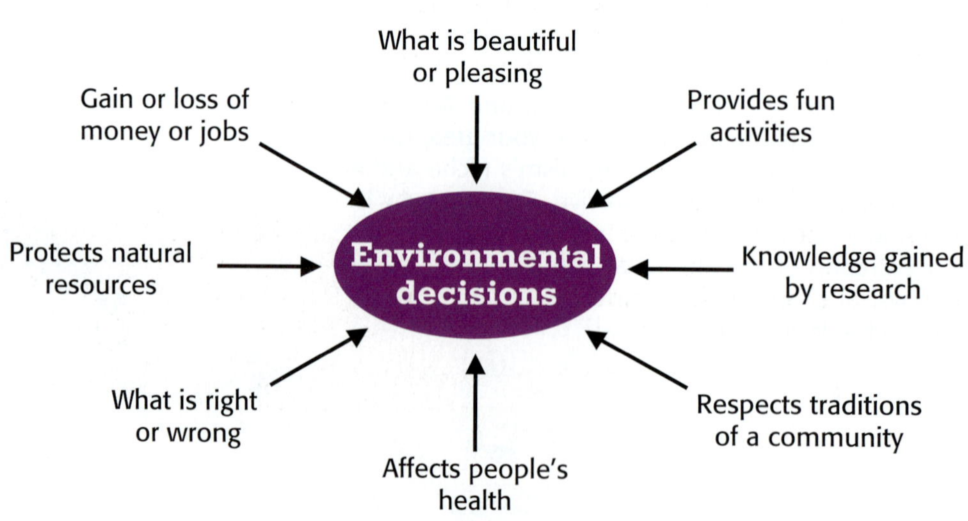

4. Consequences are the things that result from a certain course of action. Create a table similar to the one below. Use your table to organize your thoughts about consequences related to your environmental issue. List your values at the top. Fill in each space with the consequences for each value.

Consequences Table				
	Values			
Consequences				
Positive short-term consequences				
Negative short-term consequences		DO NOT WRITE IN BOOK		
Positive long-term consequences				
Negative long-term consequences				

5. Thoroughly consider all of the consequences you have recorded in your table. Evaluate how important each consequence is. Make a decision about what course of action you would choose on the issue.

Analyze the Results

1. In your evaluation, did you consider short-term consequences or long-term consequences to be more important? Why?
2. Which value or values had the greatest influence on your final decision? Explain your reasoning.

Communicating Your Data

Compare your table with your classmates' tables. Did you all make the same decision about a similar issue? If not, form teams, and organize a formal classroom debate of a specific environmental issue.

Skills Practice Lab

Enzymes in Action

You know how important enzymes are in the process of digestion. This lab will help you see enzymes at work. Hydrogen peroxide is continuously produced by your cells. If it is not quickly broken down, hydrogen peroxide will kill your cells. Luckily, your cells contain an enzyme that converts hydrogen peroxide into two nonpoisonous substances. This enzyme is also present in the cells of beef liver. In this lab, you will observe the action of this enzyme on hydrogen peroxide.

MATERIALS

- beef liver, 1 cm cubes (3)
- gloves, protective
- graduated cylinder, 10 mL
- hydrogen peroxide, fresh (4 mL)
- mortar and pestle (or fork and watch glass)
- plate, small
- spatula
- test tube (3)
- test-tube rack
- tweezers
- water

SAFETY

Procedure

1. Draw a data table similar to the one below. Be sure to leave enough space to write your observations.

Data Table		
Size and condition of liver	**Experimental liquid**	**Observations**
1 cm cube beef liver	2 mL water	
1 cm cube beef liver	2 mL hydrogen peroxide	DO NOT WRITE IN BOOK
1 cm cube beef liver (mashed)	2 mL hydrogen peroxide	

2. Get three equal-sized pieces of beef liver from your teacher, and use your forceps to place them on your plate.
3. Pour 2 mL of water into a test tube labeled "Water and liver."
4. Using the tweezers, carefully place one piece of liver in the test tube. Record your observations in your data table.
5. Pour 2 mL of hydrogen peroxide into a second test tube labeled "Liver and hydrogen peroxide."
   Caution: Do not splash hydrogen peroxide on your skin. If you do get hydrogen peroxide on your skin, rinse the affected area with running water immediately, and tell your teacher.
6. Using the tweezers, carefully place one piece of liver in the test tube. Record your observations of the second test tube in your data table.
7. Pour another 2 mL of hydrogen peroxide into a third test tube labeled "Ground liver and hydrogen peroxide."
8. Using a mortar and pestle (or fork and watch glass), carefully grind the third piece of liver.
9. Using the spatula, scrape the ground liver into the third test tube. Record your observations of the third test tube in your data table.

Analyze the Results

1. What was the purpose of putting the first piece of liver in water? Why was this a necessary step?
2. Describe the difference you observed between the liver and the ground liver when each was placed in the hydrogen peroxide. How can you account for this difference?

Applying Your Data

Do plant cells contain enzymes that break down hydrogen peroxide? Try this experiment using potato cubes instead of liver to find out.

Model-Making Lab

Antibodies to the Rescue

Some cells of the immune system, called *B cells,* make antibodies that attack and kill invading viruses and microorganisms. These antibodies help make you immune to disease. Have you ever had chickenpox? If you have, your body has built up antibodies that can recognize that particular virus. Antibodies will attach themselves to the virus, tagging it for destruction. If you are exposed to the same disease again, the antibodies remember that virus. They will attack the virus even quicker and in greater number than they did the first time. This is the reason that you will probably never have chickenpox more than once.

In this activity, you will construct simple models of viruses and their antibodies. You will see how antibodies are specific for a particular virus.

MATERIALS

- craft materials, such as buttons, fabric scraps, pipe cleaners, and recycled materials
- paper, colored
- scissors
- tape (or glue)

Procedure

1. Draw the virus patterns shown on this page on a separate piece of paper, or design your own virus models from the craft supplies. Remember to design different receptors on each of your virus models.
2. Write a few sentences describing how your viruses are different.
3. Cut out the viruses, and attach them to a piece of colored paper with tape or glue.

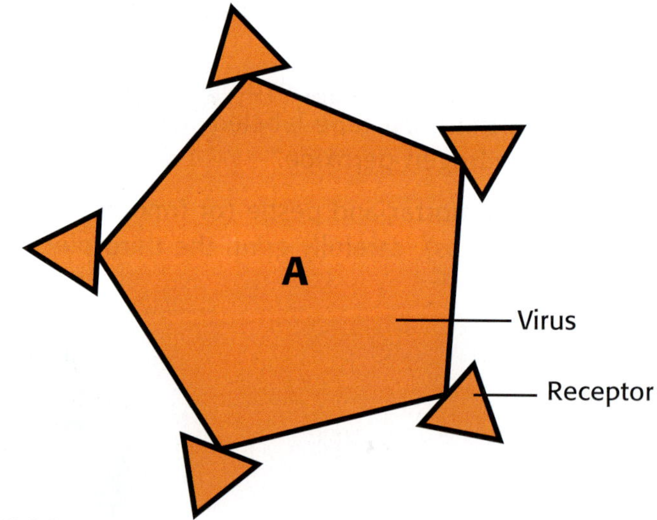

Viruses

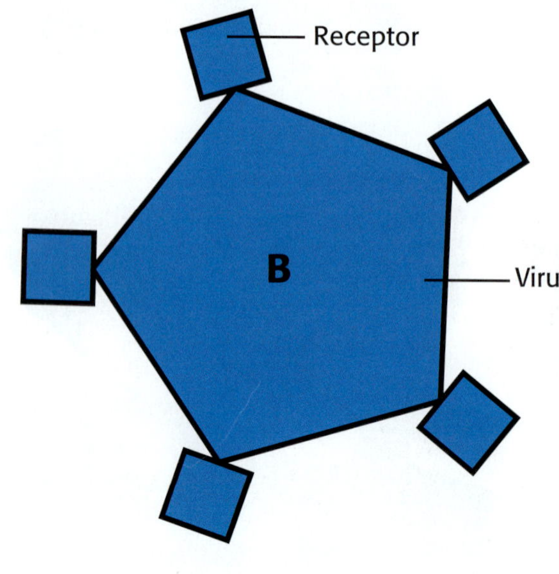

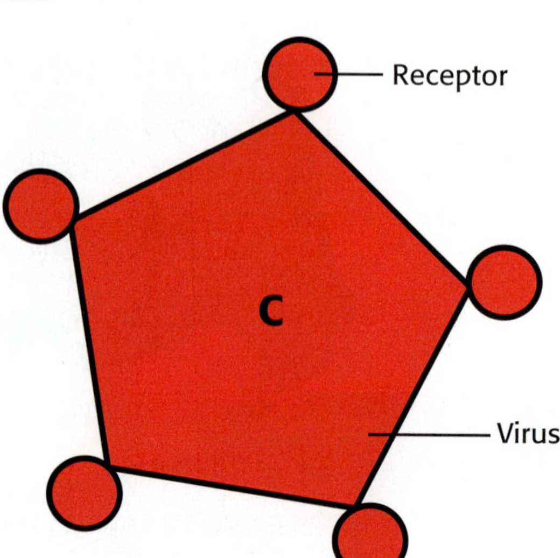

4. Select the antibodies drawn below, or design your own antibodies that will exactly fit on the receptors on your virus models. Draw or create each antibody enough times to attach one to each receptor site on the virus.

Antibodies

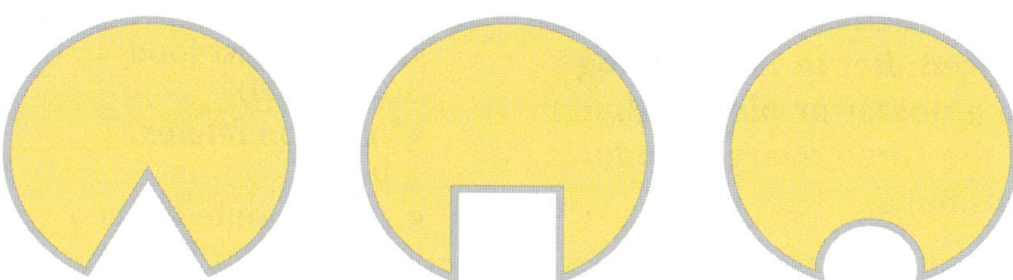

5. Cut out the antibodies you have drawn. Arrange the antibodies so that they bind to the virus at the appropriate receptor. Attach them to the virus with tape or glue.

Analyze the Results

1. Explain how an antibody "recognizes" a particular virus.
2. After the attachment of antibodies to the receptors, what would be the next step in the immune response?
3. Many vaccines use weakened copies of the virus to protect the body. Use the model of a virus and its specific antibody to explain how vaccines work.

Draw Conclusions

4. Use your model of a virus to demonstrate to the class how a receptor might change or mutate so that a vaccine would no longer be effective.

Applying Your Data

Research in the library or on the Internet to find information about the discovery of the Salk vaccine for polio. Include information on how polio affects people today.

Research in the library or on the Internet to find information and write a report about filoviruses. What do they look like? What diseases do they cause? Why are they especially dangerous? Is there an effective vaccine against any filovirus? Explain.

Skills Practice Lab

To Diet or Not to Diet

There are six main classes of foods that we need in order to keep our bodies functioning properly: water, vitamins, minerals, carbohydrates, fats, and proteins. In this activity you will investigate the importance of a well-balanced diet in maintaining a healthy body. Then you will create a poster or picture that illustrates the importance of one of the three energy-producing nutrients—carbohydrates, fats, and proteins.

MATERIALS

- crayons (or markers), assorted colors
- diet books
- menus, fast-food (optional)
- nutrition reference books
- paper, white unlined

Procedure

1. Draw a table like the one below. Research in the library, on nutrition labels, in nutrition or diet books, or on the Internet to find the information you need to fill out the chart.

Nutrition Data Table

	Fats	Carbohydrates	Proteins
Found in which foods			
Functions in the body			
Consequences of deficiency			

2. Choose one of the foods you have learned about in your research, and create a poster or picture that describes its importance in a well-balanced diet.

Analyze the Results

1. Based on what you have learned in this lab, how might you change your eating habits to have a well-balanced diet? Does the nutritional value of foods concern you? Why or why not? Write down your answers, and explain your reasoning.

Communicating Your Data

Write a paragraph explaining why water is a nutrient. Analyze a typical fast-food meal, and determine its overall nutritional value.

Contents

Reading Check Answers

Chapter 1 Science in Our World

Section 1

Page 4: Science is the knowledge obtained by observing natural events and conditions in order to discover facts and formulate laws or principles that can be verified or tested.

Page 7: Society can influence technology development by identifying important problems that need technological solutions.

Page 9: A volcanologist studies volcanoes and their products, such as lava and gases.

Section 2

Page 10: a series of steps used by scientists to solve problems

Page 12: A hypothesis is testable if an experiment can be designed to test the hypothesis.

Page 14: only one

Page 16: because the scientist has learned something

Section 3

Page 19: a mathematical model

Page 20: to explain a broad range of observations, facts, and tested hypotheses, to predict what might happen, and to organize scientific thinking

Section 4

Page 22: stopwatch, graduated cylinder, meterstick, spring scale, balance, and thermometer

Page 25: the kilogram

Page 27: Safety symbols alert you to particular safety concerns or specific dangers in a lab.

Chapter 2 Rocks: Mineral Mixtures

Section 1

Page 40: Types of rocks that have been used by humans to construct buildings include granite, limestone, marble, sandstone, and slate.

Page 44: Rock within the Earth is affected by temperature and pressure.

Page 45: The minerals that a rock contains determine a rock's composition.

Page 46: Fine-grained rocks are made of small grains, such as silt or clay particles. Medium-grained rocks are made of medium-sized grains, such as sand. Coarse-grained rocks are made of large grains, such as pebbles.

Section 2

Page 49: Felsic rocks are light-colored igneous rocks rich in aluminum, potassium, silicon, and sodium. Mafic rocks are dark-colored igneous rocks rich in calcium, iron, and magnesium.

Page 51: New sea floor forms when lava that flows from fissures on the ocean floor cools and hardens.

Section 3

Page 53: Halite forms when sodium and chlorine ions in shallow bodies of water become so concentrated that halite crystallizes from solution.

Page 55: Ripple marks are the marks left by wind and water waves on lakes, seas, rivers, and sand dunes.

Section 4

Page 57: Regional metamorphism occurs when pressure builds up in rock that is buried deep below other rock formations or when large pieces of the Earth's crust collide. The increased pressure can cause thousands of square miles of rock to become deformed and chemically changed.

Page 58: An index mineral is a metamorphic mineral that forms only at certain temperatures and pressures and therefore can be used by scientists to estimate the temperature, pressure, and depth at which a rock undergoes metamorphosis.

Page 61: Deformation causes metamorphic structures, such as folds.

Chapter 3 The Rock and Fossil Record

Section 1

Page 73: Catastrophists believed that all geologic change occurs rapidly.

Page 74: A global catastrophe can cause the extinction of species.

Section 2

Page 77: Geologists use the geologic column to interpret rock sequences and to identify layers in puzzling rock sequences.

Page 79: An unconformity is a surface that represents a missing part of the geologic column.

Page 80: A disconformity is found where part of a sequence of parallel rock layers is missing. A nonconformity is found where horizontal sedimentary rock layers lie on top of an eroded surface of igneous or metamorphic rock. Angular unconformities are found between horizontal sedimentary rock layers and rock layers that have been tilted or folded.

Section 3

Page 83: A half-life is the time it takes one-half of a radioactive sample to decay.

Page 84: strontium-87

Section 4

Page 86: An organism is caught in soft, sticky tree sap, which hardens and preserves the organism.

Page 88: A mold is a cavity in rock where a plant or an animal was buried. A cast is an object created when sediment fills a mold and becomes rock.

Appendix

Page 90: To fill in missing information about changes in organisms in the fossil record, paleontologists look for similarities between fossilized organisms or between fossilized organisms and their closest living relatives.

Page 91: *Phacops* can be used to establish the age of rock layers because *Phacops* lived during a relatively short, well-defined time span and is found in rock layers throughout the world.

Section 5

Page 93: approximately 2 billion years

Page 94: The geological time scale is a scale that divides Earth's 4.6 billion–year history into distinct intervals of time.

Page 96: The Mesozoic era is known as the *Age of Reptiles* because reptiles, including the dinosaurs, were the dominant organisms on land.

Chapter 4 Energy Resources

Section 1

Page 109: A renewable resource is a natural resource that can be replaced at the same rate at which the resource is used.

Page 111: Answers may vary. Sample answer: newspapers, plastic containers, and cardboard boxes.

Section 2

Page 113: Natural gas is most often used for heating and for generating electrical energy.

Page 114: Coal was most commonly used to power trains.

Page 116: Lignite has a higher carbon content than peat does.

Page 117: Natural gas and petroleum are removed from the Earth by drilling wells into rock that contains petroleum and natural gas.

Page 118: The sulfur dioxide released from the burning coal combines with moisture in the air to produce acid rain.

Section 3

Page 121: Fusion produces few dangerous wastes.

Page 122: The energy of fossil fuels comes from the sun.

Page 124: Hydroelectric energy is renewable because water is constantly recycled.

Page 126: Geothermal power plants obtain energy from the Earth by pumping steam and hot water from wells drilled into the rock.

Chapter 5 The Restless Earth

Section 1

Page 139: The crust is the thin, outermost layer of the Earth. It is 5 km to 100 km thick and is mainly made up of the elements oxygen, silicon, and aluminum. The mantle is the layer between the crust and core. It is 2,900 km thick, is denser than the crust, and contains most of the Earth's mass. The core is the Earth's innermost layer. The core has a radius of 3,430 km and is made mostly of iron.

Page 140: The five physical layers of the Earth are the lithosphere, asthenosphere, mesosphere, outer core, and inner core.

Page 143: Although continental lithosphere is less dense than oceanic lithosphere is, continental lithosphere has a greater mass because of its greater thickness and will displace more asthenosphere than oceanic lithosphere.

Page 144: Answers may vary. A seismic wave traveling through a solid will go faster than a seismic wave traveling through a liquid.

Section 2

Page 146: Similar fossils were found on landmasses that are very far apart. The best explanation for this phenomenon is that the landmasses were once joined.

Page 149: The molten rock at mid-ocean ridges contains tiny grains of magnetic minerals. The minerals align with the Earth's magnetic field before the rock cools and hardens. When the Earth' magnetic field reverses, the orientation of the mineral grains in the rocks will also change.

Section 3

Page 151: A transform boundary forms when two tectonic plates slide past each other horizontally.

Page 152: The circulation of thermal energy causes changes in density in the asthenosphere. As rock is heated, it expands, becomes less dense, and rises. As rock cools, it contracts, becomes denser, and sinks.

Section 4

Page 154: Compression can cause rocks to be pushed into mountain ranges as tectonic plates collide at convergent boundaries. Tension can pull rocks apart as tectonic plates separate at divergent boundaries.

Page 156: In a normal fault, the hanging wall moves down. In a reverse fault, the hanging wall moves up.

Page 158: Folded mountains form when rock layers are squeezed together and pushed upward.

Section 5

Page 163: During elastic rebound, rock releases energy. Some of this energy travels as seismic waves that cause earthquakes.

Page 165: Earthquake zones are usually located along tectonic plate boundaries.

Page 167: Surface waves travel more slowly than body waves but are more destructive.

Chapter 6 Volcanoes

Section 1

Page 179: Nonexplosive eruptions are common, and they feature relatively calm flows of lava. Explosive eruptions are less common and produce large, explosive clouds of ash and gases.

Page 180: Because silica-rich magma has a high viscosity, it tends to trap gases and plug volcanic vents. This causes pressure to build up and can result in an explosive eruption.

Page 182: Volcanic bombs are large blobs of magma that harden in the air. Lapilli are small pieces of magma that harden in the air. Volcanic blocks are pieces of solid rock erupted from a volcano. Ash forms when gases in stiff magma expand rapidly and the walls of the gas bubbles shatter into tiny glasslike slivers.

Section 2

Page 184: Eruptions release large quantities of ash and gases, which can block sunlight and cause global temperatures to drop.

Page 186: Calderas form when a magma chamber partially empties and the roof overlying the chamber collapses.

Section 3

Page 189: Volcanic activity is common at tectonic plate boundaries because magma tends to form at plate boundaries.

Page 191: When a tectonic plate subducts, it becomes hotter and releases water. The water lowers the melting point of the rock above the plate, causing magma to form.

Page 192: According to one theory, a rising body of magma, called a mantle plume, causes a chain of volcanoes to form on a moving tectonic plate. According to another theory, a chain of volcanoes forms along cracks in the Earth's crust.

Chapter 7 Agents of Erosion and Deposition

Section 1

Page 205: The amount of energy released from breaking waves causes rock to break down, eventually forming sand.

Page 207: Large waves are more capable of moving large rocks on a shoreline because they have more energy than normal waves do.

Page 208: Beach material is material deposited by waves.

Section 2

Page 211: Deflation hollows form in areas where there is little vegetation.

Page 213: Dunes move in the direction of strong winds.

Section 3

Page 214: Alpine glaciers form in mountainous areas.

Page 219: A till deposit is made up of unsorted material, while stratified drift is made up of sorted material.

Section 4

Page 221: A slump is the result of a landslide in which a block of material moves downslope over a curved surface.

Page 222: A lahar is caused by the eruption of an ice-covered volcano, which melts ice and causes a hot mudflow.

Chapter 8 Climate

Section 1

Page 234: Climate is the average weather condition in an area over a long period of time. Weather is the condition of the atmosphere at a particular time.

Page 236: Locations near the equator have less seasonal variation because the tilt of the Earth does not change the amount of energy these locations receive from the sun.

Page 238: The atmosphere becomes less dense and loses its ability to absorb and hold thermal energy, at higher elevations.

Page 239: The Gulf Stream current carries warm water past Iceland, which heats the air and causes milder temperatures.

Page 240: Each biome has a different climate and different plant and animals communities.

Section 2

Page 242: You would find the tropical zone from 23.5° north latitude to 23.5° south latitude.

Page 245: Answers may vary. Sample answer: rats, lizards, snakes, and scorpions.

Section 3

Page 246: The temperate zone is located between the Tropics and the polar zone.

Page 248: Temperate deserts are cold at night because low humidity and cloudless skies allow energy to escape.

Page 251: Cities have higher temperatures than the surrounding rural areas because buildings and pavement absorb solar radiation instead of reflecting it.

Section 4

Page 253: Changes in the Earth's orbit and the tilt of the Earth's axis are the two things that Milankovitch says cause ice ages.

Page 254: Dust, ash, and smoke from volcanic eruptions block the sun's rays, which causes the Earth to cool.

Page 257: The deserts would receive less rainfall, making it harder for plants and animals in the desert to survive.

Chapter 9 Stars, Galaxies, and the Universe

Section 1

Page 268: Rigel is hotter than Betelgeuse because blue stars are hotter than red stars.

Page 270: A star's absorption spectrum indicates some of the elements that are in the star's atmosphere.

Page 272: Apparent magnitude is the brightness of a light or star.

Page 273: A light-year is the distance that light travels in 1 year.

Page 274: The actual motion of stars is hard to see because the stars are so distant.

Section 2

Page 277: Energy from gravity is not enough to power the sun, because if all of the sun's gravitational energy were released, the sun would last for only 45 million years.

Page 279: The nuclei of hydrogen atoms repel each other because they are positively charged and like charges repel each other.

Page 280: Sunspots are cooler, dark spots on the sun. Sunspots occur because when activity slows down in the convective zone, areas of the photosphere become cooler.

Section 3

Page 283: A red giant star is a star that expands and cools once it uses all of its hydrogen. As the center of a star continues to shrink a red giant star can become a red supergiant star.

Page 287: A black hole is an object that is so massive that even light cannot escape its gravity. A black hole can be detected when it gives off X rays.

Section 4

Page 288: Spiral galaxies have a bulge at the center and spiral arms. The arms of spiral galaxies are made up of gas, dust, and new stars.

Page 290: A globular cluster is a tight group of up to 1 million stars that looks like a ball. An open cluster is a group of closely grouped stars that are usually located along the spiral disk of a galaxy.

Page 291: Quasars are starlike sources of light that are extremely far away. Some scientists think that quasars may be the core of young galaxies that are in the process of forming.

Section 5

Page 293: Cosmic background radiation is radiation that is left over from the big bang. After the big bang, cosmic background radiation was distributed everywhere and filled all of space.

Page 294: One way to calculate the age of the universe is to measure the distance from Earth to various galaxies.

Page 295: If gravity stops the expansion of the universe, the universe might collapse. If the expansion of the universe continues forever, stars will age and die and the universe will eventually become cold and dark.

Chapter 10 Matter in Motion

Section 1

Page 308: A reference point is an object that appears to stay in place.

Page 310: Velocity can change by changing speed or changing direction.

Page 312: The unit for acceleration is meters per second per second (m/s^2).

Section 2

Page 315: If all of the forces act in the same direction, you must add the forces to determine the net force.

Page 316: 2 N north

Section 3

Page 319: Friction is greater between rough surfaces because rough surfaces have more microscopic hills and valleys.

Page 321: *Static* means "not moving."

Page 322: Three common lubricants are oil, grease, and wax.

Section 4

Page 325: You must exert a force to overcome the gravitational force between the object and Earth.

Page 326: Gravitational force increases as mass increases.

Page 328: The weight of an object is a measure of the gravitational force on the object.

Chapter 11 Forces and Motion

Section 1

Page 341: The acceleration due to gravity is 9.8 m/s^2.

Page 342: Air resistance will have more of an effect on the acceleration of a falling leaf.

Page 344: The word *centripetal* means "toward the center."

Page 346: Gravity gives vertical motion to an object in projectile motion.

Section 2

Page 349: When the bus is moving, both you and the bus are in motion. When the bus stops moving, no unbalanced force acts on your body, so your body continues to move forward.

Page 351: The acceleration of an object increases as the force exerted on the object increases.

Page 353: The forces in a force pair are equal in size and opposite in direction.

Page 354: Objects accelerate toward Earth because the force of gravity pulls them toward Earth.

Section 3

Page 357: When two objects collide, some or all of the momentum of each object can be transferred to the other object.

Page 358: After a collision, objects can stick together or can bounce off each other.

Chapter 12 Chemical Reactions

Section 1

Page 371: A precipitate is a solid substance that is formed in a solution.

Page 372: In a chemical reaction, the chemical bonds in the starting substances break, and then new bonds form to make new substances.

Section 2

Page 375: Ionic compounds are made up of a metal and a nonmetal.

Page 376: Reactants are the starting substances in a chemical reaction, and products are the substances that are formed.

Page 378: The coefficient is 4.

Section 3

Page 380: A synthesis reaction is a reaction in which two or more substances combine to form one new compound.

Page 381: In a decomposition reaction, a substance breaks down into simpler substances. In a synthesis reaction, two or more substances combine to form one new compound.

Page 382: In a single-displacement reaction, one element can replace another element if the replacing element is more reactive than the starting element.

Section 4

Page 385: An endothermic reaction is a chemical reaction in which energy is taken in.

Page 386: Activation energy is the energy that is needed to start a chemical reaction.

Page 388: A high concentration of reactants allows the particles to run into each other more often, so the reaction proceeds at a faster rate.

Chapter 13 Chemical Compounds

Section 1

Page 401: Ionic solutions conduct an electric current because the ions in the solution are charged and are able to move past each other easily.

Page 402: Most covalent compounds will not dissolve in water because the attraction of the water molecules to each other is much stronger than their attraction to the compound.

Section 2

Page 404: A hydronium ion forms when a hydrogen ion bonds to a water molecule in a water solution.

Page 406: Sulfuric acid is used in car batteries to conduct electric current. Hydrochloric acid is used as an algaecide in swimming pools. Nitric acid is used to make fertilizers.

Page 409: Bases can be used at home in the form of soap, oven cleaner, or antacid.

Section 3

Page 410: In a strong acid, all of the molecules of the acid break apart when the acid is dissolved in water. In a weak acid, only a few of the acid molecules break apart when the acid is dissolved in water.

Page 412: Indicators turn different colors at different pH levels. The color on the pH strip can be compared with the colors on the indicator scale to determine the pH of the solution being tested.

Section 4

Page 414: Structural formulas show how atoms in a molecule are connected.

Page 417: Proteins are made of building blocks called *amino acids.*

Page 418: Nucleic acids store genetic information and build proteins.

Chapter 14 The Nature of Sound

Section 1

Page 431: Sound waves consist of longitudinal waves carried through a medium.

Page 432: Sound needs a medium in order to travel.

Page 434: Tinnitus is caused by long-term exposure to loud sounds.

Section 2

Page 437: Frequency is the number of crests or troughs made in a given time.

Page 439: The amplitude of a sound increases as the energy of the vibrations that caused the sound increases.

Page 440: An oscilloscope turns sounds into electrical signals and graphs the signals.

Section 3

Page 443: Echolocation helps some animals find food.

Page 444: Sound wave interference can be either constructive or destructive.

Page 446: A standing wave is a pattern of vibration that looks like a wave that is standing still.

Section 4

Page 449: Musical instruments differ in the part of the instrument that vibrates and in the way that the vibrations are made.

Page 451: Music consists of sound waves that have regular patterns, and noise consists of a random mix of frequencies.

Chapter 15 The Nature of Light

Section 1

Page 463: Electric fields can be found around every charged object.

Page 464: The speed of light is about 880,000 times faster than the speed of sound.

Section 2

Page 466: The speed of a wave is determined by multiplying the wavelength and frequency of the wave.

Page 468: Radio waves carry TV signals.

Page 470: White light is the combination of visible light of all wavelengths.

Page 471: Ultraviolet light waves have shorter wavelengths and higher frequencies than visible light waves do.

Page 472: Patients are protected from X rays by special lead-lined aprons.

Section 3

Page 474: The law of reflection states that the angle of incidence equals the angle of reflection.

Page 475: Sample answer: Four light sources are a television screen, a fluorescent light in the classroom, a light bulb, and the tail of a firefly.

Page 476: You can see things outside of a beam of light because light is scattered outside of the beam.

Page 479: The amount that a wave diffracts depends on the wavelength of the wave and the size of the barrier or opening.

Page 480: Constructive interference is interference in which the resulting wave has a greater amplitude than the original waves had.

Section 4

Page 483: Sample answer: Two translucent objects are a frosted window and wax paper.

Page 484: When white light shines on a colored opaque object, some of the colors of light are absorbed and some are reflected.

Page 486: A pigment is a material that gives color to a substance by absorbing some colors of light and reflecting others.

Section 5

Page 489: Nearsightedness happens when a person's eye is too long. Farsightedness happens when a person's eye is too short.

Page 490: The three kinds of cones are red, blue, and green.

Chapter 16 It's Alive!! Or Is It?

Section 1

Page 505: Sample answer: They control their body temperature by moving from one environment to another. If they get too warm, they move to the shade. If they get too cool, they move out into the sunlight.

Page 506: making food, breaking down food, moving materials into and out of cells, and building cells

Section 2

Page 508: photosynthesis

Page 511: Simple carbohydrates are made of one sugar molecule. Complex carbohydrates are made of many sugar molecules linked together.

Page 512: Most fats are solid, and most oils are liquid.

Section 3

Page 514: Sample answer: larger size, longer life, and cell specialization

Page 515: An organ is a structure of two or more tissues working together to perform a specific function in the body.

Page 516: cell, tissue, organ, organ system

Section 4

Page 519: because they often produce many offspring and have short generation times

Page 520: Sample answer: A newly formed canyon, mountain range, or lake could divide the members of a population.

Chapter 17 The Cell in Action

Section 1

Page 533: Red blood cells would burst in pure water because water particles move from outside, where particles were dense, to inside the cell, where particles were less dense. This movement of water would cause red blood cells to fill up and burst.

Page 535: Exocytosis is the process by which a cell moves large particles to the outside of the cell.

Section 2

Page 537: Cellular respiration is a chemical process by which cells produce energy from food. Breathing supplies oxygen for cellular respiration and removes the carbon dioxide produced by cellular respiration.

Page 539: One kind of fermentation produces CO_2, and the other kind produces lactic acid.

Section 3

Page 541: No, the number of chromosomes is not always related to the complexity of organisms.

Page 542: During cytokinesis in plant cells, a cell plate is formed. During cytokinesis in animal cells, a cell plate does not form.

Section 4

Page 545: 23 chromosomes

Page 546: During meiosis, one parent cell makes four new cells.

Chapter 18 Bacteria and Viruses

Section 1

Page 562: Bacteria make up the kingdoms Eubacteria and Archaebacteria.

Page 564: Binary fission is a process of cell division in which one cell splits into two. All bacteria reproduce by binary fission.

Section 2

Page 568: Nitrogen fixing is the process by which nitrogen gas in the air is transformed into a form that plants can use.

Page 570: In genetic engineering, scientists change the genes of bacteria or other living things.

Section 3

Page 573: Viruses can be classified by shape or by the type of genetic material that they contain. Other possible answers are that viruses can be classified by life cycle or by the kind of disease that they cause.

Page 574: when a virus attacks living cells and turns them into virus factories

Chapter 19 Interactions of Living Things

Section 1

Page 589: The biosphere is the part of Earth where life exists.

Section 2

Page 591: Organisms that eat other organisms are called *consumers*.

Page 593: An energy pyramid is a diagram that shows an ecosystem's loss of energy.

Page 594: Other animals in Yellowstone National Park were affected by the disappearance of the gray wolf because the food web was interrupted. The animals that would normally be prey for the gray wolf were more plentiful. These larger populations ate more vegetation.

Section 3

Page 597: The main ways that organisms affect each other are through competition, predator and prey relationships, symbiotic relationships, and coevolution.

Page 599: Camouflage helps an organism blend in with its surroundings because of its coloring. It is harder for a predator to find a camouflaged prey.

Page 600: In a mutualistic relationship, both organisms benefit from the relationship.

Page 602: Flowers need to attract pollinators to help the flowers reproduce with other members of their species.

Chapter 20 Environmental Problems and Solutions

Section 1

Page 614: Sample answer: Hazardous waste is waste that can catch fire, wear through metal, explode, or make people sick.

Page 617: Sample answer: Exotic species are organisms that make a home for themselves in a new place.

Page 618: Point-source pollution is pollution that comes from one place. Nonpoint-source pollution is pollution that comes from many places.

Section 2

Page 620: reduce, reuse, and recycle

Page 622: Sample answer: Water is reclaimed with plants or filter-feeding animals. Then, it can be used to water crops, parks, lawns, and golf courses.

Page 625: Sample answer: The EPA is a government organization that helps protect the environment.

Chapter 21 The Digestive and Urinary Systems

Section 1

Page 639: Enzymes cut proteins into amino acids that the body can use.

Page 641: Chyme is a soupy mixture of partially digested food in the stomach.

Page 643: Bile breaks large fat droplets into very small droplets. This process allows more fat molecules to be exposed to digestive enzymes.

Page 644: Fiber keeps the stool soft and keeps material moving through the large intestine.

Section 2

Page 647: Nephrons are microscopic filters inside the kidneys.

Page 648: Diuretics are chemicals that cause the kidneys to make more urine.

Chapter 22 Understanding Diseases

Section 1

Page 661: Cooking kills dangerous bacteria or parasites living in meat, fish, and eggs.

Page 663: only when prescribed by a doctor

Section 2

Page 665: Pasteurization kills microorganisms in milk and wine.

Page 667: sanitation, food and water regulations, quarantine policies, and aseptic surgical techniques

Section 3

Page 669: Macrophages engulf, or eat, any microorganisms or viruses that enter your body.

Page 670: If a virus particle enters the body, it may pass into body cells and begin to replicate. Or it may be engulfed and broken up by macrophages.

Page 673: rheumatoid arthritis, diabetes, multiple sclerosis, and lupus

Page 674: HIV causes AIDS.

Chapter 23 Staying Healthy

Section 1

Page 687: An incomplete protein does not contain all of the essential amino acids.

Page 689: Sample answer: a peanut butter sandwich, a glass of milk, and fresh fruit and vegetable slices

Page 690: One serving of chicken noodle soup provides more than 10% of the daily recommended allowance of vitamin A and sodium.

Section 2

Page 693: Over-the-counter drugs can be bought without a prescription. Prescription drugs can be bought only with a prescription from a doctor or other medical professional.

Page 695: First-time use of cocaine can cause a heart attack or can cause a person to become addicted.

Page 696: Drug use is the proper use of a legal drug. Drug abuse is either the use of an illegal drug or the improper use of a legal drug.

Section 3

Page 699: Aerobic exercise strengthens the heart, lungs, and bones and reduces stress. Regular exercise also burns Calories and can give you more energy.

Page 701: Sample answers: Never hike or camp alone, dress for the weather, learn how to swim, wear a life jacket, and never drink unpurified water.

Page 703: CPR is a way to revive someone whose heart has stopped beating. CPR classes are available in many places in the community.

Study Skills

FoldNote Instructions

Have you ever tried to study for a test or quiz but didn't know where to start? Or have you read a chapter and found that you can remember only a few ideas? Well, FoldNotes are a fun and exciting way to help you learn and remember the ideas you encounter as you learn science!

FoldNotes are tools that you can use to organize concepts. By focusing on a few main concepts, FoldNotes help you learn and remember how the concepts fit together. They can help you see the "big picture." Below you will find instructions for building 10 different FoldNotes.

Pyramid

1. Place a sheet of paper in front of you. Fold the lower left-hand corner of the paper diagonally to the opposite edge of the paper.
2. Cut off the tab of paper created by the fold (at the top).
3. Open the paper so that it is a square. Fold the lower right-hand corner of the paper diagonally to the opposite corner to form a triangle.
4. Open the paper. The creases of the two folds will have created an X.
5. Using scissors, cut along one of the creases. Start from any corner, and stop at the center point to create two flaps. Use tape or glue to attach one of the flaps on top of the other flap.

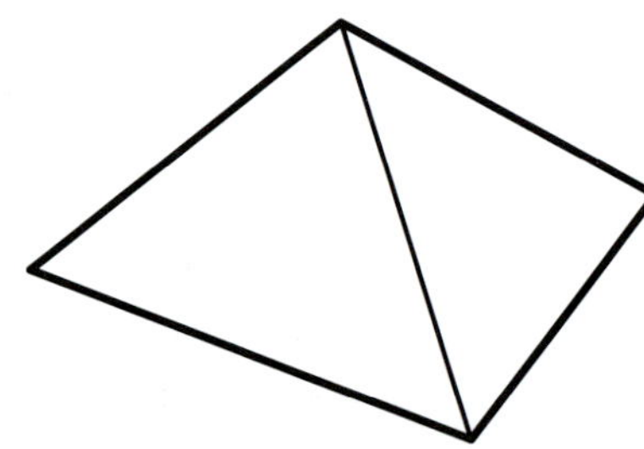

Double Door

1. Fold a sheet of paper in half from the top to the bottom. Then, unfold the paper.
2. Fold the top and bottom edges of the paper to the crease.

Booklet

1. Fold a sheet of paper in half from left to right. Then, unfold the paper.
2. Fold the sheet of paper in half again from the top to the bottom. Then, unfold the paper.
3. Refold the sheet of paper in half from left to right.
4. Fold the top and bottom edges to the center crease.
5. Completely unfold the paper.
6. Refold the paper from top to bottom.
7. Using scissors, cut a slit along the center crease of the sheet from the folded edge to the creases made in step 4. Do not cut the entire sheet in half.
8. Fold the sheet of paper in half from left to right. While holding the bottom and top edges of the paper, push the bottom and top edges together so that the center collapses at the center slit. Fold the four flaps to form a four-page book.

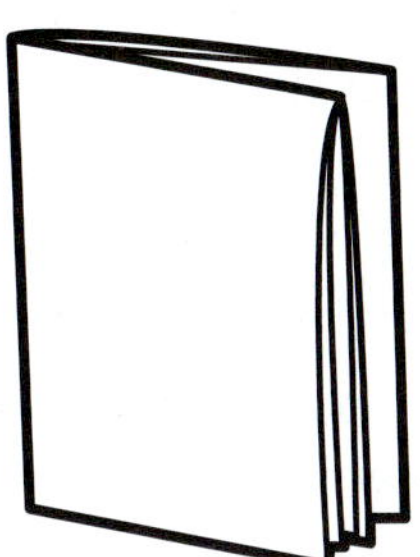

Layered Book

1. Lay one sheet of paper on top of another sheet. Slide the top sheet up so that 2 cm of the bottom sheet is showing.
2. Hold the two sheets together, fold down the top of the two sheets so that you see four 2 cm tabs along the bottom.
3. Using a stapler, staple the top of the FoldNote.

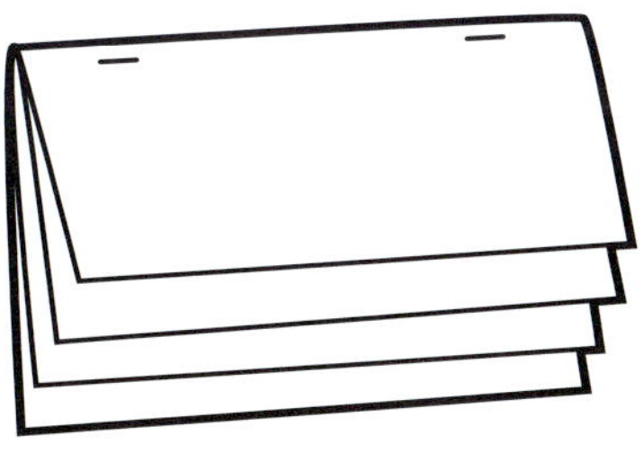

Key-Term Fold

1. Fold a sheet of lined notebook paper in half from left to right.
2. Using scissors, cut along every third line from the right edge of the paper to the center fold to make tabs.

Four-Corner Fold

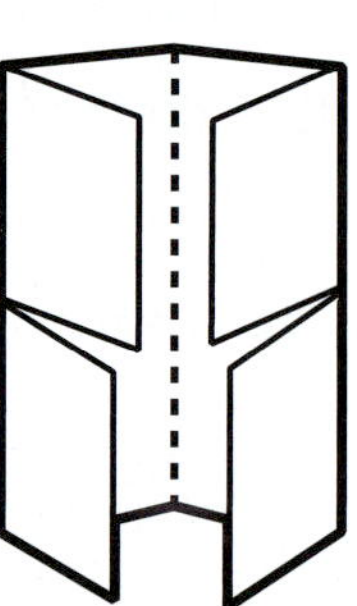

1. Fold a sheet of paper in half from left to right. Then, unfold the paper.
2. Fold each side of the paper to the crease in the center of the paper.
3. Fold the paper in half from the top to the bottom. Then, unfold the paper.
4. Using scissors, cut the top flap creases made in step 3 to form four flaps.

Three-Panel Flip Chart

1. Fold a piece of paper in half from the top to the bottom.
2. Fold the paper in thirds from side to side. Then, unfold the paper so that you can see the three sections.
3. From the top of the paper, cut along each of the vertical fold lines to the fold in the middle of the paper. You will now have three flaps.

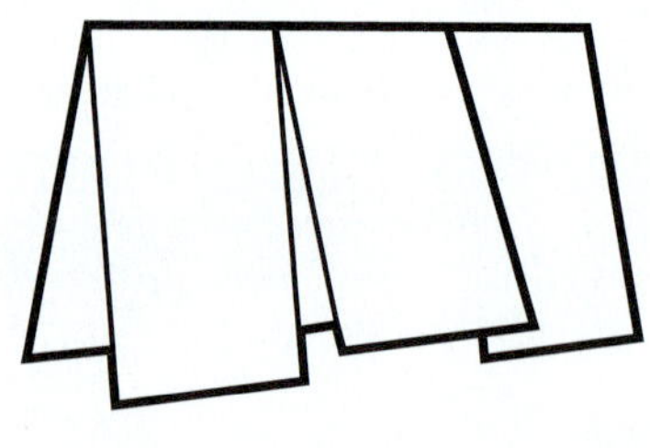

Appendix

Table Fold

1. Fold a piece of paper in half from the top to the bottom. Then, fold the paper in half again.
2. Fold the paper in thirds from side to side.
3. Unfold the paper completely. Carefully trace the fold lines by using a pen or pencil.

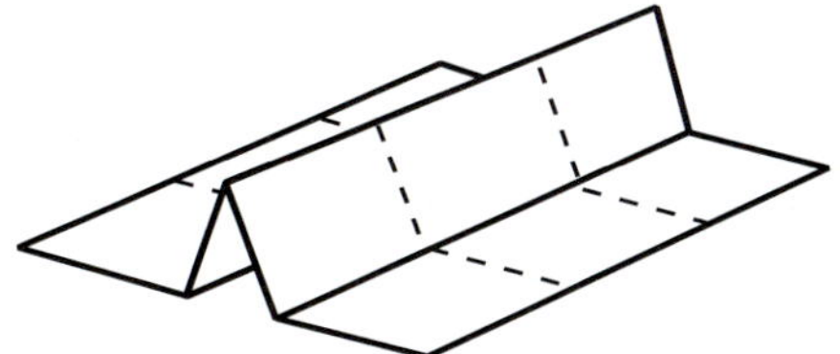

Two-Panel Flip Chart

1. Fold a piece of paper in half from the top to the bottom.
2. Fold the paper in half from side to side. Then, unfold the paper so that you can see the two sections.
3. From the top of the paper, cut along the vertical fold line to the fold in the middle of the paper. You will now have two flaps.

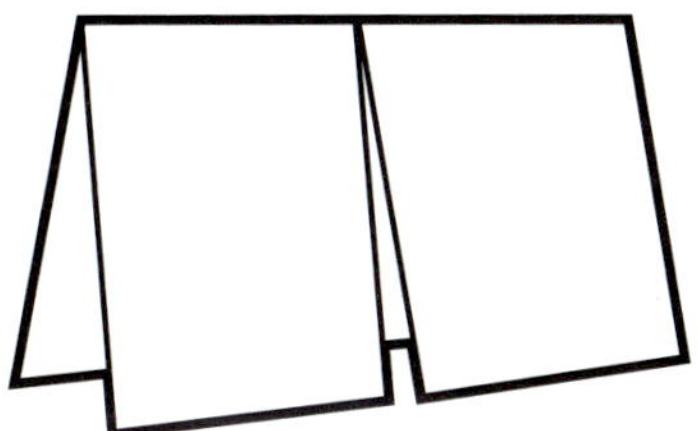

Tri-Fold

1. Fold a piece a paper in thirds from the top to the bottom.
2. Unfold the paper so that you can see the three sections. Then, turn the paper sideways so that the three sections form vertical columns.
3. Trace the fold lines by using a pen or pencil. Label the columns "Know," "Want," and "Learn."

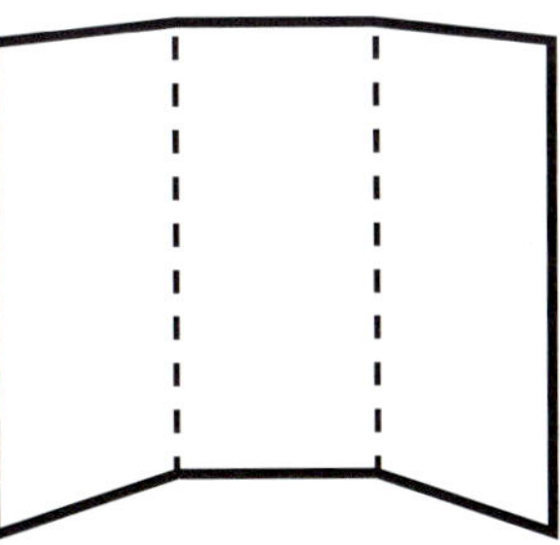

Graphic Organizer Instructions

Have you ever wished that you could "draw out" the many concepts you learn in your science class? Sometimes, being able to *see* how concepts are related really helps you remember what you've learned. Graphic Organizers do just that! They give you a way to draw or map out concepts.

All you need to make a Graphic Organizer is a piece of paper and a pencil. Below you will find instructions for four different Graphic Organizers designed to help you organize the concepts you'll learn in this book.

Spider Map

1. Draw a diagram like the one shown. In the circle, write the main topic.
2. From the circle, draw legs to represent different categories of the main topic. You can have as many categories as you want.
3. From the category legs, draw horizontal lines. As you read the chapter, write details about each category on the horizontal lines.

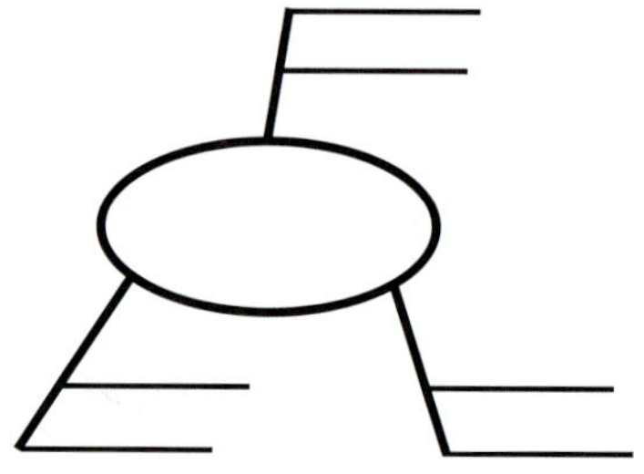

Comparison Table

1. Draw a chart like the one shown. Your chart can have as many columns and rows as you want.
2. In the top row, write the topics that you want to compare.
3. In the left column, write characteristics of the topics that you want to compare. As you read the chapter, fill in the characteristics for each topic in the appropriate boxes.

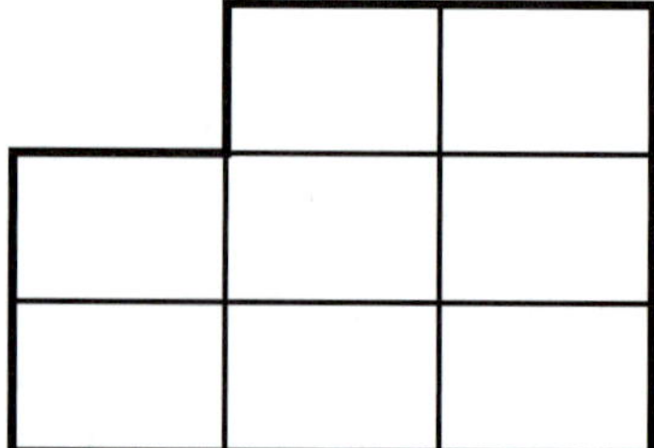

Appendix

Chain-of-Events-Chart

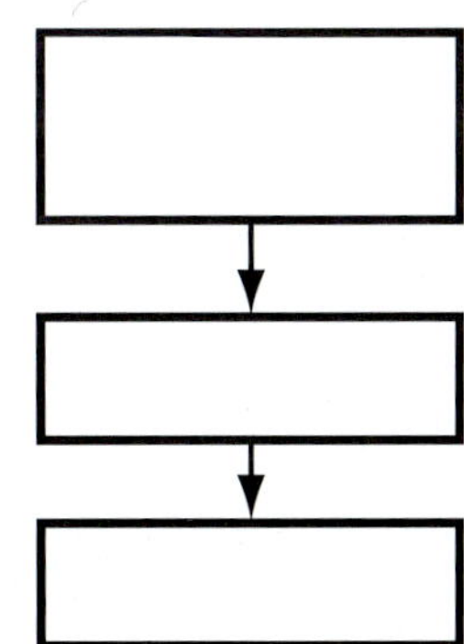

1. Draw a box. In the box, write the first step of a process or the first event of a timeline.
2. Under the box, draw another box, and use an arrow to connect the two boxes. In the second box, write the next step of the process or the next event in the timeline.
3. Continue adding boxes until the process or timeline is finished.

Concept Map

1. Draw a circle in the center of a piece of paper. Write the main idea of the chapter in the center of the circle.
2. From the circle, draw other circles. In those circles, write characteristics of the main idea. Draw arrows from the center circle to the circles that contain the characteristics.
3. From each circle that contains a characteristic, draw other circles. In those circles, write specific details about the characteristic. Draw arrows from each circle that contains a characteristic to the circles that contain specific details. You may draw as many circles as you want.

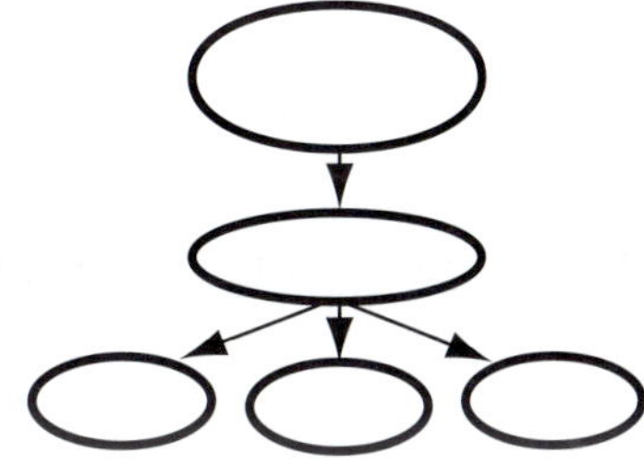

SI Measurement

The International System of Units, or SI, is the standard system of measurement used by many scientists. Using the same standards of measurement makes it easier for scientists to communicate with one another.

SI works by combining prefixes and base units. Each base unit can be used with different prefixes to define smaller and larger quantities. The table below lists common SI prefixes.

SI Prefixes			
Prefix	**Symbol**	**Factor**	**Example**
kilo-	k	1,000	kilogram, 1 kg = 1,000 g
hecto-	h	100	hectoliter, 1 hL = 100 L
deka-	da	10	dekameter, 1 dam = 10 m
		1	meter, liter, gram
deci-	d	0.1	decigram, 1 dg = 0.1 g
centi-	c	0.01	centimeter, 1 cm = 0.01 m
milli-	m	0.001	milliliter, 1 mL = 0.001 L
micro-	μ	0.000 001	micrometer, 1 μm = 0.000 001 m

SI Conversion Table		
SI units	**From SI to English**	**From English to SI**
Length		
kilometer (km) = 1,000 m	1 km = 0.621 mi	1 mi = 1.609 km
meter (m) = 100 cm	1 m = 3.281 ft	1 ft = 0.305 m
centimeter (cm) = 0.01 m	1 cm = 0.394 in.	1 in. = 2.540 cm
millimeter (mm) = 0.001 m	1 mm = 0.039 in.	
micrometer (μm) = 0.000 001 m		
nanometer (nm) = 0.000 000 001 m		
Area		
square kilometer (km^2) = 100 hectares	1 km^2 = 0.386 mi^2	1 mi^2 = 2.590 km^2
hectare (ha) = 10,000 m^2	1 ha = 2.471 acres	1 acre = 0.405 ha
square meter (m^2) = 10,000 cm^2	1 m^2 = 10.764 ft^2	1 ft^2 = 0.093 m^2
square centimeter (cm^2) = 100 mm^2	1 cm^2 = 0.155 $in.^2$	1 $in.^2$ = 6.452 cm^2
Volume		
liter (L) = 1,000 mL = 1 dm^3	1 L = 1.057 fl qt	1 fl qt = 0.946 L
milliliter (mL) = 0.001 L = 1 cm^3	1 mL = 0.034 fl oz	1 fl oz = 29.574 mL
microliter (μL) = 0.000 001 L		
Mass		
kilogram (kg) = 1,000 g	1 kg = 2.205 lb	1 lb = 0.454 kg
gram (g) = 1,000 mg	1 g = 0.035 oz	1 oz = 28.350 g
milligram (mg) = 0.001 g		
microgram (μg) = 0.000 001 g		

Temperature Scales

Temperature can be expressed by using three different scales: Fahrenheit, Celsius, and Kelvin. The SI unit for temperature is the kelvin (K). Although 0 K is much colder than 0°C, a change of 1 K is equal to a change of 1°C.

Three Temperature Scales

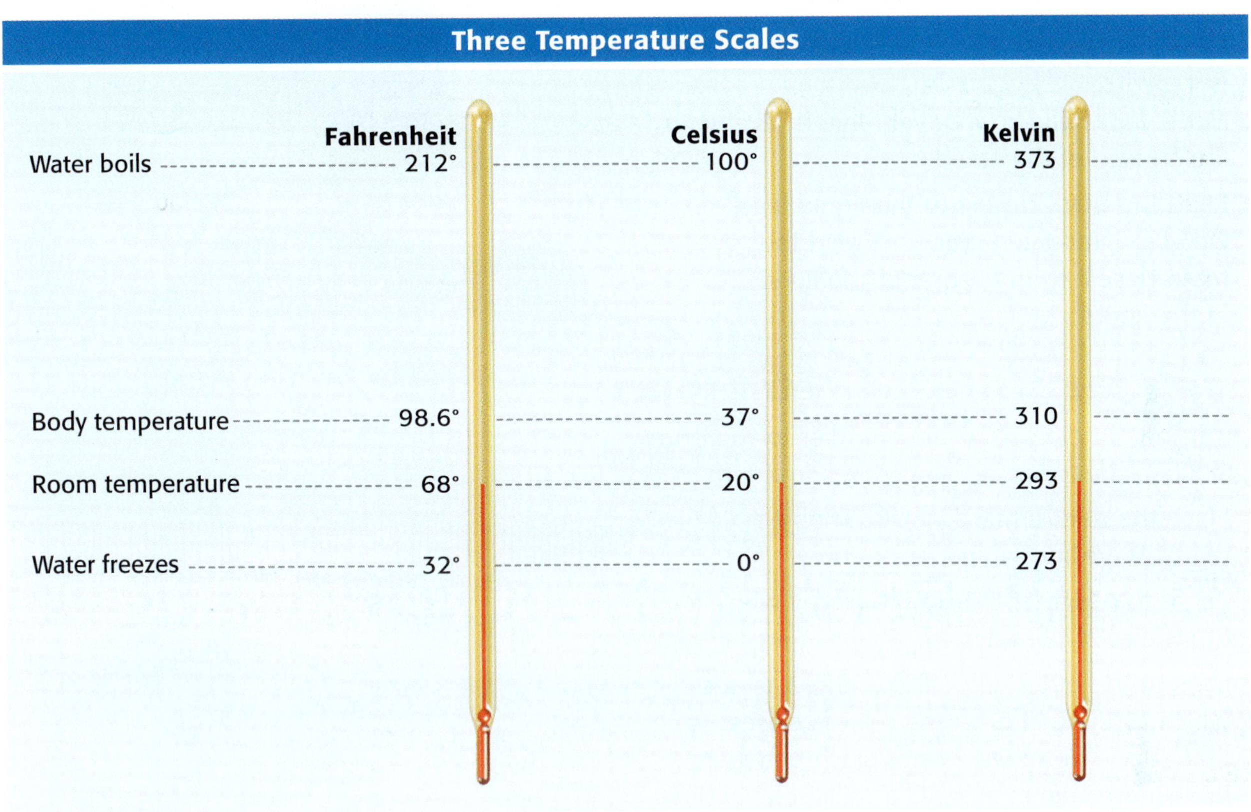

Temperature Conversions Table

To convert	Use this equation:	Example
Celsius to Fahrenheit °C → °F	$°F = \left(\frac{9}{5} \times °C\right) + 32$	Convert 45°C to °F. $°F = \left(\frac{9}{5} \times 45°C\right) + 32 = 113°F$
Fahrenheit to Celsius °F → °C	$°C = \frac{5}{9} \times (°F - 32)$	Convert 68°F to °C. $°C = \frac{5}{9} \times (68°F - 32) = 20°C$
Celsius to Kelvin °C → K	$K = °C + 273$	Convert 45°C to K. $K = 45°C + 273 = 318\ K$
Kelvin to Celsius K → °C	$°C = K - 273$	Convert 32 K to °C. $°C = 32K - 273 = -241°C$

Measuring Skills

Using a Graduated Cylinder

When using a graduated cylinder to measure volume, keep the following procedures in mind:

1. Place the cylinder on a flat, level surface before measuring liquid.
2. Move your head so that your eye is level with the surface of the liquid.
3. Read the mark closest to the liquid level. On glass graduated cylinders, read the mark closest to the center of the curve in the liquid's surface.

Using a Meterstick or Metric Ruler

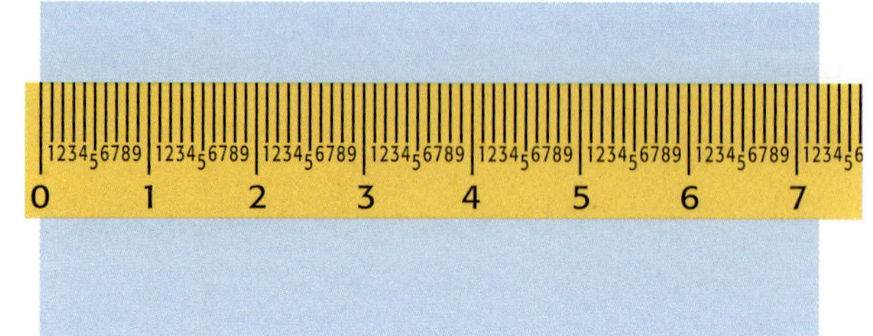

When using a meterstick or metric ruler to measure length, keep the following procedures in mind:

1. Place the ruler firmly against the object that you are measuring.
2. Align one edge of the object exactly with the 0 end of the ruler.
3. Look at the other edge of the object to see which of the marks on the ruler is closest to that edge. (Note: Each small slash between the centimeters represents a millimeter, which is one-tenth of a centimeter.)

Using a Triple-Beam Balance

When using a triple-beam balance to measure mass, keep the following procedures in mind:

1. Make sure the balance is on a level surface.
2. Place all of the countermasses at 0. Adjust the balancing knob until the pointer rests at 0.
3. Place the object you wish to measure on the pan. **Caution:** Do not place hot objects or chemicals directly on the balance pan.
4. Move the largest countermass along the beam to the right until it is at the last notch that does not tip the balance. Follow the same procedure with the next-largest countermass. Then, move the smallest countermass until the pointer rests at 0.
5. Add the readings from the three beams together to determine the mass of the object.
6. When determining the mass of crystals or powders, first find the mass of a piece of filter paper. Then, add the crystals or powder to the paper, and remeasure. The actual mass of the crystals or powder is the total mass minus the mass of the paper. When finding the mass of liquids, first find the mass of the empty container. Then, find the combined mass of the liquid and container. The mass of the liquid is the total mass minus the mass of the container.

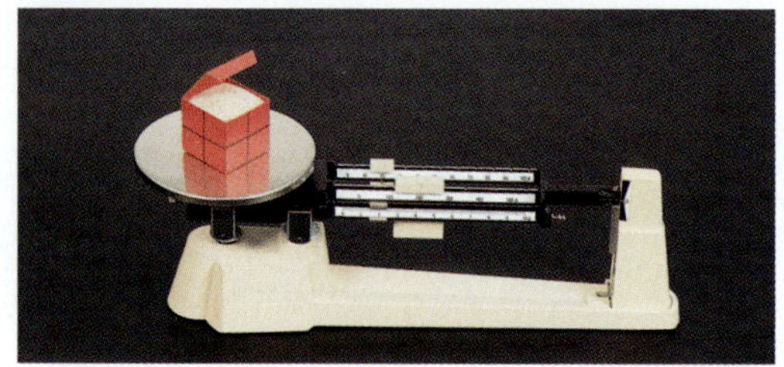

Scientific Methods

The ways in which scientists answer questions and solve problems are called **scientific methods.** The same steps are often used by scientists as they look for answers. However, there is more than one way to use these steps. Scientists may use all of the steps or just some of the steps during an investigation. They may even repeat some of the steps. The goal of using scientific methods is to come up with reliable answers and solutions.

Six Steps of Scientific Methods

Good questions come from careful **observations.** You make observations by using your senses to gather information. Sometimes, you may use instruments, such as microscopes and telescopes, to extend the range of your senses. As you observe the natural world, you will discover that you have many more questions than answers. These questions drive investigations.

Questions beginning with *what, why, how,* and *when* are important in focusing an investigation. Here is an example of a question that could lead to an investigation.

Question: How does acid rain affect plant growth?

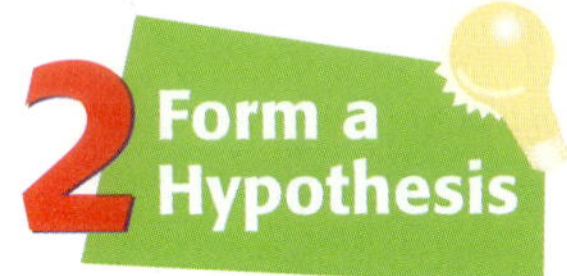

After you ask a question, you need to form a **hypothesis.** A hypothesis is a clear statement of what you expect the answer to your question to be. Your hypothesis will represent your best "educated guess" based on what you have observed and what you already know. A good hypothesis is testable. Otherwise, the investigation can go no further. Here is a hypothesis based on the question, "How does acid rain affect plant growth?"

Hypothesis: Acid rain slows plant growth.

The hypothesis can lead to predictions. A prediction is what you think the outcome of your experiment or data collection will be. Predictions are usually stated in an if-then format. Here is a sample prediction for the hypothesis that acid rain slows plant growth.

Prediction: If a plant is watered with only acid rain (which has a pH of 4), then the plant will grow at half its normal rate.

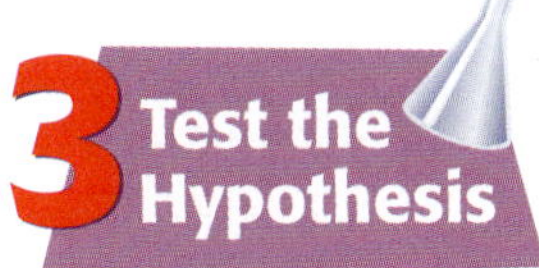

After you have formed a hypothesis and made a prediction, your hypothesis should be tested. One way to test a hypothesis is with a controlled experiment. A **controlled experiment** tests only one factor at a time. In an experiment to test the effect of acid rain on plant growth, the **control group** would be watered with normal rain water. The **experimental group** would be watered with acid rain. All of the plants should receive the same amount of sunlight and water each day. The air temperature should be the same for all groups. However, the acidity of the water will be a variable. In fact, any factor that is different from one group to another is a **variable.** If your hypothesis is correct, then the acidity of the water and plant growth are *dependant variables.* The amount a plant grows is dependent on the acidity of the water. However, the amount of water each plant receives and the amount of sunlight each plant receives are *independent variables.* Either of these factors could change without affecting the other factor.

Sometimes, the nature of an investigation makes a controlled experiment impossible. For example, the Earth's core is surrounded by thousands of meters of rock. Under such circumstances, a hypothesis may be tested by making detailed observations.

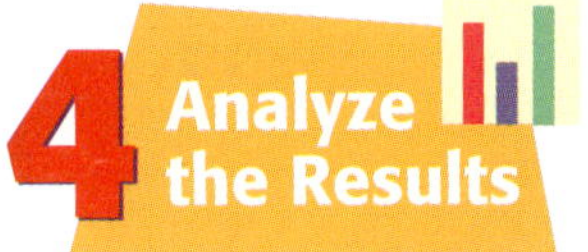

After you have completed your experiments, made your observations, and collected your data, you must analyze all the information you have gathered. Tables and graphs are often used in this step to organize the data.

After analyzing your data, you can determine if your results support your hypothesis. If your hypothesis is supported, you (or others) might want to repeat the observations or experiments to verify your results. If your hypothesis is not supported by the data, you may have to check your procedure for errors. You may even have to reject your hypothesis and make a new one. If you cannot draw a conclusion from your results, you may have to try the investigation again or carry out further observations or experiments.

After any scientific investigation, you should report your results. By preparing a written or oral report, you let others know what you have learned. They may repeat your investigation to see if they get the same results. Your report may even lead to another question and then to another investigation.

Scientific Methods in Action

Scientific methods contain loops in which several steps may be repeated over and over again. In some cases, certain steps are unnecessary. Thus, there is not a "straight line" of steps. For example, sometimes scientists find that testing one hypothesis raises new questions and new hypotheses to be tested. And sometimes, testing the hypothesis leads directly to a conclusion. Furthermore, the steps in scientific methods are not always used in the same order. Follow the steps in the diagram, and see how many different directions scientific methods can take you.

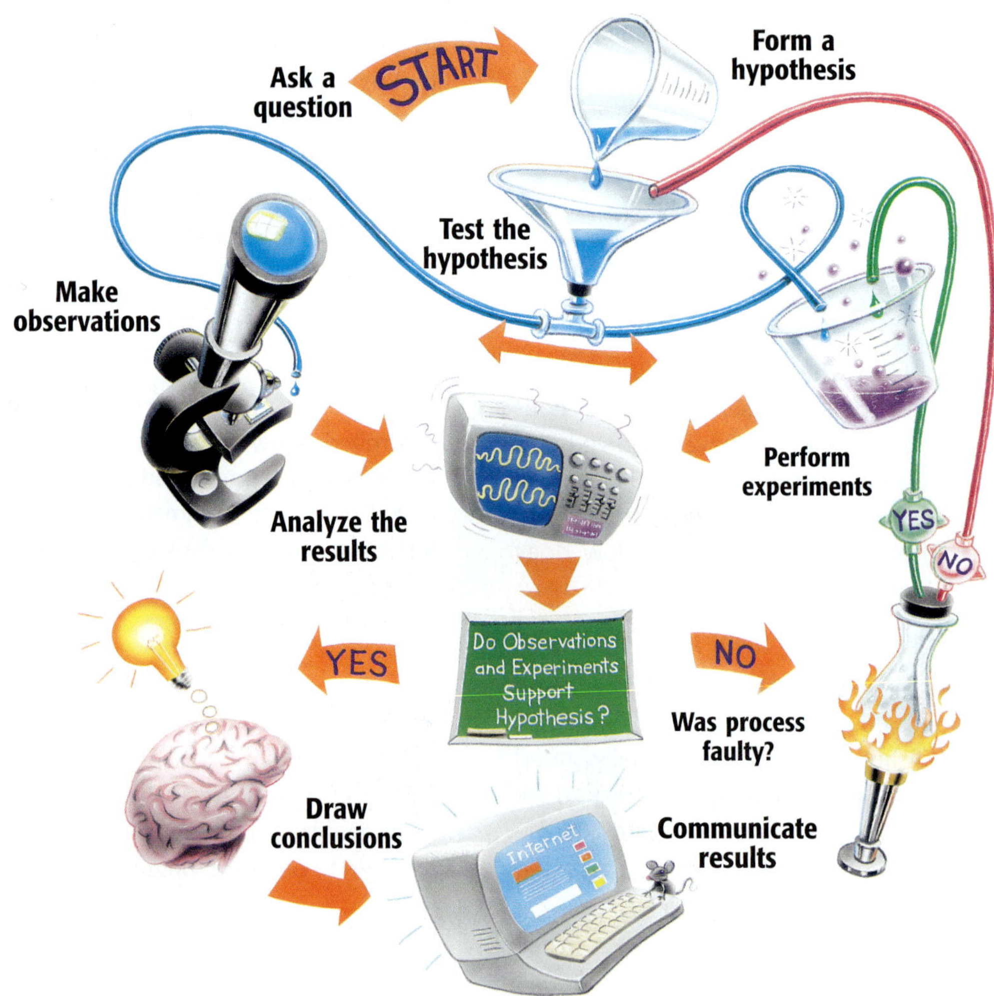

Making Charts and Graphs

Pie Charts

A pie chart shows how each group of data relates to all of the data. Each part of the circle forming the chart represents a category of the data. The entire circle represents all of the data. For example, a biologist studying a hardwood forest in Wisconsin found that there were five different types of trees. The data table at right summarizes the biologist's findings.

Wisconsin Hardwood Trees	
Type of tree	**Number found**
Oak	600
Maple	750
Beech	300
Birch	1,200
Hickory	150
Total	3,000

How to Make a Pie Chart

1. To make a pie chart of these data, first find the percentage of each type of tree. Divide the number of trees of each type by the total number of trees, and multiply by 100.

$$\frac{600 \text{ oak}}{3{,}000 \text{ trees}} \times 100 = 20\%$$

$$\frac{750 \text{ maple}}{3{,}000 \text{ trees}} \times 100 = 25\%$$

$$\frac{300 \text{ beech}}{3{,}000 \text{ trees}} \times 100 = 10\%$$

$$\frac{1{,}200 \text{ birch}}{3{,}000 \text{ trees}} \times 100 = 40\%$$

$$\frac{150 \text{ hickory}}{3{,}000 \text{ trees}} \times 100 = 5\%$$

2. Now, determine the size of the wedges that make up the pie chart. Multiply each percentage by 360°. Remember that a circle contains 360°.

$20\% \times 360° = 72°$ $\quad 25\% \times 360° = 90°$

$10\% \times 360° = 36°$ $\quad 40\% \times 360° = 144°$

$5\% \times 360° = 18°$

3. Check that the sum of the percentages is 100 and the sum of the degrees is 360.

$20\% + 25\% + 10\% + 40\% + 5\% = 100\%$

$72° + 90° + 36° + 144° + 18° = 360°$

4. Use a compass to draw a circle and mark the center of the circle.

5. Then, use a protractor to draw angles of 72°, 90°, 36°, 144°, and 18° in the circle.

6. Finally, label each part of the chart, and choose an appropriate title.

A Community of Wisconsin Hardwood Trees

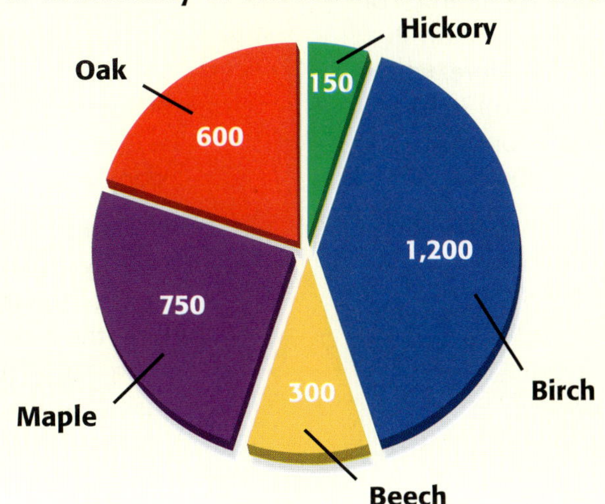

Appendix

Line Graphs

Line graphs are most often used to demonstrate continuous change. For example, Mr. Smith's students analyzed the population records for their hometown, Appleton, between 1900 and 2000. Examine the data at right.

Because the year and the population change, they are the *variables*. The population is determined by, or dependent on, the year. Therefore, the population is called the **dependent variable,** and the year is called the **independent variable.** Each set of data is called a **data pair.** To prepare a line graph, you must first organize data pairs into a table like the one at right.

Population of Appleton, 1900–2000

Year	Population
1900	1,800
1920	2,500
1940	3,200
1960	3,900
1980	4,600
2000	5,300

How to Make a Line Graph

1. Place the independent variable along the horizontal (*x*) axis. Place the dependent variable along the vertical (*y*) axis.
2. Label the *x*-axis "Year" and the *y*-axis "Population." Look at your largest and smallest values for the population. For the *y*-axis, determine a scale that will provide enough space to show these values. You must use the same scale for the entire length of the axis. Next, find an appropriate scale for the *x*-axis.
3. Choose reasonable starting points for each axis.
4. Plot the data pairs as accurately as possible.
5. Choose a title that accurately represents the data.

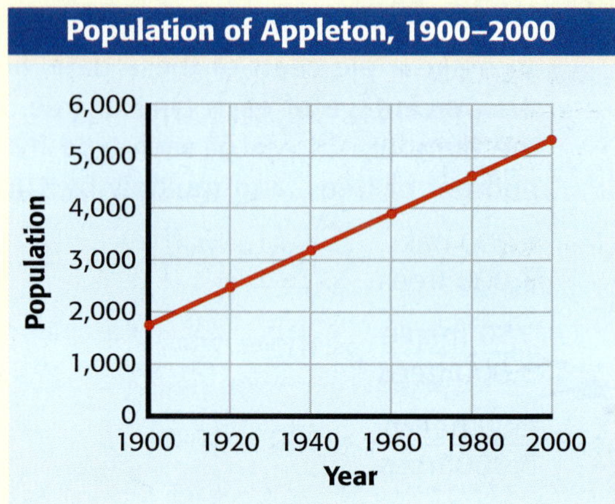

How to Determine Slope

Slope is the ratio of the change in the *y*-value to the change in the *x*-value, or "rise over run."

1. Choose two points on the line graph. For example, the population of Appleton in 2000 was 5,300 people. Therefore, you can define point *a* as (2000, 5,300). In 1900, the population was 1,800 people. You can define point *b* as (1900, 1,800).
2. Find the change in the *y*-value. (*y* at point *a*) − (*y* at point *b*) = 5,300 people − 1,800 people = 3,500 people
3. Find the change in the *x*-value. (*x* at point *a*) − (*x* at point *b*) = 2000 − 1900 = 100 years
4. Calculate the slope of the graph by dividing the change in *y* by the change in *x*.

$$slope = \frac{change\ in\ y}{change\ in\ x}$$

$$slope = \frac{3{,}500 \text{ people}}{100 \text{ years}}$$

$$slope = 35 \text{ people per year}$$

In this example, the population in Appleton increased by a fixed amount each year. The graph of these data is a straight line. Therefore, the relationship is **linear.** When the graph of a set of data is not a straight line, the relationship is **nonlinear.**

Appendix

Using Algebra to Determine Slope

The equation in step 4 may also be arranged to be

$y = kx$

where y represents the change in the y-value, k represents the slope, and x represents the change in the x-value.

$$slope = \frac{change\ in\ y}{change\ in\ x}$$

$$k = \frac{y}{x}$$

$$k \times x = \frac{y \times x}{x}$$

$$kx = y$$

Bar Graphs

Bar graphs are used to demonstrate change that is not continuous. These graphs can be used to indicate trends when the data cover a long period of time. A meteorologist gathered the precipitation data shown here for Hartford, Connecticut, for April 1–15, 1996, and used a bar graph to represent the data.

Precipitation in Hartford, Connecticut April 1–15, 1996

Date	Precipitation (cm)	Date	Precipitation (cm)
April 1	0.5	April 9	0.25
April 2	1.25	April 10	0.0
April 3	0.0	April 11	1.0
April 4	0.0	April 12	0.0
April 5	0.0	April 13	0.25
April 6	0.0	April 14	0.0
April 7	0.0	April 15	6.50
April 8	1.75		

How to Make a Bar Graph

1. Use an appropriate scale and a reasonable starting point for each axis.
2. Label the axes, and plot the data.
3. Choose a title that accurately represents the data.

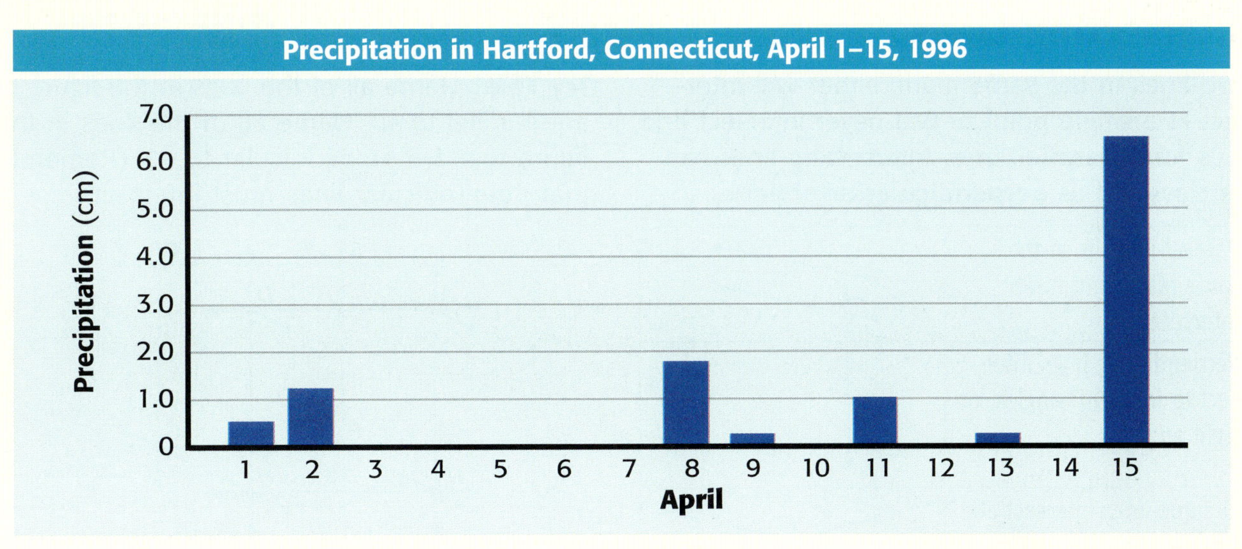

Appendix

Math Refresher

Science requires an understanding of many math concepts. The following pages will help you review some important math skills.

Binary Numbers

Computers use a **binary,** or base-2, system to represent numbers. In the conventional (base-10) number system, each place value is 10 times the value of the place value to the right, and 10 digits (0 through 9) are used.

5,102 = (**5** × 1,000) + (**1** × 100) + (**0** × 10) + (**2** × 1)

10^3	10^2	10^1	10^0
Thousands	**Hundreds**	**Tens**	**Ones**

In the binary number system, each place value is 2 times the value of the place value to the right, and only 2 digits—0 and 1—are used. The base-10 number 13 would be written as 1101 in the binary system.

1101 = (**1** × 8) + (**1** × 4) + (**0** × 2) + (**1** × 1)

2^3	2^2	2^1	2^0
Eights	**Fours**	**Twos**	**Ones**

By using only 2 digits, binary numbers can be transmitted and stored by switching an electric circuit or a magnetic particle "on" (= 1) or "off" (= 0).

Example 1: Find the base-10 value for the binary number 10111.

Step 1: Write each place value in base-10.
10111 = (**1** × 16) + (**0** × 8) + (**1** × 4) + (**1** × 2) + (**1** × 1)
= 16 + 4 + 2 + 1

Step 2: Find the sum of the place values.
16 + 4 + 2 + 1 = 23

The number 10111 in the binary system is equal to 23 in the base-10 system.

Example 2: Find the binary value for the base-10 number 29.

Step 1: Write the base-10 number as a sum of powers of 2. Start with the highest power of 2 that is not more than the base-10 number.
29 = (**1** × 16) + (**1** × 8) + (**1** × 4) + (**0** × 2) + (**1** × 1)
= 11101

Try This! Write 10101 and 10001 as base-10 numbers. Write 13 and 16 as binary numbers.

Intersecting Lines

Two lines in the same plane either will intersect at a single point or will never intersect and thus form **parallel** lines. Intersecting lines can be classified as **perpendicular** or **oblique.**

Parallel lines lie in the same plane and never intersect.	
Perpendicular lines intersect to form 90° angles, or right angles.	
Oblique lines intersect at angles other than 90°.	

Try This! Name all of the lines in the figure that are parallel to $\overleftrightarrow{AD}$. Name all of the lines in the figure that are perpendicular to $\overleftrightarrow{FG}$. (Remember that perpendicular lines must intersect.)

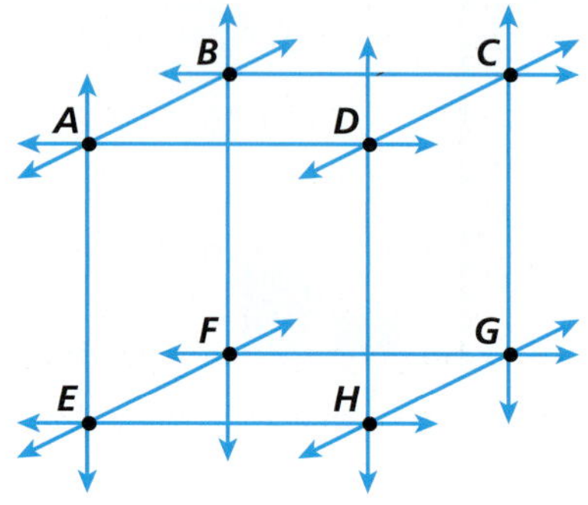

Negative Numbers

Although many items can be counted or measured by using only positive numbers, certain quantities are best represented as **negative numbers,** or numbers that are less than 0. Some temperatures and elevations are negative numbers. Losses or decreases in quantities are often expressed as negative numbers.

On a number line, negative numbers are to the left of 0.

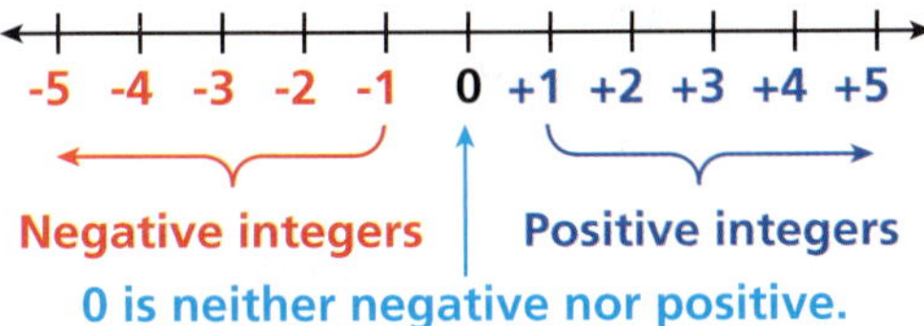

Example: Use a negative number to represent each situation.

A: 17°C below the freezing point of water

The freezing point of water is 0°C.
17° below this temperature would be $^{-}17$°C.

B: a loss of 5 L of a liquid

Losses can be represented by negative numbers: $^{-}5$ L

Try This! Use a negative number to represent each situation.

A. 45 m below sea level
B. removing 5 g from a balance

Probability

The **probability** of an event can be estimated by comparing the frequency with which an event occurs to the total number of observations *(experimental probability).*

Example 1: Will has hit 40 free throws out of 50 attempts. What is the probability that Will can make his next free throw?

Step 1: Determine the number of times the desired event occurs and the total number of events. There were 40 free throws made in 50 observations.

Step 2: Calculate the probability.

$$P = \frac{40}{50} = \frac{4}{5} = 0.8\text{, or }80\%$$

Will has an 80% chance of making his next free throw.

When all possible outcomes of a trial are equally likely, the probability can be determined without observation by calculating the total number of possible outcomes *(theoretical probability).*

Example 2: Zoe rolls a fair die. What is the probability that she will roll an odd number?

Step 1: Determine the number of times the desired event could occur and the total number of possible outcomes. On a fair die, rolling any given number on the die is equally likely. There are three possible outcomes that will be an odd number and six total possible outcomes.

Step 2: Calculate the probability.

$$P = \frac{3}{6} = \frac{1}{2} = 0.5\text{, or }50\%$$

Zoe has a 50% chance of rolling an odd number.

Try This 1! Aaron has 20 hits in 80 at-bats. What is the probability that he will get a hit during his next at-bat?

Try This 2! What is the probability that you will roll a 6 on a die?

Misleading Graphs

When data are shown on a graph, it is possible to distort the graph so that differences in the data are either exaggerated or diminished. This distortion can be done by changing the scale of either axis of the graph. Sometimes, if the smallest data points are not close to 0, the graph will have a "broken" axis. However, a broken axis can change the appearance of a graph so that the graph is misleading.

Example: In the following graphs, Graph A has a broken axis, and Graph B does not. The graphs look very different even though both show the same data. Graph A exaggerates the differences in the data.

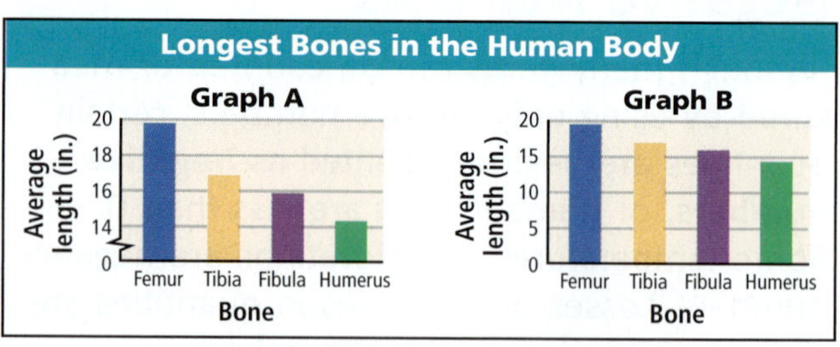

Try This! How is this graph misleading?

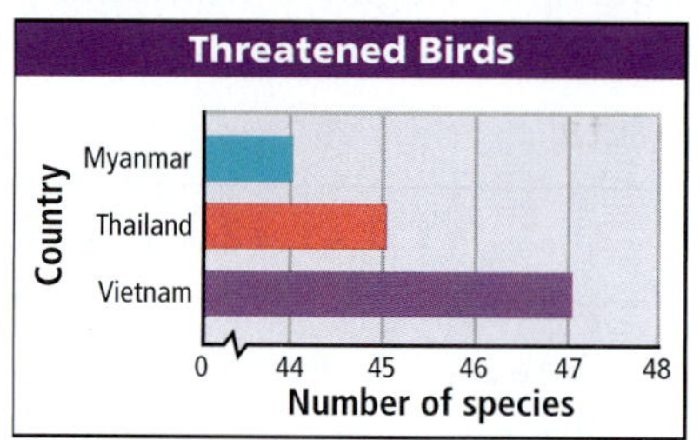

Rates of Change

A car driving down the highway will travel a distance that depends on the speed of the car and the time traveled. At a constant speed, the distance traveled will increase a fixed amount for every unit of time traveled. The formula *distance* = *rate* × *time* ($d = rt$) tells how far an object travels, d, at a certain rate, r, over a certain time, t.

Example: How many miles does a car travel in 4 h at 65 mi/h?

Step 1: Substitute into the equation $d = rt$.

$$d = 65 \text{ mi/h}(4 \text{ h})$$

Step 2: Solve the equation.

$$d = 260 \text{ mi}$$

The car traveled 260 mi.

Try This! Teresa's car can travel 25 mi on a gallon of gasoline. How far can Teresa travel on 15 gal of gasoline? Write an equation using d for distance traveled, g for the amount of gas, and r for the mi/gal that her car can travel.

Sampling of Populations

When collecting data about a group, you must collect data on each member of the group in order to have completely accurate information. Often, this is not possible, so you must study a part of the group called a *sample*. The sample must be chosen so that it represents the entire population well.

Example: For a preelection survey, you (1) telephone every tenth registered voter or (2) set up a survey on a Web page. Which sample is more representative of the population?

The first sample is; more people have phones than have Internet access, and you are interested in people who vote.

Try This! To assess the quality of 1,000 widgets as they are manufactured, you (1) test every fiftieth widget or (2) test the first and last widget.

Appendix

Perimeter, Area, and Volume

Perimeter is the distance around the outside of a closed geometric figure. It can be determined by adding the lengths of all of the sides.

Area is a measure of how much surface a figure has.

The area of a rectangle is the product of the length and the width.

Example 1: Find the perimeter and area of the rectangle.

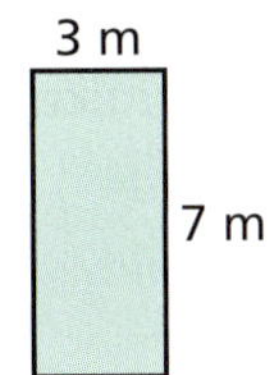

Perimeter: 3 m + 7 m + 3 m + 7 m = 20 m

$$\begin{aligned} \textit{area} &= \textit{length} \times \textit{width} \\ &= (3\text{ m})\ (7\text{ m}) \\ &= 21\text{ m}^2 \end{aligned}$$

The perimeter of a circle is called the *circumference.* The circumference of a circle is the product of π and the diameter of the circle.

The area of a circle is the product of π and the square of the radius. The radius is half the diameter of the circle.

Example 2: Find the circumference and area of the circle to the nearest tenth. Use 3.14 for π.

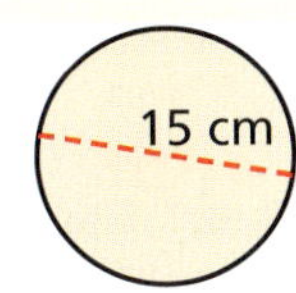

$$\text{circumference} = \pi d$$
$$15\pi = 15(3.14) = 47.1\text{ cm}$$
$$\textit{area} = \pi r^2$$
$$\pi(7.5)^2 = 56.25\pi = 176.6\text{ cm}^2$$

The area of a triangle is one-half the product of the base and height.

Example 3: Find the area of the triangle.

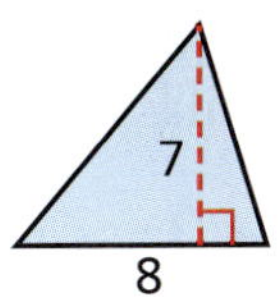

$$\begin{aligned} \textit{area} &= \tfrac{1}{2}\ \textit{base} \times \textit{height} \\ &= \tfrac{1}{2}\ (8)(7) \\ &= 28\text{ units}^2 \end{aligned}$$

Volume is the number of cubic units in a three-dimensional figure. The volume of a rectangular prism or cube is the product of the length, width, and height.

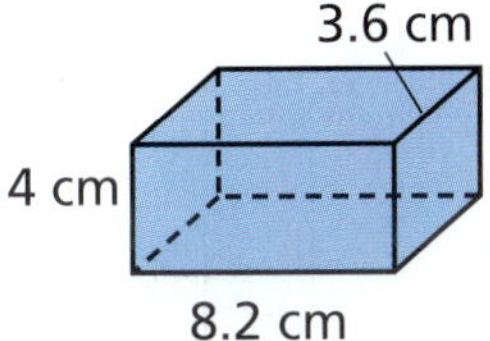

Example 4: Find the volume of the figure.

$$\begin{aligned} \textit{volume} &= \textit{length} \times \textit{width} \times \textit{height} \\ &= (8.2)(3.6)(4) \\ &= 118.1\text{ cm}^3 \end{aligned}$$

Try This! Find the perimeter or circumference and the area of each figure.

1. 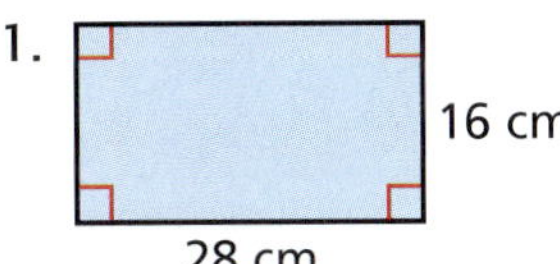

2. 3.4 m

Find the volume of the figure.

3. 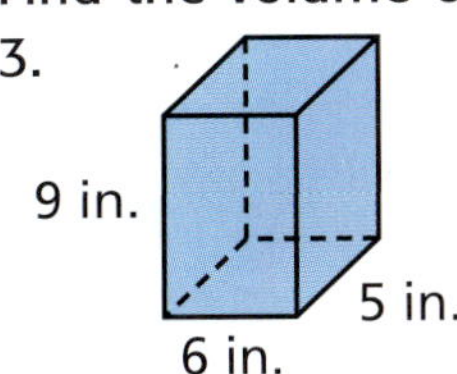

Appendix

Periodic Table of the Elements

Each square on the table includes an element's name, chemical symbol, atomic number, and atomic mass.

The color of the chemical symbol indicates the physical state at room temperature. Carbon is a solid.

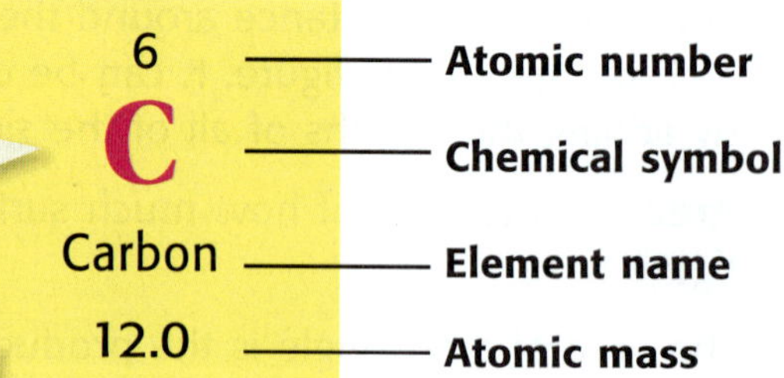

The background color indicates the type of element. Carbon is a nonmetal.

Background
- Metals
- Metalloids
- Nonmetals

Chemical symbol
- Solid
- Liquid
- Gas

	Group 1	Group 2	Group 3	Group 4	Group 5	Group 6	Group 7	Group 8	Group 9
Period 1	1 H Hydrogen 1.0								
Period 2	3 Li Lithium 6.9	4 Be Beryllium 9.0							
Period 3	11 Na Sodium 23.0	12 Mg Magnesium 24.3							
Period 4	19 K Potassium 39.1	20 Ca Calcium 40.1	21 Sc Scandium 45.0	22 Ti Titanium 47.9	23 V Vanadium 50.9	24 Cr Chromium 52.0	25 Mn Manganese 54.9	26 Fe Iron 55.8	27 Co Cobalt 58.9
Period 5	37 Rb Rubidium 85.5	38 Sr Strontium 87.6	39 Y Yttrium 88.9	40 Zr Zirconium 91.2	41 Nb Niobium 92.9	42 Mo Molybdenum 95.9	43 Tc Technetium (98)	44 Ru Ruthenium 101.1	45 Rh Rhodium 102.9
Period 6	55 Cs Cesium 132.9	56 Ba Barium 137.3	57 La Lanthanum 138.9	72 Hf Hafnium 178.5	73 Ta Tantalum 180.9	74 W Tungsten 183.8	75 Re Rhenium 186.2	76 Os Osmium 190.2	77 Ir Iridium 192.2
Period 7	87 Fr Francium (223)	88 Ra Radium (226)	89 Ac Actinium (227)	104 Rf Rutherfordium (261)	105 Db Dubnium (262)	106 Sg Seaborgium (263)	107 Bh Bohrium (264)	108 Hs Hassium (265)†	109 Mt Meitnerium (268)†

† Estimated from currently available IUPAC data.

A row of elements is called a *period*.

A column of elements is called a *group* or *family*.

Values in parentheses are of the most stable isotope of the element.

Lanthanides	58 Ce Cerium 140.1	59 Pr Praseodymium 140.9	60 Nd Neodymium 144.2	61 Pm Promethium (145)	62 Sm Samarium 150.4
Actinides	90 Th Thorium 232.0	91 Pa Protactinium 231.0	92 U Uranium 238.0	93 Np Neptunium (237)	94 Pu Plutonium (244)

These elements are placed below the table to allow the table to be narrower.

Appendix

Topic: **Periodic Table**
Go To: **go.hrw.com**
Keyword: **HNO PERIODIC**
Visit the HRW Web site for updates on the periodic table.

This zigzag line reminds you where the metals, nonmetals, and metalloids are.

Group 10	Group 11	Group 12	Group 13	Group 14	Group 15	Group 16	Group 17	Group 18
								2 **He** Helium 4.0
			5 **B** Boron 10.8	6 **C** Carbon 12.0	7 **N** Nitrogen 14.0	8 **O** Oxygen 16.0	9 **F** Fluorine 19.0	10 **Ne** Neon 20.2
			13 **Al** Aluminum 27.0	14 **Si** Silicon 28.1	15 **P** Phosphorus 31.0	16 **S** Sulfur 32.1	17 **Cl** Chlorine 35.5	18 **Ar** Argon 39.9
28 **Ni** Nickel 58.7	29 **Cu** Copper 63.5	30 **Zn** Zinc 65.4	31 **Ga** Gallium 69.7	32 **Ge** Germanium 72.6	33 **As** Arsenic 74.9	34 **Se** Selenium 79.0	35 **Br** Bromine 79.9	36 **Kr** Krypton 83.8
46 **Pd** Palladium 106.4	47 **Ag** Silver 107.9	48 **Cd** Cadmium 112.4	49 **In** Indium 114.8	50 **Sn** Tin 118.7	51 **Sb** Antimony 121.8	52 **Te** Tellurium 127.6	53 **I** Iodine 126.9	54 **Xe** Xenon 131.3
78 **Pt** Platinum 195.1	79 **Au** Gold 197.0	80 **Hg** Mercury 200.6	81 **Tl** Thallium 204.4	82 **Pb** Lead 207.2	83 **Bi** Bismuth 209.0	84 **Po** Polonium (209)	85 **At** Astatine (210)	86 **Rn** Radon (222)
110 **Ds** Darmstadtium (269)†	111 **Uuu** Unununium (272)†	112 **Uub** Ununbium (277)†		114 **Uuq** Ununquadium (285)†				

† The names and three-letter symbols of elements are temporary. They are based on the atomic numbers of the elements. Official names and symbols will be approved by an international committee of scientists.

63 **Eu** Europium 152.0	64 **Gd** Gadolinium 157.2	65 **Tb** Terbium 158.9	66 **Dy** Dysprosium 162.5	67 **Ho** Holmium 164.9	68 **Er** Erbium 167.3	69 **Tm** Thulium 168.9	70 **Yb** Ytterbium 173.0	71 **Lu** Lutetium 175.0
95 **Am** Americium (243)	96 **Cm** Curium (247)	97 **Bk** Berkelium (247)	98 **Cf** Californium (251)	99 **Es** Einsteinium (252)	100 **Fm** Fermium (257)	101 **Md** Mendelevium (258)	102 **No** Nobelium (259)	103 **Lr** Lawrencium (262)

Physical Science Refresher

Atoms and Elements

Every object in the universe is made up of particles of some kind of matter. **Matter** is anything that takes up space and has mass. All matter is made up of elements. An **element** is a substance that cannot be separated into simpler components by ordinary chemical means. This is because each element consists of only one kind of atom. An **atom** is the smallest unit of an element that has all of the properties of that element.

Atomic Structure

Atoms are made up of small particles called subatomic particles. The three major types of subatomic particles are **electrons, protons,** and **neutrons.** Electrons have a negative electric charge, protons have a positive charge, and neutrons have no electric charge. The protons and neutrons are packed close to one another to form the **nucleus.** The protons give the nucleus a positive charge. Electrons are most likely to be found in regions around the nucleus called **electron clouds.** The negatively charged electrons are attracted to the positively charged nucleus. An atom may have several energy levels in which electrons are located.

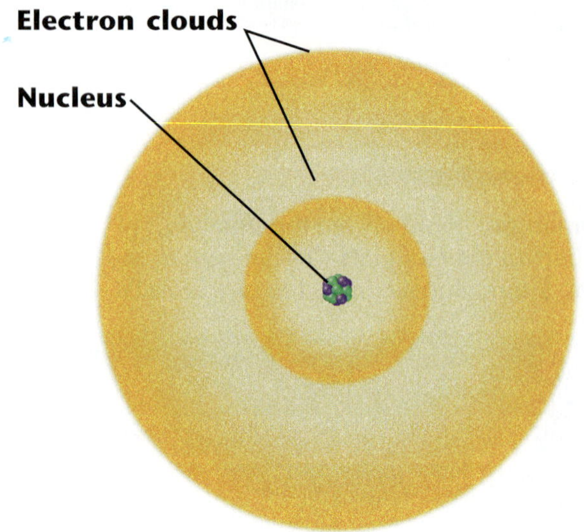

Atomic Number

To help in the identification of elements, scientists have assigned an **atomic number** to each kind of atom. The atomic number is the number of protons in the atom. Atoms with the same number of protons are all the same kind of element. In an uncharged, or electrically neutral, atom there are an equal number of protons and electrons. Therefore, the atomic number equals the number of electrons in an uncharged atom. The number of neutrons, however, can vary for a given element. Atoms of the same element that have different numbers of neutrons are called **isotopes.**

Periodic Table of the Elements

In the periodic table, the elements are arranged from left to right in order of increasing atomic number. Each element in the table is in a separate box. An uncharged atom of each element has one more electron and one more proton than an uncharged atom of the element to its left. Each horizontal row of the table is called a **period.** Changes in chemical properties of elements across a period correspond to changes in the electron arrangements of their atoms. Each vertical column of the table, known as a **group,** lists elements with similar properties. The elements in a group have similar chemical properties because their atoms have the same number of electrons in their outer energy level. For example, the elements helium, neon, argon, krypton, xenon, and radon all have similar properties and are known as the noble gases.

Molecules and Compounds

When two or more elements are joined chemically, the resulting substance is called a **compound.** A compound is a new substance with properties different from those of the elements that compose it. For example, water, H_2O, is a compound formed when hydrogen (H) and oxygen (O) combine. The smallest complete unit of a compound that has the properties of that compound is called a **molecule.** A chemical formula indicates the elements in a compound. It also indicates the relative number of atoms of each element present. The chemical formula for water is H_2O, which indicates that each water molecule consists of two atoms of hydrogen and one atom of oxygen. The subscript number after the symbol for an element indicates how many atoms of that element are in a single molecule of the compound.

Acids, Bases, and pH

An ion is an atom or group of atoms that has an electric charge because it has lost or gained one or more electrons. When an acid, such as hydrochloric acid, HCl, is mixed with water, it separates into ions. An **acid** is a compound that produces hydrogen ions, H+, in water. The hydrogen ions then combine with a water molecule to form a hydronium ion, H_3O^+. A **base,** on the other hand, is a substance that produces hydroxide ions, OH^-, in water.

To determine whether a solution is acidic or basic, scientists use pH. The **pH** is a measure of the hydronium ion concentration in a solution. The pH scale ranges from 0 to 14. The middle point, pH = 7, is neutral, neither acidic nor basic. Acids have a pH less than 7; bases have a pH greater than 7. The lower the number is, the more acidic the solution. The higher the number is, the more basic the solution.

Chemical Equations

A chemical reaction occurs when a chemical change takes place. (In a chemical change, new substances with new properties are formed.) A chemical equation is a useful way of describing a chemical reaction by means of chemical formulas. The equation indicates what substances react and what the products are. For example, when carbon and oxygen combine, they can form carbon dioxide. The equation for the reaction is as follows: $C + O_2 \rightarrow CO_2$.

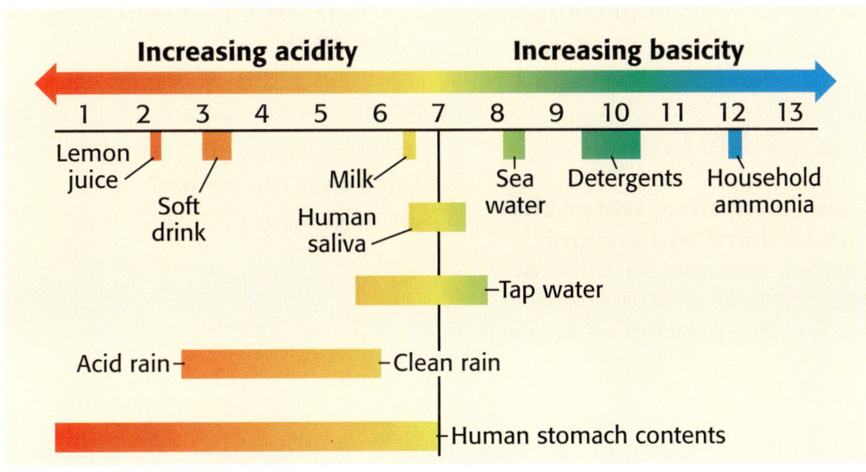

Physical Science Laws and Equations

Law of Conservation of Energy

The law of conservation of energy states that energy can be neither created nor destroyed.

The total amount of energy in a closed system is always the same. Energy can be changed from one form to another, but all of the different forms of energy in a system always add up to the same total amount of energy no matter how many energy conversions occur.

Law of Universal Gravitation

The law of universal gravitation states that all objects in the universe attract each other by a force called *gravity*. The size of the force depends on the masses of the objects and the distance between objects.

The first part of the law explains why a bowling ball is much harder to lift than a table-tennis ball. Because the bowling ball has a much larger mass than the table-tennis ball does, the amount of gravity between the Earth and the bowling ball is greater than the amount of gravity between the Earth and the table-tennis ball.

The second part of the law explains why a satellite can remain in orbit around the Earth. The satellite is carefully placed at a distance great enough to prevent the Earth's gravity from immediately pulling the satellite down but small enough to prevent the satellite from completely escaping the Earth's gravity and wandering off into space.

Newton's Laws of Motion

Newton's first law of motion states that an object at rest remains at rest and an object in motion remains in motion at constant speed and in a straight line unless acted on by an unbalanced force.

The first part of the law explains why a football will remain on a tee until it is kicked off or until a gust of wind blows it off.

The second part of the law explains why a bike rider will continue moving forward after the bike comes to an abrupt stop. Gravity and the friction of the sidewalk will eventually stop the rider.

Newton's second law of motion states that the acceleration of an object depends on the mass of the object and the amount of force applied.

The first part of the law explains why the acceleration of a 4 kg bowling ball will be greater than the acceleration of a 6 kg bowling ball if the same force is applied to both.

The second part of the law explains why the acceleration of a bowling ball will be larger if a larger force is applied to the bowling ball.

The relationship of acceleration (a) to mass (m) and force (F) can be expressed mathematically by the following equation:

$$acceleration = \frac{force}{mass}, \text{ or } a = \frac{F}{m}$$

This equation is often rearranged to the form

$$force = mass \times acceleration$$

or

$$F = m \times a$$

Newton's third law of motion states that whenever one object exerts a force on a second object, the second object exerts an equal and opposite force on the first.

This law explains that a runner is able to move forward because of the equal and opposite force that the ground exerts on the runner's foot after each step.

Appendix

Useful Equations

Average speed

$$\textit{average speed} = \frac{\textit{total distance}}{\textit{total time}}$$

Example: A bicycle messenger traveled a distance of 136 km in 8 h. What was the messenger's average speed?

$$\frac{136 \text{ km}}{8 \text{ h}} = 17 \text{ km/h}$$

The messenger's average speed was **17 km/h.**

Average acceleration

$$\textit{average acceleration} = \frac{\textit{final velocity} - \textit{starting velocity}}{\textit{time it takes to change velocity}}$$

Example: Calculate the average acceleration of an Olympic 100 m dash sprinter who reaches a velocity of 20 m/s south at the finish line. The race was in a straight line and lasted 10 s.

$$\frac{20 \text{ m/s} - 0 \text{ m/s}}{10 \text{ s}} = 2 \text{ m/s/s}$$

The sprinter's average acceleration is **2 m/s/s south.**

Net force

Forces in the Same Direction

When forces are in the same direction, add the forces together to determine the net force.

Example: Calculate the net force on a stalled car that is being pushed by two people. One person is pushing with a force of 13 N northwest, and the other person is pushing with a force of 8 N in the same direction.

$$13 \text{ N} + 8 \text{ N} = 21 \text{ N}$$

The net force is **21 N northwest.**

Forces in Opposite Directions

When forces are in opposite directions, subtract the smaller force from the larger force to determine the net force. The net force will be in the direction of the larger force.

Net force (continued)

Example: Calculate the net force on a rope that is being pulled on each end. One person is pulling on one end of the rope with a force of 12 N south. Another person is pulling on the opposite end of the rope with a force of 7 N north.

$$12 \text{ N} - 7 \text{ N} = 5 \text{ N}$$

The net force is **5 N south.**

Density

$$\textit{density} = \frac{\textit{mass}}{\textit{volume}}$$

Example: Calculate the density of a sponge that has a mass of 10 g and a volume of 40 cm^3.

$$\frac{10 \text{ g}}{40 \text{ cm}^3} = \frac{0.25\text{g}}{\text{cm}^3}$$

The density of the sponge is **0.25 g/cm^3.**

Pressure

Pressure is the force exerted over a given area. The SI unit for pressure is the pascal, whose symbol is Pa.

$$\textit{pressure} = \frac{\textit{force}}{\textit{area}}$$

Example: Calculate the pressure of the air in a soccer ball if the air exerts a force of 10 N over an area of 0.5 m^2.

$$\textit{pressure} = \frac{10 \text{ N}}{0.5 \text{ m}^2} = \frac{20 \text{ N}}{\text{m}^2} = 20 \text{ Pa}$$

The pressure of the air inside the soccer ball is **20 Pa.**

Concentration

$$\textit{concentration} = \frac{\textit{mass of solute}}{\textit{volume of solvent}}$$

Example: Calculate the concentration of a solution in which 10 g of sugar is dissolved in 125 mL of water.

$$\frac{10 \text{ g of sugar}}{125 \text{ mL of water}} = \frac{0.08 \text{ g}}{\text{mL}}$$

The concentration of this solution is **0.08 g/mL.**

Properties of Common Minerals

	Mineral	Color	Luster	Streak	Hardness
Silicate Minerals	**Beryl**	deep green, pink, white, bluish green, or yellow	vitreous	white	7.5–8
	Chlorite	green	vitreous to pearly	pale green	2–2.5
	Garnet	green, red, brown, black	vitreous	white	6.5–7.5
	Hornblende	dark green, brown, or black	vitreous	none	5–6
	Muscovite	colorless, silvery white, or brown	vitreous or pearly	white	2–2.5
	Olivine	olive green, yellow	vitreous	white or none	6.5–7
	Orthoclase	colorless, white, pink, or other colors	vitreous	white or none	6
	Plagioclase	colorless, white, yellow, pink, green	vitreous	white	6
	Quartz	colorless or white; any color when not pure	vitreous or waxy	white or none	7
Nonsilicate Minerals	**Native Elements**				
	Copper	copper-red	metallic	copper-red	2.5–3
	Diamond	pale yellow or colorless	adamantine	none	10
	Graphite	black to gray	submetallic	black	1–2
	Carbonates				
	Aragonite	colorless, white, or pale yellow	vitreous	white	3.5–4
	Calcite	colorless or white to tan	vitreous	white	3
	Halides				
	Fluorite	light green, yellow, purple, bluish green, or other colors	vitreous	none	4
	Halite	white	vitreous	white	2.0–2.5
	Oxides				
	Hematite	reddish brown to black	metallic to earthy	dark red to red-brown	5.6–6.5
	Magnetite	iron-black	metallic	black	5.5–6.5
	Sulfates				
	Anhydrite	colorless, bluish, or violet	vitreous to pearly	white	3–3.5
	Gypsum	white, pink, gray, or colorless	vitreous, pearly, or silky	white	2.0
	Sulfides				
	Galena	lead-gray	metallic	lead-gray to black	2.5–2.8
	Pyrite	brassy yellow	metallic	greenish, brownish, or black	6–6.5

Density (g/cm^3)	Cleavage, Fracture, Special Properties	Common Uses
2.6–2.8	1 cleavage direction; irregular fracture; some varieties fluoresce in ultraviolet light	gemstones, ore of the metal beryllium
2.6–3.3	1 cleavage direction; irregular fracture	
4.2	no cleavage; conchoidal to splintery fracture	gemstones, abrasives
3.0–3.4	2 cleavage directions; hackly to splintery fracture	
2.7–3	1 cleavage direction; irregular fracture	electrical insulation, wallpaper, fireproofing material, lubricant
3.2–3.3	no cleavage; conchoidal fracture	gemstones, casting
2.6	2 cleavage directions; irregular fracture	porcelain
2.6–2.7	2 cleavage directions; irregular fracture	ceramics
2.6	no cleavage; conchoidal fracture	gemstones, concrete, glass, porcelain, sandpaper, lenses
8.9	no cleavage; hackly fracture	wiring, brass, bronze, coins
3.5	4 cleavage directions; irregular to conchoidal fracture	gemstones, drilling
2.3	1 cleavage direction; irregular fracture	pencils, paints, lubricants, batteries
2.95	2 cleavage directions; irregular fracture; reacts with hydrochloric acid	no important industrial uses
2.7	3 cleavage directions; irregular fracture; reacts with weak acid; double refraction	cements, soil conditioner, whitewash, construction materials
3.0–3.3	4 cleavage directions; irregular fracture; some varieties fluoresce	hydrofluoric acid, steel, glass, fiberglass, pottery, enamel
2.1–2.2	3 cleavage directions; splintery to conchoidal fracture; salty taste	tanning hides, salting icy roads, food preservation
5.2–5.3	no cleavage; splintery fracture; magnetic when heated	iron ore for steel, pigments
5.2	no cleavage; splintery fracture; magnetic	iron ore
3.0	3 cleavage directions; conchoidal to splintery fracture	soil conditioner, sulfuric acid
2.3	3 cleavage directions; conchoidal to splintery fracture	plaster of Paris, wallboard, soil conditioner
7.4–7.6	3 cleavage directions; irregular fracture	batteries, paints
5	no cleavage; conchoidal to splintery fracture	sulfuric acid

Using the Microscope

Parts of the Compound Light Microscope

- The **ocular lens** magnifies the image 10×.
- The **low-power objective** magnifies the image 10×.
- The **high-power objective** magnifies the image either 40× or 43×.
- The **revolving nosepiece** holds the objectives and can be turned to change from one magnification to the other.
- The **body tube** maintains the correct distance between the ocular lens and objectives.
- The **coarse-adjustment knob** moves the body tube up and down to allow focusing of the image.
- The **fine-adjustment knob** moves the body tube slightly to bring the image into sharper focus. It is usually located in the center of the coarse-adjustment knob.
- The **stage** supports a slide.
- **Stage clips** hold the slide in place for viewing.
- The **diaphragm** controls the amount of light coming through the stage.
- The light source provides a **light** for viewing the slide.
- The **arm** supports the body tube.
- The **base** supports the microscope.

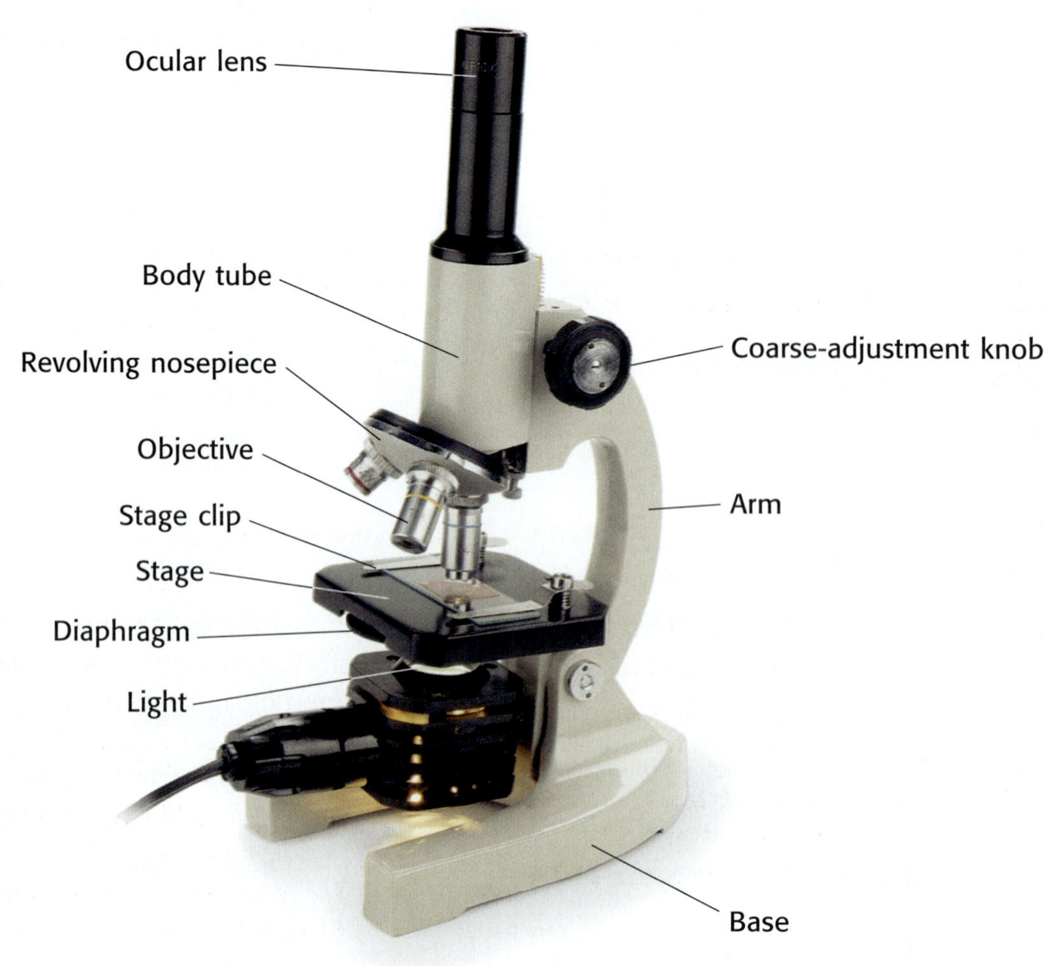

Proper Use of the Compound Light Microscope

1. Use both hands to carry the microscope to your lab table. Place one hand beneath the base, and use the other hand to hold the arm of the microscope. Hold the microscope close to your body while carrying it to your lab table.
2. Place the microscope on the lab table at least 5 cm from the edge of the table.
3. Check to see what type of light source is used by your microscope. If the microscope has a lamp, plug it in and make sure that the cord is out of the way. If the microscope has a mirror, adjust the mirror to reflect light through the hole in the stage. **Caution:** If your microscope has a mirror, do not use direct sunlight as a light source. Direct sunlight can damage your eyes.
4. Always begin work with the low-power objective in line with the body tube. Adjust the revolving nosepiece.
5. Place a prepared slide over the hole in the stage. Secure the slide with the stage clips.
6. Look through the ocular lens. Move the diaphragm to adjust the amount of light coming through the stage.
7. Look at the stage from eye level. Slowly turn the coarse adjustment to lower the objective until the objective almost touches the slide. Do not allow the objective to touch the slide.
8. Look through the ocular lens. Turn the coarse adjustment to raise the low-power objective until the image is in focus. Always focus by raising the objective away from the slide. Never focus the objective downward. Use the fine adjustment to sharpen the focus. Keep both eyes open while viewing a slide.
9. Make sure that the image is exactly in the center of your field of vision. Then, switch to the high-power objective. Focus the image by using only the fine adjustment. Never use the coarse adjustment at high power.
10. When you are finished using the microscope, remove the slide. Clean the ocular lens and objectives with lens paper. Return the microscope to its storage area. Remember to use both hands when carrying the microscope.

Making a Wet Mount

1. Use lens paper to clean a glass slide and a coverslip.
2. Place the specimen that you wish to observe in the center of the slide.
3. Using a medicine dropper, place one drop of water on the specimen.
4. Hold the coverslip at the edge of the water and at a 45° angle to the slide. Make sure that the water runs along the edge of the coverslip.
5. Lower the coverslip slowly to avoid trapping air bubbles.
6. Water might evaporate from the slide as you work. Add more water to keep the specimen fresh. Place the tip of the medicine dropper next to the edge of the coverslip. Add a drop of water. (You can also use this method to add stain or solutions to a wet mount.) Remove excess water from the slide by using the corner of a paper towel as a blotter. Do not lift the coverslip to add or remove water.

Glossary

A

abiotic describes the nonliving part of the environment, including water, rocks, light, and temperature (586)

abrasion the grinding and wearing away of rock surfaces through the mechanical action of other rock or sand particles (211)

absolute dating any method of measuring the age of an event or object in years (82)

absolute magnitude the brightness that a star would have at a distance of 32.6 light-years from Earth (272)

absorption in optics, the transfer of light energy to particles of matter (476)

acceleration (ak SEL uhr AY shuhn) the rate at which velocity changes over time; an object accelerates if its speed, direction, or both change (311)

acid any compound that increases the number of hydronium ions when dissolved in water (404)

acid precipitation rain, sleet, or snow that contains a high concentration of acids (118)

activation energy the minimum amount of energy required to start a chemical reaction (386)

active transport the movement of substances across the cell membrane that requires the cell to use energy (534)

addiction a dependence on a substance, such as alcohol or drugs (693)

aerobic exercise physical exercise intended to increase the activity of the heart and lungs to promote the body's use of oxygen (699)

alcoholism a disorder in which a person repeatedly drinks alcoholic beverages in an amount that interferes with the person's health and activities (694)

allergy a reaction to a harmless or common substance by the body's immune system (673)

antibiotic medicine used to kill bacteria and other microorganisms (570)

antibody a protein made by B cells that binds to a specific antigen (669)

apparent magnitude the brightness of a star as seen from the Earth (272)

area a measure of the size of a surface or a region (24)

asexual reproduction reproduction that does not involve the union of sex cells and in which a single parent produces offspring that are genetically identical to the parent (506)

asthenosphere the soft layer of the mantle on which the tectonic plates move (140)

ATP **a**denosine **tri**phosphate, a molecule that acts as the main energy source for cell processes (512)

autoimmune disease a disease in which the immune system attacks the organism's own cells (673)

B

base any compound that increases the number of hydroxide ions when dissolved in water (407)

B cell a white blood cell that makes antibodies (669)

beach an area of the shoreline made up of material deposited by waves (208)

big bang theory the theory that states that the universe began with a tremendous explosion about 13.7 billion years ago (293)

binary fission (BIE nuh ree FISH uhn) a form of asexual reproduction in single-celled organisms by which one cell divides into two cells of the same size (564)

biodiversity the number and variety of organisms in a given area during a specific period of time (618)

biomass organic matter that can be a source of energy; the total mass of the organisms in a given area (125)

biome a large region characterized by a specific type of climate and certain types of plant and animal communities (240)

bioremediation (BIE oh ri MEE dee AY shuhn) the biological treatment of hazardous waste by living organisms (569)

biosphere the part of Earth where life exists (589)

biotic describes living factors in the environment (586)

black hole an object so massive and dense that even light cannot escape its gravity (287)

C

caldera a large, semicircular depression that forms when the magma chamber below a volcano partially empties and causes the ground above to sink (186)

cancer a tumor in which the cells begin dividing at an uncontrolled rate and become invasive (674)

carbohydrate a class of energy-giving nutrients that includes sugars, starches, and fiber; contains carbon, hydrogen, and oxygen (416, 686, 511)

carnivore an organism that eats animals (591)

carrying capacity the largest population that an environment can support at any given time (597)

cast a type of fossil that forms when sediments fill in the cavity left by a decomposed organism (88)

catalyst (KAT uh LIST) a substance that changes the rate of a chemical reaction without being used up or changed very much (389)

catastrophism a principle that states that geologic change occurs suddenly (73)

cell in biology, the smallest unit that can perform all life processes; cells are covered by a membrane and contain DNA and cytoplasm (504)

cell cycle the life cycle of a cell (540)

cellular respiration the process by which cells use oxygen to produce energy from food (537)

chemical bond an interaction that holds atoms or ions together (400)

chemical energy the energy released when a chemical compound reacts to produce new compounds (122)

chemical equation a representation of a chemical reaction that uses symbols to show the relationship between the reactants and the products (376)

chemical formula a combination of chemical symbols and numbers to represent a substance (374)

chemical reaction the process by which one or more substances change to produce one or more different substances (370)

chromosome in a eukaryotic cell, one of the structures in the nucleus that are made up of DNA and protein; in a prokaryotic cell, the main ring of DNA (540)

climate the average weather conditions in an area over a long period of time (234)

coal a fossil fuel that forms underground from partially decomposed plant material (114)

coevolution the evolution of two species that is due to mutual influence, often in a way that makes the relationship more beneficial to both species (601)

commensalism a relationship between two organisms in which one organism benefits and the other is unaffected (600)

community all of the populations of species that live in the same habitat and interact with each other (588)

composition the chemical makeup of a rock; describes either the minerals or other materials in the rock (45)

compression stress that occurs when forces act to squeeze an object (154)

conservation (KAHN suhr VAY shuhn) the preservation and wise use of natural resources (620)

consumer an organism that eats other organisms or organic matter (509)

continental drift the hypothesis that states that the continents once formed a single landmass, broke up, and drifted to their present locations (146)

controlled experiment an experiment that tests only one factor at a time by using a comparison of a control group with an experimental group (14)

convergent boundary the boundary formed by the collision of two lithospheric plates (151)

core the central part of the Earth below the mantle (139)

cosmology the study of the origin, properties, processes, and evolution of the universe (292)

covalent compound a chemical compound that is formed by the sharing of electrons (402)

crater a funnel-shaped pit near the top of the central vent of a volcano (186)

creep the slow downhill movement of weathered rock material (223)

crust the thin and solid outermost layer of the Earth above the mantle (138)

cytokinesis the division of the cytoplasm of a cell (542)

D

decibel the most common unit used to measure loudness (symbol, dB) (440)

decomposer an organism that gets energy by breaking down the remains of dead organisms or animal wastes and consuming or absorbing the nutrients (509)

decomposition reaction a reaction in which a single compound breaks down to form two or more simpler substances (381)

deflation a form of wind erosion in which fine, dry soil particles are blown away (211)

deformation the bending, tilting, and breaking of the Earth's crust; the change in the shape of rock in response to stress (163)

density the ratio of the mass of a substance to the volume of the substance (26)

deposition the process in which material is laid down (41)

diffraction a change in the direction of a wave when the wave finds an obstacle or an edge, such as an opening (479)

diffusion (di FYOO zhuhn) the movement of particles from regions of higher density to regions of lower density (532)

digestive system the organs that break down food so that it can be used by the body (638)

divergent boundary the boundary between two tectonic plates that are moving away from each other (151)

Doppler effect an observed change in the frequency of a wave when the source or observer is moving (438)

double-displacement reaction a reaction in which a gas, a solid precipitate, or a molecular compound forms from the exchange of ions between two compounds (383)

drug any substance that causes a change in a person's physical or psychological state (692)

dune a mound of wind-deposited sand that keeps its shape even though it moves (212)

E

echo a reflected sound wave (442)

echolocation the process of using reflected sound waves to find objects; used by animals such as bats (443)

ecology the study of the interactions of living organisms with one another and with their environment (586)

ecosystem a community of organisms and their abiotic, or nonliving, environment (589)

elastic rebound the sudden return of elastically deformed rock to its undeformed shape (163)

electromagnetic spectrum all of the frequencies or wavelengths of electromagnetic radiation (466)

electromagnetic wave a wave that consists of electric and magnetic fields that vibrate at right angles to each other (462)

elevation the height of an object above sea level (238)

endocytosis (EN doh sie TOH sis) the process by which a cell membrane surrounds a particle and encloses the particle in a vesicle to bring the particle into the cell (534)

endospore (EN doh SPAWR) a thick-walled protective spore that forms inside a bacterial cell and resists harsh conditions (565)

endothermic reaction a chemical reaction that requires heat (385)

energy pyramid a triangular diagram that shows an ecosystem's loss of energy, which results as energy passes through the ecosystem's food chain (593)

eon (EE AHN) the largest division of geologic time (95)

epoch (EP uhk) a subdivision of a geologic period (95)

era a unit of geologic time that includes two or more periods (95)

erosion the process by which wind, water, ice, or gravity transports soil and sediment from one location to another (41)

esophagus (i SAHF uh guhs) a long, straight tube that connects the pharynx to the stomach (640)

exocytosis (EK soh sie TOH sis) the process in which a cell releases a particle by enclosing the particle in a vesicle that then moves to the cell surface and fuses with the cell membrane (535)

exothermic reaction a chemical reaction in which heat is released to the surroundings (384)

extinction the death of every member of a species (95)

extrusive igneous rock rock that forms as a result of volcanic activity at or near the Earth's surface (51)

F

farsightedness a condition in which the lens of the eye focuses distant objects behind rather than on the retina (489)

fat an energy-storage nutrient that helps the body store some vitamins (687)

fault a break in a body of rock along which one block slides relative to another (156)

fermentation the breakdown of food without the use of oxygen (539)

folding the bending of rock layers due to stress (155)

foliated describes the texture of metamorphic rock in which the mineral grains are arranged in planes or bands (59)

food chain the pathway of energy transfer through various stages as a result of the feeding patterns of a series of organisms (592)

food web a diagram that shows the feeding relationships between organisms in an ecosystem (592)

force a push or a pull exerted on an object in order to change the motion of the object; force has size and direction (314)

fossil the remains or physical evidence of an organism preserved by geological processes (86)

fossil fuel a nonrenewable energy resource formed from the remains of organisms that lived long ago (112)

free fall the motion of a body when only the force of gravity is acting on the body (343)

friction a force that opposes motion between two surfaces that are in contact (318)

function the special, normal, or proper activity of an organ or part (517)

G

galaxy a collection of stars, dust, and gas bound together by gravity (288)

gallbladder a sac-shaped organ that stores bile produced by the liver (643)

gasohol a mixture of gasoline and alcohol that is used as a fuel (125)

generation time the period between the birth of one generation and the birth of the next generation (519)

geologic column an arrangement of rock layers in which the oldest rocks are at the bottom (77)

geologic time scale the standard method used to divide the Earth's long natural history into manageable parts (94)

geothermal energy the energy produced by heat within the Earth (126)

glacial drift the rock material carried and deposited by glaciers (218)

glacier a large mass of moving ice (214)

global warming a gradual increase in average global temperature (256)

globular cluster a tight group of stars that looks like a ball and contains up to 1 million stars (290)

gravity a force of attraction between objects that is due to their masses (324)

greenhouse effect the warming of the surface and lower atmosphere of Earth that occurs when water vapor, carbon dioxide, and other gases absorb and reradiate thermal energy (256)

H

half-life the time needed for half of a sample of a radioactive substance to undergo radioactive decay (83)

herbivore an organism that eats only plants (591)

heredity the passing of genetic traits from parent to offspring (506)

homeostasis (HOH mee OH STAY sis) the maintenance of a constant internal state in a changing environment (505)

homologous chromosomes (hoh MAHL uh guhs KROH muh SOHMZ) chromosomes that have the same sequence of genes and the same structure (541)

host an organism from which a parasite takes food or shelter (572)

hot spot a volcanically active area of Earth's surface far from a tectonic plate boundary (192)

H-R diagram **H**ertzsprung-**R**ussell diagram, a graph that shows the relationship between a star's surface temperature and absolute magnitude (284)

hydrocarbon an organic compound composed only of carbon and hydrogen (415)

hydroelectric energy electrical energy produced by falling water (124)

hygiene the science of health and ways to preserve health (698)

hypothesis (hie PAHTH uh sis) an explanation that is based on prior scientific research or observations and that can be tested (12)

I

ice age a long period of climate cooling during which ice sheets cover large areas of Earth's surface; also known as a *glacial period* (252)

immune system the cells and tissues that recognize and attack foreign substances in the body (669)

immunity the ability to resist an infectious disease (666)

index fossil a fossil that is found in the rock layers of only one geologic age and that is used to establish the age of the rock layers (90)

indicator a compound that can reversibly change color depending on conditions such as pH (405)

inertia (in UHR shuh) the tendency of an object to resist being moved or, if the object is moving, to resist a change in speed or direction until an outside force acts on the object (350)

infectious disease a disease that is caused by a pathogen and that can be spread from one individual to another (660)

inhibitor a substance that slows down or stops a chemical reaction (388)

interference the combination of two or more waves that results in a single wave (444, 480)

intrusive igneous rock rock formed from the cooling and solidification of magma beneath the Earth's surface (50)

ionic compound a compound made of oppositely charged ions (400)

isotope an atom that has the same number of protons (or the same atomic number) as other atoms of the same element do but that has a different number of neutrons (and thus a different atomic mass) (82)

K

kidney one of the pair of organs that filter water and wastes from the blood and that excrete products as urine (647)

L

landslide the sudden movement of rock and soil down a slope (221)

large intestine the wider and shorter portion of the intestine that removes water from mostly digested food and that turns the waste into semisolid feces, or stool (644)

latitude the distance north or south from the equator; expressed in degrees (235)

lava plateau a wide, flat landform that results from repeated nonexplosive eruptions of lava that spread over a large area (187)

law a summary of many experimental results and observations; a law tells how things work (21)

law of conservation of energy the law that states that energy cannot be created or destroyed but can be changed from one form to another (385)

law of conservation of mass the law that states that mass cannot be created or destroyed in ordinary chemical and physical changes (377)

light-year the distance that light travels in one year; about 9.46 trillion kilometers (273)

lipid a type of biochemical that does not dissolve in water; fats and steroids are lipids (417, 512)

lithosphere the solid, outer layer of the Earth that consists of the crust and the rigid upper part of the mantle (140)

liver the largest organ in the body; it makes bile, stores and filters blood, and stores excess sugars as glycogen (643)

loess (LOH ES) very fertile sediments of quartz, feldspar, hornblende, mica, and clay deposited by the wind (212)

M

macrophage (MAK roh FAYJ) an immune system cell that engulfs pathogens and other materials (669)

magma chamber the body of molten rock that feeds a volcano (180)

main sequence the location on the H-R diagram where most stars lie; it has a diagonal pattern from the lower right (low temperature and luminosity) to the upper left (high temperature and luminosity) (285)

malnutrition a disorder of nutrition that results when a person does not consume enough of each of the nutrients that are needed by the human body (690)

mantle the layer of rock between the Earth's crust and core (139)

mass a measure of the amount of matter in an object (25, 328)

mass movement a movement of a section of land down a slope (220)

medium a physical environment in which phenomena occur (432)

meiosis (mie OH sis) a process in cell division during which the number of chromosomes decreases to half the original number by two divisions of the nucleus, which results in the production of sex cells (gametes or spores) (545)

memory B cell a B cell that responds to an antigen more strongly when the body is reinfected with an antigen than it does during its first encounter with the antigen (672)

mesosphere the strong, lower part of the mantle between the asthenosphere and the outer core (141)

metabolism (muh TAB uh LIZ uhm) the sum of all chemical processes that occur in an organism (506)

meter the basic unit of length in the SI (symbol, m) (24)

microclimate the climate of a small area (250)

mineral a class of nutrients that are chemical elements that are needed for certain body processes (688)

mitosis in eukaryotic cells, a process of cell division that forms two new nuclei, each of which has the same number of chromosomes (541)

model a pattern, plan, representation, or description designed to show the structure or workings of an object, system, or concept (18)

mold a mark or cavity made in a sedimentary surface by a shell or other body (88)

momentum (moh MEN tuhm) a quantity defined as the product of the mass and velocity of an object (356)

motion an object's change in position relative to a reference point (308)

mudflow the flow of a mass of mud or rock and soil mixed with a large amount of water (222)

mutualism (MYOO choo uhl IZ uhm) a relationship between two species in which both species benefit (600)

N

narcotic a drug that is derived from opium and that relieves pain and induces sleep; examples include heroine, morphine, and codeine (695)

natural gas a mixture of gaseous hydrocarbons located under the surface of the Earth, often near petroleum deposits; used as a fuel (113)

natural resource any natural material that is used by humans, such as water, petroleum, minerals, forests, and animals (108)

nearsightedness a condition in which the lens of the eye focuses distant objects in front of rather than on the retina (488)

nebula a large cloud of gas and dust in interstellar space; a region in space where stars are born or where stars explode at the end of their lives (290)

nephron the unit in the kidney that filters blood (647)

net force the combination of all of the forces acting on an object (315)

neutralization reaction (NOO truhl i ZA shuhn ree AK shuhn) the reaction of an acid and a base to form a neutral solution of water and a salt (411)

neutron star a star that has collapsed under gravity to the point that the electrons and protons have smashed together to form neutrons (286)

newton the SI unit for force (symbol, N) (314)

nicotine (NIK uh TEEN) a toxic, addictive chemical that is found in tobacco and that is one of the major contributors to the harmful effects of smoking (694)

noise a sound that consists of a random mix of frequencies (451)

nonfoliated describes the texture of metamorphic rock in which the mineral grains are not arranged in planes or bands (60)

noninfectious disease a disease that cannot spread from one individual to another (660)

nonrenewable resource a resource that forms at a rate that is much slower than the rate at which it is consumed (109, 616)

nuclear energy the energy released by a fission or fusion reaction; the binding energy of the atomic nucleus (120)

nuclear fusion the combination of the nuclei of small atoms to form a larger nucleus; releases energy (278)

nucleic acid (noo KLEE ik AS id) a molecule made up of subunits called *nucleotides* (418, 513)

nutrient a substance in food that provides energy or helps form body tissues and that is necessary for life and growth (686)

O

omnivore an organism that eats both plants and animals (591)

opaque (oh PAYK) describes an object that is not transparent or translucent (483)

open cluster a group of stars that are close together relative to surrounding stars (290)

organ a collection of tissues that carry out a specialized function of the body (515)

organic compound a covalently bonded compound that contains carbon (414)

organism a living thing; anything that can carry out life processes independently (516)

organ system a group of organs that work together to perform body functions (516)

osmosis (ahs MOH sis) the diffusion of water through a semipermeable membrane (533)

overpopulation the presence of too many individuals in an area for the available resources (617)

P

paleontology the scientific study of fossils (74)

pancreas the organ that lies behind the stomach and that makes digestive enzymes and hormones that regulate sugar levels (642)

parallax an apparent shift in the position of an object when viewed from different locations (273)

parasitism (PAR uh SIET IZ uhm) a relationship between two species in which one species, the parasite, benefits from the other species, the host, which is harmed (601)

passive transport the movement of substances across a cell membrane without the use of energy by the cell (534)

pasteurization the process of using heat to kill microorganisms in foods by heating the food to a certain temperature for a specified period of time (665)

pathogen a virus, microorganism, or other organism that causes disease (660)

pathogenic bacteria (PATH uh JEN ik bak TIR ee uh) bacteria that cause disease (570)

pedigree a diagram that shows the occurrence of a genetic trait in several generations of a family (550)

period a unit of geologic time into which eras are divided (95)

petroleum a liquid mixture of complex hydrocarbon compounds; used widely as a fuel source (113)

pH a value that is used to express the acidity or basicity (alkalinity) of a system (411)

phospholipid (FAHS foh LIP id) a lipid that contains phosphorus and that is a structural component in cell membranes (512)

photosynthesis (FOHT oh SIN thuh sis) the process by which plants, algae, and some bacteria use sunlight, carbon dioxide, and water to make food (536)

pigment a substance that gives another substance or a mixture its color (487)

pitch a measure of how high or low a sound is perceived to be, depending on the frequency of the sound wave (437)

plate tectonics the theory that explains how large pieces of the Earth's outermost layer, called *tectonic plates,* move and change shape (150)

polar zone the North or South Pole and the surrounding region (249)

pollution an unwanted change in the environment caused by substances or forms of energy (614)

population a group of organisms of the same species that live in a specific geographical area (588)

precipitate (pree SIP uh TAYT) a solid that is produced as a result of a chemical reaction in solution (371)

predator an organism that eats all or part of another organism (598)

prevailing winds winds that blow mainly from one direction during a given period (237)

prey an organism that is killed and eaten by another organism (598)

producer an organism that can make its own food by using energy from its surroundings (509)

product a substance that forms in a chemical reaction (376)

projectile motion (proh JEK tuhl MOH shuhn) the curved path that an object follows when thrown, launched, or otherwise projected near the surface of Earth (345)

prokaryote (pro KAR ee OHT) an organism that consists of a single cell that does not have a nucleus (563)

protein a molecule that is made up of amino acids and that is needed to build and repair body structures and to regulate processes in the body (510)

pulsar a rapidly spinning neutron star that emits rapid pulses of radio and optical energy (286)

P wave a seismic wave that causes particles of rock to move in a back-and-forth direction (166)

Q

quasar a very luminous, starlike object that generates energy at a high rate; quasars are thought to be the most distant objects in the universe (291)

R

radiation the transfer of energy as electromagnetic waves (463)

radioactive decay the process in which a radioactive isotope tends to break down into a stable isotope of the same element or another element (82)

radiometric dating a method of determining the age of an object by estimating the relative percentages of a radioactive (parent) isotope and a stable (daughter) (83)

reactant (ree AK tuhnt) a substance or molecule that participates in a chemical reaction (376)

recycling the process of recovering valuable or useful materials from waste or scrap; the process of reusing some items (111, 623)

red giant a large, reddish star late in its life cycle (283)

reflection the bouncing back of a ray of light, sound, or heat when the ray hits a surface that it does not go through (474)

refraction the bending of a wave as the wave passes between two substances in which the speed of the wave differs (477)

relative dating any method of determining whether an event or object is older or younger than other events or objects (76)

renewable resource a natural resource that can be replaced at the same rate at which the resource is consumed (109, 616)

resonance a phenomenon that occurs when two objects naturally vibrate at the same frequency; the sound produced by one object causes the other object to vibrate (446)

rift zone an area of deep cracks that forms between two tectonic plates that are pulling away from each other (190)

rock a naturally occurring solid mixture of one or more minerals or organic matter (40)

rock cycle the series of processes in which a rock forms, changes from one type to another, is destroyed, and forms again by geological processes (40)

rock fall the rapid mass movement of rock down a steep slope or cliff (221)

S

salt an ionic compound that forms when a metal atom replaces the hydrogen of an acid (413)

saltation the movement of sand or other sediments by short jumps and bounces that is caused by wind or water (210)

scattering an interaction of light with matter that causes light to change its energy, direction of motion, or both (476)

science the knowledge obtained by observing natural events and conditions in order to discover facts and formulate laws or principles that can be verified or tested (4)

scientific methods a series of steps followed to solve problems (10)

sea-floor spreading the process by which new oceanic lithosphere forms as magma rises toward the surface and solidifies (148)

seismic wave a wave of energy that travels through the Earth and away from an earthquake in all directions (166)

seismology (siez MAHL uh jee) the study of earthquakes (162)

sex chromosome one of the pair of chromosomes that determine the sex of an individual (549)

sexual reproduction reproduction in which sex cells from two parents unite to produce offspring that share traits from both parents (506)

shoreline the boundary between land and a body of water (204)

single-displacement reaction a reaction in which one element takes the place of another element in a compound (381)

small intestine the organ between the stomach and the large intestine where most of the breakdown of food happens and most of the nutrients from food are absorbed (642)

smog photochemical haze that forms when sunlight acts on industrial pollutants and burning fuels (118)

solar energy the energy received by the Earth from the sun in the form of radiation (122)

sonic boom the explosive sound heard when a shock wave from an object traveling faster than the speed of sound reaches a person's ears (445)

sound quality the result of the blending of several pitches through interference (448)

sound wave a longitudinal wave that is caused by vibrations and that travels through a material medium (431)

speciation (SPEE shee AY shuhn) the formation of new species as a result of evolution (520)

spectrum the band of colors produced when white light passes through a prism (269)

speed the distance traveled divided by the time interval during which the motion occurred (309)

standing wave a pattern of vibration that simulates a wave that is standing still (446)

stimulus anything that causes a reaction or change in an organism or any part of an organism (505)

stomach the saclike, digestive organ between the esophagus and the small intestine that breaks down food by the action of muscles, enzymes, and acids (641)

strata layers of rock (singular, *stratum*) (52)

stratification the process in which sedimentary rocks are arranged in layers (55)

stratified drift a glacial deposit that has been sorted and layered by the action of streams or meltwater (219)

stress a physical or mental response to pressure (700)

structure the arrangement of parts in an organism (517)

subsidence (suhb SIED'ns) the sinking of regions of the Earth's crust to lower elevations (160)

sunspot a dark area of the photosphere of the sun that is cooler than the surrounding areas and that has a strong magnetic field (280)

supernova a gigantic explosion in which a massive star collapses and throws its outer layers into space (286)

superposition a principle that states that younger rocks lie above older rocks if the layers have not been disturbed (76)

surface current a horizontal movement of ocean water that is caused by wind and that occurs at or near the ocean's surface (239)

S wave a seismic wave that causes particles of rock to move in a side-to-side direction (166)

symbiosis a relationship in which two different organisms live in close association with each other (600)

synthesis reaction (SIN thuh sis ree AK shuhn) a reaction in which two or more substances combine to form a new compound (380)

T

T cell an immune system cell that coordinates the immune system and attacks many infected cells (669)

tectonic plate a block of lithosphere that consists of the crust and the rigid, outermost part of the mantle (142)

temperate zone the climate zone between the Tropics and the polar zone (246)

temperature a measure of how hot (or cold) something is; specifically, a measure of the average kinetic energy of the particles in an object (26)

tension stress that occurs when forces act to stretch an object (154)

terminal velocity the constant velocity of a falling object when the force of air resistance is equal in magnitude and opposite in direction to the force of gravity (342)

texture the quality of a rock that is based on the sizes, shapes, and positions of the rock's grains (46)

theory an explanation that ties together many hypotheses and observations (20)

till unsorted rock material that is deposited directly by a melting glacier (218)

tissue a group of similar cells that perform a common function (515)

trace fossil a fossilized mark that is formed in soft sediment by the movement of an animal (88)

transform boundary the boundary between tectonic plates that are sliding past each other horizontally (151)

translucent (trans LOO suhnt) describes matter that transmits light but that does not transmit an image (483)

transmission the passing of light or other form of energy through matter (482)

transparent describes matter that allows light to pass through with little interference (483)

tropical zone the region that surrounds the equator and that extends from about 23° north latitude to 23° south latitude (242)

U

unconformity a break in the geologic record created when rock layers are eroded or when sediment is not deposited for a long period of time (79)

uniformitarianism a principle that states that geologic processes that occurred in the past can be explained by current geologic processes (73)

uplift the rising of regions of the Earth's crust to higher elevations (160)

urinary system the organs that make, store, and eliminate urine (646)

V

vaccine a substance prepared from killed or weakened pathogens or from genetic material and that is introduced into a body to produce immunity (666)

variable a factor that changes in an experiment in order to test a hypothesis (14)

velocity (vuh LAHS uh tee) the speed of an object in a particular direction (310)

vent an opening at the surface of the Earth through which volcanic material passes (180)

virus a microscopic particle that gets inside a cell and often destroys the cell (572)

vitamin a class of nutrients that contain carbon and that are needed in small amounts to maintain health and allow growth (688)

volcano a vent or fissure in the Earth's surface through which magma and gases are expelled (178)

volume a measure of the size of a body or region in three-dimensional space (25)

W

weather the short-term state of the atmosphere, including temperature, humidity, precipitation, wind, and visibility (234)

weight a measure of the gravitational force exerted on an object; its value can change with the location of the object in the universe (328)

white dwarf a small, hot, dim star that is the leftover center of an old star (283)

wind power the use of a windmill to drive an electric generator (123)

Spanish Glossary

A

abiotic/abiótico término que describe la parte sin vida del ambiente, incluyendo el agua, las rocas, la luz y la temperatura (586)

abrasion/abrasión proceso por el cual las superficies de las rocas se muelen o desgastan por medio de la acción mecánica de otras rocas y partículas de arena (211)

absolute dating/datación absoluta cualquier método que sirve para determinar la edad de un suceso u objeto en años (82)

absolute magnitude/magnitud absoluta el brillo que una estrella tendría a una distancia de 32.6 años luz de la Tierra (272)

absorption/absorción en la óptica, la transferencia de energía luminosa a las partículas de materia (476)

acceleration/aceleración la tasa a la que la velocidad cambia con el tiempo; un objeto acelera si su rapidez cambia, si su dirección cambia, o si tanto su rapidez como su dirección cambian (311)

acid/ácido cualquier compuesto que aumenta el número de iones de hidrógeno cuando se disuelve en agua (404)

acid precipitation/precipitación ácida lluvia, aguanieve o nieve que contiene una alta concentración de ácidos (118)

activation energy/energía de activación la cantidad mínima de energía que se requiere para iniciar una reacción química (386)

active transport/transporte activo el movimiento de substancias a través de la membrana celular que requiere que la célula gaste energía (534)

addiction/adicción dependencia de substancia, tal como el alcohol u otra droga (693)

aerobic exercise/ejercicio aeróbico ejercicio físico cuyo objetivo es aumentar la actividad del corazón y los pulmones para hacer que el cuerpo use más oxígeno (699)

alcoholism/alcoholismo un trastorno en el cual una persona consume bebidas alcohólicas repetidamente en una cantidad tal que interfiere con su salud y sus actividades (694)

allergy/alergia una reacción del sistema inmunológico del cuerpo a una substancia inofensiva o común (673)

antibiotic/antibiótico medicina utilizada para matar bacterias y otros microorganismos (570)

antibody/anticuerpo una proteína producida por las células B que se une a un antígeno específico (669)

apparent magnitude/magnitud aparente el brillo de una estrella como se percibe desde la Tierra (272)

area/área una medida del tamaño de una superficie o región (24)

asexual reproduction/reproducción asexual reproducción que no involucra la unión de células sexuales, en la que un solo progenitor produce descendencia que es genéticamente igual al progenitor (506)

asthenosphere/astenosfera la capa blanda del manto sobre la que se mueven las placas tectónicas (140)

ATP/ATP adenosín trifosfato, una molécula orgánica que funciona como la fuente principal de energía para los procesos celulares (512)

autoimmune disease/enfermedad autoinmune una enfermedad en la que el sistema inmunológico ataca las células del propio organismo (673)

B

base/base cualquier compuesto que aumenta el número de iones de hidróxido cuando se disuelve en agua (407)

B cell/célula B un glóbulo blanco de la sangre que fabrica anticuerpos (669)

beach/playa un área de la costa formada por materiales depositados por las olas (208)

big bang theory/teoría del Big Bang la teoría que establece que el universo comenzó con una tremenda explosión hace aproximadamente 13.7 mil millones de años (293)

binary fission/fisión binaria una forma de reproducción asexual de los organismos unicelulares, por medio de la cual la célula se divide en dos células del mismo tamaño (564)

biodiversity/biodiversidad el número y la variedad de organismos que se encuentran en un área determinada durante un período específico de tiempo (618)

biomass/biomasa materia orgánica que puede ser una fuente de energía; la masa total de los organismos en un área determinada (125)

biome/bioma una región extensa caracterizada por un tipo de clima específico y ciertos tipos de comunidades de plantas y animales (240)

bioremediation/bioremediación el tratamiento biológico de desechos peligrosos por medio de organismos vivos (569)

biosphere/biosfera la parte de la Tierra donde existe la vida (589)

biotic/biótico término que describe los factores vivientes del ambiente (586)

black hole/hoyo negro un objeto tan masivo y denso que ni siquiera la luz puede salir de su campo gravitacional (287)

C

caldera/caldera una depresión grande y semicircular que se forma cuando se vacía parcialmente la cámara de magma que hay debajo de un volcán, lo cual hace que el suelo se hunda (186)

cancer/cáncer un tumor en el cual las células comienzan a dividirse a una tasa incontrolable y se vuelven invasivas (674)

carbohydrate/carbohidrato una clase de nutrientes que proporcionan energía; incluye los azúcares, los almidones y las fibras; contiene carbono, hidrógeno y oxígeno (416, 511, 686)

carnivore/carnívoro un organismo que se alimenta de animales (591)

carrying capacity/capacidad de carga la población más grande que un ambiente puede sostener en cualquier momento dado (597)

cast/molde un tipo de fósil que se forma cuando un organismo descompuesto deja una cavidad que es llenada por sedimentos (88)

catalyst/catalizador una substancia que cambia la tasa de una reacción química sin consumirse ni cambiar demasiado (389)

catastrophism/catastrofismo un principio que establece que los cambios geológicos ocurren súbitamente (73)

cell/célula en biología, la unidad más pequeña que puede realizar todos los procesos vitales; las células están cubiertas por una membrana y tienen ADN y citoplasma (504)

cell cycle/ciclo celular el ciclo de vida de una célula (540)

cellular respiration/respiración celular el proceso por medio del cual las células utilizan oxígeno para producir energía a partir de los alimentos (537)

chemical bond/enlace químico una interacción que mantiene unidos los átomos o los iones (400)

chemical energy/energía química la energía que se libera cuando un compuesto químico reacciona para producir nuevos compuestos (122)

chemical equation/ecuación química una representación de una reacción química que usa símbolos para mostrar la relación entre los reactivos y los productos (376)

chemical formula/fórmula química una combinación de símbolos químicos y números que se usan para representar una substancia (374)

chemical reaction/reacción química el proceso por medio del cual una o más substancia cambian para producir una o más substancias distintas (370)

chromosome/cromosoma en una célula eucariótica, una de las estructuras del núcleo que está hecha de ADN y proteína; en una célula procariótica, el anillo principal de ADN (540)

climate/clima las condiciones promedio del tiempo en un área durante un largo período de tiempo (234)

coal/carbón un combustible fósil que se forma en el subsuelo a partir de materiales vegetales parcialmente descompuestos (114)

coevolution/coevolución la evolución de dos especies que se debe a su influencia mutua, a menudo de un modo que hace que la relación sea más beneficiosa para ambas (601)

commensalism/comensalismo una relación entre dos organismos en la que uno se beneficia y el otro no es afectado (600)

community/comunidad todas las poblaciones de especies que viven en el mismo hábitat e interactúan entre sí (588)

composition/composición la constitución química de una roca; describe los minerales u otros materiales presentes en ella (45)

compression/compresión estrés que se produce cuando distintas fuerzas actúan para estrechar un objeto (154)

conservation/conservación la preservación y el uso inteligente de los recursos naturales (620)

consumer/consumidor un organismo que se alimenta de otros organismos o de materia orgánica (509)

continental drift/deriva continental la hipótesis que establece que alguna vez los continentes formaron una sola masa de tierra, se dividieron y se fueron a la deriva hasta terminar en sus ubicaciones actuales (146)

controlled experiment/experimento controlado un experimento que prueba sólo un factor a la vez, comparando un grupo de control con un grupo experimental (14)

Spanish Glossary

convergent boundary/límite convergente el límite que se forma debido al choque de dos placas de la litosfera (151)

core/núcleo la parte central de la Tierra, debajo del manto (139)

cosmology/cosmología el estudio del origen, propiedades, procesos y evolución del universo (292)

covalent compound/compuesto covalente un compuesto químico que se forma al compartir electrones (402)

crater/cráter una depresión con forma de embudo que se encuentra cerca de la parte superior de la chimenea central de un volcán (186)

creep/arrastre el movimiento lento y descendente de materiales rocosos desgastados (223)

crust/corteza la capa externa, delgada y sólida de la Tierra, que se encuentra sobre el manto (138)

cytokinesis/citoquinesis la división del citoplasma de una célula (542)

D

decibel/decibel la unidad más común que se usa para medir el volumen del sonido (símbolo: dB) (440)

decomposer/descomponedor un organismo que, para obtener energía, desintegra los restos de organismos muertos o los desechos de animales y consume o absorbe los nutrientes (509)

decomposition reaction/reacción de descomposición una reacción en la que un solo compuesto se descompone para formar dos o más substancias más simples (381)

deflation/deflación una forma de erosión del viento en la que se mueven partículas de suelo finas y secas (211)

deformation/deformación el proceso de doblar, inclinar y romper la corteza de la Tierra; el cambio en la forma de una roca en respuesta a la tensión (163)

density/densidad la relación entre la masa de una substancia y su volumen (26)

deposition/deposición el proceso por medio del cual un material se deposita (41)

diffraction/difracción un cambio en la dirección de una onda cuando ésta se encuentra con un obstáculo o un borde, tal como una abertura (479)

diffusion/difusión el movimiento de partículas de regiones de mayor densidad a regiones de menor densidad (532)

digestive system/aparato digestivo los órganos que descomponen la comida de modo que el cuerpo la pueda usar (638)

divergent boundary/límite divergente el límite entre dos placas tectónicas que se están separando una de la otra (151)

Doppler effect/efecto Doppler un cambio que se observa en la frecuencia de una onda cuando la fuente o el observador está en movimiento (438)

double-displacement reaction/reacción de doble desplazamiento una reacción en la que se forma un gas, un precipitado sólido o un compuesto molecular a partir del intercambio de iones entre dos compuestos (383)

drug/droga cualquier substancia que produce un cambio en el estado físico o psicológico de una persona (692)

dune/duna un montículo de arena depositada por el viendo que conserva su forma incluso cuando se mueve (212)

E

echo/eco una onda de sonido reflejada (442)

echolocation/ecolocación el proceso de usar ondas de sonido reflejadas para buscar objetos; utilizado por animales tales como los murciélagos (443)

ecology/ecología el estudio de las interacciones de los seres vivos entre sí mismos y entre sí mismos y su ambiente (586)

ecosystem/ecosistema una comunidad de organismos y su ambiente abiótico o no vivo (589)

elastic rebound/rebote elástico ocurre cuando una roca deformada elásticamente vuelve súbitamente a su forma no deformada (163)

electromagnetic spectrum/espectro electromagnético todas las frecuencias o longitudes de onda de la radiación electromagnética (466)

electromagnetic wave/onda electromagnética una onda que está formada por campos eléctricos y magnéticos que vibran formando un ángulo recto unos con otros (462)

elevation/elevación la altura de un objeto sobre el nivel del mar (238)

endocytosis/endocitosis el proceso por medio del cual la membrana celular rodea una partícula y la encierra en una vesícula para llevarla al interior de la célula (534)

endospore/endospora una espora protectiva que tiene una pared gruesa, se forma dentro de una célula bacteriana y resiste condiciones adversas (565)

endothermic reaction/reacción endotérmica una reacción química que necesita calor (385)

energy pyramid/pirámide de energía un diagrama triangular que muestra la pérdida de energía en un ecosistema, producida a medida que la energía pasa a través de la cadena alimenticia del ecosistema (593)

eon/eón la mayor división del tiempo geológico (95)

epoch/época una subdivisión de un período geológico (95)

era/era una unidad de tiempo geológico que incluye dos o más períodos (95)

erosion/erosión el proceso por medio del cual el viento, el agua, el hielo o la gravedad transporta tierra y sedimentos de un lugar a otro (41)

esophagus/esófago un conducto largo y recto que conecta la faringe con el estómago (640)

exocytosis/exocitosis el proceso por medio del cual una célula libera una partícula encerrándola en una vesícula que luego se traslada a la superficie de la célula y se fusiona con la membrana celular (535)

exothermic reaction/reacción exotérmica una reacción química en la que se libera calor a los alrededores (384)

extinction/extinción la muerte de todos los miembros de una especie (95)

extrusive igneous rock/roca ígnea extrusiva una roca que se forma como resultado de la actividad volcánica en la superficie de la Tierra o cerca de ella (51)

F

farsightedness/hipermetropía condición en la que el cristalino del ojo enfoca los objetos lejanos detrás de la retina en lugar de en ella (489)

fat/grasa un nutriente que almacena energía y ayuda al cuerpo a almacenar algunas vitaminas (687)

fault/falla una grieta en un cuerpo rocoso a lo largo de la cual un bloque se desliza respecto a otro (156)

fermentation/fermentación la descomposición de los alimentos sin utilizar oxígeno (537)

folding/plegamiento fenómeno que ocurre cuando las capas de roca se doblan debido a la compresión (155)

foliated/foliada término que describe la textura de una roca metamórfica en la que los granos de mineral están ordenados en planos o bandas (59)

food chain/cadena alimenticia la vía de transferencia de energía través de varias etapas, que ocurre como resultado de los patrones de alimentación de una serie de organismos (592)

food web/red alimenticia un diagrama que muestra las relaciones de alimentación entre los organismos de un ecosistema (592)

force/fuerza una acción de empuje o atracción que se ejerce sobre un objeto con el fin de cambiar su movimiento; la fuerza tiene magnitud y dirección (314)

fossil/fósil los restos o las pruebas físicas de un organismo preservados por los procesos geológicos (86)

fossil fuel/combustible fósil un recurso energético no renovable formado a partir de los restos de organismos que vivieron hace mucho tiempo (112)

free fall/caída libre el movimiento de un cuerpo cuando la única fuerza que actúa sobre él es la fuerza de gravedad (343)

friction/fricción una fuerza que se opone al movimiento entre dos superficies que están en contacto (318)

function/función la actividad especial, normal o adecuada de un órgano o parte (517)

G

galaxy/galaxia un conjunto de estrellas, polvo y gas unidos por la gravedad (288)

gallbladder/vesícula biliar un órgano que tiene la forma de una bolsa y que almacena la bilis producida por el hígado (643)

gasohol/gasohol una mezcla de gasolina y alcohol que se usa como combustible (125)

generation time/tiempo de generación el período entre el nacimiento de una generación y el nacimiento de la siguiente generación (519)

geologic column/columna geológica un arreglo de las capas de roca en el que las rocas más antiguas están al fondo (77)

geologic time scale/escala de tiempo geológico el método estándar que se usa para dividir la larga historia natural de la Tierra en partes razonables (94)

geothermal energy/energía geotérmica la energía producida por el calor del interior de la Tierra (126)

glacial drift/deriva glacial el material rocoso que es transportado y depositado por los glaciares (218)

glacier/glaciar una masa grande de hielo en movimiento (214)

global warming/calentamiento global un aumento gradual de la temperatura global promedio (256)

globular cluster/cúmulo globular un grupo compacto de estrellas que parece una bola y contiene hasta un millón de estrellas (290)

gravity/gravedad una fuerza de atracción entre dos objetos debido a sus masas (324)

greenhouse effect/efecto de invernadero el calentamiento de la superficie y de la parte más baja de la atmósfera, el cual se produce cuando el vapor de agua, el dióxido de carbono y otros gases absorben y vuelven a irradiar la energía térmica (256)

H

half-life/vida media el tiempo que tarda la mitad de la muestra de una substancia radiactiva en desintegrarse por desintegración radiactiva (83)

herbivore/herbívoro un organismo que sólo come plantas (591)

heredity/herencia la transmisión de caracteres genéticos de padres a hijos (506)

homeostasis/homeostasis la capacidad de mantener un estado interno constante en un ambiente en cambio (505)

homologous chromosomes/cromosomas homólogos cromosomas con la misma secuencia de genes y la misma estructura (541)

host/huésped el organismo del cual un parásito obtiene alimento y refugio (572)

hot spot/mancha caliente un área volcánicamente activa de la superficie de la Tierra que se encuentra lejos de un límite entre placas tectónicas (192)

H-R diagram/diagrama H-R diagrama de Hertzsprung-Russell; una gráfica que muestra la relación entre la temperatura de la superficie de una estrella y su magnitud absoluta (284)

hydrocarbon/hidrocarburo un compuesto orgánico compuesto únicamente por carbono e hidrogeno (415)

hydroelectric energy/energía hidroeléctrica energía eléctrica producida por agua en caída (124)

hygiene/higiene la ciencia de la salud y las formas de preservar la salud (698)

hypothesis/hipótesis una explicación que se basa en observaciones o investigaciones científicas previas y que se puede probar (12)

I

ice age/edad de hielo un largo período de tiempo frío durante el cual grandes áreas de la superficie terrestre están cubiertas por capas de hielo; también conocido como período glacial (252)

immune system/sistema inmunológico las células y tejidos que reconocen y atacan substancias extrañas en el cuerpo (669)

immunity/inmunidad la capacidad de resistir una enfermedad infecciosa (666)

index fossil/fósil guía un fósil que se encuentra en las capas de roca de una sola era geológica y que se usa para establecer la edad de las capas de roca (90)

indicator/indicador un compuesto que puede cambiar de color de forma reversible dependiendo de condiciones tales como el pH (405)

inertia/inercia la tendencia de un objeto a no moverse o, si el objeto se está moviendo, la tendencia a resistir un cambio en su rapidez o dirección hasta que una fuerza externa actúe en el objeto (350)

infectious disease/enfermedad infecciosa una enfermedad que es causada por un patógeno y que puede transmitirse de un individuo a otro (660)

inhibitor/inhibidor una substancia que desacelera o detiene una reacción química (388)

interference/interferencia la combinación de dos o más ondas que resulta en una sola onda (444, 480)

intrusive igneous rock/roca ígnea intrusiva una roca formada a partir del enfriamiento y solidificación del magma debajo de la superficie terrestre (50)

ionic compound/compuesto iónico un compuesto formado por iones con cargas opuestas (400)

isotope/isótopo un átomo que tiene el mismo número de protones (o el mismo número atómico) que otros átomos del mismo elemento, pero que tiene un número diferente de neutrones (y, por lo tanto, otra masa atómica) (82)

K

kidney/riñón uno de los dos órganos que filtran el agua y los desechos de la sangre y excretan productos en fomra de orina (647)

L

landslide/derrumbamiento el movimiento súbito hacia abajo de rocas y suelo por una pendiente (221)

large intestine/intestino grueso la porción más ancha y más corta del intestino, que elimina el agua de los alimentos casi totalmente digeridos y convierte los desechos en heces semisólidas o excremento (644)

latitude/latitud la distancia hacia el norte o hacia el sur del ecuador; se expresa en grados (235)

lava plateau/meseta de lava un accidente geográfico amplio y plano que se forma debido a repetidas erupciones no explosivas de lava que se expanden por un área extensa (187)

law/ley un resumen de muchos resultados y observaciones experimentales; una ley dice cómo funcionan las cosas (21)

law of conservation of energy/ley de la conservación de la energía la ley que establece que la energía ni se crea ni se destruye, sólo se transforma de una forma a otra (385)

law of conservation of mass/ley de la conservación de la masa la ley que establece que la masa no se crea ni se destruye por cambios químicos o físicos comunes (377)

light-year/año luz la distancia que viaja la luz en un año; aproximadamente 9.46 trillones de kilómetros (273)

lipid/lípido un tipo de substancia bioquímica que no se disuelve en agua; las grasas y los esteroides son lípidos (417, 512)

lithosphere/litosfera la capa externa y sólida de la Tierra que está formada por la corteza y la parte superior y rígida del manto (140)

liver/hígado el órgano más grande del cuerpo; produce bilis, almacena y filtra la sangre, y almacena el exceso de azúcares en forma de glucógeno (643)

loess/loess sedimentos muy fértiles de cuarzo, feldespato, horneblenda, mica y arcilla depositados por el viento (212)

M

macrophage/macrófago una célula del sistema inmunológico que envuelve a los patógenos y otros materiales (669)

magma chamber/cámara de magma la masa de roca fundida que alimenta un volcán (180)

main sequence/secuencia principal la ubicación en el diagrama H-R donde se encuentran la mayoría de las estrellas; tiene un patrón diagonal de la parte inferior derecha (baja temperatura y luminosidad) a la parte superior izquierda (alta temperatura y luminosidad) (285)

malnutrition/desnutrición un trastorno de nutrición que resulta cuando una persona no consume una cantidad suficiente de cada nutriente que el cuerpo humano necesita (690)

mantle/manto la capa de roca que se encuentra entre la corteza terrestre y el núcleo (139)

mass/masa una medida de la cantidad de materia que tiene un objeto (25, 328)

mass movement/movimiento masivo un movimiento hacia abajo de una sección de terreno por una pendiente (220)

medium/medio un ambiente físico en el que ocurren fenómenos (432)

meiosis/meiosis un proceso de división celular durante el cual el número de cromosomas disminuye a la mitad del número original por medio de dos divisiones del núcleo, lo cual resulta en la producción de células sexuales (gametos o esporas) (545)

memory B cell/célula B de memoria una célula B que responde con mayor eficacia a un antígeno cuando el cuerpo vuelve a infectarse con él que cuando lo encuentra por primera vez (672)

mesosphere/mesosfera la parte fuerte e inferior del manto que se encuentra entre la astenosfera y el núcleo externo (141, 193)

metabolism/metabolismo la suma de todos los procesos químicos que ocurren en un organismo (506)

meter/metro la unidad fundamental de longitud en el sistema internacional de unidades (símbolo: m) (24)

microclimate/microclima el clima de un área pequeña (250)

mineral/mineral una clase de nutrientes que son elementos químicos necesarios para ciertos procesos del cuerpo (688)

mitosis/mitosis en las células eucarióticas, un proceso de división celular que forma dos núcleos nuevos, cada uno de los cuales posee el mismo número de cromosomas (541)

model/modelo un diseño, plan, representación o descripción cuyo objetivo es mostrar la estructura o funcionamiento de un objeto, sistema o concepto (18)

mold/molde una marca o cavidad hecha en una superficie sedimentaria por una concha u otro cuerpo (88)

momentum/momento una cantidad que se define como el producto de la masa de un objeto por su velocidad (356)

motion/movimiento el cambio en la posición de un objeto respecto a un punto de referencia (308)

mudflow/flujo de lodo el flujo de una masa de lodo o roca y suelo mezclados con una gran cantidad de agua (222)

mutualism/mutualismo una relación entre dos especies en la que ambas se benefician (600)

N

narcotic/narcótico una droga que proviene del opio, la cual alivia el dolor e induce el sueño; entre los ejemplos se encuentran la heroína, morfina y codeína (695)

natural gas/gas natural una mezcla de hidrocarburos gaseosos que se encuentran debajo de la superficie de la Tierra, normalmente cerca de los depósitos de petróleo, y los cuales se usan como combustible (113)

natural resource/recurso natural cualquier material natural que es utilizado por los seres humanos, como agua, petróleo, minerales, bosques y animales (108)

nearsightedness/miopía condición en la que el cristalino del ojo enfoca los objetos lejanos delante de la retina en lugar de en ella (488)

nebula/nebulosa una nube grande de gas y polvo en el espacio interestelar; una región en el espacio donde las estrellas nacen o donde explotan al final de su vida (290)

nephron/nefrona la unidad del riñón que filtra la sangre (647)

net force/fuerza neta la combinación de todas las fuerzas que actúan sobre un objeto (315)

neutralization reaction/reacción de neutralización la reacción de un ácido y una base que forma una solución neutra de agua y una sal (411)

neutron star/estrella de neutrones una estrella que se ha colapsado debido a la gravedad hasta el punto en que los electrones y protones han chocado unos contra otros para formar neutrones (286)

newton/newton la unidad de fuerza del sistema internacional de unidades (símbolo: N) (314)

nicotine/nicotina una substancia química tóxica y adictiva que se encuentra en el tabaco y que es una de las principales causas de los efectos dañinos de fumar (694)

noise/ruido un sonido que está constituido por una mezcla aleatoria de frecuencias (451)

nonfoliated/no foliada término que describe la textura de una roca metamórfica en la que los granos de mineral no están ordenados en planos ni bandas (60)

noninfectious disease/enfermedad no infecciosa una enfermedad que no se contagia de una persona a otra (660)

nonrenewable resource/recurso no renovable un recurso que se forma a una tasa que es mucho más lenta que la tasa a la que se consume (109, 616)

nuclear energy/energía nuclear la energía liberada por una reacción de fisión o fusión; la energía de enlace del núcleo atómico (120)

nuclear fusion/fusión nuclear combinación de los núcleos de átomos pequeños para formar un núcleo más grande; libera energía (278)

nucleic acid/ácido nucleico una molécula formada por subunidades llamadas *nucleótidos* (418, 513)

nutrient/nutriente una substancia de los alimentos que proporciona energía o ayuda a formar tejidos corporales y que es necesaria para la vida y el crecimiento (686)

O

omnivore/omnívoro un organismo que come tanto plantas como animales (591)

opaque/opaco término que describe un objeto que no es transparente ni translúcido (483)

open cluster/conglomerado abierto un grupo de estrellas que se encuentran juntas respecto a las estrellas que las rodean (290)

organ/órgano un conjunto de tejidos que desempeñan una función especializada en el cuerpo (515)

organic compound/compuesto orgánico un compuesto enlazado de manera covalente que contiene carbono (414)

organism/organismo un ser vivo; cualquier cosa que pueda llevar a cabo procesos vitales independientemente (516)

organ system/aparato (o sistema) de órganos un grupo de órganos que trabajan en conjunto para desempeñar funciones corporales (516)

osmosis/ósmosis la difusión del agua a través de una membrana semipermeable (533)

overpopulation/sobrepoblación la presencia de demasiados individuos en un área para los recursos disponibles (617)

P

paleontology/paleontología el estudio científico de los fósiles (74)

pancreas/páncreas el órgano que se encuentra detrás del estómago y que produce las enzimas digestivas y las hormonas que regulan los niveles de azúcar (642)

parallax/paralaje un cambio aparente en la posición de un objeto cuando se ve desde lugares distintos (273)

parasitism/parasitismo una relación entre dos especies en la que una, el parásito, se beneficia de la otra, el huésped, que resulta perjudicada (601)

passive transport/transporte pasivo el movimiento de substancias a través de una membrana celular sin que la célula tenga que usar energía (534)

pasteurization/pasteurización una manera de usar el calor para matar microorganismos en los alimentos, calentando los alimentos a una cierta temperatura durante un período específico de tiempocome (665)

pathogen/patógeno un virus, microorganismo u otra substancia que causa enfermedades (660)

pathogenic bacteria/bacteria patogénica bacteria que causa una enfermedad (570)

pedigree/pedigrí un diagrama que muestra la incidencia de un carácter genético en varias generaciones de una familia (550)

period/período una unidad de tiempo geológico en la que se dividen las eras (95)

petroleum/petróleo una mezcla líquida de compuestos hidrocarburos complejos; se usa ampliamente como una fuente de combustible (113)

pH/pH un valor que expresa la acidez o la basicidad (alcalinidad) de un sistema (411)

phospholipid/fosfolípido un lípido que contiene fósforo y que es un componente estructural de la membrana celular (512)

photosynthesis/fotosíntesis el proceso por medio del cual las plantas, las algas y algunas bacterias utilizan la luz solar, el dióxido de carbono y el agua para producir alimento (536)

pigment/pigmento una substancia que le da color a otra substancia o mezcla (487)

pitch/altura tonal una medida de qué tan agudo o grave se percibe un sonido, dependiendo de la frecuencia de la onda sonora (437)

plate tectonics/tectónica de placas la teoría que explica cómo se mueven y cambian de forma las placas tectónicas, que son grandes porciones de la capa más externa de la Tierra (150)

polar zone/zona polar el Polo Norte y el Polo Sur y la región circundante (249)

pollution/contaminación un cambio indeseable en el ambiente producido por substancias dañinas, desechos, gases, ruidos o radiación (614)

population/población un grupo de organismos de la misma especie que viven en un área geográfica específica (588)

precipitate/precipitado un sólido que se produce como resultado de una reacción química en una solución (371)

predator/depredador un organismo que se alimenta de otro organismo o de parte de él (598)

prevailing winds/vientos prevalecientes vientos que soplan principalmente de una dirección durante un período de tiempo determinado (237)

prey/presa un organismo al que otro organismo mata para alimentarse de él (598)

producer/productor un organismo que puede elaborar sus propios alimentos utilizando la energía de su entorno (509)

product/producto una substancia que se forma en una reacción química (376)

projectile motion/movimiento proyectil la trayectoria curva que sigue un objeto cuando es aventado, lanzado o proyectado de cualquier otra manera cerca de la superficie de la Tierra (345)

prokaryote/procariote un organismo que está formado por una sola célula y que no tiene núcleo (563)

protein/proteína una molécula formada por aminoácidos que es necesaria para construir y reparar estructuras corporales y para regular procesos del cuerpo (510)

pulsar/pulsar una estrella de neutrones que gira rápidamente y emite pulsaciones rápidas de energía radioeléctrica y óptica (286)

P wave/onda P una onda sísmica que hace que las partículas de roca se muevan en una dirección de atrás hacia delante (166)

Q

quasar/cuasar un objeto muy luminoso, parecido a una estrella, que genera energía a una gran velocidad; se piensa que los cuásares son los objetos más distantes del universo (291)

R

radiation/radiación la transferencia de energía en forma de ondas electromagnéticas (463)

radioactive decay/desintegración radiactiva el proceso por medio del cual un isótopo radiactivo tiende a desintegrarse y formar un isótopo estable del mismo elemento o de otro elemento (82)

radiometric dating/datación radiométrica un método para determinar la edad de un objeto estimando los porcentajes relativos de un isótopo radiactivo (precursor) y un isótopo estable (hijo) (83)

reactant/reactivo una substancia o molécula que participa en una reacción química (376)

recycling/reciclar el proceso de recuperar materiales valiosos o útiles de los desechos o de la basura; el proceso de reutilizar algunas cosas (111, 623)

red giant/gigante roja una estrella grande de color rojizo que se encuentra en una etapa avanzada de su vida (283)

reflection/reflexión el rebote de un rayo de luz, sonido o calor cuando el rayo golpea una superficie pero no la atraviesa (474)

refraction/refracción el curvamiento de una onda cuando ésta pasa entre dos substancias en las que su velocidad difiere (477)

relative dating/datación relativa cualquier método que se utiliza para determinar si un acontecimiento u objeto es más viejo o más joven que otros acontecimientos u objetos (76)

renewable resource/recurso renovable un recurso natural que puede reemplazarse a la misma tasa a la que se consume (109, 616)

resonance/resonancia un fenómeno que ocurre cuando dos objetos vibran naturalmente a la misma frecuencia; el sonido producido por un objeto hace que el otro objeto vibre (446)

rift zone/zona de rift un área de grietas profundas que se forma entre dos placas tectónicas que se están alejando una de la otra (190)

rock/roca una mezcla sólida de uno o más minerales o de materia orgánica que se produce de forma natural (40)

rock cycle/ciclo de las rocas la serie de procesos por medio de los cuales una roca se forma, cambia de un tipo a otro, se destruye y se forma nuevamente por procesos geológicos (40)

rock fall/desprendimiento de rocas el movimiento rápido y masivo de rocas por una pendiente empinada o un precipicio (221)

S

salt/sal un compuesto iónico que se forma cuando un átomo de un metal reemplaza el hidrógeno de un ácido (413)

saltation/saltación el movimiento de la arena u otros sedimentos por medio de saltos pequeños y rebotes debido al viento o al agua (210)

scattering/dispersión una interacción de la luz con la materia que hace que la luz cambie su energía, la dirección del movimiento o ambas (476)

science/ciencia el conocimiento que se obtiene por medio de la observación natural de acontecimientos y condiciones con el fin de descubrir hechos y formular leyes o principios que puedan ser verificados o probados (4)

scientific methods/métodos científicos una serie de pasos que se siguen para solucionar problemas (10)

sea-floor spreading/expansión del suelo marino el proceso por medio del cual se forma nueva litosfera oceánica a medida que el magma se eleva hacia la superficie y se solidifica (148)

seismic wave/onda sísmica una onda de energía que viaja a través de la Tierra y se aleja de un terremoto en todas direcciones (166)

seismology/sismología el estudio de los terremotos (162)

sex chromosome/cromosoma sexual uno de los dos cromosomas que determinan el sexo de un individuo (549)

sexual reproduction/reproducción sexual reproducción en la que se unen las células sexuales de los dos progenitores para producir descendencia que comparte caracteres de ambos progenitores (506)

shoreline/orilla el límite entre la tierra y una masa de agua (204)

single-displacement reaction/reacción de sustitución simple una reacción en la que un elemento toma el lugar de otro elemento en un compuesto (381)

small intestine/intestino delgado el órgano que se encuentra entre el estómago y el intestino grueso en el cual se produce la mayor parte de la descomposición de los alimentos y se absorben la mayoría de los nutrientes (642)

smog/esmog bruma fotoquímica que se forma cuando la luz solar actúa sobre contaminantes industriales y combustibles (118)

solar energy/energía solar la energía que la Tierra recibe del Sol en forma de radiación (122)

sonic boom/estampido sónico el sonido explosivo que se escucha cuando la onda de choque de un objeto que se desplaza a una velocidad superior a la de la luz llega a los oídos de una persona (445)

sound quality/calidad del sonido el resultado de la combinación de varios tonos por medio de la interferencia (448)

sound wave/onda de sonido una onda longitudinal que se origina debido a vibraciones y que se desplaza a través de un medio material (431)

speciation/especiación la formación de especies nuevas como resultado de la evolución (520)

spectrum/espectro la banda de colores que se produce cuando la luz blanca pasa a través de un prisma (269)

speed/rapidez la distancia que un objeto se desplaza dividida entre el intervalo de tiempo durante el cual ocurrió el movimiento (309)

standing wave/onda estacionaria un patrón de vibración que simula una onda que está parada (446)

stimulus/estímulo cualquier cosa que causa una reacción o cambio en un organismo o cualquier parte de un organismo (505)

stomach/estómago el órgano digestivo con forma de bolsa ubicado entre el esófago y el intestino delgado, que descompone los alimentos por la acción de músculos, enzimas y ácidos (641)

strata/estratos capas de roca (52)

stratification/estratificación el proceso por medio del cual las rocas sedimentarias se acomodan en capas (55)

stratified drift/deriva estratificada un depósito glacial que ha formado capas debido a la acción de los arroyos o de las aguas de ablación (219)

stress/estrés una respuesta física o mental a la presión (700)

structure/estructura el orden y distribución de las partes de un organismo (516)

subsidence/hundimiento del terreno el hundimiento de regiones de la corteza terrestre a elevaciones más bajas (160)

sunspot/mancha solar un área oscura en la fotosfera del Sol que es más fría que las áreas que la rodean y que tiene un campo magnético fuerte (280)

supernova/supernova una explosión gigantesca en la que una estrella masiva se colapsa y lanza sus capas externas hacia el espacio (286)

superposition/superposición un principio que establece que las rocas más jóvenes se encontrarán sobre las rocas más viejas si las capas no han sido alteradas (76)

surface current/corriente superficial un movimiento horizontal del agua del océano que es producido por el viento y que ocurre en la superficie del océano o cerca de ella (239)

S wave/onda S una onda sísmica que hace que las partículas de roca se muevan en una dirección de lado a lado (166)

symbiosis/simbiosis una relación en la que dos organismos diferentes viven estrechamente asociados uno con el otro (600)

synthesis reaction/reacción de síntesis una reacción en la que dos o más sustancias se combinan para formar un compuesto nuevo (380)

T

T cell/célula T una célula del sistema inmunológico que coordina el sistema inmunológico y ataca a muchas células infectadas (669)

tectonic plate/placa tectónica un bloque de litosfera formado por la corteza y la parte rígida y más externa del manto (142)

temperate zone/zona templada la zona climática ubicada entre los trópicos y la zona polar (246)

temperature/temperatura una medida de qué tan caliente (o frío) está algo; específicamente, una medida de la energía cinética promedio de las partículas de un objeto (26)

tension/tensión estrés que se produce cuando distintas fuerzas actúan para estirar un objeto (154)

terminal velocity/velocidad terminal la velocidad constante de un objeto en caída cuando la fuerza de resistencia del aire es igual en magnitud y opuesta en dirección a la fuerza de gravedad (342)

texture/textura la cualidad de una roca que se basa en el tamaño, la forma y la posición de los granos que la forman (46)

theory/teoría una explicación que relaciona muchas hipótesis y observaciones (20)

till/arcilla glaciárica material rocoso desordenado que deposita directamente un glaciar que se está derritiendo (218)

tissue/tejido un grupo de células similares que llevan a cabo una función común (515)

trace fossil/fósil traza una marca fosilizada que se forma en un sedimento blando debido al movimiento de un animal (88)

transform boundary/límite de transformación el límite entre placas tectónicas que se están deslizando horizontalmente una sobre otra (151)

translucent/traslúcido término que describe la materia que transmite luz, pero que no transmite una imagen (483)

transmission/transmisión el paso de la luz u otra forma de energía a través de la materia (482)

transparent/transparente término que describe materia que permite el paso de la luz con poca interferencia (483)

tropical zone/zona tropical la región que rodea el ecuador y se extiende desde aproximadamente 23° de latitud norte hasta 23° de latitud sur (242)

U

unconformity/disconformidad una ruptura en el registro geológico, creada cuando las capas de roca se erosionan o cuando el sedimento no se deposita durante un largo período de tiempo (79)

uniformitarianism/uniformitarianismo un principio que establece que es posible explicar los procesos geológicos que ocurrieron en el pasado en función de los procesos geológicos actuales (73)

uplift/levantamiento la elevación de regiones de la corteza terrestre a elevaciones más altas (160)

urinary system/sistema urinario los órganos que producen, almacenan y eliminan la orina (646)

V

vaccine/vacuna la administración a seres humanos o animales de organismos que han sido tratados para inducir una respuesta inmunológica (666)

variable/variable un factor que se modifica en un experimento con el fin de probar una hipótesis (14)

velocity/velocidad la rapidez de un objeto en una dirección dada (310)

vent/chimenea una abertura en la superficie de la Tierra a través de la cual pasa material volcánico (180)

virus/virus una partícula microscópica que se introduce en una célula y a menudo la destruye (572)

vitamin/vitamina una clase de nutrientes que contiene carbono y que es necesaria en pequeñas cantidades para mantener la salud y permitir el crecimiento (688)

volcano/volcán una chimenea o fisura en la superficie de la Tierra a través de la cual se expulsan magma y gases (178)

volume/volumen una medida del tamaño de un cuerpo o región en un espacio de tres dimensiones (25)

W

weather/tiempo el estado de la atmósfera a corto plazo que incluye la temperatura, la humedad, la precipitación, el viento y la visibilidad (234)

weight/peso una medida de la fuerza gravitacional ejercida sobre un objeto; su valor puede cambiar en función de la ubicación del objeto en el universo (328)

white dwarf/enana blanca una estrella pequeña, caliente y tenue que es el centro sobrante de una estrella vieja (283)

wind power/potencia eólica el uso de un molino de viento para hacer funcionar un generador eléctrico (123)

Index

Boldface page numbers refer to illustrative material, such as figures, tables, margin elements, photographs, and illustrations.

Index

B

C

Index

D

E

G

H

I

Index

M

N

O

P

Index

Q

R

S

Index

T

U

V

Index

W

Z

Index

Acknowledgments

continued from page ii

Academic Reviewers

continued

John Brockhaus, Ph.D.
Professor of Geospatial Information Science and Director of Geospatial Information Science Program
Department of Geography and Environmental Engineering
United States Military Academy
West Point, New York

Howard L. Brooks, Ph.D.
Professor of Physics and Astronomy
Department of Physics and Astronomy
DePauw University
Greencastle, Indiana

Dan Bruton, Ph.D.
Associate Professor
Department of Physics and Astronomy
Stephen F. Austin State University
Nacogdoches, Texas

Wesley N. Colley, Ph.D.
Lecturer
Department of Astronomy
University of Virginia
Charlottesville, Virginia

Joe W. Crim, Ph.D.
Professor and Head of Cellular Biology
Department of Cellular Biology
University of Georgia
Athens, Georgia

Roger J. Cuffey, Ph.D.
Professor of Paleontology
Department of Geosciences
Pennsylvania State University
University Park, Pennsylvania

Jim Denbow, Ph.D.
Associate Professor of Archaeology
Department of Anthropology and Archaeology
The University of Texas at Austin
Austin, Texas

William E. Dunscombe
Chairman
Biology Department
Union County College
Cranford, New Jersey

Simonetta Frittelli, Ph.D.
Associate Professor
Department of Physics
Duquesne University
Pittsburgh, Pennsylvania

Linda K. Gaul
Epidemiologist
Texas Department of Health
Austin, Texas

David Haig, Ph.D.
Professor of Biology
Organismic and Evolutionary Biology
Harvard University
Cambridge, Massachusetts

David S. Hall, Ph.D.
Assistant Professor of Physics
Department of Physics
Amherst College
Amherst, Massachusetts

Richard N. Hey, Ph.D.
Professor of Geophysics
Department of Geophysics and Planetology
University of Hawaii at Manoa
Honolulu, Hawaii

Ken Hon, Ph.D.
Associate Professor of Volcanology
Geology Department
University of Hawaii at Hilo
Hilo, Hawaii

Susan Hough, Ph.D.
Scientist
United States Geological Survey (USGS)
Pasadena, California

William H. Ingham, Ph.D.
Professor of Physics
James Madison University
Harrisonburg, Virginia

Steven A. Jennings, Ph.D.
Associate Professor
Geography and Environmental Studies
University of Colorado at Colorado Springs
Colorado Springs, Colorado

Ping H. Johnson, M.D., Ph.D., CHES
Assistant Professor of Health Education
Department of Health, Physical Education and Sport Science
Kennesaw State University
Kennesaw, Georgia

Linda Jones
Program Manager
Texas Department of Public Health
Austin, Texas

Joel S. Leventhal, Ph.D.
Emeritus Scientist
United States Geological Survey (USGS)
Lakewood, Colorado

Mark Mattson, Ph.D.
Assistant Professor
Physics Department
James Madison University
Harrisonburg, Virginia

Nancy L. McQueen, Ph.D.
Professor of Microbiology
Department of Biological Sciences
California State University, Los Angeles
Los Angeles, California

Madeline Micceri Mignone, Ph.D.
Assistant Professor
Natural Science
Dominican College
Orangeburg, New York

Richard F. Niedziela, Ph.D.
Assistant Professor of Chemistry
Department of Chemistry
DePaul University
Chicago, Illinois

Eva Oberdoerster, Ph.D.
Lecturer
Department of Biology
Southern Methodist University
Dallas, Texas

Sten Odenwald, Ph.D.
Astronomer
NASA Goddard Space Flight Center and Raytheon ITSS
Greenbelt, Maryland

Kenneth H. Rubin, Ph.D.
Associate Professor
Department of Geology and Geophysics
University of Hawaii at Manoa
Honolulu, Hawaii

Laurie Santos, Ph.D.
Assistant Professor
Department of Psychology
Yale University
New Haven, Connecticut

Fred Seaman, Ph.D.
Retired Research Associate
College of Pharmacy
The University of Texas at Austin
Austin, Texas

Patrick K. Schoff, Ph.D.
Research Associate
Natural Resources Research Institute
University of Minnesota—Duluth
Duluth, Minnesota

Colin D. Sumrall, Ph.D.
Lecturer of Paleontology
Earth and Planetary Sciences
The University of Tennessee
Knoxville, Tennessee

Richard S. Treptow, Ph.D.
Professor of Chemistry
Department of Chemistry and Physics
Chicago State University
Chicago, Illinois

Peter W. Weigand, Ph.D.
Professor Emeritus
Department of Geological Sciences
California State University
Northridge, California

Dale Wheeler
Assistant Professor of Chemistry
A. R. Smith Department of Chemistry
Appalachian State University
Boone, North Carolina

Ross Whitwam, Ph.D.
Assistant Professor of Biology
Division of Science and Mathematics
Mississippi University for Women
Columbus, Mississippi

Lab Testing

Barry L. Bishop
Science Teacher and Department Chair
San Rafael Junior High School
Ferron, Utah

Paul Boyle
Science Teacher
Perry Heights Middle School
Evansville, Indiana

Yvonne Brannum
Science Teacher and Department Chair
Hine Junior High School
Washington, D.C.

Daniel Bugenhagen
Science Teacher and Department Chair
Yutan Jr.–Sr. High
Yutan, Nebraska

Gladys Cherniak
Science Teacher
St. Paul's Episcopal School
Mobile, Alabama

James Chin
Science Teacher
Frank A. Day Middle School
Newtonville, Massachusetts

Alonda Droege
Biology Teacher
Evergreen High School
Seattle, Washington

Vicky Farland
Science Teacher and Department Chair
Centennial Middle School
Yuma, Arizona

Laura Fleet
Science Teacher
Alice B. Landrum Middle School
Ponte Verde Beach, Florida

Jennifer Ford
Science Teacher and Department Chair
North Ridge Middle School
North Richland Hills, Texas

Susan Gorman
Science Teacher
North Ridge Middle School
North Richland Hills, Texas

C. John Graves
Science Teacher
Monforton Middle School
Bozeman, Montana

Janel Guse
Science Teacher and Department Chair
West Central Middle School
Hartford, South Dakota

Dennis Hanson
Science Teacher and Department Chair
Big Bear Middle School
Big Bear Lake, California

Tracy Jahn
Science Teacher
Berkshire Junior-Senior High School
Canaan, New York

M. R. Penny Kisiah
Science Teacher and Department Chair
Fairview Middle School
Tallahassee, Florida

Kathy LaRoe
Science Teacher
East Valley Middle School
East Helena, Montana

Jason P. Marsh
Biology Teacher
Montevideo High School and Montevideo Country School
Montevideo, Minnesota

Edith C. McAlanis
Science Teacher and Department Chair
Socorro Middle School
El Paso, Texas

Kevin McCurdy, Ph.D.
Science Teacher
Elmwood Junior High School
Rogers, Arkansas

Kathy McKee
Science Teacher
Hoyt Middle School
Des Moines, Iowa

Dwight Patton
Science Teacher
Carrol T. Welch Middle School
Horizon City, Texas

Terry J. Rakes
Science Teacher
Elmwood Junior High School
Rogers, Arkansas

Elizabeth Rustad
Science Teacher
Higley School District
Gilbert, Arizona

Debra A. Sampson
Science Teacher
Booker T. Washington Middle School
Elgin, Texas

Rodney A. Sandefur
Science Teacher
Naturita Middle School
Naturita, Colorado

Helen Schiller
Instructional Coach
Greenville County Schools
Greenville, South Carolina

Bert J. Sherwood
Science Teacher
Socorro Middle School
El Paso, Texas

David M. Sparks
Science Teacher
Redwater Junior High School
Redwater, Texas

Larry Tackett
Science Teacher and Department Chair
R. H. Terrell Junior High School
Washington, D.C.

Ivora Washington
Science Teacher and Department Chair
Hyattsville Middle School
Washington, D.C.

Elsie N. Waynes
Science Teacher and Department Chair
R. H. Terrell Junior High School
Washington, D.C.

Gordon Zibelman
Science Teacher
Drexel Hill Middle School
Drexel Hill, Pennsylvania

Teacher Reviewers

Barbara Gavin Akre
Teacher of Biology, Anatomy-Physiology, and Life Science
Duluth Independent School District
Duluth, Minnesota

Laura Buchanan
Science Teacher and Department Chairperson
Corkran Middle School
Glen Burnie, Maryland

Sarah Carver
Science Teacher
Jackson Creek Middle School
Bloomington, Indiana

Robin K. Clanton
Science Department Head
Berrien Middle School
Nashville, Georgia

Hilary Cochran
Science Teacher
Indian Crest Junior High School
Souderton, Pennsylvania

Karen Dietrich, S.S.J., Ph.D.
Principal and Biology Instructor
Mount Saint Joseph Academy
Flourtown, Pennsylvania

Randy Dye, M.S.
Middle School Science Department Head
Earth Science
Wood Middle School
Waynesville School District #6, Missouri

Meredith Hanson
Science Teacher
Westside Middle School
Rocky Face, Georgia

James Kerr
Oklahoma Teacher of the Year 2002–2003
Oklahoma State Department of Education
Union Public Schools
Tulsa, Oklahoma

Laura Kitselman
Science Teacher and Coordinator
Loudoun Country Day School
Leesburg, Virginia

Debra S. Kogelman, MAed.
Science Teacher
University of Chicago Laboratory Schools
Chicago, Illinois

Deborah L. Kronsteiner
Teacher
Science Department
Spring Grove Area Middle School
Spring Grove, Pennsylvania

Jennifer L. Lamkie
Science Teacher
Thomas Jefferson Middle School
Edison, New Jersey

Stacy Loeak
Science Teacher and Department Chair
Baker Middle School
Columbus, Georgia

Augie Maldonado
Science Teacher
Grisham Middle School
Round Rock, Texas

Maureen Martin
Science Teacher
Jackson Creek Middle School
Bloomington, Indiana

Jean Pletchette
Health Educator
Winterset Community Schools
Winterset, Iowa

Thomas Lee Reed
Science Teacher
Rising Starr Middle School
Fayetteville, Georgia

Shannon Ripple
Science Teacher
Science Department
Canyon Vista Middle School
Round Rock, Texas

Susan H. Robinson
Science Teacher
Oglethorpe County Middle School
Lexington, Georgia

Elizabeth Rustad
Science Teacher
Higley School District
Gilbert, Arizona

Helen Schiller
Instructional Coach
Greenville County Schools
Greenville, South Carolina

Mark Schnably
Science Instructor
Thomas Jefferson Middle School
Winston-Salem, North Carolina

Stephanie Snowden
Science Teacher
Canyon Vista Middle School
Round Rock, Texas

Marci L. Stadiem
Department Head
Science Department
Cascade Middle School, Highline School District
Seattle, Washington

Bruce A. Starek
Department Chairperson, Science Teacher
Baker Middle School
Michigan City, Indiana

Martha Tedrow
Science Teacher
Thomas Jefferson Middle School
Winston-Salem, North Carolina

Martha B. Trisler
Science Teacher
Rising Starr Middle School
Fayetteville, Georgia

Florence Vaughan
Science Teacher
University of Chicago Laboratory Schools
Chicago, Illinois

Angie Williams
Teacher
Riversprings Middle School
Crawfordville, Florida

Roberta Young
Science Teacher
Gunn Junior High School
Arlington, Texas

Answer Checking

Hatim Belyamani
Austin, Texas

John A. Benner
Austin, Texas

Catherine Podeszwa
Duluth, Minnesota

Staff Credits

Editorial

Leigh Ann García, *Executive Editor*
Kelly Rizk, *Senior Editor*
David Westerberg, *Senior Editor*
Laura Zapanta, *Senior Editor*

Editorial Development Team

Karin Akre
Monica Brown
Jen Driscoll
Shari Husain
Michael Mazza
Karl Pallmeyer
Laura Prescott
Bill Rader
Jim Ratcliffe
Dennis Rathnaw
Betsy Roll
Kenneth Shepardson

Copyeditors

Dawn Marie Spinozza, *Copyediting Manager*
Simon Key
Jane A. Kirschman
Kira J. Watkins

Editorial Support Staff

Debbie Starr, *Managing Editor*
Kristina Bigelow
Suzanne Krejci
Shannon Oehler

Online Products

Bob Tucek, *Executive Editor*
Wesley M. Bain

Design

Book Design

Kay Selke, *Director of Book Design*
Lisa Woods, *Page Designer*
Holly Whittaker, *Project Administrator*

Media Design

Richard Metzger, *Design Director*
Chris Smith, *Developmental Designer*

Image Acquisitions

Curtis Riker, *Director*
Jeannie Taylor, *Photo Research Manager*
Diana Goetting, *Senior Photo Researcher*
Elaine Tate, *Art Buyer Supervisor*
Angela Boehm, *Senior Art Buyer*

Publishing Services

Carol Martin, *Director*

Graphic Services

Bruce Bond, *Director*
Jeff Bowers, *Graphic Services Manager*
Katrina Gnader, *Graphics Specialist*
Cathy Murphy, *Senior Graphics Specialist*
Nanda Patel, *Graphics Specialist*
JoAnn Stringer, *Senior Graphics Specialist II*

Technology Services

Laura Likon, *Director*
Juan Baquera, *Technology Services Manager*
Lana Kaupp, *Senior Technology Services Analyst*
Margaret Sanchez, *Senior Technology Services Analyst*
Sara Buller, *Technology Services Analyst*
Patty Zepeda, *Technology Services Analyst*
Jeff Robinson, *Ancillary Design Manager*

New Media

Armin Gutzmer, *Director*
Melanie Baccus, *New Media Coordinator*
Lydia Doty, *Senior Project Manager*
Cathy Kuhles, *Technical Assistant*
Marsh Flournoy, *Quality Assurance Analyst*
Tara F. Ross, *Senior Project Manager*

Design New Media

Ed Blake, *Director*
Kimberly Cammerata, *Design Manager*
Michael Rinella, *Senior Designer*

Production

Eddie Dawson, *Production Manager*
Sherry Sprague, *Project Manager*
Suzanne Brooks, *Production Coordinator*

Teacher Edition

Alicia Sullivan
David Hernandez
April Litz

Manufacturing and Inventory

Jevara Jackson
Ivania Quant Lee
Wilonda Ieans

Ancillary Development and Production

General Learning Communications, Northbrook, Illinois

Credits

Abbreviations used: (t) top, (c) center, (b) bottom, (l) left, (r) right, (bkgd) background

PHOTOGRAPHY

Front Cover (tl) Mike Powell/Getty images; (bl) Daryl Benson/Masterfile; (r) Andrew Syred/Getty Images; (DNA strand) David Mack/Science Photo Library

Skills Practice Lab Teens Sam Dudgeon/HRW

Connection to Astrology Corbis Images; **Connection to Biology** David M. Phillips/Visuals Unlimited; **Connection to Chemistry** Digital Image copyright © 2005 PhotoDisc; **Connection to Environment** Digital Image copyright © 2005 PhotoDisc; **Connection to Geology** Letraset Phototone; **Connection to Language Arts** Digital Image copyright © 2005 PhotoDisc; **Connection to Meteorology** Digital Image copyright © 2005 PhotoDisc; **Connection to Oceanography** © ICONOTEC; **Connection to Physics** Digital Image copyright © 2005 PhotoDisc

Table of Contents iii (t), Sam Dudgeon/HRW; iii (b), NASA; iv (t), Howard B. Bluestein; iv (bl), Tom Pantages Photography; v (t), E. R. Degginger/Color-Pic, Inc.; v (green), Dr. E.R. Degginger/Bruce Coleman Inc.; v (purple), Mark A. Schneider/Photo Researchers, Inc.; v, CORBIS Images/HRW; vi, Laurent Gillieron/Keystone/AP/Wide World Photos; vi (b), The G.R. "Dick" Roberts Photo Library; viii (t), National Geographic Image Collection/Robert W. Madden; viii (b), Bob Krueger/Photo Researchers, Inc.; ix (t), Glenn M. Oliver/Visuals Unlimited; ix (b) Tom Bean/CORBIS; x (t), Stuart Westmorland/CORBIS; xi (t), Goddard Space Flight Center Scientific Visualization Studio/NASA; xi (b), NASA; xii (tl), Index Stock; xii (c), MSFC/NASA; xii (bl), Peter Van Steen/HRW; xiii (t), Bill & Sally Fletcher/Tom Stack & Associates; xiii (b), NASA/TSADO/Tom Stack & Associates; xiv (t), NASA/Peter Arnold, Inc.; xv; Sam Dudgeon/HRW; xvi, Victoria Smith/HRW; xviii, xix, xx, xxii, Victoria Smith/HRW; xxvi, Sam Dudgeon/HRW; xxvii (t), John Langford/HRW; xxvii (b), xxviii (t, bl), Sam Dudgeon/HRW; xxviii (bl), Stephanie Morris/HRW; xxix (tl), Sam Dudgeon/HRW; xxix (tr), Jana Birchum/HRW; xxix (b), Sam Dudgeon/HRW

Chapter One 2-3 Craig Line/AP/Wide World Photos; 4 Peter Van Steen/HRW Photo; 5 (t) Peter Van Steen/HRW Photo; 5 (b) Sam Dudgeon/HRW Photo; 6 (b) Peter Van Steen/HRW Photo; 6 (t) Hank Morgan/Photo Researchers, Inc.; 7 Dale Miquelle/National Geographic Society Image Collection; 8 (b) John Langford/HRW Photo; 8 (t) NC: Science VU/PNNL/Visuals Unlimited; 9 Jeremy Bishop/Science Photo Library/Photo Researchers, Inc.; 11 (t) Peter Van Steen/HRW Photo; 12 Sam Dudgeon/HRW Photo; 14 John Mitchell/Photo Researchers; 16 Sam Dudgeon/HRW Photo; 17 (t) John Mitchell/Photo Researchers; 18 (l) © Fujifotos/The Image Works; 18 (r) © Fujifotos/The Image Works; 20 Art by Christopher Sloan/Photograph by Mark Thiessen both National Geographic_Image Collection/© National Geographic Image Collection; 24 David Austen/Publishers Network, Inc.; 25 (l) Peter Van Steen/HRW Photo; 25 (r) Peter Van Steen/HRW Photo; 26 (tl) Tony Freeman/PhotoEdit; 26 (tr) Victoria Smith/HRW; 26 (bl) Corbis Images; 34 (l), Craig Fugii/©1988 The Seattle Times; 35 (r), Bettman/CORBIS; 35 (bl), Layne Kennedy/CORBIS

Unit One 36 (tl), Science Photo Library/Photo Researchers, Inc; 36 (c), Francois Gohier; 36 (bl), © UPI/ Bettmann/CORBIS; 36 (br), Thomas Laird/Peter Arnold, Inc; 37 (tl), Science VU/Visuals Unlimited; 37 (tr), SuperStock; 37 (cl), AP/Wide World Photos; 37 (cr), NASA/Image State; 37 (br), File/AP/Wide World Photos

Chapter Two 38-39, Tom Till; 40 (bl), Michael Melford/Getty Images/The Image Bank; 40 (br), Joseph Sohm; Visions of America/CORBIS; 41, CORBIS Images/HRW; 44 (t), Joyce Photographics/Photo Researchers, Inc.; 44 (l), Pat Lanza/Bruce Coleman Inc.; 44 (r), Sam Dudgeon/HRW ; 44 (b), James Watt/Animals Animals/Earth Scenes; 44 (l), Pat Lanza/Bruce Coleman Inc.; 45, (granite), Pat Lanza/Bruce Coleman Inc.; 45, (mica), E. R. Degginger/Color-Pic, Inc.; 45, (aragonite), Breck P. Kent; 45, (limestone), Breck P. Kent; 45, (calcite), Mark Schneider/Visuals Unlimited; 45, (feldspar), Mark Schneider/Visuals Unlimited; 45, (quartz), Digital Image copyright © 2005 PhotoDisc; 46 (tl), Sam Dudgeon/HRW; 46 (tc), Dorling Kindersley; 46 (tr, br), Breck P. Kent; 46 (bl), E. R. Degginger/Color-Pic, Inc.; 47, Joseph Sohm; Visions of America/CORBIS; 48 (l), E. R. Degginger/Color-Pic, Inc.; 49 (tr, tl, bl), Breck P. Kent; 49 (br), Victoria Smith/HRW; 51, J.D. Griggs/USGS; 52, CORBIS Images/HRW; 53, (conglomerate), Breck P. Kent; 53, (siltstone), Sam Dudgeon/HRW; 53, (sandstone), Joyce Photographics/Photo Researchers, Inc.; 53, (shale), Sam Dudgeon/HRW; 54 (tl), Stephen Frink/Corbis; 54 (br), Breck P. Kent; 54 (bc), David Muench/CORBIS; 55, Franklin P. OSF/Animals Animals/Earth Scenes; 56, George Wuerthner; 58, (calcite), Dane S. Johnson/Visuals Unlimited; 58, (quartz), Carlyn Iverson/Absolute Science Illustration and Photography; 58, (hematite), Breck P. Kent; 58, (garnet), Breck P. Kent/Animals Animals/Earth Scenes; 58, (chlorite), Sam Dudgeon/HRW; 58, (mica), Tom Pantages; 59, (shale), Ken Karp/HRW; 59, (slate), Sam Dudgeon/HRW; 59, (phyllite), Sam Dudgeon/HRW; 59, (gneiss), Breck P. Kent; 59, (schist), Sam Dudgeon/HRW; 60 (tl), E. R. Degginger/Color-Pic, Inc.; 60 (bl), Ray Simmons/Photo Researchers, Inc; 60 (tr), The Natural History Museum, London; 60 (br), Breck P. Kent; 61, Jim Wark/Airphoto; 63 (t), Sam Dudgeon/HRW; 63 (b), James Tallon; 68 (l), Wolfgang Kaehler/CORBIS; 68 (tr), Dr. David Kring/Science Photo Library/Photo Researchers, Inc.; 69 (r), James Miller/Courtesy Robert Folk, Department of Geological Sciences, University of Texas at Austin; 69 (l), Dr. Philppa Uwins, Whistler Research PTY/SPL/Photo Researchers, Inc.

Chapter Three 70, National Geographic Image Collection/Jonathan Blair, Courtesy Hessian Regional Museum, Darmstadt, Germany; 73, GeoScience Features Picture Library; 75, Museum of Northern Arizona; 76 (l), Sam Dudgeon/HRW; 76 (r), Andy Christiansen/HRW; 78 (tl), Fletcher & Baylis/Photo Researchers, Inc.; 78 (tr), Ken M. Johns/Photo Researchers, Inc.; 78 (bl), Glenn M. Oliver/Visuals Unlimited; 78 (br), Francois Gohier/Photo Researchers, Inc.; 83, Sam Dudgeon/HRW; 84, Tom Till/DRK Photo; 85, Courtesy Charles S. Tucek/University of Arizona at Tucson; 86, Howard Grey/Getty Images/Stone; 87, Francis Latreille/Nova Productions/AP/Wide World Photos; 88 (b), The G.R. "Dick" Roberts Photo Library; 88 (t), © Louie Psihoyos/psihoyos.com; 89 (l), Brian Exton; 89 (r), Chip Clark/Smithsonian; 90 (l), Thomas R. Taylor/Photo Researchers, Inc.; 92, James L. Amos/CORBIS; 93 (tl), Tom Till Photography; 93 (fish), Tom Bean/CORBIS; 93 (leaf), James L. Amos/CORBIS; 93 (turtle), Layne Kennedy/CORBIS; 93 (fly), Ken Lucas/Visuals Unlimited; 95, Chip Clark/Smithsonian; 96 (t), Neg. no. 5793 Courtesy Dept. of Library Services., American Museum of Natural History; 96 (b), Neg. no. 5799 Courtesy Department of Library Services., American Museum of Natural History; 177, Neg. no. 5801 Courtesy Department of Library Services, American Museum of Natural History; 98, Jonathan Blair/CORBIS; 100 (b), The G.R. "Dick" Roberts Photo Library; 101 (fly), Ken Lucas/Visuals Unlimited; 104 (tl), Beth A. Keiser/AP/Wide World Photos; 104 (tr), Jonathan Blair/CORBIS; 105, Courtesy Kevin C. May

Chapter Four 106-107 (inset), Novovitch/Liaison/Getty Images; 108 (tc), Andy Christiansen/HRW; 108 (tl), John Blaustein/Liaison/Getty Images; 108 (tr), Mark Lewis/Getty Images/Stone; 109 (tl), James Randklev/Getty Images/Stone; 109 (b), Ed Malles/Liaison/Newsmakers/Getty Images; 109 (tr), Myrleen Furgusson Cate/PhotoEdit; 110, Victoria Smith/HRW; 112, Data courtesy Marc Imhoff of NASA/GSFC and Christopher Elvidge of NOAA/NGDC. Image by Craig Mayhew and Robert Simmon, NASA/GSFC.; 113 (b), John Zoiner; 113 (t), Mark Green/Getty Images/Taxi; 114, John Zoiner; 116, 2, Paolo Koch/Photo Researchers, Inc.; 116, 1, Horst Schafer/Peter Arnold, Inc.; 116, 3, Brian Parker/Tom Stack & Associates; 116, 4, C. Kuhn/Getty Images/The Image Bank; 117 (br), Alberto Incrocci/Getty Images/The Image Bank; 118 (inset), ©1994 NYC Parks Photo Archive/Fundamental Photographs; 118 (tl), © 1994 Kristen Brochmann/Fundamental Photographs; 118, Martin Harvey; 121 (tr), Tom Myers/Photo Researchers, Inc; 122 (t), Laurent Gillieron/Keystone/AP/Wide World Photos; 123 (b), Terry W. Eggers/CORBIS; 124 (t), Craig Sands/National Geographic Image Collection/Getty Images; 124 (b), Caio Coronel/Reuters/NewsCom; 125, G.R. Roberts Photo Library; 127, Laurent Gillieron/Keystone/AP/Wide World Photos; 129, Sam Dudgeon/HRW; 130, HRW; 131, Martin Harvey; 134 (t), Junko Kimura/Getty Images; 134 (b), STR/AP/Wide World Photos; 135 (t), Courtesy of Los Alamos National Laboratories; 135 (b), Corbis Images

Chapter Five 136-137, James Balog/Getty Images/Stone; 139 (t), James Wall/Animals Animals/Earth Scenes; 142, Bruce C. Heezen and Marie Tharp; 153 (tc), ESA/CE/Eurocontrol/Science Photo Library/Photo Researchers, Inc.; 153 (tr), NASA; 154 (bl, br), Peter Van Steen/HRW; 155 (bc), Visuals Unlimited/SylvesterAllred; 155 (br), G.R. Roberts Photo Library; 157 (tl), Tom Bean; 157 (tr), Landform Slides; 158, Jay Dickman/CORBIS; 159 (b), Michele & Tom Grimm Photography; 160, Y. Arthus-B./Peter Arnold, Inc.; 161, Peter Van Steen/HRW; 163, Roger Ressmeyer/CORBIS; 167 Michael S. Yamashita/CORBIS; 169, Sam Dudgeon/HRW; 174 (bl), NASA/Science Photo Library/Photo Researchers, Inc.; 174 (c), Ron Miller/Fran Heyl Associates; 174 (tr), Photo by S. Thorarinsson/Solar-Filma/Sun Film-15/3/courtesy of Edward T. Baker, Pacific Marine Environmental Laboratory, NOAA; 175 (r), Bettman/CORBIS

Chapter Six 176-177, Carl Shaneff/Pacific Stock; 178 (bl), National Geographic Image Collection/Robert W. Madden; 178 (br), Ken Sakamoto/Black Star; 179 (b), Breck P. Kent/Animals Animals/Earth Scenes; 179, Joyce Warren/USGS Photo Library; 181 (tl), Tui De Roy/Minden Pictures; 181 (bl), B. Murton/Southampton Oceanography Centre/Science Photo Library/Photo Researchers, Inc.; 181 (tr), Visuals Unlimited/ Martin Miller; 181 (br), Buddy Mays/CORBIS; 182 (tr), Tom Bean/DRK Photo; 182 (tl), Francois Gohier/Photo Researchers, Inc.; 182, (tlc), Visuals Unlimited Inc./Glenn Oliver; 182, (tlb), E. R. Degginger/Color-Pic, Inc.; 183 (tr), Alberto Garcia/SABA/ CORBIS; 183, Robert W. Madden/National Geographic Society; 184, Images & Volcans/Photo Researchers, Inc.; 185 (br), SuperStock; 185 (cr), SuperStock; 185 (tr), Roger Ressmeyer/CORBIS; 186 (tl), Yann Arthus-Bertrand/CORBIS; 187, Joseph Sohm; ChromoSohm Inc./CORBIS; 192 (bl), Robert McGimsey/USGS Alaska Volcano Observatory; 196 (tr), Alberto Garcia/SABA/CORBIS; 200 (bl), CORBIS; 200 (tr), Photo courtesy of Alan V. Morgan, Department of Earth Sciences, University of Waterloo; 200 (tc), © Sigurgeir Jonasson; Frank Lane Picture Agency/CORBIS; 201 (bl), Courtesy Christina Neal; 201 (r), Courtesy Alaska Volcano Observatory

Chapter Seven 203, John Kuntz/Reuters NewMedia Inc./CORBIS; 204, Aaron Chang/ Corbis Stock Market; 205 (t), Tom Bean; 205 (b), CORBIS Images/HRW; 206 (tc), The G.R. "Dick" Roberts Photo Library; 206 (br), Jeff Foott/DRK Photo; 206 (bl), CORBIS Images/HRW; 207 (tl), Breck P. Kent; 207 (tr), John S. Shelton ; 208 (t), Don Herbert/Getty Images/Taxi; 208 (b), Jonathan Weston/ImageState; 208 (t), SuperStock; 209, InterNetwork Media/Getty Images; 211, Jonathan Blair/CORBIS; 212, Telegraph Colour Library/FPG International/Getty Images/Taxi; 214, Tom Bean/ CORBIS; 216 (b), Getty Images/Stone; 216 (t), Visuals Unlimited/Glenn M. Oliver; 218, 219, Tom Bean; 220 (l), Sam Dudgeon/HRW; 220 (r), Sam Dudgeon/HRW Photo; 221 (b), Sebastian d'Souza/AFP/CORBIS; 221 (t), Jacques Jangoux/Getty Images/Stone; 222 (t), Jebb Harris/Orange County Register/SABA/CORBIS; 222 (b), Mike Yamashita/Woodfin Camp & Associates; 223, Visuals Unlimited/John D. Cunningham; 224, Sam Dudgeon/HRW; 226, Tom Bean/CORBIS; 227, Aaron Chang/ Corbis Stock Market; 230 (bl), Geological Survey of Canada, Photo #2002-581, Photographer Dr. Rejean Couture; 230, Charles H. Stites/The Lost Squadron Museum; 230 (t), Louis Sapienza/The Lost Squadron Museum; 231 (r), ©National Geographic Image Collection/ Marla Stenzel; 231 (l), Martin Mejia/AP/Wide World Photos

Chapter Eight 232-233, Steve Bloom Images; 234 (bkgd), Tom Van Sant, Geosphere Project/Planetary Visions/Science Photo Library/Photo Researchers, Inc.; 234 (tl), G.R. Roberts Photo Library; 234 (tr), Index Stock; 234 (c), Yva Momatiuk & John Eastcott; 234 (bl), Gary Retherford/Photo Researchers, Inc.; 234 (br), SuperStock; 235 (tr), CALLER-TIMES/AP/Wide World Photos; 235 (tc), Doug Mills/AP/Wide World Photos; 237 (b), Tom Van Sant, Geosphere Project/Planetary Visions/Science Photo Library/ Photo Researchers, Inc.; 238 (bl), Larry Ulrich Photography; 238 (br), Paul Wakefield/ Getty Images/Stone; 241, Index Stock; 242 (br), Tom Van Sant/Geosphere Project, Santa Monica/Science Photo Library/Photo Researchers, Inc.; 243 (tl), Carlos Navajas/ Getty Images/The Image Bank; 243 (tr), Michael Fogden/Bruce Coleman, Inc.; 244, Nadine Zuber/Photo Researchers, Inc.; 245, Larry Ulrich Photography; 246 (br), Tom Van Sant/Geosphere Project, Santa Monica/Science Photo Library/Photo Researchers, Inc.; 247 (b), Tom Bean/Getty Images/Stone; 247 (t), CORBIS Images/HRW; 248 (b), Steven Simpson/Getty Images/FPG International; 248 (t), Fred Hirschmann; 249 (b), Harry Walker/Alaska Stock; 249 (tr), Tom Van Sant/Geosphere Project, Santa Monica/ Science Photo Library/Photo Researchers, Inc.; 250, SuperStock; 245 (br), Roger Werth/Woodfin Camp & Associates; 255, D. Van Ravenswaay/Photo Researchers, Inc.; 260, Gunter Ziesler/Peter Arnold, Inc.; 261, SuperStock; 264, Roger Ressmeyer/ CORBIS; 264 (b), Terry Brandt/Grant Heilman Photography, Inc.; 265 (t), Courtesy of The University of Michigan

Chapter Nine 266-267, NASA; 268 (bl), Phil Degginger/Color-Pic, Inc.; 268 (br), John Sanford/Astrostock; 269, Sam Dudgeon/HRW; 271, Roger Ressmeyer/CORBIS; 272, Andre Gallant/Getty Images/The Image Bank; 280, NASA/Mark Marten/Photo Researchers, Inc. ; 281, NASA/TSADO/Tom Stack & Associates; 282, V. Bujarrabal (OAN, Spain), WFPC2, HST, ESA/ NASA ; 283, Royal Observatory, Edinburgh/SPL/ Photo Researchers, Inc.; 286 (br), Dr. Christopher Burrows, ESA/STScl/NASA; 286, blt Anglo-Australian Telescope Board; 286 (bl), Anglo-Australian Telescope Board; 287, V. Bujarrabal (OAN, Spain), WFPC2, HST, ESA/ NASA ; 288, Bill & Sally Fletcher/Tom Stack & Associates; 289 (br), Dennis Di Cicco/Peter Arnold, Inc.; 289 (bl), David Malin/Anglo-Australian Observatory; 290 (bl), I M House/Getty Images/Stone; 290 (br), Bill &Sally Fletcher/Tom Stack & Associates; 290 (bc), Jerry Lodriguss/Photo Researchers, Inc; 291, NASA/CXC/Smithsonian Astrophysical Observatory; 296, Sam Dudgeon/HRW; 297, John Sanford/Photo Researchers, Inc.; 302 (bl), NASA; 302 (tr), Jon Morse (University of Colorado)/NASA; 303 (r), The Open University; 303 (bkgd), Dutlev Van Ravenswaay/SPL/Photo Researchers, Inc.

Unit Two 304 (c), W.A. Mozart at the age of 7: oil on canvas, 1763, by P.A. Lorenzoni/The Granger Collection; 304 (bl), Photo Researchers, Inc.; 304 (t), © SPL/ Photo Researchers, Inc.; 305 (tc), The Vittoria, colored line engraving, 16th century/ The Granger Collection; 305 (tr), Stock Montage, Inc.; 305 (cl), Getty Images; 305 (cr), Underwood & Underwood/Corbis-Bettmann; 305 (br), © AFP/CORBIS

Chapter Ten 306-307 (all), © AFP/CORBIS; 308 (all), © SuperStock; 310 (bl), Robert Ginn/PhotoEdit; 312 (t), Sergio Purtell/Foca; 313 (tr), Digital Image copyright © 2005 PhotoDisc; 314 (b), Michelle Bridwell/HRW; 315 (b), Michelle Bridwell/HRW; 315 (tr), © Roger Ressmeyer/CORBIS; 316 (t), Daniel Schaefer/HRW; 316 (bl), Sam Dudgeon/ HRW; 317 (tr), age fotostock/Fabio Cardoso; 320 (bl, br), Michelle Bridwell/HRW; 320 (inset), Stephanie Morris/HRW; 322 (br), © Annie Griffiths Belt/CORBIS; 323 (tr), Sam Dudgeon/HRW; 133 (cr), Victoria Smith/HRW; 324 (br), NASA; 329 (tr), Digital Image copyright © 2005 PhotoDisc; 330 (bl), Sam Dudgeon/HRW; 331 (b), Sam Dudgeon/HRW; 332 (br), © Roger Ressmeyer/CORBIS; 332 (tl), Digital Image copy-right © 2005 PhotoDisc 333 (br), Sam Dudgeon/HRW; 336 (tl, c), Sam Dudgeon/ HRW; 336 (tr), Justin Sullivan/Getty Images; 337 (bl), Allsport Concepts/Getty Images; 337 (cr), Courtesy Dartmouth University

Chapter Eleven 338-339 (all), NASA; 339 (br), NASA; 340 (bl), Richard Megna/ Fundamental Photographs; 342 (cl), Toby Rankin/Masterfile; 343 (tr), James Sugar/ Black Star; 343 (bl), NASA; 345 (bl), Michelle Bridwell/Frontera Fotos; 345 (br), Image copyright © 2005 PhotoDisc, Inc.; 346 (tc), Richard Megna/Fundamental Photographs; 347 (tr), Toby Rankin/Masterfile; 348 (b), John Langford/HRW; 350 (br), Mavournea Hay/HRW; 350 (bc), Michelle Bridwell/Frontera Fotos; 351 (all), Victoria Smith/HRW; 352 (all), Image copyright © 2005 PhotoDisc, Inc.; 353 (b), David Madison; 354 (tc), Gerard Lacz/Animals Animals/Earth Scenes; 354 (tr), Sam Dudgeon/HRW; 354 (tr), Image copyright © 2005 PhotoDisc, Inc.; 354 (tl), NASA; 355 (br), Lance Schriner/HRW; 355 (tr), Victoria Smith/HRW; 357 (all), Michelle Bridwell/HRW; 358 (br), Zigy Kaluzny/Getty Images; 358 (bl), © SuperStock; 359 (cl), Michelle Bridwell/HRW; 360 (bl), Image ©2001 PhotoDisc, Inc.; 361 (all), Sam Dudgeon/HRW; 362 (tc), Gerard Lacz/Animals Animals/Earth Scenes; 363 (all), Sam Dudgeon/HRW; 364 (tl), AP Photo/Martyn Hayhow; 364 (tr), Junko Kimura/Getty Images/NewsCom; 365 (tr), Steve Okamoto; 365 (br), Lee Schwabe

Chapter Twelve 368-369 (all), Corbis Images; 370 (bl), Rob Matheson/The Stock Market; 370 (br), Sam Dudgeon/HRW; 371 (cl, cr), Richard Megna/Fundamental Photographs, New York; 371 (br), Scott Van Osdol/HRW; 371 (bl), J.T. Wright/Bruce Coleman Inc./Picture Quest; 372 (all), Charlie Winters; 373 (br), Charlie Winters/ HRW; 376 (tl), John Langford/HRW; 376 (bl), Richard Haynes/HRW; 377 (tr), Charles D. Winters/Photo Researchers, Inc.; 377 (tc), John Langford/HRW; 377 (tl), © Ingram Publishing; 382 (tl), Peticolas/Megna/Fundamental Photographs; 382 (tr), Richard Megna/Fundamental Photographs; 384 (bl), Victoria Smith/HRW; 384 (bc), Peter Van Steen/HRW; 384 (br), © Tom Stewart/The Stock Market; 385 (br), © David Stoecklein/CORBIS; 386 (t), Michael Newman/PhotoEdit; 387 (cr), Richard Megna/ Fundamental Photographs; 388 (t), Sam Dudgeon/HRW; 389 (tr), Dorling Kindersley Limited courtesy of the Science Museum, London/CORBIS; 389 (bl), Victoria Smith/ HRW; 390 (b), Victoria Smith/HRW; 392 (tr), Richard Megna/Fundamental Photographs; 393 (cr), Richard Megna/Fundamental Photographs; 393 (br), Rob Matheson/The Stock Market; 396 (tr), Tony Freeman/PhotoEdit; 396 (tl), Henry Bargas/Amarillo Globe-News/AP/Wide World Photos; 397 (all), Bob Parker/Austin Fire Investigation

Chapter Thirteen 398-399 (all), © Dr. Dennis Kunkel/Visuals Unlimited; 400 (bl), © Andrew Syred/Getty Images; 401 (all), Richard Megna/Fundamental Photographs; 402 (b), Victoria Smith/HRW; 403 (tr), Richard Megna/Fundamental Photographs; 404 (br), Jack Newkirk/HRW; 405 (br), Charles D. Winters/Timeframe Photography, Inc.; 405 (tl, tr), Peter Van Steen/HRW ; 406 (br), Tom Tracy/The Stock Shop/ Medichrome ; 407 (tc), Victoria Smith/HRW; 407 (tr), © Peter Cade/Getty Images; 407 (tl), © Bob Thomason/Getty Images; 408 (all), Peter Van Steen/HRW ; 409 (tr), Peter Van Steen/HRW ; 412 (bl), Digital Image copyright © 2005 PhotoDisc; 412 (tl), Victoria Smith/HRW; 412 (tc, tr), Scott Van Osdol/HRW; 413 (tr), Miro Vinton/Stock Boston/PictureQuest; 415 (tl), Sam Dudgeon/HRW; 415 (tc), John Langford/HRW; 415 (tr), Charles D. Winters/Timeframe Photography, Inc.; 416 (tc), Digital Image copyright © 2005 PhotoDisc; 417 (bl), Sam Dudgeon/HRW; 418 (tl), Hans Reinhard/ Bruce Coleman, Inc. ; 419 (tr), CORBIS Images/HRW; 420 (b), Sam Dudgeon/HRW; 422 (tr), Peter Van Steen/HRW ; 423 (all), Digital Image copyright © 2005 PhotoDisc; 426 (tr), Dan Loh/AP/Wide World Photos; 427 (tl), Sygma; 427 (tr), Nicole Guglielmo; 427 (bl), Corbis Images

Chapter Fourteen 428-429 (all), © Flip Nicklin/Minden Pictures; 431 (tl), John Langford/HRW; 432 (tr), Sam Dudgeon/HRW; 434 (tr), Sam Dudgeon/HRW; 435 (tr), Mary Kate Denny/PhotoEdit; 436 (all), Archive Photos; 438 (t), John Langford/HRW; 439 (bl), John Langford/HRW; 441 (tr), Charles D. Winters; 443 (t), © Stephen Dalton/Photo Researchers, Inc.; 444 (tl), Matt Meadows/Photo Researchers, Inc.; 446 (all), Richard Megna/Fundamental Photographs; 447 (tr), Sam Dudgeon/HRW; 448 (bc), Sam Dudgeon/HRW; 449 (bl), Digital Image copyright © 2005 EyeWire ; 449 (br), John Langford/HRW; 450 (tr, tl), Digital Image copyright © 2005 EyeWire ; 450 (bc), Bob Daemmrich/HRW; 452 (bl), Richard Megna/Fundamental Photographs; 453 (br), Sam Dudgeon/HRW; 454 (tl), © Flip Nicklin/Minden Pictures; 454 (bc), Sam Dudgeon/HRW; 455 (tc), © Ross Harrison Koty/Getty Images; 455 (cl), Dick Luria/ Photo Researchers, Inc.; 455 (br), John Langford/HRW; 459 (all), Victoria Smith/HRW

Chapter Fifteen 460-461 (all), Matt Meadows/Peter Arnold, Inc.; 463 (tl), Charlie Winters/Photo Researchers, Inc.; 463 (tr), Richard Megna/Fundamental Photographs; 464 (t), © A.T. Willett/Getty Images; 465 (tr), © Detlev Van Ravenswaay/Photo Researchers, Inc.; 466 (bc), Sam Dudgeon/HRW; 466 (br, bl), John Langford/HRW; 467 (bcr), Hugh Turvey/Science Photo Library/Photo Researchers, Inc.; 467 (br), Blair Seitz/Photo Researchers, Inc.; 467 (bcl), Leonide Principe/Photo Researchers, Inc.; 467 (tc, tr), Sam Dudgeon/HRW; 469 (br), © Tony Mcconnell/Photo Researchers, Inc.; 469 (tl), © Najlah Feanny/CORBIS SABA; 470 (t), © Cameron Davidson/Getty Images; 471 (cr), © Sinclair Stammers/SPL/Photo Researchers, Inc.; 472 (br), © Michael English/Custom Medical Stock Photo; 473 (tr), Hugh Turvey/Science Photo Library/Photo Researchers, Inc.; 475 (br), © Darwin Dale/Photo Researchers, Inc.; 476 (bl), Sovfoto/Eastfoto; 477 (bl), Richard Megna/Fundamental Photographs; 479 (br), Ken Kay/Fundamental Photographs; 481 (cr), Ken Kay/Fundamental Photographs; 482 (br), Stephanie Morris/HRW; 483 (all), John Langford/HRW; 484 (tl), Image copyright ©1998 PhotoDisc, Inc.; 484 (tr), Renee Lynn/Davis/Lynn Images; 484 (bl), Robert Wolf/HRW; 485 (tl), Leonard Lessin/Peter Arnold, Inc.; 486 (br), Sam Dudgeon/HRW; 487 (t), Index Stock Photography, Inc.; 487 (cr), Peter Van Steen/HRW; 488 (tl), © Digital Vision Ltd.; 490 (tr), Courtesy www.vischeck.com (program)/©Digital Vision Ltd. (frogs); 491 (tr), © Yoav Levy/Phototake; 492 (tr), Sam Dudgeon/HRW; 494 (br), Matt Meadows/Peter Arnold, Inc.; 494 (tl), Image copyright © 2005 PhotoDisc, Inc.; 495 (cr), Charles D. Winters/Photo Researchers, Inc.; 495 (bcr), © Mark E. Gibson; 495 (br), Richard Megna/Fundamental Photographs; 498 (tl), Dr. E. R. Degginger; 498 (tr), courtesy of the Raytheon Company; 499 (cr), © Underwood & Underwood/CORBIS

Unit Three 500 (tl), O.S.F./Animals Animals; 500 (cl), Hulton Archive/Getty Images; 500 (bl), Digital Image copyright © 2005 PhotoDisc; 500-501 (br & bl), Peter Veit/DRK Photo; 501 (cl), University of Pennsylvania/Hulton Getty; 501 (t), National Portrait Gallery, Smithsonian Institution/Art Resource; 501 (br), © National Geographic Image Collection/O. Louis Mazzatenta; 501 (cr), Digital Image copyright © 2005 PhotoDisc

Chapter Sixteen 502-503 (t), Rick Friedman/Blackstar Publishing/Picture Quest; 504 (r), Visuals Unlimited/Science Visuals Unlimited; 504 (l), Wolfgang Kaehler Photography; 505 (l), David M. Dennis/Tom Stack and Associates; 505 (r), David M. Dennis/Tom Stack and Associates; 506 (l), Visuals Unlimited/Stanley Flegler; 506 (r), James M. McCann/Photo Researchers, Inc. ; 508 (b), Wolfgang Bayer; 509 (t), Visuals Unlimited/Rob Simpson ; 509 (b), © Alex Kerstitch/Visuals Unlimited, Inc.; 510 (l), William J. Hebert/Stone; 510 (c), SuperStock; 510 (r), Kevin Schafer/Peter Arnold, Inc.; 511 Peter Dean/Grant Heilman Photography; 518 (b), Getty Images/Stone; 521 (r), Zig Leszczynski/Animals Animals/Earth Scenes; 521 (l), Gary Mezaros/Visuals Unlimited; 523 Peter Van Steen/HRW; 524 David M. Dennis/Tom Stack and Associates; 525 (tc), Victoria Smith/HRW; 525 (c), Victoria Smith/HRW; 525 (bc), Victoria Smith/HRW; 525 (t), © Wolfgang Kaehler/Liaison International/Getty News Images; 525 (b), © Alex Kerstitch/Visuals Unlimited, Inc.; 528 (b), Chip East/Reuters/NewsCom; 529 (r), Courtesy Janis Davis-Street/NASA; 529 (l), NASA

Chapter Seventeen 530-531 © Michael & Patricia Fogden/CORBIS; 532 Sam Dudgeon/HRW; 534 (br), Photo Researchers; 535 (tr), Birgit H. Satir; 536 (l), Runk/Schoenberger/Grant Heilman; 537 (r), John Langford/HRW Photo; 539 Corbis Images; 540 CNRI/Science Photo Library/Photo Researchers, Inc. ; 541 (t), L. Willatt, East Anglian Regional Genetics Service/Science Photo Library/Photo Researchers, Inc.; 541 (b), Biophoto Associates/Photo Researchers; 541 (b), Visuals Unlimited/R. Calentine; 541 (cl), Ed Reschke/Peter Arnold, Inc.; 541 (c), Ed Reschke/Peter Arnold, Inc.; 541 (cr), Ed Reschke/Peter Arnold, Inc.; 543 (cl), Ed Reschke/Peter Arnold, Inc.; 543 (c), Biology Media/Photo Researchers, Inc.; 543 (cr), Biology Media/Photo Researchers, Inc.; 544 (br), Biophoto Associates/Photo Researchers, Inc.; 544 (b), Phototake/CNRI/Phototake NYC; 549 (b), © Rob vanNostrand; 550 (b), © ImageState; 551 © ImageState; 552 Sam Dudgeon/HRW; 553 Sam Dudgeon/HRW; 555 (cl), Biophoto Associates/Science Source/Photo Researchers; 555 (cr), Biophoto Associates/Science Source/Photo Researchers; 555 (br), John Langford/HRW Photo; 558 (l), Lee D. Simons/Science Souce/Photo Researchers; 559 (tr), Courtesy Dr. Jarrel Yakel; 559 (tr), David McCarthy/SPL/Photo Researchers, Inc.

Chapter Eighteen 560 (t), Digital Image copyright © 2005 PhotoDisc Green; 560-561 (t), CAMR/A.B. Dowsett/Science Photo Library/Photo Researchers, Inc.; 562 (b), Robert Yin/Corbis; 562 (inset), Dr. Norman R. Pace and Dr. Esther R. Angert; 563 (c), Visuals Unlimited/David M. Phillips; 563 (l), Fran Heyl Associates; 563 (r), CNRI/Science Photo Library/Photo Researchers; 564 (b), Institut Pasteur/CNRI/Phototake; 565 (t), Heather Angel; 565 (bl), SuperStock; 565 (bc), SuperStock; 565 (br), © David M. Phillips/Visuals Unlimited; 566 (br), SuperStock; 566 (bl), Dr. Kari Lounatmaa/Science Photo Library/Photo Researchers, Inc.; 567 (t), Robert Yin/Corbis; 569 (tr), Bio-Logic Remediation LTD; 569 (b), Peter Van Steen/HRW; 570 (b), © Aaron Haupt/Photo Researchers, Inc.; 570 (t), © Carmela Leszczynski/Animals Animals/Earth Scenes; 571 (t), Visuals Unlimited/Sherman Thomson; 572 (b), E.O.S./Gelderblom/Photo Researchers; 573 (tl), Visuals Unlimited/Hans Gelderblom; 573 (tr), Visuals Unlimited/K. G. Murti; 573 (bl), Dr. O. Bradfute/Peter Arnold; 573 (br), Oliver Meckes/MPI-Tubingen/Photo Researchers; 575 (tr), Omni Photo Communications, Inc./Index Stock Imagery, Inc.; 576 (b), Victoria Smith/HRW; 578 (t), © Carmela Leszczynski/Animals Animals/Earth Scenes; 579 (b), Digital Image copyright © 2005 PhotoDisc Green; 582 (l), HRW Photo composite; 582 (r), ©CORBIS; 583 Courtesy of Annie O'Neal, Laytonville Middle School

Chapter Nineteen 584-585 © Roine Magnusson/Getty Images/The Image Bank; 585 (r), © David M. Phillips/Visuals Unlimited; 593 (t), © George H. H. Huey/CORBIS; 593 (c), © D. Robert & Lorri Franz/CORBIS; 593 (b), © Jason Brindel Photography/Alamy Photos; 594 (t), Laguna Photo/Liaison International/Getty News Images; 594 (b), Jeff Lepore/Photo Researchers; 596 Jeff Foott/AUSCAPE; 597 © Ross Hamilton/Getty Images/Stone; 598 (l), Visuals Unlimited/Gerald & Buff Corsi; 598 (cr), Hans Pfletschinger/Peter Arnold; 599 (r), W. Peckover/Academy of Natural Sciences Philadelphia/VIREO; 599 (l), Leroy Simon/Visuals Unlimited; 600 (b), Ed Robinson/Tom Stack & Associates; 600 (tl), © Telegraph Color Library/Getty Images/FPG International; 600 (tr), OSF/Peter Parks/Animals Animals Earth Scenes; 601 (t), © Gay Bumgarner/Getty Images/Stone; 601 (b), Carol Hughes/Bruce Coleman; 602 (tl), CSIRO Wildlife & Ecology; 602 (r), © Rick & Nora Bowers/Visuals Unlimited; 603 Leroy Simon/Visuals Unlimited; 604 Sam Dudgeon/HRW; 605 Sam Dudgeon/HRW; 605 Sam Dudgeon/HRW; 606 Leroy Simon/Visuals Unlimited; 610 (r), © National Geographic Image Collection/Darlyne Murawski; 610 (l), Digital Image copyright © 2005 PhotoDisc; 611 (r), Photo from the Dept. of Communication Services, North Carolina State University; 611 (l), Digital Image copyright © 2005 Artville

Chapter Twenty 612-613 Martin Harvey/NHPA; 614 Larry Lefever/Grant Heilman Photography; 615 (t), J. Roche/Peter Arnold, Inc.; 615 (b), NASA; 616 (b), © Jacques Jangoux/Getty Images/Stone; 617 REUTERS/Jonathan Searle/NewsCom; 618 © Rex Ziak/Getty Images/Stone; 619 REUTERS/Jonathan Searle/NewsCom; 620 (l), Peter Van Steen/HRW; 620 (c), Peter Van Steen/HRW; 620 (r), Peter Van Steen/HRW; 621 (t), Argonne National Laboratory; 621 (b), Digital Image copyright © 2005 PhotoDisc; 622 (b), PhotoEdit; 622 (t), Kay Park-Rec Corp.; 623 (t), Peter Van Steen/HRW; 623 (b), Martin Bond/Science Photo Library/Photo Researchers; 624 (b), K. W. Fink/Bruce Coleman; 624 (t), © Sindre Ellingsen/Alamy Photos; 625 Stephen J. Krasemann/DRK Photo; 626 (bl), Sam Dudgeon/HRW; 626 (t), © Will & Deni McIntyre/Getty Images/Stone; 626 (br), Stephen J. Krasemann/DRK Photo; 627 (t), K. W. Fink/Bruce Coleman; 628 Peter Van Steen/HRW; 629 (tl), Tom Bean/DRK Photo; 629 (tr), Darrell Gulin/DRK Photo; 630 (l), Peter Van Steen/HRW; 630 (r), Peter Van Steen/HRW; 631 Larry Lefever/Grant Heilman Photography; 634 (l), Art Wolfe; 634 (r), © Toru Yamanaka/AFP/CORBIS; 635 (b), Gamma-Liaison/Getty News Images; 635 (t), Huntsville Times

Chapter Twenty One 636-637 © ISM/Phototake; 645 (t), Victoria Smith/HRW; 648 Getty Images/The Image Bank; 649 Stephen J. Krasemann/DRK Photo; 650 (b), Sam Dudgeon/HRW; 656 (l), J.H. Robinson/Photo Researchers; 656 (r), REUTERS/David Gray/NewsCom; 657 (t), Peter Van Steen/HRW

Chapter Twenty Two 658-659 ©Andrew Syred / Photo Researchers, Inc., Inc.; 660 (br), CNRI/Science Photo Library/Photo Researchers; 660 (bl), Tektoff-RM/CNRI/Science Photo Library/Photo Researchers; 661 (t), Kent Wood/Photo Researchers; 662 Dr. Jeremy Burgess/Photo Researchers, Inc.; 664 Erich Lessing/ Art Resource; 665 (t)Reuters NewMedia Inc./ CORBIS; 666 (t) Michelle Bridwell/PhotoEdit; 666 (b) Bettman/CORBIS; 667 © Jim Richardson/CORBIS; 668 (b), Peter Van Steen/HRW; 672 (t), John Langford/HRW Photo; 673 (b), Clinical Radiology Dept., Salisbury District Hospital/Science Photo Library/Photo Researchers; 673 (t), SuperStock; 674 (b), Photo Lennart Nilsson/Albert Bonniers Forlag AB; 674 (tl), Dr. A. Liepins/Science Photo Library/Photo Researchers; 674 (tr), Dr. A. Liepins/Science Photo Library/Photo Researchers; 676 Sam Dudgeon/HRW; 679 (t), Peter Van Steen/HRW ; 682 (l), E. R. Degginger/Bruce Coleman; 682 (r), Chris Rogers/Index Stock Imagery, Inc.; 683 (t), Peter Van Steen/HRW; 683 (b), Corbis

Chapter Twenty Three 684-685 © Arthur Tilley/Getty Images/Taxi; 686 Peter Van Steen/HRW; 687 (b), Peter Van Steen/HRW; 687 (t), Sam Dudgeon/HRW; 688 (c), Image Copyright ©2004 PhotoDisc, Inc./HRW; 688 (t), Image Copyright ©2004 PhotoDisc, Inc./HRW; 688 (bl), CORBIS Images/HRW; 688 (br), CORBIS Images/HRW; 689 © John Kelly/Getty Images/Stone; 690 John Burwell/FoodPix; 691 Peter Van Steen/HRW; 692 Peter Van Steen/HRW; 693 (b), Peter Van Steen/HRW; 693 (t), ©1999 Steven Foster; 694 (tl), E. Dirksen/Photo Researchers; 694 (b), Spencer Grant/Photo Researchers, Inc.; 694 (tr), Dr. Andrew P. Evans/Indiana University; 696 Jeff Greenberg/PhotoEdit; 697 Mike Siluk/The Image Works; 698 Sam Dudgeon/HRW; 699 (t), © Rob Van Petten/Getty Images/The Image Bank; 699 (b), Peter Van Steen/HRW; 700 Sam Dudgeon/HRW; 702 (b), Peter Van Steen/HRW; 702 (t), © Mug Shots/CORBIS; 703 Peter Van Steen/HRW; 704 Digital Image copyright © 2005 PhotoDisc; 706 (t), © John Kelly/Getty Images/Stone; 706 (b), Peter Van Steen/HRW; 707 (t), Peter Van Steen/HRW; 707 (b), Peter Van Steen/HRW; 710 (l), Brian Hagiwara/FoodPix; 711 (r), Courtesy Russell Selger; 711 (l), © Eyebyte/Alamy Photos

Lab Book TOC 712 (l, tr, br), Sam Dudgeon/HRW; 712 (c), Scott Van Osdol/HRW; 714, 715 (hematite, br), Sam Dudgeon/HRW; 716 (all), Andy Christiansen/HRW; 717, 719, Sam Dudgeon/HRW; 720, Tom Bean; 721, 722, Sam Dudgeon/HRW; 729, Sam Dudgeon/HRW; 730, Andy Christiansen/HRW; 731, 732 Sam Dudgeon/HRW; 733 (all), Sam Dudgeon/HRW; 734 (br), Sam Dudgeon/HRW; 736 (c), Sam Dudgeon/HRW; 741 (all), Sam Dudgeon/HRW; 742 (br), Sam Dudgeon/HRW; 743 (b), Sam Dudgeon/HRW; 744 (br), Rob Boudreau/Getty Images; 745 (br), Victoria Smith/HRW; 746 (cr), John Langford/HRW; 748 (br), Sam Dudgeon/HRW; 749 (br), HRW Photo; 750 (r), Sam Dudgeon/HRW; 752 (br), Sam Dudgeon/HRW; 763 Sam Dudgeon/HRW; 765 (b), Sam Dudgeon/HRW; 765 (c), Sam Dudgeon/HRW; 766 Sam Dudgeon/HRW; 772 (tr), Peter Van Steen/HRW; 772 (c), Peter Van Steen/HRW; 772 (br), Peter Van Steen/HRW; 793 Peter Van Steen/HRW; 796 Sam Dudgeon/HRW; 797 Sam Dudgeon/HRW; 801 Peter Van Steen/HRW; 767, Sam Dudgeon/HRW; 779 (tr), Sam Dudgeon/HRW; 830 CENCO